The Economics of
Climate C
The Stern Revie

NICHOLAS STERN

Cabinet Office – HM Treasury

There is now clear scientific evidence that emissions from economic activity, particularly the burning of fossil fuels for energy, are causing changes to the Earth's climate. A sound understanding of the economics of climate change is needed in order to underpin an effective global response to this challenge.

The Stern Review is an accessible, independent, and comprehensive analysis of the economic aspects of this crucial issue. It has been conducted by Nicholas Stern, Head of the UK Government Economic Service, and a former Chief Economist of the World Bank.

The Review considers all aspects of the issue, including the nature of the economics and the science; the impact of climate change on growth and development in both rich and poor countries; the economics of cutting emissions and stabilising greenhouse gas emissions in the atmosphere; the components of policy on both mitigation and adaptation; and the challenges of achieving sustained international collective action. *The Review* will help to promote a greater understanding of the impact and effectiveness of national and international policies and arrangements in reducing emissions in a cost-effective way, and promoting a dynamic, equitable and sustainable global economy.

The Economics of Climate Change: The Stern Review will be invaluable for anyone interested in the economics and policy implications of climate change, and students, economists, scientists and policy makers involved in all aspects of climate change.

Sir Nicholas Stern is Adviser to the UK Government on the Economics of Climate Change and Development, reporting to the Prime Minister. As well as being Head of the *Stern Review on the Economics of Climate Change*, he is Head of the Government Economic Service, and previously Second Permanent Secretary to Her Majesty's Treasury and Director of Policy and Research for the Prime Minister's Commission for Africa. He is also a former Chief Economist for the World Bank and Special Counsellor to the President of the European Bank for Reconstruction and Development. His research and publications have focused on economic development and growth, economic theory, tax reform, public policy and the role of the state and economies in transition. He is a Fellow of the British Academy and a Foreign Honorary Member of the American Academy of Arts and Sciences. His most recent book is *Growth and Empowerment: Making Development Happen* (2005, MIT Press).

Comments on *The Economics of Climate Change: The Stern Review*

'Nicholas Stern's review will be seen as a landmark in the struggle against climate change. It gives a stark warning, but also offers hope. It proves comprehensively that tackling climate change is a pro-growth strategy. The economic benefits of strong early action easily outweigh the costs. The framework for action laid out by the Review is both ambitious and realistic. Climate change is a global problem and Stern's conclusions are a wake-up call not just for the UK, but for every country in the world.'
Rt Hon Tony Blair MP, UK Prime Minister

'*The Stern Review* on the *Economics of Climate Change* is the most comprehensive analysis yet, not only of the challenges, but also of the opportunities from climate change. Stern makes clear that climate change is a global challenge that demands a global solution. Above all, environmental policy is economic policy. It is my hope that this Review is discussed and understood as widely as possible, throughout the world, not just by Governments, but also by business leaders, NGOs, international institutions and society as a whole.'
Rt Hon Gordon Brown MP, UK Chancellor of the Exchequer

'*The Stern Review* shows us, with utmost clarity, while allowing fully for all the uncertainties, what global warming is going to mean; and what can and should be done to reduce it. It provides numbers for the economic impact, and for the necessary economic policies. It deserves the widest circulation. I wish it the greatest possible impa   Governments have a clear and immediate duty to accept the challenge it represent.
James Mirrle , recipient of the Nobel Prize for Economics, 1996

'The stark prospects of climate change and its mounting economic and human costs are clearly brought out in this searching investigation. What is particularly striking is the identification of ways and means of sharply minimizing these penalties through acting right now, rather than waiting for our lives to be overrun by rapidly advancing adversities. The world would be foolish to neglect this strong but strictly time-bound practical message.'
Amartya Sen, recipient of the Nobel Prize for Economics, 1998

'*The Stern Review* of *The Economics of Climate Change* provides the most thorough and rigorous analysis to date of the costs and risks of climate change, and the costs and risks of reducing emissions. It makes clear that the question is not whether we can afford to act, but whether we can afford not to act. ... And it provides a comprehensive agenda – one which is economically and politically feasible – behind which the entire world can unite in addressing this most important threat to our future well being.'
Joseph Stiglitz, recipient of the Nobel Prize for Economics, 2001

' ... the world is waiting for a calm, reasonable, carefully argued approach to climate change: Nick Stern and his team have produced one.'
Robert M. Solow , recipient of the Nobel Prize for Economics, 1987

'I very much welcome *The Stern Review*, which provides a much needed critical economic analysis of the issues associated with climate change ...'
Paul Wolfowitz, President of the World Bank

'*The Stern Review* of *The Economics of Climate Change* is a vital step forward in securing an effective global policy on climate change. Led by one of the world's top economists, the *Stern Review* shows convincingly that the benefits of early global action to mitigate climate change will be far lower than the costs. The report establishes realistic guidelines for action ... . *The Stern Review* will play an important role in helping the world to agree on a sensible post-Kyoto policy.'
Professor Jeffrey D. Sachs, Director of the Earth Institute at Columbia University and Special Advisor to UN Secretary General

'*The Economics of Climate Change* sends a very important and timely message: that the benefits of strong, early action on climate change outweigh the costs. ... Congratulations to Sir Nick Stern and his team for producing a landmark review which I have no doubt will strengthen the political will to change of governments around the world.'
Claude Mandil, Executive Director of the International Energy Agency

'The scientific evidence of global warming is overwhelming but some commentators and lobby groups have continued to oppose offsetting actions on economic and competitiveness grounds. This comprehensive and authoritative report demolishes their arguments, explaining clearly the complex economics of climate change. It makes plain that we can cut emissions radically at a cost to the economy far less than the economic and human welfare costs which climate change could impose.'
Adair Turner, Former Director of UK Confederation of British Industry and Economic Advisor to the Sustainable Development Commission

'When the history of the world's response to climate change is written, the *Stern Review* will be recognized as a turning point. ... Sir Nicholas and his team have provided important intellectual leadership as humanity engages with its greatest challenge. ... While the details will be debated, the main thrust of the report is clear and compelling – the expected benefits of tackling climate change far outweigh the expected costs.'
Cameron Hepburn, Elizabeth Wordsworth Junior Research Fellow in Economics, Oxford University

'Pay now to fix global warming or risk a worldwide economic depression later ... The [Stern] report moves economic discussion of how humanity should deal with global warming to center stage ...'
USA Today

'The overwhelming message of ... [the] Stern review on the economics of climate change is that it is now time to move on from arguing about statistics to taking drastic action at an international level. ... Even if Stern is only half right then ... the consequence of doing nothing is still so dreadful that it ought not to be contemplated.'
The Guardian

'[The report's] basic point seems unassailable: failure to act now will exact much greater penalties later on ... If people and industries are made to pay heavily for the privilege, they will inevitable be driven to develop cleaner fuels, cars and factories...'
The New York Times

'The Stern review makes two invaluable contributions. The first is that it recasts environmentalism as economics ... Stern's second serious contribution is to provide a formula for durable environmentalism, one which binds business and government.'
The Times

'The [Stern] report argues that environmentalism and economic growth can go hand in hand in the battle against global warming ... The report by Sir Nicholas Stern, a senior government economist, represents a huge contrast to the U.S. government's wait-and-see policies.'
Chicago Tribune

'... a comprehensive overview of the threat posed by climate change – and how we should respond to it. ... Sir Nicholas Stern spells out a bleak vision of a future gripped by violent storms, rising sea-levels, crippling droughts and economic chaos unless urgent action is taken to tackle global warming. ... a heavyweight review ... Sir Nicholas Stern's review of the economic impact of global warming is a watershed. The former World Bank chief economist has put a price-tag on saving the planet. ... Sir Nicholas is a sober and respected economist, which makes his findings all the more chilling.'
The Daily Telegraph

'Future generations may come to regard the apocalyptic report by Sir Nicholas Stern ... as the turning point in combating global warming, or as the missed opportunity. ... what Sir Nicholas Stern has done with his report on the economics of climate change is remarkable; he has ripped up the last excuse for inaction. ... one wouldn't want to exaggerate, but it does feel like one of those moments that are truly historic ... the first really comprehensive review of the economics of climate change. For nearly 20 years if has been the *science* of climate change that has made all the headlines ... We've heard a thousand calls to action, to stop global warming happening. But what would that *cost* the world? And what would doing nothing cost us? ... now Sir Nicholas Stern and his team have come up with concrete numbers.'
The Independent

The Economics of
Climate Change
The Stern Review

NICHOLAS STERN

Cabinet Office – HM Treasury

ON THE ECONOMICS
OF CLIMATE CHANGE

CAMBRIDGE
UNIVERSITY PRESS

CAMBRIDGE UNIVERSITY PRESS
Cambridge, New York, Melbourne, Madrid, Cape Town, Singapore, São Paulo

Cambridge University Press
The Edinburgh Building, Cambridge CB2 8RU, UK

Published in the United States of America by Cambridge University Press, New York

www.cambridge.org
Information on this title: www.cambridge.org/9780521700801

The Stern Review Report © Crown copyright 2006

The Economics of Climate Change: The Stern Review © Cambridge University Press 2007

First published 2007
Reprinted 2007

Printed in the United Kingdom at the University Press, Cambridge

A catalogue record for this publication is available from the British Library

ISBN-13 978-0-521-70080-1 paperback

Printed on 100% recycled paper

Contents

Preface

This Review was announced by the Chancellor of the Exchequer in July 2005. The Review set out to provide a report to the Prime Minister and Chancellor by Autumn 2006 assessing:

- the economics of moving to a low-carbon global economy, focusing on the medium to long-term perspective, and drawing implications for the timescales for action, and the choice of policies and institutions;
- the potential of different approaches for adaptation to changes in the climate; and
- specific lessons for the UK, in the context of its existing climate change goals.

The terms of reference for the Review included a requirement to consult broadly with stakeholders and to examine the evidence on:

- the implications for energy demand and emissions of the prospects for economic growth over the coming decades, including the composition and energy intensity of growth in developed and developing countries;
- the economic, social and environmental consequences of climate change in both developed and developing countries, taking into account the risks of increased climate volatility and major irreversible impacts, and the climatic interaction with other air pollutants, as well as possible actions to adapt to the changing climate and the costs associated with them;
- the costs and benefits of actions to reduce the net global balance of greenhouse gas emissions from energy use and other sources, including the role of land-use changes and forestry, taking into account the potential impact of technological advances on future costs; and
- the impact and effectiveness of national and international policies and arrangements in reducing net emissions in a cost-effective way and promoting a dynamic, equitable and sustainable global economy, including distributional effects and impacts on incentives for investment in cleaner technologies.

Overall approach to the Review

We have taken a broad view of the economics required to understand the challenges of climate change. Wherever possible, we have based our Review on gathering and structuring existing research material.

Submissions to the Review were invited from 10 October 2005 to 15 January 2006. Sir Nicholas Stern set out his initial views on the approach to the Review in the Oxonia lecture on 31 January 2006, and invited further responses to this lecture up to 17 March 2006.

During the Review, Sir Nicholas and members of the team visited a number of key countries and institutions, including Brazil, Canada, China, the European Commission, France, Germany, India, Japan, Mexico, Norway, Russia, South Africa and the USA. These visits and work in the UK have included a wide range of interactions, including with economists, scientists, policy-makers, business and NGOs.

The report also draws on the analysis prepared for the International Energy Agency publications "Energy Technology Perspectives" and "World Energy Outlook 2006".

There is a solid basis in the literature for the principles underlying our analysis. The scientific literature on the impacts of climate change is evolving rapidly, and the economic modelling has yet to reflect the full range of the new evidence.

In some areas, we found that existing literature did not provide answers. In these cases, we have conducted some of our own research, within the constraints allowed by our timetable and resources. We also commissioned some papers and analysis to feed into the Review. A full list of commissioned work and links to the papers are at www.sternreview.org.uk

Acknowledgements

The team was led by Siobhan Peters. Team members included Vicki Bakhshi, Alex Bowen, Catherine Cameron, Sebastian Catovsky, Di Crane, Sophie Cruickshank, Simon Dietz, Nicola Edmondson, Su-Lin Garbett, Lorraine Hamid, Gideon Hoffman, Daniel Ingram, Ben Jones, Nicola Patmore, Helene Radcliffe, Raj Sathiyarajah, Michelle Stock, Chris Taylor, Tamsin Vernon, Hannah Wanjie, and Dimitri Zenghelis.

We are very grateful to the following organisations for their invaluable contributions throughout the course of the Review: Vicky Pope and all those who have helped us at the Hadley Centre for Climate Prediction; Claude Mandil, Fatih Birol and their team at the International Energy Agency; Francois Bourguignon, Katherine Sierra, Ken Chomitz, Maureen Cropper, Ian Noble and all those who have lent their support at the World Bank; the OECD, EBRD, IADB, and UNEP; Rajendra Pachauri, Bert Metz, Martin Parry and others at the IPCC; Chatham House; as well as Martin Rees and the Royal Society.

Many government departments and public bodies have supported our work, with resources, ideas and expertise. We are indebted to them. They include: HM Treasury, Cabinet Office, Department for Environment Food and Rural Affairs, Department of Trade and Industry, Department for International Development, Department for Transport, Foreign and Commonwealth Office, and the Office of Science and Innovation. We are also grateful for support and assistance from the Bank of England and the Economic and Social Research Council, and for advice from the Environment Agency and Carbon Trust.

We owe thanks to the academics and researchers with whom we have worked closely throughout the Review. A special mention goes to Dennis Anderson who contributed greatly to our understanding of the costs of energy technologies and of technology policy, and has provided invaluable support and advice to the team. Special thanks too to Halsey Rogers and to Tony Robinson who worked with us to edit drafts of the Review. And we are very grateful to: Neil Adger, Sudhir Anand, Nigel Arnell, Terry Barker, John Broome, Andy Challinor, Paul Collier, Sam Fankhauser, Michael Grubb, Roger Guesnerie, Cameron Hepburn, Dieter Helm, Claude Henry, Chris Hope, Paul Johnson, Paul Klemperer, Robert May, David Newbery, Robert Nicholls, Peter Sinclair, Julia Slingo, Max Tse, Rachel Warren and Adrian Wood.

Throughout our work we have learned greatly from academics and researchers who have advised us, including: Philippe Aghion, Shardul Agrawala, Edward Anderson, Tony Atkinson, The Auto Project (Vance

Wagner and Alex Whitworth), Paul Baer, Philip Bagnoli, Hewson Baltzell, Scott Barrett, Marcel Berk, Richard Betts, Ken Binmore, Victor Blinov, Christopher Bliss, Katharine Blundell, Severin Borenstein, Jean-Paul Bouttes, Richard Boyd, Alan Budd, Frances Cairncross, Daniel Cullenward, Larry Dale, Victor Danilov-Daniliyan, Amy Davidsen, Angus Deaton, Michelden Elzen, Richard Eckaus, Jae Edmonds, Jorgen Elmeskov, Paul Epstein, Gunnar Eskeland, Alexander Farrell, Brian Fender, Anthony Fisher, Meredith Fowley, Jeffrey Frankel, Jose Garibaldi, Leila Gohar, Maryanne Grieg-Gran, Bronwyn Hall, Jim Hall, Stephane Hallegate, Kate Hampton, Michael Hanemann, Bill Hare, Geoffrey Heal, Merylyn Hedger, Molly Hellmuth, David Henderson, David Hendry, Marc Henry, Margaret Hiller, Niklas Hoehne, Bjart Holtsmark, Brian Hoskins, Jean-Charles Hourcade, Jo Hossell, Alistair Hunt, Saleem Huq, Mark Jaccard, Sarah Joy, Jiang Kejun, Ian Johnson, Tom Jones, Dale Jorgenson, Paul Joskow, Kassim Kulindwa, Daniel Kammen, Jonathan Köhler, Paul Krugman, Sari Kovats, Klaus Lackner, John Lawton, Tim Lenton, Li Junfeng, Lin Erda, Richard Lindzen, Björn Lomborg, Gordon MacKerron, Joaquim Oliveira Martins, Warwick McKibbin, Malte Meinshausen, Robert Mendelsohn, Evan Mills, Vladimir Milov, James Mirrlees, Richard Morgenstern, Mu Haoming, Robert Muir-Wood, Justin Mundy, Gustavo Nagy, Nebojša Nakicenovic, Karsten Neuhoff, Greg Nimmet, J.C Nkomo, William Nordhaus, David Norse, Anthony Nyong, Pan Jiahua, John Parsons, Cedric Philibert, Robert Pindyck, William Pizer, Oleg Pluzhnikov, Jonathon Porritt, Lant Pritchett, John Reilly, Richard Richels, David Roland-Holst, Nick Rowley, Cynthia Rosenzweig, Joyashree Roy, Jeffrey Sachs, Mark Salmon, Alan Sanstad, Mark Schankerman, John Schellnhueber, Michael Schlesinger, Ken Schomitz, Amartya Sen, Robert Sherman, Keith Shine, P. R. Shukla, Brian Smith, Leonard Smith, Robert Socolow, David Stainforth, Robert Stavins, David Stephenson, Joe Stiglitz, Peter Stone, Roger Street, Josuè Tanaka, Evgeniy Sokolov, Robert Solow, James Sweeney, Richard Tol, Asbjorn Torvanger, Laurence Tubiana, David Vaughnan, Steven Ward, Paul Watkiss, Jim Watson, Martin Weitzman, Hege Westskog, John Weyant, Tony White, Gary Yohe, Ernesto Zedillo, Zhang Anhua, Zhang Qun, Zhao Xingshu, and Zou Ji.

We are grateful to the leaders, officials, academics, NGO staff and business people who assisted us during our visits to: Brazil, Canada, China, the European Commission, France, Germany, Iceland, India, Japan, Mexico, Norway, Russia, South Africa and the USA.

And thanks to the numerous business leaders and representatives who have advised us, including, in particular, John Browne, Paul Golby, Jane Milne, Vincent de Rivaz, James Smith, Adair Turner, and the Corporate Leaders Group.

Also to the NGOs that have offered advice and help including: Christian Aid, The Climate Group, Friends of the Earth, Global Cool, Green Alliance, Greenpeace, IIED, IPPR, New Economics Foundation, Oxfam, Practical Action, RSPB, Stop Climate Chaos, Tearfund, Women's Institute, and WWF UK.

Finally, thanks also go to Australian Antarctic Division for permission to use the picture for the logo and to David Barnett, for designing the logo.

Introduction

The economics of climate change is shaped by the science. That is what dictates the structure of the economic analysis and policies; therefore we start with the science.

Human-induced climate change is caused by the emissions of carbon dioxide and other greenhouse gases (GHGs) that have accumulated in the atmosphere mainly over the past 100 years.

The scientific evidence that climate change is a serious and urgent issue is now compelling. It warrants strong action to reduce greenhouse gas emissions around the world to reduce the risk of very damaging and potentially irreversible impacts on ecosystems, societies and economies. With good policies the costs of action need not be prohibitive and would be much smaller than the damage averted.

Reversing the trend to higher global temperatures requires an urgent, world-wide shift towards a low-carbon economy. Delay makes the problem much more difficult and action to deal with it much more costly. Managing that transition effectively and efficiently poses ethical and economic challenges, but also opportunities, which this Review sets out to explore.

Economics has much to say about assessing and managing the risks of climate change, and about how to design national and international responses for both the reduction of emissions and adaptation to the impacts that we can no longer avoid. If economics is used to design cost-effective policies, then taking action to tackle climate change will enable societies' potential for well-being to increase much faster in the long run than without action; we can be 'green' and grow. Indeed, if we are not 'green', we will eventually undermine growth, however measured.

This Review takes an international perspective on the economics of climate change. Climate change is a global issue that requires a global response. The science tells us that emissions have the same effects from wherever they arise. The implication for the economics is that this is clearly and unambiguously an international collective action problem with all the attendant difficulties of generating coherent action and of avoiding free riding. It is a problem requiring international cooperation and leadership.

Our approach emphasises a number of key themes, which will feature throughout.

- We use a consistent approach towards **uncertainty.** The science of climate change is reliable, and the direction is clear. But we do not know precisely when and where particular impacts will occur. Uncertainty about impacts strengthens the argument for mitigation: this Review is about the economics of the management of very large risks.

- We focus on a quantitative understanding of **risk**, assisted by recent advances in the science that have begun to assign probabilities to the relationships between emissions and changes in the climate system, and to those between the climate and the natural environment.

- We take a systematic approach to the treatment of inter- and intra-generational **equity** in our analysis, informed by a consideration of what various ethical perspectives imply in the context of climate change. Inaction now risks great damage to the prospects of future generations, and particularly to the poorest amongst them. A coherent economic analysis of policy requires that we be explicit about the effects.

Economists describe human-induced climate change as an 'externality' and the global climate as a 'public good'. Those who create greenhouse gas emissions as they generate electricity, power their factories, flare off gases, cut down forests, fly in planes, heat their homes or drive their cars do not have to pay for the costs of the climate change that results from their contribution to the accumulation of those gases in the atmosphere.

But climate change has a number of features that together distinguish it from other externalities. It is global in its causes and consequences; the impacts of climate change are persistent and develop over the long run; there are uncertainties that prevent precise quantification of the economic impacts; and there is a serious risk of major, irreversible change with non-marginal economic effects.

This analysis leads us to *five sets of questions* that shape Parts 2 to 6 of the Review.

- What is our understanding of the risks of the impacts of climate change, their costs, and on whom they fall?

- What are the options for reducing greenhouse-gas emissions, and what do they cost? What does this mean for the economics of the choice of paths to stabilisation for the world? What are the economic opportunities generated by action on reducing emissions and adopting new technologies?

- For mitigation of climate change, what kind of incentive structures and policies will be most effective, efficient and equitable? What are the implications for the public finances?

- For adaptation, what approaches are appropriate and how should they be financed?

- How can approaches to both mitigation and adaptation work at an international level?

Summary of Conclusions

There is still time to avoid the worst impacts of climate change, if we take strong action now.

The scientific evidence is now overwhelming: climate change is a serious global threat, and it demands an urgent global response.

This Review has assessed a wide range of evidence on the impacts of climate change and on the economic costs, and has used a number of different techniques to assess costs and risks. From all of these perspectives, the evidence gathered by the Review leads to a simple conclusion: the benefits of strong and early action far outweigh the economic costs of not acting.

Climate change will affect the basic elements of life for people around the world – access to water, food production, health, and the environment. Hundreds of millions of people could suffer hunger, water shortages and coastal flooding as the world warms.

Using the results from formal economic models, the Review estimates that if we don't act, the overall costs and risks of climate change will be equivalent to losing at least 5% of global GDP each year, now and forever. If a wider range of risks and impacts is taken into account, the estimates of damage could rise to 20% of GDP or more.

In contrast, the costs of action – reducing greenhouse gas emissions to avoid the worst impacts of climate change – can be limited to around 1% of global GDP each year.

The investment that takes place in the next 10-20 years will have a profound effect on the climate in the second half of this century and in the next. Our actions now and over the coming decades could create risks of major disruption to economic and social activity, on a scale similar to those associated with the great wars and the economic depression of the first half of the 20th century. And it will be difficult or impossible to reverse these changes.

So prompt and strong action is clearly warranted. Because climate change is a global problem, the response to it must be international. It must be based on a shared vision of long-term goals and agreement on frameworks that will accelerate action over the next decade, and it must build on mutually reinforcing approaches at national, regional and international level.

Climate change could have very serious impacts on growth and development.

If no action is taken to reduce emissions, the concentration of green-house gases in the atmosphere could reach double its pre-industrial level as early as 2035, virtually committing us to a global average temperature rise of over 2°C. In the longer term, there would be more than a 50% chance that the temperature rise would exceed 5°C. This rise would be very dangerous indeed; it is equivalent to the change in average temperatures from the last ice age to today. Such a radical change in the physical geography of the world must lead to major changes in the human geography – where people live and how they live their lives.

Even at more moderate levels of warming, all the evidence – from detailed studies of regional and sectoral impacts of changing weather patterns through to economic models of the global effects – shows that climate change will have serious impacts on world output, on human life and on the environment.

All countries will be affected. The most vulnerable – the poorest countries and populations – will suffer earliest and most, even though they have contributed least to the causes of climate change. The costs of extreme weather, including floods, droughts and storms, are already rising, including for rich countries.

Adaptation to climate change – that is, taking steps to build resilience and minimise costs – is essential. It is no longer possible to prevent the climate change that will take place over the next two to three decades, but it is still possible to protect our societies and economies from its impacts to some extent – for example, by providing better information, improved planning and more climate-resilient crops and infrastructure. Adaptation will cost tens of billions of dollars a year in developing countries alone, and will put still further pressure on already scarce resources. Adaptation efforts, particularly in developing countries, should be accelerated.

The costs of stabilising the climate are significant but manageable; delay would be dangerous and much more costly.

The risks of the worst impacts of climate change can be substantially reduced if greenhouse gas levels in the atmosphere can be stabilised between 450 and 550 ppm CO_2 equivalent (CO_2e). The current level is 430 ppm CO_2e today, and it is rising at more than 2 ppm each year. Stabilisation in this range would require emissions to be at least 25% below current levels by 2050, and perhaps much more.

Ultimately, stabilisation – at whatever level – requires that annual emissions be brought down to more than 80% below current levels.

This is a major challenge, but sustained long-term action can achieve it at costs that are low in comparison to the risks of inaction. Central estimates of the annual costs of achieving stabilisation between 500 and 550 ppm CO_2e are around 1% of global GDP, if we start to take strong action now.

Costs could be even lower than that if there are major gains in efficiency, or if the strong co-benefits, for example from reduced air pollution, are measured. Costs will be higher if innovation in low-carbon technologies is slower than expected, or if policy-makers fail to make the most of economic instruments that allow emissions to be reduced whenever, wherever and however it is cheapest to do so.

It would already be very difficult and costly to aim to stabilise at 450 ppm CO_2e. If we delay, the opportunity to stabilise at 500-550 ppm CO_2e may slip away.

Action on climate change is required across all countries, and it need not cap the aspirations for growth of rich or poor countries.

The costs of taking action are not evenly distributed across sectors or around the world. Even if the rich world takes on responsibility for absolute cuts in emissions of 60-80% by 2050, developing countries must take significant action too. But developing countries should not be required to bear the full costs of this action alone, and they will not have to. Carbon markets in rich countries are already beginning to deliver flows of finance to support low-carbon development, including through the Clean Development Mechanism. A transformation of these flows is now required to support action on the scale required.

Action on climate change will also create significant business opportunities, as new markets are created in low-carbon energy technologies and other low-carbon goods and services. These markets could grow to be worth hundreds of billions of dollars each year, and employment in these sectors will expand accordingly.

The world does not need to choose between averting climate change and promoting growth and development. Changes in energy technologies and in the structure of economies have created opportunities to decouple growth from greenhouse gas emissions. Indeed, ignoring climate change will eventually damage economic growth.

Tackling climate change is the pro-growth strategy for the longer term, and it can be done in a way that does not cap the aspirations for growth of rich or poor countries.

A range of options exists to cut emissions; strong, deliberate policy action is required to motivate their take-up.

Emissions can be cut through increased energy efficiency, changes in demand, and through adoption of clean power, heat and transport technologies. The power sector around the world would need to be at least 60% decarbonised by 2050 for atmospheric concentrations to stabilise at or below 550 ppm CO_2e, and deep emissions cuts will also be required in the transport sector.

Even with very strong expansion of the use of renewable energy and other low-carbon energy sources, fossil fuels could still make up over half of global energy supply in 2050. Coal will continue to be important in the energy mix around the world, including in fast-growing economies. Extensive carbon capture and storage will be necessary to allow the continued use of fossil fuels without damage to the atmosphere.

Cuts in non-energy emissions, such as those resulting from deforestation and from agricultural and industrial processes, are also essential.

With strong, deliberate policy choices, it is possible to reduce emissions in both developed and developing economies on the scale necessary for stabilisation in the required range while continuing to grow.

Climate change is the greatest market failure the world has ever seen, and it interacts with other market imperfections. Three elements of policy are required for an effective global response. The first is the pricing of carbon, implemented through tax, trading or regulation. The second is policy to support innovation and the deployment of low-carbon technologies. And the third is action to remove barriers to energy efficiency, and to inform, educate and persuade individuals about what they can do to respond to climate change.

Climate change demands an international response, based on a shared understanding of long-term goals and agreement on frameworks for action.

Many countries and regions are taking action already: the EU, California and China are among those with the most ambitious policies that will reduce greenhouse gas emissions. The UN Framework Convention on Climate Change and the Kyoto Protocol provide a basis for international co-operation, along with a range of partnerships and other approaches. But more ambitious action is now required around the world.

Countries facing diverse circumstances will use different approaches to make their contribution to tackling climate change. But action by individual countries is not enough. Each country, however large, is just a part of the problem. It is essential to create a shared international vision of long-term goals, and to build the international frameworks that will help each country to play its part in meeting these common goals.

Key elements of future international frameworks should include:

- *Emissions trading:* Expanding and linking the growing number of emissions trading schemes around the world is a powerful way to promote cost-effective reductions in emissions and to bring forward action in developing countries: strong targets in rich countries could drive flows amounting to tens of billions of dollars each year to support the transition to low-carbon development paths.
- *Technology cooperation:* Informal co-ordination as well as formal agreements can boost the effectiveness of investments in innovation around the world. Globally, support for energy R&D should at least double, and support for the deployment of new low-carbon technologies should increase up to five-fold. International co-operation on product standards is a powerful way to boost energy efficiency.
- *Action to reduce deforestation:* The loss of natural forests around the world contributes more to global emissions each year than the transport sector. Curbing deforestation is a highly cost-effective way to reduce emissions; large-scale international pilot programmes to explore the best ways to do this could get underway very quickly.

- *Adaptation:* The poorest countries are most vulnerable to climate change. It is essential that climate change be fully integrated into development policy, and that rich countries honour their pledges to increase support through overseas development assistance. International funding should also support improved regional information on climate change impacts, and research into new crop varieties that will be more resilient to drought and flood.

I Climate Change – Our Approach

Part I of the Review considers the nature of the scientific evidence for climate change, and the nature of the economic analysis required by the structure of the problem which follows from the science.

The first half of the Review examines the evidence on the economic impacts of climate change itself, and explores the economics of stabilising greenhouse gas concentrations in the atmosphere. The second half of the Review considers the complex policy challenges involved in managing the transition to a low-carbon economy and in ensuring that societies can adapt to the consequences of climate change that can no longer be avoided.

The Review takes an international perspective. Climate change is global in its causes and consequences, and the response requires international collective action. Working together is essential to respond to the scale of the challenge. An effective, efficient and equitable collective response to climate change will require deeper international co-operation in areas including the creation of price signals and markets for carbon, scientific research, infrastructure investment, and economic development.

Climate change presents a unique challenge for economics: it is the greatest example of market failure we have ever seen. The economic analysis must be global, deal with long time horizons, have the economics of risk and uncertainty at its core, and examine the possibility of major, non-marginal change. Analysing climate change requires ideas and techniques from most of the important areas of economics, including many recent advances.

Part I is structured as follows:

- **Chapter 1** examines the latest scientific evidence on climate change. The basic physics and chemistry of the scientific understanding begins in the 19th century when Fourier, Tyndall and Arrhenius laid the foundations. But we must also draw on the very latest science which allows a much more explicit analysis of risk than was possible five years ago.
- **Chapter 2** considers how economic theory can help us analyse the relationship between climate change and the divergent paths for growth and development that will result from 'business as usual' approaches and from strong action to reduce emissions. We look at the range of theories required and explain some of the technical foundations necessary for the economics that the scientific analysis dictates.
- **The technical annex to Chapter 2** addresses the complex issues involved in the comparison of alternative paths and their implications for individuals in different places and generations. Building on Chapter 2, we explore the ethical issues concerning the aggregation of the welfare of individuals across time, place and uncertain outcomes. This annex also provides a technical explanation of the approach to discounting used throughout the Review, and in particular in our own analysis of the costs of climate-change impacts.

1

1 The Science of Climate Change: Scale of the Environment Challenge

KEY MESSAGES

An overwhelming body of scientific evidence now clearly indicates that **climate change is a serious and urgent issue**. The Earth's climate is rapidly changing, mainly as a result of increases in greenhouse gases caused by human activities.

Most climate models show that **a doubling of pre-industrial levels of greenhouse gases is very likely to commit the Earth to a rise of between 2–5°C in global mean temperatures**. This level of greenhouse gases will probably be reached between 2030 and 2060. A warming of 5°C on a global scale would be far outside the experience of human civilisation and comparable to the difference between temperatures during the last ice age and today. Several new studies suggest up to a 20% chance that warming could be greater than 5°C.

If annual greenhouse gas emissions remained at the current level, concentrations would be more than treble pre-industrial levels by 2100, committing the world to 3–10°C warming, based on the latest climate projections.

Some impacts of climate change itself may amplify warming further by triggering the release of additional greenhouse gases. This creates a real risk of even higher temperature changes.

- Higher temperatures cause plants and soils to soak up less carbon from the atmosphere and cause permafrost to thaw, potentially releasing large quantities of methane.

- Analysis of warming events in the distant past indicates that such feedbacks could amplify warming by an additional 1–2°C by the end of the century.

Warming is very likely to intensify the water cycle, reinforcing existing patterns of water scarcity and abundance and increasing the risk of droughts and floods.

Rainfall is likely to increase at high latitudes, while regions with Mediterranean-like climates in both hemispheres will experience significant reductions in rainfall. Preliminary estimates suggest that the fraction of land area in extreme drought at any one time will increase from 1% to 30% by the end of this century. In other regions, warmer air and warmer oceans are likely to drive more intense storms, particularly hurricanes and typhoons.

As the world warms, the risk of abrupt and large-scale changes in the climate system will rise.

- Changes in the distribution of heat around the world are likely to disrupt ocean and atmospheric circulations, leading to large and possibly abrupt shifts in regional weather patterns.

- If the Greenland or West Antarctic Ice Sheets began to melt irreversibly, the rate of sea level rise could more than double, committing the world to an eventual sea level rise of 5–12 m over several centuries.

The body of evidence and the growing quantitative assessment of risks are now sufficient to give clear and strong guidance to economists and policy-makers in shaping a response.

1.1   Introduction

Understanding the scientific evidence for the human influence on climate is an essential starting point for the economics, both for establishing that there is indeed a problem to be tackled and for comprehending its risk and scale. It is the science that dictates the type of economics and where the analyses should focus, for example, on the economics of risk, the nature of public goods or how to deal with externalities, growth and development and intra- and inter-generational equity. The relevance of these concepts, and others, is discussed in Chapter 2.

This chapter begins by describing the changes observed in the Earth's system, examining briefly the debate over the attribution of these changes to human activities. It is a debate that, after more than a decade of research and discussion, has reached the conclusion there is no other plausible explanation for the observed warming for at least the past 50 years. The question of precisely how much the world will warm in the future is still an area of active research. The Third Assessment Report (TAR) of the Intergovernmental Panel on Climate Change (IPCC)[1] in 2001 was the last comprehensive assessment of the state of the science. This chapter uses the 2001 report as a base and builds on it with more recent studies that embody a more explicit treatment of risk. These studies support the broad conclusions of that report, but demonstrate a sizeable probability that the sensitivity of the climate to greenhouse gases is greater than previously thought. Scientists have also begun to quantify the effects of feedbacks with the natural carbon cycle, for example, exploring how warming may affect the rate of absorption of carbon dioxide by forests and soils. These types of feedbacks are predicted to further amplify warming, but are not typically included in climate models to date. The final section of this chapter provides a starting point for Part II, by exploring what basic science reveals about how warming will affect people around the world.

1.2   The Earth's climate is changing

An overwhelming body of scientific evidence indicates that the Earth's climate is rapidly changing, predominantly as a result of increases in greenhouse gases caused by human activities.

Human activities are changing the composition of the atmosphere and its properties. Since pre-industrial times (around 1750), carbon dioxide concentrations have increased by just over one third from 280 parts per million (ppm) to 380 ppm today (Figure 1.1), predominantly as a result of burning fossil fuels, deforestation, and other changes in land-use.[2] This has been accompanied by rising concentrations of other greenhouse gases, particularly methane and nitrous oxide.

There is compelling evidence that the rising levels of greenhouse gases will have a warming effect on the climate through increasing the amount of infrared

---

[1] The fourth assessment is due in 2007. The scientific advances since the TAR are discussed in Schellnhuber *et al.* (2006).

[2] The human origin of the accumulation of carbon dioxide in the atmosphere is demonstrated through, for example, the isotope composition and hemispheric gradient of atmospheric carbon dioxide (IPCC 2001a).

radiation (heat energy) trapped by the atmosphere: "the greenhouse effect" (Figure 1.2). In total, the warming effect due to all (Kyoto) greenhouse gases emitted by human activities is now equivalent to around 430 ppm of carbon dioxide (hereafter, CO_2 equivalent or CO_2e)[3] (Figure 1.1) and rising at around 2.3 ppm per year[4]. Current levels of greenhouse gases are higher now than at any time in at least the past 650,000 years.[5]

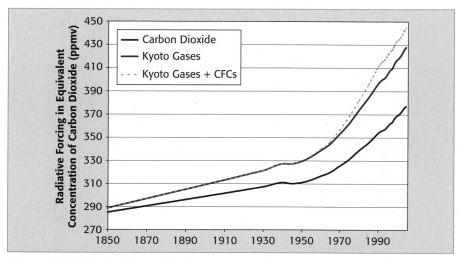

Figure 1.1 Rising levels of greenhouse gases

The figure shows the warming effect of greenhouse gases (the 'radiative forcing') in terms of the equivalent concentration of carbon dioxide (a quantity known as the CO_2 equivalent). The blue line shows the value for carbon dioxide only. The red line is the value for the six Kyoto greenhouse gases (carbon dioxide, methane, nitrous oxide, PFCs, HFCs and SF_6)[6] and the grey line includes CFCs (regulated under the Montreal Protocol). The uncertainty on each of these is up to 10%[7]. The rate of annual increase in greenhouse gas levels is variable year-on-year, but is increasing.

Source: Dr L Gohar and Prof K Shine, Dept. of Meteorology, University of Reading

---

[3] In this Review, the total radiative effect of greenhouse gases is quoted in terms of the equivalent concentration (in ppm) of carbon dioxide and will include the six Kyoto greenhouse gases. It will not include other human influences on the radiation budget of the atmosphere, such as ozone, land properties (i.e. albedo), aerosols or the non-greenhouse gas effects of aircraft unless otherwise stated, because the radiative forcing of these substances is less certain, their effects have a shorter timescale and they are unlikely to form a substantial component of the radiative forcing at equilibrium (they will be substantially decreasing over the timescale of stabilisation). The definition excludes greenhouse gases controlled under the Montreal Protocol (e.g. CFCs). Note however, that such effects are included in future temperature projections. The CO_2 equivalence here measures only the instantaneous radiative effect of greenhouse gases in the atmosphere and ignores the lifetimes of the gases in the atmosphere (i.e. their future effect).

[4] The 1980–2004 average, based on data provided by Prof K Shine and Dr L Gohar, Dept. of Meteorology, University of Reading.

[5] Siegenthaler *et al.* (2005) using data from ice cores. The same research groups recently presented analyses at the 2006 conference of the European Geosciences Union, which suggest that carbon dioxide levels are unprecedented for 800,000 years.

[6] Kyoto greenhouse gases are the six main greenhouse gases covered by the targets set out in the Kyoto Protocol.

[7] Based on the error on the radiative forcing (in CO_2 equivalent) of all long-lived greenhouse gases from Figure 6.6, IPCC (2001b).

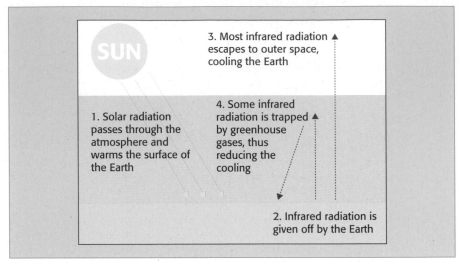

Figure 1.2 The Greenhouse Effect

Source: Based on DEFRA (2005)

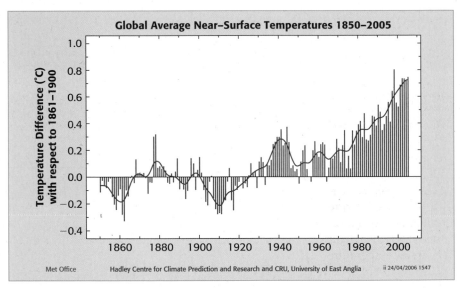

Figure 1.3 The Earth has warmed 0.7°C since around 1900.

The figure above shows the change in global average near-surface temperature from 1850 to 2005. The individual annual averages are shown as red bars and the blue line is the smoothed trend. The temperatures are shown relative to the average over 1861–1900.

Source: Brohan et al. (2006)

As anticipated by scientists, global mean surface temperatures have risen over the past century. The Earth has warmed by 0.7°C since around 1900 (Figure 1.3). Global mean temperature is referred to throughout the Review and is used as a rough index of the scale of climate change. This measure is an average over both space (globally across the land-surface air, up to about 1.5 m above the ground, and sea-surface temperature to around 1 m depth) and time (an annual mean over a defined time period). All temperatures are given relative to pre-industrial, unless otherwise stated. As discussed later in this chapter, this warming does not occur evenly across the planet.

Over the past 30 years, global temperatures have risen rapidly and continuously at around 0.2°C per decade, bringing the global mean temperature to what is probably at or near the warmest level reached in the current interglacial period, which began around 12,000 years ago[8]. All of the ten warmest years on record have occurred since 1990. The first signs of changes can be seen in many physical and biological systems, for example many species have been moving poleward by 6 km on average each decade for the past 30–40 years. Another sign is changing seasonal events, such as flowering and egg laying, which have been occurring 2–3 days earlier each decade in many Northern Hemisphere temperate regions.[9]

The IPCC concluded in 2001 that there is new and stronger evidence that most of the warming observed over at least the past 50 years is attributable to human activities.[10] Their confidence is based on several decades of active debate and effort to scrutinise the detail of the evidence and to investigate a broad range of hypotheses.

Over the past few decades, there has been considerable debate over whether the trend in global mean temperatures can be attributed to human activities. Attributing trends to a single influence is difficult to establish unequivocally because the climate system can often respond in unexpected ways to external influences and has a strong natural variability. For example, Box 1.1 briefly describes the debate over whether the observed increase in temperatures over the last century is beyond that expected from natural variability alone throughout the last Millennium.

BOX 1.1 The "Hockey Stick" Debate.

Much discussion has focused on whether the current trend in rising global temperatures is unprecedented or within the range expected from natural variations. This is commonly referred to as the "Hockey Stick" debate as it discusses the validity of figures that show sustained temperatures for around 1000 years and then a sharp increase since around 1800 (for example, Mann *et al.* 1999, shown as a purple line in the figure below).

Some have interpreted the "Hockey Stick" as definitive proof of the human influence on climate. However, others have suggested that the data and methodologies used to produce this type of figure are questionable (e.g. von Storch *et al.* 2004), because widespread, accurate temperature records are only available for the past 150 years. Much of the temperature record is recreated from a range of 'proxy' sources such as tree rings, historical records, ice cores, lake sediments and corals.

Climate change arguments do not rest on "proving" that the warming trend is unprecedented over the past Millennium. Whether or not this debate is now settled, this is only one in a number of lines of evidence for human induced climate change. The key conclusion, that the build-up of greenhouse gases in the atmosphere will lead to several degrees of warming, rests on the laws of physics and chemistry and a broad range of evidence beyond one particular graph.

[8] Hansen et al. (2006)
[9] Parmesan and Yohe (2003) and Root et al. (2005) have correlated a shift in timing and distribution of 130 different plant and animal species with observed climate change.
[10] IPCC (2001a) – this key conclusion has been supported in the Joint Statement of Science Academies in 2005 and a report from the US Climate Change Science Programme (2006).

Reconstruction of annual temperature changes in the Northern Hemisphere for the past millennium using a range of proxy indicators by several authors. The figure suggests that the sharp increase in global temperatures since around 1850 has been unprecedented over the past millennium. Source: IDAG (2005)

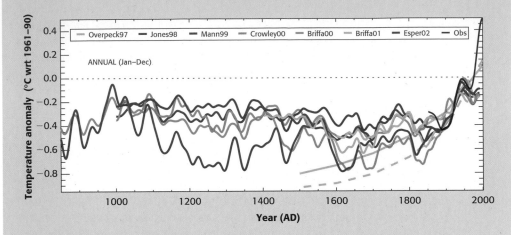

Recent research, for example from the Ad hoc detection and attribution group (IDAG), uses a wider range of proxy data to support the broad conclusion that the rate and scale of 20th century warming is greater than in the past 1000 years (at least for the Northern Hemisphere). Based on this kind of analysis, the US National Research Council (2006)[11] concluded that there is a high level of confidence that the global mean surface temperature during the past few decades is higher than at any time over the preceding four centuries. But there is less confidence beyond this. However, they state that in some regions the warming is unambiguously shown to be unprecedented over the past millennium.

Much of the debate over the attribution of climate change has now been settled as new evidence has emerged to reconcile outstanding issues. It is now clear that, while natural factors, such as changes in solar intensity and volcanic eruptions, can explain much of the trend in global temperatures in the early nineteenth century, the rising levels of greenhouse gases provide the only plausible explanation for the observed trend for at least the past 50 years. Over this period, the sustained globally averaged warming contrasts strongly with the slight cooling expected from natural factors alone. Recent modelling by the Hadley Centre and other research institutes supports this. These models show that the observed trends in temperatures at the surface and in the oceans[12], as well as the spatial distribution of warming[13], cannot be replicated without the inclusion of both human and natural effects.

Taking into account the rising levels of aerosols, which cool the atmosphere,[14] and the observed heat uptake by the oceans, the calculated warming effect of greenhouse gases is more than enough to explain the observed temperature rise.

[11] National Research Council (2006) – a report requested by the US Congress
[12] Barnett et al. (2005a)
[13] For example, Ad hoc detection and attribution group (2005)
[14] Aerosols are tiny particles in the atmosphere also created by human activities (e.g. sulphate aerosol emitted by many industrial processes). They have several effects on the atmosphere, one of which is to reflect solar radiation and therefore, cool the surface. This effect is thought to have offset some of the warming effect of greenhouse gases, but the exact amount is uncertain.

1.3 Linking Greenhouse Gases and Temperature

The causal link between greenhouse gases concentrations and global temperatures is well established, founded on principles established by scientists in the nineteenth century.

The greenhouse effect is a natural process that keeps the Earth's surface around 30°C warmer than it would be otherwise. Without this effect, the Earth would be too cold to support life. Current understanding of the greenhouse effect has its roots in the simple calculations laid out in the nineteenth century by scientists such as Fourier, Tyndall and Arrhenius[15]. Fourier realised in the 1820s that the atmosphere was more permeable to incoming solar radiation than outgoing infrared radiation and therefore trapped heat. Thirty years later, Tyndall identified the types of molecules (known as greenhouse gases), chiefly carbon dioxide and water vapour, which create the heat-trapping effect. Arrhenius took this a step further showing that doubling the concentration of carbon dioxide in the atmosphere would lead to significant changes in surface temperatures.

Since Fourier, Tyndall and Arrhenius made their first estimates, scientists have improved their understanding of how greenhouse gases absorb radiation, allowing them to make more accurate calculations of the links between greenhouse gas concentrations and temperatures. For example, it is now well established that the warming effect of carbon dioxide rises approximately logarithmically with its concentration in the atmosphere[16]. From simple energy-balance calculations, the direct warming effect of a doubling of carbon dioxide concentrations would lead to an average surface warming of around 1°C.

But the atmosphere is much more complicated than these simple models suggest. The resulting warming will in fact be much greater than 1°C because of the interaction between feedbacks in the atmosphere that act to amplify or dampen the direct warming (Figure 1.4). The main positive feedback comes from water vapour, a very powerful greenhouse gas itself. Evidence shows that, as expected from basic physics, a warmer atmosphere holds more water vapour and traps more heat, amplifying the initial warming.[17]

Using climate models that follow basic physical laws, scientists can now assess the likely range of warming for a given level of greenhouse gases in the atmosphere.

It is currently impossible to pinpoint the exact change in temperature that will be associated with a level of greenhouse gases. Nevertheless, increasingly sophisticated climate models are able to capture some of the chaotic nature of the climate, allowing scientists to develop a greater understanding of the many

---

[15] For example, Pearce (2003), Pierrehumbert (2004)

[16] i.e. the incremental increase in radiative forcing due to an increase in concentration (from pre-industrial) will fall to around half of the initial increase when concentrations reach around 600 ppm, a quarter at 1200 ppm and an eighth at 2400 ppm. Note that other greenhouse gases, such as methane and nitrous oxide, have a linear relationship.

[17] It has been suggested that water vapour could act as a negative feedback on warming, on the basis that the upper atmosphere would dry out as it warms (Lindzen 2005). Re-analysis of satellite measurements published last year indicated that in fact the opposite is happening (Soden *et al.* 2005). Over the past two decades, the air in the upper troposphere has become wetter, not drier, countering Lindzen's theory and confirming that water vapour is having a *positive* feedback effect on global warming. This positive feedback is a major driver of the indirect warming effects from greenhouse gases.

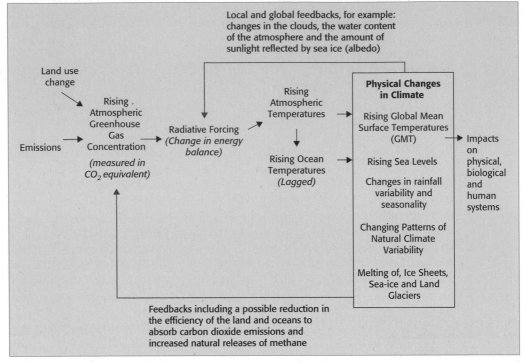

Figure 1.4 The link between greenhouse gases and climate change.

complex interactions within the system and estimate how changing greenhouse gas levels will affect the climate. Climate models use the laws of nature to simulate the radiative balance and flows of energy and materials. These models are vastly different from those generally used in economic analyses, which rely predominantly on curve fitting. Climate models cover multiple dimensions, from temperature at different heights in the atmosphere, to wind speeds and snow cover. Also, climate models are tested for their ability to reproduce past climate variations across several dimensions, and to simulate aspects of present climate that they have not been specifically tuned to fit.

The accuracy of climate predictions is limited by computing power. This, for example, restricts the scale of detail of models, meaning that small-scale processes must be included through highly simplified calculations. It is important to continue the active research and development of more powerful climate models to reduce the remaining uncertainties in climate projections.

The sensitivity of mean surface temperatures to greenhouse gas levels is benchmarked against the warming expected for a doubling of carbon dioxide levels from pre-industrial (roughly equivalent to 550 ppm CO_2e). This is called the "climate sensitivity" and is an important quantity in accessing the economics of climate change. By comparing predictions of different state-of-the-art climate models, the IPCC TAR concluded that the likely range of climate sensitivity is 1.5°–4.5°C. This range is much larger than the 1°C direct warming effect expected from a doubling of carbon dioxide concentrations, thus emphasising the importance of feedbacks within the atmosphere. For illustration, using this range of sensitivities, if green-house gas levels could be stabilised at today's levels (430 ppm CO_2e), global mean temperatures would eventually rise to around 1–3°C above pre-industrial (up to 2°C more than today)[18]. This is not the same as the "warming commitment" today from past emissions, which includes the current levels of aerosols in the atmosphere (discussed later in this chapter).

Results from new risk based assessments suggest there is a significant chance that the climate system is more sensitive than was originally thought.

Since 2001, a number of studies have used both observations and modelling to explore the full range of climate sensitivities that appear realistic given current knowledge (Box 1.2). This new evidence is important in two ways: firstly, the conclusions are broadly consistent with the IPCC TAR, but indicate that higher climate sensitivities cannot be excluded; and secondly, it allows a more explicit treatment of risk. For example, eleven recent studies suggest only between a 0% and 2% chance that the climate sensitivity is less than 1°C, but between a 2% and 20% chance that climate sensitivity is greater than 5°C[19]. These sensitivities imply that there is up to a one-in-five chance that the world would experience a warming

BOX 1.2 Recent advances in estimating climate sensitivity

Climate sensitivity remains an area of active research. Recently, new approaches have used climate models and observations to develop a better understanding of climate sensitivity.

- Several studies have estimated climate sensitivity by benchmarking climate models against the observed warming trend of the 20th century, e.g. Forest et al. (2006) and Knutti et al. (2002).

- Building on this work, modellers have systematically varied a range of uncertain parameters in more complex climate models (such as those controlling cloud behaviour) and run ensembles of these models, e.g. Murphy *et al.* (2004) and Stainforth et al. (2005). The outputs are then checked against observational data, and the more plausible outcomes (judged by their representation of current climate) are weighted more highly in the probability distributions produced.

- Some studies, e.g. Annan & Hargreaves (2006), have used statistical techniques to estimate climate sensitivity through combining several observational datasets (such as the 20th century warming, cooling following volcanic eruptions, warming after last glacial maximum).

These studies provide an important first attempt to apply a probabilistic framework to climate projections. Their outcome is a series of probability distribution functions (PDFs) that aim to capture some of the uncertainty in current estimates. Meinshausen (2006) brings together the results of eleven recent studies (below). The red and blue lines are probability distributions based on the IPCC TAR (Wigley and Raper (2001)) and recent Hadley Centre ensemble work (Murphy *et al.* (2004)), respectively. These two distributions lie close to the centre of the results from the eleven studies.

[18] Calculated using method shown in Meinshausen (2006).
[19] Meinshausen (2006)

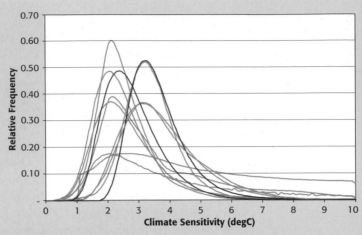

Source: Reproduced from Meinshausen (2006)

The distributions share the characteristic of a long tail that stretches up to high tempera-tures. This is primarily because of uncertainty over clouds[20] and the cooling effect of aerosols. For example, if cloud properties are sensitive to climate change, they could create an important addition feedback. Similarly, if the cooling effect of aerosols is large it will have offset a substantial part of past warming due to greenhouse gases, making high climate sen-sitivity compatible with the observed warming.

in excess of 3°C above pre-industrial even if greenhouse gas concentrations were stabilised at today's level of 430 ppm CO_2e.

In the future, climate change itself could trigger additional increases in greenhouse gases in the atmosphere, further amplifying warming. These potentially powerful feedbacks are less well understood and only beginning to be quantified.

Climate change projections must also take into account the strong possibility that climate change itself may accelerate future warming by reducing natural absorption and releasing stores of carbon dioxide and methane. These feedbacks are not incorporated into most climate models to date because their effects are only just beginning to be understood and quantified.

Rising temperatures and changes in rainfall patterns are expected to weaken the ability of the Earth's natural sinks to absorb carbon dioxide (Box 1.3), causing a larger fraction of human emissions to accumulate in the atmosphere. While this finding is not new, until recently the effect was not quantified. New models, which explicitly include interactions between carbon sinks and climate, suggest that by 2100, greenhouse gas concentrations will be 20–200 ppm higher than they would have otherwise been, amplifying warming by 0.1–1.5°C.[21] Some models

[20] An increase in low clouds would have a negative feedback effect, as they have little effect on infrared radiation but block sunlight, causing a local cooling. Conversely, an increase in high clouds would trap more infrared radiation, amplifying warming.
[21] Friedlingstein *et al.* (2006)

predict future reductions in tropical rainforests, particularly the Amazon, also releasing more carbon into the atmosphere[22]. Chapter 8 discusses the implications of weakened carbon sinks for stabilising greenhouse gas concentrations.

Widespread thawing of permafrost regions is likely to add to the extra warming caused by weakening of carbon sinks. Large quantities of methane (and carbon dioxide) could be released from the thawing of permafrost and frozen peat bogs. One estimate, for example, suggests that if all the carbon accumulated in peat alone since the last ice age were released into the atmosphere, this would raise greenhouse gas levels by 200 ppm CO_2e.[23] Additional emissions may be seen from warming tropical wetlands, but this is more uncertain. Together, wetlands and frozen lands store more carbon than has been released already by human activities since industrialisation began. Substantial thawing of permafrost has already begun in some areas; methane emissions have increased by 60% in northern Siberia since the mid-1970s[24]. Studies of the overall scale and timing of future releases are scarce, but initial estimates suggest that methane emissions (currently 15% of all emissions in terms of CO_2 equivalent[25]) may increase by around 50% by 2100 (Box 1.3).

Preliminary estimates suggest that these "positive feedbacks" could lead to an addition rise in temperatures of 1–2°C by 2100.

Recent studies have used information from past ice ages to estimate how much extra warming would be produced by such feedbacks. Warming following previous ice ages triggered the release of carbon dioxide and methane from the land and oceans, raising temperatures by more than that expected from solar effects alone. If present day climate change triggered feedbacks of a similar size, temperatures in 2100 would be 1–2°C higher than expected from the direct warming caused by greenhouse gases.[26]

There are still many unanswered questions about these positive feedbacks between the atmosphere, land and ocean. The combined effect of high climate sensitivity and carbon cycle feedbacks is only beginning to be explored, but first indications are that this could lead to far higher temperature increases than are currently anticipated (discussed in chapter 6). It remains unclear whether warming could initiate a self-perpetuating effect that would lead to a much larger temperature rise or even runaway warming, or if some unknown feedback could reduce the sensitivity substantially[27]. Further research is urgently required to quantify the combined effects of these types of feedbacks.

---

[22] Cox et al. (2000) with the Hadley Centre model and Scholze et al (2006) with several models.

[23] Gorham et al. (1991)

[24] Walter et al. (2006)

[25] Emissions measured in CO_2 equivalent are weighted by their global warming potential (see chapter 8).

[26] These estimates come from recent papers by Torn and Harte (2006) and Scheffer et al. (2006), which estimate the scale of positive feedbacks from release of carbon dioxide and methane from past natural climate change episodes, e.g. Little Ice Age and previous inter-glacial period, into current climate models.

[27] One study to date has examined this question and suggested that a run away effect is unlikely, at least for the land-carbon sink (Cox et al. 2006). It remains unclear how the risk of run-away climate change would change with the inclusion of other feedbacks.

BOX 1.3 Changes in the earth system that could amplify global warming

Weakening of Natural Land-Carbon Sinks: Initially, higher levels of carbon dioxide in the atmosphere will act as a fertiliser for plants, increasing forest growth and the amount of carbon absorbed by the land. A warmer climate will increasingly offset this effect through an increase in plant and soil respiration (increasing release of carbon from the land). Recent modelling suggests that net absorption may initially increase because of the carbon fertilisation effects (chapter 3). But, by the end of this century it will reduce significantly as a result of increased respiration and limits to plant growth (nutrient and water availability).[28]

Weakening of Natural Ocean-Carbon Sinks: The amount of carbon dioxide absorbed by the oceans is likely to weaken in the future through a number of chemical, biological and physical changes. For example, chemical uptake processes may be exhausted, warming surface waters will reduce the rate of absorption and CO_2 absorbing organisms are likely to be damaged by ocean acidification[29]. Most carbon cycle models agree that climate change will weaken the ocean sink, but suggest that this would be a smaller effect than the weakening of the land sink[30].

Release of Methane from Peat Deposits, Wetlands and Thawing Permafrost: Thawing permafrost and the warming and drying of wetland areas could release methane (and carbon dioxide) to the atmosphere in the future. Models suggest that up to 90% of the upper layer of permafrost will thaw by 2100.[31] These regions contain a substantial store of carbon. One set of estimates suggests that wetlands store equivalent to around 1600 $GtCO_2$e (where Gt is one billion tonnes) and permafrost soils store a further 1500 $GtCO_2$e[32]. Together these stores comprise more than double the total cumulative emissions from fossil fuel burning so far. Recent measurements show a 10–15% increase in the area of thaw lakes in northern and western Siberia. In northern Siberia, methane emissions from thaw lakes are estimated to have increased by 60% since the mid 1970's[33]. It remains unclear at what rate methane would be released in the future. Preliminary estimates indicate that, in total, methane emissions each year from thawing permafrost and wetlands could increase by around 4–10 $GtCO_2$e, more than 50% of current methane emissions and equivalent to 10–25% of current man-made emissions.[34]

Release of Methane from Hydrate Stores: An immense quantity of methane (equivalent to tens of thousands of $GtCO_2$, twice as much as in coal, oil and gas reserves) may also be trapped under the oceans in the form of gas hydrates. These exist in regions sufficiently cold and under enough high pressures to keep them stable. There is considerable uncertainty whether these deposits will be affected by climate change at all. However, if ocean warming penetrated deeply enough to destabilise even a small amount of this methane and release it to the atmosphere, it would lead to a rapid increase in warming.[35] Estimates of the size of potential releases are scarce, but are of a similar scale to those from wetlands and permafrost.

---

[28] Friedlingstein *et al.* (2006) found that all eleven climate models that explicitly include carbon cycle feedbacks showed a weakening of carbon sinks.

[29] Orr *et al.* (2005)

[30] Friedlingstein *et al.* (2006)

[31] Lawrence and Slater (2005), based on IPCC A2 Scenario

[32] Summarised in Davidson and Janssens (2006) (wetlands) and Archer (2005) (permafrost) – CO_2 equivalent emissions (chapter 7).

[33] Walter et al. (2006) and Smith et al. (2005)

[34] Estimates of potential methane emissions from thawing permafrost range around 2–4$GtCO_2$/yr. Wetlands emit equivalent to 2–6 $GtCO_2$/yr and studies project that this may rise by up to 80%. Davidson & Janssens (2006), Gedney et al. (2004) and Archer (2005).

[35] Hadley Centre (2005)

1.4   Current Projections

Additional warming is already in the pipeline due to past and present emissions.

The full warming effect of past emissions is yet to be realised. Observations show that the oceans have taken up around 84% of the total heating of the Earth's system over the last 40 years[36]. If global emissions were stopped today, some of this heat would be exchanged with the atmosphere as the system came back into equilibrium, causing an additional warming. Climate models project that the world is committed to a further warming of 0.5°–1°C over several decades due to past emissions[37]. This warming is smaller than the warming expected if concentrations were stabilised at 430 ppm CO_2e, because atmospheric aerosols mask a proportion of the current warming effect of greenhouse gases. Aerosols remain in the atmosphere for only a few weeks and are not expected to be present in significant levels at stabilisation[38].

If annual emissions continued at today's levels, greenhouse gas levels would be close to double pre-industrial levels by the middle of the century. If this concentration were sustained, temperatures are projected to eventually rise by 2–5°C or even higher.

Projections of future warming depend on projections of global emissions (discussed in chapter 7). If annual emissions were to remain at today's levels, greenhouse gas levels would reach close to 550 ppm CO_2e by 2050[39]. Using the lower and upper 90% confidence bounds based on the IPCC TAR range and recent research from the Hadley Centre, this would commit the world to a warming of around 2–5°C (Table 1.1). As demonstrated in Box 1.2, these two climate sensitivity distributions lie close to the centre of recent projections and are used throughout this Review to give illustrative temperature projections. Positive feedbacks, such as methane emissions from permafrost, could drive temperatures even higher.

Near the middle of this range of warming (around 2–3°C above today), the Earth would reach a temperature not seen since the middle Pliocene around 3 million years ago[40]. This level of warming on a global scale is far outside the experience of human civilisation.

However, these are conservative estimates of the expected warming, because in the absence of an effective climate policy, changes in land use and the growth in population and energy consumption around the world will drive greenhouse gas emissions far higher than today. This would lead greenhouse gas levels to attain higher levels than suggested above. The IPCC projects that without intervention greenhouse gas levels will rise to 550–700 ppm CO_2e by 2050 and 650–1200 ppm CO_2e by 2100[41]. These projections and others are discussed in Chapter 7, which concludes that, without mitigation, greenhouse gas levels are likely to be towards the upper end of these ranges. If greenhouse gas levels were to reach 1000 ppm, more than treble pre-industrial levels, the Earth would be committed to around a

---

[36] Barnett et al. (2005a) and Levitus *et al.* (2005)

[37] Wigley (2005) and Meehl *et al.* (2005) look at the amount of warming "in the pipeline" using different techniques.

[38] In many countries, aerosol levels have already been reduced by regulation because of their negative health effects.

[39] For example, 45 years at 2.5 ppm/yr gives 112.5 ppm. Added to the current level, this gives 542.5 ppm in 2050.

[40] Hansen *et al.* (2006)

[41] Based on the IPCC TAR central radiative forcing projections for the six illustrative SRES scenarios (IPCC 2001b).

Table 1.1 Temperature projections at stabilisation

Meinshausen (2006) used climate sensitivity estimates from eleven recent studies to estimate the range of equilibrium temperature changes expected at stabilisation. The table below gives the equilibrium temperature projections using the 5–95% climate sensitivity ranges based on the IPCC TAR (Wigley and Raper (2001)), Hadley Centre (Murphy *et al.* 2004) and the range over all eleven studies. Note that the temperature changes expected prior to equilibrium, for example in 2100, would be lower.

Stabilisation level (ppm CO₂ equivalent)	Temperature increase at equilibrium relative to pre-industrial (°C)		
	IPCC TAR 2001 (Wigley and Raper)	Hadley Centre Ensemble	Eleven Studies
400	0.8–2.4	1.3–2.8	0.6–4.9
450	1.0–3.1	1.7–3.7	0.8–6.4
500	1.3–3.8	2.0–4.5	1.0–7.9
550	1.5–4.4	2.4–5.3	1.2–9.1
650	1.8–5.5	2.9–6.6	1.5–11.4
750	2.2–6.4	3.4–7.7	1.7–13.3
1000	2.8–8.3	4.4–9.9	2.2–17.1

3–10°C of warming or more, even without considering the risk of positive feed-backs (Table 1.1).

1.5 Large Scale Changes and Regional Impacts

This chapter has so far considered only the expected changes in global average surface temperatures. However, this can often mask both the variability in temperature changes across the earth's surface and changes in extremes. In addition, the impacts on people will be felt mainly through water, driven by shifts in regional weather patterns, particularly rainfall and extreme events (more detail in Part II).

In general, higher latitudes and continental regions will experience temperature increases significantly greater than the global average.

Future warming will occur unevenly and will be superimposed on existing temperature patterns. Today, the tropics are around 15°C warmer than the mid-latitudes and more than 25°C warmer than the high latitudes. In future, the smallest temperature increases will generally occur over the oceans and some tropical coastal regions. The largest temperature increases are expected in the high latitudes (particularly around the poles), where melting snow and sea ice will reduce the reflectivity of the surface, leading to a greater than average warming. For a global average warming of around 4°C, the oceans and coasts generally warm by around 3°C, the mid-latitudes warm by more than 5°C and the poles by around 8°C.

The risk of heat waves is expected to increase (Figure 1.5). For example, new modelling work by the Hadley Centre shows that the summer of 2003 was Europe's hottest for 500 years and that human-induced climate change has already more than doubled the chance of a summer as hot as 2003 in Europe

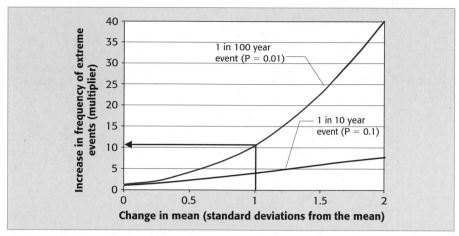

Figure 1.5 Rising probability of heatwaves

There will be more extreme heat days (relative to today) and fewer very cold days, as the distribution of temperatures shifts upwards. The figure below illustrates the change in frequency of a one-in-ten (blue) and one-in-one-hundred (red) year event. The black arrow shows that if the mean temperature increases by one standard deviation (equal to, for example, only 1°C for summer temperatures in parts of Europe), then the probability of today's one-in-one-hundred year event (such as a severe heatwave) will increase ten-fold. This result assumes that the shape of the temperature distribution will remain constant. However, in many areas, the drying of land is expected to skew the distribution towards higher temperatures, further increasing the frequency of temperature extremes[44].

Source: Based on Wigley (1985) assuming normally distributed events.

occurring.[42] By 2050, under a relatively high emissions scenario, the temperatures experienced during the heatwave of 2003 could be an average summer. The rise in heatwave frequency will be felt most severely in cities, where temperatures are further amplified by the urban heat island effect.

Changes in rainfall patterns and extreme weather events will lead to more severe impacts on people than that caused by warming alone.

Warming will change rainfall patterns, partly because warmer air holds more moisture, and also because the uneven distribution of warming around the world will lead to shifts in large-scale weather regimes. Most climate models predict increases in rainfall at high latitudes, while changes in circulation patterns are expected to cause a drying of the subtropics, with northern Africa and the Mediterranean experiencing significant reductions in rainfall. There is more uncertainty about changes in rainfall in the tropics (Figure 1.6), mainly because of complicated interactions between climate change and natural cycles like the El Niño, which dominate climate in the tropics.[43] For example, an El Niño event

[42] According to Stott *et al.* (2004), climate change has increased the chance of the 2003 European heatwave occurring by between 2 and 8 times. In 2003, temperatures were 2.3°C warmer than the long-term average.

[43] In an El Niño year (around once every 3–7 years), the pattern of tropical sea surface temperatures changes, with the eastern Pacific warming significantly. This radically alters large-scale atmospheric circulations across the globe, and causes rainfall patterns to shift, with some regions experiencing flooding and others severe droughts. As the world warms, many models suggest that the East Pacific may warm more intensely than the West Pacific, mimicking the pattern of an El Niño, although significant uncertainties remain. Models do not yet agree on the nature of changes in the frequency or intensity of the El Niño (Collins and the CMIP Modelling Groups 2005).

[44] Schär C *et al.* (2004)

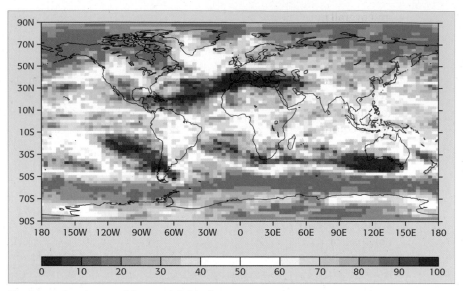

Figure 1.6 Consistency of future rainfall estimates

The figure above indicates the percentage of models (out of a total of 23) that predict that annual rainfall will increase by 2100 (for a warming of around 3.5°C above pre-industrial). Blue shading indicates that most models (>75%) show an increase in annual rainfall, while red shading indicates that most models show a decrease in rainfall. Lightly shaded areas are where models show inconsistent results. The figure shows only the direction of change and gives no information about its scale. In general, there is agreement between most of the models that high latitudes will see increases in rainfall, while much of the subtropics will see reductions in rainfall. Changes in rainfall in the tropics are still uncertain.

Source: Climate Directorate of the National Centre for Atmospheric Science, University of Reading

with strong warming in the central Pacific can cause the Indian monsoon to switch into a "dry mode", characterised by significant reductions in rainfall leading to severe droughts. These delicate interactions could cause abrupt shifts in rainfall patterns. This is an area that urgently needs more research because of the potential effect on billions of people, especially in South and East Asia (more detail in Part II).

Greater evaporation and more intense rainfall will increase the risk of droughts and flooding in areas already at risk.[45] It could also increase the size of areas at risk; one recent study, the first of its kind, estimates that the fraction of land area in moderate drought at any one time will increase from 25% at present to 50% by the 2090s, and the fraction in extreme drought from 3% to 30%[46].

Hurricanes and other storms are likely to become more intense in a warmer, more energised world, as the water cycle intensifies, but changes to their location

---

[45] Huntington (2006) reviewed more than 50 peer-reviewed studies and found that many aspects of the global water cycle have intensified in the past 50 years, including rainfall and evaporation. Modelling work by Wetherald & Manabe (2002) confirms that warming will increase rates of both precipitation and evaporation.
[46] Burke, Brown and Christidis (2006) using one model under a high emissions scenario. Other climate models are needed to verify these results. The study uses one commonly used drought index: The Palmer Drought Severity Index (PDSI). This uses temperature and rainfall data to formulate a measure of 'dryness'. Other drought indices do not show such large changes.

and overall numbers[47] remain less certain. There is growing evidence the expected increases in hurricane severity are already occurring, above and beyond any natural decadal cycles. Recent work suggests that the frequency of very intense hurricanes and typhoons (Category 4 and 5) in the Atlantic Basin has doubled since the 1970s as a result of rising sea-surface temperatures.[48] This remains an active area of scientific debate[49]. In higher latitudes, some models show a general shift in winter storm tracks towards the poles.[50] In Australia, this could lead to water scarcity as the country relies on winter storms to supply water[51].

Climate change could weaken the Atlantic Thermohaline Circulation, partially offsetting warming in both Europe and eastern North America, or in an extreme case causing a significant cooling.

The warming effect of greenhouse gases has the potential to trigger abrupt, large-scale and irreversible changes in the climate system. One example is a possible collapse of the North Atlantic Thermohaline Circulation (THC). In the North Atlantic, the Gulf Stream and North Atlantic drift (important currents of the North Atlantic THC) have a significant warming effect on the climates of Europe and parts of North America. The THC may be weakened, as the upper ocean warms and/or if more fresh water (from melting glaciers and increased rainfall) is laid over the salty seawater.[52] No complex climate models currently predict a complete collapse. Instead, these models point towards a weakening of up to half by the end of the century[53]. Any sustained weakening of the THC is likely to have a cooling effect on the climates of Europe and eastern North America, but this would only offset a portion of the regional warming due to greenhouse gases. A recent study using direct ocean measurements (the first of its kind) suggests that part of the THC may already have weakened by up to 30% in the past few decades, but the significance of this is not yet known.[54] The potential for abrupt, large-scale changes in climate requires further research.

Sea levels will continue to rise, with very large increases if the Greenland Ice Sheet starts to melt irreversibly or the West Antarctic Ice Sheet (WAIS) collapses.

Sea levels will respond more slowly than temperatures to changing greenhouse gas concentrations. Sea levels are currently rising globally at around 3 mm per year and the rise has been accelerating[55]. According to the IPCC TAR, sea levels are projected to rise by 9–88 cm by 2100, mainly due to expansion of the warmer oceans and melting glaciers on land.[56] However, because warming only penetrates the oceans very slowly, sea levels will continue to rise substantially more

---

[47] For example, Lambert and Fyfe (2006) and Fyfe (2003)

[48] Emanuel (2005); Webster et al. (2005)

[49] Pielke (2005); Landsea (2005)

[50] For example, Geng and Sugi (2003); Bengtsson, Hodges and Roeckner (2006)

[51] Hope (2006)

[52] Summarised in Schlesinger *et al.* (2006)

[53] Wood *et al.* (2006). Complex climate models project a weakening of between 0% and 50% by the end of the century.

[54] Bryden *et al.* (2005). It is unclear whether the weakening is part of a natural cycle or the start of a downward trend.

[55] Church and White (2006)

[56] IPCC (2001b). This range covers several sources of uncertainty, including emissions, climate sensitivity and ocean responses

over several centuries. On past emissions alone, the world has built up a substantial commitment to sea level rise. One study estimates an existing commitment of between 0.1 and 1.1 metres over 400 years.[57]

BOX 1.4 Ice sheets and sea level rise

Melting ice sheets are already contributing a small amount to sea level rise. Most of recent and current global sea level rise results from the thermal expansion of the ocean with a contribution from glacier melt. As global temperatures rise, the likelihood of substantial contributions from melting ice sheets increases, but the scale and timing remain highly uncertain. While some models project that the net contribution from ice sheets will remain close to zero or negative over the coming century, recent observations suggest that the Greenland and West Antarctic ice sheets may be more vulnerable to rising temperatures than is projected by current climate models:

• **Greenland Ice Sheet.** Measurements of the Greenland ice sheet have shown a slight inland growth,[58] but significant melting and an acceleration of ice flows near the coast,[59] greater than predicted by models. Melt water is seeping down through the crevices of the melting ice, lubricating glaciers and accelerating their movement to the ocean. Some models suggest that as local temperatures exceed 3–4.5°C (equivalent to a global increase of around 2–3°C) above pre-industrial,[60] the surface temperature of the ice sheet will become too warm to allow recovery from summertime melting and the ice sheet will begin to melt irreversibly. During the last interglacial period, around 125,000 years ago when Greenland temperatures reached around 4–5°C above the present[61], melting of ice in the Arctic contributed several metres to sea level rise.

• **Collapse of the West Antarctic Ice Sheet.**[62] In 2002, instabilities in the Larsen Ice Shelf led to the collapse of a section of the shelf the size of Rhode Island (Larsen B – over 3200 km^2 – and 200 m thick) from the Antarctic Peninsula. The collapse has been associated with a sustained warming and resulting rapid thinning of Larsen B at a rate of just under 20 cm per year[63]. A similar rapid rate of thinning has now been observed on other parts of the WAIS around Amundsen Bay (this area alone contains enough water to raise sea levels by 1.5 m)[64]. Rivers of ice on the ice-sheet have been accelerating towards the ocean. It is possible that ocean warming and the acceleration of ice flows will destabilise the ice sheet and cause a runaway discharge into the oceans. Uncertainties over the dynamics of the ice sheet are so great that there are few estimates of critical thresholds for collapse. One study gives temperatures between 2°C and 5°C, but these remain disputed.

[57] Wigley (2005). The uncertainty reflects a range of climate sensitivities, aerosol forcings and melt-rates.

[58] For example, Zwally et al. 2006 and Johannessen et al. 2005

[59] For example, Hanna et al. 2005 and Rignot and Kanagaratnam 2006

[60] Lower and higher estimates based on Huybrechts and de Wolde (1999) and Gregory and Huybrechts (2006), respectively.

[61] North Greenland Ice Core Project (2004). The warm temperatures in the Northern Hemisphere during the previous interglacial reflected a maximum in the cycle of warming from the Sun due to the orbital position of the Earth. In the future, Greenland is expected to experience some of the largest temperature changes. A 4–5°C greenhouse warming of Greenland would correspond to a global mean temperature rise of around 3°C (Gregory and Huybrechts (2006)).

[62] Rapley (2006)

[63] Shepherd et al. 2003. The collapse of Larsen B followed the collapse in 1995 of the smaller Larsen A ice shelf.

[64] Zwally et al. (2006)

As global temperatures continue to rise, so do the risks of additional sea level contributions from large-scale melting or collapse of ice sheets. If the Greenland and West Antarctic ice sheets began to melt irreversibly, the world would be committed to substantial increases in sea level in the range 5–12 m over a timescale of centuries to millennia.[65] The immediate effect would be a potential doubling of the rate of sea level rise: 1–3 mm per year from Greenland and as high as 5 mm per year from the WAIS.[66] For illustration, if these higher rates were reached by the end of this century, the upper range of global sea level rise projections would exceed 1 m by 2100. Both of these ice sheets are already showing signs of vulnerability, with ice discharge accelerating over large areas, but the thresholds at which large-scale changes are triggered remain uncertain (Box 1.4).

1.6 Conclusions

Climate change is a serious and urgent issue. While climate change and climate modelling are subject to inherent uncertainties, it is clear that human activities have a powerful role in influencing the climate and the risks and scale of impacts in the future. All the science implies a strong likelihood that, if emissions continue unabated, the world will experience a radical transformation of its climate. Part II goes on to discuss the profound implications that this will have for our way of life.

The science provides clear guidance for the analysis of the economics and policy. The following chapter examines the implications of the science for the structuring of the economics.

References

The Third Assessment Report of the IPCC gives the most comprehensive assessment of the science of climate change up to 2001 (IPCC 2001a,b). The summary for policymakers gives a good introduction to the more in-depth analyses of the three working groups. Maslin (2004) provides a more narrative description of climate change, including an overview of the history. Schellnhuber (2006) gives a good summary of the evolution of the science from early 2001 to 2005, including articles describing temperature projections based on new estimates of climate sensitivity (e.g. Meinshausen (2006)), positive feedbacks in the carbon cycle (e.g. Cox *et al.* (2006)) and several articles on the impacts of climate change.

Annan, J.D. and J.C. Hargreaves (2006): 'Using multiple observationally-based constraints to estimate climate sensitivity', Geophysical Research Letters **33**: L06704

Archer, D. (2005): 'Methane hydrates and anthropogenic climate change', Reviews of Geophysics, submitted, available from http://geosci.uchicago.edu/~archer/reprints/archer.ms.clathrates.pdf

---

[65] Based on 7m and 5m from the Greenland and West Antarctic ice sheets, respectively. Rapley (2006) and Wood *et al.* (2006)

[66] Huybrechts and DeWolde (1999) simulated the melting of the Greenland Ice Sheet for a local temperature rise of 3°C and 5.5°C. These scenarios led to a contribution to sea level rise of 1m and 3m over 1000 years (1 mm/yr and 3 mm/yr), respectively. Possible contributions from the West Antarctic Ice Sheet (WAIS) remain highly uncertain. In an expert survey reported by Vaughan and Spouge (2002), most glaciologists agree that collapse might be possible on a thousand-year timescale (5 mm/yr), but that this contribution is unlikely to be seen in this century. Few scientists considered that collapse might occur on a century timescale.

Barnett T.P., D.W. Pierce, K.M. AchutaRao et al. (2005): 'Penetration of human-induced warming into the world's oceans', Science **309:** 284–287

Bengtsson, L., K. Hodges and E. Roeckner (2006): 'Storm tracks and climate change', Journal of Climate, in press.

Brohan. P., J.J. Kennedy, I. Harris, et al. (2006): 'Uncertainty estimates in regional and global observed temperature changes: a new dataset from 1850'. Journal of Geophysical Research, **111,** D12106, doi: 10.1029/2005JD006548

Burke, E.J., S.J. Brown and N. Christidis (2006): 'Modelling the Recent Evolution of Global Drought and Projections for the Twenty-First Century with the Hadley Centre Climate Model', Journal of Hydrometeorology, **7**(5):1113–1125

Bryden, H.L., H.R. Longworth and S.A. Cunningham (2005): 'Slowing of the Atlantic meridional overturning circulation at 25°N', Nature **438:** 655–657

Church, J.A., and N.J. White (2006): 'A 20th century acceleration in global sea-level rise', Geophysical Research Letters, **33,** L01602, doi: 10.1029/2005GL024826.

Collins, M. and the CMIP Modelling Group (2005): 'El Nino – or La Nina-like climate change? Climate Dynamics' **24:** 89–104

Cox, P.M., R.A. Betts, C.D. Jones, et al. (2000): 'Acceleration of global warming due to carbon-cycle feedbacks in a coupled climate model', Nature **408:** 184–187

Cox P.M., C. Huntingford and C.D. Jones (2006): 'Conditions for Sink-to-Source Transitions and Runaway Feedbacks from the Land Carbon Cycle', in Avoiding dangerous climate change, H.J. Schellnhuber et al. (eds.), Cambridge: Cambridge University Press, pp.155–163.

Davidson and Janssens (2006): 'Temperature sensitivity of soil carbon decomposition and feedbacks to climate change', Nature **440:** 165–173

DEFRA (2005): 'Climate change and the greenhouse effect: a briefing from the Hadley Centre', available from http://www.metoffice.com/research/hadleycentre/pubs/brochures/2005/climate_greenhouse.pdf

Emanuel, K. (2005): 'Increased destructiveness of tropical cyclones over the past 30 years', Nature **436:** 686–688

Forest, C.E., P.H. Stone and A.P. Sokolov (2006): 'Estimates PDFs of climate system properties including natural and anthropogenic forcings', Geophysical Research Letters **33:** L01705, doi: 10.1029/2005GL023977

Friedlingstein, P., P. Cox, R. Betts et al. (2006): 'Climate-carbon cycle feedback analysis: results from C4MIP model intercomparison', Journal of Climate, **19:** 3337–3353

Fyfe, J.C. (2003): 'Extratropical southern hemisphere cyclones: Harbingers of climate change?' Journal of Climate **16,** 2802–2805

Gedney, N., P.M. Cox and C. Huntingford (2004): 'Climate feedback from wetland methane emissions', Geophysical Research Letters **31** (20): L20503

Geng, Q.Z. and M. Sugi (2003): 'Possible changes in extratropical cyclone activity due to enhanced greenhouse gases and aerosols – Study with a high resolution AGCM'. Journal of Climate **16:** 2262–2274.

Gorham, E. (1991): 'Northern Peatlands: Role in the Carbon Cycle and Probable Responses to Climatic Warming', Ecological Applications **1:** 182–195, doi: 10.2307/1941811

Gregory, J. and P. Huybrechts (2006): 'Ice sheet contributions to future sea level change', Phil Trans Royal Soc A **364:** 1709–1731, doi: 10.1098/rsta.2006.1796

Hadley Centre (2005): 'Stabilising climate to avoid dangerous climate change – a summary of relevant research at the Hadley Centre', available from http://www.metoffice.com/research/hadleycentre/pubs/brochures

Hanna, E., P. Huybrechts, and I. Janssens, et al. (2005): 'Runoff and mass balance of the Greenland ice sheet: 1958–2003'. Journal of Geophysical Research **110,** D13108, doi: 10.1029/2004JD005641

Hansen, J., M. Sato, R. Ruedy, et al. (2006): 'Global temperature change, Proceedings of the National Academy', **103:** 14288–14293

Hope, P.K. (2006): 'Projected future changes in synoptic systems influencing southwest Western Australia. Climate Dynamics' **26:** 765–780, doi: 10.1007/s00382-006-0116-x

Huntington, T.G. (2006): 'Evidence for intensification of the global water cycle: review and synthesis', Journal of Hydrology **319:** 1–13

Huybrechts, P. and J. de Wolde (1999): 'The dynamic response of the Greenland Ice Sheet and Antarctic ice sheets to a multiple century climatic-warming', Journal of Climate **12:** 2169–2188

International ad hoc detection group (2005): 'Detecting and attributing external influences on the climate system: a review of recent advances', Journal of Climate **18:** 1291–1314

Intergovernmental Panel on Climate Change (2001a): 'Climate change 2001: summary for policymakers, A contribution of Working Groups I, II and III to the Third Assessment Report of the

Intergovernmental Panel on Climate Change' [Watson RT, and the Core Writing Team (eds.)], Cambridge: Cambridge University Press.

Intergovernmental Panel on Climate Change (2001b): 'Climate change 2001: the scientific basis. Contribution of Working Group I to the Third Assessment Report of the Intergovernmental Panel on Climate Change' [Houghton JT, Ding Y, Griggs DJ, et al. (eds.)], Cambridge: Cambridge University Press.

Johannessen, O.M., K. Khvorostovsky, M.W. Miles et al. (2005): 'Recent ice-sheet growth in the interior of Greenland'. Science **310**: 1013–1016

Knutti, R., T.F. Stocker, F. Joos and G-K Plattner (2002): 'Constraints on radiative forcing and future climate change from observations and climate model ensembles', Nature **416**: 719–723

Lawrence D.M. and A.G. Slater (2005): 'A projection of severe near-surface permafrost degradation during the 21st century', Geophysical Research Letters **32**: L24401

Lambert S.J. and J.C. Fyfe (2006): 'Changes in winter cyclone frequencies and strengths simulated in enhanced greenhouse warming experiments: results from models participating in the IPCC diagnostic exercise', Climate Dynamics **1432**, 0894

Landsea, C. (2005): 'Atlantic hurricanes and global warming', Nature **438**, E11–E12

Levitus, S.J., J. Antonov and T. Boyer (2005): 'Warming of the world ocean 1955–2003', Geophysical Research Letters **32**: L02604, doi:10.1029/2004GL021592

Lindzen, R.S., M-D Chou and A.Y. Hou (2001): 'Does the earth have an adaptive infrared iris?' Bulletin of the American Meteorological Society **82**: 417–432

Mann, M.E., R.S. Bradley and M.K. Hughes (1999): 'Northern hemisphere temperatures during the past millennium: inferences, uncertainties, and limitations', Geophysical Research Letters, 26, 759–762.

Maslin, M. (2004): 'Global warming: a very short introduction', New York: Oxford University Press,

Meehl, G.A., W.M. Washington, W.D. Collins et al. (2005): 'How much more global warming and sea level rise?' Science **307**:1769–1772

Meinshausen, M. (2006): 'What does a 2°C target mean for greenhouse gas concentrations? A brief analysis based on multi-gas emission pathways and several climate sensitivity uncertainty estimates', Avoiding dangerous climate change, in H.J. Schellnhuber et al. (eds.), Cambridge: Cambridge University Press, pp.265–280.

Murphy, J.M., D.M.H. Sexton D.N. Barnett et al. (2004): 'Quantification of modelling uncertainties in a large ensemble of climate change simulations', Nature **430**: 768–772

National Research Council (2006): 'Surface temperature reconstructions for the past 2,000 years', available from http://www.nap.edu/catalog/11676.html

North Greenland Ice Core Project (2004): 'High-resolution record of Northern Hemisphere climate extending into the last interglacial maximum', Nature **431**: 147–151

Orr, J.C., V.J. Fabry, O. Aumont et al. (2005): 'Anthropogenic ocean acidification over the twenty-first century and its impact on calcifying organisms', Nature **437**: 681–686

Parmesan, C. and G. Yohe (2003): 'A globally coherent fingerprint of climate change impacts across natural systems', Nature **421**: 37–42

Pearce, F. (2003): 'Land of the midnight sums', New Scientist **177**: 2379

Pierrehumbert, R.T. (2004): 'Warming the world', Nature **432**: 677, doi: 10.1038/432677a

Pielke, R. (2005): 'Meteorology: Are there trends in hurricane destruction?' Nature **438**: E11

Rapley, C. (2006): 'The Antarctic ice sheet and sea level rise', in Avoiding dangerous climate change, Schellnhuber HJ (ed.), Cambridge: Cambridge University Press, pp. 25–28.

Rignot, E. and P. Kanagaratnam (2006): 'Changes in the velocity structure of the Greenland ice sheet. Science' **311**: 986–990

Root, T.L., D.P. MacMynowski, M.D. Mastrandrea and S.H. Schneider (2005): 'Human-modified temperatures induce species changes: combined attribution', Proceedings of the National Academy of Sciences **102**: 7465–7469

Schär, C., P.L. Vidale, D. Lüthi, et al. (2004): 'The role of increasing temperature variability in European summer heatwaves', Nature **427**: 332–336, doi: 10.1038/nature02300

Scheffer, M., V. Brovkin and P. Cox (2006): 'Positive feedback between global warming and the atmospheric CO2 concentration inferred from past climate change'. Geophysical Research Letters 33, L10702

Schellnhuber, H.J., W. Cramer, N. Nakicenovic et al. (eds.) (2006): 'Avoiding dangerous climate change', Cambridge: Cambridge University Press.

Schlesinger, M.E., J. Yin, G. Yohe, et al. (2006): 'Assessing the risk of a collapse of the Atlantic Thermohaline Circulation', in Avoiding dangerous climate change, H.J. Schellnhuber et al. (eds.) Cambridge: Cambridge University Press, pp. 37–47.

Scholze, M., K. Wolfgang, N. Arnell and C. Prentice (2006): 'A climate-change risk analysis for world ecosystems', Proceedings of the National Academy of Sciences **103**: 13116–13120

Siegenthaler U, Stocker TF, Monnin E, et al. (2005) Stable carbon cycle-climate relationship during the late Pleistocene, Science **310**: 1313–1317

Shepherd, A., D. Wingham, T. Payne and P. Skvarca (2003): 'Larsen Ice Shelf has progressively thinned', Science **302**: pp 856–859, doi: 10.1126/science.1089768

Smith, L.C., Y. Sheng, G.M. MacDonald, et al. (2005): 'Disappearing artic lakes'. Science, **308**: 1429

Soden, B.J., D.L. Jackson, V. Ramaswamy, et al. (2005): 'The radiative signature of upper tropospheric moistening', Science, **310**: 841–844

Stainforth, D., T. Aina, C. Christensen, et al. (2005): 'Uncertainty in predictions of the climate response to rising levels of greenhouse gases', Nature **433**: 403–406

Stott P.A., D.A. Stone and M.R. Allen (2004): 'Human contribution to the European heatwave of 2003', Nature **432**: 610–614

Torn, M.S. and J. Harte (2006): 'Missing feedbacks, asymmetric uncertainties, and underestimation of future warming', Geophysical Research Letters **33**: L10703

US Climate Change Programme (2006): 'Temperature trends in the lower atmosphere: steps for understanding and reconciling differences', available from http://www.climatescience.gov/Library/sap/sap1-1/finalreport

Vaughan, D.G. and J.R. Spouge (2002): 'Risk estimation of collapse of the West Antarctic ice sheet', Climatic Change **52**: 65–91

von Storch, H., E. Zorita, J.M. Jones, et al. (2004): Reconstructing past climate from noisy data, Science **306**: 679–682

Walter, K.M., S.A. Zimov, J.P. Chanton, et al. (2006): 'Methane bubbling from Siberian thaw lakes as a positive feedback to climate warming', Nature **443**: 71–75

Webster, P.J., G.J. Holland, J.A. Curry and H-R Chang (2005): 'Changes in tropical cyclone number, duration, and intensity in a warming environment', Science **309**,1844–1846

Wetherald, R.T. and S. Manabe (2002): 'Simulation of hydrologic changes associated with global warming', Journal of Geophysical Research **107**: 4379

Wigley, T.M.L. (1985): 'Impact of extreme events', Nature **316**: 106–107

Wigley, T.M.L. and S.C.B. Raper (2001): 'Interpretation of high projections for global-mean warming', Science **293**: 451–454

Wigley, T.M.L. (2005): 'The climate change commitment', Science **307**: 1766–1769

Wood, R., M. Collins, J. Gregory, et al. (2006): 'Towards a risk assessment for shutdown of the Atlantic Thermohaline Circulation', in Avoiding dangerous climate change, H.J Schellnhuber et al. (eds.), Cambridge: Cambridge University Press, pp. 49–54.

Zwally, H.J., M.B. Giovinetto, J. Li, et al. (2006): 'Mass changes of the Greenland and Antarctic ice sheets and shelves and contributions to sea-level rise: 1992– 2002', Journal of Glaciology **51**: 509–527

2   Economics, Ethics and Climate Change

KEY MESSAGES

Climate change is a result of the externality associated with greenhouse-gas emissions – it entails costs that are not paid for by those who create the emissions.

It has a number of features that together distinguish it from other externalities:

- It is **global** in its causes and consequences;
- The impacts of climate change are **long-term and persistent;**
- **Uncertainties and risks** in the economic impacts are pervasive.
- There is a serious risk of major, irreversible change with **non-marginal economic effects**.

These features shape the economic analysis: it must be **global**, deal with **long** time horizons, have the economics of **risk and uncertainty** at its core, and examine the possibility of major, **non-marginal** changes.

The **impacts of climate change are very broad ranging and interact with other market failures and economic dynamics, giving rise to many complex policy problems**. Ideas and techniques from most of the important areas of economics, including many recent advances, have to be deployed to analyse them.

The **breadth, magnitude and nature of impacts imply that several ethical perspectives, such as those focusing on welfare, equity and justice, freedoms and rights, are relevant.** Most of these perspectives imply that the outcomes of climate-change policy are to be understood in terms of impacts on consumption, health, education and the environment over time but different ethical perspectives may point to different policy recommendations.

Questions of intra- and inter-generational equity are central. Climate change will have serious impacts within the lifetime of most of those alive today. Future generations will be even more strongly affected, yet they lack representation in present-day decisions.

Standard externality and cost-benefit approaches have their usefulness for analysing climate change, but, as they are methods focused on evaluating marginal changes, and generally abstract from dynamics and risk, they can only be starting points for further work.

Standard treatments of discounting are valuable for analysing marginal projects but are inappropriate for non-marginal comparisons of paths; the approach to discounting must meet the challenge of assessing and comparing paths that have very different trajectories and involve very long-term and large inter-generational impacts. We must go back to the first principles from which the standard marginal results are derived.

The severity of the likely consequences and the application of the above analytical approaches form the basis of powerful arguments, developed in the Review, in favour of strong and urgent global action to reduce greenhouse-gas emissions, and of major action to adapt to the consequences that now cannot be avoided.

2.1   Introduction

The science described in the previous chapter drives the economics that is required for the analysis of policy. This chapter introduces the conceptual frameworks that we will use to examine the economics of climate change. It explores, in Section 2.2, the distinctive features of the externalities associated with greenhouse-gas emissions and draws attention to some of the difficulties associated with a simplistic application of the standard theory of externalities to this problem. Section 2.3 introduces a variety of ethical approaches and relates them to the global and long-term nature of the impacts (the discussion is extended in the appendix to the chapter). Section 2.4 examines some specifics of intertemporal allocation, including discounting (some further technical details are provided in the appendix to the chapter). Sections 2.5 and 2.6 consider how economic analysis can get to grips with a problem that is uncertain and involves a serious risk of large losses of wellbeing, due to deaths, extinctions of species and heavy economic costs, rather than the marginal changes more commonly considered in economics. For most of economic policy, the underlying ethical assumptions are of great importance, and this applies particularly for climate change: that is why they are given special attention in this chapter.

The economics introduced in this chapter applies, in principle, to the whole Review but the analysis of Sections 2.2 to 2.6 is of special relevance to Parts II and III, which look at impacts and at the economics of mitigation – assessing how much action is necessary to reduce greenhouse-gas emissions. Parts IV, V, VI of this report are devoted to the analysis of policy to promote mitigation and adaptation. The detailed, and often difficult, economics of public policy and collective action that are involved in these analyses are introduced in the sections themselves and we provided only brief coverage in Sections 2.7 and 2.8. In the former section, we refer briefly to the modern public economics of carbon taxation, trading and regulation and of the promotion of research, development and deployment, including the problems of various forms of market imperfection affecting innovation. It also covers an analysis of the role of 'responsible behaviour' and how public understanding of this notion might be influenced by public policy. Section 2.8 explores some of the difficulties of building and sustaining global collective action in response to the global challenge of climate change.

In these ways, this chapter lays the analytical foundations for much of the economics required by the challenge of climate change and which is put to work in the course of the analysis presented in this Review.

The subject demands analysis across an enormous range of issues and requires all the tools of economics we can muster – and indeed some we wish we had. In setting out some of these tools, some of the economic analysis of this chapter is inevitably technical, even though the more mathematical material has been banished to an appendix. Some readers less interested in the technical underpinnings of the analysis may wish to skim the more formal analytical material. Nevertheless, it is important to set out some of the analytical instruments at the beginning of the Review, since they underpin the analysis of risk, equity and allocation over time that must lie at the heart of a serious analysis of the economics of climate change.

2.2   Understanding the market failures that lead to climate change

Climate change results from greenhouse-gas emissions associated with economic activities including energy, industry, transport and land use.

In common with many other environmental problems, human-induced climate change is at its most basic level an externality. Those who produce greenhouse-gas emissions are bringing about climate change, thereby imposing costs on the world and on future generations, but they do not face directly, neither via markets nor in other ways, the full consequences of the costs of their actions.

Much economic activity involves the emission of greenhouse gases (GHGs). As GHGs accumulate in the atmosphere, temperatures increase, and the climatic changes that result impose costs (and some benefits) on society. However, the full costs of GHG emissions, in terms of climate change, are not immediately – indeed they are unlikely ever to be – borne by the emitter, so they face little or no economic incentive to reduce emissions. Similarly, emitters do not have to compensate those who lose out because of climate change.[1] In this sense, human-induced climate change is an externality, one that is not 'corrected' through any institution or market,[2] unless policy intervenes.

The climate is a public good: those who fail to pay for it cannot be excluded from enjoying its benefits and one person's enjoyment of the climate does not diminish the capacity of others to enjoy it too.[3] Markets do not automatically provide the right type and quantity of public goods, because in the absence of public policy there are limited or no returns to private investors for doing so: in this case, markets for relevant goods and services (energy, land use, innovation, etc) do not reflect the consequences of different consumption and investment choices for the climate. Thus, climate change is an example of market failure involving externalities and public goods.[4] Given the magnitude and nature of the effects initially described in the previous chapter and taken forward in Parts II and III, it has profound implications for economic growth and development. All in all, it must be regarded as market failure on the greatest scale the world has seen.

The basic theory of externalities and public goods is the starting point for most economic analyses of climate change and this Review is no exception. The starting point embodies the basic insights of Pigou, Meade, Samuelson and Coase

---

[1]   Symmetrically, those who benefit from climate change do not have to *reward* emitters.
[2]   Pigou (1912).
[3]   Samuelson (1954).
[4]   Formally, in economic theory, public goods are a special case of externalities where the effects of the latter are independent of the identity of the emitters or origin of the externalities.

(see Part IV). But the special features of this particular externality demand, as we shall see, that the economic analysis go much further.

The science of climate change means that this is a very different form of externality from the types commonly analysed.

Climate change has special features that, together, pose particular challenges for the standard economic theory of externalities. There are four distinct issues that will be considered in turn in the sections below.

- Climate change is an externality that is global in both its causes and consequences. The incremental impact of a tonne of GHG on climate change is independent of where in the world it is emitted (unlike other negative impacts such as air pollution and its cost to public health), because GHGs diffuse in the atmosphere and because local climatic changes depend on the global climate system. While different countries produce different volumes the marginal damage of an extra unit is independent of whether it comes from the UK or Australia.
- The impacts of climate change are persistent and develop over time. Once in the atmosphere, some GHGs stay there for hundreds of years. Furthermore, the climate system is slow to respond to increases in atmospheric GHG concentrations and there are yet more lags in the environmental, economic and social response to climate change. The effects of GHGs are being experienced now and will continue to work their way through in the very long term.
- The uncertainties are considerable, both about the potential size, type and timing of impacts and about the costs of combating climate change; hence the framework used must be able to handle risk and uncertainty.
- The impacts are likely to have a significant effect on the global economy if action is not taken to prevent climate change, so the analysis has to consider potentially non-marginal changes to societies, not merely small changes amenable to ordinary project appraisal.

These features shape much of the detailed economic analysis throughout this Review. We illustrate with just one example, an important one, which shows how the dynamic nature of the accumulation of GHGs over time affects one of the standard analytical workhorses of the economics of externalities and the environment. It is common to present policy towards climate change in terms of the social cost of carbon on the margin (SCC) and the marginal abatement cost (MAC). The former is the total damage from now into the indefinite future of emitting an extra unit of GHGs now – the science says that GHGs (particularly CO_2) stay in the atmosphere for a very long time. Thus, in its simplest form, the nature of the problem is that the stock of gases in the atmosphere increases with the net flow of GHG emissions in this period, and thus decreases with abatement. Therefore, on the one hand, the SCC curve slopes downwards with increasing abatement in any given period, assuming that the lower the stock at any point in the future, the less the marginal damage. On the other hand, the MAC curve slopes upwards with increasing abatement, if it is more costly on the margin to do more abatement as abatement increases in the given period. The optimum level of abatement must satisfy the condition that the MAC equals the SCC. If, for example, the SCC were bigger than the MAC, the social gain from one extra unit of abatement would be less than the cost and it would be better to do a little more. We call the optimum level this period x^*_0.

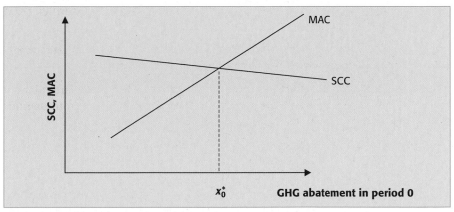

Figure 2.1 The optimum degree of abatement in a given period

In the figure, the SCC and the MAC are drawn as functions of emissions in this period, call it period 0. As drawn, the SCC curve is fairly flat and downward sloping, since extra emissions this period do not affect the total stock very much, but nevertheless extra abatement now implies a slightly lower stock in the future. The MAC curve rises, since we assume that, as abatement increases in this period, the marginal cost goes up. The optimum path for abatement is where $x_0^*, x_1^*, x_2^*, \ldots x_t^*, \ldots$ are all set optimally for each period $0, 1, 2, t, \ldots$ into the indefinite future, and the SCC curve is drawn for each period on the assumption that all future periods are set optimally.

It should be clear that the SCC curve this period depends on future emissions: if we revised upwards our specified assumptions on future emissions, the whole SCC curve would shift upwards, and so would the optimum abatement level in this period, x_0^*. Thus, if we are thinking about an optimum path over time, rather than simply an optimum emission for this period, we must recognise that the SCC curve for any given period depends on the future stock and thus on the future path of emissions. **We cannot sensibly calculate an SCC without assuming that future emissions and stocks follow some specified path. For different specified paths, the SCC will be different.** For example, it will be much higher on a 'business as usual' path (BAU) than it will be on a path that cuts emissions strongly and eventually stabilises concentrations. It is remarkable how often SCC calculations are vague on this crucial point (see Chapter 13 for a further discussion). Thus we must be very careful how we use a diagram that is pervasive in the economics of climate change – see Figure 2.1.

A number of important points follow from this, in addition to the basic one that an SCC curve cannot be drawn, nor an SCC calculated, without specific assumptions on future paths. First, if the SCC rises over time along the specified path then, for optimality, so too must the MAC. It is very likely that the SCC *will* rise over time, since stocks of GHGs will rise as further emissions take place, up to the point where stabilisation is reached. Thus the MAC at the optimum rises and the intersection of the MAC and SCC curves will imply successively greater abatement. This is true even though the whole MAC curve is likely to be lower for any particular degree of abatement in the future because learning will have taken place.

Figure 2.2 is thus perhaps more helpful than Figure 2.1 in sketching the nature of the solution to the problem. The position of the schedule in the left-hand side panel depends on the stabilisation target chosen for the atmospheric concentration of greenhouse gases, which in turn depends upon how the expected present values (in

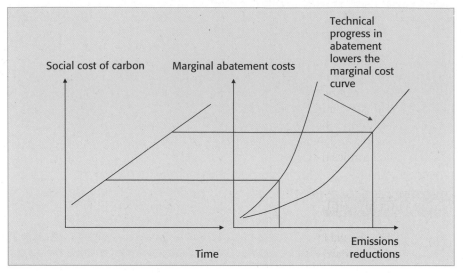

Figure 2.2 How the path for the social cost of carbon drives the extent of abatement

terms of discounted utility) of costs and benefits of mitigation through time change as the stabilisation level changes. Hence the choice of stabilisation target implies a view about what is likely to happen to abatement costs over time. The right-hand panel shows the shifts in the MAC curve expected at the time the stabilisation target is chosen.

This illustrates how important it is that the dynamics of the problem are considered. The conclusion that the MAC rises along an optimum path does *not* automatically follow from an analysis that simply shifts the SCC curve upwards over time (with higher stocks) and shifts the MAC down over time (with learning), without linking to the full dynamic optimisation. That optimisation takes account of the known future fall in costs in determining the whole path for the SCC. We are simply assuming that this fall in costs could not be of a magnitude to make it optimum for stocks to fall, that is, for emissions to be less than the Earth system's equilibrium capacity to absorb greenhouse gases from the atmosphere.

This analysis raises the second point, about the role of uncertainty. In the above argument, there is no consideration of uncertainty. If that vital element is now introduced, the argument becomes more complex. It has to be asked whether the resolution of uncertainty in any period would lead to a revision of views about the future probability distributions for abatement costs and climate-change damages. If, for example, there is unexpected good news that abatement is likely to be much cheaper than previously thought, then a lower stabilisation target and more abatement over time than originally planned would become appropriate. This would reduce the SCC from where it would otherwise have been. However, one surprisingly good period for costs does not necessarily imply that future periods will be just as good. In Figure 2.2, persistently faster technical progress than expected (as opposed to random fluctuations of the MAC around its expected value) would lead to a downward revision of the stabilisation target and hence a downward shift in the schedule in the left-hand panel.

Dynamics and uncertainty are explored further in Chapters 13 and 14, while analyses involving risk are taken further in Sections 2.5 and 2.6 and in Chapter 6.

This important example shows how important it is to integrate the scientific features of the externality into the economics and shows further that there are difficult conceptual and technical questions to be tackled. The analysis must cover a very broad range, including the economics of: growth and development; industry; innovation and technological change; institutions; the international economy; demography and migration; public finance; information and uncertainty; and the economics of risk and equity; and environmental and public economics throughout.

2.3   Ethics, welfare and economic policy

The special features of the climate-change externality pose difficult questions for the standard welfare-economic approach to policy.

Chapter 1 shows that the effects of climate change are global, intertemporal and highly inequitable. The inequity of climate change is examined further in Part II. Generally, poor countries, and poor people in any given country, suffer the most, notwithstanding that the rich countries are responsible for the bulk of past emissions. These features of climate change, together with the fact that they have an impact on many dimensions of human well-being, force us to look carefully at the underlying ethical judgements and presumptions which underpin, often implicitly, the standard framework of policy analysis. Indeed, it is important to consider a broader range of ethical arguments and frameworks than is standard in economics, both because there are many ways of looking at the ethics of policy towards climate change, and, also, because in so doing we can learn something about how to apply the more standard economic approach. There is a growing literature on the ethics of climate change: analysis of policy cannot avoid grappling directly with the difficult issues that arise. These ethical frameworks are discussed more formally in the technical appendix to this chapter; the discussion here is only summary[5].

The underlying ethics of basic welfare economics, which underpins much of the standard analysis of public policy, focuses on the consequences of policy for the consumption of goods and services by individuals in a community. These goods and services are generated by labour, past saving, knowledge and natural resources. The perspective sees individuals as having utility, or welfare, arising from this consumption.

In this approach, the objective is to work out the policies that would be set by a decision-maker acting on behalf of the community and whose role it is to improve, or maximise, overall social welfare. This social welfare depends on the welfare of each individual in the community. When goods and services are defined in a broad way, they can include, for example, education, health and goods appearing at different dates and in different circumstances. Thus the theory covers time and uncertainty. And, to the extent that individuals value the

---

[5] Particularly important contributions on ethics are those of Beckerman and Pasek (2001), Broome (1992, 1994, 2004, 2005), Gardiner (2004) and Müller (2006). We are very grateful to John Broome for his advice and guidance, but he is not responsible for the views expressed here.

environment, that too is part of the analysis. Many goods or services, including education, health and the environment, perform a dual role: individuals directly value them and they are inputs into the use or acquisition of other consumption goods. In the jargon, they are both goals and instruments.

The standard economic theory then focuses on flows of goods or services over time and their distribution across individuals. The list of goods or services should include consumption (usually monetary or the equivalent), education, health and the environment. These are usually the areas focused upon in cross-country comparisons of living standards, such as, for example, in the *World Development Indicators* of the World Bank, the *Human Development Report* of the UNDP, and the *Millennium Development Goals* (MDGs) agreed at the UN at the turn of the millennium. 'Stocks' of wealth, infrastructure, the natural environment and so on enter into the analysis in terms of their influence on flows. Through these choices of data for central attention and through the choice of goals, the international community has identified a strong and shared view on the key dimensions of human well-being.

Those choices of data and goals can be derived from a number of different ethical perspectives (see, for example, Sen (1999)). Most ethical frameworks generally used in the analyses of economic policy have some relevance for the economics of climate change and there are some – for example, those involving stewardship and sustainability – that are particularly focused on environmental issues.

The ethical framework of standard welfare economics looks first only at the consequences of actions (an approach often described as 'consequentialism') and then assesses consequences in terms of impacts on 'utility' (an approach often described as 'welfarism', as in Sen (1999), Chapter 3 and the appendix to this chapter). This standard welfare-economic approach has no room, for example, for ethical dimensions concerning the processes by which outcomes are reached. Some different notions of ethics, including those based on concepts of rights, justice and freedoms, do consider process. Others, such as sustainability, and stewardship, emphasise particular aspects of the consequences of decisions for others and for the future, as explained in the technical appendix.

Nevertheless, the consequences on which most of these notions would focus for each generation often have strong similarities: above all, with respect to the attention they pay to consumption, education, health and the environment.

And all the perspectives would take into account the distribution of outcomes within and across generations, together with the risks involved in different actions, now and over time. Hence the Review focuses on the implications of action or inaction on climate change for these four dimensions.

How the implications for these four dimensions are assessed, will, of course, vary according to the ethical position adopted. How policy-makers aggregate over consequences (i) within generations, (ii) over time, and (iii) according to risk will be crucial to policy design and choice. Aggregation requires being quantitative in comparing consequences of different kinds and for different people. The Review pays special attention to all three forms of aggregation. In arriving at decisions, or a view, it is not, however, always necessary to derive a single number that gives full quantitative content and appropriate weight to all the dimensions and elements involved (see below).

Climate change is an externality that is global in both its causes and consequences. Both involve deep inequalities that are relevant for policy.

The incremental impact of a tonne of GHG is independent of where in the world it is emitted. But the volume of GHGs emitted globally is not uniform. Historically, rich countries have produced the majority of GHG emissions. Though all countries are affected by climate change, they are affected in different ways and to different extents. Developing countries will be particularly badly hit, for three reasons: their geography; their stronger dependence on agriculture; and because with their fewer resources comes greater vulnerability. There is therefore a double inequity in climate change: the rich countries have special responsibility for where the world is now, and thus for the consequences which flow from this difficult starting point, whereas poor countries will be particularly badly hit.

   The standard welfare-economics framework has a single criterion, and implicitly, a single governmental decision-maker. It can be useful in providing a benchmark for what a 'good' global policy would look like. But the global nature of climate change implies that the simple economic theory with one jurisdiction, one decision-maker, and one social welfare function cannot be taken literally. Instead, it is necessary to model how different players or countries will interact (see Section 2.8 below and Pt VI) and to ask ethical questions about how people in one country or region should react to the impacts of their actions on those in another. This raises questions of how the welfare of people with very different standards of living should be assessed and combined in forming judgements on policy.

There are particular challenges in valuing social welfare across countries at different stages of development and across different income or consumption levels.

The ethical question of how consequences for people in very different circumstances should be aggregated must be faced directly. For the sake of simplicity and clarity, we shall adopt the perspective of the 'social welfare function' approach, as explained in Box 2.1.

BOX 2.1 The 'social welfare function' approach to 'adding up' the wellbeing of different people.

The stripped-down approach that we shall adopt when we attempt to assess the potential costs of climate change uses the standard framework of welfare economics. The objective of policy is taken to be the maximisation of the sum across individuals of social utilities of consumption. Thus, in this framework, aggregation of impacts across individuals using social value judgements is assumed to be possible. In particular, we consider consumption as involving a broad range of goods and services that includes education, health and the environment. The relationship between the measure of social wellbeing – the sum of social utilities in this argument – and the goods and services consumed by each household, on which it depends, is called the social welfare function.

   In drawing up a social welfare function, we have to make explicit value judgements about the distribution of consumption across individuals – how much difference should it make, for example, if a given loss of consumption opportunities affects a rich person rather than a poor person, or someone today rather than in a hundred years' time?[6] Aggregating social utility across individuals to come up with a measure of social welfare has its problems. Different value judgements can lead to different rankings of possible outcomes, and deciding what values should be applied is difficult in democratic societies[7]. It is not always consistent with ethical perspectives based on rights and freedoms. But the approach has the virtue of clarity and simplicity, making it easy to test the sensitivity of the policy choice that emerges to the value judgements made. It is fairly standard in the economics of applied policy problems and allows for a consistent treatment of aggregation within and across generations and for uncertainty. The social welfare function's treatment of income differences can be calibrated by simple thought experiments. For example, suppose the decision-maker is considering two possible policy outcomes. In the second outcome, a poor person receives an income $X more than in the first, but a rich person receives $Y less; how much bigger than X would Y have to be for the decision-maker to decide that the second outcome is worse than the first?

Aggregation across education, health, income and environment raises profound difficulties, particularly when comparisons are made across individuals. Some common currency or 'numeraire' is necessary: the most common way of expressing an aggregate measure of wellbeing is in terms of real income. That immediately raises the challenge of expressing health (including mortality) and environmental quality in terms of income. There have been many attempts to do just that. These should not be lightly dismissed, since nations often decide how much to allocate to, for example, accident and emergency services or environmental protection in the knowledge that a little extra money saves lives and improves the environment. Indeed, individuals make similar choices in their own lives.

Nevertheless, there are significant difficulties inherent in the valuation of health and the environment, many of which are magnified across countries where major differences in income affect individuals' willingness and ability to pay for them. For example, a very poor person may not be 'willing-to-pay' very much money to insure her life, whereas a rich person may be prepared to pay a very large sum. Can it be right to conclude that a poor person's life or health is therefore less valuable?[8] It is surely within the realms of sensible discourse to think of the consequences of different strategies simultaneously in terms of income, lives and the environment: that is the approach we adopt where possible. At some points (such as in Chapter 6), however, we present models from the literature that do embody estimates of the monetary equivalent of the impacts of climate change on broader dimensions of welfare (although generally in these contexts increments in income are valued differently at different levels in income – see Box 2.1). Such exercises should be viewed with some circumspection.

---

[6]   Effectively, in putting it this way, we resist the interpretation that this is a strict utilitarian sum of 'actual utility'. On some of the difficulties and attractions of consequentialism, welfarism, utilitarianism and other approaches, see e.g. Sen and Williams (1982) and Sen (1999).

[7]   The difficulties of this type of aggregation using democratic methods have been examined by Kenneth Arrow (1951, 1963) using his famous 'impossibility theorem'. It has been examined in a series of studies by Amartya Sen (see, for example, Sen (1970, 1986 and 1999)).

[8]   Notice however that if the valuation of life in money terms in country A is twice that of country B, where income in A is twice that in B, we may choose to value increases in income in A half as much as for B (see Box 2.1 and Chapter 6). In that case, extra mortality would be valued in the same way for both countries.

2.4   The long-run impacts of climate change: evaluation over time and discounting

The effects of GHGs emitted today will be felt for a very long time. That makes some form of evaluation or aggregation across generations unavoidable. The ethical decisions on, and approaches to, this issue have major consequences for the assessment of policy.

The approach we adopt here is similar to that for assessing impacts that fall on different people or nations, and in some respects continues the discussion of ethics in the preceding section. When we do this formally, we work in terms of sums of utilities of consumption. Again there is a problem of calibrating the social welfare function for this purpose but, as with aggregating across people with different incomes at a moment in time, one can use a series of 'thought experiments' to help (see Box 2.1).

Typically, in the application of the theory of welfare economics to project and policy appraisal, an increment in future consumption is held to be worth less than an increment in present consumption, for two reasons. First, if consumption grows, people are better off in the future than they are now and an extra unit of consumption is generally taken to be worth less, the richer people are. Second, it is sometimes suggested that people prefer to have good things earlier rather than later – 'pure time preference' – based presumably in some part on an assessment of the chances of being alive to enjoy consumption later and in some part 'impatience'.

Yet assessing impacts over a very long time period emphasises the problem that future generations are not fully represented in current discussion. Thus we have to ask how they should be represented in the views and decisions of current generations. This throws the second rationale for 'discounting' future consumption mentioned above – pure time preference – into question. We take a simple approach in this Review: if a future generation will be present, we suppose that it has the same claim on our ethical attention as the current one.

Thus, while we do allow, for example, for the possibility that, say, a meteorite might obliterate the world, and for the possibility that future generations might be richer (or poorer), we treat the *welfare* of future generations on a par with our own. It is, of course, possible that people actually do place less value on the welfare of future generations, simply on the grounds that they are more distant in time. But it is hard to see any ethical justification for this. It raises logical difficulties, too. The discussion of the issue of pure time preference has a long and distinguished history in economics, particularly among those economists with a strong interest and involvement in philosophy[9]. It has produced some powerful assertions. Ramsey (1928, p.543) described pure time discounting as 'ethically indefensible and [arising] merely from the weakness of the imagination'. Pigou (1932, pp 24–25) referred to it as implying that 'our telescopic faculty is defective'. Harrod (1948, pp 37–40) described it as a 'human infirmity' and 'a polite expression for rapacity and the conquest of reason by passion'. Solow (1974, p.9) said 'we ought to act as if the social rate of time preference were zero (though we would

---

[9]   See Dasgupta (1974), Anand and Sen (2000), for a technical discussion of these issues, and further references and quotes beyond those here. And see Broome (1991) and (2004) for an extended discussion. We are grateful to Sudhir Anand and John Broome for discussions of these issues.

simultaneously discount future *consumption* if we expected the future to be richer than the present)'. Anand and Sen (2000) take a similar view. So does Cline (1992) in his analysis of the economics of global warming. The appendix to this chapter explores these issues in more technical detail, and includes references to one or two dissenting views.

However, we must emphasise that the approach we adopt, aggregating utility of consumption, does take directly into account the possibility that future generations may be richer or poorer, the first rationale for discounting above. Uncertainty about future prospects plays an important role in the analysis of the Review. How well off we may be when a cost or benefit arrives does matter to its evaluation, as does the probability of the occurrence of costs and benefits. Those issues, *per se*, are not reasons for discounting (other than the case of uncertainty about existence).

A formal discussion of discounting inevitably becomes mathematically technical, as one must be explicit about growth paths and intertemporal allocations. The simple techniques of comparing future incomes or consumption with those occurring now using discount rates (other than for 'pure time preference') is not valid for comparing across paths that are very different. Further, where comparisons are for marginal decisions and the use of discount rates is valid, then, for a number of reasons, particularly uncertainty, discount rates may fall over time. A formal discussion is provided in the appendix to this chapter: the results are summarised in Box 2.2.

BOX 2.2   Discounting

Discounting, as generally used in economics, is a technique relevant for marginal perturbations around a given growth path. A discount rate that is common across projects can be used only for assessing projects that involve perturbations around a path and not for comparing across very different paths.

With marginal perturbations, the key concept is the discount factor: the value of an increment in consumption at a time in the future relative to now. The discount factor will generally depend on the consumption level in the future relative to that now, i.e. on growth, and on the social utility or welfare function used to evaluate consumption (see Box 2.1).

The discount rate is the rate of fall of the discount factor. There is no presumption that it is constant over time, as it depends on the way in which consumption grows over time.

- If consumption falls along a path, the discount rate can be negative.

- If inequality rises over time, this would work to reduce the discount rate, for the social welfare functions typically used.

- If uncertainty rises as outcomes further into the future are contemplated, this would work to reduce the discount rate, with the welfare functions typically used. Quantification of this effect requires specification of the form of uncertainty, and how it changes, and of the utility function.

With many goods and many households, there will be many discount rates. For example, if conventional consumption is growing but the environment is deteriorating, then the discount rate for consumption would be positive but for the environment it would be negative. Similarly, if the consumption of one group is rising but another is falling, the discount rate would be positive for the former but negative for the latter.

Taking the analysis of this section and that of the appendix to this chapter together with the discussion of ethics earlier in this chapter, it can be seen that the standard welfare framework is highly relevant as a theoretical basis for assessing strategies and projects in the context of climate change. However, the implications of that theory are very different from those of the techniques often used in cost-benefit analysis. For example, a single constant discount rate would generally be unacceptable for dealing with the long-run, global, non-marginal impacts of climate change.

For further discussion of discounting, and references to the relevant literature, see the technical annex to this chapter.

This approach to discounting and the ethics from which it is derived is of great importance for the analysis of climate change. That is why we have devoted space to it at the beginning of our Review. **If little or no value were placed on prospects for the long-run future, then climate change would be seen as much less of a problem. If, however, one thinks about the ethics in terms of most standard ethical frameworks, there is every reason to take these prospects very seriously.**

2.5   Risk and Uncertainty

The risks and uncertainties around the costs and benefits of climate policy are large; hence the analytical framework should be able to handle risk and uncertainty explicitly.

For the moment, we do not make a distinction between risk and uncertainty, but the distinction is important and we return to it below. Uncertainty affects every link in the chain from emissions of GHGs through to their impacts. There are uncertainties associated, for example, with future rates of economic growth, with the volume of emissions that will follow, with the increases in temperature resulting from emissions, with the impacts of these temperature increases and so on. Similarly, there are uncertainties associated with the economic response to policy measures, and hence about how much it will cost to reduce GHG emissions.

Our treatment of uncertainty follows a similar approach to that for evaluation or aggregation over space and time. Where we embody uncertainty formally in our models, we add utilities over possible states of the world that might result from climate change, weighting by the probability of those states. This yields what is known as 'expected' utility.

This is essentially the extension of the social utility approach to an uncertain or 'stochastic' environment. As in a certain or 'deterministic' environment, it has its ethical difficulties, but it has the virtues of transparency, clarity, and consistency. Again, it is fairly standard in applied economics.

The basis of such probabilities should be up-to-date knowledge from science and economics. This amounts to a 'subjective' probability approach.[10] It is a pragmatic

---

[10] Often called a 'Bayesian' approach, after Thomas Bayes, the 18th century mathematician. However, the application of Bayes' ideas to a subjective theory of probability was made in the 20th century. See Ramsey (1931).

response to the fact that many of the 'true' uncertainties around climate-change policy cannot themselves be observed and quantified precisely, as they can be in many engineering problems, for example.

The standard expected-utility framework involves aversion to risk and, in this narrow sense, a 'precautionary principle'.

This approach to uncertainty, combined with the assumption that the social marginal utility of income declines as income rises, implies that society will be willing to pay a premium (insurance) to avoid a simple actuarially fair gamble where potential losses and gains are large. As Parts II and III show, potential losses from climate change are large and the costs of avoidance (the insurance premium involved in mitigation), we argue, seem modest by comparison.

The analytical approach incorporates aspects of insurance, caution and precaution directly, and does not therefore require a separate 'precautionary principle' to be imposed as an extra ethical criterion.

More modern theories embodying a distinction between uncertainty and risk suggest an explicit 'precautionary principle' beyond that following from standard expected-utility theory.

The distinction between uncertainty and risk is an old one, going back at least to Knight (1921) and Keynes (1921). In their analysis, risk applied when one could make some assessment of probabilities and uncertainty when one does not have the ability to assess probabilities. In a fascinating paper, Claude Henry (2006) puts these ideas to work on problems in science and links them to modern theories of behaviour towards risk. He uses two important examples to illustrate the relevance of a precautionary principle in the presence of uncertainty. The first is the link between bovine spongiform encephalopathy (BSE) in cows and Creutzfeld-Jacob Disease (CJD) in humans and the second, the link between asbestos and lung disease.

For the first, UK scientists asserted for some time that there could be no link because of 'a barrier between species'. However in 1991 scientists in Bristol succeeded in inoculating a cat with BSE and the hypothesis of 'a barrier' was destroyed. Around the same time, a scientist, Stanley Prusiner, identified protein mutations that could form the basis of a link. These results did not establish probabilities but they destroyed 'certainty'. By introducing uncertainty, the finding opened up the possibility of applying a precautionary principle.

For the second, a possible link between asbestos and lung disease was suggested as early as 1898 by health inspectors in the UK, and in 1911 on a more scientific basis after experiments on rats. Again the work was not of a kind to establish probabilities but provided grounds for precaution. Unfortunately, industry lobbying prevented a ban on asbestos and the delay of fifty years led to considerable loss of life. Application of the precautionary principle could have saved lives.

Henry refers to recent work by Maccheroni et al (2005) and Klibanoff et al (2005) that formalises this type of argument,[11] giving, in effect, a formal description of the precautionary principle. In this formalisation, there are a number of possible probability distributions over outcomes that could follow from some

---

[11] See also Chichilnisky (2000)

action. But the decision-maker, who is trying to choose which action to take, does not know which of these distributions is more or less likely for any given action. It can be shown under formal but reasonable assumptions[12] that she would act as if she chooses the action that maximises a weighted average of the worst expected utility and the best expected utility, where best and worst are calculated by comparing expected utilities using the different probability distributions. The weight placed on the worst outcome would be influenced by concern of the individual about the magnitude of associated threats, or pessimism, and possibly any hunch about which probability might be more or less plausible. It is an explicit embodiment of 'aversion to uncertainty', sometimes called 'aversion to ambiguity', and is an expression of the 'precautionary principle'. It is different from and additional to the idea of 'aversion to risk' associated with and derived from expected utility.

The ability to work with probability distributions in the analysis of climate change was demonstrated in Chapter 1. But there is genuine uncertainty over which of these distributions should apply. In particular, the science and economics are particularly sparse precisely where the stakes are highest – at the high temperatures we now know may be possible. Uncertainty over probability distributions is precisely the situation we confront in the modelling of Chapter 6. As Claude Henry puts it in the conclusion to his 2006 paper, 'uncertainty should not be inflated and invoked as an alibi for inaction'. We now have a theory that can describe how to act.

2.6   Non-marginal policy decisions

There is a serious risk that, without action to prevent climate change, its impacts will be large relative to the global economy, much more so than for most other environmental problems.

The impacts of climate change on economies and societies worldwide could be large relative to the global economy. Specifically, it cannot be assumed that the global economy, net of the costs of climate change, will grow at a certain rate in the future, regardless of whether nations follow a 'business as usual' path or choose together to reduce GHG emissions. In this sense, the decision is not a marginal one.

The issues are represented schematically in Figure 2.3, which compares two paths, one with mitigation and one without. We should note that, in this diagram, there is uncertainty around each path, which should be analysed using the approaches of the preceding section. This is crucial to the analysis in much of the Review. Income on the 'path with mitigation' is below that on the path without ('business as usual') for the earlier time period, because costs of mitigation are incurred. Later, as the damages from climate change accumulate, growth on the 'path without mitigation' will slow and income will fall below the level on the other path. The analysis of Part III attempts to quantify these effects and finds that the 'greener' path (with mitigation) allows growth to continue but, on the path without mitigation, income will suffer. The analysis requires formal comparison between

---

[12] Essentially the axioms are similar to those of the standard Von Neumann-Morgenstern theorem deriving expected utility except the dependence axiom is relaxed slightly. See Gollier (2001), for example, for a description of the Von Neumann-Morgenstern approach.

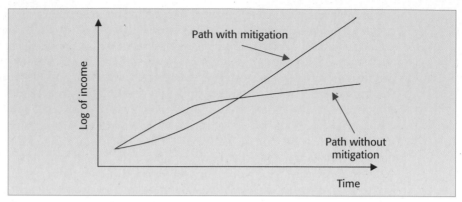

Figure 2.3 Conceptual approach to comparing divergent growth paths over the long term

paths and Part III shows that the losses from mitigation in the near future are strongly outweighed by the later gains in averted damage.

2.7 The public policy of promoting mitigation

Having established the importance of strong mitigation in Parts II and III of the Review, Part IV is devoted to policy to bring it about. The basic theory of externalities identifies the source of the economic problem in untaxed or unpriced emissions of GHGs.

The externality requires a price for emissions: that is the first task of mitigation policy.

The first requirement is therefore to introduce taxes or prices for GHGs. The Pigou treatment of externalities points to taxes based on the marginal damages caused by carbon emissions. In the diagram shown in Figure 2.1, the appropriate tax would be equal to the social cost of carbon at the point where it is equal to the marginal abatement cost. Faced with this tax, the emitters would choose the appropriate level of abatement.

However, the modern theory of risk indicates that long-term quantity targets may be the right direction for policy, with trading within those targets or regular revision of taxes to keep on course towards the long-run objective (see Chapter 14). Given the long-run nature of many of the relevant decisions, whichever policies are chosen, credibility and predictability of policy will be crucial to effectiveness.

The second task of mitigation policy is to promote research, development and deployment.

However, the inevitable absence of total credibility for GHG pricing policy decades into the future may inhibit investment in emission reduction, particularly the development of new technologies. Action on climate change requires urgency, and there are generally obstacles, due to inadequate property rights, preventing investors reaping the full return to new ideas. Specifically, there are spillovers in learning (another externality), associated with the development and adoption of new low-emission technologies that can affect how much emissions are reduced. Thus the economics of mitigating climate change involves understanding the processes of innovation.

The spillovers occur in a number of ways. A firm is unlikely to be able to appropriate all the benefits, largely because knowledge has some characteristics of a public good. In particular, once new information has been created, it can be virtually costless to copy. This allows a competitor with access to the information to capture the benefits without undertaking the research and development (R&D). Patents are commonly used to reduce this problem. In addition, there are typically 'adoptive externalities' to other firms that arise from the processes whereby technology costs fall as a result of increasing adoption. These spillovers are likely to be particularly important in the case of low-emission technologies that can help to mitigate climate change, as Chapter 16 explains.

Other interacting barriers or problem that are relevant include

- asymmetric and inadequate information – for example, about energy-efficiency measures
- policy-induced uncertainties – such as uncertainty about the implicit price of carbon in the future
- moral hazard or 'gaming' – for example firms might rush to make carbon-emitting investments to avoid the possibility of more stringent regulation in the future
- perverse regulatory incentives – such as the incentive to establish a high baseline of emissions in regimes where carbon quotas are 'grandfathered'
- the endogenous price dynamics of exhaustible natural resources – and the risk that fossil-fuel prices could fall in response to strong climate-change policy, threatening to undermine it.[13]

These issues involve many of the most interesting theoretical questions studied by economists in recent years in industrial, regulatory and natural resource economics.

There are important challenges for public policy to promote mitigation beyond the two tasks just described. That is the subject of Chapter 16. These include regulation and standards and deepening public understanding of responsible behaviour.

Standards and regulation can provide powerful and effective policies to promote action on mitigation.

The learning process for new technologies is uncertain. There are probably important scale effects in this process due to experience or learning-by-doing and the externalities of learning-by-watching. In these circumstances, standards for emissions, for example, can provide a clear sense of direction and reduced uncertainty for investors, allowing these economies of scale to be realised.

In other circumstances, particularly concerning energy efficiency, there will be market imperfections, for example due to the nature of landlord-tenant relations in property, that may inhibit adaptation of beneficial investments or technologies. In these circumstances, regulation can produce results more efficient than those that are available from other instruments alone.

Information, education and public discussion can play a powerful role in shaping understanding of reasonable behaviour.

---

[13] The economic theory of exhaustible natural resources is expounded by Dasgupta and Heal (1979). A seminal reference is Hotelling (1931). See, also, Ulph and Ulph (1994), Sinclair (1992) and Sinclair (1994).

Economists tend to put most of weight in public-policy analyses and recommendations on market instruments to which firms and households respond. And there are excellent reasons for this – firms and households know more about their own circumstances and can respond strongly to incentives. But the standard 'sticks and carrots' of this line of argument do not constitute the whole story.

Chapter 17 argues that changing attitudes is indeed likely to be a crucial part of a policy package. But it raises ethical difficulties: who has the right or authority to attempt to change preferences or attitudes? We shall adopt the approach of John Stuart Mill and others who have emphasised 'government by discussion' as the way in which individuals can come to decisions individually and collectively as to the ethical and other justifications of different approaches to policy.

2.8   International action for mitigation and adaptation

The principles of public policy for mitigation elaborated so far do not take very explicit account of the international nature of the challenge. This is a global problem and mitigation is a global public good. This means that it is, from some perspectives, 'an international game' and the theory of games does indeed provide powerful insights. The challenge is to promote and sustain international collective action in a context where 'free-riding' is a serious problem. Adaptation, like mitigation, raises strong and difficult international issues of responsibility and equity, and also has some elements of the problem of providing public goods.

Aspects of adaptation to climate change also have some of the characteristics of public goods and require public policy intervention.

Concerns about the provision of public goods affect policy to guide adaptation to the adverse impacts of climate change. This is the subject of Part V of the Review. Compared with efforts to reduce emissions, adaptation provides immediate, local benefits for which there is some degree of private return. Nevertheless, efficient adaptation to climate change is also hindered by market failures, notably inadequate information on future climate change and positive externalities in the provision of adaptation (where the social return remains higher than the return that will be captured by private investors). These market failures may limit the amount of adaptation undertaken – even where it would be cost-effective.

The ethics of adaptation imply strong support from the rich countries to the most vulnerable.

The poorest in society are likely to have the least capacity to adapt, partly because of resource constraints on upfront investment in adaptive capacity. Given that the greatest need for adaptation will be in low-income countries, overcoming financial constraints is also a key objective. This will involve transfers from rich countries to poor countries. The argument is strongly reinforced by the historical responsibility of rich countries for the bulk of accumulated stocks of GHGs. Poor countries are suffering and will suffer from climate change generated in the past by consumption and growth in rich countries.

Action on climate change that is up to the scale of the challenge requires countries to participate voluntarily in a sustained, coordinated, international effort.

Climate change shares some characteristics with other environmental challenges linked to the management of common international resources, including the protection of the ozone layer and the depletion of fisheries. Crucially, there is no global single authority with the legal, moral, practical or other capacity to manage the climate resource.

This is particularly challenging, because, as Chapter 8 makes clear, no one country, region or sector alone can achieve the reductions in GHG emissions required to stabilise atmospheric concentrations of GHGs at the necessary level. In addition, there are significant gains to co-operating across borders, for example in undertaking emission reductions in the most cost-effective way. The economics and science point to the need for emitters to face a common price of emissions at the margin. And, although adaptation to climate change will often deliver some local reduction in its impact, those countries most vulnerable to climate change are particularly short of the resources to invest in adaptation. Hence international collective action on both mitigation and adaptation is required, and Part VI of the Review discusses the challenges and options.

Economic tools such as game theory, as well as insights from international relations, can aid the understanding of how different countries, with differing incentives, preferences and cost structures, can reach agreement. The problem of free-riding on the actions of others is severe. International collective action on any issue rests on the voluntary co-operation of sovereign states. Economic analysis suggests that multilateral regimes succeed when they are able to define the gain to co-operation, share it equitably and can sustain co-operation in ways that overcome incentives for free-riding.

Our response to climate change as a world is about the choices we make about development, growth, the kind of society we want to live in, and the opportunities it affords this and future generations. The challenge requires focusing on outcomes that promote wealth, consumption, health, reduced mortality and greater social justice.

The empirical analysis of impacts and costs, together with the ethical frameworks we have examined, points to strong action to mitigate GHG emissions. And, given the responsibility of the rich countries for the bulk of the current stock of GHGs, and the poverty and vulnerability of developing countries that would be hardest hit, the analysis suggests that rich countries should bear the major responsibility for providing the resources for adjustment, at least for the next few years. The reasons for strong action by the rich countries are similar to those for aid:

- the moral consequences which flow from a recognition of a common humanity of deep poverty;
- the desire to build a more collaborative, inclusive and better world;
- common interest in the climate and in avoiding dislocation;
- historical responsibility.

2.9   Conclusions

Much of the economics we have begun to describe here and that is put to use in the subsequent parts of this Review is not simple. But the structure of this economics is essentially dictated by the structure of the science. And we have seen that it is not possible to provide a coherent and serious account of the economics of climate change without close attention to the ethics underlying economic policy raised by the challenges of climate change.

The economics of climate change is as broad ranging, deep and complicated as any other area of economics. Indeed, it combines most of the difficulties of other areas of economics. It is unavoidably technical in places. It is the task of this Review to explore the economics of climate change in the depth that is possible given the current state of economic and scientific knowledge. And it should already be clear that much more research is necessary. In many ways, the science has progressed further than the economics.

The scope and depth of the subject require us to put the tools of economics to work across the whole range of the subject. Indeed they point to the importance of tools we wish we had. Nevertheless, the economics can be very powerful in pointing us towards important policy conclusions, as we have already begun to see in this chapter. The urgency of the problems established by the science points to the urgency of translating what we can already show with the economic analysis into concrete policy actions. In doing so, the international dimension must be at centre stage.

References

This chapter touches on many important topics in ethics, social choice and economic theory. On social choice theory, Sen (1986) is a thorough guide, while Sen (1970) is a classic analysis of collective choice and social welfare. Broome (2004) examines the complexities of assessing inter-generational welfare. Henry (2006) considers the distinction between risk aversion and aversion to ambiguity or Knightian uncertainty. Gollier (2001) also raises many important issues concerning discounting of the future. Dasgupta and Heal (1979) investigate the economics of exhaustible resources and its implications for price theory, many of which are not intuitive.

Anand, S and A.K. Sen (2000): 'Human development and economic sustainability', World Development, 28(12): 2029–2049
Arrow, K. (1951): 'Individual Values and Social Choice', New York: Wiley.
Arrow, K. (1963): 'Individual Values and Social Choice', 2nd edition, New York: Wiley.
Beckerman, W. and J. Pasek (2001): 'Justice, Posterity and the Environment', Oxford: Oxford University Press.
Broome, J. (1992): 'Counting the Cost of Global Warming', Cambridge: The White Horse Press.
Broome, J. (1994): 'Discounting the future', Philosophy and Public Affairs 23: 128–56
Broome, J. (2004): 'Weighing lives', Oxford: Oxford University Press.
Broome, J. (2005): 'Should we value population?', The Journal of Political Philosophy. 13(4): 399–413
Chichilnisky, G. (2000): 'An axiomatic approach to choice under uncertainty with catastrophic risk', Resource and Energy Economics, 22: 221–31
Cline, W.R. (1992): 'The Economics of Global Warming', Washington D.C.: Institute for International Economics
Dasgupta, P. S. (1974): 'On some alternative criteria for justice between generations', Journal of Public Economics, 3(4): 405–423
Dasgupta, P. S. and G. M. Heal (1979): 'Economic theory and exhaustible resources', Cambridge: Cambridge University Press.
Fankhauser, S. (1998): 'The costs of adapting to climate change', Global Environment Facility Working Paper 16, Washington, DC: Global Environment Facility.
Gardiner, S. (2004): 'Ethics and global climate change', Ethics 114: 555–600
Gollier, C. (2001): 'The Economics of Risk and Time', Cambridge, MA: MIT Press.
Harrod, R.F. (1948): 'Towards a Dynamic Economics', London: Macmillan.
Henry, C. (2006): 'Decision–Making Under Scientific, Political and Economic Uncertainty', Cahier no. DDX-06-12, Chaire Developpement Durable, Paris; Laboratoire d'Economètrie de l'Ecole Polytechnique.

Hills, J. and K. Stewart (2005): 'A More Equal Society: New Labour, Poverty, Inequality and Exclusion', Bristol: Policy Press.

Hotelling, H. (1931): 'The economics of exhaustible resources', Journal of Political Economy **39**(2): 137–175

Keynes, J.M. (1921): 'A Treatise on Probability', London: Macmillan.

Klibanoff, P., M. Marinacci and S.Mukerji (2005): 'A smooth model of decision-making under ambiguity' Econometrica, **73**: 1849–1892

Knight, F. (1921): 'Risk, Uncertainty and Profit', New York: Kelly.

Maccheroni, F., Marinacci, M., and A. Rustichini (2005): 'Ambiguity aversion', robustness and the variational representation of preferences', available from http://web.econ.unito.it/gma/fabio/mmr-r.pdf

Mueller, B. (2006): 'Adaptation Funding and the World Bank Investment Framework Initiative'. Background Report prepared for the Gleneagles Dialogue Government Working Groups Mexico, 7–9 June 2006.

Pigou, A. C. (1912): 'Wealth and Welfare', London: Macmillan.

Pigou, A. C. (1932): 'The Economics of Welfare', (4th ed), London: Macmillan.

Ramsey, F.P. (1928): 'A Mathematical Theory of Saving', Economics Journal, **38** (December): 543–559

Ramsey, F.P. (1931): 'Foundations of Mathematics and Other Logical Essays', New York: Harcourt Brace.

Samuelson, P. (1954): 'The pure theory of public expenditure'. Review of Economics and Statistics **36**(4): 387–389

Sen, A.K. (1970): 'Collective Choice and Social Welfare', San Francisco: Holden-Day.

Sen, A.K. (1986): 'Social Choice Theory', Handbook of Mathematical Economics, K.J. Arrow and M. Intriligator, Vol3. Amsterdam: North-Holland.

Sen, A.K. (1999): 'The Possibility of Social Choice', American Economic Review **89**(3): 349–378

Sen, A. and B. Williams (eds.) (1982): 'Utilitarianism and Beyond', Cambridge: Cambridge University Press/Maison des Sciences de l'Homme.

Sinclair, P. J. N. (1992): 'High does nothing and rising is worse: carbon taxes should keep declining to cut harmful emissions', The Manchester School of Economic and Social Studies, **60**(1): 41–52.

Sinclair, P. J. N. (1994): 'On the optimum trend of fossil fuel taxation'. Oxford Economic Papers **46**: 869–877.

Solow, R.M. (1974): 'The economics of resources or the resources of economics'. American Economic Review, **64**(2): 1–14.

Ulph, A. and D. Ulph (1994). 'The optimal time path of a carbon tax'. Oxford Economic Papers **46**: 857–868

2A Ethical Frameworks and Intertemporal Equity

2A.1 Ethical frameworks for climate change

The 'consequentialist' and 'welfarist' approach – the assessment of a policy in terms of its consequences for individual welfare – that is embodied in standard welfare economics is highly relevant to the ethics of climate change.

In Section 2.3, we described the standard approach to ethics in welfare economics i.e. the evaluation of actions in terms of their consequences for consumption by individuals of goods and services. We emphasised that 'goods and services' in consumption were multi-dimensional and should be interpreted broadly. In this appendix we examine that approach in a little more detail and compare it with different ethical perspectives of relevance to the economics of climate change.

For many applications of the standard theory, the community is defined as the nation-state and the decision-maker is interpreted as the government. Indeed this is often seen as sufficiently obvious as to go unstated. This is not, of course, intended to deny the complexities and pressures of political systems: the results of this approach should be seen as an ethical benchmark rather than a descriptive model of how political decisions are actually taken.

Nevertheless, questions such as 'what do individuals value', 'what should be their relation to decisions and decision-making', 'what is the decision-making process' and 'who are the decision-makers' arise immediately and strongly in the ethical analysis of climate change. These questions take us immediately to different perspectives on ethics.

Economics, together with the other social sciences, has in fact embraced a much broader perspective on the objectives of policy than that of standard welfare-economic analysis. Amartya Sen[1], for example, has focused on the capabilities and freedoms of individuals to live a life they have reason to value, rather than narrowly on the bundles of goods and services they consume. His focus is on opportunities and the processes that create them, rather than on outcomes only. Similar emphases come from discussions of equity[2] (with its focus on opportunity), empowerment[3], or social inclusion[4].

While such perspectives are indeed different, in practice many of the indicators arising from them would overlap strongly with the areas of focus in the Millennium Development Goals (MDGs) and other indicators commonly used by international institutions. Indeed, the MDGs were the outcome of analyses and discussions which themselves embraced a range of ethical approaches.

---

[1] Sen (1999).
[2] e.g. *World Development Report 2006*.
[3] e.g. Stern *et al.* (2005).
[4] Atkinson and Hills (1998), Atkinson *et al.* (2002), Hills and Stewart (2005).

Impacts of climate change on future generations and other nations raise very firmly questions of rights. Protection from harm done by others lies at the heart of many philosophical approaches to liberty, freedom and justice.[5]

Protection from harm is also expressed in many legal structures round the world in terms of legal responsibility for damage to the property or well-being of others. This is often applied whether or not the individual or firm was knowingly doing harm. A clear example is asbestos, whose use was not prohibited[6] when it was placed in buildings with the worthy purpose of protecting against the spread of fire. Nevertheless insurance companies are still today paying large sums as compensation for its consequences.

This is a version of the 'polluter pays' principle that is derived from notions of rights, although, as we saw, for example, in the discussion of Fig. 2.1 above, it also arises from an efficiency perspective within the standard economic framework. If this interpretation of rights were applied to climate change, it would place at least a moral, if not a legal, responsibility on those groups or nations whose past consumption has led to climate change.

Looking at the moral responsibilities of this generation, many would argue that future generations have the right to enjoy a world whose climate has not been transformed in a way that makes human life much more difficult; or that current generations across the world have the right to be protected from environmental damage inflicted by the consumption and production patterns of others.

The notions of the right to climate protection or climate security of future generations and of shared responsibilities in a common world can be combined to assert that, collectively, we have the right only to emit some very small amount of GHGs, equal for all, and that no-one has the right to emit beyond that level without incurring the duty to compensate. We are therefore obliged to pay for the right to emit above that common level. This can be seen as one argument in favour of the 'contract and converge' proposition, whereby 'large emitters' should contract emissions and all individuals in the world should either converge to a common (low) level or pay for the excess (and those below that level could sell rights).

There are problems with this approach, however. One is that this right, while it might seem natural to some, is essentially asserted. It is not clear why a common humanity in a shared world automatically implies that there are equal rights to emit GHGs (however low). Equality of rights, for example to basic education and health, or to common treatment in voting, can be related to notions of capabilities, empowerment, or the ability to participate in a society. Further, they have very powerful consequences in terms of law, policy and structures of society. How does the 'right to emit' stand in relation to these rights? Rights are of great importance in ethics but they should be argued rather than merely asserted. More pragmatically, as we shall examine in Part VI of this report, action on climate change requires international agreement and this is not a proposition likely to gain the approval necessary for it to be widely adopted.

---

[5] See, for example, Shue (1999) on the 'no-harm principle' in the context of climate change and Gardiner (2004) for a link with John Rawls' theory of justice. From the point of view of jurisprudence, and for a discussion of links with notions of retribution, see Hart (1968).

[6] As Henry (2006) argued, the possibility of harmful effects had been discovered around 100 years ago, but this would not necessarily be generally known by those whose used it.

A concept related to the idea of the rights of future generations is that of sustainable development: future generations should have a right to a standard of living no lower than the current one.

In other words, the current generation does not have the right to consume or damage the environment and the planet in a way that gives its successor worse life chances than it itself enjoyed. The life chances of the next generation, it is understood here, are assessed assuming that it behaves in a sustainable way, as defined here, in relation to its own successor generation[7].

Expressed in this form, however, the principle need not imply that the whole natural environment and endowment of resources should be preserved by this generation for the next generation in a form exactly as received from the previous generation. The capital stock passed on to the next generation consists of many things, mostly in the form of stocks covering, for example, education, health, capital equipment, buildings, natural resources, and the natural environment. The standard of living available to the next generation depends on this whole collection of stocks. A decline in one of them, say copper, might be compensated by another stock, say education or infrastructure, which has increased.

On the other hand, it seems quite clear that, at a basic level, the global environmental and ecological system, which provides us with life support functions such as stable and tolerable climatic conditions, cannot be substituted. The relation between emissions of GHGs and the risks to these functions is examined in detail in the Review, particularly Part II. The commitment of Article 2 of the United Nations Framework Convention on Climate Change (UNFCCC) to 'achieve stabilisation of greenhouse gas concentrations at a level that would prevent dangerous anthropogenic [i.e. human-induced] interference with the climate system' can be interpreted as just such a sustainability rule.

The notion of 'stewardship' can be seen as a special form of sustainability. It points to particular aspects of the world, which should themselves be passed on in a state at least as good as that inherited from the previous generation.

Examples might be historic buildings, particular pieces of countryside, such as National Parks, or even whole ecosystems such as tracts of primary tropical rainforest. This involves a particular interpretation of the responsibilities of the current generation in terms of a limit on its rights to property. Essentially, in this approach each generation has the responsibility of stewardship. Some would see the climate in this way, since it shapes so much of all the natural environment and is not straightforwardly substitutable with other capital. Others[8] might ask still more basic questions as to how we ought to live, particularly in relation to nature.

These different notions of ethics emphasise different aspects of the consequences of decisions for others and for the future. Nevertheless, the list of consequences on which they would focus for each generation are similar: above all consumption, education, health and environment.

And all the perspectives would take into account the distribution of outcomes within and across generations, together with the risks involved in different actions,

---

[7] A valuable summary of the analytic background and foundations of sustainability is given by Anand and Sen (2000). See also Solow (1974).
[8] Jamieson (1992).

now and over time. Hence in the Review we shall focus our analysis on the implications of action or inaction on climate change for these four dimensions.

How the implications on these four dimensions are assessed, will, of course, vary according to the ethical position adopted. How and whether, in making assessments, we attempt to aggregate over consequences (i) within generations, (ii) over time, and (iii) according to risk will be crucial to policy design and choice. When we do aggregate explicitly we have to be quantitative in comparing consequences of different kinds and for different people. We shall be paying special attention to all three forms of aggregation. Aggregation across dimensions poses different kinds of questions and problems, as was discussed in Section 2.3 above.

2A.2   Intertemporal appraisals and discounting[9]

Introduction: the underlying welfare framework for appraisal and cost-benefit analysis

Different strategies for climate change will yield different patterns of consumption over time. We assume that a choice between strategies will depend on their consequences for households now and in the future (see Chapter 2 and 2A.1 above, for a brief discussion of 'consequentialism'). The households to be included and examined in this weighting of consequences will depend on the perspective of those making the judgements: we assume here that the assessment is done from the perspective of the world as a whole. Narrower perspectives would include, for example, only those households associated with a particular country or region and would follow similar reasoning except that net benefits would be assessed for a narrower group. If all the perspectives are from narrow groups, one country, or just the next one or two generations, it is likely that little action would be taken on global warming. As is emphasised throughout this Review, this is a global and long-run issue.

An analysis of how to carry out an intertemporal assessment of consequences of strategies or actions is inevitable if somewhat formal: usually there would be first a modelling of the consequences, second an aggregation of the consequences into overall welfare indicators for households, and third an aggregation across households within generations, across generations and across uncertain outcomes. We focus here on the second and third elements, particularly the third.

We can compare the consequences of different strategies and actions by thinking of overall welfare, W, calculated across households (and generations) as a function of the welfare of these households, where we write welfare of household h as u^h. The joint specification of W and u^h constitutes a set of value judgements which will guide the assessment of consequences. We think of h as ranging across households now and in the future and can allow (via specification of W and u^h) for the possibility that a household does not live forever. Then, if we are

---

[9]   This section has benefited from discussions with Cameron Hepburn and Paul Klemperer, although they are not responsible for the views expressed here. See also Hepburn (2006).

comparing a strategy indexed by the number 1 with that indexed by zero we will prefer strategy 1 if

$$W^1 > W^0 \tag{1}$$

where W^1 is evaluated across the path 1 with its consequences for all households now and in the future, and similarly W^0.

In the above, the two strategies can yield very different patterns of outcomes across individuals and over time – they can differ in a non-marginal way. There is, however, a major part of economic theory that works in terms of a marginal change, for example an investment project. Then we can write, where W^1 is welfare in the world with the project and W^0 is welfare in the world without the project,

$$\Delta W \equiv W^1 - W^0 = \sum_h \frac{\partial W}{\partial u^h} \Delta u^h \tag{2}$$

where Δu^h is the change in household welfare for h as a result of the project. Calculating Δu^h will then depend on the structure of the economic model and the characteristics of the project. This is the theory of cost-benefit analysis set out clearly by James Meade (1955) and explored in some detail by Drèze and Stern (1987) and (1990) for imperfect economies.

As we have argued, strategies on climate change cannot be reduced to marginal comparisons, so we have to examine W^1 and W^0 (for different strategies) and, for many climate change questions, we must compare the two without using the very special case of marginal comparisons as in equation (2).

Nevertheless there will be investment projects which can be considered as small variations around a particular path e.g. a new technique in electricity generation. In this case marginal analysis can be appropriate. In this context we can think about comparing benefits occurring at different points in time, in terms of how we should value small changes around a particular path. This leads to the subject of discounting and how we value marginal benefits which are similar in nature but which occur at different points in time. We must emphasise very strongly that these valuations occur with respect to variations around a particular path. If the path is shifted, so too are the marginal valuations and thus discount factors and rates (see below).

An investment carried out now may yield returns which are dependent on which strategy, and thus which growth paths, might be followed. If we are uncertain about these strategies, for example, we do not know whether the world would follow a strong mitigation strategy or not, then we should evaluate the project for each of the relevant scenarios arising from the strategies. Each of these evaluations would then be relative to a different growth path. The next step would not be straightforward. We could aggregate across the scenarios or growth paths using probabilities and relative values of social marginal utilities relevant for the different paths (i.e. we would have to compare the numeraire used for each path) but only if we are in a position to assign probabilities. Further, a related discussion over strategies may be going on at the same time as the projection evaluation.

Discounting: a very simple case

Discounting and discount rates have been controversial in environmental economics and the economics of climate change, because a high rate of discounting of the future will favour avoiding the costs of reducing emissions now, since the gains from a safer and better climate in the future are a long way off and heavily discounted (and vice versa for low discount rates). Our first and crucial point has been made already: discounting is in general a marginal approach where the evaluation of marginal changes depends on the path under consideration. If the two paths are very different, a marginal/discounting approach for comparing the two is unacceptable in logic – we have to go back to an evaluation of the underlying W for each path.

The discounting approach is, however, relevant for small changes around a given path and, since some of the literature has been somewhat confused on the issue and because it brings out some important issues relevant for this Review, we provide a brief description of the main principles here. To do this, we narrow down the relevant determinants of utility to just consumption at each point in time and take a very special additive form of W. Thus we think of the overall objective as the sum (or integral) across all households and all time of the utility of consumption. In order to establish principles as clearly as possible, we simplify still further to write

$$W = \int_0^\infty u(c)e^{-\delta t}dt \qquad\qquad (3)$$

We assume here that there is just one individual at each point in time (or a group of identical individuals) and that the utility or valuation function is unchanging over time. We introduce population and its change later in the discussion.

In Chapter 2 we argued, following distinguished economists from Frank Ramsey in the 1920s to Amartya Sen and Robert Solow more recently, the only sound ethical basis for placing less value on the utility (as opposed to consumption) of future generations was the uncertainty over whether or not the world will exist, or whether those generations will all be present. Thus we should interpret the factor $e^{-\delta t}$ in (3) as the probability that the world exists at that time. In fact this is exactly the probability of survival that would apply if the destruction of the world was the first event in a Poisson process with parameter δ (i.e. the probability of an event occurring in a small time interval Δt is $\delta\Delta t$). Of course, there are other possible stochastic processes that could be used to model this probability of survival, in which case the probability would take a different form. The probability reduces at rate δ. With or without the stochastic interpretation here, δ is sometimes called 'the pure time discount rate.' We discuss possible parameter values below.

The key concept for discounting is the marginal valuation of an extra unit of consumption at time t, or *discount factor*, which we denote by λ. We can normalise utility so that the value of λ at time 0 along the path under consideration is 1. We are considering a project which perturbs consumption over time around this particular path. Then, following the basic criterion, equation 2, for marginal changes we have to sum the net incremental benefits accruing at each point in time, weighting those accruing at time t by λ. Thus, from the basic marginal criteria (2), in the special case (3), we accept the project if

$$\Delta W = \int_0^\infty \lambda \Delta c\, dt > 0 \qquad\qquad (4)$$

where λ and c are each evaluated at time t, Δc is the perturbation to consumption at time t arising from the project and λ is the marginal utility of consumption where

$$\lambda = u'(c)e^{-\delta t} \tag{5}$$

If, for example, we have to invest to gain benefits then Δc will be negative for early time periods and positive later.

The rate of fall of the discount factor is the *discount rate*, which we denote by ρ. These definitions and the special form of λ as in (5) are in the context of the very strong simplifications used. Under uncertainty or with many goods or with many individuals, there will be a number of relevant concepts of discount factors and discount rates.

The discount factors and rates depend on the numeraire that is chosen for the calculation. Here it is consumption and we examine how the present value of a unit of consumption changes over time. If there are many goods, households, or uses of revenue we must be explicit about choice of numeraire. There will, in principle, be different discount factors and rates associated with different choices of numeraire see below.

Even in this very special case, there is no reason to assume the discount rate is constant. On the contrary, it will depend on the underlying pattern of consumption for the path being examined; remember that λ is essentially the discounted marginal utility of consumption along the path.

Let us simplify further and assume the very special 'isoelastic' function for utility

$$u(c) = \frac{c^{1-\eta}}{1-\eta} \tag{6}$$

(where, for $\eta = 1$, u(c) = log c). Then

$$\lambda = c^{-\eta}e^{-\delta t} \tag{7}$$

and the discount rate ρ, defined as $-\dot{\lambda}/\lambda$, is given by

$$\rho = \eta\frac{\dot{c}}{c} + \delta \tag{8}$$

To work out the discount rate in this very simple formulation we must consider three things. The first is η which is the elasticity of the marginal utility of consumption. [10] In this context it is essentially a value judgement. If, for example $\eta = 1$, then we would value an increment in consumption occurring when utility was 2c as half as valuable as if it occurred when consumption was c. The second is $\dot{c}/c$, the growth rate of consumption along the path: this is a specification of the path itself or the scenario or forecast of the path of consumption as we look to the future. The third is δ, the pure time discount rate, which generates, as discussed, a probability of existence of $e^{-\delta t}$ at time t (thus δ is the rate of fall of this probability).

---

[10] See e.g. Stern (1977), Pearce and Ulph (1999) or HM Treasury (2003) for a discussion of some of the issues.

Table 2A.1

	Probability of human race surviving 10 years	Not surviving 10 years	Probability of human race surviving 100 years	Not surviving 100 years
$\delta = 0.1$	0.990	0.010	0.905	0.095
0.5	0.951	0.049	0.607	0.393
1.0	0.905	0.095	0.368	0.632
1.5	0.861	0.139	0.223	0.777

The advantage of (8) as an expression for the discount rate is that it is very simple and we can discuss its value in terms of the three elements above. The Treasury's Green Book (2003) focuses on projects or programmes that have only a marginal effect relative to the overall growth path and thus uses the expression (8) for the discount rate. The disadvantage of (8) is that it depends on the very specific assumptions involved in simplifying the social welfare function into the form (3).

There is, however, one aspect of the argument that will be important for us in the analysis that follows in the Review and that is the appropriate pure time discount rate. We have argued that it should be present for a particular reason, i.e. uncertainty about existence of future generations arising from some possible shock which is exogenous to the issues and choices under examination (we used the metaphor of the meteorite).

But what then would be appropriate levels for δ? That is not an easy question, but the consequences for the probability for existence of different δs can illuminate — see Table 2A.1.

For $\delta = 0.1$ per cent, there is an almost 10% chance of extinction by the end of a century. That itself seems high – indeed if this were true, and had been true in the past, it would be remarkable that the human race had lasted this long. Nevertheless, that is the case we shall focus on later in the Review, arguing that there is a weak case for still higher levels.[11] Using $\delta = 1.5$ per cent, for example, i.e. 0.015, the probability of the human race being extinct by the end of a century would be as high as 78%, indeed there would be a probability of extinction in the next decade of 14%. That seems implausibly, indeed unacceptably, high as a description of the chances of extinction.

However, we should examine other interpretations of 'extinction'. We have expressed survival or extinction of the human race as either one or the other and have used the metaphor of the devastating meteorite. There are also possibilities of partial extinction by some exogenous or man-made force which has little to do with climate change. Nuclear war would be one possibility or a devastating outbreak of some disease which 'took out' a significant fraction of the world's population.

In the context of *project uncertainty*, rather different issues arise. Individual projects can and do collapse for various reasons and in modelling this type of process we might indeed consider values of δ rather higher than shown in this table. This type of issue is relevant for the assessment of public sector projects, see, for example, HM Treasury (2003), the Green Book.

---

[11] See also Hepburn (2006).

A different perspective on the pure time preference rate comes from Arrow (1995). He argues that one problem with the absence of pure time discounting is that it gives an implausibly high optimum saving rate using the utility functions as described above, in a particular model where output is proportional to capital. If $\delta = 0$ then one can show that the optimum savings rate in such a model[12] is $1/\eta$; for η between 1 and 1.5 this looks very high. From a discussion of 'plausible' saving rates he suggests a δ of 1%. The problem with Arrow's argument is, first, that there are other aspects influencing optimum saving in possible models that could lower the optimum saving rate and, second, that his way of 'solving' the 'over-saving' complication is very ad hoc. Thus the argument is not convincing.

Arrow does in his article draw the very important distinction between the 'prescriptive' and the 'descriptive' approach to judgements of how to 'weigh the welfare' of future generations – a distinction due to Nordhaus (see Samuelson and Nordhaus (2005)). He, like the authors described in Chapter 2 on this issue, is very clear that this should be seen as a prescriptive or ethical issue rather than one which depends on the revealed preference of individuals in allocating their own consumption and wealth (the descriptive approach). The allocation an individual makes in her own lifetime may well reflect the possibility of her death and the probability that she will survive a hundred years may indeed be very small. But this intertemporal allocation by the individual has only limited relevance for the long-run ethical question associated with climate change.

There is nevertheless an interesting question here of combining short-term and long-term discounting. If a project's costs and benefits affect only this generation then it is reasonable to argue that the revealed relative valuations across periods has strong relevance (as it does across goods). On the other hand, as we have emphasised allocation across generations and centuries is an ethical issue for which the arguments for low pure time discount rates are strong.

Further, we should emphasise that using a low δ does not imply a low discount rate. From (8) we see, e.g., that if η were, say, 1.5, and c/c were 2.5% the discount rate would be, for $\delta = 0$, 3.75%. Growing consumption *is* a reason for discounting. Similarly if consumption were falling the discount rate would negative.

As the table shows the issue of pure time discounting is important. **If the ethical judgement is that future generations count very little *regardless of their consumption level* then investments with mainly long-run pay-offs would not be favoured. In other words, if you care little about future generations you will care little about climate change. As we have argued, that is not a position which has much foundation in ethics and which many would find unacceptable.**

Beyond the very simple case

We examine in summary form the key simplifying assumptions associated with the formulation giving equations (3) and (8) above, and ask how the form and time pattern of the various discount factors and discount rates might change when these assumptions are relaxed.

---

[12] This uses the optimality condition that the discount rate (as in (8)) should be equal to the marginal product of capital.

Case 1   Changing population

With population N at time t and total consumption of C, we may write the social welfare function to generalise (3) as

$$W = \int_0^\infty Nu(C/N)e^{-\delta t}dt \tag{9}$$

In words, we add, over time, the utility of consumption per head times the number of people with that consumption: i.e. we simply add across people in this generation, just as in (3) we added across time; we abstract here from inequality within the generation (see below). Then the social marginal utility of an increment in *total* consumption at time t is again given by (5) where c is now C/N consumption per head. Thus the expression (8) for the discount rate is unchanged. We should emphasise here that expression (9) is the appropriate form for the welfare function where population is exogenous. In other words we know that there will be N people at time t. Where population is endogenous some difficult ethical issues arise – see, for example, Dasgupta (2001) and Broome (2004, 2005).

Case 2   Inequality within generations

Suppose group i has consumption C_i and population N_i. We write the utility of consumption at time t as

$$\sum_i N_i u(C_i / N_i)e^{-\delta t} \tag{10}$$

and integrate this over time: in the same spirit as for (9), we are adding utility across sub-groups in this generation. Then we have, replacing (5), where c_i is consumption per head for group i,

$$\lambda_i = u'(c_i)e^{-\delta t} \tag{11}$$

as the discount factor for weighting increments of consumption to group i. Note that in principle the probability of extinction could vary across groups, thus making δ_i dependent on i.

An increment in *aggregate consumption* can be evaluated only if we specify how it is distributed. Let us assume a unit increment is distributed across groups in proportions α_i. Then

$$\lambda = \sum_i \alpha_i u'(c_i)e^{-\delta t} \tag{12}$$

For some cases α_i may depend on c_i, for example, if the increment were distributed just as total consumption, so that $\alpha_i = C_i/C$ where C is total consumption. In this case, the direction of movement of the discount rate will depend on the form of the utility function. For example, in this last case, if $\eta = 1$, the discount rate would be unaffected by changing inequality.

If $\alpha_i = 1/N$ this is essentially 'expected utility' for a 'utility function' given by $u'(\ )$. Hence the Atkinson theorem (1970) tells us that if $\{c_i\}$ becomes more unequal[13] then λ will rise and the discount rate will fall if u' is convex (and vice versa if it is concave). The convexity of $u'(\ )$ is essentially the condition that the third derivative of u is positive: all the isoelastic utility functions considered here satisfy this condition[14].

For α_i 'tilted' towards the bottom end of the income distribution, the rise is reinforced. Conversely, it is muted or reversed if α_i is 'tilted' towards the top end of the income distribution. For example, where $\alpha_i = 1$ for the poorest subset of households, then λ will rise where rising inequality makes the poorest worse off. But where $\alpha_N = 1$ for the richest household, λ will fall if rising income inequality makes the richest better off. Note that in the above specification the contribution of individual i to overall social welfare depends only on the consumption of that individual. Thus we are assuming away consumption externalities such as envy.

Case 3   Uncertainty over the growth path

We cannot forecast, for a given set of policies, future growth with certainty. In this case, we have to replace the right-hand side of (5) in the expression for λ by its expectation. This then gives us an expression similar to (12), where we can now interpret α_i as the probability of having consumption in period t, denoted as p_i in equation (13). We would expect uncertainty to grow over time in the sense that the dispersion would increase. Under the same assumptions, i.e. convexity of u', as for the increasing inequality case, this increasing dispersion would reduce the discount rate over time. Increased uncertainty (see Rothschild and Stiglitz (1976) and also Gollier (2001)) increases λ if u' is convex since λ is essentially expected utility with u' as the utility function.

$$\lambda = \sum_i p_i u'(c_i) e^{-\delta t} \tag{13}$$

Figure 2A.1 shows a simple example of how the discount factor falls as consumption increases over time, when the utility function takes the simple form given in equation (6). The chart plots the discount factor along a range of growth paths for consumption; along each path, the growth rate of consumption is constant, ranging from 0 per cent to 6 per cent per year. The value of δ is taken to be 0.1 per cent and of η 1.05. The paths with the lowest growth rates of consumption are the ones towards the top of the chart, along which the discount factor declines at the slowest rate. Figure 2A.2 shows the average discount rate over time corresponding to the discount factor given by equation (13), assuming that all the paths are equally likely. This falls over time. For further discussion of declining discount rates, see Hepburn (2006).

---

[13] This property can be defined via distribution functions and Lorenz curves. It is also called second-order stochastic domination or Lorenz-dominance: see e.g. Gollier (2001), Atkinson (1970) and Rothschild and Stiglitz (1970).

[14] Applying the same theory to the utility function shows that total utility will be lower under greater inequality for a concave utility function.

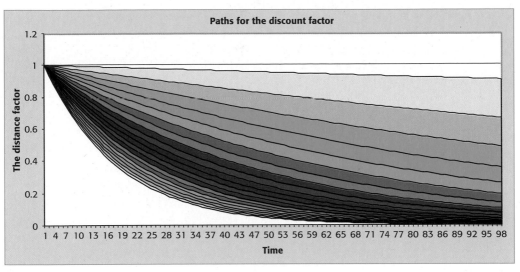

Figure 2A.1

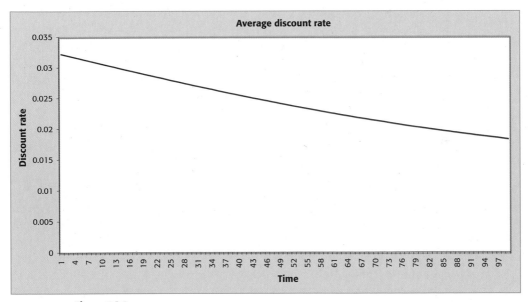

Figure 2A.2

Further complications

The above treatment has kept things very simple and focused on a case with one consumption good and one type of consumer, and says little about markets.

Where there are *many goods*, and *different types of household* and *market imperfections* we have to go back to the basic marginal criterion specified in (2) and evaluate Δu^h for each household taking into account these complications: for a discussion, see Drèze and Stern (1990). There will generally be a different discount rate for each good and for each consumer. One can, however, work in terms of a discount rate for aggregate (shadow) public revenue.

A case of particular relevance in this context would be where utility depended on both current consumption and the natural environment. Then it is highly likely that the relative 'price' of consumption and the environment (in terms of willingness-to-pay) will change over time. The changing price should be explicit and the discount rate used will differ according to whether consumption or the environment is numeraire (see below on Arrow (1966)).

Growing benefits in a growing economy: convergence of integrals

We examine the special case (4) of the basic marginal criteria (2). The convergence of the integral requires λ to fall faster than the net benefits Δc are rising. Without convergence, it will appear from (4) that the project has infinite value. Suppose consumption grows at rate g and the net benefits at g. From (8) and (4) we have that for convergence we need, in the limit into the distant future,

$$\eta g + \delta > \hat{g} \tag{14}$$

If, for example, g and $\hat{g}$ are the same (benefits are proportional to consumption) then for convergence we need, in the limit,

$$\delta > (1 - \eta)g \tag{15}$$

Where $\eta \geqslant 1$ and $\delta > 0$, this will be satisfied. But for $\eta < 1$ there can be problems. Given that infinite aggregate net benefits are implausible this could be interpreted as an argument for a high η or more precisely a high limiting value of η. We have so far assumed that η is constant (the isoelastic case) but it could, however, in principle be higher for very high c. As we have indicated, arguments for a high δ should be conducted on a separate basis concerning the probability of existence, and we have, in this context, argued for a low value of δ.

Market rates, capital market imperfections and intergenerational welfare

Some may object that the discount rates which would arise from (8), e.g. 3-4% or lower, may not directly reflect market interest rates[15]. Further, it may be argued, market interest rates give the terms under which individuals actually do make intertemporal allocations and thus these market rates reflect individual marginal rates of substitution between goods now and in the future. Thus, in this argument, market rates should be used as discount rates.

There are a number of reasons why this argument may be misleading, including capital market imperfections and myopia. And the argument begs the question of which of the many different market rates of interest and return might be relevant. In this context, however, we would emphasise, as argued in Chapter 2,

---

[15] However, these values are not far away from real long-run returns on government bonds.

that the decisions at issue for the long-run analyses concern allocation *across generations* rather than within. One can confront these only by looking carefully at the ethical issues themselves. The intertemporal valuations of individuals over their own lifetimes, as we have argued, is not the same issue. They do not constitute a market-revealed preference of the trade-offs at stake here.

This is not the place for a detailed analysis of market imperfections, 'crowding out of investment' and discount rates. The reader may wish to consult Drèze and Stern (1987 and 1990) and some of the references therein, in particular Arrow (1966). An intuitive expression of the Arrow argument is as follows. The issue concerns the relative value of two forms of income, call this relative value μ. These different forms of income can be, for example, consumption, investment or government income. If μ is constant over time, then the discount rates, whether we work in terms of consumption, investment or government income, should be the same. The reason is that the difference between the two discount rates for the two forms of income is simply the rate of change of μ (since $\mu = \lambda^A / \lambda^B$ where λ are the discount factors for incomes type A and B respectively). The reason that μ is not unity arises from various market imperfections and constraints on the tax system (otherwise the government would shift resources so that λ^A is equal to λ^B). And if the intensities of these imperfections and constraints are unchanged over time, then μ will be constant and the relevant discount rates will be equal.

2A.3   Conclusions

Discounting is a technique relevant for marginal perturbations around a given growth path. Where the strategies being compared involve very different paths, then discounting can be used only for assessing projects which involve perturbations around a path and not for comparing across paths. There will be important decisions for which marginal analysis is appropriate, including, for example, technological choices to sustain given paths of emission reduction. We must emphasise, however, that, as with any marginal analysis, the marginal valuations will depend on the paths under consideration. Which path or paths are relevant will depend on the overall strategies adopted.

Within the case of marginal perturbations, the key concept is the discount factor, i.e. the present value of the numeraire good: here the discount factor is the relative value of an increment in consumption at a time in the future relative to now. The discount factor will generally depend on the consumption level in the future relative to that now, i.e. on the growth path, and on the social utility or welfare function used to evaluate consumption. The discount rate is the rate of fall of the discount factor. It depends on the way in which consumption grows over time. If consumption falls along a path then the discount rate can be negative. There is no presumption that it is constant over time.

- If inequality rises over time then this would work to reduce the discount rate, for the welfare functions commonly used.
- If uncertainty rises as we go into the future, this would work to reduce the discount rate, for the welfare functions commonly used. Quantification of this effect requires specification of the form of uncertainty, and how it changes, and of the utility function.

- With many goods and many households there will be many discount rates. For example, if conventional consumption is growing but the environment is deteriorating, then the discount rate for consumption would be positive but for the environment it would be negative. Similarly, if the consumption of one group is rising but another is falling, then the discount rate would be positive for the former but negative for the latter.

Taken together with our discussion of ethics, we see that the standard welfare framework is highly relevant as a theoretical basis for assessing strategies and projects in the context of climate change. However, the implications of that theory are very different from those of the techniques often used in cost-benefit analysis. For example, a single constant discount rate would generally be unacceptable.

Whether we are considering marginal or non-marginal changes or strategies the 'pure time discount rate' is of great importance for a long-run challenge such as climate change. The argument in the chapter and in the appendix and that of many other economists and philosophers who have examined these long-run, ethical issues, is that 'pure time discounting' is relevant only to account for the exogenous possibility of extinction. From this perspective, it should be small. On the other hand, those who would put little weight on the future (regardless of how living standards develop) would similarly show little concern for the problem of climate change.

References

Anand, S and A.K. Sen (2000): 'Human development and economic sustainability'. World Development, 28(12): 2029–2049

Arrow, K.J. (1966): 'Discounting and public investment criteria', in Water Research, A.V. Kneese and S.C. Smith, (eds.), John Hopkins University Press.

Arrow, K.J. (1995): 'Inter-generational equity and the rate of discount in long-term social investment' paper at IEA World Congress (December), available at www.econ.stanford.adu/faculty/workp/swp97005.htm

Atkinson, A. B. (1970): 'On the measurement of inequality', Journal of Economic Theory, 2(3): 244–263

Atkinson, A.B. and J. Hills (1998): 'Social exclusion, poverty and unemployment', Exclusion, Employment and Opportunity, A.B. Atkinson and J. Hills, London: London School of Economics. CASE Paper 4.

Atkinson, A.B., B. Cantillon, E. Marlier and B.Nolan (2002): 'Social Indicators:The EU and Social Inclusion'. Oxford: Oxford University Press.

Broome, J. (2004): 'Weighing Lives', Oxford: Oxford University Press.

Broome, J. (2005): 'Should we value population', The Journal of Political Philosophy: 13(4): 399–413

Dasgupta, P. (2001):'Human Well-Being and the Natural Environment', Oxford: Oxford University Press.

Drèze, J.P. and N.H. Stern (1987): 'The Theory of Cost-Benefit Analysis' in (ed.) Auerback and Feldstein, Handbook of Public Economics, Vol 12, North Holland.

Drèze, J.P. and N.H. Stern (1990): 'Policy reform, shadow prices and market prices', Journal of Public Economics, 42: 1–45

Gardiner, S. (2004): 'Ethics and global climate change', Ethics 114: 555–600

Gollier, C. (2001): 'The Economics of Risk and Time', Cambridge, MA: MIT Press.

Hart, H.L.A. (1968): 'Punishment and Responsibility', Oxford: Oxford University Press.

Hepburn, C. (2006): 'Discounting climate change damages', Overview for the Stern Review, St Hugh's College, Oxford.

Henry, C. (2006): 'Decision–Making Under Scientific, Political and Economic Uncertainty', Cahier no. DDX-06-12, Chaire Developpement Durable, Paris; Laboratoire d'Econométrie de l'Ecole Polytechnique.

Hills, J. and K. Stewart (2005) 'A More Equal Society: New Labour, Poverty, Inequality and Exclusion'. Bristol: Policy Press.

HM Treasury (2003): 'The Green Book – Appraisal and Evaluation in Central Government', HM Treasury, London.

Jamieson, D. (1992): 'Ethics, public policy and global warming', Science, Technology and Human Values 17:139–53. Reprinted in Jamieson, D. (2003), Morality's Progress, Oxford: Oxford University Press.

Meade, J.E. (1955): ''Trade and Welfare: A mathematical supplement', Oxford: Oxford University Press.

Pearce, D.W. and A. Ulph (1999): 'A social discount rate for the United Kingdom'. Economics and the Environment: Essays in Ecological Economics and Sustainable Development, D.W. Pearce, Cheltenham: Edward Elgar.

Rothschild, M.D. and J.E. Stiglitz (1970): 'Increasing Risk: I. A Definition', Journal of Economic Theory, September 2 pp. 255–243.

Samuelson, P.A., and W.D. Nordhaus (2005): 'Economics', 18th edition, New York: McGraw Hill.

Shue, H. (1999): 'Bequeathing Hazards: Security Rights and Property Rights of Future Humans', in M. Dore and T. Mount, 'Global Environmental Economics: Equity and the Limits to Markets', Oxford: Blackwell: 38–53.

Sen, A (1999) : 'Development as Freedom', Anchor Books, London.

Solow, R.M. (1974): 'Intergenerational equity and exhaustible resources', Review of Economic Studies Symposium: 29–46.

Stern, N.H. (1977): 'Welfare weights and the elasticity of the marginal valuation of income'. Published as Chapter 8 in M. Artis and R. Nobay. Proceedings of the AUTE Edinburgh Meeting of April 1976. Basil Blackwell.

Stern, N.H., J-J Dethier and F.H. Rogers (2005): 'Growth and Empowerment: Making Development Happen'. Munich Lectures in Economics. Massachusetts: MIT Press.

World Development Report (2006) :'Equity and Development', World Bank, Washington DC.

II Impacts of Climate Change on Growth and Development

Part II considers how climate change will affect people's lives, the environment and the prospects for growth and development in different parts of the world. All three dimensions are fundamental to understanding how climate change will affect our future.

These effects will not be felt evenly across the globe. Although some parts of the world would benefit from modest rises in temperature, at higher temperature increases, most countries will suffer heavily and global growth will be affected adversely. For some of the poorest countries there is a real risk of being pushed into a downwards spiral of increasing vulnerability and poverty.

Average global temperature increases of only 1-2°C (above pre-industrial levels) could commit 15-40 percent of species to extinction. As temperatures rise above 2-3°C, as will very probably happen in the latter part of this century, so the risk of abrupt and large-scale damage increases, and the costs associated with climate change – across the three dimensions of mortality, ecosystems and income – are likely to rise more steeply. In mathematical terms, the global damage function is convex.

No region would be left untouched by changes of this magnitude, though developing countries would be affected especially adversely. This applies particularly to the poorest people within the large populations of both sub-Saharan Africa, and South Asia. By 2100, in South Asia and sub-Saharan Africa, up to 145 - 220 million additional people could fall below the $2-a-day poverty line, and every year an additional 165,000 - 250,000 children could die compared with a world without climate change.

Modelling work undertaken by the Review suggests that the risks and costs of climate change over the next two centuries could be equivalent to an average reduction in global per capita consumption of at least 5%, now and forever. The estimated damages would be much higher if non-market impacts, the possibility of climate-carbon feedbacks, and distributional issues were taken into account.

Part II is structured as follows:

- **Chapter 3** begins by exploring how climate change will affect people around the world, including the potential implications for access to food, water stress, health and well being, and the environment.
- **Chapter 4** considers the implications of climate change for developing countries. It explains why developing countries are so vulnerable to climate change – a volatile mix of geographic location, existing vulnerability and, linked to this, limited ability to deal with the pressures that climate change will create.

- **Chapter 5** focuses on the implications for developed countries. Some regions will benefit from temperature rises of up to 1 to 2°C, but the balance of impacts will become increasingly negative as temperature rises.
- **Chapter 6** aims to pull together the existing modelling work that has been done to estimate the monetary costs of climate change, and also sets out the detail of modelling work undertaken by the Review.

3 How Climate Change will Affect People Around the World

KEY MESSAGES

Climate change threatens the basic elements of life for people around the world – access to water, food, health, and use of land and the environment. On current trends, average global temperatures could rise by 2 - 3°C within the next fifty years or so,[1] leading to many severe impacts, often mediated by water, including more frequent droughts and floods (Table 3.1).

- **Melting glaciers** will increase flood risk during the wet season and strongly reduce dry-season water supplies to one-sixth of the world's population, predominantly in the Indian sub-continent, parts of China, and the Andes in South America.

- **Declining crop yields**, especially in Africa, are likely to leave hundreds of millions without the ability to produce or purchase sufficient food – particularly if the carbon fertilisation effect is weaker than previously thought, as some recent studies suggest. At mid to high latitudes, crop yields may increase for moderate temperature rises (2 – 3°C), but then decline with greater amounts of warming.

- **Ocean acidification**, a direct result of rising carbon dioxide levels, will have major effects on marine ecosystems, with possible adverse consequences on fish stocks.

- **Rising sea levels** will result in tens to hundreds of millions more people flooded each year with a warming of 3 to 4°C. There will be serious risks and increasing pressures for coastal protection in South East Asia (Bangladesh and Vietnam), small islands in the Caribbean and the Pacific, and large coastal cities, such as Tokyo, Shanghai, Hong Kong, Mumbai, Calcutta, Karachi, Buenos Aires, St Petersburg, New York, Miami and London.

- Climate change will increase worldwide deaths from **malnutrition and heat stress**. Vector-borne diseases such as malaria and dengue fever could become more widespread if effective control measures are not in place. In higher latitudes, cold-related deaths will decrease.

- By the middle of the century, 200 million more people may become **permanently displaced** due to rising sea levels, heavier floods, and more intense droughts, according to one estimate.

- **Ecosystems** will be particularly vulnerable to climate change, with one study estimating that around 15 – 40% of species face extinction with 2°C of warming. Strong drying over the Amazon, as predicted by some climate models, would result in dieback of the forest with the highest biodiversity on the planet.

[1] All changes in global mean temperature are expressed relative to pre-industrial levels (1750 - 1850). A temperature rise of 1°C represents the range 0.5 – 1.5°C, a temperature rise of 2°C represents the range 1.5 – 2.5°C etc.

Table 3.1 Highlights of possible climate impacts discussed in this chapter

Temp rise (°C)	Water	Food	Health	Land	Environment	Abrupt and Large-Scale Impacts
1°C	Small glaciers in the Andes disappear completely, threatening water supplies for 50 million people	Modest increases in cereal yields in temperate regions	At least 300,000 people each year die from climate-related diseases (predominantly diarrhoea, malaria, and malnutrition) Reduction in winter mortality in higher latitudes (Northern Europe, USA)	Permafrost thawing damages buildings and roads in parts of Canada and Russia	At least 10% of land species facing extinction (according to one estimate) 80% bleaching of coral reefs, including Great Barrier Reef	Atlantic Thermohaline Circulation starts to weaken
2°C	Potentially 20 – 30% decrease in water availability in some vulnerable regions, e.g. Southern Africa and Mediterranean	Sharp declines in crop yield in tropical regions (5 – 10% in Africa)	40 – 60 million more people exposed to malaria in Africa	Up to 10 million more people affected by coastal flooding each year	15 – 40% of species facing extinction (according to one estimate) High risk of extinction of Arctic species, including polar bear and caribou	Potential for Greenland ice sheet to begin melting irreversibly, accelerating sea level rise and committing world to an eventual 7 m sea level rise
3°C	In Southern Europe, serious droughts occur every once every 10 years	150 – 550 additional millions at risk of hunger (if carbon fertilisation weak)	1 – 3 million more people die from malnutrition (if carbon fertilisation weak)	1 – 170 million more people affected by coastal flooding each year	20 – 50% of species facing extinction (according to one estimate), including 25 – 60% mammals, 30 – 40% birds and	Rising risk of abrupt changes to atmospheric circulations, e.g. the monsoon

	1 – 4 billion more people suffer water shortages, while 1 – 5 billion gain water, which may increase flood risk	Agricultural yields in higher latitudes likely to peak		15 – 70% butterflies in South Africa Onset of Amazon forest collapse (some models only)	Rising risk of collapse of West Antarctic Ice Sheet Rising risk of collapse of Atlantic Thermohaline Circulation
4°C	Potentially 30 – 50% decrease in water availability in Southern Africa and Mediterranean	Agricultural yields decline by 15 – 35% in Africa, and entire regions out of parts of Australia	Up to 80 million more people exposed to malaria in Africa	7 – 300 million more people affected by coastal flooding each year	Loss of around half Arctic tundra Around half of all the world's nature reserves cannot fulfill objectives
5°C	Possible disappearance of large glaciers in Himalayas, affecting one-quarter of China's population and hundreds of millions in India	Continued increase in ocean acidity seriously disrupting marine ecosystems and possibly fish stocks		Sea level rise threatens small islands, low-lying coastal areas (Florida) and major world cities such as New York, London, and Tokyo	
More than 5°C	The latest science suggests that the Earth's average temperature will rise by even more than 5 or 6°C if emissions continue to grow and positive feedbacks amplify the warming effect of greenhouse gases (e.g. release of carbon dioxide or methane from permafrost). This level of global temperature rise would be equivalent to the amount of warming that occurred between the last age and today – and is likely to lead to major disruption and large-scale movement of population. Such "socially contingent" effects could be catastrophic, but are currently very hard to capture with current models as temperatures would be so far outside human experience.				

Note: This table shows illustrative impacts at different degrees of warming. Some of the uncertainty is captured in the ranges shown, but there will be additional uncertainties about the exact size of impacts (more detail in Box 3.2). Temperatures represent increases relative to pre-industrial levels. At each temperature, the impacts are expressed for a 1°C band around the central temperature, e.g. 1°C represents the range 0.5 – 1.5°C etc. Numbers of people affected at different temperatures assume population and GDP scenarios for the 2080s from the Intergovernmental Panel on Climate Change (IPCC). Figures generally assume adaptation at the level of an individual or firm, but not economy-wide adaptations due to policy intervention (covered in Part V).

The consequences of climate change will become disproportionately more damaging with increased warming. Higher temperatures will increase the chance of triggering abrupt and large-scale changes that lead to regional disruption, migration and conflict.

● Warming may induce **sudden shifts in regional weather patterns** like the monsoons or the El Niño. Such changes would have severe consequences for water availability and flooding in tropical regions and threaten the livelihoods of billions.

● **Melting or collapse of ice sheets** would raise sea levels and eventually threaten at least 4 million Km^2 of land, which today is home to 5% of the world's population.

3.1   Introduction

This chapter examines the increasingly serious impacts on people as the world warms.

Climate change is a serious and urgent issue. The Earth has already warmed by 0.7°C since around 1900 and is committed to further warming over coming decades simply due to past emissions (Chapter 1). On current trends, average global temperatures could rise by 2 - 3°C within the next fifty years or so, with several degrees more in the pipeline by the end of the century if emissions continue to grow (Figure 3.1; Chapters 7 and 8).

This chapter examines how the physical changes in climate outlined in Chapter 1 affect the essential components of lives and livelihoods of people around the world - water supply, food production, human health, availability of land, and ecosystems. It looks in particular at how these impacts intensify with increasing amounts of warming. The latest science suggests that the Earth's average temperature will rise by even more than 5 or 6°C if feedbacks amplify the warming effect of greenhouse gases through the release of carbon dioxide from soils or methane from permafrost (Chapter 1). Throughout the chapter, changes in global mean temperature are expressed relative to pre-industrial levels (1750 - 1850).

The chapter builds up a comprehensive picture of impacts by incorporating two effects that are not usually included in existing studies (extreme events and threshold effects at higher temperatures). In general, impact studies have focused predominantly on changes in average conditions and rarely examine the consequences of increased variability and more extreme weather. In addition, almost all impact studies have only considered global temperature rises up to 4 or 5°C and therefore do not take account of threshold effects that could be triggered by temperatures higher than 5 or 6°C (Chapter 1).

● **Extreme weather events.** Climate change is likely to increase the costs imposed by extreme weather, both by shifting the probability distribution upwards (more heatwaves, but fewer cold-snaps) and by intensifying the water cycle, so that severe floods, droughts and storms occur more often (Chapter 1).[2] Even if

---

[2] "Extreme events" occur when a climate variable (e.g. temperature or rainfall) exceeds a particular threshold, e.g. two standard deviations from the mean.

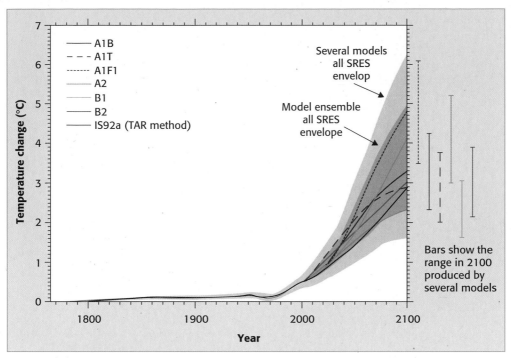

Figure 3.1 Temperature projections for the 21st century

Notes: The graph shows predicted temperature changes through to 2100 relative to pre-industrial levels. Nine illustrative emissions scenarios are shown with the different coloured lines. Blue shading represents uncertainty between the seven different climate models used. Coloured bars show the full range of climate uncertainty in 2100 for each emissions scenario based on the models with highest and lowest climate sensitivity. Updated projections will be available in the Fourth Assessment Report of the Intergovernmental Panel of Climate Change (IPCC) in 2007. These are likely to incorporate some of the newer results that have emerged from probabilistic climate simulations and climate models including carbon cycle feedbacks, such as the Hadley Centre's (more details in Chapter 1).

Source: IPCC (2001)

the shape of the distribution of temperatures does not change, an upward shift in the distribution as a whole will disproportionately increase the probability of exceeding damaging temperature thresholds.[3] Changes in the variability of climate in the future are more uncertain, but could have very significant impacts on lives and livelihoods. For example, India's economy and social infrastructure are finely tuned to the remarkable stability of the monsoon, with the result that fluctuations in the strength of the monsoon both year-to-year and within a

---

[3] In looking at the effects on crop yields of severe weather during the Little Ice Age, Prof Martin Parry (1978) argued that the frequency of extreme events would change dramatically as a result of even a small change in the mean climate and that the probability of two successive extremes is even more sensitive to small changes in the mean. Often a single extreme event is easy to withstand, but a second in succession could be far more devastating. In a follow-up paper, Tom Wigley (1985) demonstrated these effects on extremes mathematically.

single season can lead to significant flooding or drought, with significant repercussions for the economy (see Box 3.5 later).[4]

- **Non-linear changes and threshold effects at higher temperatures (convexity).**
The impacts of climate change will become increasingly severe at higher temperatures, particularly because of rising risks of triggering abrupt and large-
scale changes, such as melting of the Greenland ice sheet or loss of the Amazon
forest. Few studies have examined the shape of the damage function at higher
temperatures, even though the latest science suggests that temperatures are 5
or 6°C or higher are plausible because of feedbacks that amplify warming
(Chapter 1). For some sectors, damages may increase much faster than temperatures rise, so that the damage curve becomes convex - the consequences of
moving from 4 to 5°C are much greater than the consequences of moving from
2 to 3°C. For example, hurricane damages increase as a cube (or more) of wind-
speed, which itself scales closely with sea temperatures (Chapter 1 and Section
3.6). Theory suggests impacts in several key sectors will increase strongly at
higher temperatures, although there is not enough direct quantitative evidence
on the impacts at higher temperatures (Box 3.1).

The combined effect of impacts across several sectors could be very damaging and
further amplify the consequences of climate change. Little work has been done to
quantify these interactions, but the potential consequences could be substantial.
For example, in some tropical regions, the combined effect of loss of native pollinators, greater risks of pest outbreaks, reduced water supply, and greater incidence of heatwaves could lead to much greater declines in food production than
through the individual effects themselves (see Table 3.2 later in chapter).

The consequences of climate change will depend on how the physical impacts
interact with socio-economic factors. Population movement and growth will
often exacerbate the impacts by increasing society's exposure to environmental
stresses (for example, more people living by the coast) and reducing the amount
of resource available per person (for example, less food per person and causing
greater food shortages).[5] In contrast, economic growth often reduces vulnerability to climate change (for example, better nutrition or health care; Chapter 4) and
increases society's ability to adapt to the impacts (for example, availability of
technology to make crops more drought-tolerant; Chapter 20). This chapter
focuses on studies that in general calculate impacts by superimposing climate
change onto a future world that has developed economically and socially and
comparing it to the same future world without climate change (Box 3.2 for further
details). Most of the studies generally assume adaptation at the level of an individual or firm, but not economy-wide adaptations due to policy intervention
(covered in Part V).

Building on the analyses presented in this chapter, Chapters 4 and 5 trace the
physical impacts through to examine the consequences for economic growth and
social progress in developing and developed countries. Chapter 6 brings together
evidence on the aggregate impacts of climate change, including updated projections from the PAGE2002 model that incorporate the risk of abrupt climate change.

---

[4] Based on a technical paper prepared for the Stern Review by Challinor *et al.* (2006b)
[5] This will also depend on efficiency of use as well.

BOX 3.1 The types of relationship between rising damages and sectoral impacts

Basic physical and biological principles indicate that impacts in many sectors will become disproportionately more severe with rising temperatures. Some of these effects are summarised below, but are covered in detail in the relevant section of the chapter. Empirical support for these relationships is lacking. Hitz and Smith (2004) reviewed studies that examined the nature of the relationship between the impacts of climate change and increasing global temperatures. They found increasingly adverse impacts for several climate-sensitive sectors but were not able to determine if the increase was linear or exponential. For other sectors like water and energy where there was a mix of costs and benefits they found no consistent relationship with temperature.

Type of effect	Sector [location of source]	Proposed Functional Form		Basis
Climate system	Water [Chapter 1]	Exponential $y = e^x$		The Clausius-Clapeyron equation shows that the water holding capacity of air increases exponentially with temperature. This means that the water cycle will intensify, leading to more severe floods and droughts. There will also be more energy to drive storms and hurricanes.
	Extreme temperatures (threshold effects) [Chapter 1]	Convex curve (i.e. gradient increases with temperature)		Because of the shape of the normal distribution, a small increase in the mean dramatically increases the frequency of an extreme event.
Physical impacts	Agricultural production [Section 3.3]	Inverse parabolic ("hill function") $y = -x^2$		In cooler regions, low levels of warming may improve conditions for crop growth (extended growing season and new areas opened up for production), but further warming will have increasingly negative impacts as critical temperature thresholds are crossed more often. Tropical regions may already be past the peak. The shape and location of the curve depend on crop.

Type of effect	Sector [location of source]	Proposed Functional Form		Basis
	Heat-related human mortality [Section 3.4]	U-shaped		Sharp increase in mortality once human temperature tolerances are exceeded (heatwaves and cold-snaps). Initially mortality will be reduced by warming in cold regions.
	Storm damage [Section 3.6]	Cubic $y = x^3$		Infrastructure damage increases as a cube of wind-speed
Human response	Costs of coastal protection [Section 3.5]	Parabolic $y = x^2$		Costs of sea-wall construction increase as a square of defence height

BOX 3.2 Assumptions and scenarios used in impact studies

This chapter bases much of its detailed analysis on a series of papers prepared by Prof. Martin Parry and colleagues ("FastTrack"), one of the few that clearly sets out the assumptions used and explores different sources of uncertainty.[6]

Climate change scenarios. Climate models produce different regional patterns of temperature and rainfall (especially). The original "FastTrack" studies were based on outputs of the Hadley Centre climate model. However, in some cases the analyses have been updated to examine sensitivity to a range of different climate models.[7] Other science uncertainties, such as the link between greenhouse gas concentrations and global temperatures, were not directly examined by the work (more detail in Chapter 1).

Socio-economic scenarios. The studies carefully separated out the effects of climate change from socio-economic effects, such as growing wealth or population size. In these studies, population and GDP per capita grew on the basis of four socio-economic pathways, as described by the IPCC (see table below).[8] The effects of climate change were calculated by comparing a future world with and without climate change (but with socio-economic development in every case). Changing socio-economic factors alongside climate may be crucial because: (1) a growing population will increase society's exposure to stress from malnutrition, water shortages and coastal flooding, while (2) growing wealth will reduce vulnerability to climate change, for example by developing crops that are more drought-tolerant. Other impact studies superimpose climate change in a future world where population and GDP remain constant at today's levels. These studies are perhaps less realistic, but still provide a useful indication of the scale of the impacts and may be easier to interpret.

Summary characteristics of IPCC socio-economic scenarios (numbers in brackets for 2100)

IPCC Scenarios	A1 FI	A2	B1	B2
Name	World Markets	National Enterprise	Global Sustainability	Local Stewardship
Population growth	Low (7 billion)	High (15 billion)	Low (7 billion)	Medium (10 billion)
World GDP growth[9]	Very high, 3.5% p.a. ($550 trillion)	Medium, 2% p.a. ($243 trillion)	High, 2.75% p.a. ($328 trillion)	Medium 2% p.a. ($235 trillion)
Degree of convergence: ratio of GDP per capita in rich vs. poor countries[10]	High (1.6)	Low (4.2)	High (1.8)	Medium (3.0)
Emissions	High	Medium High	Low	Medium Low

[6] Special Issue of Global Environmental Change, Volume 14, April 2004 - further details on the new analysis are available from Warren *et al.* (2006). Risk and uncertainty are often used interchangeably, but in a formal sense, risk covers situations when the probabilities are known and uncertainty when the probabilities are not known.
[7] See, for example, Arnell (2006a)
[8] IPCC (2000)
[9] In 1990 US $
[10] Problematic as based on Market Exchange Rates

Adaptation assumptions. Clarity over adaptation is critical for work on the impacts of climate change, because large amounts of adaptation would reduce the overall damages caused by climate change (net of costs of adaptation). Within the literature, the picture remains mixed: some studies assume no adaptation, many studies assume individual (or "autonomous") adaptation, while other studies assume an "efficient" adaptation response where the costs of adaptation plus the costs of residual damages are minimised over time.[11] Unless otherwise stated, the results presented assume adaptation at the level of an individual or firm ("autonomous"), but not economy-wide. Such adaptations are likely to occur gradually as the impacts are felt but that require little policy intervention (more details in Part V). This provides the "policy neutral" baseline for analysing the relative costs and benefits of adaptation and mitigation.

3.2   Water

People will feel the impact of climate change most strongly through changes in the distribution of water around the world and its seasonal and annual variability.

Water is an essential resource for all life and a requirement for good health and sanitation. It is a critical input for almost all production and essential for sustainable growth and poverty reduction.[12] The location of water around the world is a critical determinant of livelihoods. Globally, around 70% of all freshwater supply is used for irrigating crops and providing food. 22% is used for manufacturing and energy (cooling power stations and producing hydro-electric power), while only 8% is used directly by households and businesses for drinking, sanitation, and recreation.[13]

Climate change will alter patterns of water availability by intensifying the water cycle.[14] Droughts and floods will become more severe in many areas. There will be more rain at high latitudes, less rain in the dry subtropics, and uncertain but probably substantial changes in tropical areas.[15] Hotter land surface temperatures induce more powerful evaporation and hence more intense rainfall, with increased risk of flash flooding.

Differences in water availability between regions will become increasingly pronounced. Areas that are already relatively dry, such as the Mediterranean basin and parts of Southern Africa and South America, are likely to experience further

---

[11] For example, many integrated assessment models – details in Chapter 7
[12]   Grey and Sadoff (2006) make a strong case for water resources being at the heart of economic growth and development. They show how in the late 19th and early 20th centuries, industrialised countries invested heavily in water infrastructure and institutions to facilitate strong economic growth. In least developed economies, climate variability and extremes are often quite marked, while the capacity to manage water is generally more limited.
[13] World Water Development Report (2006)
[14] Further detail in Chapter 1 - rising temperatures increase the water holding capacity of the air, so that more water will evaporate from the land in dry areas of the world. But where it rains, the water will fall in more intense bursts.
[15] At the same time, rising carbon dioxide levels will cause plants to use less water (a consequence of the carbon fertilisation effect – see Box 3.4 later) and this could increase water availability in some areas. Gedney *et al.* (2006) found that suppression of plant transpiration due to the direct effects of carbon dioxide on the closure of plant stomata (the pores on the leaves of plants) could explain a significant amount of the increase in global continental runoff over the 20th century.

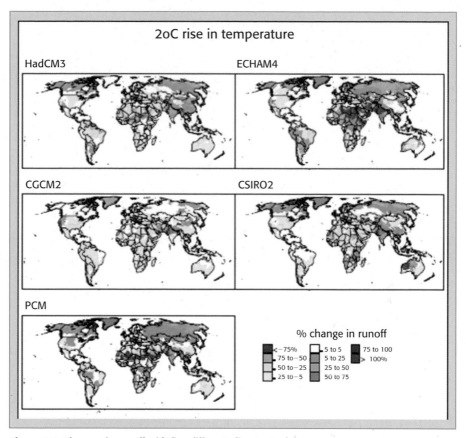

Figure 3.2 Changes in runoff with five different climate models

Source: Warren *et al.* (2006) analysing data from Arnell (2004) and Arnell (2006a)

Note: Runoff refers to the amount of water that flows over the land surface. Typically this water flows in channels such as streams and rivers, but may also flow over the land surface directly. It provides a measure of potential water availability (see Box 3.3).

decreases in water availability, for example several (but not all) climate models predict up to 30% decrease in annual runoff in these regions for a 2°C global temperature rise (Figure 3.2) and 40 – 50% for 4°C.[16] In contrast, South Asia and parts of Northern Europe and Russia are likely to experience increases in water availability (runoff), for example a 10 – 20% increase for a 2°C temperature rise and slightly greater increases for 4°C, according to several climate models.

These changes in the annual volume of water each region receives mask another critical element of climate change – its impact on year-to-year and seasonal variability. An increase in annual river flows is not necessarily beneficial, particularly in highly seasonal climates, because: (1) there may not be sufficient storage to hold the extra water for use during the dry season,[17] and (2) rivers may flood

---

[16] From Arnell (2006a); runoff, the amount of water that flows over the land surface, not only represents potential changes in water availability to people, but also provides a useful indication of whether communities will need to invest in infrastructure to help manage patterns of water supply (more details in Box 3.3).

[17] Arnell (2006a)

more frequently.[18] In dry regions, where runoff one-year-in-ten can be less than 20% of the average annual amount, understanding the impacts of climate change on variability of water supplies is perhaps even more crucial. One recent study from the Hadley Centre predicts that the proportion of land area experiencing severe droughts at any one time will increase from around 10% today to 40% for a warming of 3 to 4°C, and the proportion of land area experiencing extreme droughts will increase from 3% to 30%.[19] In Southern Europe, serious droughts may occur every 10 years with a 3°C rise in global temperatures instead of every 100 years if today's climate persisted.[20]

As the water cycle intensifies, billions of people will lose or gain water. Some risk becoming newly or further water stressed, while others see increases in water availability. Seasonal and annual variability in water supply will determine the consequences for people through floods or droughts.

Around one-third of today's global population live in countries experiencing moderate to high water stress, and 1.1 billion people lack access to safe water (Box 3.3 for an explanation of water stress). Water stress is a useful indicator of water availability but does not necessarily reflect access to safe water. Even without climate change, population growth by itself may result in several billion more people living in areas of more limited water availability.

The effects of rising temperatures against a background of a growing population are likely to cause changes in the water status of billions of people. According to one study, temperature rises of 2°C will result in 1 – 4 billion people experiencing growing water shortages, predominantly in Africa, the Middle East, Southern Europe, and parts of South and Central America (Figure 3.3).[21] In these regions, water management is already crucial for their growth and development. Considerably more effort and expense will be required on top of existing practices to meet people's demand for water. At the same time, 1 – 5 billion people, mostly in South and East Asia, may receive more water.[22] However, much of the extra water will come during the wet season and will only be useful for alleviating shortages in the dry season if storage could be created (at a cost). The additional water could also give rise to more serious flooding during the wet season.

Melting glaciers and loss of mountain snow will increase flood risk during the wet season and threaten dry-season water supplies to one-sixth of the world's population (over one billion people today).

Climate change will have serious consequences for people who depend heavily on glacier meltwater to maintain supplies during the dry season, including large

---

[18] Milly *et al.* (2002)
[19] Burke *et al.* (2006) using the Hadley Centre climate model (HadCM3). Drought was assessed with the Palmer Drought Severity Index (PDSI), with severe and extreme droughts classed as PDSI of less than 3.3 and 4.0, respectively.
[20] Lehner *et al.* (2001)
[21] Warren *et al.* (2006) have prepared these results, based on the original analysis of Arnell (2004) for the 2080s. The results are based on hydrology models driven by monthly data from five different climate models. The results do not include adaptation and thus only represent "potential water stress".
[22] The large ranges come about from differences in the predictions of the five different climate models – particularly for tropical areas where the impacts are uncertain due to the dominant influence of the El Niño and the monsoon and the difficulty of predicting interactions with climate change.

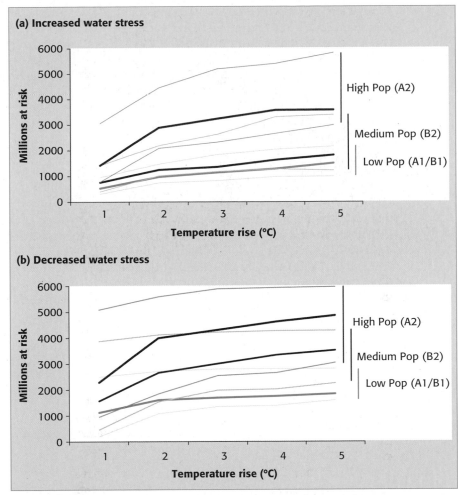

Figure 3.3 Changes in millions at risk of water stress with increasing global temperature

Source: Warren *et al.* (2006) analysing data from Arnell (2004)

Note: Lines represent different population futures for the 2080s: green – low population (7 billion), blue – medium population (10 billion), red – high population (15 billion). The thick lines show the average based on six climate models, and the thin lines the upper and lower bounds. "Millions at risk of water stress" is defined as a threshold when a population has less than 1000 m^3 per person per day (more details in Box 3.3). "Increased stress" includes people becoming water stressed who would not have been and those whose water stress worsens because of climate change. "Decreased stress" includes people who cease to become water stressed because of climate change and those whose water stress situation improves (if not to take them out of water stress completely). These aggregate figures mask the importance of annual and seasonal variability in water supply and the potential role of water management to reduce stress, but often at considerable cost.

parts of the Indian sub-continent, over quarter of a billion people in China, and tens of millions in the Andes.[23] Initially, water flows may increase in the spring as the glacier melts more rapidly. This may increase the risk of damaging glacial lake outburst floods, especially in the Himalayas,[24] and also lead to shortages later in the year. In the long run dry-season water will disappear permanently once the glacier has completely melted. Parts of the developed world that rely on mountain snowmelt (Western USA, Canadian prairies, Western Europe) will also have their summer water supply affected, unless storage capacity is increased to capture the "early water".

In the Himalaya-Hindu Kush region, meltwater from glaciers feeds seven of Asia's largest rivers, including 70% of the summer flow in the Ganges, which provides water to around 500 million people. In China, 23% of the population (250 million people) lives in the western region that depends principally on glacier meltwater. Virtually all glaciers are showing substantial melting in China, where spring stream-flows have advanced by nearly one month since records began. In the tropical Andes in South America, the area covered by glaciers has been reduced by nearly one-quarter in the past 30 years. Some small glaciers are likely to disappear completely in the next decade given current trends.[25] Many large cities such as La Paz, Lima and Quito and up to 40% of agriculture in Andean valleys rely on glacier meltwater supplies. Up to 50 million people in this region will be affected by loss of dry-season water.[26]

BOX 3.3 Meaning of water stress metrics

Water is essential for human existence and all other forms of life. Over half of the world's drinking water is taken directly from rivers or reservoirs (natural or man-made), and the rest from groundwater. Water supply is determined by runoff – the amount of water that flows over the land surface. Typically this water flows in channels such as streams and rivers, but may also flow over the land surface directly.

Water stress is a useful indicator of water availability but does not necessarily reflect access to safe water. The availability of water resources in a watershed can be calculated by dividing long-term average annual runoff (or "renewable resource") by the number of people living in the watershed.[27] A country experiences *water scarcity (or "severe water stress")* when supply is below $1000\,\text{m}^3$ per person per year and *absolute scarcity (or "extreme water stress")* when supply is below $500\,\text{m}^3$ per person per year. The thresholds are based loosely

---

[23] Barnett *et al.* (2005) have comprehensively reviewed the glacier/water supply impacts. There are 1 billion people in snowmelt regions today, and potentially 1.5 billion by 2050. In a warmer world, runoff from snowmelt will occur earlier in the spring or winter, leading to reduced flows in the summer and autumn when additional supplies will be most needed.

[24] Nepal is particularly vulnerable to glacial lake outburst floods – catastrophic discharges of large volumes of water following the breach of the natural dams that contain glacial lakes (described in Agrawala *et al.* 2005). The most significant flood occurred in 1985. A surge of water and debris up to 15 m high flooded down the Bhote Koshi and Dudh Koshi rivers. At its peak the discharge was $2000\,\text{m}^3/\text{s}$, up to four times greater than the maximum monsoon flood level. The flood destroyed the almost-completed Namche Small Hydro Project (cost $1 billion), 14 bridges, many major roads and vast tracts of arable land.

[25] Reported in Coudrain *et al.* (2005)

[26] Nagy *et al.* (2006)

[27] Based on the work of Falkenmark *et al.* 1989, water availability per person per year is the most frequently used measure of water resource availability. The UN has widely adopted this measure, for which data are readily available. The next most frequently used measure is the ratio of withdrawals to availability, but this requires reliable estimates of actual and, most crucially, future withdrawals.

on average annual estimates of water requirements in the household, agricultural, industrial and energy sectors, and the needs of the environment.

For comparative purposes, the basic water requirement for personal human needs, excluding that used directly for growing food, is around 50 Litres (L) per person per day or 18.25 m³ per person per year which includes allowances for drinking (2 – 5 L per person per day), sanitation (20 L per person per day), bathing (15 L per person per day), and food preparation (10 L per person per day). This does *not* include any allowance for growing food, industrial uses or the environment, which constitute the bulk of the use (see next point).[28]

The threshold for water scarcity is considerably higher than the basic water requirement for three reasons:

- Much of the water available to communities is used for purposes other than direct human consumption. Globally, the largest user of water is irrigated agriculture, representing 70% of present freshwater withdrawals. Industry accounts for 22% through manufacturing and cooling of thermoelectric power generation, although much of this is returned to the water system but at higher temperature. Domestic, municipal and service industry use accounts for just 8% of global water use. The proportions of water used in each sector can vary considerably by country. For example in Europe, water used for domestic, municipal and service industries is a very high proportion of total demand. Agriculture in large parts of Asia and Africa is rain-fed and does not rely on irrigation and storage infrastructure.

- Not all river flows are available for use (some flows occur during floods, and some is used by ecosystems). On average, approximately 30% of river flows occur as non-captured flood flows, and freshwater ecosystem use ranges between 20 and 50% of average flows. Taken together, 50 - 80% of average flow is unavailable to humans, meaning that a threshold of 1000 m³ per person per year of average flows translates into 200 to 50 m³/person/year *available* flows.

- The 1000 m³ per person per year is an annual average and does not reflect year-to-year variability. In dry regions, runoff one-year-in-ten can be less than one-fifth the average, so that less than 200 m³ would be available per person even before other uses are taken into account.

Water availability per person is only one indicator of potential exposure to stress. Some "stressed" watersheds will have effective management systems and water pricing in place to provide adequate supplies (e.g. through storage), while other watersheds with more than 1000 m³ per person per year may experience severe water shortages because of lack of access to water.

Source: Prepared with assistance from Prof Nigel Arnell, Tyndall Centre and University of Southampton

3.3 Food

In tropical regions, even small amounts of warming will lead to declines in yield. In higher latitudes, crop yields may increase initially for moderate increases in

---

[28] Based on work of Gleick (1996). Actual usage varies considerably, depending on water availability, price, and cultural preferences (domestic consumption in UK is around 170 L per person per day; in large parts of Africa it is less than 20 L per person per day).

temperature but then fall. Higher temperatures will lead to substantial declines in cereal production around the world, particularly if the carbon fertilisation effect is smaller than previously thought, as some recent studies suggest.

Food production will be particularly sensitive to climate change, because crop yields depend in large part on prevailing climate conditions (temperature and rainfall patterns). Agriculture currently accounts for 24% of world output, employs 22% of the global population, and occupies 40% of the land area. 75% of the poorest people in the world (the one billion people who live on less than $1 a day) live in rural areas and rely on agriculture for their livelihood.[29]

Low levels of warming in mid to high latitudes (US, Europe, Australia, Siberia and some parts of China) may improve the conditions for crop growth by extending the growing season[30] and/or opening up new areas for agriculture. Further warming will have increasingly negative impacts – the classic "hill function" (refer back to Box 3.1) - as damaging temperature thresholds are reached more often and water shortages limit growth in regions such as Southern Europe and Western USA.[31] High temperature episodes can reduce yields by up to half if they coincide with a critical phase in the crop cycle like flowering (Figure 3.4).[32]

The impacts of climate change on agriculture depend crucially on the size of the "carbon fertilisation" effect (Box 3.4). Carbon dioxide is a basic building block for plant growth. Rising concentrations in the atmosphere may enhance the initial benefits of warming and even offset reductions in yield due to heat and water stress. Work based on the original predictions for the carbon fertilisation effect suggests that yields of several cereals (wheat and rice in particular) will increase for 2 or 3°C of warming globally, according to some models, but then start to fall once temperatures reach 3 or 4°C.[33] Maize shows greater declines in yield with rising temperatures because its different physiology makes it less responsive to the direct effects of rising carbon dioxide. Correspondingly, world cereal production only falls marginally (1 – 2%) for warming up to 4°C (Box 3.4).[34] But the latest analysis from crops grown in more realistic field conditions suggests that the effect is likely to be no

[29] FAO World Agriculture report (Bruinsma 2003 ed.)
[30] Plants also develop faster at warmer temperatures such that the duration from seedling emergence to crop harvest becomes shorter as temperatures warm, allowing less time for plant growth. This effect varies with both species and cultivar. With appropriate selection of cultivar, effective use of the extended growing season can be made.
[31] Previous crop studies use a quadratic functional form, where yields are increasing in temperature up to an "optimal" level when further temperature increases become harmful (for example Mendelsohn *et al.* 1994). A crucial implicit assumption behind the quadratic functional form is symmetry around the optimum: temperature deviations above and below the "optimal" level give equivalent yield reductions. However, recent studies (e.g. Schlenker and Roberts 2006) have shown that the relationship is highly asymmetric, where temperature increases above the "optimal" level are much more harmful than comparable deviations below it. This has strong implications for climate change, as continued temperature increases can result in accelerating yield reductions.
[32] Evidence reviewed in Slingo *et al.* (2005); Ciais *et al.* (2005)
[33] The impacts depend crucially on the distribution of warming over land (Chapter 1). In general, higher latitudes and continental regions will experience temperature increases significantly greater than the global average. For a global average warming of around 4°C, the oceans and coasts generally warm by around 3°C, the mid-latitudes warm by more than 5°C and the poles by around 8°C.
[34] Warren *et al.* (2006) have prepared this analysis, based on the original work of Parry *et al.* (2004). More detail on method and assumptions are set out in Box 3.4. Production declines less than yields with increasing temperature because more land area at higher latitudes becomes more suitable for agriculture.

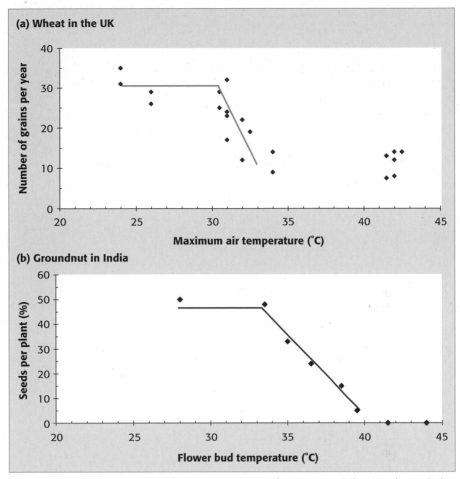

Figure 3.4 Yield loss caused by high temperature in a cool-season crop (wheat) and a tropical crop (groundnut)

Source [Figure3.4(a)]: Wheeler *et al.* (1996)

Source [Figure3.4(b)]: Vara Prasad *et al.* (2001)

Notes: Figures show how indicators of crop yield (y-axis) change with increases in daily maximum temperature during flowering (x-axis). In both cases, crops show sharp declines in yield at a threshold maximum temperature.

more than half that typically included in crop models.[35] When a weak carbon fertilisation effect is used, worldwide cereal production declines by 5% for a 2°C rise in temperature and 10% for a 4°C rise. By 4°C, entire regions may be too hot and dry to grow crops, including parts of Australia. Agricultural collapse across large areas of the world is possible at even higher temperatures (5 or 6°C) but clear empirical evidence is still limited.

---

[35] New analysis by Long *et al.* (2006) showed that the high-end estimates (25 – 30%) were largely based on studies of crops grown in greenhouses or field chambers, while analysis of studies of crops grown in near-field conditions suggest that the benefits of carbon dioxide may be significantly less, e.g. no more than half.

BOX 3.4 Agriculture and the carbon fertilisation effect

Carbon dioxide is a basic building block for crop growth. Rising concentrations in the atmosphere will have benefits on agriculture – both by stimulating photosynthesis and decreasing water requirements (by adjusting the size of the pores in the leaves). But the extent to which crops respond depends on their physiology and other prevailing conditions (water availability, nutrient availability, pests and diseases).

Until recently, research suggested that the positive benefits of increasing carbon dioxide concentrations might compensate for the negative effects of rising mean temperatures (namely shorter growing season and reduced yields). Most crop models have been based on hundreds of experiments in greenhouses and field-chambers dating back decades, which suggest that crop yields will increase by 20 – 30% at 550 ppm carbon dioxide. Even maize, which uses a different system for photosynthesis and does not respond to the direct effects of carbon dioxide, shows increases of 18 – 25% in greenhouse conditions due to improved efficiency of water use. But new analysis by Long et al. (2006) showed that the high-end estimates were largely based on studies of crops grown in greenhouses or field chambers, whereas analysis of studies of crops grown in near-field conditions suggest that the benefits of carbon dioxide may be significantly less – an 8 – 15% increase in yield for a doubling of carbon dioxide for responsive species (wheat, rice, soybean) and no significant increase for non-responsive species (maize, sorghum).

These new findings may have very significant consequences for current predictions about impacts of climate change on agriculture. Parry et al. (2004) examined the impacts of increasing global temperatures on cereal production and found that significant global declines in productivity could occur if the carbon fertilisation is small (figures below). Regardless of the strength of the carbon fertilisation effect, higher temperatures are likely to become increasingly damaging to crops, as droughts intensify and critical temperature thresholds for crop production are reached more often.

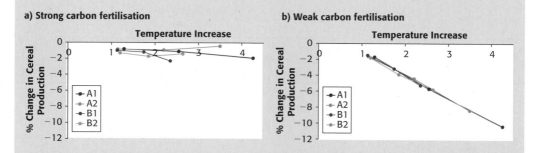

Source: Warren et al. (2006) analysing data from Parry et al. (2004)

Note: Percent changes in production are relative to what they would be in a future with no climate change but with socio-economic development. Lines represent different socio-economic scenarios developed by the IPCC. The results are based on crop models driven by monthly data from the Hadley Centre climate model, which shows greater declines in yield than two other climate models (GISS, GFDL) – see comparison in Figure 3.5. The research did not take account of the impacts of extremes, which could be significant (Box 3.5). The work assumed mostly farm-level adaptation in developing countries, but some economy-wide adaptation in developed countries (details in Figure 3.5).

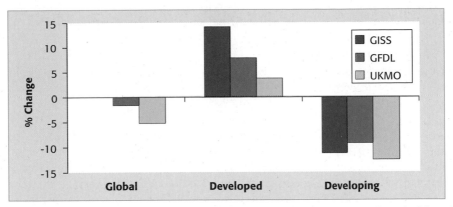

Figure 3.5 Change in cereal production in developed and developing countries for a doubling of carbon dioxide levels (equivalent to around 3°C of warming in models used) simulated with three climate models (GISS, GFDL and UKMO Hadley Centre)

Source: Parry *et al.* (2005) analysing data from Rosenzweig and Parry (1994)

Note: Percent changes in production are relative to what they would be in a future with no climate change. Overall changes are relatively robust to different model outputs, but regional patterns differ depending on the model's rainfall patterns – more details in Fischer et al. (2005). The work assumed mostly farm-level adaptation in developing countries but some economy-wide adaptation in developed countries. The work also assumed a strong carbon fertilisation effect - 15 – 25% increase in yield for a doubling of carbon dioxide levels for responsive crops (wheat, rice, soybean) and a 5 – 10% increase for non-responsive crops (maize). These are about twice as high as the latest field-based studies suggest – see Box 3.4 for more detail.

While agriculture in higher-latitude developed countries is likely to benefit from moderate warming (2 – 3°C), even small amounts of climate change in tropical regions will lead to declines in yield. Here crops are already close to critical temperature thresholds[36] and many countries have limited capacity to make economy-wide adjustments to farming patterns (Figure 3.5). The impacts will be strongest across Africa and Western Asia (including the Middle East), where yields of the predominant regional crops may fall by 25 – 35% (weak carbon fertilisation) or 15 – 20% (strong carbon fertilisation) once temperatures reach 3 or 4°C. Maize-based agriculture in tropical regions, such as parts of Africa and Central America, is likely to suffer substantial declines, because maize has a different physiology to most crops and is less responsive to the direct effects of rising carbon dioxide.[37]

Many of the effects of climate change on agriculture will depend on the degree of adaptation (see Part V), which itself will be determined by income levels, market structure, and farming type, such as rain-fed or irrigated.[38] Studies that take a more optimistic view of adaptation and assume that a substantial amount of land at higher latitudes becomes suitable for production find more positive effects of

[36] The optimum temperature for crop growth is typically around 25 – 30°C, while the lethal temperature is usually around 40°C.
[37] Other staple crops in Africa (millet and sorghum) are also relatively unresponsive to the carbon fertilisation effect. They all show a small positive response because they require less water to grow.
[38] Types of adaptation discussed by Parry *et al.* (2005)

Table 3.2 Climate change will have a wide range of effects on the environment, which could have knock-on consequences for food production. The <u>combined</u> effect of several factors could be very damaging.

Loss of essential species	Climate change will affect species' distributions and abundance (see Section 3.7), which in turn will threaten the viability of species that are essential for sustained agricultural outputs, including native pollinators for crops and soil organisms that maintain the productivity and fertility of land. Pollination is essential for the reproduction of many wild flowers and crops and its economic value worldwide has been estimated at $30 - 60 billion.
Increased incidence of flooding	Flood losses to US corn production from waterlogging could double in the next thirty years, causing additional damages totalling an estimated $3 billion per year (Rosenzweig *et al.* 2002).
Forest and crop fires	The 2003 European heatwave and drought led to severe wildfires across Portugal, Spain and France, resulting in total losses in forestry and agriculture of $15 billion (Munich Re 2004).
Climate-induced outbreaks of pests and diseases	The northward spread of Bluetongue virus in Europe, a devastating disease of sheep, has been linked to increased persistence of the virus in warmer winters and the northward expansion of the midge vector (Purse *et al.* 2005).
Rising surface ozone	Fossil fuel burning increases concentrations of nitrogen oxide in the atmosphere, which increase levels of ozone at the surface in the presence of sunlight and rising temperatures. Ozone is toxic to plants at concentrations as low as 30 ppb (parts per billion), but these effects are rarely included in future predictions. Many rural areas in continental Europe and Midwestern USA are forecast to see increases in average ozone concentrations of around 20% by the middle of the century, even though peak episodes may decline (Long *et al.* 2006).

climate change on yield.[39] But the transition costs are often ignored and the movement of population required to make this form of adaptation a reality could be very disruptive. At the same time, many existing estimates do not include the impacts of short-term weather events, such as floods, droughts and heatwaves. These have only recently been incorporated into crop models, but are likely to have additional negative impacts on crop production (Table 3.2). Expansion of agricultural land at the expense of natural vegetation may itself exert additional effects on local climates with tropical deforestation leading to rainfall reductions because of less moisture being returned to the atmosphere once trees are removed.[40]

Declining crop yields are likely to leave hundreds of millions without the ability to produce or purchase sufficient food, particularly in the poorest parts of the world.

Around 800 million people are currently at risk of hunger (~ 12% of world's population),[41] and malnutrition causes around 4 million deaths annually, almost half

---

[39] For example Fischer *et al.* (2005)

[40] These effects are not yet routinely considered in climate models or impacts studies (Betts 2005).

[41] According to Parry *et al.* (2004) people at risk of hunger are defined as the population with an income insufficient either to produce or procure their food requirements, estimated by FAO based on energy requirements deduced from an understanding of human physiology (1.2 – 1.4 times basal metabolic rate as minimum maintenance requirement to avoid undernourishment).

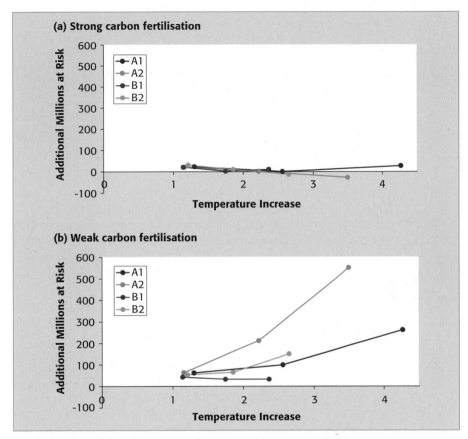

Figure 3.6 Changes in millions at risk of hunger with increasing global temperature

Source: Warren *et al.* (2006) analysing data from Parry *et al.* (2004)

Note: Lines represent different socio-economic growth paths and emissions scenarios for the 2080s developed by the IPCC (details in Box 3.2). People at risk of hunger are defined as the population with an income insufficient either to produce or procure their food requirements, estimated by the Food and Agriculture Organisation (FAO) based on energy requirements deduced from an understanding of human physiology (1.2 – 1.4 times basal metabolic rate as minimum maintenance requirement to avoid undernourishment). There are currently around 800 million people malnourished based on this definition. The IIASA Basic Linked System (BLS) world food trade model was used to examine impacts of changes in crop yields on food distribution and hunger around the world, determined both by regional agricultural production and GDP per capita (a measure of purchasing power for any additional food required). The model assumes economic growth in different regions following the IPCC scenarios. "Strong carbon fertilisation" refers to runs where the fertilisation effect was about twice as high as the latest field-based studies suggest (see Box 3.4 and Long *et al.* 2006), while "weak carbon fertilisation" includes a minimal amount.

in Africa.[42] According to one study, temperature rises of 2 to 3°C will increase the people at risk of hunger, potentially by 30 - 200 million (if the carbon fertilisation effect is small) (Figure 3.6).[43] Once temperatures increase by 3°C, 250 - 550 million

---

[42] Links between changes in income and mortality are explored in Chapter 5.

[43] Warren *et al.* (2006) have prepared these results, based on the original analysis of Parry *et al.* (2004) (more details in Box 3.6). These figures assume future socio-economic development, but no carbon fertilisation effect. There is likely to be some positive effect of rising levels of carbon dioxide (if not as much as assumed by most studies).

additional people may be at risk – over half in Africa and Western Asia, where (1) the declines in yield are greatest, (2) dependence on agriculture highest, and (3) purchasing power most limited. If crop responses to carbon dioxide are stronger, the effects of warming on risk of hunger will be considerably smaller. But at even higher temperatures, the impacts are likely to be damaging regardless of the carbon fertilisation effect, as large parts of the world become too hot or too dry for agricultural production, such as parts of Africa and even Western Australia.

Ocean acidification, a direct result of rising carbon dioxide levels, will have major effects on marine ecosystems, with possible adverse consequences on fish stocks.

For fisheries, information on the likely impacts of climate change is very limited – a major gap in knowledge considering that about one billion people worldwide (one-sixth of the world's population) rely on fish as their primary source of animal protein. While higher ocean temperatures may increase growth rates of some fish, reduced nutrient supplies due to warming may limit growth.

Ocean acidification is likely to be particularly damaging. The oceans have become more acidic in the past 200 years, because of chemical changes caused by increasing amounts of carbon dioxide dissolving in seawater.[44] If global emissions continue to rise on current trends, ocean acidity is likely to increase further, with pH declining by an additional 0.15 units if carbon dioxide levels double (to 560 ppm) relative to pre-industrial and an additional 0.3 units if carbon dioxide levels treble (to 840 ppm).[45] Changes on this scale have not been experienced for hundreds of thousands of years and are occurring at an extremely rapid rate. Increasing ocean acidity makes it harder for many ocean creatures to form shells and skeletons from calcium carbonate. These chemical changes have the potential to disrupt marine ecosystems irreversibly - at the very least halting the growth of corals, which provide important nursery grounds for commercial fish, and damaging molluscs and certain types of plankton at the base of the food chain. Plankton and marine snails are critical to sustaining species such as salmon, mackerel and baleen whales, and such changes are expected to have serious but as-yet-unquantified wider impacts.

3.4   Health

Climate change will increase worldwide deaths from malnutrition and heat stress. Vector-borne diseases such as malaria and dengue fever could become more widespread if effective control measures are not in place. In higher latitudes, cold-related deaths will decrease.

Climate-sensitive aspects of human health make up a significant proportion of the global disease burden and may grow in importance.[46] The health of the

---

[44] Turley *et al.* (2006) – Ocean pH has changed by 0.1 pH unit over the last 200 yrs. As pH is on a log scale, this corresponds to a 30% increase in the hydrogen ion concentration, the main component of acidity.

[45] Royal Society (2005) – a drop of 0.15 pH units corresponds to a 40% increase in the hydrogen ion concentration, the main component of acidity. A drop of 0.3 pH units corresponds to a doubling of hydrogen ion concentration.

[46] Comprehensively reviewed by Patz *et al.* (2005)

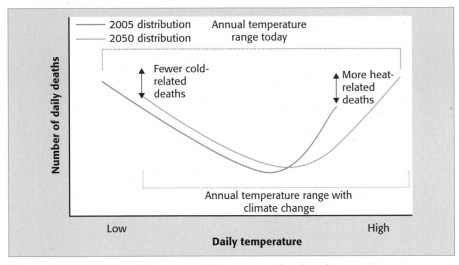

Figure 3.7 Stylised U-shaped human mortality curves as a function of temperature.

Source: Redrawn from McMichael *et al.* (2006).

Note: The blue line shows a stylised version of today's distribution of daily temperatures through the year, and the purple line shows a future distribution shifted to the right because of climate change. Deaths increase sharply at both ends of the distribution, because Heatwaves and cold snaps that exceed thresholds for human temperature tolerance become more frequent. With climate change, there will be more heatwaves (in tropical areas or continental cities) but fewer cold snaps (in higher latitudes). The overall shape of the curve is not yet clearly characterised but is crucial because it determines the net effects of decreased deaths from the cold and increased deaths from heatwaves. These costs and benefits will not be evenly distributed around the world.

world's population has improved remarkably over the past 50 years, although striking disparities remain.[47] Slum populations in urban areas are particularly exposed to disease, suffering from poor air quality and heat stress, and with limited access to clean water.

In some tropical areas, temperatures may already be at the limit of human tolerance. Peak temperatures in the Indo-Gangetic Plain often already exceed 45°C before the arrival of the monsoon.[48] In contrast, in northern latitudes (Europe, Russia, Canada, United States), global warming may imply fewer deaths overall, because more people are saved from cold-related death in the winter than succumb to heat-related death in the summer (Figure 3.7; more detail in Chapter 5).[49] In cities heatwaves will become increasingly dangerous, as regional warming together with the urban heat island effect (where cities concentrate and retain heat) leads to extreme temperatures and more dangerous air pollution incidents (see Box 6.4 in Chapter 5).

Climate change will amplify health disparities between rich and poor parts of the world. The World Health Organisation (WHO) estimates that climate change since the 1970s is already responsible for over 150,000 deaths each year through increasing incidence of diarrhoea, malaria and malnutrition, predominantly in

---

[47] Average life expectancy at birth has increased by 20 years since the 1960s. But in parts of Africa life expectancy has fallen in the past 20 years because of the HIV/AIDS pandemic (McMichael *et al.* 2004).

[48] De *et al.* (2005)

[49] See Tol (2002) for indicative figures for different OECD regions

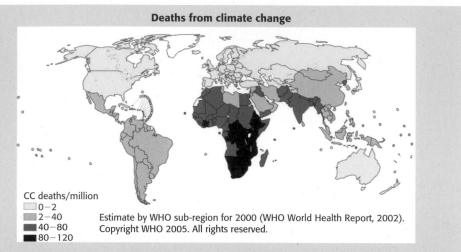

Deaths from climate change

CC deaths/million
- 0–2
- 2–40
- 40–80
- 80–120

Estimate by WHO sub-region for 2000 (WHO World Health Report, 2002).

Figure 3.8 WHO estimates of extra deaths (per million people) from climate change in 2000

Disease/Illness	Annual Deaths	Climate change component (death / % total)
Diarrhoeal diseases	2.0 million	47,000 / 2%
Malaria	1.1 million	27,000 / 2%
Malnutrition	3.7 million	77,000 / 2%
Cardiovascular disease	17.5 million	Total heat/cold data not provided
HIV/AIDS	2.8 million	No climate change element
Cancer	7.6 million	No climate change element

Source: WHO (2006) based on data from McMichael *et al.* (2004). The numbers are expected to at least double to 300,000 deaths each year by 2030.

Africa and other developing regions (Figure 3.8).[50] Just a 1°C increase in global temperature above pre-industrial could double annual deaths from climate change to at least 300,000 according to the WHO.[51] These figures do not account for any reductions in cold-related deaths, which could be substantial.[52] At higher temperatures, death rates will increase sharply, for example millions more people dying from malnutrition each year.[53] Climate change will also affect health via other diseases not included in the WHO modelling.[54]

[50] Based on detailed analysis by McMichael *et al.* (2004), using existing quantitative studies of climate-health relationships and the UK Hadley Centre GCM (business as usual emissions) to estimate relative changes in a range of climate-sensitive outcomes, including diarrhoea, malaria, dengue fever and malnutrition. Changes in heat- and cold-related deaths were not included in the aggregate estimates of mortality. Climate change contributes 2% to today's climate disease burden (6.8 million deaths annually) and 0.3% to today's total global disease burden.

[51] Projections from Patz *et al.* (2005)

[52] See, for example, Tol (2002) and Bosello *et al.* (2006)

[53] As described earlier, today 800 million people are at risk of hunger and around 4 million of those die from malnutrition each year. Once temperatures increase by 3°C, 200 – 600 million additional people could be at risk (with little carbon fertilisation effect), suggesting 1 – 3 million more dying each year from malnutrition, assuming that the ratio of risk of hunger to mortality from malnutrition remains the same. This ratio will of course change with income status – see Chapter 4 for more detail.

[54] The impacts on human development mediated through changes in income are explored in Chapter 4.

The distribution and abundance of disease vectors are closely linked to temperature and rainfall patterns, and will therefore be very sensitive to changes in regional climate in a warmer world. Changes to mosquito distributions and abundance will have profound impacts on malaria prevalence in affected areas. This will be particularly significant in Africa, where 450 million people are exposed to malaria today, of whom around 1 million die each year. According to one study, a 2°C rise in temperature may lead to 40 – 60 million more people exposed to malaria in Africa (9 – 14% increase on present-day), increasing to 70 – 80 million (16 – 19%) at higher temperatures, assuming no change to malaria control efforts.[55] Much of the increase will occur in Sub-Saharan Africa, including East Africa. Some studies suggest that malaria will decrease in parts of West Africa, e.g. taking 25 – 50 million people out of an exposed region, because of reductions in rainfall.[56] Changes in future exposure depend on the success of national and international malaria programmes. Such adaptations are not taken into account in the estimates presented, but the effectiveness of such programmes remains variable.[57] Climate change will also increase the global population exposed to dengue fever, predominantly in the developing world, e.g. 5 – 6 billion people exposed with a 4°C temperature rise compared with 3.5 billion people exposed with no climate change.[58]

Health will be further affected by changes in the water cycle. Droughts and floods are harbingers of disease, as well as causing death from dehydration or drowning.[59] Prolonged droughts will fuel forest fires that release respiratory pollutants, while floods foster growth of infectious fungal spores, create new breeding sites for disease vectors such as mosquitoes, and trigger outbreaks of water-borne diseases like cholera. In the aftermath of Hurricane Mitch in 1998, Honduras recorded an additional 30,000 cases of malaria and 1,000 cases of dengue fever. The toxic moulds left in New Orleans in the wake of Hurricane Katrina continue to create health problems for its population, for example the so-called "Katrina cough".

[55] Calculations from Warren *et al.* (2006) based on research from Tanser *et al.* (2003), using one of only two models which has been validated directly to account for the observed effect of climate variables on vector and parasite population biology. They assume no increase in population size in the future or change in vulnerability (through effective treatment/prophylaxis). While this assumption of no change in control efforts is not realistic, the results illustrate the potential scale of the problem. The study used the Hadley Centre climate model to estimate regional temperature and rainfall patterns; other models produce different rainfall patterns and therefore may result in different regional patterns for malaria.

[56] Calculations from Warren *et al.* (2006) based on research from Van Lieshout *et al.* (2004), who take into account future population projections and used the Hadley Centre climate model. Similar to Tanser *et al.* (2003), they use the Hadley Centre model and find an increase in malaria exposure in Sub-Saharan Africa, but with slightly fewer people affected (50 million rather than 80 million for a 4°C temperature rise) because of different assumptions about rainfall thresholds.

[57] Malaria in Africa is particularly difficult to control because of the large numbers of mosquitoes spreading the disease, their effectiveness as transmitting the disease, and increasing drug resistance problems. Alternatives can be very effective, but are often much more expensive (WHO 2005).

[58] Hales *et al.* (2002) used a vector-specific model coupled to outputs of two climate models. Their estimates take account of projected population growth to the 2080s, but not any control measures.

[59] Reviewed in Epstein and Mills (2005)

3.5   Land

Sea level rise will increase coastal flooding, raise costs of coastal protection, lead to loss of wetlands and coastal erosion, and increase saltwater intrusion into surface and groundwater.

Warming from the last century has already committed the world to rising seas for many centuries to come. Further warming this century will increase this commitment.[60] Rising sea levels will increase the amount of land lost and people displaced due to permanent inundation, while the costs of sea walls will rise approximately as a square of the required height. Coastal areas are amongst the most densely populated areas in the world and support several important ecosystems on which local communities depend. Critical infrastructure is often concentrated around coastlines, including oil refineries, nuclear power stations, port and industrial facilities.[61]

Currently, more than 200 million people live in coastal floodplains around the world, with 2 million Km^2 of land and $1 trillion worth of assets less than 1-m elevation above current sea level. One-quarter of Bangladesh's population (~35 million people) lives within the coastal floodplain.[62] Many of the world's major cities (22 of the top 50) are at risk of flooding from coastal surges, including Tokyo, Shanghai, Hong Kong, Mumbai, Calcutta, Karachi, Buenos Aires, St Petersburg, New York, Miami and London.[63] In almost every case, the city relies on costly flood defences for protection. Even if protected, these cities would lie below sea level with a residual risk of flooding like New Orleans today.

The homes of tens of millions more people are likely to be affected by flooding from coastal storm surges with rising sea levels. People in South and East Asia will be most vulnerable, along with those living on the coast of Africa and on small islands.

Sea level rises will lead to large increases in the number of people whose homes are flooded (Figure 3.9).[64] According to one study that assumes protection levels rise in line with GDP per capita, between 7 – 70 million and 20 – 300 million additional people will be flooded each year by 3 to 4°C of warming causing 20 – 80 cm of sea level rise (low and high population growth assumptions respectively).[65] Upgrading coastal defences further could partially offset these impacts, but would require substantial capital investment and ongoing maintenance. At higher levels of warming and increased rates of sea level rise, the risks will become increasingly serious (more on melting polar ice sheets in Section 3.8).

---

[60] More detail in Chapter 1

[61] See Chapter 6 for discussion of implications for global trade

[62] Ali (2000)

[63] Munich Re (2005)

[64] Increased storm intensity could cause similar impacts and will exacerbate the effects of sea level rise – these effects are not included in the impact estimates provided here (see Chapter 6).

[65] Warren *et al.* (2006) have prepared these results, based on the original analysis of Nicholls (2004), Nicholls and Tol (2006) and Nicholls and Lowe (2006) for impacts of sea level rise on populations in 2080s with and without climate change. More details on method are set out in Figure 3.8. "Average annual people flooded" refers to the average annual number of people who experience episodic flooding by storm surge, including the influence of any coastal protection. In some low-lying areas without protection,

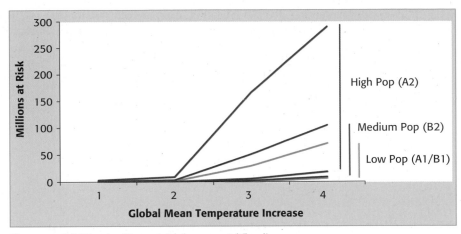

Figure 3.9 Additional millions at risk from coastal flooding

Source: Warren *et al.* (2006) analysing data from Nicholls (2004), Nicholls and Tol (2006) and Nicholls and Lowe (2006)

Notes: Figure shows increase in number of people flooded by storm surge on average each year in the 2080s for different levels of global temperature rise (relative to pre-industrial levels). Results assume that flood defences are upgraded in phase with GDP per capita, but ignoring sea level rise itself. Lines represent different socio-economic futures for the 2080s based on a range of population and growth paths taken from the IPCC: green – A1/B1 low population (7 billion), red – B2 medium population (10 billion), blue – A2 high population (15 billion) (details of population and GDP per capita for each scenario set out in Box 3.2). A richer more populous country will be able to spend more on flood defences, but will have a greater number of people at risk. The impacts are shown for the "transient sea level rise" associated with reaching a particular level of warming, but do not include the consequences of the additional sea level rise that the world would be committed to for a given level of warming (0 – 15 cm for 1°C, 10 – 30 cm for 2°C, 20 – 50 cm for 3°C, 35 – 80 cm for 4°C; more details in Chapter 1). The ranges cover the uncertainties in climate modelling and how much sea level rises for a given change in temperature (based on IPCC Third Assessment Report data from 2001, which may be revised in the Fourth Assessment due in 2007).

South and East Asia will be most vulnerable because of their large coastal populations in low-lying areas, such as Vietnam, Bangladesh and parts of China (Shanghai) and India. Millions will also be at risk around the coastline of Africa, particularly in the Nile Delta and along the west coast. Small island states in the Caribbean, and in the Indian and Pacific Oceans (e.g. Micronesia and French Polynesia, the Maldives, Tuvalu) are acutely threatened, because of their high concentrations of development along the coast. In the Caribbean, more than half the population lives within 1.5 Km of the shoreline.

Some estimates suggest that 150 - 200 million people may become permanently displaced by the middle of the century due to rising sea levels, more frequent floods, and more intense droughts.

Today, almost as many people are forced to leave their homes because of environmental disasters and natural resource scarcity as flee political oppression, religious persecution and ethnic troubles (25 million compared with 27 million).[66] Estimates in this area, however, are still problematic. Norman Myers uses conservative

---

[66] International Federation of Red Cross and Red Crescent Societies (2001)

assumptions and calculates that climate change could lead to as many as 150 - 200 million environmental refugees by the middle of the century (2% of projected population).[67] This estimate has not been rigorously tested, but it remains in line with the evidence presented throughout this chapter that climate change will lead to hundreds of millions more people without sufficient water or food to survive or threatened by dangerous floods and increased disease. People may also be driven to migrate within a region – Chapter 5 looks in detail at a possible climate-induced shift in population and economic activity from southern regions to northern regions of Europe and the USA.

3.6   Infrastructure

Damage to infrastructure from storms will increase substantially from only small increases in event intensity. Changes in soil conditions (from droughts or permafrost melting) will influence the stability of buildings.

By increasing the amount of energy available to fuel storms (Chapter 1), climate change is likely to increase the intensity of storms. Infrastructure damage costs will increase substantially from even small increases in sea temperatures because: (1) peak wind speeds of tropical storms are a strongly exponential function of temperature, increasing by about 15 - 20% for a 3°C increase in tropical sea surface temperatures;[68] and (2) damage costs typically scale as the cube of windspeed or more (Figure 3.10).[69] Storms and associated flooding are already the most costly natural disaster today, making up almost 90% of the total losses from natural catastrophes in 2005 ($184 billion from windstorms alone, particularly

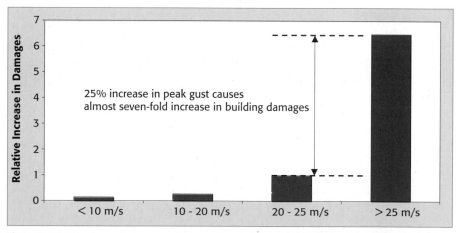

Figure 3.10 Damage costs increase disproportionately for small increases in peak wind speed

Source: IAG (2005)

[67] Myers and Kent (1995)
[68] Emanuel (1987)
[69] In fact Nordhaus (2006) found that economic damages from hurricanes rise as the ninth power of maximum wind-speed, perhaps as a result of threshold effects, such as water overtopping storm levees.

hurricanes and typhoons).[70] A large proportion of the financial losses fall in the developed world, because of the high value and large amount of infrastructure at risk (more details in Chapter 5).

High latitude regions are already experiencing the effects of warming on previously frozen soil. Thawing weakens soil conditions and causes subsidence of buildings and infrastructure. Climate change is likely to lead to significant damage to buildings and roads in settlements in Canada and parts of Russia currently built on permafrost.[71] The Quinghai-Tibet Railway, planned to run over 500 Km of permafrost, is designed with a complex and costly insulation and cooling system to prevent thawing of the permafrost layer (more details in Chapter 20). However, most of the existing infrastructure is not so well designed to cope with permafrost thawing and land instability.

3.7   Environment

Climate change is likely to occur too rapidly for many species to adapt. One study estimates that around 15 – 40% of species face extinction with 2°C of warming. Strong drying over the Amazon, as predicted by some climate models, would result in dieback of forest with the highest biodiversity on the planet.

The warming of the 20th century has already directly affected ecosystems. Over the past 40 years, species have been moving polewards by 6 Km on average per decade, and seasonal events, such as flowering or egg-laying, have been occurring several days earlier each decade.[72] Coral bleaching has become increasingly prevalent since the 1980s. Arctic and mountain ecosystems are acutely vulnerable – polar bears, caribou and white spruce have all experienced recent declines.[73] Climate change has already contributed to the extinction of over 1% of the world's amphibian species from tropical mountains.[74]

Ecosystems will be highly sensitive to climate change (Table 3.1). For many species, the rate of warming will be too rapid to withstand. Many species will have to migrate across fragmented landscapes to stay within their "climate envelope" (at rates that many will not be able to achieve). Migration becomes more difficult with faster rates of warming. In some cases, the "climate envelope" of a species may move beyond reach, for example moving above the tops of mountains or beyond coastlines. Conservation reserves may find their local climates becoming less amenable to the native species. Other pressures from human activities, including land-use change, harvesting/hunting, pollution and transport of alien species around the world, have already had a dramatic effect on species and will make it even harder for species to cope with further warming. Since 1500, 245 extinctions have been recorded across most major species groups, including mammals, birds, reptiles, amphibians, and trees. A further 800 known species in these groups are threatened with extinction.[75]

---

[70] Munich Re (2006)
[71] Nelson (2003)
[72] Root *et al.* (2005); Parmesan and Yohe (2003)
[73] Arctic Climate Impacts Assessment (2004)
[74] Pounds *et al.* (2006)
[75] Ricketts *et al.* (2005)

A warming world will accelerate species extinctions and has the potential to lead to the irreversible loss of many species around the world, with most kinds of animals and plants affected (see below). Rising levels of carbon dioxide have some direct impacts on ecosystems and biodiversity,[76] but increases in temperature and changes in rainfall will have even more profound effects. Vulnerable ecosystems are likely to disappear almost completely at even quite moderate levels of warming.[77] The Arctic will be particularly hard hit, since many of its species, including polar bears and seals, will be very sensitive to the rapid warming predicted and substantial loss of sea ice (more detail in Chapter 5).[78]

- **1°C warming.** At least 10% of land species could be facing extinction, according to one study.[79] Coral reef bleaching will become much more frequent, with slow recovery, particularly in the southern Indian Ocean, Great Barrier Reef and the Caribbean.[80] Tropical mountain habitats are very species rich and are likely to lose many species as suitable habitat disappears.
- **2°C warming.** Around 15 – 40% of land species could be facing extinction, with most major species groups affected, including 25 – 60% of mammals in South Africa and 15 – 25% of butterflies in Australia. Coral reefs are expected to bleach annually in many areas, with most never recovering, affecting tens of millions of people that rely on coral reefs for their livelihood or food supply.[81] This level of warming is expected to lead to the loss of vast areas of tundra and forest – almost half the low tundra and about one-quarter of the cool conifer forest according to one study.[82]
- **3°C warming.** Around 20 – 50% of land species could be facing extinction. Thousands of species may be lost in biodiversity hotspots around the world,

---

[76] For example, fast-growing tropical tree species show greater growth enhancements with increased carbon dioxide concentrations than slower-growing species and could gain a dominant competitive advantage in tropical forests in the future (Körner 2004).

[77] Reviewed in detail in Hare (2006). These figures are likely to underestimate the impacts of climate change, because many of the most severe effects are likely to come from interactions with factors not taken into account in these calculations, including land use change and habitat fragmentation/loss, spread of invasive species, new pests and diseases, and loss of pollinators. In addition, ecosystem assessments rarely consider the rate of temperature change. It is likely that rates of change exceeding 0.05 – 0.1°C per decade (regional temperature) are more than most ecosystems can withstand, because species cannot migrate polewards fast enough (further details in Warren 2006).

[78] According to the Arctic Climate Impacts Assessment (2004), Arctic ecosystems will be strongly affected by climate change as temperatures here are rising at close to double the global average.

[79] Thomas *et al.* (2004a) – these (and subsequent) estimates of extinction risk are based on calculations of decreases in the availability of areas with suitable climate conditions for species into the future. As suitable areas to support a certain level of biodiversity disappear, species become "committed to extinction" when the average rate of recruitment of adults into the population is less than the average rate of adult mortality. There is likely to be a lag in response depending on the life span of the species in question – short-lived species rapidly disappear from an area while long-lived species can survive as adults for several years. There is a great deal of uncertainty inherent in such estimates of extinction risk (Pearson and Dawson 2003) and alternative modelling approaches have been shown to yield different estimates (Thuiller *et al.* 2004, Pearson *et al.* 2006). However, other studies looking at climate suitability also predict high levels of extinction, for example McClean *et al.* (2005) predict that 25 – 40% of African plan species will lose all suitable climate area with 3°C of warming globally.

[80] Coral bleaching describes the process that occurs when the tiny brightly coloured organisms that feed the main coral (through photosynthesis) leave the skeleton because they become heat-stressed. Bleached corals have significantly higher rates of mortality.

[81] Donner *et al.* (2005)

[82] Leemans and Eichkout (2004)

e.g. over 40% of endemic species in some biodiversity hotspots such as African national parks and Queensland rain forest.[83] Large areas of coastal wetlands will be permanently lost because of sea level rise (up to one-quarter according to some estimates), with acute risks in the Mediterranean, the USA and South East Asia. Mangroves and coral reefs are at particular risk from rapid sea level rise (more than 5 mm per year) and their loss would remove natural coastal defences in many regions. Strong drying over the Amazon, according to some climate models, would result in dieback of forest with the highest biodiversity on the planet.[84]

Temperatures could rise by more than 4 or 5°C if emissions continue unabated, but the full range of consequences at this level of warming have not been clearly articulated to date. Nevertheless, a basic understanding of ecological processes leads quickly to the conclusion that many of the ecosystem effects will become compounded with increased levels of warming, particularly since small shifts in the composition of ecosystems or the timing of biological events will have knock-on effects through the food-chain (e.g. loss of pollinators or food supply).[85]

3.8   Non-linear changes and threshold effects

Warming will increase the chance of triggering abrupt and large-scale changes.

Human civilisation has lived through a relatively stable climate. But the climate system has behaved erratically in the past.[86] The chaotic nature of the climate system means that even relatively small amounts of warming can become amplified, leading to major shifts as the system adjusts to balance the new conditions. Abrupt and large-scale changes could potentially destabilise regions and increase regional conflict – for example shutdown of Atlantic Thermohaline Circulation (more details in Chapter 5).[87] While there is still uncertainty over the possible triggers for such changes, the latest science indicates that the risk is more serious than once thought (Table 3.3).[88] Some temperature triggers, like 3 or 4°C of warming, could be reached this century if warming occurs quite rapidly.

Melting/collapse of polar ice sheets would accelerate sea level rise and eventually lead to substantial loss of land, affecting around 5% of the global population.

The impacts of sea level rise in the long term depend critically on changes in both the Greenland and West Antarctic ice sheets. As temperatures rise, the world risks crossing a threshold level of warming beyond which melting or collapse of these polar ice sheets would be irreversible. This would commit the world to increases in sea level of around 5 to 12-m over coming centuries to millennia,

---

[83] Malcolm *et al.* (2006)

[84] This effect has been found with the Hadley Centre model (Cox *et al.* 2000) and several other climate models (Scholze *et al.* 2006).

[85] Visser and Both (2005); Both *et al.* (2006) report declines of 90% in pied flycatcher populations in the Netherlands in areas where caterpillar numbers have been peaking two weeks earlier due to warming, which means there is little food when the flycatcher eggs hatch.

[86] For example, Rial *et al.* (2004)

[87] As set out in a Pentagon commissioned report by Schwartz and Randall (2004)

[88] Schellnhuber (2006)

Table 3.3 Potential temperature triggers for large-scale and abrupt changes in climate system

Phenomenon	Global Temperature Rise (above pre-industrial)	Relative Confidence*	References
Shifts in regional weather regimes (e.g. changes in monsoons or the El Niño)	Uncertain (although some changes are expected)	Medium	Hoskins (2003)
Onset of irreversible melting of Greenland	2-3°C	Medium	Lowe *et al.* (2006)
Substantial melting threatening the stability of the West Antarctic Ice Sheet	>2-5°C	Low	Oppenheimer (2005)
Weakening of North Atlantic Thermohaline Circulation	Gradual weakening from present	High	Wood *et al.* (2006)
Complete collapse of North Atlantic Thermohaline Circulation	>3-5°C	Low	O'Neill and Oppenheimer (2002)

Source: Adapted from Schneider and Lane (2006)

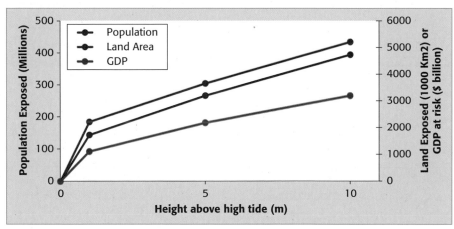

Figure 3.11 Global flood exposure from major sea level rise (based on present conditions)
Source: Anthoff *et al.* (2006)

much greater than from thermal expansion alone, and significantly accelerate the rate of increase (Chapter 1). A substantial area of land and a large number of people would be put at risk from permanent inundation and coastal surges. Currently, around 5% of the world's population (around 270 million people) and $2 trillion worth of GDP would be threatened by a 5-m rise (Figure 3.11). The most vulnerable regions are South and East Asia, which could lose 15% of their land area (an area over three times the size of the UK). Many major world cities would likely have to be abandoned unless costly flood defences were constructed.[89]

[89] Nicholls *et al.* (2004)

Warming may induce sudden shifts in regional weather patterns that have severe consequences for water availability in tropical regions.

The strongly non-linear nature of weather systems, like the Asian and African monsoons, and patterns of variability, such as the El Niño (chapter 1), suggests that they may be particularly vulnerable to abrupt shifts. For example, recent evidence shows that an El Niño with strong warming in the central Pacific can cause the Indian monsoon to switch into a dry state, leading to severe droughts.[90] Currently, this type of shift is a temporary occurrence, but in the past, there is evidence that climate changes have caused such shifts to persist for many decades. For example, cold periods in the North Atlantic since the last ice age, such as a 2.5°C regional cooling during the Little Ice Age, led to an abrupt weakening of the Asian summer monsoon.[91] If such abrupt shifts were replicated in the future, they could have a very severe effect on the livelihoods of hundreds of millions of people (Box 3.5). The impacts would be strongest in the tropics, where such weather systems are a key driver of rainfall patterns. However, the confidence in projections of future changes is relatively low. Currently, several climate models predict that in the future average rainfall patterns will look more like an El Niño.[92] This could mean a significant shift in weather in many parts of the world, with areas that are normally wet perhaps rapidly becoming dryer. In the long term, it may be possible to adapt to such changes, but the short-term impacts would be highly disruptive. For example, the strong El Niño in 1997/98 had severe impacts around the Indian and Pacific oceans, causing flooding and droughts that led to thousands of deaths and several billion dollars of damage.

Extreme high temperatures will occur more often, increasing human mortality during the dry pre-monsoon months and damaging crops.[93] Critical temperatures, above which damage to crops increases rapidly, are likely to be exceeded more frequently. A recent study predicts up to a 70% reduction in crop yields by the end of this century under these conditions, assuming no adaptation.[94]

BOX 3.5 Possible impacts of an abrupt change in Asian monsoon reliability

Any changes in rainfall patterns of the Asian monsoon would severely affects the lives of millions of people across southern Asia. Summer monsoon rains play a crucial role for agricultural and industrial production throughout South and East Asia. In India, for example, summer monsoon rains provide 75 – 90% of the annual rainfall.

Models suggest that climate change will bring a warmer, wetter monsoon by the end of the century.[95] This could increase water availability for around two billion people in South and East Asia.[96] However, the increased runoff would probably increase flood risk, particularly because models predict that rain will fall in more intense bursts. Without adaptation this could have devastating impacts. For example, over 1000 people died when Mumbai

---

[90] Kumar *et al.* (2006)
[91] Gupta *et al.* (2003)
[92] Collins and the CMIP Modelling Groups (2005)
[93] Defra (2005)
[94] Challinor *et al.* (2006a)
[95] Reviewed in detail in a report prepared for the Stern Review by Challinor *et al.* (2006b)
[96] This is a result from Arnell (2006b), who superimposed rainfall and temperature changes from past extreme monsoon years (average over five driest and five wettest years) on today's mean summer climate to understand consequences for water availability.

was devastated by flash floods from extremely heavy rainfall in August 2005.[97] A record-breaking one-metre of rain fell in just 24 hours and parts of Mumbai were flooded to a depth of 3 metres. Schools, banks, the stock exchange, and the airport all had to be closed. Hundreds of cases of dysentery and cholera were recorded as a result of contaminated water, and medical supplies were limited because of damages to storage warehouses.

But it is changes in the timing and variability of rainfall, both within the wet season and between years that are likely to have the most significant impacts on lives and livelihoods. A year-to-year fluctuation of just 10% in average rainfall can lead to food and water shortages. Confidence in projections of future rainfall variability is relatively low; however, this represents the difference between steady, predictable rainfall and a destructive cycle of flooding and drought. Most models predict a modest increase in year-to-year variability but to differing degrees. At the heart of this are the projections of what will happen to El Niño. Changes in variability within the wet season are more uncertain, but also vital to livelihoods. For example, in 2002, the monsoon rains failed during July, resulting in a seasonal rainfall deficit of 20%. This caused a massive loss of agricultural production, leading to severe hardship for hundreds of millions of people.

3.9   Conclusion

Climate change will have increasingly severe impacts on people around the world, with a growing risk of abrupt and large-scale changes at higher temperatures.

This chapter has outlined the main mechanisms through which physical changes in climate will affect the lives and livelihoods of people around the world. A warmer world with a more intense water cycle and rising sea levels will influence many key determinants of wealth and wellbeing, including water supply, food production, human health, availability of land, and the environment. While there may be some initial benefits in higher latitudes for moderate levels of warming (1 – 2°C), the impacts will become increasingly severe at higher temperatures (3, 4 or 5°C). While there is some evidence in individual sectors for disproportionate increases in damages with increasing temperatures, such as heat stress (Box 3.1), the most powerful consequences will arise when interactions between sectors magnify the effects of rising temperatures. For example, infrastructure damage will rise sharply in a warmer world, because of the combined effects of increasing potency of storms from warmer ocean waters and the increasing vulnerability of infrastructure to rising windspeeds. At the same time, the science is becoming stronger, suggesting that higher temperatures will bring a growing risk of abrupt and large-scale changes in the climate system, such as melting of the Greenland Ice Sheet or sudden shift in the pattern of monsoon rains. Such changes are still hard to predict, but their consequences could be potentially catastrophic, with the risk of large-scale movement of populations and global insecurity. Chapter 6 brings this disparate material together to examine the full costs in aggregate.

While modelling efforts are still limited, they provide a powerful tool for taking a comprehensive look at the impacts of climate change. At the same time, it is the

[97] Described in detail in Munich Re (2006)

underlying detail, as described in this and the next two chapters, rather than the aggregate models that should be the primary focus. It is not possible in aggregate models to bring out the key elements of the effects, much is lost in aggregation, and the particular model structure can have their own characteristics. What matters is the magnitude of the risks of different kind for different people and the fact that they rise so sharply as temperatures move upwards.

Chapters 4 and 5 pick up this story. The poorest will be hit earliest and most severely. In many developing countries, even small amounts of warming will lead to declines in agricultural production because crops are already close to critical temperature thresholds. The human consequences will be most serious and widespread in Sub-Saharan Africa, where millions more will die from malnutrition, diarrhoea, malaria and dengue fever, unless effective control measures are in place. There will be acute risks all over the world – from the Inuits in the Arctic to the inhabitants of small islands in the Caribbean and Pacific. Developed countries may experience some initial benefits from warming, such as longer growing seasons for crops, less winter mortality, and reduced heating demands. These are likely to be short-lived and counteracted at higher temperatures by sharp increases in damaging extreme events such as hurricanes, floods, and heatwaves.

References

Dr Rachel Warren and colleagues from the Tyndall Centre (Warren *et al.* 2006) have prepared a detailed technical report for the Stern Review that looks at how the impacts of climate change vary with rising temperatures (working paper available from http://www.tyndall.ac.uk). The analysis drew heavily on a series of papers, known as "FastTrack", prepared by Prof. Martin Parry and colleagues in a Special Issue of Global Environmental Change (introduced in Parry 2004). These studies are among the few that use a consistent set of climate and socio-economic scenarios and explore different sources of risks and uncertainty (more details in Box 3.2). The work built on previous analyses by Grassl *et al.* (2003), Hare (2006) and Warren (2006). Sam Hitz and Joel Smith also analysed the "FastTrack" work, amongst others, in a special report for the OECD, focusing in particular on the functional form of the impacts with rising temperatures (Hitz and Smith 2004). They found increasingly adverse impacts for several climate-sensitive sectors but were not able to determine if the increase was linear or exponential (more details in Box 3.1). Prof. Richard Tol (2002) carried out a detailed study to examine both the costs and the benefits of climate change at different levels of global temperature rise in key economic sectors – agriculture, forestry, natural ecosystems, sea level rise, human mortality, energy consumption, and water resources. He found that some developed countries show net economic benefits for low levels of warming (1 – 2°C) because of reduced winter heating and cold-related deaths, and increased agricultural productivity due to carbon fertilisation. The book "Avoiding dangerous climate change" (edited by Schellnhuber 2006) currently provides the most up-to-date assessment of the full range of impacts of climate change, particularly the risk of abrupt and large-scale changes. The Fourth Assessment Report of the IPCC is expected to be published in 2007 and will provide the most comprehensive picture of the latest science.

Agrawala, S., S. Gigli, V. Raksakulthai et al. (2005): 'Climate change and natural resource management: key themes from case studies', in Bridge over troubled water: linking climate change and development, S. Agrawala (ed.), Paris: OECD, pp. 85 – 131

Ali, A. (2000): 'Vulnerability of Bangladesh coastal region to climate change with adaptation options', Dhaka: Bangladesh Space Research and Remote Sensing Organisation (SPARRSO), available from http://www.survas.mdx.ac.uk/pdfs/3anwaral.pdf

Anthoff, D., R. Nicholls, R.S.J. Tol, and A.T. Vafeidis (2006): 'Global and regional exposure to large rises in sea-level', Research report prepared for the Stern Review, Tyndall Centre Working Paper 96, Norwich: Tyndall Centre, available from http://www.tyndall.ac.uk/publications/working_papers/twp96.pdf

Arctic Climate Impacts Assessment (2004): 'Impacts of a warming Arctic', Cambridge: Cambridge University Press.

Arnell, N.W. (2004): 'Climate change and global water resources: SRES emissions and socio-economic scenarios', Global Environmental Change **14**: 31 – 52

Arnell, N.W. (2006a): 'Climate change and water resources', in Avoiding dangerous climate change, H.J. Schellnhuber (eds.), Cambridge: Cambridge University Press, pp. 167 – 176

Arnell, N.W. (2006b): 'Global impacts of abrupt climate change: an initial assessment', Report commissioned by the Stern Review, available from http://www.sternreview.org.uk

Barnett, T.P., J.C, Adam, and D.P. Lettenmaier (2005): 'Potential impacts of a warming climate on water availability in snow-dominated regions', Nature **438**: 303-309

Betts, R.A. (2005): 'Integrated approaches to climate-crop modelling: needs and challenges'. Philosophical Transactions of the Royal Society B **360**: 2049-2065

Bosello, F., R. Roson, R.S.J. Tol et al. (2006): 'Economy-wide estimates of the implications of climate change: human health', Ecological Economics **58**: 579-591

Both, C., S.Bouwhius, C.M. Lessells, and M.E. Visser (2006): 'Climate change and population declines in a long-distance migratory bird', Nature **441**: 81 – 83

Bruinsma, J. (ed.) (2003): 'World agriculture: towards 2015/2030', FAO, London: Earthscan

Burke, E.J., S.J. Brown, and N. Christidis (2006): 'Modelling the recent evolution of global drought and projections for the twenty-first century with the Hadley Centre climate model', Journal of Hydrometeorology, **7**: 1113 – 1125

Challinor, A.J., T.R. Wheeler, et al. (2006a): 'Adaptation of crops to climate change through genotypic responses to mean and extreme temperatures, Agriculture, Ecosystems and Environment', (in press)

Challinor, A.J., J. Slingo, A. Turner, and T. Wheeler (2006b): 'Indian Monsoon, Contribution to Stern Review', available from http://www.sternreview.org.uk

Ciais, P., M. Reichstein, N. Viovy, F et al. (2005): 'Europe-wide reduction in primary productivity caused by the heat and drought in 2003', Nature **437**: 529 – 533

Collins, M. and the CMIP Modelling Groups (2005): 'El Niño – or La Niño-like climate change? Climate Dynamics' **24**: 89-104

Coudrain, A., B. Francou et al. (2005): 'Glacier shrinkage in the Andes and consequences for water resources', Hydrological Sciences Journal **50**: 925 – 932

Cox P.M., R.A. Betts, C.D. Jones, et al. (2000): 'Acceleration of global warming due to carbon-cycle feedbacks in a coupled climate model', Nature **408**: 184-187

Defra (Department for Environment, Food and Rural Affairs) (2005): 'Investigating the impacts of climate change in India', available from http://www.defra.gov.uk/environment/climate-change.htm

De, U.S., R.K. Dube and G.S. Prakasa Rao (2005): 'Extreme weather events over India in the last 100 years', Journal of the Indian Geophysical Union **9**: 173 – 187

Donner, S.D, W.J. Skirving, C.M. Little et al. (2005): 'Global assessment of coral bleaching and required rates of adaptation under climate change', Global Change Biology **11**: 2251 – 2265

Emanuel, K.A. (1987): 'The dependence of hurricane intensity on climate', Nature **326**: 483 – 485

Epstein, P.R. and E. Mills (2005): 'Climate change futures: health, ecological and economic dimensions', Center for Health and the Global Environment, Harvard Medical School, Cambridge, MA: Harvard University, available from http://www.climatechangefutures.org

Falkenmark, M., J. Lunquist, and C. Widstrand (1989): 'Macro-scale water scarcity requires micro-scale approaches', Natural Resources Forum **13**: 258 – 267

Fischer, G., M. Shah, F.N. Tubiello, and H. van Velhuizen (2005): 'Socio-economic and climate change impacts on agriculture', Philosophical Transactions of the Royal Society B **360**: 2067 – 2083

Gedney, N., P.M. Cox, R.A. Betts, et al. (2006) 'Detection of a direct carbon dioxide effect in continental river runoff records', Nature **439**: 835 – 838

Gleick, P.H. (1996): 'Basic water requirements for human activities: meeting basic needs', Water International **21**: 83 – 92

Grassl, H., J. Kokott, M. Kulessa, et al. (2003): 'Climate protection strategies for the 21st century: Kyoto and beyond', Special Report for German Advisory Council on Global Change (WBGU), Berlin: WBGU, available from http://www.wbgu.de/wbgu_sn2003_voll_engl.html

Grey, D., and C. Sadoff (2006): 'Water for growth and development, in Thematic documents of the IV World Water Forum', Mexico City: Comisíon Nacional del Agua

Gupta, A., D.M. Anderson, and J.T. Overpeck (2003): 'Abrupt changes in the Asian southwest monsoon during the Holocene and their links to the North Atlantic Ocean', Nature **421**: 354 – 356

Hales, S., N. de Wet, J. Macdonald, and A. Woodward (2002): 'Potential effect of population and climate changes on global distribution of dengue fever: an empirical model', The Lancet **360**: 830 – 834

Hare, B. (2006): 'Relationship between increases in global mean temperature and impacts on ecosystems, food production, water and socio-economic systems', in Avoiding dangerous climate change, H.J. Schellnhuber (eds.), Cambridge: Cambridge University Press, pp. 177 – 185

Hitz S. and J. Smith (2004): 'Estimating Global Impacts from Climate Change', pp. 31 – 82 in Morlot JC, Agrawala S (eds.), The benefits of climate change policies, Paris: OECD

Hoskins, B.J. (2003): 'Atmospheric processes and observations', Philosophical Transactions of the Royal Society B **361**:1945-1960

Insurance Australia Group (2005): 'Evidence to the Stern Review', Insurance Australia Group, Australia, available from http://www.sternreview.org.uk

International Federation of Red Cross and Red Crescent Societies (2001): 'World disasters report 2001', Geneva: International Federation of Red Cross and Red Crescent Societies, available from http://www.ifrc.org/publicat/wdr2001

Intergovernmental Panel on Climate Change (2000:) 'Emissions Scenarios. Special Report of the Intergovernmental Panel on Climate Change', [Nakicenovic N and Swart R (eds.)], Cambridge: Cambridge University Press

Intergovernmental Panel on Climate Change (2001): 'Climate Change 2001: Synthesis Report', Contribution of Working Groups I, II and III to the Third Assessment Report of the Intergovernmental Panel on Climate Change' [Watson R.T. and Core Team (Eds.)], Cambridge: Cambridge University Press

Körner, Ch. (2004): 'Through enhanced tree dynamics carbon dioxide enrichment may cause tropical forests to lose carbon', Philosophical Transactions of the Royal Society London B **359**: 493 – 498

Kumar, K.K., B. Rajagopalan, M. Hoerling et al. (2006): 'Unravelling the mystery of indian monsoon failure during the El Niño', Science **314**: 115-119, doi: 10.1126/science.1131152

Leemans, R. and B. Eickhout (2004): 'Another reason for concern: regional and global impacts on ecosystems for different levels of climate change', Global and Environmental Change **14**: 219 – 228

Lehner B, Henrichs T, Döll P, Alcamo J (2001): 'EuroWasser: Model-Based Assessment of European Water Resources and Hydrology in the Face of Global Change'. World Water Series 5. Centre for Environmental Systems Research, Kassel: University of Kassel, available from http://www.usf.uni-kassel.de/usf/archiv/dokumente/kwws/kwws.5.en.htm

Long, S.P., E.A. Ainsworth, A.D.B. Leakey et al. (2006): 'Food for thought: lower-than-expected crop yield stimulation with rising CO_2 concentrations', Science **312**: 1918 – 1921

Malcolm, J.R., C.R. Liu, R.P. Neilson, et al. (2006): Global warming and extinctions of endemic species from biodiversity hotspots, Conservation Biology **20**: 538 – 548

McClean, C.J., J.C. Lovett, W. Kuper, et al. (2005): 'African plant diversity and climate change', Annals of the Missouri Botanical Garden **92**: 139 – 152

McMichael, A., D. Campbell-Lendrum, S. Kovats, et al. (2004): 'Global climate change', in M.J. Ezzati, et al. (eds.), Comparative quantification of health risks: global and regional burden of disease due to selected major risk factors, Geneva: World Health Organisation, pp. 1543 – 1649

McMichael, A.J., R.E. Woodruff, and S. Hales (2006): 'Climate change and human health: present and future risks', The Lancet **367**: 859-869

Mendelsohn, R., W.D. Nordhaus, and D. Shaw (1994): 'The impact of global warming on agriculture: a Ricardian analysis', American Economic Review **84**: 753 – 771

Milly P.C.D., R.T. Wetherald, K.A. Dunne, and T.L. Delworth (2002): 'Increasing risk of great floods in a changing climate', Nature **415**: 514-517

Munich Re (2004): 'Annual review: natural catastrophes 2003', Munich Re Group: Munich

Munich Re (2005): 'Megacities – megarisks.' Munich Re Group: Munich

Munich Re (2006): 'Annual review: natural catastrophes 2005', Munich Re Group: Munich

Myers, N., and Kent, J. (1995): 'Environmental exodus: an emergent crisis in the global arena, Washington, DC: The Climate Institute

Nagy, G. et al. (2006): 'Understanding the potential impact of climate change in Latin America and the Caribbean', Report prepared for the Stern Review, available from http://www.sternreview.org.uk

Nelson, F.E. (2003): '(Un)frozen in Time', Science **299**: 1673 – 1675

Nicholls R.J. (2004): 'Coastal flooding and wetland loss in the 21st century: changes under the SRES climate and socio-economic scenarios', Global Environmental Change **14**: 69 – 86

Nicholls R.J., J.A. Lowe (2006): 'Climate stabilisation and impacts of sea-level rise', in Avoiding dangerous climate change, H.J. Schellnhuber (eds.), Cambridge: Cambridge University Press, pp. 195 – 201

Nicholls, R.J., R.S.J. Tol, and N. Vafeidis (2004): 'Global estimates of the impact of a collapse of the West Antarctic Ice Sheet', Report for the Atlantis project, available from http://www.uni-hamburg.de/Wiss/FB/15/Sustainability/atlantis.htm

Nicholls, R.J. and R.S.J. Tol, (2006): 'Impacts and responses to sea level rise: global analysis of the SRES scenarios over 21st century', Philosophical Transactions of the Royal Society B **364**: 1073 – 1095

Nordhaus, W.D. (2006): 'The economics of hurricanes in the United States, Prepared for Snowmass Workshop on Abrupt Climate Change', Snowmass: Annual Meetings of the American Economic Association, available from http://nordhaus.econ.yale.edu/hurricanes.pdf

O'Neill, B.C., and M. Oppenheimer (2002): 'Climate change – dangerous climate impacts and the Kyoto protocol', Science 296: 1971-1972

Oppenheimer, M. (2005): 'Defining dangerous anthropogenic interference: the role of science, the limits of science', Risk Analysis 25: 1399–1408

Parmesan, C. and G. Yohe (2003): 'A globally coherent fingerprint of climate change impacts across natural systems', Nature 421: 37 – 42

Parry, M.L. (1978): 'Climatic change, agriculture and settlement', Folkestone, UK: Dawson and Sons

Parry, M.L (2004): 'Global impacts of climate change under the SRES scenarios', Global Environmental Change 14: 1

Parry, M.L, C. Rosenzweig, A. Iglesias et al. (2004): 'Effects of climate change on global food production under SRES emissions and socio-economic scenarios', Global Environmental Change 14: 53 – 67

Parry, M.L, C. Rosenzweig, and M. Livermore (2005): 'Climate change, global food supply and risk of hunger', Philosophical Transactions of the Royal Society B 360: 2125 – 2136

Patz, J.A., D. Campbell-Lendrum, T. Holloway, and J.A. Foley (2005): 'Impact of regional climate change on human health', Nature 438: 310 – 317

Pearson, R.G, W. Thuiller, M.B. Araújo, et al. (2006): 'Model-based uncertainty in species' range prediction'. Journal of Biogeography, 33: 1704–1711, doi: 10.1111/j.1365-2699.2006.01460.x

Pearson, R.G., and T.P. Dawson (2003): 'Predicting the impacts of climate change on the distribution of species: are bioclimate envelope models useful?' Global Ecology and Biogeography 12: 361 – 371

Pounds, J.A., M.R. Bustamante, and L.A. Coloma (2006): 'Widespread amphibian extinctions from epidemic disease driven by global warming', Nature 439: 161 – 167

Purse, B.V., P.S. Mellor, D.J. Rogers, et al. (2005): 'Climate change and the recent emergence of bluetongue in Europe', Nature Reviews Microbiology 3: 171–181

Rial, J.A., R.A. Pielke, M. Beniston et al. (2004): 'Non-linearities, feedbacks and critical thresholds within the earth's climate system', Climatic Change 65: 11–38

Ricketts, T.H., E. Dinerstein, T. Boucher, et al. (2005): 'Pinpointing and preventing imminent extinctions', Proceedings of the National Academy of Sciences 102: 18497 – 18501

Root, T.L., D.P. MacMynowski, M.D. Mastrandrea, and S.H. Schneider (2005): 'Human-modified temperatures induce species changes', Proceedings of the National Academy of Sciences 102: 7465 – 7469

Rosenzweig, C. and M.L. Parry (1994): 'Potential impacts of climate change on world food supply', Nature 367: 133 – 138

Rosenzweig, C, F.N. Tubiello, R. Goldberg, et al. (2002): 'Increased crop damage in the US from excess precipitation under climate change', Global Environmental Change 12: 197 – 202

Royal Society (2005): 'Ocean acidification due to increasing atmospheric carbon dioxide', London: Royal Society, available from http://www.royalsoc.ac.uk

Schellnhuber, H.J. (ed.) (2006): 'Avoiding dangerous climate change, Cambridge: Cambridge University Press

Schlenker, W. and M.J. Roberts (2006): 'Nonlinear effects of weather on corn yields', Review of Agricultural Economics 28: in press

Schneider, S.H. and J. Lane (2006): 'An overview of 'dangerous' climate change', in Avoiding dangerous climate change, H.J. Schellnhuber (eds.), Cambridge: Cambridge University Press, pp. 7 – 24

Scholze, M., W. Knorr, N.W. Arnell, and I.C. Prentice (2006): 'A climate-change risk analysis for world ecosystems', Proceedings of the National Academy of Sciences, DOI: 10.1073/pnas.0601816103

Schwartz, P. and D. Randall (2004): 'An abrupt climate change scenario and its implications for United States national security', Report prepared by the Global Business Network (GBN) for the Department of Defense, San Francisco: GBN, available from 26231" |http://www.gbn.com/ArticleDisplayServlet.srv?aid=26231

Slingo, J.M., A.J. Challinor, B.J. Hoskins, and T.R. Wheeler (2005): 'Introduction: food crops in a changing climate', Philosophical Transactions of the Royal Society B 360: 1983 – 1989

Tanser, F.C., B. Sharp, and D. le Seur (2003): 'Potential effect of climate change on malaria transmission in Africa', The Lancet 362: 1792 – 1798

Thomas, C.D., A. Cameron, R.E. Green et al. (2004a): 'Extinction risk from climate change', Nature 427: 145 – 148

Thomas C.D., S.E. Williams, A. Cameron, et al. (2004b): 'Reply (brief communication)', Nature 430: doi:10.1038/nature02719

Thuiller, W., M.B. Araújo, R.G. Pearson, et al. (2004): 'Uncertainty in predictions of extinction risk', Nature 430: doi:10.1038/nature02716

Tol, R.S.J. (2002): 'Estimates of the damage costs of climate change'. Part I: Benchmark estimates, Environmental and Resource Economics **21**: 47 – 73

Turley, C., J.C. Blackford, S. Widdicombe, et al. (2006): 'Reviewing the impact of increased atmospheric CO_2 on oceanic pH and the marine ecosystem', in Avoiding dangerous climate change, H.J. Schellnhuber (eds.), Cambridge: Cambridge University Press, pp. 65 – 70

Van Lieshout, M., R.S. Kovats, M.T. Livermore, and P. Martens (2004): 'Climate change and malaria: analysis of the SRES climate and socio-economic scenarios', Global Environmental Change **14**: 87 – 99

Vara Prasad, P.V., P.Q. Craufurd, V.G. Kakani, et al. (2001): 'Influence of high temperature on fruit-set and pollen germination in peanuts', Australian Journal of Plant Physiology **28**: 233

Visser, M.E. and C. Both (2005): 'Shifts in phenology due to global climate change: the need for a yardstick', Proceedings of the Royal Society B **272**: 2561 – 2569

Warren, R. (2006): 'Impacts of global climate change at different annual mean global temperature increases', in Avoiding dangerous climate change, H.J. Schellnhuber (eds.), Cambridge: Cambridge University Press, pp. 93 – 131

Warren, R., N. Arnell, R. Nicholls, P. Levy, and J. Price (2006): 'Understanding the regional impacts of climate change', Research report prepared for the Stern Review, Tyndall Centre Working Paper 90, Norwich: Tyndall Centre, available from http://www.tyndall.ac.uk/publications/working_papers/twp90.pdf

Wheeler, T.R., G.R. Batts, R.H. Ellis, et al. (1996): 'Growth and yield of winter wheat (Triticum aestivum) crops in response to CO_2 and temperature', Journal of Agricultural Science **127**: 37 – 48

World Health Organisation (2005): 'Susceptibility of Plasmodium falciparum to antimalarial drugs – report on global monitoring 1996 – 2004', Geneva: World Health Organisation.

World Health Organisation (2006): WHO/UNEP Health and Environment Linkages Initiative (website), WHO, Geneva, available from http://www.who.int/heli/risks/climate/climate-change

Wigley, T.M.L. (1985): 'Impact of extreme events', Nature **316**: 106 – 107

World Bank (2004): 'Towards a water-secure Kenya', Water Resources Sector Memorandum, Washington, DC: World Bank

World Water Development Report (2006): 'Water: a shared responsibility, World Water Assessment Programme', New York: United Nations, available from http://www.unesco.org/water/wwap

4 Implications of Climate Change for Development

KEY MESSAGES

Climate change poses a real threat to the developing world. Unchecked it will become a major obstacle to continued poverty reduction.

Developing countries are especially vulnerable to climate change because of their geographic exposure, low incomes, and greater reliance on climate sensitive sectors such as agriculture. Ethiopia, for example, already has far greater hydrological variability than North America but less than 1% of the artificial water storage capacity per capita. **Together these mean that impacts on developing countries are proportionally greater and the ability to adapt smaller.**

Many developing countries are already struggling to cope with their current climate. For low-income countries, major natural disasters today can cost an average of 5% of GDP.

For example:

Health and agricultural incomes will be under particular threat from climate change.

- Falling farm incomes will increase poverty and reduce the ability of households to invest in a better future and force them to use up meagre savings just to survive.

- Millions of people will potentially be at risk of climate-driven heat stress, flooding, malnutrition and water related and vector borne diseases. For example, dengue transmission in South America may increase by 2 to 5 fold by the 2050s.

- The cost of climate change in India and South East Asia could be as high as a 9–13% loss in GDP by 2100 compared with what could have been achieved in a world without climate change. There could be up to an additional 145–220 million people living on less than $2 a day and an additional 165 000 to 250 000 child deaths per year in South Asia and sub-Saharan Africa by 2100 due to income losses alone.

Severe deterioration in the local climate could lead to mass migration and conflict in some parts of the developing world, especially as another 2–3 billion people are added to the developing world's population in the next few decades. For example:

- Rising sea levels, advancing desertification and other climate-driven changes could drive millions of people to migrate. More than a fifth of Bangladesh could be under water with a 1 m rise in sea levels – a possibility by the end of the century.

- Drought and other climate-related shocks risk sparking conflict and violence, with West Africa and the Nile Basin particularly vulnerable given their high water interdependence.

These risks place an even greater premium on fostering growth and development to reduce the vulnerability of developing countries to some of the now inevitable impacts of climate change.

Little can now be done to change the likely adverse effects that developing countries will face in the next few decades, and so some adaptation will be essential. Strong and early mitigation is the only way to avoid some of the very severe impacts that could occur in the second half of this century.

4.1   Introduction

While all regions will eventually feel the effects of climate change, it will have a disproportionately harmful effect on developing countries – and in particular poor communities who are already living at or close to the margins of survival. Changes in the climate will amplify the existing challenges posed by tropical geography, a heavy dependence on agriculture, rapid population growth and poverty. The world is already likely to fall short of the Millennium Development Goals for 2015 in many countries (see Box 4.1 for the Goals). Climate change threatens the long-term sustainability of development progress.[1]

BOX 4.1 Millennium Development Goals

In September 2000, 189 countries signed the United Nations Millennium Declaration. In so doing, they agreed on the fundamental dimensions of development, translated into an international blueprint for poverty reduction. This is encapsulated by the Millennium Development Goals that are focused on a target date of 2015:

● Halve extreme poverty and hunger

● Achieve universal primary education

● Empower women and promote equality between women and men

● Reduce under five mortality by two thirds

● Reduce maternal mortality by three-quarters

● Reverse the spread of diseases, especially HIV/AIDS and malaria

● Ensure environmental sustainability

● Create a global partnership for development, with targets for aid, trade and debt relief

But it is important to recognise that the scale of future climate impacts will vary between regions, countries and people. The last 30 years or so has already seen strong advances in many developing countries on income, health and education. Those developing countries that continue to experience rapid growth will be much better placed to deal with the consequences of climate change. Other

---

[1] The physical effects of climate change are predicted to become progressively more significant by the 2050s with a 2 to 3°C warming, as explained in Chapter 3.

areas, predominantly low-income countries, where growth is stagnating may find their vulnerability increases.

The challenge now is to limit the damage, both by mitigation and adaptation. It is vital therefore to understand just how, and how much, climate change is likely to slow development progress. The chapter begins by examining the processes by which climate change impacts will be felt in developing countries. Section 4.2 considers what it is about the starting position of these countries that makes them vulnerable to the physical changes set out in Chapter 3. Understanding why developing countries are especially vulnerable is critical to understanding how best to improve their ability to deal with climate change (discussed in Chapter 20). Sections 4.3 and 4.4 move on to consider the consequences of a changing climate on health, income and growth. The first part of the analysis draws on evidence from past and current exposure to climate variability to show how vulnerable groups are affected by a hostile climate. The second summarises key regional impacts from climate change impact studies. Section 4.5 explores the potential effects of climate change on future growth and income levels, which in turn affect the numbers of people living below poverty thresholds as well as the child mortality rate. The chapter concludes with Section 4.6 reviewing the possible consequences for migration, displacement and risk of conflict resulting from the socio-economic and environmental pressures of climate change.

4.2   The vulnerability of developing countries to a changing climate

Developing countries are especially vulnerable to the physical impacts of climate change because of their exposure to an already fragile environment, an economic structure that is highly sensitive to an adverse and changing climate, and low incomes that constrain their ability to adapt.

The effects of climate change on economies and societies will vary greatly over the world. The circumstances of each country – its initial climate, socio-economic conditions, and growth prospects – will shape the scale of the social, economic and environmental impacts of climate change. Vulnerability to climate change can be classified as: *exposure* to changes in the climate, *sensitivity* – the degree to which a system is affected by or responsive to climate stimuli,[2] and *adaptive capacity* – the ability to prepare for, respond to and tackle the effects of climate change. This is illustrated in Figure 4.1. Developing countries generally score poorly on all three criteria. This section provides a brief overview of some of the key vulnerabilities facing many developing countries. Unless these vulnerabilities are overcome they are likely to increase the risk and scale of damaging impacts posed by climate change.

Exposure: The geography of many developing countries leaves them especially vulnerable to climate change.

Geographical exposure plays an important role in determining a country's growth and development prospects. Many developing countries are located in tropical areas. As a result, they already endure climate extremes (such as those

---

[2]   IPCC (2001). The classification of *sensitivity* is similar to *susceptibility* to climate change, the degree to which a system is open, liable, or sensitive to climate stimuli.

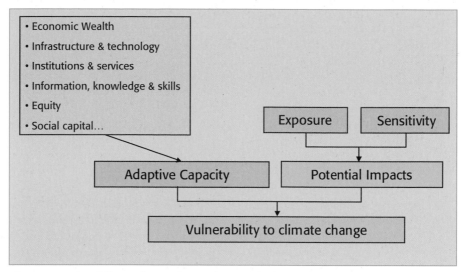

Figure 4.1 Vulnerability to climate change: the IPCC Third Assessment Report

Source: Redrawn from Ionescu et al (2005)

that accompany the monsoon and El Niño and La Niña cycles), intra and inter-annual variability in rainfall,[3] and very high temperatures. India, for example, experienced peak temperatures of between 45°C and 49°C during the pre-monsoon months of 2003.[4] Geographical conditions have been identified as important contributors to lower levels of growth in developing countries. If rainfall – that arrives only in a single season in many tropical areas – fails for example, a country will be left dry for over a year with powerful implications for their agricultural sector. This occurred in India in 2002 when the monsoon rains failed, resulting in a seasonal rainfall deficit of 19% and causing large losses of agricultural production and a drop of over 3% in India's GDP.[5] Recent analysis has led Nordhaus to conclude that "tropical geography has a substantial negative impact on output density and output per capita compared to temperate regions".[6] Sachs, similarly, argues that poor soils, the presence of pests and parasites, higher crop respiration rates due to warmer temperatures, and difficulty in water availability and control explain much of the tropical disadvantage in agriculture.[7] Climate change is predicted to make these conditions even more challenging, with the range of possible physical impacts set out in Chapter 3. Even slight variations in the climate can have very large costs in developing countries as many places are already close to the upper

---

[3] Intra-annual variability refers to rainfall concentrated in a single season, whilst interannual variability refers to large differences in the annual total of rainfall. The latter may be driven by phenomena such as the El Niño-Southern Oscillation (ENSO) or longer-term climate shifts such as those that caused the ongoing drought in the African Sahel. Brown and Lall (2006)

[4] De et al (2005)

[5] Challinor et al (2006). The scale of losses in the agricultural sector is indicated by the fact that this sector contributed just over one fifth of GDP at the time.

[6] Nordhaus (2006). Approximately 20% of the difference in per capita output between tropical Africa and two industrial regions is attributed to geography according to Nordhaus' model and analysis.

[7] Sachs (2001)

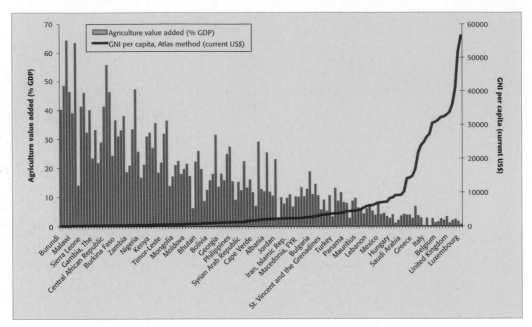

Figure 4.2 The share of agriculture in GDP and per capita income in 2004

Source: Updated from an earlier version by Tol et al (2004) using data from World Bank (World Development Indicators for 2004) for all countries for which such data are available. Countries are ranked by per capita income.

temperature tolerance of activities such as crop production. Put another way, climate change will have a disproportionately damaging impact on developing countries due, in part at least, to factors such as their location in low latitudes, the amount and variability of rainfall they receive, and the fact that they are "already too hot".[8]

Sensitivity: Developing economies are very sensitive to the direct impacts of climate change given their heavy dependence on agriculture and ecosystems, rapid population growth and concentration of millions of people in slum and squatter settlements, and low health levels.

Dependence on agriculture   Agriculture and related activities are crucial to many developing countries, in particular for low income or semi-subsistence economies. The rural sector contributes 21% of GDP in India, for example, rising to 39% in a country like Malawi,[9] whilst 61% and 64% of people in South Asia and sub-Saharan Africa, respectively are employed in the rural sector.[10] This concentration of economic activities in the rural sector – and in some cases around just a few commodities – is associated with low levels of income, as illustrated in Figure 4.2.[11] The concentration of activities in one sector also limits flexibility to switch to less climate-sensitive activities such as manufacturing and services. The agricultural sector is one of the most at risk to the damaging impacts of climate

---

[8] Mendelsohn et al (2006)
[9] World Bank (2006a) using 2004 data
[10] ILO (2005). The employment figures are given as a share of total employment, 2005.
[11] For example, the Central African Republic derives more than 50% of its export earnings from cotton alone (1997/99). Commission for Africa (2005)

change – and indeed current extreme climate variability – in developing countries, as discussed in Chapter 3.

Dependence on vulnerable ecosystems All humans depend on the services provided by natural systems. However, environmental assets and the services they provide are especially important for poor people, ranging from the provision of subsistence products and market income, to food security and health services.[12] Poor people are consequently highly sensitive to the degradation and destruction of these natural assets and systems by climate change. For example, dieback of large areas of forest – some climate models show strong drying over the Amazon if global temperature increases by more than 2°C for example – would affect many of the one billion or more people who depend to varying degrees on forests for their livelihoods (Table 4.1).[13]

Population growth and rapid urbanisation Over the next few decades, another 2–3 billion people will be added to the world's population, virtually all of them in developing countries.[14] This will add to the existing strain on natural resources and the social fabric already felt in many poor countries, and expose an ever greater number of people to the effects of climate change. Greater effort is required

Table 4.1 Direct roles of forests in household livelihood strategies

Poverty aspects	Function	Description
Safety net	Insurance	Food and cash income in periods of unexpected food and income shortfall
Support current consumption	Gap-filling	Regular (seasonal, for example) shortfall of food and income
	Regular subsistence uses	Fuelwood, wild meat, medicinal plants, and so on
	Low-return cash activities	A wide range of extractive or "soft management" activities, normally in economies with low market integration
Poverty reduction	Diversified forest strategies	Forest activities that are maintained in economies with high market integration
	Specialised forest strategies	Forest activities that form the majority of the cash income in local economies with high market integration
	Payment for environmental services	Direct transfers to local communities from off-site beneficiaries

Source: Vedeld et al (2004). Classification based on Arnold (2001), Kaimowitz (2002), Angelsen and Wunder (2003), and Belcher, Ruiz-perez, and Achdiawan (2003)

[12] Natural medicines, for example, are often the only source of medicine for poor people and can help reduce national costs of supplying medical provisions in developing countries. The ratio of traditional healers to western-trained doctors is approximately 150:1 in some African countries for example. UNEP-WCMC (2006)

[13] Vedeld et al (2004). This effect on the Amazon has been found with the Hadley Centre model, as reported in Cox *et al.* (2000), and several other climate models (Scholze *et al.* 2006) as discussed in Chapter 3.

[14] World Bank (2003b)

to encourage lower rates of population growth. Development on the MDG dimensions (in particular income, the education of women, and reproductive health) is the most powerful and sustainable way to approach population growth.[15]

Developing countries are also undergoing rapid urbanisation, and the trend is set to continue as populations grow. The number of people living in cities in developing countries is predicted to rise from 43% in 2005 to 56% by 2030.[16] In Africa, for example, the 500 km coast between Accra and the Niger delta will likely become a continuous urban megalopolis with more than 50 million people by 2020.[17] It does not follow from this that policies to slow urbanisation are desirable. Urbanisation is closely linked to economic growth and it can provide opportunities for reducing poverty and decreasing vulnerability to climate change.[18] Nonetheless, many of those migrating to cities live in poor conditions on marginal land as city planners and public services have been unable to keep pace with the influx of people.[19] These areas are particularly vulnerable because of their limited access to clean water, sanitation and location in low quality housing and on lands that are already prone to flooding and other environmental hazards. In Latin America, for example, where urbanisation has gone far further than in Africa or Asia, more and more people are likely to be forced to locate in cheaper, hazard-prone areas such as floodplains or steep slopes.

Food insecurity, malnutrition and health   Approximately 40% of the population of sub-Saharan Africa is undernourished, largely because of the poor diet and severe and repeated infections that afflict poor people.[20] Even if the Millennium Development Goals are met, more than 400 million people could be suffering from chronic hunger in 2015.[21] Malnutrition is a health outcome in itself, but it also lowers natural resistance to infectious diseases by weakening the immune system. This is a challenge today – malnutrition was associated with 54% of child deaths in developing countries in 2001 (10.8 million children), as illustrated in Figure 4.3. Climate change will potentially exacerbate this vulnerability as a greater number of malaria carrying mosquitoes move into previously uninfected areas. This is likely to generate higher morbidity and mortality rates among people suffering from malnutrition than among food-secure people.

Adaptive capacity: People will adapt to changes in the climate as far as their resources and knowledge allow. But developing countries lack the infrastructure (most notably in the area of water supply and management), financial means, and access to public services that would otherwise help them adapt.

Poor water-related infrastructure and management   Developing countries are highly dependent on water – the most climate-sensitive economic resource – for

---

[15] Stern et al (2005)

[16] World Population Prospects (2004); and World Urbanization Prospects (2005)

[17] Hewawasam (2002)

[18] For example, proximity and economies of scale enable cost-effective and efficient targeting and provision of basic infrastructure and services.

[19] Approximately 72% of Africa's urban inhabitants now live in slums and squatter settlements for example. Commission for Africa (2005).

[20] WHO (2005). Poverty impacts a person's standard of living, the environmental conditions in which they live, and their ability to meet basic needs such as food, housing and health care that in turn affects their level of nutrition.

[21] One of the MDGs is to halve, between 1990 and 2015, the proportion of people who suffer from hunger. In 2002 there were 815 million hungry people in the developing world, 9 million less than in 1990. UN (2005)

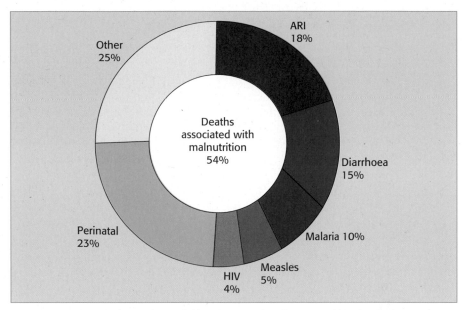

Figure 4.3 Proportional mortality in children younger than five years old in developing countries

Source: WHO (2005)

Note: Acute Respiratory Infection (ARI)

their growth and development. Water is a key input to agriculture, industry, energy and transport and is essential for domestic purposes. Irrigation and effective water management will be very important in helping to reduce and manage the effects of climate change on agriculture.[22] But many developing countries have low investment in irrigation systems, dams, and ground water. For example, Ethiopia has less than 1% of the artificial water storage capacity per capita of North America, despite having to manage far greater hydrological variability.[23] Many developing countries do not have enough water storage to manage annual water demand based on the current average seasonal rainfall cycle, as illustrated in Table 4.2. This will become an even greater bind with a future, less predictable cycle.

In addition, inappropriate water pricing and subsidised electricity tariffs that encourage the excessive use of groundwater pumping (for agricultural use, for example) also increase vulnerability to changing climatic conditions. For example, 104 of Mexico's 653 aquifers (that provide half the water consumed in the country) drain faster than they can replenish themselves, with 60% of the withdrawals being for irrigation.[24] Similarly, water tables are falling in some drought-affected districts of Pakistan by up to 3 m per year, with water now available only at depths of 200–300 meterd.[25] The consequences of inadequate investment in water-related

[22] Irrigation plays an important role in improving returns from land, with studies identifying an increase in cropping intensity of 30% with the use of irrigation (Commission for Africa, 2005). Similarly, effective water management enables water to be stored for multiple uses, increases the reliability of water services, reduces peak flows and increases off-peak flows, and reduces the risk of water-related shocks and damage. World Bank (2006b).

[23] World Bank (2006c)

[24] International Commission on Irrigation and Drainage (2005)

[25] Roy (2006)

Table 4.2 Investment in water storage in developing countries

The seasonal storage index (SSI) indicates the volume of storage needed to satisfy annual water demand based on the average seasonal rainfall cycle (calculated as the volume needed to transfer water from wet months to dry months). The countries listed below need water during dry seasons and have water available to be captured during wet seasons. The 'Hard Water' column represents water storage requirements. Surface water reservoir development or groundwater development could provide additional storage. Some developing countries will also require 'soft water' (with water needs in excess of the volume that can be captured from internal renewable water resources) through increasing the efficiency of water use. South Asia faces problems of seasonal and inter-annual deficits, requiring both seasonal and inter-annual storage, and 'soft' water.[26]

	Seasonal Storage Index (km3)	SSI as % of Annual Volume	% Hard Water (of total)	Current Storage (% of SSI)	GDP per capita ($, 2003)
India	356.6	21%	17%	76%	555
Bangladesh	62.28	41%	40%	33%	385
Ethiopia	40.99	10%	100%	8%	91
Nepal	29.86	47%	100%	0%	233
Vietnam	27.64	10%	100%	3%	471
North Korea	23.32	45%	100%	0%	494
Senegal	22.3	40%	100%	7%	641
Malawi	18.98	34%	100%	0%	158
Algeria	6.6	6%	100%	91%	2,049
Tanzania	5.5	1%	33%	76%	271
El Salvador	5.45	37%	100%	59%	2,302
Haiti	3.73	25%	79%	0%	300
Guinea	3.71	2%	100%	51%	424
Eritrea	2.75	11%	15%	3%	305
Burundi	2.64	19%	27%	0%	86
Albania	2.64	23%	100%	21%	1,915
Guinea-Bissau	2.48	11%	100%	0%	208
Sierra Leone	2.21	3%	100%	0%	197
The Gambia	2.14	56%	100%	0%	224
Rwanda	1.38	9%	3%	0%	185
Mauritania	1.34	2%	100%	66%	381
Swaziland	0.98	15%	100%	59%	1,653
Bhutan	0.4	1%	13%	0%	303

Source: Brown and Lall (2006)

[26] Brown and Lall (2006)

infrastructure and poor management are important given that most climate change impacts are mediated through water (as discussed in Chapter 3).

Low incomes and underdeveloped financial markets   In many developing countries the capacity of poor people to withstand extreme weather events such as a drought is constrained both by low income levels and by limited access to credit, loans or insurance (in terms of access and affordability).[27] These constraints are likely to become worse as wet and dry seasons become increasingly difficult to predict with climate change.[28] This is often exacerbated by weak social safety nets that leave the poorest people very vulnerable to climate shocks. At the national level, many low-income countries have limited financial reserves to cushion the economy against natural disasters,[29] coupled with underdeveloped financial markets and weak links to world financial markets that limit the ability to diversify risk or obtain or reallocate financial resources. Less than 1% of the total losses from natural disasters, for example, were insured in low-income countries during the period 1985 to 1999.[30]

Poor public services   Inadequate resources and poor governance (including corruption) often result in poor provision of public services. Early warning systems for extreme weather conditions, education programmes raising awareness of climate change, and preventive measures and control programmes for diseases spread by vectors or caused by poor nutrition are examples of public services that would help to manage and cope with the effects of climate change but currently receive weak support and attention in developing countries.

Implications of different growth pathways for future vulnerability

Some parts of the developing world may look very different by the end of the century. If development progress is strong, then much of Asia and Latin America may be middle income or above, with substantial progress also being made in Africa. Growth and development should equip these countries to better manage the effects of climate change, and possibly avoid some of the most adverse impacts. For example, if there are more resources to build protection against rising sea levels, and economies become more diversified. But the extent to which these countries will be able to cope with climate change will depend on the scale of future impacts, and hence the action taken today to curb greenhouse gas emissions.

Further, the speed and type of climate change experienced over the next few decades will in part determine the ability of developing countries to develop and grow. Climate change is likely to lead to an increase in extreme weather

---

[27] An estimated 2.5 billion low income people globally do not have access to bank accounts, with less than 20% of people in many African countries having access compared to 90–95% of people in the developed world (CGAP, 2004). Poor people are typically constrained by their lack of collateral to offer lenders, unclear property rights, insufficient information to enable lenders to judge credit risk, volatile incomes, and lack of financial literacy, among other things.

[28] The incomes of poor people will become less predictable, making them less able to guarantee the returns that are needed to pay back loans, while insurers will face higher risks and losses making them even less willing to cover those most in need.

[29] IMF (2003)

[30] Freeman et al (2002)

events.[31] Evidence (discussed below) shows that extreme climate variability can set back growth and development prospects in the poorest countries. If climatic shocks do become more intense and frequent before these countries have been able to reduce their vulnerability, their long-term growth potential could be affected. And some developing countries are already exposed to the damaging impacts of climate change that, in extreme cases such as Tuvalu, have already constrained their long-term development prospects.

4.3   Direct implications of climate change for health, livelihoods and growth: what can be learnt from natural disasters?

The impact of climate change on poor countries is likely to be severe through both the effects of extreme weather events and a longer-term decline in the environment. The impact of previous extreme weather events provides an insight into the potential consequences of climate change.

Many developing countries are already struggling to cope with their current climate. Both the economic costs of natural disasters and their frequency have increased dramatically in the recent past. Global losses from weather related disasters amounted to a total of around $83 billion during the 1970s increasing to a total of around $440 billion in the 1990s, with the number of 'great natural catastrophe' events increasing from 29 to 74 between those decades.[32] The financial costs of extreme weather events represent a greater proportion of GDP loss in developing countries, even if the absolute costs are more in developed countries given the higher monetary value of infrastructure.[33] And over 96% of all disaster related deaths worldwide in recent years have occurred in developing countries. Climatic shocks can – and do – cause setbacks to economic and social development in developing countries. The IMF, for example, estimates costs of over 5% of GDP per large disaster on average in low-income countries between 1997 and 2001.[34]

Climate change will exacerbate the existing vulnerability of developing countries to an often difficult and changing climate. This section focuses on those

---

[31] For example, a recent study from the Hadley Centre shows that the proportion of land experiencing extreme droughts is predicted to increase from 3% today to 30% for a warming of around 4°C, and severe droughts at any one time will increase from 10% today to 40% (discussed in Chapters 1 and 3).

[32] Data extracted from Munich Re (2004). These figures are calculated on the basis of the occurrence and consequences of 'great natural disasters'. This definition is in line with that used by the United Nations and includes those events that over-stretch the ability of the affected regions to help themselves. As a rule, this is the case when there are thousands of fatalities, when hundreds of thousands of people are made homeless or when the overall losses and/or insured losses reach exceptional orders of magnitude. While increases in wealth and population growth account for a proportion of this increase, it cannot explain it all (see Chapter 5 for more details). The losses are given in constant 2003 values.

[33] The true cost of disasters for developing countries is also often undervalued. Much of the data on the costs of natural disasters is compiled by reinsurance companies and focused on economic losses rather than livelihood losses, and is unlikely to capture the effect of slow-onset and small-scale disasters and the impact these have on households. Furthermore, the assessments typically do not capture the cumulative economic losses as they are based on snapshots in time. Benson and Clay (2004)

[34] IMF (2003)

aspects that will likely feel the largest impacts: health, livelihoods and growth. The analysis draws on evidence from past and current exposure to climate variability to demonstrate the mechanisms at work.

Despite some beneficial effects in colder regions, climate change is expected to worsen health outcomes substantially.

Climate change will alter the distribution and incidence of climate-related health impacts, ranging from a reduction in cold-related deaths to greater mortality and illness associated with heat stress, droughts and floods. Equally, climate change will contribute to the geographical extension or contraction of potential outbreaks of vector borne diseases such as malaria. As noted in Chapter 3, if there is no change in malaria control efforts an additional 40 to 60 million people in Africa could be exposed to malaria with a 2°C rise in temperature, increasing to 70 to 80 million at 3–4°C.[35] Though some regions, such as parts of West Africa, may experience a reduction in exposure to vector borne diseases (see Chapter 3), previously unaffected regions may not have appropriate health systems to cope with and control malaria outbreaks. Poor people in slums are at significant risk of the effects of high temperature, water shortage, and increased prevalence of malaria and cholera given their limited or no access to clean water, exposure to stagnant water and poor sanitation and malnutrition. In Delhi, for example, gastroenteritis cases increased by 25% during a recent heat wave as slum dwellers had to drink contaminated water.[36]

The additional heath risks will not only cost lives, but also increase poverty. Malnutrition, for example, reduces peoples' capacity to work and affects a child's mental development and educational achievements with life-long effects. The drought in Zimbabwe in 2000, for example, is estimated to have contributed to a loss of 7–12% of lifetime earnings for the children who suffered from malnutrition.[37] And managing the consequences of these health impacts can in itself lead to further impoverishment. Households face higher personal health expenditures through clinic fees, anti-malarial drugs and burials, for example. This was seen in the case of Vietnam where rising health expenditures were found to have pushed about 3.5% of the population into absolute poverty in both 1993 and 1998.[38] The effects can be macroeconomic in scale: malaria is estimated to have reduced growth in the most-affected countries by 1.3% per year.[39]

Falling agricultural output and deteriorating conditions in rural areas caused by climate change will directly increase poverty of households in poor countries.

Current experience of extreme weather events underlines how devastating droughts and floods can be for household incomes. For example:

- In North-Eastern Ethiopia drought induced losses in crop and livestock between 1998–2000 were estimated at $266 per household – greater than the

[35] Warren et al (2006)
[36] Huq and Reid (2005)
[37] Alderman et al (2003)
[38] Wagstaff and van Doorslaer (2003)
[39] These results were estimated after controlling for initial poverty, economic policy, tropical location and life expectancy (using different time frames). Sachs and Gallup (2001)

annual average cash income for more than 75% of households in the study region;[40]

- In Ecuador the 1997–98 El Niño contributed to a loss of harvest and rise in unemployment that together increased poverty incidence by 10 percentage points in the affected municipalities.[41]

These direct impacts are often compounded by the rising cost of food and loss of environmental assets that would otherwise help to provide a safety net for poor people. Following the drought in Zimbabwe in 1991–92, for example, food prices increased by 72%[42].

These risks may increase with climate change if people remain highly exposed to the agricultural sector and have limited resources to invest in water management or crop development. As discussed in Chapter 1, climate change is likely to result in more heatwaves, droughts, and severe floods. In addition to these short-term shocks in output, climate change also risks a long-term decline in agricultural productivity in tropical regions. As Chapter 3 notes, yields of the key crops across Africa and Western Asia may fall by between 15% to 35% or 5% to 20% (assuming a weak or high carbon fertilisation respectively) once temperatures reach 3 or 4°C. Such a decline in productivity would pose a real challenge for the poorest countries, especially those already facing water scarcity. In sub-Saharan Africa, for example, only 4% of arable land is currently irrigated. The effects of climate change may constrain the long-term feasibility of future large-scale investment in irrigation.[43] For example, some extreme scenarios suggest that by 2100 the Nile could face a decrease in flow of up to 75%,[44] with normal irrigation practices having been found to cease when annual flow is reduced by more than 20%.[45]

Strategies to manage the risks and impacts of an adverse climate can lock people into long-term poverty traps.

The survival strategies adopted by poor people to cope with a changing climate may damage their long-term development prospects. Equally, if there is a risk of more frequent extreme weather events, then households may also have shorter periods in which to recover. Both will increase the possibility of households being pushed into a poverty-trap (as illustrated in Figure 4.4).[46] There are two aspects to this:

- *Risk-managing strategies*: Poor, risk-averse households may switch to low risk crops. In India, for example, poor households have been found to allocate a larger share of land to safer traditional varieties of rice and castor than to riskier but high-return varieties. This response in itself can reduce the average income of these people. For example, households in Tanzania that allocated

[40] Carter et al (2004)
[41] Vos et al (1999)
[42] IMF (2003). This was largely due to the higher price of food that had to be imported following a drought induced reduction in agricultural output, as described in Box 4.2, coupled with an increase in inflation to 46%.
[43] Commission for Africa (2005)
[44] Strzepek et al (2001)
[45] Cited in Nkomo et al (2006)
[46] This refers to a minimum asset threshold beyond which people are unable to build up their productive assets, educate their children and improve their economic position over time. Carter et al (2005)

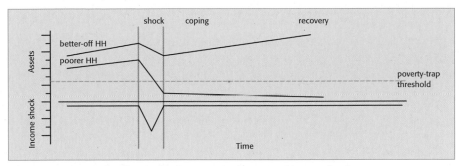

Figure 4.4 Impact of a climate shock on asset trajectory and income levels

This diagram illustrates: a) the period of shock itself (e.g. hurricanes or drought), b) the coping period in which households deal with the immediate losses created by the shock, and c) the recovery period where a household will try to rebuild the assets they have lost as a result of the climate shock or through the coping strategy they adopted.

Source: Carter et al (2005)

more of their land to sweet potatoes (a low return, low risk crop) were found to have a lower return per adult.[47]

- *Risk-coping strategies*: Poor households may also be forced to sell their only assets such as cattle and land. This can compromise their long-term development prospects as they are unable to educate their children, or raise levels of income over time. Following the 1991–92 droughts in Zimbabwe, many households had to sell their goats that were intended as a form of savings to pay, for example, for secondary education.[48] These risks will be amplified as prices of assets change relative to the price of food and other necessities as poor households respond to drought in a similar way.[49] Alternatively, to try and avoid permanent destitution households may decide to reduce their current consumption levels to protect their assets or prevent further depletion of their asset stock. This strategy can have long-term effects on health and human capital.[50] Reductions in consumption levels during a drought in Zimbabwe, for example, led to permanent and irreversible growth losses among children – losses that would reduce their future educational and economic achievement.[51]

Climate change and extreme variability could cut the revenue and increase the spending of nations, worsening their budget situation.

---

[47] Dercon (2003). Households with an average livestock holding in Tanzania were found to allocate 20% less of their land to sweet potatoes than a household with no liquid assets, with the return per adult of the wealthiest group being 25% higher for the crop portfolio compared to the poorest quintile.

[48] Hicks (1993)

[49] A household survey in eight peasant associations in Ethiopia found that distressed sales of livestock following the drought in 1999 sold for less than 50% of the normal price. Carter et al (2004)

[50] People can be pushed below a critical nutritional level whereby no productive activity is possible, with little scope for recovery given dependence on their own labour following the loss or depletion of their physical assets. Dasgupta and Ray (1986)

[51] Hoddinott (2004)

BOX 4.2 Economic Impacts of Drought in Zimbabwe, 1991–92

In late 1991 to early 1992, Zimbabwe was hit by a severe drought. This resulted in a range of impacts including: a fall in production of maize, cotton and sugarcane by 83%, 72% and 61% respectively; the death/slaughter of more than 23% of the national herd; water shortages that led to the deterioration in quality and price of Zimbabwean tobacco; and reduction in hydro-electricity generation that affected industry and the mineral export sector.

The direct impacts of the drought contributed to a doubling of the current account deficit from 6% to 12% of GDP between 1991 and 1992 and an increase in external debt from 36% of GDP in 1991 to 60% in 1992 and 75% by 1995. Government revenues fell in 1992–93 due to drought-induced loss of incomes, slowdown in non-food imports and slow-down in the private sector. Current expenditures increased by 2 percentage points of GDP in 1992–93 due predominantly to drought-related emergency outlays. Government expenditures on health and education were reduced as a share of the budget, in particular for primary education. By the end of 1992, real GDP had fallen by 9% and inflation increased to 46% with food prices having increased by 72%.

At the time of the drought the country was one of the better educated and more functional of states in sub-Saharan Africa. The more recent difficulties with governance, mismanagement and inflation, for example, were not anywhere near as problematic at the time of the drought.

Source: IMF (2003)

Dealing with climate change and extreme variability will also place a strain on government budgets, as illustrated in the case of Zimbabwe following the drought of 1991–92 (Box 4.2). The severity of the effect on government revenues will in part depend on the concentration of activity in climate-sensitive sectors. For example, the drought in southern Africa in 1991–92 resulted in a fall in income of over 8% in Malawi where agriculture contributed 45% of GDP at the time, but only 2% of GDP in South Africa where just 5% of GDP was obtained from agriculture.[52] Climate change will also necessitate an increase in spending at the national level to deal with the aftermath of extreme weather events and the consequences of a long-term decline in food and water supplies. For example, the logistical costs alone of importing cereal into drought affected southern African countries in 1991–92 were $500 million.[53] In some cases the expenditure requirements may be beyond the government's capacity. This was the case following Hurricane Mitch in 1998 where the Honduras government with a GNP of $850 per capita faced reconstruction costs equivalent to $1250 per capita.[54]

When governments face financial constraints, their response to the impacts of climate change and extreme variability – ranging from expenditure switching to additional financing through increasing debt levels – can itself amplify the negative effect on the growth and development of the economy. For example, if key investments to raise economic performance are deferred indefinitely.[55] In reality

[52] IMF (2003); World Bank (2006a)
[53] Benson and Clay (2004). Similarly the climatically less severe 1994–95 drought involved costs of $1 billion in cereal losses (due to higher prices in a tighter international cereal market).
[54] ODI (2005)
[55] IMF (2003)

Official Development Assistance (ODA) will often step in to help fill this financing gap, as was the case in Honduras following Hurricane Mitch. However these emergency funds are rarely additional and often reallocated funds or existing commitments within multi-year country programmes brought forward.

The experience of past extreme weather events and episodes testifies to the damaging effect that an adverse climate can have on social and economic prospects in developing countries. If climate change increases the frequency and severity of these events, as the science suggests, the costs on developing countries will grow significantly unless considerable effort is made today to reduce their vulnerability and exposure. And coupled with this will be a longer-term decline in the environment that will have to be managed. This will exert greater pressure still on resources and declines in the productivity and output of climate sensitive sectors.

4.4   What do global climate change models predict for developing countries?

Climate models predict a range of impacts on developing countries from a decrease in agricultural output and food security to a loss of vital river flows. The impacts are predominantly negative.

Evidence from the past and current extreme climate variability demonstrates the effect that a hostile climate can have on development. Looking now to the future, this section summarises some of the key findings from climate change impact studies undertaken by academics from particular developing regions to contribute to the Stern Review. These reports can be found on the Stern Review website (www.sternreview.org.uk). These summaries are not intended to be comprehensive but are rather more to highlight the key areas where climate change will be seen.

South Asia[56]

- India's economy and societal infrastructures are vulnerable to even small changes in monsoon rainfall. Climate change may increase the intensity of heavy rainfall events (the Mumbai floods of 2005 may be an example)[57] while the number of rainy days may decrease. Floods could become more extreme as a result with droughts remaining just as likely. Temperatures will increase for all months. Consequently, during the dry pre-monsoon months of April and May, the incidence of extreme heat is likely to increase, leading to greater mortality.
- Changes in the intensity of rainfall events, and the active/break cycles of the monsoon – combined with an increased risk of critical temperatures being exceeded more frequently – could significantly change crop yields. For example, mean yields for some crops in northern India could be reduced by up to 70% by

[56]   Information based largely on Challinor et al (2006). See also Roy (2006)
[57]   As ever it is difficult to attribute an outside event to climate change but the evidence is strong that the severity of such events is likely to increase.
[58]   Challinor et al (2006). 70% was the maximum reduction in yield that came from the study, in northern regions. Reductions in the 30–60% range were found over much of India. Strictly speaking these results are for groundnut only, although many annual crops are expected to behave similarly. The study was based on an SRES A2 scenario. The values assume no adaptation.

2100.[58] This is set against a background of a rapidly rising population that will need an additional 5 million tons of food production per year just to keep pace with the predicted increase in population to about 1.5 billion by 2030.

- Meltwater from Himalayan glaciers and snowfields currently supply up to 85% of the dry season flow of the great rivers of the Northern Indian Plain. This could be reduced to about 30% of its current contribution over the next 50 years, if forecasts of climate change and glacial retreat are realised. This will have major implications for water management and irrigated crop production, as well as introducing additional hazards to highland communities through increasingly unstable terrain.[59]

Sub-Saharan Africa[60]

- Africa will be under severe pressure from climate change. Many vulnerable regions, embracing millions of people, are likely to be adversely affected by climate change, including the mixed arid-semiarid systems in the Sahel, arid-semiarid rangeland systems in parts of eastern Africa, the systems in the Great Lakes region of eastern Africa, the coastal regions of eastern Africa, and many of the drier zones of southern Africa.[61]
- Between 250–550 million additional people may be at risk of hunger with a temperature increase of 3°C, with more than half of these people concentrated in Africa and Western Asia.[62] And there are risks of higher temperatures still. Climate change is also predicted to decrease in size and/or shift the areas of suitable climate for 81–97% of Africa's plant species, with 25–42% predicted to lose their entire area by 2085.[63]
- Tens of millions of additional people could be at risk of malaria by the 2080s.[64] Previously unsuitable areas for malaria in Zimbabwe could become suitable for transmission with slight temperature and precipitations variations, while in South Africa the area suitable for malaria may double with 7.8 million people at risk by 2100.[65]
- Water pressures may be intensified as rainfall becomes more erratic, glaciers retreat and rivers dry up. While there is much uncertainty about the flow of

[59] Challinor et al (2006)
[60] Information based largely on Nkomo et al (2006)
[61] Thornton et al. (2006). The regions at risk of climate change were identified by looking at the possibility of losses in length of growing period that was used as an integrator of changing temperatures and rainfall to 2050. This was projected by downscaling the outputs from several coupled Atmosphere-Ocean General Circulation Models for four different scenarios of the future using the SRES scenarios of the IPCC. Several different combinations of GCM and scenario were used. The vulnerability indicator was derived from the weighted sum of the following four components: 1) public health expenditure and food security issues; 2) human diseases and governance; 3) Human Poverty Index and internal renewable water resources; and 4) market access and soil degradation. Thornton et al. (2006)
[62] Cited in Warren et al. (2006) based on the original analysis of Parry et al (2004). These figures assume future socio-economic development but no carbon fertilisation effect, as discussed in Chapter 3.
[63] McClean et al (2005). This is estimated using the Hadley Centre third generation coupled ocean-atmosphere General Circulation Model.
[64] van Lieshout et al (2004)
[65] Republic of South Africa (2000) cited in Nkomo et al (2006)

the Nile, for example, several models suggest a decrease in river flow with nine recent climate scenario impacts ranging from no change to more than 75% reduction in flows by 2100.[66] This will have a significant impact on the millions of people that have competing claims on its supplies.

- Many large cities in Africa that lie on or very close to the coast could suffer severe damages from sea level rise. According to national communications to the UNFCCC, a 1 meter sea-level rise (a possibility by the end of the century) could result in the complete submergence of the capital city of Gambia, and losses of more than $470 million in Kenya for damage to three crops alone (mangoes, cashew nuts and coconuts).[67]

Latin America[68]

- Countries in Latin American and the Caribbean are significantly affected by climate variability and extremes, particularly the ENSO events.[69] The region's economy is strongly dependent on natural resources linked to the climate, and patterns of income distribution and poverty exacerbate the impacts of climate change for specific sub-regions, countries and populations.
- Living conditions and livelihood opportunities for millions of people may be affected by climate change. By 2055 subsistence farmers' maize production (the main source of food security) in the Andean countries and Central America could fall by around 15% on average, for example, based on projections of HadCM2.[70] The potential die-back, or even collapse, of the Amazon rainforest (discussed in Chapter 3) presents a great threat to the region. The Amazonian forests are home to around 1 million people of 400 different indigenous groups, and provide a source of income and medical and pharmaceutical supplies to millions more.
- Climate change could contribute to a 70% rise in the projected number of people with severe difficulties in accessing safe water by 2025. About 40 million people may be at risk of water supply for human consumption, hydro-power and agriculture in 2020, rising to 50 million in 2050 through the predicted melting of tropical Andean glaciers between 2010 and 2050. The cities of Quito, Lima and La Paz are likely to be most affected. Dengue transmission is likely to increase by 2 to 5 fold by the 2050s in most areas of South America and new transmission areas are likely to appear in the southern half of the continent and at higher elevations.

China[71]

- There is significant variation in climatic patterns across China's regions including arid, temperate and mountainous regions. The average surface air

---

[66] Strzepek et al (2001)
[67] Gambia (2003) and Republic of Kenya (2002) cited in Nkomo et al (2006)
[68] Information based on Nagy et al (2006)
[69] El Nino-Southern Oscillation events (as discussed in Chapter 1).
[70] Jones and Thornton (2003), cited in Nagy et al (2006)
[71] Information based on Erda and Ji (2006)

temperature in China has increased by between 0.5 and 0.8°C over the 20th century with increases more marked in North China and Tibetan Plateau compared to southern regions. Temperature rise will lead to temperate zones in China moving north as well as an extension of arid regions. Cities such as Shangai are expected to experience an increase in the frequency and severity of heat waves causing significant discomfort to fast growing urban populations.

- Overall water scarcity is a critical problem in China with existing water shortages, particularly in the north (exacerbated by economic and population growth). Climate change is expected to increase water scarcity in northern provinces such as Ningxia, Gansu, Shanxi and Jilin province increasing frequency of drought. An increase in average rainfall in southern provinces such as Fujian, Zhejiang and Jiangxi is anticipated over the next 50 to 100 years leading to more instances of flooding. From 1988 to 2004, China experienced economic losses from drought and flood equating to 1.2% and 0.8% of GDP respectively.

- Climate change is expected to have mixed effects on agricultural output and productivity across different regions with impacts closely related to changes in water availability. On average, irrigated land productivity is expected to decrease between 1.5% to 7% and rain fed land by between 1.1% to 12.6% (under rain-fed conditions) between 2020s and 2080s under HadCM2, CGCM1 and ECHAM4 scenarios in China.[72] Overall a net decrease in agricultural production is anticipated with seven provinces in the north and northwest of China particularly vulnerable (accounting for 25% of total arable land and 14% of China's total agricultural output by value).[73]

Middle East and North Africa

- The region is already very short of fresh water and faces difficulty meeting the needs of fast-growing populations. Most if not all the region may be adversely affected by changing rainfall patterns as a result of climate change. An additional 155 to 600 million people may be suffering an increase in water stress in North Africa with a 3°C rise in temperature according to one study.[74] Yemen is particularly at risk given its low income levels, rapidly growing populations and acute water shortages today. Competition for water within the region and across its borders may grow, carrying the risk of conflict.

- Reduced water availability combined with higher temperatures will reduce agricultural productivity and in some areas may make crops unsustainable. Maize yields in North Africa, for example, could fall by between 15–25% with a 3°C rise in temperature according to one recent report.[74]

- Some parts of the region – notably the Nile Delta and the Gulf coast of the Arabian peninsula – are in addition vulnerable to flooding from rising sea levels which could lead to loss of agricultural land and/or threats to coastal cities and critical revenue-generating infrastructure; others are vulnerable to increased desertification.

---

[72] Tang Guoping et al (2000)
[73] NBSC (2005)
[74] Warren et al (2006)

Climate change poses a wide range of potentially very severe threats to developing countries. Understanding the impact of climate change on developing countries – at both a regional and national level – is essential. This will enable a better understanding of the scale of threat and urgency of mitigation action, but will also help countries to prepare for some of the now inevitable impacts of climate change. To date, however, analysis undertaken in developing countries of potential threats and impacts has been very limited. Many climate changes are on the way and foresight and action will be crucial if damages to development progress are to be managed by both the private and by public sectors. Further work is required on studying the impacts of climate change on developing countries at a national, regional and global level.

4.5   Impact of climate change on economic growth prospects and implications for incomes and health

Over time, there is a real risk that climate change will have adverse implications for growth and development. This section looks at how income levels and growth have been affected by extreme climate variability and then moves on to summarise illustrative modelling work undertaken as part of the review. If climate change results in lower output and growth levels than would otherwise be the case, there will be implications for poverty levels. But income levels also affect health, and mortality rates will be likely to rise above what they would otherwise have been, in addition to any effects from the physical environment such as via vector borne diseases. The previous section reviewed a range of projected direct climate impacts on factors affecting lives and livelihoods that recent research has highlighted. This section provides an analysis of their possible impacts on income and health.

Extreme weather events can – and do – affect growth rates in developing countries. Climate change presents a greater threat still.

The output of an economy in a given year depends on labour, environmental quality and capital available in that year (illustrated, for example, in Box 5.1 of Chapter 5). All three will be affected by climate change – be it through the damaging effects on the health and productivity of the labour force, the loss and damage to agriculture and infrastructure, or lower quality investment and capital. As the output and factors of production of an economy are repeatedly affected, so growth prospects will change. This will be particularly true for poorer economies with a stronger focus on agriculture and with less ability to diversify their economies. Falling agricultural output and productivity will have consequences across the economy.[75]

The effects of current extreme climate variability demonstrate the potential impact a changing climate can have on output and growth. Changes in the hydrological cycle can be especially damaging. Too much rainfall can inundate transport, for example, limiting trade potential and communication. It has been estimated

---

[75] Increased agricultural productivity has been identified as a key factor in reducing poverty and inequality. This is based on work undertaken by Bourguignon and Morrisson (1998) using data from a broad sample of developing countries in the early 1970s and mid 1980s. Evidence from Zambia, for example, suggests that an extra $1.5 of income is generated in other businesses for every $1 of farm income. Hazel and Hojjati (1995). Similarly, Block and Timer (1994) estimated an agricultural multiplier in Kenya of 1.64 versus a non-agricultural multiplier of 1.23 in Kenya.

that the 2000 floods in West Bengal, for example, destroyed 450 km of rail track and 30 bridges and culverts, and adversely affected 1739 km of district roads, 1173 km of state highways and 328 km of national highways.[76] Too little rainfall will affect crop production but also reduce the flow of surface water that could provide irrigation and hydroelectricity production. The La Niña drought in Kenya, for example, caused damage to the country amounting to 16% of GDP in each of 1998–99 and 1999–2000 financial years, with 26% of these damages due to hydropower losses and 58% due to shortfalls in industrial production.[77]

Economy-wide, multi-market models that incorporate historical hydrological variability project that hydrological variability may cut average annual GDP growth rates in Ethiopia by up to 38% and increase poverty rates by 25%.[78] These models capture the impacts of both deficit and excess rainfall on agricultural and non-agricultural sectors. As climate change increases the variability of rainfall, the scale of these growth impacts could rise significantly.

Slower growth could cause an increase in poverty and child mortality relative to a world without climate change, as found by illustrative modelling work undertaken by and for the Stern Review.

The Stern Review has used the PAGE2002 model (an integrated assessment model that takes account of a wide range of risks and uncertainties) to assess how climate change may affect output and growth in the future.[79] Integrated assessment models can be useful vehicles for exploring the kinds of costs that might follow from climate change. However, these are highly aggregative and simplified models and, as such, the results should be seen as illustrative only.

Under a baseline-climate-change scenario[80] the mean cost of climate change in India and South East Asia, and in Africa and the Middle East is predicted by PAGE2002[81] to be around a 2.5% and 1.9% loss in GDP by 2100, compared with what could have been achieved in a world without climate change. Under a high-climate-change scenario,[82] the mean cost of climate change is predicted by PAGE2002 to be equal to losses of 3.5% in GDP in India and South East Asia, and 2.7% in Africa and the Middle East by 2100.

---

[76] Cited in Roy (2006)

[77] World Bank (2006c)

[78] World Bank (2006c). The model shows growth projections dropping 38% when historical levels of hydrological variability are assumed, relative to the same model's results when average annual rainfall is assumed in all years. Hydrological variability includes drought, floods and normal variability of 20% around the mean.

[79] The estimates used in this analysis are based on the impact of climate change on market sectors such as agriculture. PAGE2002 allows examination of either market impacts only (as used here to ensure no double counting of poverty impacts) or market plus non-market impacts. These estimates and further details on the PAGE2002 model are given in Chapter 6.

[80] The baseline-climate-change scenario is based largely on scientific evidence in the Third Assessment Report of the IPCC, in which global mean temperature increases to 3.9°C in 2100 (see Chapter 6 for more detail).

[81] Using the IPCC A2 SRES baseline.

[82] In the high-climate-change scenario, global mean temperature increases to 4.3°C in 2100. The high-climate-change scenario is designed to explore the impacts that may be seen if the level of temperature change is pushed to higher levels through positive feedbacks in the climate system, as suggested by recent studies (see Chapter 1 and Chapter 6 for more detail).

There are good reasons, however, for giving more emphasis to the higher (95th percentile) impacts predicted in these scenarios, as the model is unlikely to capture the full range of costs to developing countries. In particular:

- The poorest people will be hit the hardest by climate change, an effect for which the highly aggregated models do not allow;
- There are specific effects, such as possible loss of Nile waters and the cumulative effects of extreme weather events (as discussed above), that aggregated global and regional models do not capture;
- This is a long-term story. If emissions continue unabated, temperatures will rise to much higher levels in the next century, committing these regions to far greater impacts (as discussed in Chapters 3 and 6), including the risks associated with mass migration and conflict discussed in the next section.

At the 95th percentile, and under the baseline-climate-change scenario, the projections rise to a 9% loss in GDP in India and South East Asia, and a 7% loss in Africa and the Middle East by 2100. And under the high-climate-change scenario, the costs of climate change rise significantly to losses of 13% and 10% in GDP respectively (again at the 95th percentile).

Given the strong correlation between growth and poverty reduction (see Box 4.3), a climate-driven reduction in GDP would increase the number of people below the $2 a day poverty line by 2100, and raise the child mortality rate compared with a world without climate change. This is illustrated below by modelling work undertaken for the Stern Review.[83] This analysis assumes reductions in poverty and child mortality are driven primarily by GDP growth.[84] As with the PAGE2002 model itself, projections that extend so far into the future should be treated with caution, but are useful for illustrative purposes. The projections summarised below focus only on income effects.

BOX 4.3 Relationship between growth and development

Countries with higher overall growth rates tend to have higher growth in incomes of poor people. Poverty is estimated to decline on average by 2% for a 1 percentage point rise in economic growth across countries.[85] Kraay estimates that growth accounts for about 70% of the variation in poverty over the short run (as measured by a $1 a day poverty line). As the time horizon lengthens, that proportion increases to above 95%.[86] East Asia has grown rapidly (5.8% in the 80s and 6.3% in the 90s) and has seen the fastest fall in poverty in human history. An annual growth of more than 7% will be needed to halve severe poverty in Africa by 2015 (and a 5% annual growth is required just to keep the number of poor people from rising).[87] There is also a close relationship between growth and many non-income indicators of development, ranging from under-five mortality to educational attainment and peace and security. Income-earning opportunities provide citizens with a vested interest in avoiding conflict, and security allows governments to invest in productive assets and social expenditures rather than defence.

[83] Using the IPCC A2 SRES baseline
[84] Other factors – such as changes in income distribution – that may also affect poverty levels or child mortality are assumed to be constant.
[85] Ravallion (2001)
[86] Kraay (2005)
[87] World Bank (2000)

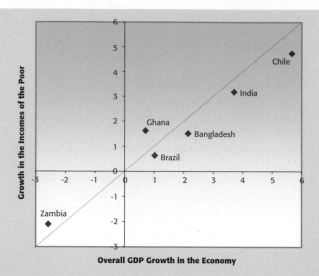

Overall GDP Growth in the Economy

Source: World Bank (2003c)

While growth is clearly an important contributor to poverty reduction, much depends on how the benefits of this growth are distributed and the extent to which the additional resources generated are used to fund public services such as healthcare and education. Poor people benefit the most from economic growth when it occurs in those parts of the economy that offer higher returns for poor people's assets.

Poverty projections

By 2100 climate change could cause up to an additional 145 million people to be living on less than $2 a day in South Asia and sub-Saharan Africa because of GDP losses alone at the 95th percentile of the baseline-climate-change scenario and runs, or 35 million people at the mean of these runs. Under the high-climate-change scenario at the 95th percentile, up to an additional 220 million people could be living on less than $2 a day by 2100 in South Asia and sub-Saharan Africa because of GDP losses alone. The effects at the mean of the distribution are smaller but still significant: up to an additional 50 million people living on less than $2 a day per year. This is illustrated in Box 4.4 below.

If growth proceeds faster than predicted, then the overall numbers of people living on below $2 a day will be less, while if it is slower there will be more people pushed into poverty. These calculations should be viewed as indicative of the risks.

Child mortality projections

There is a well-studied relationship between reduced income and child mortality. Falling income and GDP levels from what could have been achieved in a world without climate change will slow the improvement of child (and adult) health in

BOX 4.4 Potential impact of climate change on additional people living on less than $2 a day in South Asia and sub-Saharan Africa

These projections are calculated using the formulae for the poverty headcount used in World Bank calculations,[88] population forecasts, and the assumptions that average household income grows at 0.8 times the rate of GDP per capita[89] and distribution of income remains constant.

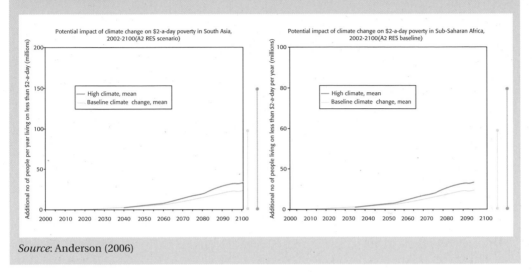

Source: Anderson (2006)

developing countries.[90] Lower per capita expenditures at both a public and private level are likely on goods that improve health, such as safe water, food and basic sanitation. Previous econometric studies have reported a range of values for the income elasticity of infant and child mortality, the vast majority falling between –0.3 and –0.7. Taking an elasticity of 0.4, for example, implies that a 5% fall in GDP from what could have been achieved in a world without climate change will lead to a 2% increase in infant mortality.[91] This analysis uses a value of -0.5 for the elasticity of the child mortality rate (deaths per 1,000 births) with respect to per capita income, the mid-point of this range.[92]

[88] The formulae express the level of poverty as a function of the poverty line, average household income and the distribution of income. The $2 poverty line is used throughout.

[89] This figure is obtained from a cross-country regression of rates of growth in mean household expenditure per capita on GDP per capita. Ravaillion (2003)

[90] It is important to note that income alone does not determine health outcomes. Efficient public programmes and access to education for women are also important factors, for example. Furthermore, the way in which GDP per capita changes (for example if there is a change in the distribution of income that coincides with the change in national income) can affect the impact it has on health.

[91] Analysis demonstrates the health effects today of slowing or negative per capita growth. For example, in 1990 over 900,000 infant deaths would have been prevented had developing countries been able to maintain the same rate of growth in the 1980s as in the period 1960–80 (assuming an elasticity of -0.4), rather than the slow or negative growth they in fact experienced. The effects were particularly significant in Africa and Latin America, where growth was lower by 2.5% on average. Pritchett and Summers (1993).

[92] The elasticity is assumed to be a constant across countries and over time, consistent with econometric evidence (such as Kakwani (1993)). However, the average elasticity of child mortality with respect to GDP over a period of time will typically not be the same as the actual elasticity that applies on a year-to-year basis, even if the latter is assumed constant, because of compounding.

Using the illustrative output and growth scenarios generated by PAGE2002, climate change could cause an additional 40,000 (mean) to 165,000 (95th percentile) child deaths per year in South Asia and sub-Saharan Africa through GDP losses alone by 2100 under the baseline-climate-change scenario. Under the high-climate-change scenario, climate change could cause an additional 60,000 (mean) to 250,000 (95th percentile) child deaths per year by 2100 in South Asia and sub-Saharan Africa through GDP losses alone and compared with a world without climate change. This is illustrated in Box 4.5 below.

BOX 4.5 Potential impact of climate change on additional child deaths per year in South Asia and sub-Saharan Africa

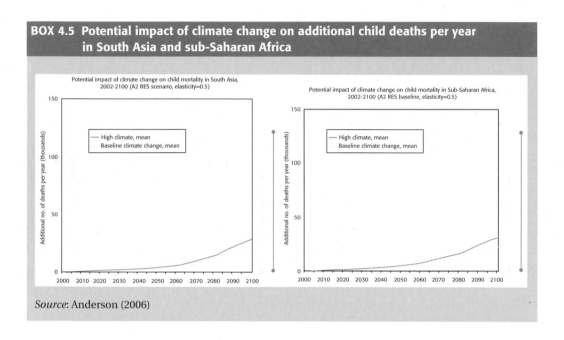

Source: Anderson (2006)

The above projections pick up the pure income effect of climate change on poverty and child mortality through its dampening effect on GDP. They do *not* include the millions of people that will be exposed to heat stress or malaria, or risk losing their jobs, assets and livelihoods through extreme weather events, for example, as discussed in Sections 4.3 and 4.4. This analysis and projections are simply illustrative of possible risks associated with a loss in income through climate change.

4.6   Population movement and risk of conflict

Greater resource scarcity, desertification, risks of droughts and floods, and rising sea levels could drive many millions of people to migrate – a last-resort adaptation for individuals, but one that could be very costly to them and the world.

The impacts of climate change, coupled with population growth in developing countries, will exert significant pressure for cross-border and internal population displacement and movement. There is already evidence of the pressure that an adverse climate can impose for migration. Approximately 7 million people migrated in order to obtain relief food out of the 80 million considered to be semi-starving in sub-Saharan Africa, primarily due to environmental factors.[93] Millions of people could be compelled to move between countries and regions in order to seek new sources of water and food if these fall below critical thresholds. Rising sea levels may force others to move out of low-lying coastal zones. For example, if sea levels rise by 1 metre (a possible scenario by the end of the century, Chapter 3) and no dyke enforcement measures are taken, more than one-fifth of Bangladesh may be under water.[94] And atolls and small islands are at particular risk of displacement with the added danger of complete abandonment. As one indication of this, the government of Tuvalu have already begun negotiating migration rights to New Zealand in the event of serious climate change impacts.[95]

The total number of people at risk of displacement or migration in developing countries is very large. This ranges from the millions of people at risk of malnutrition and lack of clean water to those currently living in flood plains. Worldwide, nearly 200 million people today live in coastal flood zones that are at risk; in South Asia alone, the number exceeds 60 million people.[96] Coupled with this there are potentially between 30 to 200 million people at risk of hunger with temperature rises of 2 to 3°C – rising to 250 to 550 million people with a 3°C warming;[97] and between 0.7 to 4.4 billion people who will experience growing water shortages with a temperature rise of 2°C,[98] as discussed in Chapter 3.

The exact number of people who will actually be displaced or forced to migrate will depend on the level of investment, planning and resources at a government's disposal to defend these areas or provide access to public services and food aid. The Thames Barrier, for example, protects large parts of London, while flood defences and pumped drainage enable large areas of Shanghai and Tokyo to be below normal tides. Protection is expensive, however, particularly relative to income levels in developing countries. A project to construct 8,000 km of river dykes in Bangladesh, a country with a GNI of $61billion, is costing $10 billion for example. These high costs will discourage governments from investing. Defensive investments must be made early to be effective, but they may be politically unpopular if they divert large amounts of money from programmes with more immediate impact such as infrastructure, health and education.

Drought and other climate-related shocks may spark conflict and violence, as they have done already in many parts of Africa.

The effects of climate change – particularly when coupled with rapid population growth, and existing economic, political, ethnic or religious tensions – could be a

[93] Myers (2005)

[94] Nicholls (1995) and Anwar (2000/2001)

[95] Barnett and Adger (2003)

[96] Warren et al. (2006) analysing data from Nicholls (2004), Nicholls and Tol (2006) and Nicholls and Lowe (2006). This is calculated on the basis of the number of people that are exposed each year to storm surge elevation that has a one in a thousand year chance of occurring. These odds and the numbers explored could be rising rapidly. This has already been demonstrated in the case of heat waves in Southern Europe where the chance of having a summer as hot as in 2003 that in the past would be expected to occur once every 1000 years, will be commonplace by the middle of the century due to climate change, as discussed in Chapter 5.

[97] Warren et al. (2006) based on the original analysis of Parry et al. (2004). As noted in Chapter 3, these figures assume future socio-economic development, but no carbon fertilisation effect.

[98] Warren et al. (2006) based on the original analysis of Arnell (2004) for the 2080s. As noted in Chapter 3, the results only represent "potential water stress" as they do not include adaptation.

contributory factor in both national and cross-border conflicts in some developing countries. There are two more aspects to this:

- Long-term climate deterioration (such as rising temperatures and sea levels) will exacerbate the competition for resources and may contribute to forced dislocation and migration that can generate destabilising pressures and tensions in neighbouring areas.
- Increased climate variability (such as periods of intense rain to prolonged dry periods) can result in adverse growth shocks and cause higher risks of conflict as work opportunities are reduced, making recruitment into rebel groups much easier. Support for this relationship has been provided by empirical work in Africa, using rainfall shocks as an instrument for growth shocks.[99]

Adverse climatic conditions already make societies more prone to violence and conflict across the developing world, both internally and cross-border. Long periods of drought in the 1970s and 1980s in Sudan's Northern Darfur State, for example, resulted in deep, widespread poverty and, along with many other factors such as a breakdown in methods of coping with drought, has been identified by some studies as a contributor to the current crisis there.[100] While climate change could contribute to the risk of conflict, however, it is very unlikely to be the single driving factor. Empirical evidence shows that a changing and hostile climate has resulted in tension and conflict in some countries but not others. The risk of climate change sparking conflict is far greater if other factors such as poor governance and political instability, ethnic tensions and, in the case of declining water availability, high water interdependence are already present. In light of this, West Africa, the Nile Basin and Central Asia have been identified as regions potentially at risk of future tension and conflict. Box 4.6 indicates areas vulnerable to future tension and past conflicts where an adverse climate has played an important role.

BOX 4.6 Future risks and past conflicts

Future risks

- *West Africa*: While there is still much uncertainty surrounding the future changes in rainfall in this part of the world, the region is already exposed to declining average annual rainfall (ranging from 10% in the wet tropical zone to more than 30% in the Sahelian zone since the early 1970s) and falling discharge in major river systems of between 40 to 60% on average. Changes of this magnitude already give some indication of the magnitude of risks in the future given that we have only seen 0.7°C increase and 3 or 4°C more could be on the way in the next 100 to 150 years. The implications of this are amplified by both the high water interdependence in the region – 17 countries share 25 transboundary watercourses – and plans by many of the countries to invest in large dams that will both increase water withdrawals and change natural water allocation patterns between riparian countries.[101] The region faces a serious risk of water-related conflict in the future if cooperative mechanisms are not agreed.[102]

---

[99] Miguel et al (2004), Collier and Hoeffler (2002), Hendrix and Glaser (2005) and Levy et al (2005)
[100] University for Peace Africa Programme (2005)
[101] For example, there are 20 plans in place to build large dams along the Niger River alone.
[102] Niasse (2005)

- *The Nile*: Ten countries share the Nile.[103] While Egypt is water scarce and almost entirely dependent on water originating from the upstream Nile basin countries, approximately 70% of the Nile's waters flow from the Ethiopian highlands. Climate change threatens an increase in competition for water in the region, compounded by rapid population growth that will increase demand for water. The population of the ten Nile countries is projected to increase from 280 million in 2000 to 860 million by 2050. A recent study by Strzepek et al. (discussed above) found a propensity for lower Nile flows in 8 out of 8 climate scenarios, with impacts ranging from no change to a roughly 40% reduction in flows by 2025 to over 60% by 2050 in 3 of the flow scenarios.[104] Regional cooperation will be critical to avoid future climate-driven conflict and tension in the region.

Past conflicts

- *National conflict*: Drought in Mali in the 1970s and 1980s damaged the pastoral livelihoods of the semi-nomadic Tuareg. This resulted in many people having to seek refuge in camps or urban areas where they experienced social and economic marginalisation or migrated to other countries. On their return to Mali, these people faced unemployment and marginalisation which, coupled with the lack of social support networks for returning migrants, continuing drought, and competition for resources between nomadic and settles peoples (among many other things), helped create the conditions for the 'Second Tuareg Rebellion' in 1990. A similar scenario has played out in the Horn of Africa,[105] and may now be replicating itself in northern Nigeria, where low rainfall combined with land-use pressures have reduced the productivity of grazing lands, and herders are responding by migrating southward into farm areas.[106]

- *Cross-border conflict*: Following repeated droughts in the Senegal River Basin in the 1970s–80s, the Senegal River Basin Development Authority was created by Mali, Mauritania and Senegal with the mandate of developing and implementing a major water infrastructure programme. Following the commissioning and completion of agreed dams, conflict erupted between Senegal and Mauritania when the river started to recede from adjacent floodplains. The dispute and tension escalated with hundreds of Senegalese residents being killed in Mauritania and a curfew imposed by both Governments such that 75,000 Senegalese and 150,000 Mauritanians were repatriated by June 1989. Diplomatic relationships between the two countries were restored in 1992, but a virtual wall has effectively been erected along the river.[107] Drought has also caused conflict between Ugandan and Kenyan pastoralists, and has led Ethiopian troops to move up north to stop the Somalis crossing the border in search of pasture and water for their livestock.[108] Similarly, extreme weather events in 2000 that affected approximately 3 million people in Bangladesh resulted in migration and violence as tribal people in North India clashed with emigrating Bangladeshis.[109]

[103] Ethiopia, the Sudan, Egypt, Kenya, Uganda, Burundi, Tanzania, Rwanda, the Democratic Republic of Congo and Eritrea.
[104] Strzepek et al (2001). While there is general agreement regarding an increase in temperature with climate change that will lead to greater losses to evaporation, there is more uncertainty regarding the direction and magnitude of future changes in rainfall. This is due to large differences in climate model rainfall predictions.
[105] Meier and Bond (2005)
[106] AIACC (2005)
[107] Niasse (2005)
[108] Christian Aid (2006)
[109] Tanzler et al (2002)

4.7   Implications of Climate Change on other Aspects of Development

Practically all development aspirations could eventually be challenged by climate change. Education and gender goals, for example, may be adversely affected by climate change, in turn further amplifying vulnerability to the impacts of climate change (as discussed in Box 4.7). Limited research has been undertaken to date on the impact of climate change on these important aspects of development. This merits much greater attention going forward.

BOX 4.7 Impact of Climate Change on Education and Gender Equality

Education

Climatic disasters can threaten educational infrastructure making it physically impossible for children to attend school. For example in 1998 Hurricane Mitch destroyed 25% of Honduras' schools.[110] Education levels may also decline through climate-induced changes in income and health conditions. Schooling will become less affordable and accessible, especially for girls, as income, assets and employment opportunities are affected by climate change. Children will need to help more with household tasks or prematurely engage in paid employment leaving less time for schooling. Deteriorating health conditions will also affect both a child's learning abilities and school attendance, and the supply of teachers. Children will be deprived of the long-term benefits of education and be more vulnerable to the effects of climate change. Better-educated farmers, for example, absorb new information quickly, use unfamiliar inputs, and are more willing to innovate. An additional year of education has been associated with an annual increase in farm output of between 2 to 5%.[111]

Gender equality

Gender inequalities will likely worsen with climate change. Workloads and responsibilities such as collecting water, fuel and food will grow and become more time consuming in light of greater resource scarcity. This will allow less time for education or participation in market-based work. A particular burden will be imposed on those households that are short of labour, further exacerbated if the men migrate in times of extreme stress leaving women vulnerable to impoverishment, forced marriage, labour exploitation and trafficking.[112] Women are 'over-represented' in agriculture and the informal economy, sectors that will be hardest hit by climate change. This exposure is coupled with a low capacity to adapt given their unequal access to resources such as credit and transport. Women are also particularly vulnerable to the effects of natural disasters with women and children accounting for more than 75% of displaced persons following natural disasters.[113]

[110] ODI (2005)
[111] This takes into account farm size, inputs, hours worked etc. This is drawing on evidence from Malaysia, Ghana and Peru. Information drawn from Birdsall (1992)
[112] Chew and Ramdas (2005)
[113] Chew and Ramdas (2005)

4.8 Conclusion

The impacts of climate change will exacerbate poverty – in particular through its effects on health, income and future growth prospects. Equally, poverty makes developing countries more vulnerable to the impacts of climate change. This chapter has discussed some of the specific risks faced by developing countries. However it is the sum of the parts that creates perhaps the greatest concern. Poor households and governments may, for example, face falling food and water supplies and rising sea levels that will increase poverty directly, while also facing greater risk of illness and death through malaria or the effects of extreme weather events. These impacts may be compounded if governments have limited – or reduced – financial resources to manage these impacts, and to invest in building resilience against the future impacts of climate change. An important priority for future research will be to identify the type and scale of climate change impacts on developing countries and to understand more deeply the nature of these compounding, aggregated effects.

The threats posed by climate change increase the urgency of promoting growth and development today. This is key to reducing the vulnerability of developing countries to some of the now inevitable climate change, and enabling them to better manage these impacts. But adaptation can only mute the effects and there are limits to what it can achieve.

Unchecked climate change could radically alter the prospects for growth and development in some of the poorest countries. This underlines the urgency of strong and early action to reduce greenhouse gas emissions. This is discussed further in Part III of the report.

References

Abramovitz, J. N. (2001): 'Averting unnatural disasters' in State of the World 2001: A Worldwatch Institute Report on Progress Toward a Sustainable Society, Brown L. R., C. Flavin, and H. French, New York: W. W. Norton & Company.

AIACC (2005): 'For Whom the Bell Tolls: Vulnerabilities in a changing climate' Washington, DC: AICC Working Papers.

Alderman, H., J. Hoddinott, and B. Kinsey (2003): 'Long term consequences of early childhood malnutrition' FCND Discussion Paper 168, Washington, DC: International Food Policy Research Institute (IFPRI).

Anderson, E. (2006): 'Potential impacts of climate change on $2 a day poverty and child mortality in Sub-Saharan Africa and South Asia'. Mimeo, Overseas Development Institute, London.

Angelsen, A. and S. Wunder (2003): Exploring the forest–poverty link: key concepts, issues and research implications. Occasional Paper No. 40. Bogor, Indonesia: Center for International Forestry Research.

Anwar, A. (2000/2001): 'Vulnerability of Bangladesh coastal region to climate change with adaptation options', Bangladesh: SPARRSO.

Arnell, N.W. (2004): 'Climate change and global water resources: SRES emissions and socio-economic scenarios', Global Environmental Change 14: 31 52

Arnold, J. E. M. (2001): 'Forestry, poverty and aid'. Occasional Paper No 33. Bogor, Indonesia: Center for International Forestry Research.

Auffret, P. (2003): 'High consumption volatility: the impact of natural disasters', World Bank Working Paper 2962, Washington, DC: World Bank.

Barnett, J. and W.N. Adger (2003): 'Climate dangers and atoll countries', Climatic Change 61: 321–337

Belcher, B., M. Ruiz-Perez, and R. Achdiawan (2003): 'Global Patterns and Trends in Non-Timber Forest Product Use.' International Conference on Rural Livelihoods, Forest and Biodiversity. Bonn, Germany, 20–23 May 2003.

Benson, C. and E. Clay (2004): 'Understanding the economic and financial impacts of natural disasters', Washington, DC: World Bank.

Birdsall, N. (1992): 'Social development is economic development' Presentation to delegates of Social Committee, United Nations General Assembly, October 19, D3934.

Block, S. and C.P. Timmer (1994): 'Agriculture and economic growth: conceptual issues and the Kenyan experience', Cambridge, MA: Harvard Institute for International Development.

Bourguignon, F. and C. Morrisson (1998): 'Inequality and development: the role of dualism', Journal of Development Economics, 57: 233–58

Brown, C. and U. Lall, (2006): 'Water and economic development: the role of interannual variability and a framework for resilience', Working Paper, Columbia University, NY: International Research Institute for Climate Prediction.

Carter, M.R., P.D. Little, T. Mogues and W. Negatu (2005): 'The long-term impacts of short-term shocks: poverty traps and environmental disasters in Ethiopia and Honduras', Wisconsin: BASIS.

Carter, M.R., P.D. Little, T. Mogues and W. Negatu (2004): 'Shock, sensitivity and resilience: tracking the economic impacts of environmental disaster on assets in Ethiopia and Honduras'. Wisconsin: BASIS.

CGAP (2004): 'Financial Institutions with a 'double bottom line': implications for the future of microfinance' Occasional Paper No 8.

Challinor, A., J. Slingo, A. Turner, and T. Wheeler (2006): 'Indian monsoon: contribution to chapter 4 of Stern Review', University of Reading.

Chew, L. and K.N. Ramdas (2005): 'Caught in the storm: impact of natural disasters on women'. San Fransisco, CA: Global Fund for Women.

Christian Aid Report (2006): 'The climate of poverty: facts, fears and hope'. May 2006, London: Christian Aid.

Collier, P. and Hoeffler, A. (2001): 'Greed and grievance in civil war'. Washington, DC: World Bank

Collier, P. and A. Hoeffler (2002): 'On the incidence of civil war in Africa'. Journal of Conflict Resolution 46 (1): 13–28

Commission for Africa (2005): 'Our common interest – report of the Commission for Africa', London: Penguin.

Conway, D. (2005): 'From headwater tributaries to international river: Observing and adapting to climate variability and change in the Nile Basin', Global Environmental Change, 2005, 99–114.

UN (2005): 'The Millennium Development Goals Report'. New York: United Nations Department of Public Information.

Cox, P.M., R.A. Betts, C.D. Jones, et al. (2000): 'Acceleration of global warming due to carbon-cycle feedbacks in a coupled climate model', Nature 408: 184–187

Dasgupta, P. and D. Ray (1986): Inequality as a determinant of malnutrition and unemployment: theory', Economic Journal, 90, December: 1011–34

De, U.S., R.K. Dube and G.S. Prakasa Rao, (2005): 'Extreme weather events over India in the last 100 years' J. Ind Geophys. Union (July 2005) 9 (3): 173–187

Dercon, S. (2003): 'Poverty traps and development: the equity-efficiency trade-off revisited'. Paper prepared for the Conference on Growth, Inequality and Poverty, organised by the Agence francaise de development and the European Development Research Network (EUDN).

Erda, L. and Ji, Z. et al (2006): 'Climate Change Impacts and its Economics in China', Report prepared for the Stern Review, available from http://www.sternreview.org.uk.

Freeman, P. K., L.A. Martin, R. Mechler, et al. (2002): 'Catastrophes and development: integrating natural catastrophes into development planning,' World Bank DMF Paper 1, Washington, DC: World Bank.

Gambia (2003): 'First national communication of the republic of the Gambia to the United Nations Framework Convention on Climate Change'.

Guoping, T., L. Xiubin, G. Fischer, and S. Prieler (2000): Climate change and its impacts on China's agriculture. Acta Geographic Sinica, vol.55 (2), p129–138.

Hallegatte, S. and J.-C. Hourcade (2006): 'Why economic dynamics matters in the assessment of climate change damages: illustration of extreme events', Ecological Economics (forthcoming).

Hazell, P. and B. Hojjati (1995): 'Farm–nonfarm linkages in Zambia'. Journal of African Economies 4 (3): 406–435

Hendrix, C. S., and S.M. Glaser (2005): 'Trends and triggers: climate change and civil conflict in sub-Saharan Africa, human security and climate change: an international workshop', 21–23 June 2005.

Hewawasam, I. (2002): 'Managing the marine and coastal environment of sub-Saharan Africa. Strategic directions for sustainable development'. Washington, DC: World Bank.

Hicks, D. (1993): 'An evaluation of the Zimbabwe drought relief programme 1992/93: the roles of household level response and decentralised decision-making', Harare: World Food Programme.

Hoddinott, J. (2004): 'Shocks and their consequences across and within households in rural Zimbabwe', Wisconsin: BASIS CRSP.

Huq, S. and H. Reid (2005): 'Millennium Development Goals' in Tiempo: A bulletin on climate and development, Issue 54, January 2005.

International Labour Office (2005): 'Global Employment Trends Model', Geneva: ILO.

International Monetary Fund (2003): 'Fund assistance for countries facing exogenous shocks'. Prepared by the Policy Development and Review Department (In consultation with the Area, Finance, and Fiscal Affairs Departments), Washington, DC: IMF.

International Commission on Irrigation and Drainage (2005): 'Water policy issues of Mexico', Delhi: ICID.

Ionescu, C., R.J.T. Klein, J. Hinkel, et al. (2005): 'Towards a formal framework of vulnerability to climate change', NeWater Working Paper 2, Osnabrueck: NeWater.

Intergovernmental Panel on Climate Change (2001): 'Climate Change 2001: Synthesis Report', Contribution of Working Groups I, II and III to the Third Assessment Report of the Intergovernmental Panel on Climate Change' [Watson R.T. and Core Team (Eds.)], Cambridge: Cambridge University Press.

Jones P G., and P. K. Thornton (2003): 'The potential impacts of climate change on maize production in Africa and Latin America in 2055'. Global Environmental Change 13: 51–59.

Kaimowitz, D. (2002): Forest and rural livelihoods in developing countries. Bogor, Indonesia: Center for International Forestry Research.

Kakwani, N. (1993): 'Performance in living standards: an international comparison'. Journal of Development Economics 41(2): 307–336

Kraay, A. (2005): 'When is growth pro-poor? Evidence from a Panel of Countries.' Journal of Development Economics, 80, June: 198–227

Levy, M.L., C. Thorkelson, C. Vörösmarty, et al. (2005): 'Freshwater availability anomalies and outbreak of internal war: results from a global spatial time series analysis'. Oslo: Human Security and Climate Change International Workshop.

McClean, C. J., J. C. Lovett, W. Kuper, et al. (2005): 'African plant diversity and climate change', Annals of the Missouri Botanical Garden 92 (2): 139–152

Meier, P. and D. Bond (2005): 'The influence of environmental factors on conflict in the horn'. Journal of Political Geography (in review).

Mendelsohn, R., A. Dinar, and L. Williams. (2006): 'The distributional impact of climate change on rich and poor countries', Environment and Development Economics 2006 11: 1–20

Miguel, E., S. Satyanath, and E. Sergenti (2004): 'Economic shocks and civil conflict: an instrumental variables approach', Journal of Political Economy 112

Munich Re (2004): 'Topics Geo Annual review: Natural catastrophes' Munich: Munich Re Group.

Myers, N. (2005) 'Environmental Refugees: An Emergent Security Issue,' 13th Economic forum Prague, 23–27 May

Nagy, G. et al. (2006): 'Understanding the potential impact of climate change in Latin America and the Caribbean', Report prepared for the Stern Review, available from http://www.sternreview.org.uk

NBSC (National Bureau of Statistics of China) (2005): 'China Agricultural Yearbook', 2005, Beijing: Statistic Press of China, (in Chinese).

Niasse, M. (2005): 'Climate-induced water conflict risks in West Africa: recognising and coping with increasing climate impacts on shares watercourses' available at http://www.gechs.org/activities/holmen/Niasse.pdf

Nicholls, R.J. (1995): 'Synthesis of vulnerability analysis studies', in Proceedings of the World Coast Conference 1993, P. Beukenkamp et al. (eds.), Noordwijk, The Netherlands,1–5 November 1993, Coastal Zone Management Centre Publication 4, National Institute for Coastal and Marine Management: The Hague: 181–216

Nicholls R.J. (2004): 'Coastal flooding and wetland loss in the 21st century: changes under the SRES climate and socio-economic scenarios', Global Environmental Change 14: 69–86

Nicholls, R.J. and J.A. Lowe (2006): 'Climate stabilisation and impacts of sea-level rise', in Avoiding dangerous climate change, H.J. Schellnhuber (eds.), Cambridge: Cambridge University Press, pp. 195–201.

Nicholls, R.J. and R.S.J. Tol, (2006): 'Impacts and responses to sea level rise: global analysis of the SRES scenarios over 21st century', Philosophical Transactions of the Royal Society B 364: 1073–1095

Nkomo, J.C., A. Nyong and K. Kulindwa (2006): 'The Impacts of Climate Change in Africa', Report prepared for the Stern Review, available from http://www.sternreview.org.uk

Nordhaus, W. (2006): 'Geography and macroeconomics: New data and new findings', PNAS, 103(10): 3510–3517

Overseas Development Institute (2005): 'Aftershocks: Natural Disaster Risk and Economic Development Policy', London: ODI.

Parry, M.L, C. Rosenzweig, A. Iglesias et al. (2004): 'Effects of climate change on global food production under SRES emissions and socio-economic scenarios', Global Environmental Change 14: 53 – 67

Population Division of the Department of Economic and Social Affairs of the United Nations Secretariat (2004): 'World Population Prospects: The 2004 Revision'. Washington, DC: World Bank.

Pritchett, L. and L.H. Summers, (1993): 'Wealthier is healthier'. World Bank, WPS 1150, Washington, DC: World Bank.

Ravallion, M. (2001): 'Growth, inequality and poverty: looking beyond averages' Working Paper no 2558, Washington, DC: World Bank.

Ravallion, M. (2003): 'Measuring aggregate welfare in developing countries: how well do national accounts and surveys agree?' Review of Economics and Statistics, 85 (3): 645–652

Republic of Kenya (2002): 'First national communication of the Republic of Kenya to the United Nations Framework Convention on Climate Change,' Ministry of Finance and Natural Resources.

Republic of South Africa (2000): 'South Africa initial national communication to the United Nations Framework Convention on Climate Change.' Pretoria.

Roy, J. (2006): 'The economics of climate change: a review of studies in the context of South Asia with a special focus on India', Report prepared for the Stern Review, available from http://www.sternreview.org.uk

Sachs, J. D., and J. L. Gallup, (2001): 'The economic burden of malaria', CMH Working Paper Series paper No WG1: 10: Commission on Macroeconomics and Health, Washington, DC: World Health Organisation.

Sachs, J.D., (2001): 'Tropical Underdevelopment'. Working Paper w8119, Cambridge, MA: National Bureau of Economic Research.

Scholze, M., W. Knorr, N.W. Arnell and I.C. Prentice (2006): 'A climate-change risk analysis for world ecosystems', Proceedings of the National Academy of Sciences, DOI: 10.1073/pnas.0601816103

Stern, N, J-J-. Dethier, and F.H. Rogers (2005): 'Growth and Empowerment – making development happen', Cambridge, MA: MIT Press.

Strzepek, K., D.N. Yates, G. Yohe, et al. (2001): 'Constructing not 'implausible' climate and economic scenarios for Egypt'. Integrated Assessment 2, 139–157

Tanzler, D., A. Carius, and S. Oberthus (2002): 'Climate change and conflict prevention' report on behalf of the German Federal Ministry for the Environment, Nature Conservation and Nuclear Safety.

Thornton, P.K., P.G. Jones, T. Owiyo et al. (2006): 'Mapping climate vulnerability and poverty in Africa.' Report to the Department for International Development, ILRI, PO Box 30709, Nairobi 00100, Kenya. pp 171.

Tol, R.S.J., T.E. Downing, O.J. Kuik and J.B. Smith (2004): 'Distributional Aspects of Climate Change Impacts', Global Environmental Change, 14 (3), 259–272

United Nations (2005): 'The Millennium Development Goals Report 2005' available from http://millenniumindicators.un.org/unsd/mi/pdf/MDG%20Book.pdf

UNEP-WCMC (2006): 'Biodiversity and poverty reduction: the importance of biodiversity for ecosystem services', Cambridge: UNEP.

University for Peace Africa Programme (2005): 'Environmental degradation as a cause of conflict in Darfur', Switzerland: University for Peace.

Vedeld, P., A. Angelsen, E. Sjaasrad and G. Berg (2004): 'Counting on the environment: Forest income and the rural poor', Environmental Economics Series No. 98, Washington DC: World Bank.

Vincent, K. (2004): 'Creating an index of social vulnerability to climate change for Africa'. Tyndall Centre for Climate Change Research, Working Paper 54, Norwich: Tyndall Centre.

van Lieshout, M., R.S. Kovats, M.T.J. Livermore and P. Martens, (2004): 'Climate change and malaria: analysis of the SRES climate and socio-economic scenarios', Global Environmental Change 14: 87–99.

Vos R., M. Velasco, and E. De Labatisda. (1999): 'Economic and social effects of El Niño in Ecuador'. Washington, D.C: Inter-American Development Bank.

Wagstaff, A., and E. van Doorslaer. (2003): 'Paying for health care: quantifying fairness, catastrophe, and impoverishment with applications to Vietnam, 1993–98.' Health Economics 12 (11): 921–33

Warren, R., N. Arnell, R. Nicholls, et al. (2006): 'Understanding the regional impacts of climate change', Research report prepared for the Stern Review, Tyndall Centre Working Paper 90, Norwich, UK, available from http://www.tyndall.ac.uk/publications/working_papers/working_papers.shtml

World Health Organisation (2005): 'Malnutrition: Quantifying the health impact at national and local levels'. Environmental burden of diseases series, No. 12, Washington DC: WHO.

World Bank (2000): 'Can Africa Claim the 21st Century,' Washington DC: World Bank

World Bank (2003a): 'World Development Indicators,' Washington, DC: World Bank.

World Bank (2003b): 'World Development Report,' Washington, DC: World Bank.

World Bank (2003c): 'Global Poverty Monitoring Database,' Washington, DC: World Bank.

World Bank (2004): 'World Development Indicators,' Washington, DC: World Bank.

World Bank (2006a): 'The Little Data Book,' Washington, DC: World Bank.

World Bank (2006b) 'Water for growth and development.' D. Grey and C. W. Sadoff in Thematic Documents of the IV World Water Forum. Comision Nacional del Agua: Mexico City. 2006, Washington, DC: World Bank.

World Bank (2006c) 'Managing water resources to maximize sustainable growth: A country water resources assistance strategy for ethiopia', Washington, DC: World Bank.

World Bank (2006d) 'Financing Health in Low-Income Countries'

World Urbanization Prospects: 'The 2005 Revision', available from http://esa.un.org/unup.

5 Costs of Climate Change in Developed Countries

KEY MESSAGES

Climate change will have some positive effects for a few developed countries for moderate amounts of warming, but will become very damaging at the higher temperatures that threaten the world in the second half of this century.

- In higher latitude regions, such as Canada, Russia and Scandinavia, climate change could bring net benefits up to 2 or 3°C through higher agricultural yields, lower winter mortality, lower heating requirements, and a potential boost to tourism. But these regions will also experience the most rapid rates of warming with serious consequences for biodiversity and local livelihoods.

- Developed countries in lower latitudes will be more vulnerable. Regions where water is already scarce will face serious difficulties and rising costs. Recent studies suggest a 2°C rise in global temperatures may lead to a 20% reduction in water availability and crop yields in southern Europe and a more erratic water supply in California, as the mountain snowpack melts by 25 – 40%.

- In the USA, one study predicts a mix of costs and benefits initially ($\pm$1% GDP), but then declines in GDP even in the most optimistic scenarios once global temperatures exceed 3°C.

- The poorest will be the most vulnerable. People on lower incomes are more likely to live in poor-quality housing in higher-risk areas and have fewer financial resources to cope with climate change, including lack of comprehensive insurance cover.

The costs of extreme weather events, such as storms, floods, droughts, and heatwaves, will increase rapidly at higher temperatures, potentially counteracting some of the early benefits of climate change. Costs of extreme weather alone could reach 0.5 – 1% of world GDP by the middle of the century, and will keep rising as the world warms.

- Damage from hurricanes and typhoons will increase substantially from even small increases in storm severity, because they scale as the cube of windspeed or more. A 5 – 10% increase in hurricane windspeed is predicted to approximately double annual damages, resulting in total losses of 0.13% of GDP each year on average in the USA alone.

- The costs of flooding in Europe are likely to increase, unless flood management is strengthened in line with the rising risk. In the UK, annual flood losses could increase from around 0.1% of GDP today to 0.2 – 0.4% of GDP once global temperature increases reach 3 to 4°C.

- Heatwaves like 2003 in Europe, when 35,000 people died and agricultural losses reached $15 billion, will be commonplace by the middle of the century.

At higher temperatures, developed economies face a growing risk of large-scale shocks.

- Extreme weather events could affect trade and global financial markets through disruptions to communications and more volatile costs of insurance and capital.

- Major areas of the world could be devastated by the social and economic consequences of very high temperatures. As history shows, this could lead to large-scale and disruptive population movement and trigger regional conflict.

5.1   Introduction

While the most serious impacts of climate change will fall on the poorest countries, the developed world will be far from immune.

On the whole, developed countries will be less vulnerable to climate change because:[1]

- A smaller proportion of their economy is in sectors such as agriculture that are most sensitive to climate.
- They are located in cooler higher latitudes and therefore further from critical temperature thresholds for humans and crops. Higher latitudes are expected to warm faster than lower latitudes, but this effect is small compared with the initial difference in temperatures between regions.
- Adaptive capacity is higher. Richer countries have more resources to invest in adaptation, more flexible economies, and more liquid financial markets to increase resilience to climate change.

Nevertheless, the advances in the science over the last few years have shown that there are now significant risks of temperatures much higher than the 2 or 3°C that were the focus of analytical discourse up to a few years ago. The potential damages with temperature increases of 4 to 5°C and higher are likely to be very severe for all countries, rich and poor.

This chapter examines the potential costs and opportunities of climate change in developed countries, with a particular focus on the consequences for wealth and output. The analysis suggests that, while there may be benefits in some sectors for 1 or 2°C of warming, climate change will have increasingly negative effects on developed countries as the world warms, even under the most optimistic assumptions. In particular, at higher temperatures (4 or 5°C), the impacts will become disproportionately more damaging (Chapter 3). Extreme weather events (storms, floods, droughts and heatwaves) are likely to intensify in many cases. The risks of large-scale and abrupt impacts will increase significantly, such as melting/collapse of ice-sheets or shutdown of the thermohaline circulation (Gulf Stream). Large-scale shocks and financial contagion originating from poorer countries who are more vulnerable to climate change (Chapter 4) will also pose growing risks for rich countries, with increasing pressures for large-scale migration and political instability.

---

[1]   Tol *et al.* (2004) set out these arguments in some detail and with great clarity.

5.2   Impacts on wealth and output

Climate change will have some positive effects for a few developed countries for moderate amounts of warming, but is likely to be very damaging for the much higher temperature increases that threaten the world in the second half of this century and beyond if emissions continue to grow.

Climate change will influence economic output in the developed world via several different paths (Box 5.1), including the availability of commodities essential for economic growth, such as water, food and energy. While it will be possible to moderate increased costs through adaptation, this in itself will involve additional expenditure (Part V).

BOX 5.1 A simple production function with environmental quality

The market impacts of climate change on economic growth can be framed using a simple theoretical structure, beginning with a general production function in which the output of an economy in a given year depends on the stocks (and, implicitly, the marginal productivities) of capital, labour and environmental quality available in that year.

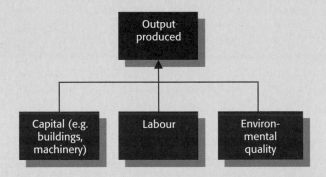

$$\text{or } Y(t) = F(K,L,E)$$

Where Y is the output of the economy in year t and is a function of capital, K, labour, L, and environmental quality, E, which together are the factors of production. In this way, environmental quality is a (natural) capital asset that provides a flow of services on which output depends.

If the net impacts of climate change are negative, then environmental quality E is reduced. This will reduce the output obtainable with a given supply of capital and labour, because output is jointly dependent on all three factors of production. In practice, either the productivity of capital and labour is directly reduced, or a portion of the output produced in a given year is destroyed that same year by climate change, for example by an extreme weather event. The opposite of this story is true if climate change brings with it net benefits, thereby increasing environmental quality.

Adaptation to climate change will be an important economic option (Part V). Adaptation will reduce losses in E and/or enhance gains in E, but it too comes at a cost relative to a world without climate change. In this case, the opportunity cost of adaptation is lost consumption or investment diverted away from adding to K.

Water   Warming will have strong impacts on water availability in the developed world. Altered patterns of rainfall and snowmelt will affect supply through changes in runoff.[2] Water availability will generally rise in higher latitude regions where rainfall becomes more intense. But regions with Mediterranean-like climates will have existing pressures on limited water resources exacerbated because of reduced rainfall and loss of snow/glacial meltwater. Population pressures and water-intensive activities, such as irrigation, already strain the water supplies in many of the regions expected to see falling supplies. Based on recent studies:

- In Southern Europe, summer water availability may fall by 20 – 30% due to warming of 2°C globally and 40 – 50% for 4°C.[3]
- The West Coast of the USA is likely to experience more erratic water supply as mountain snowpack decreases by 25 – 40% for a 2°C increase in global temperatures and 70 – 90% for 4°C.[4] The snow will melt several weeks earlier in the spring, but the supply will eventually diminish as glaciers disappear later in the century.
- In Australia (the world's driest continent) winter rainfall in the southwest and southeast is likely to decrease significantly, as storm tracks shift polewards and away from the continent itself. River flows in New South Wales, including those supplying Sydney, have been predicted to drop by 15% for a 1 – 2°C rise in temperature.[5]

Food   While agriculture is only a small component of GDP in developed countries (1 – 2% in the USA, for example), it is highly sensitive to climate change and could contribute substantially to economy-wide changes in growth.[6] In higher latitudes, such as Canada, Russia and Northern Europe, rising temperatures may initially increase production of some crops – but only if the carbon fertilisation effect is strong (still a key area of uncertainty; further details in Chapter 3) (Figure 5.1).[7] In these regions, any benefits are likely to be short-lived, as conditions begin to exceed the tolerance threshold for crops at higher temperatures. In many lower latitude regions, such as Southern Europe, Western USA, and Western Australia, increasing water shortages in regions where water is already scarce are likely to limit the carbon fertilisation effect and lead to substantial declines in crop yields. This north-south disparity in impacts was observed during the 2003 heatwave when crop yields in southern Europe dropped by 25% while they increased in northern Europe (25% in Ireland and 5% in Scandinavia).[8]

Energy   In higher latitude regions, climate change will reduce heating demands, while increasing summer cooling demands; the latter effect seems smaller in most

---

[2]   Projections for changes in rainfall patterns in developed countries are generally more reliable than those in developing countries (due to their higher latitude location).

[3]   Schröter *et al.* (2006) and Arnell (2004)

[4]   Hayhoe *et al.* (2006)

[5]   Preston and Jones (2006)

[6]   Using a general equilibrium model for the USA, Jorgenson *et al.* (2005) found that agriculture contributed 70 – 80% of the changes in GDP driven by climate change (more details later in chapter). This work did not include the costs of extreme weather, particularly infrastructure damage from hurricanes and storms.

[7]   Mendelsohn *et al.* (1994); see also Schlenker *et al.* (2005) for a recent critique of this work

[8]   COPA COGECA (2003)

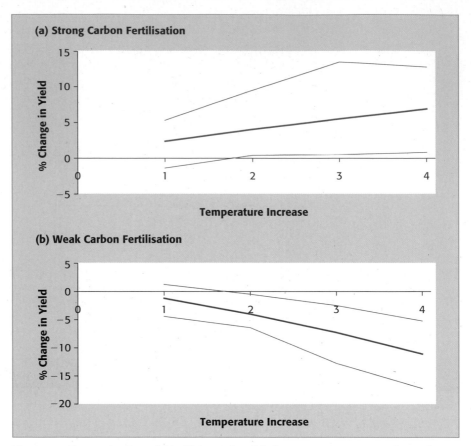

Figure 5.1 Changes in wheat yield with increasing global temperatures across North America, Europe and Australasia

Source: Warren *et al.* (2006) analysing data from Parry *et al.* (2004). More details on method in Chapter 3.

Notes: The strong carbon fertilisation runs assumed a 15 – 25% increase in yield for a doubling of carbon dioxide levels. These are about twice as high as the latest field-based studies suggest. The red line represents the average across different scenario runs developed by the IPCC, while the blue lines show the full range. Yield changes were based on monthly temperature and rainfall data from the Hadley Centre climate model. Using other climate models produces a greater increase in yield at low levels of warming. The work assumed farm-level adaptation with some economy-wide adaptation. Much larger declines in yield are expected at higher temperatures (more than 4°C), as critical thresholds for crop growth are reached. Few studies have examined the consequences of higher temperatures.

cases (Table 5.1).[9] In lower latitude regions, overall energy use is expected to increase, as incremental air-conditioning demands in the summer outstrip the reduction in heating demands in the winter. In Italy, winter energy use is predicted to fall by 20% for a warming of 3°C globally, while summer energy use rises by 30%.[10] Climate change could also disrupt energy production. During the 2003 heat

---

[9] Warren *et al.* (2006) have prepared these results, based on the original analysis of Prof Nigel Arnell (University of Southampton). Energy requirements are expressed as Heating Degree Days and Cooling Degree Days (more detail in Table 5.1).
[10] MICE (2005)

Table 5.1 Temperature-driven changes in energy requirements in the developed world

World Region	Change in Heating Degree Days	Change in Cooling Degree Days
Russia	−935	+358
Europe	−667	+310
North America	−614	+530
Australia	−277	+427

Source: Warren *et al.* (2006) analysing data from Prof Nigel Arnell, University of Southampton
Note: Regions ranked by largest net change in energy demand. Both Heating Degree Days (HDD) and Cooling Degree Days (CDD) are calculated with reference to a base temperature (B), defined as the target "comfort" temperature, and are calculated from daily temperatures T_i, summed over all days (i) in the year. In most global-scale studies, the base temperature is taken as 18°C.

$$HDD = \Sigma \, (B - T_i) \quad \text{where } T_i \text{ is less than B}$$
$$CDD = \Sigma \, (T_i - B) \quad \text{where } T_i \text{ is greater than B}$$

These changes assume: (1) no change to the target "comfort" base temperature; (2) no effects mediated through humidity; and (3) implicitly no acclimatisation or adaptation, in the sense of accepting warmer temperatures. Comfort temperatures will differ across the world, but using a fixed "base" temperature provides an index of potential changes in heating and cooling requirements in the future.

wave in Europe, for example, energy production in France's nuclear power stations fell because the river water was too hot to cool the power stations adequately. Similarly, at the height of the 2002 drought, Queensland's power stations had to reduce output considerably. In California, hydropower generation is predicted to fall by 30% for a warming of 4°C globally as storage lakes deplete.[11]

The distribution of impacts is likely to follow a strong north-south gradient – with regions such as Canada, Russia and Scandinavia experiencing some net benefits from moderate levels of warming, while low latitude regions will be more vulnerable. At higher temperatures, the risks become severe for all regions of the developed world.

Climate change will have widespread consequences across the developed world (major impacts set out in Box 5.2). The impacts will become more damaging from north to south. For example, in higher latitudes, where winter death rates are relatively high, more people are likely to be saved from cold-related death than will die from the heat in the summer.[12] In lower latitude regions, summer deaths could outstrip declines in winter deaths, leading to an overall increase in mortality.[13] Similarly, tourism may shift northwards, as cooler regions enjoy warmer

---

[11] Cayan *et al.* (2006)

[12] Department of Health (2003) study for the UK found an increase in heat-related mortality by 2,000 and decrease in cold-related mortality by 20,000 by the 2050s using the Hadley Centre climate model.

[13] Benson *et al.* (2000) report on studies in five US cities in the Mid-Atlantic region (Baltimore, Greensboro, Philadelphia, Pittsburgh and Washington DC) and find a net increase in temperature-related mortality of up to two- to three-fold by 2050 (using outputs from three global climate models). These cities see larger increases in summer heat-related mortality than some other cities in the USA.

summers, while warmer regions like southern Europe suffer increased heat wave frequency and reduce water availability. One study projected that Canada and Russia would both see a 30% increase in tourists with only 1°C of warming.[14] On the other hand, mountain regions such as the Alps or the Rockies that rely on snow for winter recreation (skiing) may experience significant declines in income. Australia's $32 billion tourism industry will suffer from almost complete bleaching of the Great Barrier Reef.[15]

This broad distribution of impacts across many sectors might stimulate a broad northward shift in economic activity and population in regions such as the North America or Europe, as southern regions begin to suffer disproportionate increases in risks to human health and extreme events, coupled with loss of competitiveness in agriculture and forestry, reduced water availability and rising energy costs.[16] There could be additional knock-on consequences for long-run growth, as changes in economic output have knock-on effects on growth and investment, capital stock, and labour (more detail in Box 5.2 for the USA and in Chapter 6 more generally).

Arctic regions will not follow this general north-south trend. Warming will occur most rapidly here - average temperatures have already risen twice as fast as in other parts of the world in recent decades.[17] For example, in Alaska and western Canada, winter temperatures have already increased by as much as 3 – 4°C in the past 50 years. Over the past 30 years, average sea ice extent has declined by 8% or nearly 1 million Km^2, an area larger than all of Norway, Sweden and Denmark combined, and the melting trend is accelerating. Over half of all the ice could have disappeared by 2100. Loss of even a small fraction of sea ice will have devastating consequences for polar bears, seals and walrus, as well as for the livelihoods of Inuits and others who rely on these animals for food. Shrinking arctic tundra will also threaten grazing animals, such as Caribou and Reindeer, and breeding habitats for millions of migratory bird species.

BOX 5.2 Summary of regional impacts of climate change

USA

- Climate change impacts in the USA will be unevenly distributed, with potential short-term benefits in the North and extensive damage possible in the South. In the short to medium term, the most costly impacts are expected from coastal flooding and extreme events. More powerful hurricanes raise risks along the eastern seaboard and Gulf of Mexico. Defensive investment could be substantial.

- Reduced snowfall and shorter winters will change snowmelt patterns – affecting water supply both along the Pacific coast and California and the farmlands of the Mississippi basin whose western tributaries are fed by snow melt.

- Impacts on overall agricultural yields should be moderate (or even positive with a strong carbon fertilisation effect) up to around 2 – 3°C given adaptation to shifting crop

---

[14] Hamilton *et al.* (2005)
[15] Preston and Jones (2006)
[16] Suggested by Pew Center study by Jorgenson *et al.* (2005)
[17] All impacts in the Arctic are clearly and comprehensively set out in the Arctic Climate Impacts Assessment (2004)

varieties and planting times. But this depends on sufficient irrigation water particularly in the southeast and Southern Great Plains. Farm production in general is expected to shift northwards. Above 3°C, total output could fall by 5 – 20% even with effective adaptation because of summer drought and high temperatures.

- The north could benefit from lower energy bills and fewer cold-related deaths as winter temperatures rise. The south will see rising summer energy use for air-conditioning and refrigeration and more heat-related deaths. This rebalance of economic activity could also induce a northward population shift.

Canada
- Canada has large areas of permafrost, forest and tundra. Melting permafrost raises the cost of protecting infrastructure and oil and gas installations from summer subsidence.

- Reduced sea-ice cover and shorter winters should increase the summer Arctic navigation period offering improved access to oil, gas and mineral resources and to isolated communities.

- But warmer summers and smaller ice packs will make life difficult for the polar bear, seal and other Arctic mammals and fish on which indigenous people depend.

- A warmer climate and carbon fertilisation could lengthen summer growing seasons and increase agricultural productivity. But thinner winter snow cover risks making winter wheat crops vulnerable.

UK
- Infrastructure damage from flooding and storms is expected to increase substantially, especially in coastal regions, although effective flood management policies are likely to keep damage in check.

- Water availability will be increasingly constrained, as runoff in summer declines, particularly in the South East where population density is increasing. Serious droughts will occur more regularly.

- Milder winters will reduce cold-related mortality rates and energy demand for heating, while heatwaves will increase heat-related mortality. Cities will become more uncomfortable in summer.

- Agricultural productivity may initially increase because of longer growing seasons and the carbon fertilisation effect but this depends on adequate water and requires changing crops and sowing times.

Mainland Europe
- Europe has large climatic variations from the Baltic to the Mediterranean and the Atlantic to the Black Sea and will be affected in a diverse fashion by climate change. The Mediterranean will see rising water stress, heat waves and forest fires. Spain, Portugal and Italy are likely to be worst affected. This could lead to a general northward shift in summer tourism, agriculture and ecosystems.

- Northern Europe could experience rising crop yields (with adaptation) and falling energy use for winter heating. But warmer summers will raise demand for air conditioning. Melting Alpine snow waters and more extreme rainfall patterns could lead to more frequent flooding in major river basins such as the Danube, Rhine and Rhone. Winter tourism will be severely affected.

- Many coastal countries across Europe are also vulnerable to rising sea levels: the Netherlands, where 70% of the population would be threatened by a 1-m sea level rise, is most at risk.

Russia

- A vast swathe of northern Russia is permafrost, apart from a short, hot summer when the surface melts to form marshy lakes. Rising temperatures will push the permafrost boundary further north and deepen the surface melt. This has big implications for future oil, gas and other investment projects. De-stabilised, shifting permafrost conditions release greenhouse gases and could lead to flooding, but also require more expensive underpinning of buildings, refineries and other infrastructure such as the Baikal Amur railway and the planned East Siberia-Pacific export oil pipeline.

- Melting of the Arctic ice cap will prolong both the northern sea and Siberian river navigation seasons but could lead to more extreme weather patterns. At higher global temperatures there is a possibility that Arctic warming could be reversed if the Gulf Stream weakens before it reaches the Barents Sea.

- Agriculture, and tree growth in the vast Siberian pine forests, should benefit from a longer, warmer growing season and the carbon fertilisation effect. But the most fertile black earth regions of Southern Russia and Ukraine could suffer from increased drought.

- Warmer winters should reduce domestic heating costs and free energy for export. But higher summer temperatures will raise air conditioning energy use.

Japan

- Japan consists of a long chain of narrow, mountainous islands on a seismic fault line, naturally subject to large climatic variations from north to south. Densely urbanised and heavily industrialised, Japan's topography, lack of raw materials, and heavy dependence on international trade, ensure that most people are concentrated in highly industrialised port cities.

- Climate change will exacerbate Japan's existing vulnerability to typhoons and coastal storms. Tokyo extends over a flat coastal plain, vulnerable both to typhoons and rising sea levels. Most other major cities are also heavily industrialised ports, with many factories, refineries, gas liquefaction and chemical plants, steel mills, shipyards, oil storage tanks and other vulnerable infrastructure.

- Agriculture, especially rice cultivation, is not significant economically but has strong cultural importance. Higher temperatures will make rice more difficult to grow in the south. Fish are another key part of a national cuisine. Fish are vulnerable to rising ocean temperatures and increased acidity.

- Major cities will be increasingly affected by the urban heat island effect. Over 40% of summer power generation is consumed by air conditioning. Rising temperatures will make a fast ageing population more vulnerable both to heat and the spread of infectious diseases such as malaria and dengue fever.

Australia

- Australia, as the world's driest continent, is particularly vulnerable to the impact of rising sea temperatures on the major Pacific and Indian Ocean currents. These determine both overall rainfall patterns and unpredictable year-to-year variations.

- At the same time the east coast – home to over 70% of the population and location for most major cities – has suffered longer droughts and declining rainfall. The 2002 drought cut farm output by 30% and shaved 1.6% off GDP. Water supply to big cities will become more difficult – Melbourne's could fall by 7 – 35% with only 2°C of warming. Water flow in the Murray–Darling system, Australia's bread basket, could fall by one quarter[18].

- Drier and hotter summers threaten the survival of the Queensland rainforest. Warmer winters and reduced snowfall endanger the habitat of mountain top fauna and flora. Rising ocean temperatures threaten the future of Australia's coral reefs and the $32 billion fishing and tourist industries. Over 60% of the Great Barrier Reef suffered coral bleaching in 2002, 10% of it permanent. Studies show ocean warming could be fatal to large tracts of reef within 40 years. Higher inland temperatures are likely to cause more bush fires.

- Tropical diseases are spreading southward. The dengue fever transmission zone could reach Brisbane and possibly Sydney with 3°C of warming.

BOX 5.3 Costs of climate change: USA case study on long-run growth impacts

Jorgenson *et al.* (2005) used a general equilibrium model to estimate the impacts of climate change on investment, the capital stock, labour and consumption in the USA for two scenarios: one "optimistic" (assuming "optimal" adaptation, a strong carbon fertilisation effect and low potential damages) and one "pessimistic" (assuming little adaptation, a weak carbon fertilisation effect and high potential damages). Recent field-based studies suggest that the carbon fertilisation effect may be about half as large as the values used in the "optimistic" case (more details in Chapter 3).

For a warming of 3°C, the study projects a net damage of 1.2% of GDP in the pessimistic case and a benefit of 1% of GDP in the optimistic case. In the optimistic case, the benefits peak at just over 2°C warming and then decline from around 3.5°C. In the pessimistic case, warming causes increasingly negative impacts on GDP. The range of outcomes encompasses other earlier estimates of the costs of climate change for the US economy, such as Mendelsohn (2001).

In both optimistic and pessimistic cases, the change was driven largely by changes in agricultural prices (70 – 80%), with a lesser contribution from changes in energy prices and mortality. In the pessimistic case, productive resources were diverted from more efficient uses to the affected sectors, leading to overall productivity losses. The end effect was a significant reduction in consumption. In the optimistic case, the reverse process occurred.

---

[18] Prepared with assistance from Nick Rowley and Josh Dowse of KINESIS Consulting, Sydney, Australia http://www.kinesis.net.au

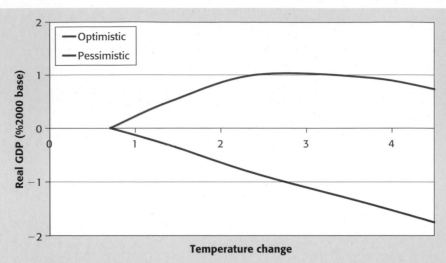

Real GDP (%2000 base) / Temperature change

The study did not take full account of the impacts of extreme weather events, which could be very significant (Section 5.4). Nordhaus (2006) shows that just a small increase in hurricane intensity (5 – 10%), which several models predict will occur 2 – 3°C of warming globally, could alone double costs of storm damage to around 0.13% GDP. The risks of higher temperatures, as the latest science suggests, could bring even greater damage costs, particularly given the very non-linear relationship between temperature and hurricane destructiveness (Chapter 3).

Source: Jorgenson *et al.* 2005

5.3   Key vulnerabilities

The poorest in developed countries will be the most vulnerable to climate change.

Low-income households will be disproportionately affected by increases in extreme weather events.[19]

- Those on lower incomes often live in higher-risk areas, marginal lands,[20] and poor quality housing. In the UK, the Environment Agency found that the most deprived 10% of the population were eight times more likely to be living in the coastal floodplain than those from the least deprived 10%.[21]
- Lower-income groups will typically have fewer financial resources to cope with climate change, including lack of comprehensive insurance cover. In New Orleans, disproportionately more people (22%) were below the poverty line in areas flooded by Hurricane Katrina than in non-flooded areas (15%) (Box 5.4a). More than half the people in flooded areas did not own a car compared with one-third in non-flooded areas.[22]
- Residents in deprived areas are likely to be less aware and worse prepared for an extreme weather event like a flood. The health impacts will be more severe for those already characterised by poor health. Across Europe, a large

[19] Environment Agency (2006), McGregor *et al.* (2006)
[20] O'Brien *et al.* (2006)
[21] Environment Agency (2003)
[22] Brookings Institution (2005)

majority of the 35,000 people who died during the 2003 heatwave were the elderly and the sick (Box 5.4b). The most deprived proportion of the population are more likely to be employed in outdoor labour and therefore have little relief from the heat at work.

5.4   Impacts of extreme events

The costs of extreme weather events, such as storms, floods, droughts, and heatwaves, will increase rapidly at higher temperatures, potentially countering some of the early benefits of climate change. Costs of extreme weather alone could reach 0.5 – 1% of world GDP by the middle of the century, and will keep rising as the world continues to warm.

The consequences of climate change in the developed world are likely to be felt earliest and most strongly through changes in extreme events - storms, floods, droughts, and heatwaves.[23] This could lead to significant infrastructure damage and faster capital depreciation, as capital-intensive infrastructure has to be replaced, or strengthened, before the end of its expected life. Increases in extreme events will be particularly costly for developed economies, which invest a considerable amount in fixed capital each year (20% of GDP or $5.5 trillion invested in gross fixed capital today). Just over one-quarter of this investment typically goes into construction ($1.5 trillion - mostly for infrastructure and buildings; more detail in Chapter 19). The long-run production losses from extreme weather could significantly amplify the immediate damage costs, particularly when there are constraints to financing reconstruction.[24]

The costs of extreme weather events are already high and rising, with annual losses of around $60 billion since the 1990s (0.2% of World GDP), and record costs of $200 billion in 2005 (more than 0.5% of World GDP).[25] New analysis based on insurance industry data has shown that weather-related catastrophe losses have increased by 2% each year since the 1970s over and above changes in wealth, inflation and population growth/movement.[26] If this trend continued or intensified with rising global temperatures, losses from extreme weather could reach 0.5 – 1% of world GDP by the middle of the century.[27] If temperatures continued to rise over the second half of the century, costs could reach several percent of GDP each year, particularly because the damages increase disproportionately at higher temperatures (convexity in damage function; Chapter 3).

[23] Described by low frequency but high impact events (e.g. more than two standard deviations from the mean)

[24] Hallegatte *et al*. (2006) define the "economic amplification ratio" as the ratio of the overall production losses from the disaster to its direct losses.

[25] 2005 prices for total losses (insured and uninsured) — analysis of data from Swiss Re and Munich Re in Mills (2005) and Epstein and Mills (2005); Munich Re (2006)

[26] Muir-Wood *et al*. (2006)

[27] Based on simple extrapolation through to the 2050s. The lower bound assumes a constant 2% increase in costs of extreme weather over and above changes in wealth and inflation. The upper band assumes that the rate of increase will increase by 1% each decade, starting at 2% today, 3% in 2015, 4% in 2025, 5% in 2035, and 6% in 2045. These values are likely underestimates: (1) they exclude "small-scale" events which have large aggregate costs, (2) they exclude data for some regions (Africa and South America), (3) they fail to capture many of the indirect economic costs, such as the impacts on oil prices arising from damages to energy infrastructure, and (4) they do not adjust for the reductions in losses that would have otherwise occurred without disaster mitigation efforts that have reduced vulnerability.

BOX 5.4 Impacts of recent extreme weather events

Extreme weather events are likely to occur with greater frequency and intensity in the future, particularly at higher temperatures.

(a) Hurricane Katrina (2005) was the costliest weather catastrophe on record, totalling $125 billion in economic losses (~1.2% of US GDP), of which around $45 billion was insured through the private market and $15 billion through the National Flood Insurance Program. More than 1,300 people died as a result of the hurricane and over one million people were displaced from their homes. By the end of August, Katrina had reached a Category 5 status (the most severe) with peak gusts of 340 km per hour, in large part driven by the exceptionally warm waters of the Gulf (1 – 3°C above the long-term average). Katrina maintained its force as it passed over the oilfields off the Louisiana coast, but dropped to a Category 3 hurricane when it hit land. New Orleans was severely damaged when the hurricane-induced 10-metre storm-surge broke through the levees and flooded several quarters (up to 1 Km inland). The Earth Policy Institute estimates that 250,000 former residents have established homes elsewhere and will not return.

Source: Munich Re (2006)

(b) European Heatwave (2003). Over a three-month period in the summer, Europe experienced exceptionally high temperatures, on average 2.3°C hotter than the long-term average. In the past, a summer as hot as 2003 would be expected to occur once every 1000 years, but climate change has already doubled the chance of such a hot summer occurring (now once every 500 years).[28] By the middle of the century, summers as hot as 2003 will be commonplace. The deaths of around 35,000 people across Europe were brought forward because of the effects of the heat (often through interactions with air pollution). Around 15,000 people died in Paris, where the urban heat island effect sustained nighttime temperatures and reduced people's tolerance for the heat the following day. In France, electricity became scarce because of a lack of water needed to cool nuclear power plants. Farming, livestock and forestry suffered damages of $15 billion from the combined effects of drought, heat stress and fire.

Source: Munich Re (2004)

Even a small increase in the intensity of hurricanes or coastal surges is likely to increase infrastructure damage substantially.

Storms are currently the costliest weather catastrophes in the developed world and they are likely to become more powerful in the future as the oceans warm and provide more energy to fuel storms. Many of the world's largest cities are at risk from severe windstorms - Miami alone has $900 billion worth of total capital stock at risk. Two recent studies have found that just a 5 – 10% rise in the intensity of major storms with a 3°C increase in global temperatures could approximately double the damage costs, resulting in total losses of 0.13% of GDP in the USA each year on average or insured losses of $100 – 150 billion in an extreme year (2004 prices).[29]

---

[28] Stott *et al.* (2004)

[29] Recent papers from Nordhaus (2006) and the Association of British Insurers (2005a) examined consequences of increased hurricane wind-speeds of 6% on loss damages, keeping socio-economic conditions and prices constant. Several climate models predict a 6% increase in storm intensity for a doubling of CO_2 concentrations (close to a 3°C temperature rise). The insurance study used existing industry catastrophe loss models validated with historic events to predict future losses. The extreme event costs are defined from an event with a 0.4% chance of occurring (1 in 250 year loss).

If temperatures increase by 4 or 5°C, the losses are likely to be substantially greater, because any further increase in storm intensity has an even larger impact on damage costs (convexity highlighted in Chapter 3). This effect will be magnified for the costs of extreme storms, which are expected to increase disproportionately more than the costs of an average storm. For example, Swiss Re recently estimated that in Europe the costs of a 100-year storm event could double by the 2080s with climate change ($50/€40 billion in the future compared with $25/€20 billion today), while average storm losses were estimated to increase by only 16 – 68% over the same period.[30]

Rising sea levels will increase the risk of damages to coastal infrastructure and accelerate capital depreciation (Box 5.5). Costs of flood defences on the coast will rise, along with insurance premiums. A Government study calculated that in the UK the average annual costs of flood damage to homes, businesses and infrastructure could increase from around 0.1% of GDP currently to 0.2 – 0.4% of GDP if global temperatures rise by 3 to 4°C.[31] Greater investment in flood protection is likely to keep damages in check. Similarly, preliminary estimates suggest that annual flood losses in Europe could rise from $10 billion today to $120 – 150 billion (€100 – 120 billion) by the end of the century.[32] If flood management is strengthened in line with the rising risk, the costs may only increase two-fold. According to one recent report, storm surge heights all along Australia's East Coast from Victoria to Cairns could rise by 25 – 30% with only a 2°C increase in global temperatures.[33]

Heatwaves like 2003 in Europe, when 35,000 people died and agricultural losses reached $15 billion, will be commonplace by the middle of the century.

People living and working in urban areas will be particularly susceptible to increases in heat-related mortality because of the interaction between regional warming, the urban heat island and air pollution (Chapter 3). In California, a warming of around 2°C relative to pre-industrial is expected to extend the heat wave season by 17 – 27 days and cause a 25 – 35% rise in high pollution days, leading to a 2 to 3-fold increase in the number of heat related deaths in urban areas.[34] In the UK, for a global temperature rise of 3°C, temperatures in London could be up to 7°C warmer than today because of the combined effect of climate change and the urban heat island effect, meaning that comfort levels will be exceeded for people at work for one-quarter of the time on average in the summer.[35] In years that are warmer than average or at higher temperatures, office buildings could become difficult to work in for large spells during the summer without additional air-conditioning. In already-dry regions, such as parts of the Mediterranean and South East England, hot summers will further increase soil drying and subsidence damage to properties that are not properly underpinned.[36]

---

[30] Heck *et al.* (2006)
[31] UK Government Foresight Programme (2004) calculations for flooding from rivers, the sea and flash-flooding in urban areas. Prof Jim Hall at the University of Newcastle has provided some additional analysis. Assumes no change in flood management policies.
[32] Research from the Association of British Insurers (2005a) extrapolated from a UK-based study of flood losses that assumed no change in flood management policies beyond existing programme. Some of the increased cost is driven by economic growth of the century and greater absolute wealth in physical assets.
[33] Preston and Jones (2006)
[34] Hayhoe *et al.* (2006)
[35] London Climate Change Partnership (2004)
[36] Association of British Insurers (2004) estimates that subsidence costs to buildings could double by the middle of the century to £600 million (2004 prices).

BOX 5.5 Costs of coastal flooding in developed country regions

1-m of sea level rise is plausible by the end of the century under rapid rates of warming (Chapter 1), particularly if one of the polar ice sheets begins to melt significantly (Greenland) or collapses (West Antarctic). This could impose significant costs on developed countries with long, exposed coastlines.

For North America, an area just under half the size of Alaska (640,000 km^2) would be lost with 1-m of sea level rise, unless defences are in place to protect the land. Much of this land will in sparsely populated areas, but a significant proportion covers the Gulf Coast and large parts of Florida. These areas will be particularly vulnerable as rising risks of tropical storms combine with rising sea levels to create sharp increases in damages from coastal surges.

In Europe, sea level rise will affect many densely populated areas. An area of 140,000 km^2 is currently within 1-m of sea level. Based on today's population and GDP, this would affect over 20 million people and put an estimated $300 billion worth of GDP at risk. The Netherlands is by far the most vulnerable European country to sea level rise, with around 25% of the population potentially flooded each year for a 1-m sea level rise.[37]

Projected costs of coastal flooding over the period 2080-2089 under two different sea level rise scenarios

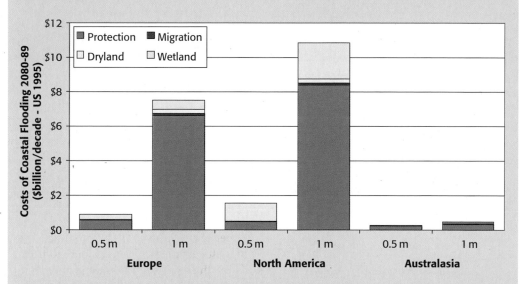

Source: Anthoff *et al.* (2006) analysing data from Nicholls and Tol (2006)

Note: Costs were calculated as net present value in US $ billion (1995 prices). Damage costs include value of dryland and wetland lost and costs of displaced people (assumed in this study to be three times average per capita income). The protection costs only include costs to protect against permanent inundation. Infrastructure damage from storm surges is not included (see additional costs in text). Discounting with a constant growth rate (2%) and a pure time preference rate of 0.1% per year increases values by around 2.5 fold (more details in Chapter 2 and technical appendix).

[37] Nicholls and Klein (2003)

5.5   Large-scale impacts and systemic shocks

Abrupt shifts in climate and rising costs of extreme weather events will affect global financial markets.

Well-developed financial markets will help richer countries moderate the impacts of climate change – for example hedging with derivatives to smooth commodity prices. Such markets help to spread the risk across different regional markets and over time, but cannot reduce the risks by themselves. In addition, they are at risk of severe disruption from climate change:

- **Physical risks.** The world's major financial centres (London, New York and Tokyo) are all located in coastal areas. The insurance industry estimates that in London alone at least $220 billion (£125 billion) of assets lie in the floodplain.[38]

- **Correlated risks.** At higher temperatures, climate change is likely to have severe impacts on many parts of the economy simultaneously. The shock may well exceed the capacity of markets and could potentially destabilise regions.[39] For example, a collapse of the Atlantic Thermohaline Circulation would have a massive effect on many parts of the economy of the countries around the Northern Atlantic Ocean and polar seas.[40] A collapse in the next few decades would lead to a decrease in temperatures across much of the northern hemisphere, with a peak cooling of around 2°C in the UK and Scandinavia. Preliminary estimates suggest that this would be accompanied by a reduction in rainfall over much of the northern hemisphere,[41] reducing agriculture productivity, water supplies and threatening ecosystems.

- **Capital constraints on insurance.** Increasing costs of extreme weather will not only raise insurance premiums - they will also increase the amount of capital that insurance companies have to hold to cover extreme losses, such as a hurricane that occurs once every 100 years (Box 5.6). The insurance industry will have to develop new financial products to gain more widespread access to international capital markets.[42] New opportunities for diversifying risk are already emerging, for example weather derivatives and catastrophe bonds, but in future these will require new risk valuation techniques to deal with the changing profile of extreme weather events. If the insurance industry looks to access additional capital from the securities and bond markets, investors are likely to demand higher rates of return for placing more capital at risk, causing a rise in the cost of capital.

- **Spillover risks to other financial sectors.**[43] Failure to raise sufficient capital could mean restrictions in insurance coverage. After seven costly hurricanes in the past two years, higher reinsurance prices have pushed up the cost of insurance coverage in the USA and contributed to decisions by some insurers to transfer more risk back to the homeowner or business, for example by raising

---

[38] Association of British Insurers (2005b)
[39] As set out in a Pentagon commissioned report by Schwartz and Randall (2004)
[40] A complete collapse of the Thermohaline Circulation is considered to be unlikely (but still plausible) this century (Chapter 1).
[41] Vellinga and Wood (2002)
[42] Salmon and Weston (2006)
[43] Mills (2005)

deductibles or cutting back on coverage in riskier areas.[44] In future, if rising weather risks cause insurance to become even less available in high-risk areas like the coast, this could be severely disruptive for other parts of the economy. Banks, for example, would be unable to offer finance where insurance is required as part of the collateral package for mortgages or loans. Lack of insurance could be particularly damaging for small and medium enterprises that will find it harder to access capital to protect against extreme events.[45]

Major areas of the world could be devastated by the social and economic consequences of very high temperatures. As history shows, this could lead to large-scale and disruptive population movement and trigger regional conflict.

BOX 5.6 Climate change and constraints on insurance capital

The insurance industry requires sufficient capital to bridge the gap between losses in an average year, which are covered by premium income, and those in an "extreme" year.[46] Today, the insurance industry holds around $120 billion to cover extreme losses from natural weather catastrophes (principally hurricanes, typhoons and winter storms).

Climate change is likely to lead to a shift in the distribution of losses towards higher values, with a greater effect at the tail.[47] Average annual losses (or expected losses) will increase by a smaller amount than the extreme losses (here shown as a 1 in 250 year event), with the result that the amount of capital that insurers are required to hold to deal with extremes increases.

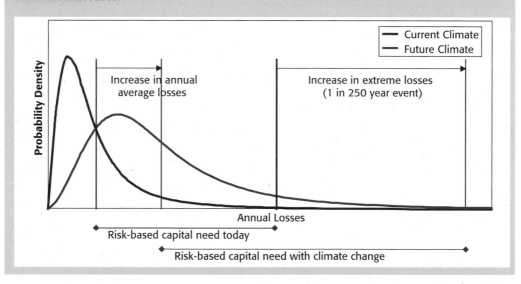

44   Mills and Lecomte (2006) provide many examples of increasing prices or withdrawing cover in the US. For example, reinsurance prices have increased by 200% in some parts of the US. Commercial customers are also being affected by the availability and affordability of insurance. Allstate insurance dropped 16,000 commercial customers in Florida in 2005, and some commercial businesses in the Gulf of Mexico are unable to find insurance at any price.
45   Crichton (2006) found that today in the UK one-third of small and medium-sized businesses had any form of business interruption cover against extreme weather.
46   "Extreme" is defined by an insurers risk appetite and regulatory requirements.
47   Heck *et al.* (2006)

If storm intensity increases by 6%, as predicted by several climate models for a doubling of carbon dioxide or a 3°C rise in temperature, this could increase insurers' capital requirements by over 90% for US hurricanes and 80% for Japanese typhoons – an additional $76 billion in today's prices.

Source: Association of British Insurers (2005a)

The impacts of climate change will be more serious for developing countries than developed countries, in part because poorer countries have more existing economic and social vulnerabilities to climate and less access to capital to invest in adaptation (Chapter 4). As the impacts become increasingly damaging at higher temperatures, the effects on the developing world may have knock-on consequences for developed economies, through disruption to global trade and security (Box 5.7), population movement and financial contagion. Climate change will affect the prices and volumes of goods traded between developed and developing countries, particularly raw materials for manufacturing and food products, with wider macroeconomic consequences.

BOX 5.7 Potential impacts of climate change on trade routes and patterns

Few studies have examined the effects of climate change on global trade patterns, but the consequences could be substantial, particularly for sea-borne trade and linked coastal manufacturing and refining activities.

Rising sea levels will demand heavy investment in flood protection around ports and the export and import related activities concentrated in and around them. Stronger storm surges, winds and heavier rainfall already point to the requirement for stronger ships and sturdier offshore oil, gas and other installations. Multi-billion dollar processing installations such as oil refineries, liquefied natural gas plants and re-gasification facilities may have to be re-located to more protected areas inland.

This would reverse decades of building steel mills, petrochemical plants and other energy-related facilities close to the deepwater ports accommodating bulk cargo vessels, super-tankers and ever larger container ships which have become the key vectors of rising global trade and just-on-time production schedules. Both increased protection and relocation inland would have significant capital and transport costs, and make imports in particular more expensive.

Rapidly rising temperatures in the polar regions will affect trade, transport and energy/resource exploitation patterns. Both Canada's putative North West passage and the Arctic sea-lanes that Russia keeps open with icebreakers could become safer and more reliable alternative transport routes. But melting permafrost risks damaging high latitude oil and gas installations, pipelines and other infrastructure, including railways, such as Russia's Baikal-Amur railway, and will also require expensive remedial investment. Stormier seas could raise the attraction of land routes from Asia to Europe, including the planned new Eurasian railway across Kazakhstan.

Any weakening of the Gulf Stream however would have a dramatic cooling impact on water temperatures in the Arctic region. At present the lingering impact of the Gulf Stream keeps Murmansk open all year as an ice-free port. Russian plans to develop the offshore Shtokman gas field and associated export facilities depend on the waterway remaining navigable. In the Middle East higher temperatures and more severe droughts will cause serious problems to both water supply and agriculture.

Table 5.2 Summary costs of extreme weather events in developed countries with moderate climate change. Costs at higher temperatures could be substantially higher.

Region	Event Type	Temperature	Costs as % GDP	Notes
Global	All extreme weather events	2°C	0.5 – 1.0% (0.1%)	Based on extrapolating and increasing current 2% rise in costs each year over and above changes in wealth
USA	Hurricane	3°C	0.13% 0.06%	Assumes a doubling of carbon dioxide leads to a 6% increase in hurricane windspeed
	Coastal Flood	1-m sea level rise	0.01 – 0.03%	Only costs of wetland loss and protection against permanent inundation
UK	Floods	3 – 4°C	0.2 – 0.4% (0.13%)	Infrastructure damage costs assuming no change in flood management to cope with rising risk
Europe	Coastal Flood	1-m sea level rise	0.01 – 0.02%	Only costs of wetland loss and protection against permanent inundation

Notes: Numbers in brackets show the costs in 2005. Temperatures are global relative to pre-industrial levels. The costs are likely to rise sharply as higher temperatures lead to even more intense extreme weather events and the risk of triggering abrupt and large-scale changes. Currently, there is little robust quantitative information for the costs at even higher temperatures (4 or 5°C), which are plausible if emissions continue to grow and feedbacks amplify the original warming effect (such as release of carbon dioxide from warming soils or release of methane from thawing permafrost).

Climate change is likely to increase migratory pressures on developed countries significantly, although the potential scale and effect are still very uncertain and require considerably more research.

- **Income gap.** Pressures for long-distance and large-scale migration is likely to grow as climate change raises existing inequalities and the relative income differential between developed and developing countries (Chapter 4). Wage differentials were a strong driver of the mass migration of 50 million people from Europe to the New World in the second half of the 19th century, alongside over-population and the resulting land hunger.[48]
- **Environmental disasters.** As temperatures rise and conditions deteriorate significantly, climate change will test the resilience of many societies around the world. Large numbers of people will be compelled to leave their home when resources drop below a critical threshold. Bangladesh, for example,

---

[48] The fundamental drivers of past, current and future world migration are clearly set out by Hatton and Williamson (2002).

faces the permanent loss of large areas of coastal land affecting 35 million people, about one-quarter of its population, while one-quarter of China's population (300 million people) could suffer from the wholesale reduction in glacial meltwater. The Irish Potato Famine is an important example from history of how a dramatic loss in basic subsistence triggered large-scale population movement.[49] The famine took hold in 1845 with the appearance of "the Blight" - a potato fungus that almost instantly destroyed the primary food source for the majority of the population. It led to the death of 1 million people and the emigration of a further 1 million, many of them to the USA.

Developed countries may become drawn into climate-induced conflicts in regions that are hardest hit by the impacts (Chapter 4), particularly as the world becomes increasingly interconnected politically and socially. In the past, climate variability and resource management have both been important contributory factors in conflict.[50] So-called "water wars" have started because competition over water resources and the displacement of populations as a result of dam building have led to unrest.[51] Direct conflict between nation states because of water scarcity has been rare in the past, but dam building and water extraction from shared rivers has served to heighten political tensions in several regions, including the Middle East (discussed in detail in Chapter 4).

5.6   Conclusion

The costs of climate change for developed countries could reach several percent of GDP as higher temperatures lead to a sharp increase in extreme weather events and large-scale changes.

The cooler climates of many developed countries mean that small increases in temperature (2 or 3°C) may increase economic output through greater agricultural productivity, reduced winter heating bills and fewer winter deaths. But at the same time, many developed regions have existing water shortages that will be exacerbated by rising temperatures that increase evaporation and dry out land that is already dry (Southern Europe, California, South West Australia). Water shortages will increase the investment required in infrastructure, reduce agricultural output and increase infrastructure damage from subsidence.

As temperatures continue to rise, the costs of damaging storms and floods are likely to increase rapidly. Losses could potentially reach several percent of world GDP if damages increase, as expected, in a highly non-linear manner.[52] Higher

---

[49] See, for example, Woodham-Smith (1991)

[50] Brooks *et al.* (2005)

[51] Shiva (2002) describes several examples of conflict within a nation or between nations that has been exacerbated by tensions over construction of dams to manage water availability. Every river in India has become a site of major, irreconcilable water conflicts, including the Sutlej, Yamuna, Ganges, Krishna and Kaveri Rivers. The Tigris and Euphrates Rivers, the major water bodies sustaining agriculture for thousands of years in Turkey, Syria and Iraq have led to several major clashes among the three countries. The Nile, the longest river in the world, is shared by ten African countries and is another complicated site of water conflict, particularly following construction of the Aswan Dam.

[52] For example, hurricane damages scale as the cube of windspeed (or more), which itself increases exponentially with ocean temperatures.

temperatures will increase the risk of triggering abrupt and large-scale changes in the climate system. These could have a direct impact on the economies of developed countries, ranging from several metres of sea level rise following melting of Greenland ice sheet to several degrees of cooling in Northern Europe following collapse of the thermohaline circulation (considered plausible but unlikely this century). Other impacts, such as monsoon failure or loss of glacial meltwater, could have devastating effects in developing countries, particularly on food and water availability, and trigger large-scale population movement and regional conflict. These effects may exacerbate existing political tensions and could drive greater global instability.

References

The study by Jorgenson *et al.* (2004) for the Pew Center on Global Climate Change is one of the most comprehensive top-down assessments of the market impacts of climate change in a developed country (the USA). In contrast, the recent Metroeconomica (2006) study takes a bottom-up approach to calculating the costs and benefits of climate change for key sectors in the UK. Neither of these studies takes a detailed look at the costs of increased frequency or severity of extreme weather events. A new conference paper by Prof Bill Nordhaus (2006) examines the economics of hurricanes, showing that the costs may well double of the course of the century. This supports earlier work by the Association of British Insurers (2005) looking at the financial risks of climate change from hurricanes, typhoons and winter-storms. Hallegatte *et al.* (2006) develop these arguments further by setting out the potential long-run effects of changes in extreme weather on economic growth. Recent papers by Dr Evan Mills have examined the financial consequences of such changes in extreme weather for the insurance industry and wider capital markets (Mills 2005, Mills and Lecomte 2006).

Anthoff, D., R. Nicholls, R.S.J. Tol, and A.T. Vafeidis (2006): 'Global and regional exposure to large rises in sea-level: a sensitivity analysis', Research report prepared for the Stern Review, available from http://www.sternreview.org.uk
Arctic Climate Impacts Assessment (2004): 'Impacts of a warming Arctic', Cambridge: Cambridge University Press, available from http://www.acia.uaf.edu
Arnell, W.N. (2004): 'Climate change and global water resources: SRES scenarios and socio-economic scenarios'. Global Environmental Change **14**: 31-52
Association of British Insurers (2005a): 'Financial risks of climate change', London: Association of British Insurers, available from http://www.abi.org.uk/flooding
Association of British Insurers (2005b): 'Making communities sustainable: managing flood risks in the growth areas', London: Association of British Insurers, available from http://www.abi.org.uk/housing
Benson, K., P. Kocagil and J. Shortle (2000): 'Climate change and health in the Mid-Atlantic Region', Climate Research **14**: 245 – 253
Brookings Institution (2005): 'New Orleans after the storm: lessons from the past', Washington, DC: The Brookings Institution, available from http://www.brookings.edu/metro/pubs/20051012_ New Orleans.htm
Brooks, N., J. Gash, M. Hulme, et al. (2005): 'Climate stabilisation and "dangerous" climate change: a review of relevant issues', Scoping Study, London: Defra
Cayan, D., A.L. Luers, M. Hanemann, et al. (2006): 'Scenarios of climate change in California: an overview', California Climate Change Center, available from http://www.climatechange.ca.gov
COPA-COGECA (2003): 'Assessment of the impacts of the heatwave and drought of summer 2003 on agriculture and forestry', available from http://www.copa-cogeca.be
Crichton, D. (2006): 'Climate change and its effects on small businesses in the UK', London, UK: AXA Insurance, available from http://www.axa.co.uk/aboutus/corporate_publications/ climate_change.html
Department of Health (2003): 'Health effects of climate change in the UK', The Stationary Office, UK
Dlugolecki, A. (2004): 'A changing climate for insurance', London: Association of British Insurers, available from http://www.abi.org.uk/climatechange

Environment Agency (2003): 'Deprived communities experience disproportionate levels of environmental threat', Bristol: Environment Agency

Environment Agency (2006): 'Addressing environmental inequalities: flood risk', Bristol: Environment Agency

Epstein, P.R. and E. Mills (2005): 'Climate change futures: health, ecological and economic dimensions', Center for Health and the Global Environment, Harvard Medical School, Cambridge, MA: Harvard University, available from http://www.climatechangefutures.org

Hallegatte, S., J-C- Hourcade and P. Dumas (2006): 'Why economic dynamics matter in assessing climate change damages: illustration on extreme events', Ecological Economics, in press

Hamilton, J.M., D.J. Maddison and R.S.J. Tol (2005): 'Climate change and international tourism: a simulation study', Global Environmental Change 15: 253 – 266

Hatton, T.J.and J.G. Williamson (2002): 'What fundamentals drive world migration?', Working Paper 159, Cambridge, MA: National Bureau of Economic Research, available from http://www.nber.org/papers/w9159

Hayhoe, K., P. Frunhoff, S. Schneider, et al. (2006): 'Regional assessment of climate impacts on California under alternative emission scenarios', in Avoiding dangerous climate change, H.J. Schellnhuber (eds.), Cambridge: Cambridge University Press, pp. 227 – 234

Heck, P., D. Bresch and S. Tröber (2006): 'The effects of climate change: storm damage in Europe on the rise', Zurich: Swiss Re

Jorgenson, D.W., R.J. Goettle, B.H. Hurd et al. (2004): 'US market consequences of global climate change', Washington, DC: Pew Center on Global Climate Change, available from http://www.pewclimate.org/global-warming-in-depth/all_reports/marketconsequences

London Climate Change Partnership (2002): 'London's warming', London Climate Change Partnership, available from http://www.london.gov.uk/gla/publications/environment/londons_warming02.pdf

McGregor, G.R., M. Pelling and T. Wolf (2006): 'The social impacts of heat waves', Forthcoming Report to the Environment Agency

Mendelsohn, R., W.D. Nordhaus and D. Shaw (1994): 'The impact of global warming on agriculture: a Ricardian analysis', American Economic Review 84: 753 – 771

Mendelsohn, R.O. (ed.) (2001): 'Global warming and the American economy: a regional assessment of climate change impacts', Cheltenham: Edward Elgar Publishing

Metroeconomica (2006): 'Cross-regional research programme - Quantify the costs of impacts and adaptation' (GA01075), Research report for Defra, London: Defra

MICE [Modelling the Impacts of Climate Extremes] (2005): MICE Summary of Final Report, available from http://www.cru.uea.ac.uk/cru/projects/mice/html/reports.html

Mills, E., (2005): 'Insurance in a climate of change', Science 309: 1040 – 1044

Mills, E. and Lecomte E (2006): 'From risk to opportunity: how insurers can proactively and profitably manage climate change', Boston, MA: Ceres, available from http://www.ceres.org/pub/docs/Ceres_Insurance_Climate_%20Report_082206.pdf

Muir-Wood, R., S. Miller and A. Boissonade (2006): 'The search for trends in a global catalogue of normalized weather-related catastrophe losses', Climate change and disaster losses workshop, Hohenkammer: Munich Re, available from http://w3g.gkss.de/staff/storch/material/060525.hohenkammer.pdf

Munich Re (2004): 'Annual review: natural catastrophes 2003', Munich: Munich Re Group

Munich Re (2006): 'Annual review: natural catastrophes 2005', Munich: Munich Re Group

Nicholls, R.J., and R.J. Klein (2003): 'Climate change and coastal management on Europe's coast', EVA Working Paper 3, Potsdam: Potsdam Institute for Climate Impact Research

Nicholls, R.J., and R.S.J. Tol (2006): 'Impacts and responses to sea-level rise: a global analysis of the SRES scenarios over 21st century', Philosophical Transactions of the Royal Society A 364: 1073 – 1095

Nordhaus, W.D. (2006): 'The economics of hurricanes in the United States', prepared for the Snowmass Workshop on Abrupt and Catastrophic Climate Change, Snowmass, CO: Annual Meetings of the American Economic Association, available from http://nordhaus.econ.yale.edu/hurricanes.pdf

O'Brien, K., S. Eriksen, L. Sygna and L.O. Naess (2006): 'Questioning complacency: climate change impacts, vulnerability, and adaptation in Norway', Ambio 35: 50 –56

Parry, M.L., C. Rosenzweig, A. Iglesias, et al. (2004): 'Effects of climate change on global food production under SRES emissions and socio-economic scenarios', Global Environmental Change 14: 53 – 67

Preston, B.L. and R.N. Jones (2006): 'Climate change impacts on Australia and the benefits of early action to reduce global greenhouse gas emissions: a report prepared for the Australian Business Roundtable on Climate Change', Victoria: CSIRO

Salmon, M., and S. Weston (2006): 'Evidence to Stern Review', available from http://www.sternreview.org.uk

Schröter et al. (2005): Ecosystem service supply and vulnerability to global change in Europe, Science **310**: 1333:1337

Schlenker, W., W.M. Hanemann and A. Fisher (2005): 'Will US agriculture really benefit from global warming? Accounting for irrigation in the hedonic approach', American Economic Review **95**: 395 – 406

Schwartz, P. and D. Randall (2004): 'An abrupt climate change scenario and its implications for United States security', Report prepared by the Global Business Network (GBN) for the Department of Defense, San Francisco, CA: GBN, available from, http://www.gbn.com/ArticleDisplayServlet.srv?aid=26231

Shiva, V. (2002) 'Water wars', Cambridge, MA: South End Press.

Stott, P.A, D.A Stone and M.R. Allen (2004): 'Human contribution to the European heatwave of 2003', Nature **432**: 610 – 614

Tol, R.S.J., T.E. Downing, O.J. Kuik and J.B. Smith (2004): 'Distributional aspects of climate change impacts', Global Environmental Change **14**: 259 – 272

UK Government Foresight Programme (2004):' Future flooding', London: Office of Science and Technology, available from http://www.foresight.gov.uk/Previous_Projects/Flood_and_Coastal_Defence

Vellinga, M. and R.A. Wood (2002): 'Global climatic impacts of a collapse of the Atlantic thermohaline circulation', Climatic Change **54**: 251-267

Warren, R., N. Arnell, R. Nicholls, et al. (2006): 'Understanding the regional impacts of climate change', Research report prepared for the Stern Review, Tyndall Centre Working Paper 90, Norwich: Tyndall Centre, available from http://www.tyndall.ac.uk/publications/ working_papers/twp90.pdf

Woodham-Smith, C. (1991): 'The Great Hunger, 1845 – 1849', London: Penguin Books

6 Economic Modelling of Climate-Change Impacts

KEY MESSAGES

The monetary cost of climate change is now expected to be higher than many earlier studies suggested, because these studies tended not to include some of the most uncertain but potentially most damaging impacts.

Modelling the overall impact of climate change is a formidable challenge, involving forecasting over a century or more as the effects appear with long lags and are very long-lived. The limitations to our ability to model over such a time scale demand caution in interpreting results, but projections can illustrate the risks involved – and policy here is about the economics of risk and uncertainty.

Most formal modelling has used as a starting point 2–3°C warming. In this temperature range, the cost of climate change could be equivalent to around a 0–3% loss in global GDP from what could have been achieved in a world without climate change. Poor countries will suffer higher costs.

However, 'business as usual' (BAU) temperature increases may exceed 2–3°C by the end of this century. This increases the likelihood of a wider range of impacts than previously considered, more difficult to quantify, such as abrupt and large-scale climate change. With 5–6°C warming, models that include the risk of abrupt and large-scale climate change estimate a 5–10% loss in global GDP, with poor countries suffering costs in excess of 10%. The risks, however, cover a very broad range and involve the possibility of much higher losses. This underlines the importance of revisiting past estimates.

Modelling over many decades, regions and possible outcomes demands that we make distributional and ethical judgements systematically and explicitly. Attaching little weight to the future, simply because it is in the future ('pure time discounting'), would produce low estimates of cost – but if you care little for the future you will not wish to take action on climate change.

Using an Integrated Assessment Model, and with due caution about the ability to model, we estimate the total cost of BAU climate change to equate to an average reduction in global per-capita consumption of 5%, at a minimum, now and forever.

The cost of BAU would increase still further, were the model to take account of three important factors:

- First, including direct impacts on the environment and human health ('non-market' impacts) increases the total cost of BAU climate change from 5% to 11%, although valuations here raise difficult ethical and measurement issues. But this does not fully include 'socially contingent' impacts such as social and political instability, which are very difficult to measure in monetary terms;

- Second, some recent scientific evidence indicates that the climate system may be more responsive to greenhouse gas emissions than previously thought, because of the existence of amplifying feedbacks in the climate system. Our estimates indicate that the potential scale of the climate response could increase the cost of BAU climate change from 5% to 7%, or from 11% to 14% if non-market impacts are included. In fact, these may be only modest estimates of the bigger risks – the science here is still developing and broader risks are plausible;

- Third, a disproportionate burden of climate change impacts fall on poor regions of the world. Based on existing studies, giving this burden stronger relative weight could increase the cost of BAU by more than one quarter.

Putting these three additional factors together would increase the total cost of BAU climate change to the equivalent of around a 20% reduction in current per-capita consumption, now and forever. Distributional judgements, a concern with living standards beyond those elements reflected in GDP, and modern approaches to uncertainty all suggest that the appropriate estimate of damages may well lie in the upper part of the range 5–20%. Much, but not all, of that loss could be avoided through a strong mitigation policy. We argue in Part III that this can be achieved at a far lower cost.

6.1   Introduction

The cost of climate change is now expected to be larger than many earlier studies suggested.

This chapter brings together estimates from formal models of the monetary cost of climate change, including evidence on how these costs rise with increasing temperatures. It builds on and complements the evidence presented in Chapters 3, 4 and 5, which set out the effects of climate change in detail and separately considered its consequences for key indicators of development: income, health and the environment.

In estimating the costs of climate change, we build on the very valuable first round of integrated climate-change models that have come out over the past fifteen years or so. We use a model that is able to summarise cost simulations across a wide range of possible impacts – taking account of new scientific evidence – based on a theoretical framework that can deal effectively with large and uncertain climate risks many years in the future (see Section 6.4). Thus our focus is firmly on the economics of risk and uncertainty.

Our estimate of the total cost of 'business as usual' (BAU) climate change over the next two centuries equates to an average welfare loss equivalent to at least 5% of the value of global per-capita consumption, now and forever. That is a minimum in the context of this model, and there are a number of omitted features that would add substantially to this estimate. Thus the cost is shown to be higher if recent scientific findings about the responsiveness of the climate system to greenhouse gas (GHG) emissions turn out to be correct and if direct impacts on the environment and human health are taken into account. Were the model also to reflect the importance of the disproportionate burden of climate-change impacts on poor regions of the world, the cost would be higher still. Putting all these together, the cost could be equivalent to up to around 20%, now and forever.

The large uncertainties in this type of modelling and calculation should not be ignored. The model we use, although it is able to build on and go beyond previous models, nonetheless shares most of their limitations. In particular, it must rely on sparse or non-existent observational data at high temperatures and from developing regions. The possibilities of very high temperatures and abrupt and large-scale changes in the climate system are the greatest risks we face in terms of their potential impact, yet these are precisely the areas we know least about, both scientifically and economically – hence the uncertainty about the shape of the probability distributions for temperature and impacts, in particular at their upper end. Also, if the model is to quantify the full range of effects, it must place monetary values on health and the environment, which is conceptually, ethically and empirically very difficult. But, given these caveats, even at the optimistic end of the 5–20% range, 'business as usual' climate change implies the equivalent of a permanent reduction in consumption that is strikingly large.

In interpreting these results, economic models that look out over just a few years are insufficient.[1] The impacts of GHGs emitted today will still be felt well over a century from now. Uncertainty about both scientific and economic possibilities is very large and any model must be seen as illustrative. Nevertheless, getting to grips with the analysis in a serious way does require us to look forward explicitly. These models should be seen as one contribution to that discussion. They should be treated with great circumspection. There is a danger that, because they are quantitative, they will be taken too literally. They should not be. They are only one part of an argument. But they can, and do, help us to gain some understanding of the size of the risks involved, an issue that is at the heart of the economics of climate change.

Although this Review is based on a multi-dimensional view of economic and social goals, rather than a narrowly monetary one, models that can measure climate-change damage in monetary terms have an important role.

A multi-dimensional approach to development is crucial, as our discussions in Part II make clear and as is embodied, for example, in the Millennium Development Goals (MDGs). In this Chapter, we focus on three dimensions most affected by climate change: income/consumption, health, and the environment. Chapters 3 to 5 have laid out how these dimensions are affected individually. Here we consider how they might be combined in a single metric of damage[1a].

Our preference is to consider the multiple dimensions of the cost of climate change separately, examining each on its own terms. A toll in terms of lives lost gains little in eloquence when it is converted into dollars; but it loses something, from an ethical perspective, by distancing us from the human cost of climate change.

---

[1]   Cline (1992).
[1a] Ethical perspectives other than those embodied in the models below – such as the approaches based on rights and liberties, intergenerational responsibilities, and environmental stewardship discussed in Chapter 2 – also point towards focusing on the costs of climate change in terms of income/consumption, health, and environment.

Nevertheless, in this chapter the Review does engage with formal models of the monetary cost of climate change. Such models produce useful insights into the global cost of climate change. In making an analytical assessment in terms of the formal economics of risk and uncertainty, our models incorporate, systematically and transparently, the high risks that climate change is now thought to pose. Estimating those costs is essential for taking action (although we have emphasised strongly the dangers of taking them too literally). Once the aggregate cost of climate change is expressed in monetary terms, it is possible to compare this cost with the anticipated cost of mitigating and adapting to climate change. This is covered in Chapter 13, where the Review also considers other ways, beyond this modelling, of examining the case for action.

6.2   What existing models calculate and include

Modelling the monetary impacts of climate change globally is very challenging: it requires quantitative analysis of a very broad range of environmental, economic and social issues. Integrated Assessment Models (IAMs), though limited, provide a useful tool.

IAMs simulate the process of human-induced climate change, from emissions of GHGs to the socio-economic impacts of climate change (Figure 6.1). We focus on the handful of models specially designed to provide monetary estimates of climate impacts. Although the monetary cost of climate change can be presented in a number of ways, the basis is the difference between income growth with and without climate change impacts. To do this, the part of the model that simulates the impacts of climate change is in effect 'switched off' in the 'no climate change' scenario.

Income in the 'no climate change' scenario is conventionally measured in terms of GDP – the value of economic output. The difficulty is that some of the negative effects of climate change will actually lead to increases in expenditure, which increase economic output. Examples are increasing expenditure on air conditioning and flood defences. But it is correct to subtract these from GDP in the 'no climate change' scenario, because such expenditures are a cost of climate change. As a result, the measure of the monetary cost of climate change that we derive is really a measure of income loss, rather than output loss as conventionally measured by GDP.

Making such estimates is a formidable task in many ways (discussed below). It is also a computationally demanding exercise, with the result that such models must make drastic, often heroic, simplifications along all stages of the climate-change chain. What is more, large uncertainties are associated with each element in the cycle. Nevertheless, the IAMs remain the best tool available for estimating aggregate quantitative global costs and risks of climate change.

The initial focus of IAMs is on economic sectors for which prices exist or can be imputed relatively straightforwardly. These 'market' sectors include agriculture, energy use and forestry. But this market-sector approach fails to capture most direct impacts on the environment and human health, because they are not priced in markets. These important impacts – together with some other effects in agriculture and forestry that are not covered by market prices – are often described as 'non-market'.

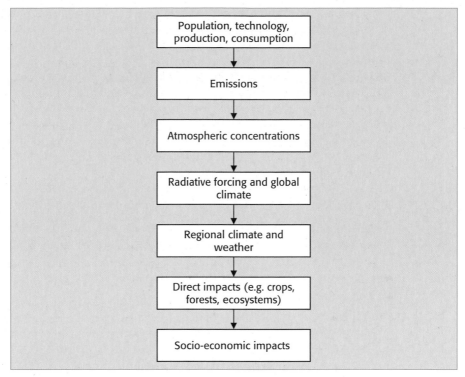

Figure 6.1 Modelling climate change from emissions to impacts.

This figure describes a simple unidirectional chain. This is a simplification as, in the real climate-human system, there will be feedbacks between many links in the chain.

Source: Hope (2005).

Economists have developed a range of techniques for calculating prices and costing non-market impacts, but the resulting estimates are problematic in terms of concept, ethical framework, and practicalities. Many would argue that it is better to present costs in human lives and environmental quality side-by-side with income and consumption, rather than trying to summarise them in monetary terms. That is indeed the approach taken across most of the Review. Nevertheless, modellers have tried to do their best to assess the full costs of climate change and the costs of avoiding it on a comparable basis, and thus make their best efforts to include 'non-market' impacts.

Estimates from the first round of IAMs laid an important foundation for later work, and their results are still valuable for informing policy. However, they were limited to snapshots of climate change at temperatures now likely to be exceeded by the end of this century.

The first round of estimates from a wide range of IAMs, presented in the IPCC's 1996 *Second Assessment Report*,[2] were based on a snapshot increase in global mean temperature. The models estimated the effects of a doubling of atmospheric

---

[2] Pearce *et al.* (1996)

CO_2 concentrations from pre-industrial levels, which was believed likely to lead to a 2.5°C mean temperature increase from pre-industrial levels. The costs of such an increase were estimated at 1.5–2.0% of world GDP, 1.0–1.5% of GDP in developed countries, and 2–9% in developing countries.

Because they took a snapshot of climate change at 2.5°C warming, these early IAM-based studies did not consider the risks associated with higher temperatures. Since then, a smaller number of models have traced the costs of climate change as temperatures increase, although their parameters are still largely calibrated on estimates of impacts with a doubling of atmospheric CO_2. These models have also covered new sectors and have looked more carefully at adaptation to climate change.

Figure 6.2 illustrates the results of three important models (whose assumptions are reported in detail in Warren *et al.* (2006)) at different global mean temperature rises:

- **The 'Mendelsohn' model**[3] estimates impacts only for five 'market' sectors: agriculture, forestry, energy, water and coastal zones. The global impact of

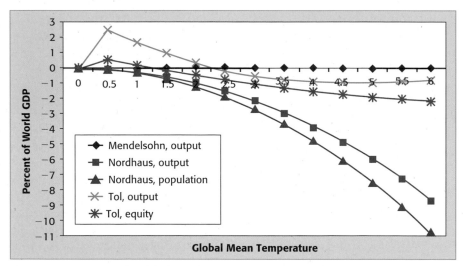

Figure 6.2 Estimates of the global impacts of climate change, as a function of global mean temperature, considered by the 2001 IPCC *Third Assessment Report*.

Source: Smith et al. (2001).

The figure above traces the global monetary cost of climate change with increases in global mean temperature above pre-industrial levels (shown on the x-axis), according to three models:

- 'Mendelsohn, output' traces the estimates of Mendelsohn *et al.* (1998), with regional monetary impact estimates aggregated to world impacts without weighting;
- 'Nordhaus, output' traces the estimates of Nordhaus and Boyer (2000), with regional monetary impact estimates aggregated to world impacts without weighting;
- 'Nordhaus, population' also traces the estimates of Nordhaus and Boyer (2000), with regional monetary impact estimates aggregated to world impacts based on regional population;
- 'Tol, output' traces the estimates of Tol (2002), with regional monetary impact estimates aggregated without weighting;
- 'Tol, equity' also traces the estimates of Tol (2002), with regional monetary impacts aggregated, to world impacts weighting by the ratio of global average per-capita income to regional average per-capita income.

---

[3] Mendelsohn *et al.* (1998)

climate change is calculated to be very small (virtually indistinguishable from the horizontal axis) and is positive for increases in global mean temperature up to about 4°C above pre-industrial levels.

- **The 'Tol' model**[4] estimates impacts for a wider range of market and non-market sectors: agriculture, forestry, water, energy, coastal zones and ecosystems, as well as mortality from vector-borne diseases, heat stress and cold stress. Costs are weighted either by output or by equity-weighted output (see below). The model estimates that initial increases in global mean temperature would actually yield net global benefits. Since these benefits accrue primarily to rich countries, the method of aggregation across countries matters for the size of the global benefits. According to the output-weighted results, global benefits peak at around 2.5% of global GDP at a warming of 0.5°C above pre-industrial. But, according to the equity-weighted results, global benefits peak at only 0.5% of global GDP (also for a 0.5°C temperature increase). Global impacts become negative beyond 1°C (equity-weighted) or 2–2.5°C (output-weighted), and they reach 0.5–2% of global GDP for higher increases in global mean temperature.

- **The 'Nordhaus' model**[5] includes a range of market and non-market impact sectors: agriculture, forestry, energy, water, construction, fisheries, outdoor recreation, coastal zones, mortality from climate-related diseases and pollution, and ecosystems. It also includes what were at the time pioneering estimates of the economic cost of catastrophic climate impacts (the small probability of losses in GDP running into tens of percentage points – see below). These catastrophic impacts drive much of the larger costs of climate change at high levels of warming. At 6°C warming, the 'Nordhaus' model estimates a global cost of between around 9–11% of global GDP, depending on whether regional impacts are aggregated by output (lower) or population (higher). The Nordhaus model also predicts that the cost of climate change will increase faster than global mean temperature, so that the aggregate loss in global GDP almost doubles as global mean temperature increases from 4°C to 6°C above pre-industrial levels. As Section 6.3 explains, this reflects the fact that higher temperatures will increase the chance of triggering abrupt and large-scale changes, such as sudden shifts in regional weather patterns like the monsoons or the El Niño phenomenon (and see Chapter 3 for a discussion of increasing marginal damages).

Models differ on whether low levels of global warming would have positive or negative global effects. But all agreed that the effects of warming above 2–3°C would reduce global welfare, and that even mild warming would harm poor countries.

These results are quite difficult to compare, because of the many differences between the models and the inputs they use, but some key points can be made:

- *Up to around 2–3°C warming*, there is disagreement about whether the global impact of climate change will be positive or negative. But, even at these levels of warming, it is clear that any benefits are temporary and confined to rich countries, with poor countries suffering significant costs. For example,

---

[4] Tol (2002)
[5] Nordhaus and Boyer (2000)

Tol estimates a cost to Africa of 4.1% of GDP for 2.5°C warming, very close to Nordhaus and Boyer's estimate of 3.9%.

- *For warming beyond 2–3°C*, the models agree that climate change will reduce global consumption. However, they disagree on the size of this cost, ranging from a very small fraction of global GDP to 10% or more. In this range too, the models agree that poor countries will suffer the highest costs, although in the Nordhaus model the estimated cost to Western Europe of 6°C warming is second only to the cost to Africa.[6]

These results depend on key modelling decisions, including how each model values the costs to poor regions and what it assumed about societies' ability to reduce costs by adapting to climate change.

Each model's results depend heavily on how it aggregates the impacts across regions, and in particular how it values costs in poor regions relative to those in rich ones. The prices of marketed goods and services, as well as the hypothetical values assigned to health and the environment, are typically higher in rich countries than in poor countries. Thus, in these models, a 10% loss in the volume of production of an economic sector is worth more in a rich country than in a poor country. Similarly, a 5% increase in mortality, if 'values of life' are based on willingness to pay, is worth more in purely monetary terms in a rich country than a poor country, because incomes are higher in the former. Many ethical observers would reject both of these statements. Thus some of the authors have used welfare or 'equity' weighting. Explicit functions to capture distributional judgements are also used in this Review – see Chapter 2 and Appendix. In summary, if aggregation is done purely on the basis of adding incomes or GDP, then very large physical impacts in poor countries will tend to be overshadowed by small impacts in rich countries.

Nordhaus and Boyer and Tol both adopt equity-weighting approaches, a step which in our view is supported by the type of ethical considerations discussed in Chapter 2 and its Appendix, as well as empirical observations of the attitudes that people actually hold towards inequality in wealth.[7,8] Mendelsohn does not use equity weights.

---

[6] The European result is driven in large part by Europe's expected willingness to pay to reduce the risk of a catastrophic event such as a significant weakening of the Atlantic thermohaline circulation – part of which keeps Western Europe warmer than its latitude would otherwise imply.

[7] Stern (1977), Pearce and Ulph (1999)

[8] Equity weights should reflect the choice of social welfare function – sometimes called the 'objective' function. This aggregates the consumption of individuals over space and time, reflecting judgements about the value of consumption enjoyed by individuals in different regions at different times (see the Appendix to Chapter 2). Here we focus on how this weighting should be carried out across regions within the present generation when considering the aggregation of small changes. The first step in calculating a weighted average change is to calculate the proportional impact of climate change on the representative individual in each region. If the utility function for an individual has constant marginal utility, the proportional impacts on per capita consumption can then be aggregated to give the proportional impact on overall social welfare by weighting them by the share of each individual's consumption in total consumption. At the regional level, this means weighting the impact on the representative individual by the region's share in global consumption (i.e. regional per-capita consumption multiplied by regional population, as a share of total global consumption). With a utility function given by the log of individual consumption, the proportional impacts on individuals should simply be added up; thus, at the regional level, the proportional impact on the representative consumer is weighted by the region's population.

Adaptation to climate change is another important factor in these models, because it has the capacity to reduce the cost of BAU climate change. The key questions are how much adaptation can be assumed without extra stimulus from policy (financial, legal and otherwise), how much will it cost, because the costs of adaptation themselves are part of the cost of climate change, and what would it achieve? Again, it is difficult to compare the models, because each treats adaptation in a different manner. In general, the models do assume that households and businesses do what they can to adapt, without extra stimulus from policy.

The 'Mendelsohn' model is most optimistic about adaptation, and – not coincidentally – it estimates the lowest cost of climate change.[9] In their method, future responses to climate change are calibrated against the relationship between output and climate that can be seen from region to region today, or that can be determined from laboratory experiments.[10] The former method models adaptation most completely. In effect, as temperatures increase, and controlling for other climate and non-climate variables, environmental and economic conditions migrate from the equator towards the poles. High-latitude regions climb a hill of rising productivity for a time as temperatures make conditions easier (e.g. for agriculture), while low-latitude regions fall further into more difficult conditions. This method encompasses a variety of ways a region can adapt, because regions can be assumed to be well adapted to their current climates. Its major drawback, however, is that it makes no provision for the costs and difficulties of transition from one climate to another or the potential movement of people. Whether these are small or large, it is, on balance, an underestimate of the cost of climate change.

A final point to keep in mind is that all three models are based on scientific evidence up to the mid- to late 1990s. Since then, new evidence has come to light, most importantly on the possibilities of higher and more rapidly increasing temperatures than envisaged then, as well as possibilities of abrupt and large-scale changes to the climate system. Section 6.3 explores the consequences of these risks at greater length.

6.3   Do the existing models fully capture the likely cost of climate change?

Existing estimates of the monetary cost of climate change, although very useful, leave many questions unanswered and omit potentially very important impacts. Taking omitted impacts into account will increase cost estimates, and probably strongly.

Understanding of the science and economics of climate change is constantly improving to overcome substantial gaps, but many remain. This is particularly true of the existing crop of IAMs, due in part to the demands of modelling and in

---

[9]   There are several reasons why the 'Mendelsohn' model estimates the lowest cost of climate change. Adaptation is likely to be one, its omission of non-market impacts and the risk of catastrophe another.

[10] That is, they estimate the relationship between production in their five market sectors and climate based on how production varies across current world climates, and control for other important determining factors.

part to their reliance on knowledge from other active areas of research. Indeed, the knowledge base on which the cost of climate change is calibrated – specialised studies of impacts on agriculture, ecosystems and so on – is particularly patchy at high temperatures.[11] In principle, the gaps that remain may lead to underestimates or overestimates of global impacts. In practice, however, most of the unresolved issues will increase damage estimates.

Existing models omit many possible impacts. Watkiss *et al.*[12] have developed a 'risk matrix' of uncertainty in projecting climate change and its impacts to illustrate the limitations of existing studies in capturing potentially important effects. Figure 6.3 presents this matrix and locates the existing models on it.

As the figure shows, most existing studies are confined to the top left part of the matrix and are thus limited to a small subset of the most well understood, but least damaging, impacts (for example, the 'Mendelsohn' model, which is also most optimistic about adaptation: see previous section). By contrast, because the impacts in the bottom right corner of the matrix are surrounded by the greatest scientific uncertainty, they have not been incorporated into IAMs. Yet it is also these paths that have the potential to inflict the greatest damage.

Extreme weather events are not fully captured in most existing IAMs;[13] the latest science suggests that extreme events will increase in frequency and severity with climate change.

Chapters 1 and 3 laid out the newer evidence that climate change will spur an increase in extreme weather events – notably floods, droughts, and storms. Experience of weather disasters in many parts of the world demonstrates that the more extreme events can have lasting economic effects, especially when they fall on an economy weakened by previous weather disasters or other shocks, or if they fall on an economy that finds it difficult to adjust quickly.[14] Thus it is very important to consider the economic impacts of variations in weather around mean trends in climate change.

However, it is at least as important to consider the climatic changes and impacts that will occur if GHG emissions lead to very substantial warming, with global mean temperatures 5–6°C above pre-industrial levels or more. High temperatures are likely to generate a hostile and extreme environment for human activity in many parts of the world. Some models capture aspects of this, because costs both in market and non-market sectors accelerate as temperatures increase.[15] At 5–6°C above pre-industrial levels, the cost of climate change on, for example, agriculture can be very high.

Further, Chapter 1 detailed emerging evidence of risks that higher temperatures will trigger massive system 'surprises', such as the melting and collapse of ice sheets and sudden shifts in regional weather patterns like the monsoons. Thus there is a danger that feedbacks could generate abrupt and large-scale changes in the climate and still further losses.

Existing IAMs largely omit these system-change effects; including them is likely to increase cost estimates significantly. Although many factors can produce

---

[11] See Hitz and Smith (2004)
[12] Watkiss *et al.* (2005)
[13] Warren *et al.* (2006)
[14] Hallegatte and Hourcade (2005) and Chapter 4.
[15] Although this depends on how rapidly costs increase in proportion to temperature.

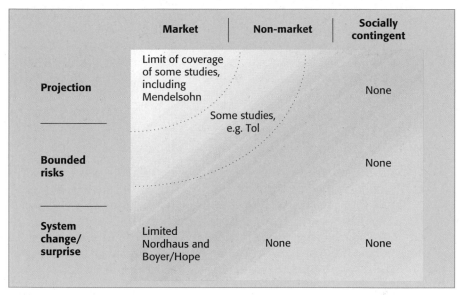

	Market	Non-market	Socially contingent
Projection	Limit of coverage of some studies, including Mendelsohn		None
		Some studies, e.g. Tol	
Bounded risks			None
System change/ surprise	Limited Nordhaus and Boyer/Hope	None	None

Figure 6.3 Coverage of existing integrated assessment studies.

Source: Watkiss, Downing *et al.* (2005).

Figure 6.3 summarises which impacts existing estimates of the monetary cost of climate change cover (by reference to the authors of the various studies) and which impacts are omitted.

The vertical axis captures uncertainty in predicting climate change, with uncertainty increasing as we go down. There are three categories:

● Projection – high confidence on the direction of these changes and bounds can be placed around their magnitude (i.e. temperature change and sea-level rise);
● Bounded risks – more uncertainty about the direction and magnitude of these changes, though reasonable bounds can be placed around them (i.e. precipitation, extreme events);
● System change and surprises – large uncertainty about the potential trigger and timing of these changes (e.g. weakening of the thermohaline circulation, collapse of the West Antarctic Ice Sheet). However, evidence on the risk of such changes is building (see Chapters 1 and 3).

The horizontal axis captures uncertainty in the economic measurement of impacts, with uncertainty increasing as we go from left to right. There are again three categories:

● 'Market' impacts – where prices exist and a valuation can be made relatively easily, such as in agriculture, energy use and forestry;
● 'Non-market' impacts – directly on human health and the environment, where market prices tend not to exist and methods are required to create them;
● 'Socially contingent' responses – large-scale, 'second-round' socio-economic responses to the impacts of climate change, such as conflict, migration and the flight of capital investment.

differences in results from model to model, it is nevertheless intuitive that the Nordhaus estimates[16], produced by the only model to include catastrophic 'system change/surprise', were the highest among the existing IAMs. For increases in global mean temperature of 5–6°C above pre-industrial levels or more, costs were estimated to approach and even exceed 10% of global GDP.

The Nordhaus method is based on polling a number of experts on the probability that a very large loss of 25% of global GDP, roughly equivalent to the effect of

---

[16] Nordhaus and Boyer (2000)

the Great Depression, will result from increases in global mean temperature of 3°C by 2090, 6°C by 2175 and 6°C by 2090. Taking account of estimated differences in regional vulnerability to catastrophic climate change, the model uses survey data to estimate people's willingness to pay to avoid the resulting risk. This approach is simple, but it takes us some way towards capturing the economic importance of complex, severe responses of the climate system.

Most existing IAMs also omit other potentially important factors – such as social and political instability and cross-sectoral impacts. And they have not yet incorporated the newest evidence on damaging warming effects.

One factor omitted at least in part from most models is 'socially contingent' responses – the possibility that climate change will not only increase the immediate costs of climate change, but also affect investment decisions, labour supply and productivity, and even social and political stability.

On the one hand, these knock-on effects could dampen the negative effect of climate change, if the economic response is to adapt, for example, by shifting production from the most climate-sensitive sectors into less climate-sensitive sectors. As mentioned, recent models have taken adaptation more fully into account.

On the other hand, knock-on effects could amplify the future consequences of today's climate change, for example if they reduce investment. This possibility has yet to be taken fully into account. In some models, baseline income is taken from outside the model, so that the impacts in any one time period do not affect growth in future periods. In other models, such as that employed by Nordhaus and Boyer,[17] the economy makes investment and saving decisions based on the level of income it starts off with and on expectations of how that income will grow in the future. Climate change reduces investment and saving, as the income available to invest and the returns to saving fall.[18]

How important might these effects be? Fankhauser and Tol[19] unpack the 'Nordhaus' estimates to show that the knock-on cost of depressed investment on the total, long-run cost of 3°C warming is at least an additional 90% over and above the immediate cost. Furthermore, substituting for a more powerful model of economic growth that is better able to explain past and present growth trends, world GDP losses are almost twice as high as they are for immediate impacts alone. These dynamic effects may be especially strong in some developing regions, where the further effect of climate change may be to precipitate instability, conflict and migration (see Chapters 3 and 4).

A second omitted factor is possible interactions between impacts in one sector and impacts in another, which past IAMs have not generally taken into account. Climate damage in one sector could multiply damage in another – for example, if water-sector impacts amplify the impacts of climate change on agriculture. The reasons for excluding these effects have to do with the modelling approach: in the

---

[17] Nordhaus and Boyer (2000)

[18] Because the Nordhaus and Boyer model simplifies the economy to one sector, it ignores the possibility that productivity will increase if production is shifted from low productivity/highly climate-sensitive sectors to high productivity/low sensitivity sectors. But a multi-sector study for the USA (Jorgensen *et al.*, 2005) indicates that such processes are negligible, at least in that region.

[19] Fankhauser and Tol (2003)

basic IAM method, impacts are characteristically enumerated on a sector-by-sector basis, and then added up to arrive at the overall economy-wide impact.

Finally, even in market sectors that the IAMs do cover well, the latest specialised impact studies suggest that IAM-based estimates may be too optimistic.[20] The underlying impacts literature on which the IAMs are based dates primarily from 2000 or earlier. Since then, many of the predictions of this literature have become more pessimistic, for example, on the possible boost from CO_2 fertilisation to agriculture (Chapter 3).

The building of the IAMs has been a valuable contribution to our understanding of possible effects. Any model must necessarily leave out much that is important and can use only the information available at the time of construction. The science has moved quickly and the economic analysis and modelling can move with it.

6.4 Calculating the global cost of climate change: an 'expected-utility' analysis

Modelling the global cost of climate change presents many challenges, including how to take account of risks of very damaging impacts, as well as uncertain changes that occur over very long periods.

A model of the monetary cost of climate change ideally should provide:

- Cost simulations across the widest range of possible impacts, taking into account the risks of the more damaging impacts that new scientific evidence suggests are possible.
- A theoretical framework that is fit for the purpose of analysing changes to economies and societies that are large, uncertain, unevenly distributed and that occur over a very long period of time.

This section begins with the first challenge, illustrating the consequences of BAU climate change in a framework that explicitly brings out risk. The second challenge is addressed later in the chapter, allowing consideration of how to value risks with different consequences, particularly the risks, however small, of very severe climate impacts.

The model we use – the PAGE2002 IAM[21] – can take account of the range of risks by allowing outcomes to vary probabilistically across many model runs, with the probabilities calibrated to the latest scientific quantitative evidence on particular risks.

The first challenge points strongly to the need for a modelling approach based on probabilities (that is, a 'stochastic' approach). The PAGE2002 (Policy Analysis of the Greenhouse Effect 2002) IAM meets this requirement by producing estimates based on 'Monte Carlo' simulation. This means that it runs each scenario many times (e.g. 1000 times), each time choosing a set of uncertain parameters randomly from pre-determined ranges of possible values. In this way, the model

---

[20] Warren *et al.* (2006)
[21] Hope (2003)

generates a probability distribution of results rather than just a single point estimate. Specifically, it yields a probability distribution of future income under climate change, where climate-driven damage and the cost of adapting to climate change are subtracted from a baseline GDP growth projection[22].

The parameter ranges used as model inputs are calibrated to the scientific and economic literatures on climate change, so that PAGE2002 in effect summarises the range of underlying research studies. So, for example, the probability distribution for the climate sensitivity parameter – which represents how temperatures will respond in equilibrium to a doubling of atmospheric carbon dioxide concentrations – captures the range of estimates across a number of peer-reviewed scientific studies. Thus, the model has in the past produced mean estimates of the global cost of climate change that are close to the centre of a range of peer-reviewed studies, including other IAMs, while also being capable of incorporating results from a wider range of studies.[23] This is a very valuable feature of the model and a key reason for its use in this study.

PAGE2002 has a number of further desirable features. It is flexible enough to include market impacts (for example, on agriculture, energy and coastal zones) and non-market impacts (direct impacts on the environment and human mortality), as well as the possibility of catastrophic climate impacts. Catastrophic impacts are modelled in a manner similar to the approach used by Nordhaus and Boyer.[24] When global mean temperature rises to high levels (an average of 5°C above pre-industrial levels), the chance of large losses in regional GDP in the range of 5–20% begins to appear. This chance increases by an average of 10% per °C rise in global mean temperature beyond 5°C.

At the same time, PAGE2002 shares many of the limitations of other formal models. It must rely on sparse or non-existent data and understanding at high temperatures and in developing regions, and it faces difficulties in valuing direct impacts on health and the environment. Moreover, like the models depicted in Figure 6.3, the PAGE2002 model does not fully cover the 'socially contingent' impacts. As a result, the estimates of catastrophic impacts may be conservative, given the damage likely at temperatures as high as 6–8°C above pre-industrial levels. Thus the results presented below should be viewed as indicative only and interpreted with great caution. Given what is excluded, they should be regarded as rather conservative estimates of costs, relative to the ability of these models to produce reliable guidance.

We present results based on different assumptions along two dimensions: first, of how fast global temperatures increase in response to GHG emissions and, second, different categories of economic impact.

To reflect the considerable uncertainty about likely probability distributions and difficulties in measuring different effects, we examine models that differ along two dimensions:

● **Response of the climate to GHG emissions**. We run the model under two different assumed levels of climatic response. The 'baseline climate' scenario is

---

[22] We follow PAGE 2002 in referring to 'GDP' but, as remarked above, it is preferable to think of a broader income concept in interpreting some of the results.
[23] Tol (2005)
[24] Nordhaus and Boyer (2000)

designed to give outputs consistent with the IPCC *Third Assessment Report* (TAR)[25]. The 'high climate' scenario adds to this a risk of there being amplifying natural feedbacks in the climate system. This is based on recent studies showing that there is a real risk of additional feedbacks, such as weakening carbon sinks and natural methane releases from wetlands and thawing permafrost. This scenario gives a higher probability of larger temperature changes. These scenarios are discussed in more detail in Box 6.1. Both climate scenarios give temperature outputs that are roughly consistent with other studies.

BOX 6.1 The PAGE2002 climate scenarios.

Baseline Climate: This is designed to give outputs consistent with the range of assumptions presented in the IPCC *Third Assessment Report* (TAR). The scenario produces a mean warming of 3.9°C relative to pre-industrial in 2100 and a 90% confidence interval of 2.4–5.8°C (see figure below) for the A2 emissions scenario used in this exercise. This is in line with the mean projection of 4.1°C given by the IPCC TAR. The IPCC does not give a probability range of temperatures. It does quote a range across several models of 3.0–5.3°C. The wider range of temperatures produced by PAGE2002 mainly reflects the wider combinations of parameters explored by the model.

High Climate: This is designed to explore the impacts that may be seen if the level of temperature change is pushed to higher levels through the action of amplifying feedbacks in the climate system. Scientists are only just beginning to quantify these effects, but these preliminary studies suggest that they will form an important part of the climate system's response to GHG emissions. No studies have yet combined ranges of climate sensitivity and feedbacks in this way, so these results should be treated as only indicative of the possible potential scale of response. The scenario includes recent estimates of two types of amplifying feedback: a weakening of natural carbon absorption and increased natural methane releases from, for example, thawing permafrost.

- **Weakened carbon sinks:** As temperatures increase, plant and soil respiration increases. Recent evidence suggests that these extra natural emissions will offset any increase in natural sink capacity due to carbon fertilisation, so that carbon sinks will be weakened overall (discussed in chapter 1). Weakening of carbon sinks is modelled as a function of temperature, based on Friedlingstein et al. (2006).

- **Increased natural methane releases:** Natural methane currently locked in wetlands and permafrost is released as temperatures rise. This is simulated using a probability distribution based on recent studies (Box 1.3)[26].

In this exercise, these feedbacks push the mean temperature change up by around 0.4°C and give a higher probability of larger temperature increases. Accordingly, the 90% confidence interval increases to 2.6–6.5°C. There is little effect on the lower bound of temperature changes, as, at this level, temperatures are not large enough to initiate a significant

---

[25] IPCC (2001)
[26] For example, the central value is based on Gedney *et al.* (2004) assuming 4.5°C temperature rise in 2100

feedback effect from the carbon cycle. The increase in the mean and upper bound are consistent with recent studies (chapter 1).

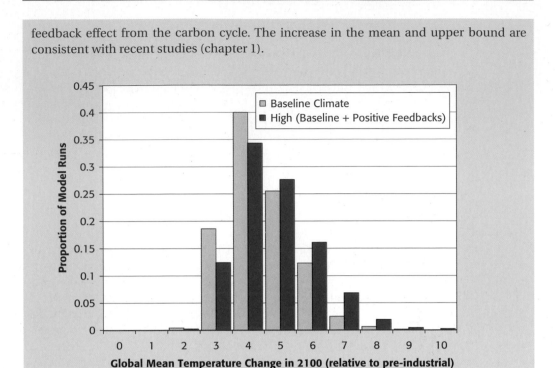

- **Categories of economic impact.** Our analyses also vary in the comprehensiveness with which they measure the impacts of climate change on the economy and on welfare. The first set of estimates includes only the impacts of 'gradual climate change' on market sectors of the economy. In other words, it takes no account of the possibility of catastrophic events that we now know may occur. The second set also includes the risk of catastrophic climate impacts at higher temperatures. Figure 6.3 illustrated that these also fall on market sectors of the economy, but are much more uncertain. Finally, the third set includes market impacts, the risk of catastrophe *and* direct, non-market impacts on human health and the environment. This chapter shall argue that attention should be focused on the second and third cases here, since there is very good reason to believe that both are relevant.

These dimensions combine to produce a 2 × 3 matrix of scenarios (Figure 6.4). For example, the lowest cost estimates would be expected to come from the scenario that (i) uses the baseline-climate scenario and (ii) considers only those impacts from gradual climate change on market sectors.

Preliminary estimates of average losses in global per-capita GDP in 2200 range from 5.3 to 13.8%, depending on the size of climate-system feedbacks and what estimates of 'non-market impacts' are included.

Estimates of losses in per-capita income over time are benchmarked against projected GDP growth in a world without climate change. The baseline-climate/

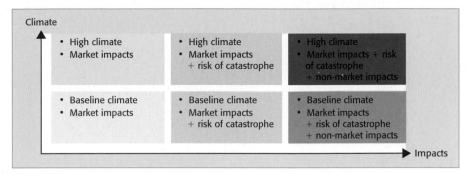

Figure 6.4 A 2 × 3 matrix of scenarios.

market-impacts scenario generates the smallest losses, where climate change reduces global per-capita GDP by, on average, 2.2% in 2200. However, as discussed in the previous section, the omission of the very real risk of abrupt and large-scale changes at high temperatures creates an unrealistic negative bias in estimates.

Figure 6.5 shows the results of scenarios including a risk of 'catastrophe'. The lower-bound estimate of the global cost of climate change in Figure 6.5 uses the baseline climate and includes both market impacts and the risk of catastrophic changes to the climate system (Figure 6.5a). In this scenario, the mean loss in global per-capita GDP is 0.2% in 2060. By 2100, it rises to 0.9%, but by 2200 it rises steeply to 5.3%.

There is a substantial dispersion of possible outcomes around the mean and, in particular, a serious risk of very high damage. The grey-shaded areas in Figure 6.5 give the range of estimates in each year taken from the 5th and 95th percentile damage estimates over the 1000 runs of the model. For the lower-bound estimate in 2100, the range is a 0.1–3% loss in global GDP per capita. By 2200, this rises to 0.6–13.4%.

Figures 6.5b to d demonstrate the loss in global GDP per capita when first, the risk of more feedbacks in the climate system is included (the high-climate scenario), and second, estimates of non-market impacts of climate change are included.

In the high-climate scenario, the losses in 2100 and 2200 are increased by around 35%. In 2200, the range of losses is increased to between 0.9% and 17.9%.

The inclusion of non-market impacts increases these estimates further still. In this Review, non-market impacts, on health and the environment, are generally considered separately to market impacts. However, if the goal is to compare the cost of climate change in monetary terms with the equivalent cost of mitigation, then excluding non-market costs is misleading. For the high-climate scenario with non-market impacts (Figure 6.5c), the mean total losses are 2.9% in 2100 and 13.8% in 2200. In 2200, the 5th and 95th percentiles increase significantly, to 2.9% to 35.2%.

These estimates still do not capture the full range of impacts. The costs of climate change could be greater still. For example, recent studies demonstrate that the climate sensitivity could be greater than the range used in the PAGE2002 climate scenarios (Chapter 1). Were this to be the case, costs would rise again. The potential impacts of higher climate sensitivity are explored speculatively in Box 6.2.

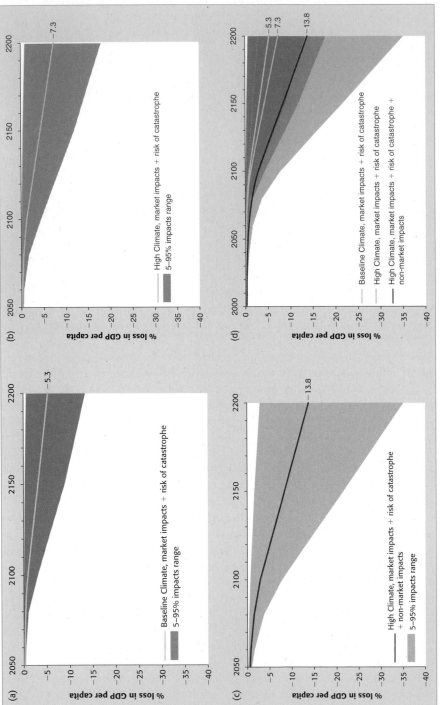

Figure 6.5a–d Losses in income per capita due to climate change over the next 200 years, according to three of our main scenarios of climate change and economic impacts. The mean loss is shown in a colour matching the scenarios of Figure 6.4. The range of estimates from the 5th to the 95th percentile is shaded grey; a. Baseline-climate scenario, with market impacts and the risk of catastrophe; b. High-climate scenario, with market impacts and the risk of catastrophe; c. High-climate scenario, with market impacts, the risk of catastrophe and non-market impacts; d. Combined scenarios.

BOX 6.2 Exploring the consequences of high climate sensitivity.

The climate scenarios described in Box 6.1 are based on a climate sensitivity (the equilibrium temperature increase following a doubling in atmospheric carbon dioxide concentrations) range of 1.5–4.5°C, as outlined in the IPCC TAR[27]. However, studies since the TAR have shown up to a 20% chance that the climate sensitivity could be greater than 5°C.

In order to explore the possible consequences of recent scientific evidence on a higher climate sensitivity, we develop a 'high+' climate scenario that combines the amplifying natural feedbacks explained in Box 6.1 with a higher probability distribution for the climate sensitivity parameter. We use the climate sensitivity distribution estimated by Murphy *et al.* (2004). This is has a 5–95% range of 2.4–5.4°C, and a mode of 3.5, with a logistic distribution (Box 1.2).

This scenario is particularly speculative, but we cannot rule out that this is the direction that further evidence might take us. Combining the high+ scenario with market impacts and the risk of catastrophe, the mean loss in global per-capita GDP is 0.4% in 2060. In 2100, it rises to 2.7%, but by 2200 it rises to 12.9%. Adding non-market impacts, the mean loss is 1.3% in 2060, 5.9% in 2100 and 24.4% in 2200.

In addition, these results reflect the aggregation of costs across the world, but aggregating simply by adding GDP across countries or regions masks the value of impacts in poor regions. A given absolute loss is more damaging for a person on lower incomes. Nordhaus and Boyer[28] and Tol[29] demonstrate that giving more weight to impacts in poor regions increases the global cost of climate change. Nordhaus and Boyer estimate that the global cost increases from 6% to 8% of GDP for 5°C warming, one quarter higher. Tol estimates that the global cost is almost twice as high for 5°C warming, if he uses welfare weights (see Section 6.2).

Only a small portion of the cost of climate change between now and 2050 can be realistically avoided, because of inertia in the climate system.

Past emissions of GHGs have already committed the world to much of the loss in global GDP per capita over the next few decades. Over this period, market impacts are likely to be relatively small. This is, in large part, because the risk of catastrophic, large-scale changes to the climate system, as well as amplifying natural feedbacks (which boost the temperature response to GHG emissions), become a bigger factor later. Non-market impacts are significant in the period to 2050, reaching around 0.5% of per-capita global GDP in 2050 in both the baseline and high-climate scenarios.

In all scenarios, the highest impacts are in Africa and the Middle East, and India and South-East Asia.

For example, in the baseline-climate scenario with all three categories of economic impact, the mean cost to India and South-East Asia is around 6% of regional GDP by 2100, compared with a global average of 2.6%.

---

[27] IPCC (2001)
[28] Nordhaus and Boyer (2000)
[29] Tol (2002)

In all scenarios, the consequences of climate change will become disproportionately more severe with increased warming.

Figure 6.6 examines the relationship between mean losses in per-capita GDP and average increases in global mean temperature produced by the baseline and high-climate scenarios. The figure makes two important points graphically:

• The first is that the climatic effects suggested by the newer scientific evidence have the potential to nudge global temperatures, and therefore impacts, to higher levels than those suggested by the IPCC TAR report. In the high scenario, global mean temperature rises to an average of nearly 4.3°C above pre-industrial levels by 2100, compared with an average of 3.9°C above pre-industrial levels in the baseline scenario. The difference between the two scenarios increases beyond 2100, because the effect of the amplifying natural feedbacks becomes more marked at higher temperatures. By 2200, the rise in global mean temperature increases to 8.6°C in the high-climate scenario, while the baseline reaches only 7.4°C. These numbers should be treated as indicative, as climate models have not yet been used to explore the high temperatures that are likely to be realised beyond 2100. They do demonstrate that, if emissions continue unabated, the climate is very likely to enter unknown territory with the potential to cause severe impacts.

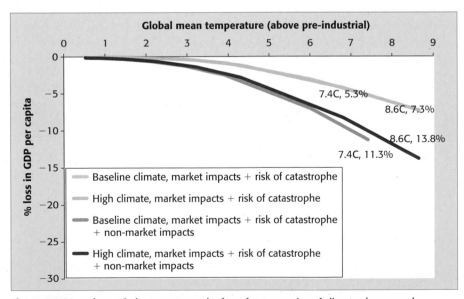

Figure 6.6 Mean losses in income per capita from four scenarios of climate change and economic impacts, plotted against average increases in global mean temperature (above pre-industrial levels).

This figure traces mean losses in per-capita GDP due to climate change as a function of increasing global mean temperature, according to four of the scenarios of climate change and economic impacts. Losses are compared to baseline growth in per-capita GDP without climate change. Because temperature is one of the probabilistic outputs of the PAGE2002 model, increases in temperature in each scenario are averaged across all 1000 runs.

- Second, scenarios that include the risk of catastrophe and non-market impacts project higher costs of climate change at any given temperature. The figure makes an additional point that the incremental cost associated with including these non-market and catastrophic impacts increases as temperatures rise, so that the wedge between the economic scenarios becomes more and more substantial.

Estimates of income effects and distribution of risks can also be used to calculate the overall welfare cost of climate change.

Whereas the first part of Section 6.4 estimated how BAU climate change would affect income, the remainder of the section tackles a still more important challenge: estimating the global welfare costs of climate change, taking explicitly into account the risks involved. Because the forecast changes are large, uncertain, and unevenly distributed, and because they occur over a very long period of time, this exercise must take on the problem of aggregating across different possible outcomes (risk), over different points in time (inter-temporal distribution), and over groups with different incomes (intra-temporal distribution). It should carry out these three types of aggregation consistently. At this stage of the analysis, we have not incorporated intra-temporal distribution.

First, the analysis requires evaluation of the significance of severe climate risks that would result in very low levels of global GDP relative to the world without climate change. In the high-climate scenario with market impacts, the risk of catastrophe and non-market impacts, for example, the 95th percentile estimate is a 35.2% loss in global per-capita GDP by 2200. This is not the statistical mean, but it is nevertheless a risk that few would want to ignore. As discussed below, such risks have a disproportionate effect on welfare calculations, because they reduce income to levels where every marginal dollar or pound has greater value. That is indeed how risk is generally treated in economics.

Second, it requires deciding how to express the future costs of BAU climate change in terms that can be compared with current levels of well-being: we have to evaluate costs occurring at different times on a common basis. The process of warming builds over many decades. In the baseline-climate scenario, 5°C warming is not predicted to occur until some time between 2100 and 2150. By then, growth in GDP will have made the world considerably richer than it is now.

To make these calculations, the model uses the standard tools of applied welfare economics, as described in Chapter 2 and its Appendix.

In these highly aggregated models, the basic approach has to be simple, but it does depend on key assumptions. It is important to lay them out transparently. First, in applying this basic welfare-economics theory to the PAGE2002 model, we follow many other studies in calculating overall social welfare (or global 'utility', to use the standard economic term) as the sum of social utilities of consumption of all individuals in the world. In practice, for this exercise, this means that we convert per-capita global GDP at each point in time into consumption[30], and

---

[30] In these calculations, we assume that some fixed proportion of income is saved for future consumption. A more sophisticated model would vary the rate of saving as a result of prospects for future consumption, as determined by the model itself.

then calculate the social utility of per-capita consumption. This is then multi-plied by global population (Box 6.3).

An approach that would better reflect the consequences of climate change on different world regions would take regional per-capita utility (e.g. for India and South-East Asia) and multiply by regional population to get 'regional utility'. Global utility would then be the sum of regional utilities[31]. Doing so was beyond the scope of this exercise, given the limited time available for analysis, but it is possible to provide some assessment of the bias from this omission. Taking this regional approach would increase the climate-change cost estimates, as illus-trated in Section 6.2, so our decision to use a simpler global aggregation approach will bias our model toward lower cost estimates.

Second, we use the assumption of diminishing marginal utility as we evaluate risks and future welfare. This standard assumption in economics, generally sup-ported by empirical evidence on behaviour and preferences, holds that the extra utility produced by additional consumption falls as the level of consumption rises. That is, an extra dollar or pound is worth more to a poor person than it is to a rich person. This assumption plays an important role in the welfare calcula-tions, in that it places greater weight on:

- Near-term consumption than on consumption in the distant future, because even with climate change, the world will be richer in the future as a result of economic growth; and
- The most severe climate impacts, because they reduce consumption to such low levels (see Chapter 2 and its Appendix for the underlying welfare economics).

Third, consumption growth is allowed to vary in the future in systematic ways. Traditionally, economic appraisal of projects and policies has taken a simplified approach to this basic welfare-economics framework. Consumption is simply assumed to grow at a certain rate in the future, with uncertainty entering the pro-jection only to the extent that there will be perturbations around this assumed path. In our case, however, climate change could substantially reduce consump-tion growth in the future, and so two probabilistic model runs with different cli-mate impacts produce different growth rates. So the simplified approach will not work here. Instead, we have to go back to the underlying theory, which implies that consumption paths must be valued separately along each of the model's many (1000, say) runs.

Fourth, in carrying out the expected-utility valuation process, we use a pure rate of time preference (or 'utility discount rate') to weight (or value) the utility of consumption at each point in the future. Thus utility in the future has a different weight simply because it is in the future.[32] This assumption is difficult to justify on ethical grounds, as discussed in Chapter 2 and its Appendix, except where we take into account the probability that individuals will be alive in the future to enjoy the projected consumption stream. In other words, if we know a future generation

---

[31] As in Nordhaus and Boyer (2000)
[32] We are not considering here the discounting of extra units of consumption in the future because consumption itself may be higher then.

will be present (that is, apart from discounting for the small chance of global annihilation), we suppose that it has the same claim on our ethical attention as the current one.

Putting all this together, we can:

- Calculate the aggregate utility of the different paths over the future by adding utilities over time, as described, and then;
- Average utility across all 1000 runs to calculate the expected utility under each scenario.

Finally, we need to decide in what terms to express the loss in expected future welfare due to climate change. If the result is to guide policy, it must be easily understandable. When we calculate social utility and aggregate over time for risk, the resulting measure might most immediately be expressed in expected 'utils', but this would not be easily understood. Instead, we introduce the idea of a 'Balanced Growth Equivalent' (hereafter BGE)[33] to calibrate welfare along a path. The BGE essentially measures the utility generated by a consumption path in terms of the consumption now that, if it grew at a constant rate, would generate the same utility.[34]

BOX 6.3   'Expected-utility' analysis of the global cost of climate change.

PAGE2002 takes baseline GDP growth from an exogenous scenario[35] and produces 1000 runs of global GDP, less the cost of climate change damage and adaptation to climate change, from 2001 to 2200. Thus we obtain a probability distribution of global income pathways net of climate change damage and adaptation costs.

We first transform this probability distribution into GDP per capita, dividing through each run by a population scenario determined exogenously.[36] Then we transform each run into global consumption per capita, taking an arbitrary, exogenous rate of saving of 20%.

We transform consumption per capita into utility:

$$U(t) = \frac{C^{1-\eta}}{1-\eta} \qquad (1)$$

where U is utility, C is consumption per capita, t is the year[37] and η is the elasticity of the marginal utility of consumption (see Appendix to Chapter 2). In our main case, we take η

---

[33] Proposed by Mirrlees and Stern (1972)

[34] Formally, the change in the BGE is a natural commodity measure of welfare that expresses changes in future consumption due to policy in terms of the percentage increase in consumption (along a steady-state growth path), now and forever, that is equal to the changes that are forecast to follow from the policy change being examined. In a one-sector growth model with natural growth α and consumption C at time t, we want to calibrate welfare from the path [C(t)]. If this is equivalent, in welfare terms, to the balanced growth path yielding consumption $\gamma e^{\alpha t}$, then γ is the BGE of [C(t)].

[35] An extrapolated version of the IPCC's A2 scenario (IPCC, 2000), characterised by annual average GDP growth of about 1.9%.

[36] Also extrapolated from the IPCC's A2 scenario. Annual average population growth is about 0.6%.

[37] In fact, the model is restricted to a subset of uneven time steps. Thus we interpolate linearly between time steps to produce an annual time series.

to be 1, in line with recent empirical estimates.[38] Further work would investigate a broader range of η, including higher values.[39] Where η is 1, the utility function is a special case:

$$U(t) = \ln C(t) \tag{2}$$

Then discounted utility (with constant population) is given by:

$$W = \int_{t=1}^{\infty} U(t)e^{-\delta t}dt \tag{3}$$

where W is social welfare and δ is the utility discount rate. The value of δ is taken to be 0.1% per annum, so that the probability of surviving beyond time T is described by a Poisson process $e^{-\delta T}$, where δ is the annual risk of catastrophe eliminating society, here 0.1%. So the probability of surviving beyond, say, 2106 is $e^{-0.001*100}$, which is 90.5%. The Appendix to Chapter 2 discusses the implications of this choice in more detail.

Where population varies exogenously over time, we would automatically weight by population. In the case of just one region (i.e. the world), this means that we integrate global utility weighted by global population over time:

$$W = \int_{t=1}^{\infty} N(t)U(t)e^{-\delta t}dt \tag{4}$$

where N is global population. Where global income data can be disaggregated, regional utility should be evaluated for consistency using similar utility functions to that used in (4).[40] For endogenous population growth, some difficult ethical issues are involved and we cannot automatically apply this criterion (see Chapter 2 and appendix).

In the PAGE2002 modelling horizon – 2001 to 2200 – we can calculate total discounted utility as the sum of discounted utility in each individual year:

$$W = \sum_{t=1}^{2200} U(t)e^{-\delta t} \tag{5}$$

We approximate utility from 2200 to infinity based on an assumed, arbitrary rate of per-capita consumption growth g, which is achieved by all paths, as well as assessing constant population. We use 1.3% per annum, which is the annual average projection from 2001 to 2200 in PAGE2002's baseline world without climate change. In other words, as a simplification, in each run the world instantaneously overcomes the problems of climate change in the year 2200 (zero damages and zero adaptation) and all runs grow at an arbitrary 1.3% into the far-off future. In this sense there is an underestimate of the costs of climate change.

Again, a special case arises where the elasticity of the marginal utility of consumption is 1:

$$W = \sum_{t=1}^{2200} N(t)\ln C(t)e^{-\delta t} + \left(\frac{N_T \ln C_T}{\delta} + \frac{N_T g}{\delta^2} \right)e^{-\delta T} \tag{6}$$

[38] See Pearce and Ulph (1999).
[39] Pearce and Ulph (1999) and Stern (1977).
[40] Nordhaus and Boyer (2000).

Expected utility is given by the mean of total discounted utility from 2001 to infinity along all 1000 runs.

Finally, we can find the balanced growth equivalent (BGE) of the discounted consumption path described in 6. This is the current level of consumption per capita (i.e. in 2001), which, growing at a constant rate g set again to 1.3% per annum, delivers the same amount of utility as in (6) for the case of $\eta = 1$.

$$W = \sum_{t=1}^{2200} N(t)\left[\frac{C_{BGE}^{1-\eta}}{1-\eta} + gt\right]e^{-\delta t} + \left(\frac{N(t)\left(\frac{(C_{BGE} + 200\,g)^{1-\eta}}{1-\eta}\right)}{\delta - g(1-\eta)}\right)e^{-\delta T} \tag{7}$$

Taking the difference between the BGE of a single consumption path with climate damage and a consumption path without it gives the costs of climate change, measured in terms of a permanent loss of consumption, now and forever. One can think of the costs measured in this way as like a tax levied on consumption now and forever, the proceeds of which are simply poured away.

We have to go beyond the simple BGE generated in this way to take account of uncertainty. Thus the BGEs calculated here calibrate the expected utility in a particular scenario (with many possible paths) in terms of the definite or certain consumption that, if it grew at a constant rate, would generate the same expected utility. One can, therefore, think of the BGE measure of climate-change costs not as a tax but as the maximum insurance premium society would be prepared to pay, on a permanent basis, to avoid the risk of climate change (if society shared the policy-maker's ethical judgements). In practice, as we shall see, society will not in fact have to pay as much as this. Thus the BGE here combines the growth idea of Mirrlees and Stern[41] with the certainty equivalence ideas in, say, Rothschild and Stiglitz[42]. The next step, if intra-temporal income distribution is taken into account explicitly, would be to combine it with the 'equally distributed equivalent' income of Atkinson[43]. Box 6.3 outlines our calculations in more detail.

The welfare costs of BAU climate change are very high. Climate change is projected to reduce average global welfare by an amount equivalent to a permanent cut in per-capita consumption of a minimum of 5%.

Table 6.1 presents results in terms of Balanced Growth Equivalents (BGEs), based on defensible values for the utility discount rate (0.1% per annum) and for the elasticity of the marginal utility of consumption (1.0) (see Chapter 2 and its

[41] Mirrlees Stern (1972)
[42] Rothschild and Stiglitz (1970)
[43] Atkinson (1970)

Table 6.1 Losses in current per-capita consumption from six scenarios of climate change and economic impacts*.

Scenario		Balanced growth equivalents: % loss in current consumption due to climate change		
Climate	Economic	Mean	5th percentile	95th percentile
Baseline climate	Market impacts	2.1	0.3	5.9
	Market impacts + risk of catastrophe	5.0	0.6	12.3
	Market impacts + risk of catastrophe + non-market impacts	10.9	2.2	27.4
High climate	Market impacts	2.5	0.3	7.5
	Market impacts + risk of catastrophe	6.9	0.9	16.5
	Market impacts + risk of catastrophe + non-market impacts	14.4	2.7	32.6

*Utility discount rate = 0.1% per annum; elasticity of marginal utility of consumption = 1.0.

Appendix for an explanation and justification). For each of our six scenarios of climate change and economic impacts, we calculate three BGEs:

- For mean total discounted utility;
- For total discounted utility along the 5th percentile run;
- For total discounted utility along the 95th percentile run.

Table 6.1 shows the results. In each case, we quote the difference between the BGEs with and without climate change – the cost of climate change – in percentage terms. These are our headline results from the modelling. The numbers express the cost of 'business as usual' (BAU) climate change over the next two centuries in terms of present per-capita consumption for each scenario as a whole and for specific paths with impacts at the low and high end of the underlying probability distributions.

The results under the different scenarios range greatly, but virtually all project that BAU climate change will have very significant costs. In our lower-bound scenario, comprising the baseline climate scenario and including both market impacts and the risk of catastrophe, the BGE of the mean outcome is 5% below the equivalent BGE without climate change, meaning that the expected welfare cost of BAU climate change between 2001 and 2200 is equivalent to a 5% loss in per-capita consumption, now and forever. The BGE of the 95th percentile run amounts to a 12.3% loss in consumption now and forever, while the BGE of the 5th percentile run amounts to a 0.6% loss.

Climate change will reduce welfare even more if non-market impacts are included, if the climatic response to rising GHG emissions takes account of feedbacks, and if regional costs are weighted using value judgements consistent with those for risk and time. Putting these three factors together would probably

increase the cost of climate change to the equivalent of a 20% cut in per-capita consumption, now and forever.

- Adding the possibility of the feedbacks involved in the high-climate scenario reduces the BGE of mean total discounted utility to 6.9% below the equivalent BGE without climate change. The BGE of the 95th percentile run is 16.5% below, while the BGE of the 5th percentile run is just 0.9% below.

- In the high-climate scenario and with all three categories of economic impact (that is, adding the non-market impact), the BGE of the mean outcome is reduced to 14.4% below the equivalent BGE without climate change. The BGE of the 95th percentile run is 32.6% below, while the BGE of the 5th percentile run is 2.7% below. If the possibility of still higher climate sensitivities is taken into account, the incremental cost might be higher still.

- Calculating the BGE cost of climate change after including value judgements for regional distribution is beyond the scope of this Review, given our limited time. But if we take as an indication of how much estimates might increase the results of Nordhaus and Boyer[44], then estimates might be one quarter higher. In addition, because their deterministic approach could not take into account the valuation of risk, there is good reason to believe that the weighting would in our model increase estimates still further (see the Appendix to Chapter 2). In total, the global cost of climate change would probably be equivalent to around a 20% reduction in the BGE compared with a world without climate change.

Finally, we should discuss where one might place the evaluation of the losses from climate change between the 5 and 20% figures. There are two types of issue. The first is the inclusion of relevant effects and the second is the presence of different possible probability distributions.

On the first, it is reasonable to include what we consider to be relevant effects. This means catastrophic events, non-market effects and distribution of impacts within a generation. We have calculated the first two of these. However, we have conceptual, ethical and practical reservations about how non-market impacts should be included, although there is no doubt they are important. We have yet to calculate the distributional effects – that is for further work – but, based on previous studies, we can hazard a guess.

The second type of issue concerns the fact that we are unsure of which probability distribution to use. This takes us back to the distinction between risk and uncertainty discussed in Chapter 2 and the Appendix. We argued there that we now have some theory to guide us. Essentially, it points to taking a weighted average of the best and worst expected utility.

The first type of issue would take the evaluation towards an overall loss in the region of 13–15% (using the 10.9% figure of Table 6.1 and scaling up by one-quarter or more for distribution). The second type of issue would lead to taking a weighted average somewhere between this figure (13 or 14%) and 20%. The weights would depend on crude judgements about likelihoods of different kinds of probability distributions, on judgements about the severity of losses in this context, and on the basic degree of cautiousness on the part of the policy-maker.

---

[44] Nordhaus and Boyer (2000)

Together, they would make up the 'aversion to ambiguity' discussed in Chapter 2 and the Appendix.

This discussion points to areas for further work in the context of this particular model: distribution within a generation and explaining different distributional judgements. Of course, there is much more to do in terms of considering different economic models – we have investigated just one – and exploring different probability distributions.

6.5   Conclusion

This Chapter has presented global cost estimates of the losses from 'business as usual' climate change. They have been expressed in terms of their equivalent permanent percentage loss in consumption. They are averages over time and risk and can be compared with percentage costs, similarly averaged over time, of mitigation – that is the subject of Part III of this Review. In the final chapter of that part, we include a discussion of how much of the losses estimated in this Chapter could be saved by mitigation. The loss estimates of this Chapter should be viewed as complementary to the discussions of the scale of the separate impacts on consumption, health, and environment that were presented in Chapters 3 to 5.

What have we learned from this exercise? Notwithstanding the limitations inherent in formal integrated models, there can be no doubt that the economic risks of a 'business as usual' approach are very severe – and probably more severe than suggested by past models. Relying on the scientific knowledge that informed the IPCC's TAR, the cost of BAU climate change over the next two centuries is equivalent to a loss of at least 5% of global per-capita consumption, now and forever. More worrying still, when the model incorporates non-market impacts and more recent scientific findings on natural feedbacks, this total average cost is pushed to 14.4%.

Cost estimates would increase still further if the model incorporated other important omitted effects. First, the welfare calculations fail to take into account distributional impacts, even though these impacts are potentially very important: poorer countries are likely to suffer the largest impacts. Second, there may be greater risks to the climate from dynamic feedbacks and from heightened climate sensitivity beyond those included here. If these are included, the total cost would be likely to be around 20% of current per-capita consumption, now and forever.

Further, there are potentially worrying 'social contingent' impacts such as migration and conflict, which have not been quantified explicitly here. If the world's physical geography is changed, so too will be its human geography.

Finally, we must close with the warning about over-literal interpretation of these results with which we began this chapter. The estimates have arisen from an attempt to add two things to the previous literature on IAM models. The first is use of recent scientific estimates of probabilities and the second is putting these probabilities to work using the economics of risk and uncertainty. The most worrying possible impacts are also among the most uncertain, given that so little is known about the risks of very high temperatures and potential dynamic instability. The exercise allows us to see what the implications of the risks, as we currently understand them, might be. The answer is that they would imply very large estimates of potential losses from climate change. They give an indication of the stakes involved in making policy on climate change. The analysis of this Chapter shows the inevitable difficulties of all these models in extrapolating over very long periods of time. We therefore urge the reader to avoid an over-literal interpretation of these results. Nevertheless, we think that they illustrate a very important point: the risks involved in a 'business as usual' approach to climate change are very large.

References

Successive IPCC assessments of the IAM literature can be found in Pearce *et al.* (1996) and Smith *et al.* (2001). Hitz and Smith (2004) provide a more recent summary, focussing on the nature of the relationship between rising temperatures and the cost of climate change. William Cline's 1992 book *The Economics of Global Warming* and William Nordhaus and Joseph Boyer's 2000 book *Warming the World* provide an important and well-structured discussion of the issues, while Hope (2005) explains Integrated Assessment Modelling in detail. Watkiss *et al.* (2005) is a valuable discussion of the uncertainties around estimating the monetary cost of climate change, while Warren et al. (2006) subject the damage functions in IAMs to critical scrutiny.

Atkinson, A. B. (1970): 'On the measurement of inequality', Journal of Economic Theory, 2(3): 244–263

Cline, W.R. (1992): 'The Economics of Global Warming'. Washington, DC: Institute for International Economics.

Fankhauser, S. and R.S.J. Tol (2003): 'On climate change and economic growth', Resource and Energy Economics 27: 1–17

Friedlingstein, P., P. Cox, R. Betts et al. (2006): 'Climate-carbon cycle feedback analysis: results from C4MIP model intercomparison', Journal of Climate, 19: 3337–3353

Gedney, N., P.M. Cox and C. Huntingford (2004): 'Climate feedback from wetland methane emissions', Geophysical Research Letters 31(20): L20503.

Hallegatte, S. and J.-C. Hourcade (2006): 'Why economic dynamics matters in the assessment of climate change damages: illustration on extreme events', Ecological Economics (forthcoming).

Hitz, S. and J.B. Smith (2004): 'Estimating global impacts from climate change', The Benefits of Climate Change Policies, J. -C. Morlot and S. Agrawala. Paris: OECD, pp 31–82.

Hope, C. (2003): 'The marginal impacts of CO2, CH4 and SF6 emissions,' Judge Institute of Management Research Paper No.2003/10, Cambridge, UK, University of Cambridge, Judge Institute of Management.

Hope, C. (2005): 'Integrated assessment models' in Helm, D. (ed.), Climate-change policy, Oxford: Oxford University Press, pp 77–98.

Intergovernmental Panel on Climate Change (2001): Climate Change 2001: The Scientific Basis. Contribution of Working Group I to the Third Assessment Report of the Intergovernmental Panel on Climate Change [Houghton JT, Ding Y, Griggs DJ, et al. (Eds.)], Cambridge: Cambridge University Press.

Jorgenson, D.W., R.J. Goettle, B.H. Hurd et al. (2004): 'US market consequences of global climate change', Washington, DC: Pew Center on Global Climate Change.

Mendelsohn, R.O., W.N. Morrison, M.E. Schlesinger and N.G. Andronova (1998): 'Country-specific market impacts of climate change', Climatic Change 45(3–4): 553–569

Mirrlees, J.A. and N.H. Stern (1972): 'Fairly good plans', Journal of Economic Theory 4(2): 268–288

Murphy J.M. et al. (2004): 'Quantification of modelling uncertainties in a large ensemble of climate change simulations', Nature 430: 768–772

Nordhaus, W.D. and J.G. Boyer (2000): 'Warming the World: the Economics of the Greenhouse Effect', Cambridge, MA: MIT Press.

Pearce, D.W. et al. (1996): 'The social costs of climate change: greenhouse damage and the benefits of control' Climate Change 1995: Economic and Social Dimensions of Climate Change, Intergovernmental Panel on Climate Change. Cambridge: Cambridge University Press pp 183–224

Pearce, D.W. and A. Ulph (1999): 'A social discount rate for the United Kingdom' Economics and the Environment: Essays in Ecological Economics and Sustainable Development, D.W. Pearce, Cheltenham: Edward Elgar.

Rothschild, M.D. and J.E. Stiglitz (1970): 'Increasing Risk: I. A Definition', Journal of Economic Theory, September 2, pp255–243.

Smith, J.B. et al. (2001): 'Vulnerability to climate change and reasons for concern: a synthesis' Climate Change 2001: Impacts, Adaptation and Vulnerability, Intergovernmental Panel on Climate Change. Cambridge: Cambridge University Press pp 913–967.

Stern, N. (1977): 'The marginal valuation of income', in Artis, M. and R. Nobay (eds.), Studies in Modern Economic Analysis, Oxford: Blackwell.

Tol, R.S.J. (2002): 'Estimates of the damage costs of climate change – part II: dynamic estimates', Environmental and Resource Economics 21: 135–160

Tol, R.S.J. (2005): 'The marginal damage costs of carbon dioxide emissions: an assessment of the uncertainties', Energy Policy 33(16): 2064–2074

Warren, R. et al. (2006): 'Spotlighting Impacts Functions in Integrated Assessment Models', Norwich, Tyndall Centre for Climate Change Research Working Paper 91.

Watkiss, P. et al. (2005): 'Methodological Approaches for Using Social Cost of Carbon Estimates in Policy Assessment, Final Report', Culham: AEA Technology Environment.

III The Economics of Stabilisation

Part III of the Review considers the economic challenges of achieving stabilisation of greenhouse gases in the atmosphere.

'Business as usual' emissions will take greenhouse gas concentrations and global temperatures way beyond the range of human experience. In the absence of action, the stock of greenhouse gases in the atmosphere could more than treble by the end of the century.

Stabilisation of concentrations will require deep emissions cuts of at least 25% by 2050, and ultimately to less than one-fifth of today's levels. The costs of achieving this will depend on a number of factors, particularly progress in bringing down the costs of technologies. Overall costs, with sensible policies, are estimated at around 1% of GDP for stabilisation levels between 500–550 ppm CO_2e.

The costs will not be evenly felt – some carbon-intensive sectors will suffer, while for others, climate change policy will create opportunities. Climate change policies may also have wider benefits where they can be designed in a way that also meets other goals.

Comparing the costs and benefits of action clearly shows that the benefits of strong, early action on climate change outweigh the costs. The current evidence suggests aiming for stabilisation somewhere within the range 450–550 ppm CO_2e. Ignoring climate change will eventually damage economic growth; tackling climate change is the pro-growth strategy.

Part III is structured as follows:

- **Chapter 7** discusses the past drivers of global emissions growth, and how these are likely to evolve in the future.
- **Chapter 8** explains what needs to happen to emissions in order to stabilise greenhouse-gas concentrations in the atmosphere, and the range of trajectories available to achieve this.
- **Chapter 9** discusses how to identify the costs of mitigation, and looks at a resource-based approach to calculating global costs.
- **Chapter 10** compares modelling approaches to calculating costs, and looks at how policy choices may influence cost.
- **Chapter 11** considers how climate change policies may affect competitiveness if they are not applied evenly worldwide.
- **Chapter 12** looks at how to take advantage of the opportunities and wider benefits arising from action on climate change.
- **Chapter 13** brings together the analysis of costs and benefits, and looks at how a global long-term goal for climate change policy can be chosen.

7 Projecting the Growth of Greenhouse-Gas Emissions

KEY MESSAGES

Greenhouse-gas concentrations in the atmosphere now stand at around 430 ppm CO_2 equivalent, compared with only 280 ppm before the Industrial Revolution. The stock is rising, driven by increasing emissions from human activities, including energy generation and land-use change.

Emissions have been driven by economic development. CO_2 emissions per head have been strongly correlated with GDP per head across time and countries. North America and Europe have produced around 70% of CO_2 emissions from energy production since 1850, while developing countries – non-Annex 1 parties under the Kyoto Protocol – account for less than one quarter of cumulative emissions.

Annual emissions are still rising. Emissions of carbon dioxide, which accounts for the largest share of greenhouse gases, grew at an average annual rate of around $2\frac{1}{2}$% between 1950 and 2000. In 2000, emissions of all greenhouse gases were around 42GtCO₂e, increasing concentrations at a rate of about 2.7 ppm CO_2e per year.

Without action to combat climate change, atmospheric concentrations of greenhouse gases will continue to rise. In a plausible 'business as usual' scenario, they will reach 550 ppm CO_2e by 2035, then increasing at $4\frac{1}{2}$ppm per year and still accelerating.

Most future emissions growth will come from today's developing countries, because of more rapid population and GDP growth than developed countries, and an increasing share of energy-intensive industries. The non-Annex 1 parties are likely to account for over three quarters of the increase in energy-related CO_2 emissions between 2004 and 2030, according to the International Energy Agency, with China alone accounting for over one third of the increase.

Total emissions are likely to increase more rapidly than emissions per head, as global population growth is likely to remain positive at least to 2050.

The relationship between economic growth and development and CO_2 emissions growth is not immutable. There are examples where changes in energy technologies, the structure of economies and the pattern of demand have reduced the responsiveness of emissions to income growth, particularly in the richest countries. Strong, deliberate policy choices will be needed, however, to decarbonise both developed and developing countries on the scale required for climate stabilisation.

Increasing scarcity of fossil fuels alone will not stop emissions growth in time. The stocks of hydrocarbons that are profitable to extract (under current policies) are more than enough to take the world to levels of CO_2 concentrations well beyond 750 ppm, with very dangerous consequences for climate-change impacts. Indeed, with business as usual, energy users are likely to switch towards more carbon-intensive coal, oil shales and synfuels, tending to *increase* rates of emissions growth. It is important to redirect energy-sector research, development and investment away from these sources towards low-carbon technologies.

Extensive carbon capture and storage would allow some continued use of fossil fuels, and help guard against the risk of fossil fuel prices falling in response to global climate-change policy, undermining its effectiveness.

7.1   Introduction

Part II showed that continuing climate change will produce harmful and ultimately dangerous impacts on the environment, the global economy and society. This chapter shows that, in the absence of deliberate policy to combat climate change, global greenhouse-gas emissions will continue to increase at a rapid rate.

Even if annual greenhouse-gas (GHG) emissions remained at the current level of 42 $GtCO_2$ equivalent[1] each year[2], the world would experience major climate change. That rate of emissions would be sufficient to take greenhouse-gas concentrations to over 650 ppm CO_2 equivalent (CO_2e) by the end of this century, likely to result eventually in a rise in the global mean temperature of at least 3°C from its pre-industrial level[3].

But annual emissions are not standing still – they are rising, at a rapid rate. If they continue to do so, then the outlook is even worse.

This chapter reviews some of the projections of emissions growth in Section 7.2, noting that, despite the uncertainties about the precise pace of increases, there is powerful evidence, robust to plausible variations in the detail of forecasts, that

---

[1] Greenhouse gases are converted to a common unit, CO_2 equivalent, which measures the amount of carbon dioxide that would produce the same global warming potential (GWP) over a given period as the total amount of the greenhouse gas in question. In 2000, 77% of the 100-year GWP of *new emissions* was from CO_2. See Table 8.1 for conversion factors for different gases. Figures for the *stock* of greenhouse gases are usually reported in terms of the amount of CO_2 that would have the equivalent effect on current radiative forcing, i.e. they focus on the GWP over one year.

[2] GHG emissions in 2000 were 42 $GtCO_2e$, WRI (2006). This does not include some emissions for which data are unavailable. For example: CO_2 emissions from soil; additional global warming effect of aviation, including the uncertain contrail effect (see Box 15.6); CFCs (for example from refrigerants in developing countries); and aerosols (for example, from the burning of biomass).

[3] Chapter 8 examines the relationship between stabilisation levels, temperatures and emissions trajectories.

with 'business as usual' emissions will reach levels at which the impacts of climate change are likely to be very dangerous. Sections 7.3 to 7.5 then look behind the head-line projections to consider the main drivers of energy-related emissions growth: economic growth, technological choices affecting carbon intensity of energy use and energy intensity of output, and population growth. This is helpful not only in understanding what underlies the projections but also in identifying the channels through which climate-change policy can work. Finally, in Section 7.6, the chapter argues that fossil fuels' increasing scarcity is not going to rein in emissions growth by itself. To the contrary, there will be a problem for climate-change policies if they induce significant falls in fossil-fuel prices. That is one reason why carbon capture and storage technology is so important.

7.2   Past greenhouse-gas emissions and current trends

57% of emissions are from burning fossil fuels in power, transport, buildings and industry; agriculture and changes in land use (particularly deforestation) produce 41% of emissions.

Total greenhouse-gas emissions were 42 $GtCO_2e$[4] in 2000[5], of which 77% were CO_2, 14% methane, 8% nitrous oxide and 1% so-called F-gases such as perfluorocarbon and sulphur hexafluoride. Sources of greenhouse-gas emissions comprise:

- Fossil-fuel combustion for energy purposes in the power, transport, buildings and industry sectors amounted to 26.1 $GtCO_2$ in 2004[6]. Combustion of coal, oil and gas in electricity and heat plants accounted for most of these emissions, followed by transport (of which three quarters is road transport), manufacturing and construction and buildings.
- Land-use change such as deforestation releases stores of CO_2 into the atmosphere.
- Methane, nitrous oxide and F-gases are produced by agriculture, waste and industrial processes. Industrial processes such as the production of cement and chemicals involve a chemical reaction that releases CO_2 and non-CO_2 emissions. Also, the process of extracting fossil fuels and making them ready for use generates CO_2 and non-CO_2 emissions (so-called fugitive emissions).

The shares are summarised in Figure 7.1 below, and emissions sources are analysed further by sector in Box 7.1 and Annexes 7.B to 7.G[7].

---

[4] WRI (2006).
[5] WRI (2006). Historical emission figures are drawn from the WRI's Climate Analysis Indicators Database (CAIT) http://cait.wri.org. Emission estimates exclude: CO_2 emissions from soil; additional global warming effect of aviation, including the uncertain cirrus cloud effect (see Box 15.6); CFCs (for example from refrigerants in developing countries); and aerosols (for example, from the burning of biomass).
[6] IEA (in press).
[7] For Annexes 7B to 7G, see www.sternreview.org.uk

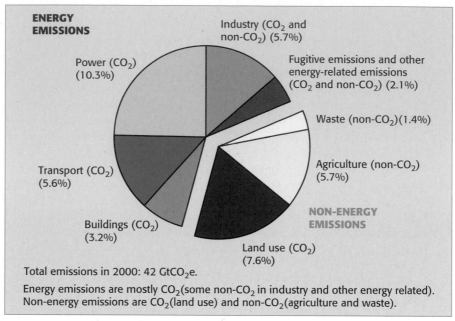

Figure 7.1 GHG emissions in 2000, by source[8]

Source: WRI (2006)

BOX 7.1 Current and projected emissions sources by sector

Power

A quarter of all global greenhouse-gas emissions come from the generation of power and heat, which is mostly used in domestic and commercial buildings, and by industry. This was the fastest growing source of emissions worldwide between 1990 and 2002, growing at a rate of 2.2% per year; developing-country emissions grew most rapidly, with emissions from Asia (including China and India), the Middle East and the transition economies doubling between 1990 and 2000.

This sector also includes emissions arising from petroleum refineries, gas works and coal mines in the transformation of fossil fuel into a form that can be used in transport, industry and buildings. Emissions from this source are likely to increase over four-fold between now and 2050 because of increased synfuel production from gas and coal, according to the IEA. Total power-sector emissions are likely to rise more than three-fold over this period. For more detail on power emissions, see Annex 7.B.

Land use

Changes in land use account for 18% of global emissions. This is driven almost entirely by emissions from deforestation. Deforestation is highly concentrated in a few countries. Currently around 30% of land-use emissions are from Indonesia and a further 20% from Brazil.

Land-use emissions are projected to fall by 2050, because it is assumed that countries stop deforestation after 85% of forest has been cleared. For more detail, see Annex 7.F.

---

[8] Emissions are presented according to the sector from which they are directly emitted, i.e. emissions are by source, as opposed to end user/activity; the difference between these classifications is discussed below.

Agriculture

Non-CO_2 emissions from agriculture amount to 14% of total GHG emissions. Of this, fertiliser use and livestock each account for one third of emissions; other sources include rice and manure management. Over half of these emissions are from developing countries. Agricultural practices such as the manner of tillage are also responsible for releasing stores of CO_2 from the soil, although there are no global estimates of this effect. Agriculture is also indirectly responsible for emissions from land-use change (agriculture is a key driver of deforestation), industry (in the production of fertiliser), and transport (in the movement of goods). Increasing demand for agricultural products, due to rising population and incomes per head, is expected to lead to continued rises in emissions from this source. For more detail on trends in agriculture emissions, see Annex 7.G.

Total non-CO_2 emissions are expected to double in the period to 2050[9].

Transport

Transport accounts for 14% of global greenhouse-gas emissions, making it the third largest source of emissions jointly with agriculture and industry. Three-quarters of these emissions are from road transport, while aviation accounts for around one eighth and rail and shipping make up the remainder. The efficiency of transport varies widely between countries, with average efficiency in the USA being around two thirds that in Europe and half that in Japan[10]. Total CO_2 emissions from transport are expected to more than double in the period to 2050, making it the second-fastest growing sector after power.

CO_2 emissions from aviation are expected to grow by over three-fold in the period to 2050, making it among the fastest growing sectors. After taking account of the additional global warming effects of aviation emissions (discussed in Box 15.8), aviation is expected to account for 5% of the total warming effect (radiative forcing) in 2050[11]. For more detail on trends in transport emissions, see annex 7.C.

Industry

Industry accounts for 14% of total direct emissions of GHG (of which 10% are CO_2 emissions from combustion of fossil fuels in manufacturing and construction and 3% are CO_2 and non-CO_2 emissions from industrial processes such as production of cement and chemicals).

Buildings

A further 8% of emissions are accounted for by direct combustion of fossil fuels and biomass in commercial and residential buildings, mostly for heating and cooking.

The contribution of the buildings and industry sectors to climate change are greater than these figures suggest, because they are also consumers of the electricity and heat produced by the power sector (as shown in Figure A below). Direct emissions from both industry and buildings are both expected to increase by around two thirds between 2000 and 2050 under BAU conditions. For more detail on industry and buildings emissions, see Annex 7.D and 7.E respectively.

---

[9] There are no projections available splitting non-CO_2 emission estimates into individual sector sources after 2020.

[10] An and Sauer (2004).

[11] For explanation of how these percentages are calculated, see Box 15.6. The transport emissions presented in Figure A and B include CO_2 emissions from aviation, but exclude the additional global warming effect of these emissions at altitude because there is no internationally agreed consensus on how to include these effects.

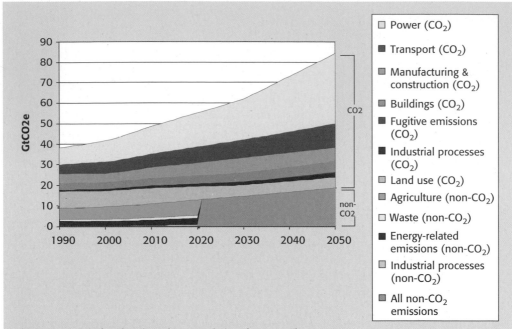

Figure A Historical and projected GHG emissions by sector (by source)

Source: WRI (2006), IEA (in press), IEA (2006), EPA (forthcoming), Houghton (2005).

GHG emissions can also be classified according to the activity associated with them. Figure B below shows the relationship between the physical source of emissions and the end use/activity associated with their production. For example, at the left-hand side of the diagram it can be seen that electricity generation leads to production of emissions at the coal, gas or oil plant; the electricity produced is then consumed by residential and commercial buildings and in a range of industries such as chemicals and aluminium.

This analysis is useful for building a detailed understanding of the drivers behind emissions growth and how emissions can be cut. For example, emissions from the power sector can be cut either by improving the efficiency and technology of the power plant, or by reducing the end-use demand for electricity.

Data sources for historical and projected GHG emissions used in this box and throughout the report:

- Historical data on all GHG emissions (1990–2002) from WRI (2006)[12].

- Fossil-fuel emissions projections (i.e. power, transport, buildings and industry CO_2 emissions) from IEA. Data for 2030 taken from IEA (in press) and data for 2050 from IEA (2006). Intermediate years calculated by extrapolation.

- Land-use emission projections were taken from Houghton (2005).

- Non-CO_2 emission projections to 2020 from EPA (forthcoming). Figures extrapolated to 2050 using IPCC SRES scenarios A1F1 and A2.

- CO_2 industrial-process and CO_2 fugitive emissions projections extrapolated at 1.8% pa (the growth rate in fossil fuel emissions anticipated by the IEA).

---

[12] Note that the estimates of energy-related CO_2 emissions in the early 1990s include approximate estimates of emissions from transition economies, which are sometimes excluded from data tables from the WRI (2006).

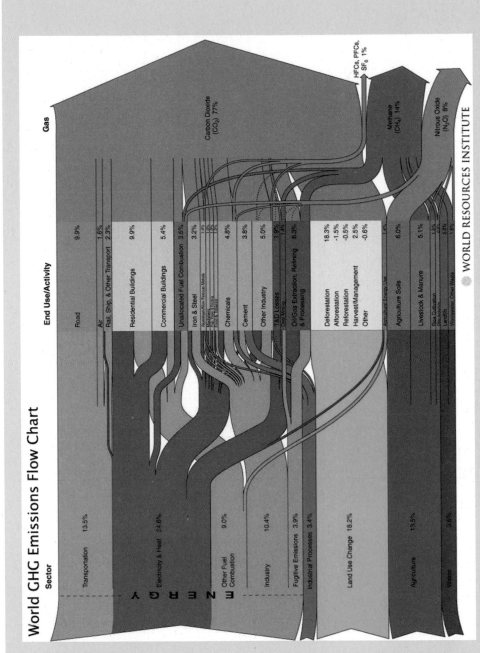

Figure B World Resources Institute mapping from sectors to greenhouse-gas emissions

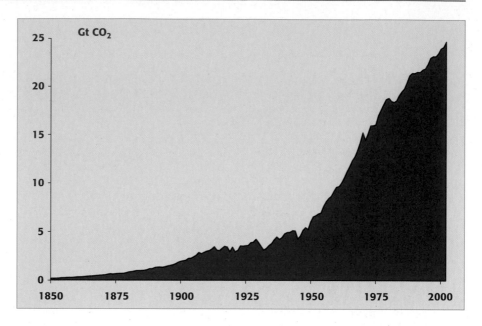

Figure 7.2 Global CO_2 emissions from fossil-fuel burning and cement over the long term

Source: Climate Analysis Indicators Tool (CAIT) Version 3.0. (Washington, DC: World Resources Institute, 2006)

Annual global greenhouse-gas emissions have been growing.

Figure 7.2 illustrates the long-run trend of energy-related CO_2 emissions[13], for which reasonable historical data exist. Between 1950 and 2002, emissions rose at an average annual rate of over 3%. Emissions from burning fossil fuels for the power and transport sectors have been increasing since the mid-nineteenth century, with a substantial acceleration in the 1950s.

The rate fell back somewhat in the three decades after 1970, but was still 1.7% on average between 1971 and 2002 (compared with an average rate of increase in energy demand of 2.0% per year). The slowdown appears to have been associated with the temporary real increases in the price of oil in the 1970s and 1980s, the sharp reduction in emissions in Eastern Europe and the former Soviet Union due to the abrupt changes in economic systems in the 1990s, and increases in energy efficiency in China following economic reforms.

The majority of emissions have come from rich countries in the past. North America and Europe have produced around 70% of the CO_2 from energy production since 1850, while developing countries – non-Annex 1 parties under the Kyoto Protocol – account for less than one quarter of cumulative emissions.

Less is known about historical trends in emissions from agriculture and changes in land use, but emissions due to land-use changes and deforestation are thought to have risen on average by around 1.5% annually between 1950 and 2000, according to the World Resources Institute.

---

[13] Including emissions from international aviation and shipping and CO_2 emissions from the industrial process of making cement.

In total, between 1990 and 2000 (the period for which comprehensive data are available), the average annual rate of growth of non-CO_2 greenhouse gases, in CO_2-equivalent terms, was 0.5% and of all GHGs together 1.2%.

Global emissions are projected to continue to rise in the absence of climate-change policies; 'business as usual' will entail continuing increases in global temperatures well beyond levels previously experienced by humankind.

Some simple arithmetic can illustrate this. The concentration of greenhouse gases in the atmosphere is currently at around 430 ppm CO_2e, adding 2–3 ppm a year. Emissions are rising. But suppose they continue to add to GHG concentrations by only 3 ppm a year. That will be sufficient to take the world to 550 ppm in 40 years and well over 700 ppm by the end of the century. Yet a stable global climate requires that the stock of greenhouse gases is constant and therefore that emissions are brought down to the level that the Earth system can naturally absorb from the atmosphere annually in the long run.

Formal projections suggest that the situation in the absence of climate-change policies is worse than in this simple example. The reference scenario[14] in the International Energy Agency (IEA)'s 2006 World Energy Outlook projects an increase of over 50% in annual global fossil fuel CO_2 emissions between 2004 and 2030, from 26 $GtCO_2$ to 40 $GtCO_2$, an annual average rate of increase of 1.7%. The reference scenario for the IEA's Energy Technology Perspectives envisages emissions of 58 $GtCO_2$ by 2050.

Developing countries will account for over three-quarters of the increase in fossil-fuel emissions to 2030, according to the World Energy Outlook, thanks to rapid economic growth rates and their growing share of many energy-intensive industries. China may account for over one third of the increase by itself, with Chinese emissions likely to overtake those of the United States by the end of this decade, driven partly by heavy use of coal.

The fastest growing sectors are driven by growth in demand for transport. The second fastest source of emissions is expected to be aviation, expected to rise about three-fold over the same period. Fugitive emissions are expected to increase over four-fold in the period to 2050, because of an increase in production of syn-fuels from gas and coal, mostly for use in the transport sector.

Other 'business as usual' (BAU) projections show similar patterns. The US Energy Information Administration is currently projecting an increase from 25 $GtCO_2$ in 2003 to 43.7 $GtCO_2$ by 2030, at an annual average rate of increase of 2.1%[15], as does the POLES model[16]. The factors responsible for the rise in energy-related missions are considered further in the sections below.

Projections of future emissions from land-use changes remain uncertain. At the current rate of deforestation, most of the top ten deforesting nations would clear their forests before 2100. Based on rates of deforestation over the past two decades, and assuming that countries stop deforestation when 85% of the forests they had in 2000 have been cut down, annual emissions will remain at around 7.5 $GtCO_2$/yr until 2012, falling to 5 $GtCO_2$/yr by 2050 and 2 $GtCO_2$/yr by 2100[17].

---

[14] The reference scenario assumes no major changes to existing policies.
[15] Different modellers may use slightly different definitions of emissions, depending on their treatment of international marine and aviation fuel bunkers and gas flaring.
[16] According to WRI (2006).
[17] Houghton (2005)

The US Environmental Protection Agency (EPA) projects an increase in agricultural emissions from 5.7 to 7.3 $GtCO_2e$ between 2000 and 2020 with business as usual. The key drivers behind agricultural emissions growth are population and income growth. While the share of emissions from the OECD and transition economies is expected to fall, the share from developing countries is expected to increase, especially in Africa and Latin America. The income elasticity of demand for meat is often high in developing countries, which will tend to raise emissions from livestock. Increases in emissions from other sources, including waste and industrial processes, are also expected.

Looking at emissions from all sources together, the IPCC Special Report on Emissions Scenarios, published in 2000, considered a wide range of possible future scenarios. Although they differ considerably, all entail substantial increases in emissions for at least the next 25 years and increases in greenhouse-gas concentrations at least until the end of the century. All but one SRES storyline envisage a concentration level well in excess of 650 ppm CO_2e by then. Academic studies also envisage steady increases. The MIT EPPA model reference projection, for example, envisages an average annual increase in CO_2 emissions of 1.26% between 1997 and 2100 (faster in the earlier years). In the rest of this Review, for the purposes of illustrating the size of the emission abatement required to achieve various CO_2e concentration levels, a BAU trajectory based on IEA, EPA, IPCC and Houghton projections has been used[18]. This is broadly representative of BAU projections in the literature and results in emissions reaching 84 $GtCO_2e$ per year, and a greenhouse-gas level of around 630 ppm CO_2e, by 2050.

Despite the differences across the emissions scenarios in the literature and the unavoidable uncertainty in making long-run projections, any plausible BAU scenario entails continuing increases in global temperatures, well beyond levels previously experienced by humankind, with the profound physical, social and economic consequences described in Part II of the Review. If, for instance, the average annual increase in greenhouse-gas emissions is 1.5%[19], concentrations will reach 550 ppm CO_2e by around 2035, by when they will be increasing at 4½ ppm per year and still accelerating.

The rest of this chapter takes a more detailed look at the drivers that lie behind these headline projections.

7.3   The determinants of energy-related CO_2 emissions

The drivers of emissions growth can be broken down into different components.

The reasons why annual emissions are projected to increase under 'business as usual' can be better understood by focusing on energy-related CO_2 emissions from

---

[18] Fossil fuel projections to 2050 are taken from IEA (2006). Non-CO_2 emission projections to 2020 are taken from EPA (forthcoming) and extrapolated forward to 2050 in a manner to be consistent with non-CO_2 emissions reached by SRES scenarios A1F1 and A2. Land use emissions to 2050 are taken from Houghton (2005). Actual estimates of CO_2 emissions from industrial processes and CO_2 fugitive emissions were taken from CAIT until 2002; henceforth, they are extrapolated at 1.8% pa (the average growth rate for fossil fuel emissions projected by IEA).
[19] This assumes that total emissions of greenhouse gases grow more slowly than emissions of CO_2. Their annual growth rate was about 0.5 percentage points lower during 1990 to 2000.

the combustion of fossil fuel, which have been more thoroughly investigated than emissions from land use, agriculture and waste[20].

The so-called Kaya identity expresses total CO_2 emissions in terms of the components of an accounting identity: the level of output (which can be further split into population growth and GDP per head); the energy intensity of that output; and the carbon intensity of energy[21]:

$$CO_2 \text{ emissions from energy} = \text{Population} \times \text{(GDP per head)} \times \text{(energy use/GDP)} \times (CO_2 \text{ emissions/energy use)}$$

Trends in each of these components can then be considered in turn. In particular, it can immediately be seen that increases in world GDP will tend to increase global emissions, unless income growth stimulates an offsetting reduction in the carbon intensity of energy use or the energy intensity of GDP.

Table 7.1 abstracts from the impact of population size and focuses on emissions per head, which are equal to the product of income per head, carbon intensity of energy and energy intensity. These are reported for the world and various countries and groupings within it. The table illustrates the wide variation in emissions per head across countries and regions, and how this variation is driven primarily by variations in income per head and, to a lesser extent, by variations in energy intensity. It also illustrates the similarity in the carbon intensity of energy across countries and regions.

Table 7.1 Key ratios for energy-related[22] CO_2 emissions in 2002

Country/ grouping	CO_2 per head (tCO_2)	GDP per head ($ppp2000)	CO_2 emissions/ energy use (tCO_2/toe)	Energy use/GDP (toe/$ppp2000 $\times 10^6$)
USA	20.4	34430	2.52	230.8
EU	9.4	23577	2.30	158.0
UK	9.6	27176	2.39	140.6
Japan	9.8	26021	2.35	155.7
China	3.0	4379	3.08	219.1
India	1.1	2555	2.05	201.3
OECD	11.7	24351	2.41	193.0
Economies in transition	7.7	7123	2.57	421.2
Non-Annex 1 parties	2.2	3870	2.48	217.8
World	4.0	7649	2.43	219.5

Source: WRI (2006).

[20] Econometric studies of past data have tended to focus on energy-related CO_2 emissions, although modellers are increasingly including non-CO_2 GHGs in their projections. See, for example, Paltsev et al. (2005).
[21] Kaya (1990)
[22] Energy-related emissions include all fossil-fuel emissions plus CO_2 emissions from industrial processes.

Some of the factors determining these ratios change only very slowly over time. Geographers have drawn attention to the empirical importance of a country's endowments of fossil fuels and availability of renewable energy sources[23], which appear to affect both the carbon intensity of energy use and energy use itself. Qatar, a Gulf oil-producing state, for example, has the highest energy use per head and the highest CO_2 emissions per head[24]. China, which uses a greater proportion of coal in its energy mix than the EU, has a relatively high figure for carbon intensity. A country's typical winter climate and population density are also important influences on the energy intensity of GDP.

But some factors are subject to change. Economists have stressed, for example, the role of the prices of different types of energy, the pace and direction of technological progress, and the structure of production in different countries in influencing carbon intensity and energy intensity[25].

Falls in the carbon intensity of energy and energy intensity of output have slowed the growth in global emissions, but total emissions have still risen, because of income and population increases.

In Table 7.2, the Kaya identity is used to break down the total growth rates of energy-related CO_2 emissions for various countries and regions over the period 1992 to 2002 into the contributions – in an accounting sense – from population growth,

Table 7.2 Annual growth rates in energy-related[26] CO_2 emissions and their components, 1992–2002 (%)

Country/grouping	CO_2 emissions (GtCO_2)	GDP per head	Carbon intensity	Energy intensity	Population
USA	1.4	1.8	0.0	−1.5	1.2
EU	0.2	1.8	−0.7	−1.2	0.3
UK	−0.4	2.4	−1.0	−2.3	0.2
Japan	0.7	0.7	−0.5	0.2	0.3
China	3.7	8.5	0.5	−6.4	0.9
India	4.3	3.9	1.1	−2.5	1.7
OECD	1.2	1.8	−0.3	−1.1	0.7
Economies in transition	−3.0	0.4	−0.6	−2.7	−0.1
Non-Annex 1 parties	3.3	3.5	0.2	−2.0	1.6
World	1.4	1.9	−0.1	−1.7	1.4

Source: WRI (2006).

[23] E.g. Neumayer (2004)
[24] Generous endowments of raw materials are not necessarily reflected in domestic consumption (e.g. South Africa and diamonds), but in the case of energy there does seem to be a significant correlation, perhaps because of the broad-based demand for energy and the tendency for local energy prices to be relatively low in energy-rich countries.
[25] E.g. Huntington (2005) and McKibbin and Stegman (2005)
[26] Energy-related emissions include all fossil-fuel emissions plus CO_2 emissions from industrial processes.

changes in the carbon intensity of energy use, changes in the energy intensity of GDP, and growth of GDP per head. It shows that, in the recent past, income growth per head has tended to raise global emissions (by 1.9% per year) whereas reductions in global carbon and energy intensity have tended to reduce them (by the same amount). Because world population has grown (by 1.4% per year), emissions have gone up.

There has been a variety of experience across countries. The EU and the economies in transition were able to reduce carbon intensity considerably during the period, but there was a significant increase in India, from a very low base. Population growth, as well as increases in GDP per head, was particularly important in developing countries. The reductions in the energy intensity of output in China, India and the economies in transition are striking. If energy intensity had fallen in China only at the speed it fell in the OECD, global emissions in 2002 would have been over 10% higher. But Table 7.1 shows that, at least in China and India, energy intensity is now below that of the United States. Economic reforms helped to reduce wasteful use of energy in many countries in the 1990s, but many of the improvements are likely to have reflected catching up with best practice, boosting the level of energy efficiency but not necessarily bringing reductions in its long-run growth rate.

7.4   The role of growth in incomes and population in driving emissions

In the absence of policies to combat climate change, CO_2 emissions are likely to rise as the global economy grows.

Historically, economic development has been associated with increased energy consumption and hence energy-related CO_2 emissions per head. Across 163 countries, from 1960 to 1999, the correlation between CO_2 emissions per head and GDP per head (expressed as natural logarithms) was nearly 0.9[27]. Similarly, one study for the United States estimated that, over the long term, a 1% rise in GDP per head leads to a 0.9% increase in emissions per head, holding other explanatory factors constant[28].

Consistent with this, emissions per head are highest in developed countries and much lower in developing countries – although developing countries are likely to be closing the gap, because of their more rapid collective growth and their increasing share of more energy-intensive industries, as shown in the example of the projection in Figure 7.3[29].

Structural shifts in economies may change the relationship between income and emissions.

---

[27] See Neumayer, (2004)

[28] See Huntington (2005). GDP per head is itself a function of many other variables, and emissions projections should in principle be based upon explicit modelling of the sources of growth; for example, the consequences for emissions will be different if growth is driven by innovations in energy technology rather than capital accumulation.

[29] Holtsmark, B (2006). McKitrick and Strazicich (2005) have pointed out that global emissions per head have behaved as a stationary series subject to structural breaks. But this does not preclude increases in global emissions per head in future, either because of structural changes within economies, or changes in the distribution of emissions across fast- and slow-growing economies, leading to further structural breaks.

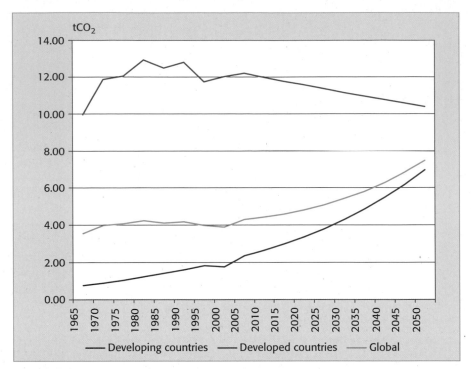

Figure 7.3 Global emissions per head: history and extrapolations
Source: Holtsmark (2006)

Structural changes in economies will have a significant impact on their emissions. In some rich countries, the shift towards a service-based economy has
helped to slow down, or even reverse, the growth in national emissions. Indeed,
emissions per head have fallen in some countries over some periods (e.g. they
peaked in the United Kingdom in 1973 and fell around 20% between then and
1984). Holtsmark's extrapolation in Figure 7.3 envisages a decline in emissions
per head for the developed world as a whole. And breaks in the relationship
between emissions per head and GDP per head have taken place, as seen in
Figure 7.4 for the USA, at income levels around $6000 per head, $12000 per head
and $22000 per head.

If it were true that the relationship between emissions and income growth
disappeared at higher income levels, emissions growth would eventually be self-
limiting, reducing the need to take action on climate change if this happened fast
enough. The observation that, at high incomes, some kinds of pollution start to
fall is often explained by invoking the 'environmental Kuznets curve' hypothesis –
see Annex 7.A. The increasing importance of the 'weightless economy' in the
developed world[30], with a rising share of spending accounted for by services, shows
how patterns of demand, and the resulting energy use, can change.

However, in the case of climate change, the hypothesis is not very convincing, for
three reasons. First, at a global level, there has been little evidence of large voluntary reductions in emissions as a result of consumers' desire to reduce emissions

[30] Quah (1996)

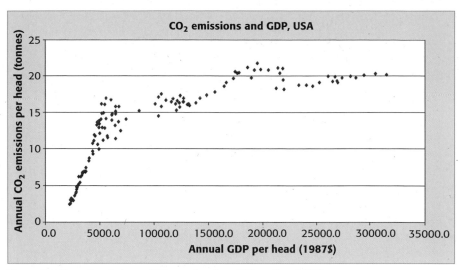

Figure 7.4 Annual emissions of CO_2 per head vs. GDP per head, USA

Source: Huntington (2005)

as they become richer. That may change as people's understanding of climate-change risks improves, but the global nature of the externality means that the incentive for uncoordinated individual action is very low. Second, the pattern seen in Figure 7.4 partly reflects the relocation of manufacturing activity to developing countries. So, at the global level, the structural shift within richer countries has less impact on total emissions. Third, demand for some carbon-intensive goods and services – such as air transport[31] – has a high income elasticity, and will continue to grow as incomes rise. Demand for car transport in many developing countries, for example, is likely to continue to increase rapidly. For these reasons, at the global level, in the absence of policy interventions, the long-run positive relationship between income growth and emissions per head is likely to persist. Breaking the link requires significant changes in preferences, relative prices of carbon-intensive goods and services and/or breaks in technological trends. But all of these are possible with appropriate policies, as Part IV of this Review argues.

Different assumptions about the definition and growth of income produce different projections for emissions, but this does not affect the conclusion that emissions are well above levels consistent with a stable climate and are likely to remain so under 'business as usual'.

Projected trajectories for CO_2 are sensitive to long-run growth projections, but the likelihood of economic growth slowing sufficiently to reverse emissions growth by itself is small. Most models assume some decline in world growth rates in the medium to long run, as poorer countries catch up and exhaust the growth possibilities from adopting best practices in production techniques. But some go further and assume that developed-country income growth per head will actually decline. There is no strong empirical basis for this assumption. Neither is the

[31] Air transport is particularly problematic given its impacts on the atmosphere over and above the simple CO_2 effect. The additional global warming effect of aviation is discussed in Box 15.8.

assumption very helpful if one wishes to assess the consequences if developed economies do manage to continue to grow at post-World War II rates.

The choice of method for converting the incomes of different countries into a common currency to allow them to be aggregated also makes some difference – see Box 7.2. But given that the growth rate of global GDP was around 2.9% per year on average between 1900 and 2000, and 3.9% between 1950 and 2000, projecting

---

BOX 7.2 Using market exchange rates or purchasing power parities in projections

There has been some controversy over how GDPs of different countries and regions should be compared for the purposes of making long-run emissions projections. Some method is required to convert data compiled in national currency terms into a common unit of account. Most emissions scenarios have used market exchange rates (MER), while others have argued for purchasing power parity (PPP) conversions. Castles and Henderson (2003) argue that "the mistaken use of MER-based comparisons, together with questionable assumptions about 'closing the gap' between rich countries and poor, have imparted an upward bias to projections of economic growth in developing countries, and hence to projections of total world emissions."

MER conversions suffer from two main problems. First, although competition tends to equalise the prices of internationally traded goods and services measured in a common currency using MERs, this is not true of non-traded goods and services. As the price of the latter relative to traded goods and services tends to be higher in rich countries than in poor ones, rich countries tend to have higher price levels converted at MERs. This phenomenon arises because the productivity differential between rich and poor countries tends to be larger for traded than non-traded goods and services (the 'Balassa-Samuelson' effect[32]). In this sense, the ratio of income per head between rich countries and poor countries is exaggerated if the comparison is intended to reflect purchasing power. Thus, the use of MERs will mean that developing countries' current GDP levels per head will be underestimated. If GDP levels per head are assumed to converge over some fixed time horizon, this means that the growth rates of the poor countries while they 'catch up' will be exaggerated. Henderson and Castles were concerned that this would lead to an over-estimate of the growth of emissions as well.

Second, MERs can be driven away from the levels that ensure the 'law of one price' for traded goods and services by movements across countries' capital accounts. Different degrees of firms' market power in different countries may also have this effect.

Instead of using MERs, one can try to use conversions based on purchasing power parity (PPP). These try to compare real incomes across countries by comparing the ability to purchase a standard basket of goods and services. But PPP exchange rates have their own problems, as explained by McKibbin et al (2004). PPP calculation requires detailed information about the prices in national currencies of many comparable goods and services. The resource costs are heavy. There are different ways of weighting individual countries' prices to obtain 'international prices' and aggregating volumes of output or expenditure. Different PPP conversions are needed for different purposes. For example, different baskets of products and PPP conversion rates are appropriate for comparing the incomes of old people across countries than for comparing the incomes of the young; similarly, different price indices need to be used for comparing industrial outputs. Data are only available for benchmark years, unlike MERs, which for many countries are available at high frequency.

---

[32] See, for example, Balassa (1964)

But efforts are under way to improve the provision of PPP data. The International Comparison Programme (ICP), launched by the World Bank when Nicholas Stern was Chief Economist, is the world's largest statistical initiative, involving 107 countries and collaboration with the OECD, Eurostat and National Statistical Offices. It produces internationally comparable price levels, economic aggregates in real terms, and Purchasing Power Parity (PPP) estimates that inform users about the relative sizes of markets, the size and structure of economies, and the relative purchasing power of currencies.

In the IPCC SRES scenarios that use MER conversions, it is not clear that the use of MERs biases upwards the projected rates of emissions growth, as the SRES calibration of the past relationship between emissions per head and GDP per head also used GDPs converted at MERs as the metric for economic activity (Holtsmark and Alfsen (2003)). Hence the scenarios are based on a lower estimate of the elasticity of emissions growth per head with respect to (the incorrectly measured) GDP growth per head. As Nakicenovic et al (2003) have argued, the use of MERs in many of the IPCC SRES scenarios is unlikely to have distorted the emissions trajectories much.

world growth to continue at between 2 and 3% per year (as in the IPCC SRES scenarios, for example) does not seem unreasonable.

Overall, the statement that, under business as usual, global emissions will be sufficient to propel greenhouse-gas concentrations to over 550 ppm CO_2e by 2050 and over 650–700 ppm by the end of this century is robust to a wide range of changes in model assumptions. It is based on a conservative assumption of constant or very slowly rising annual emissions. The proposition does not, for example, rely on convergence of growth rates of GDP per head across countries, an assumption commonly made in global projections. Cross-country growth regressions suggest that on average there has been a general tendency towards convergence of growth rates[33]. But there has been a wide range of experience over time and regions, and some signs of divergence in the 1990s[34].

Total emissions are likely to increase more rapidly than emissions per head.

The UN projects world population to increase from 6.5 billion in 2005 to 9.1 billion in 2050 in its medium variant and still to be increasing slowly then (at about 0.4% per year), despite projected falls in fertility[35]. The average annual growth rate from 2005 to 2050 is projected to be 0.75%; the UN's low and high variants give corresponding rates of 0.38% and 1.11%. Population growth rates will be higher among the developing countries, which are also likely in aggregate to have more rapid emissions growth per head. This means that emissions in the developing world will grow significantly faster than in the developed world, requiring a still sharper focus on emissions abatement in the larger economies like China, India and Brazil.

Climate change itself is also likely to have an impact on energy demand and hence emissions, but the direction of the net impact is uncertain. Warmer winters

[33] Bosworth, B, and Collins, S (2003)
[34] See McKibbin and Stegman, op. cit.; Pritchett, L (1997)
[35] Population Division of the Department of Economic and Social Affairs of the United Nations Secretariat (2005)

in higher latitudes are likely to reduce energy demand for heating[36], but the hotter summers likely in most regions are likely to increase the demand for refrigeration and air conditioning[37].

7.5 The role of technology and efficiency in breaking the link between growth and emissions

The relationship between economic development and CO_2 emissions growth is not immutable.

Historically, there have been a number of pervasive changes in energy systems, such as the decline in steam power, the spread of the internal combustion engine and electrification. The adoption of successive technologies changed the physical relationship between energy use and emissions. A number of authors have identified in several countries structural breaks in the observed relationship that are likely to have been the result of such switches[38]. Using US data, Huntington (2005) found that, after allowing for these technology shifts, the positive relationship between emissions per head and income per head has remained unchanged, casting some doubt on the scope for changes in the structure of demand to reduce emissions in the absence of deliberate policy. Also, an MIT study suggests that, since 1980, changes in US industrial structure have had little effect on energy intensity[39].

Shifts usually entailed switching from relatively low-energy-density fuels (e.g. wood, coal) to higher-energy-density ones (e.g. oil), and were driven primarily by technological developments, not income growth (although cause and effect are difficult to disentangle, and changes in the pattern of demand for goods and services may also have played a role). The energy innovations and their diffusion were largely driven by their advantages in terms of costs, convenience and suitability for powering new products (with some local environmental concerns, such as smog in London or Los Angeles, occasionally playing a part). As the discussion of technology below suggests (see Chapter 16), given the current state of knowledge, alternative technologies do not appear, on balance, to have the inherent advantages over fossil-fuel technologies (e.g. in costs, energy density or suitability for use in transport) necessary if decarbonisation were to be brought about purely by private commercial decisions. Strong policy will therefore be needed to provide the necessary incentives.

Technical progress in the energy sector and increased energy efficiency are also likely to moderate emissions growth. Figure 7.5, for instance, illustrates that the efficiency with which energy inputs are converted into useful energy services in the United States has increased seven-fold in the past century. One study has found that innovations embodied in information technology and electrical equipment capital stocks have played a key part in reducing energy intensity over the

---

[36] See Neumayer, op. cit.
[37] Asadoorian et al (2006)
[38] See, for example, Lanne and Liski (2004) and Huntington, op. cit. The former study 16 countries but use a very limited set of explanatory variables.
[39] Sue Wing and Eckaus (2004)

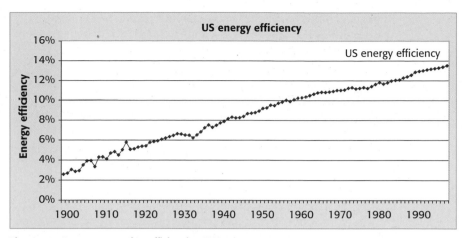

Figure 7.5 Energy conversion efficiencies, USA, 1900–1998

Source: Ayres et al (2005) and Ayres and Warr (2005) This graph shows the efficiency with which power from fossil-fuel, hydroelectric and nuclear sources is converted into useful energy services. The percentages reflect the ratio of useful work output to energy input.

long term[40]. But, in the absence of appropriate policy, incremental improvements in efficiency alone will not overwhelm the income effect. For example, a review of projections for China carried out for the Stern Review suggests that energy demand is very likely to increase substantially in 'business as usual' scenarios, despite major reductions in energy intensity[41]. And in the USA, emissions per head are projected to rise whenever income per head grows at more than 1.8% per year[42]. But the scale of potential cost-effective energy efficiency improvements, which will be explored elsewhere in this Review, indicates that energy efficiency and reductions in energy intensity constitute an important and powerful part of a wider strategy.

Chapter 9 will set out in more detail the potential for improvements in efficiency and technology; Part IV of this report will look at how policy frameworks can be designed to make this happen.

7.6   The impact of fossil-fuel scarcity on emissions growth

This chapter has argued that, without action on climate change, economic growth and development are likely to generate levels of greenhouse-gas emissions that would be very damaging. Development is likely to lead to increasing demand for fossil-fuel energy, and, without appropriate international collective action, producers and consumers will not modify their behaviour to reduce the adverse impacts. But is the increase in energy use implied actually technically feasible? In other words, are the stocks of fossil fuels in the world large enough to satisfy the

[40] Sue Wing and Eckaus (2004)
[41] Understanding China's Energy Policy: Background Paper Prepared for Stern Review on the Economics of Climate Change by the Research Centre for Sustainable Development, Chinese Academy of Social Sciences
[42] Huntington, op. cit.

demand implied by the BAU scenarios? Or will increasing scarcity drive up the relative prices of fossil fuels sufficiently to choke off demand fast enough to provide a 'laissez faire' answer to the climate-change problem?

There is enough fossil fuel in the ground to meet world consumption demand at reasonable cost until at least 2050.

To date, about 2.7 trillion barrels of oil equivalent (boe) of oil, gas and coal have been used up[43]. At least another 40 trillion boe remain in the ground, of which around 7 trillion boe can reasonably be considered economically recoverable[44]. This is comfortably enough to satisfy the BAU demand for fossil fuels in the period to 2050 (4.7 trillion boe)[45].

The IEA has looked at where the economically recoverable reserves of oil might come from in the next few decades and the associated extraction costs (see Figure 7.6). Demand for oil in the period to 2050 is expected to be 1.8 trillion boe[46]; this could be extracted at less than \$30/barrel. This alone would be enough to raise the concentration of CO_2e in the atmosphere by 50 ppm[47].

There appears to be no good reason, then, to expect large increases in real fossil-fuel prices to be necessary to bring forth supply. Yet big increases in price would be required to hold energy demand and emissions growth in check if no other method were also available. The IEA emissions projections envisage an average annual rate of increase of 1.7% to 2030. If the price elasticity of energy demand

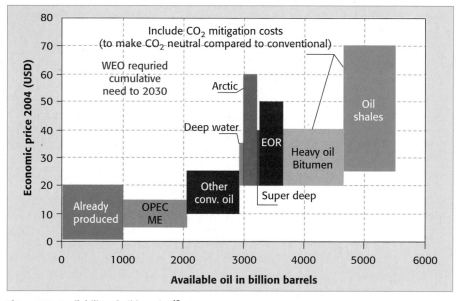

Figure 7.6 Availability of oil by price[48]

Source: International Energy Agency

[43] World Energy Council (2000)
[44] World Energy Council (2000)
[45] IEA (2006)
[46] IEA (2006)
[47] This assumes that half of CO_2 emissions are absorbed, as discussed in Chapter 1.
[48] IEA (2005)

were -0.23, an estimate in the middle of the range in the literature[49], the prices of fossil fuels would have to increase by over 7% per year in real terms merely to bring the rate of emissions growth back to zero, implying a more-than-six-fold rise in the real price of energy.

'Carbon capture and storage' technology is important, as it would allow some continued use of fossil fuels and help guard against the risk of fossil-fuel prices falling in response to global climate-change policy, undermining its effectiveness.

There are three major implications for policy. First, it is important to provide incentives to redirect research, development and investment away from the fossil fuels that are currently more difficult to extract (see Grubb (2001)). The initial costs of development provide a hurdle to the exploitation of some of the more carbon-intensive fuels like oil shales and synfuels. This obstacle can be used to help divert R,D&D efforts towards low-carbon energy resources. Second, the low resource costs of much of the remaining stock of fossil fuels have to be taken into account in climate-change policy[50]. Third, as there is a significant element of rent in the current prices of exhaustible fossil-fuel resources, particularly those of oil and natural gas, there is a danger that fossil-fuel prices could fall in response to the strengthening of climate-change policy, undermining its effectiveness[51]. Extensive carbon capture and storage would maintain the viability of fossil fuels for many uses in a manner compatible with deep cuts in emissions, and thereby help guard against this risk.

References

The World Resources Institute (2005) publication "Navigating the Numbers" provides a very good overview of global GHG emissions, by source and country. The WRI also provides a very user-friendly database in its Climate Analysis Indicators Tool. The International Energy Agency's publications provide an excellent source of information about fossil-fuel emissions and analysis of the medium-term outlook for emissions, energy demand and supply. The US Environmental Protection Agency produces estimates of historical and projected non-CO_2 emissions. Houghton (2005) is a good source of data and information on emissions due to land-use change.

The IPCC's Special Report on Emission Scenarios considers possible longer-term outlooks for emissions and discusses many of the complex issues that arise with any long-term projections. Its scenarios provide the foundation for many of the benchmark 'business as usual' scenarios used in the literature. Some of the difficult challenges posed by the need to make long-term projections have been pursued in the academic literature, for example, in the two papers co-authored by Warwick McKibbin and referenced here and the paper by Schmalensee et al (1998). There have been lively methodological exchanges, including the debates between Castles and Henderson (2003a,b), Nakicenovic et al (2003) and Holtsmark and Alfsen (2005) on how to aggregate

---

[49] See Hunt et al (2003)
[50] In calculating the costs of climate-change mitigation to the world as a whole, fossil-fuel energy should be valued at its marginal resource cost, excluding the scarcity rents, not at its market price. Some estimates of cost savings from introducing alternative energy technologies ignore this point and consequently overestimate the global cost savings.
[51] A downward shift in the demand curve for an exhaustible natural resource is likely to lead to a fall in the current and future price of the resource. In the case of resources for which the marginal extraction costs are very low, this fall could continue until the demand for the fossil fuel is restored. Pindyck (1999) found that the behaviour of oil prices has been broadly consistent with the theory of exhaustible natural resource pricing. See also Chapter 2 references on the pricing of exhaustible natural resources.

incomes across countries. A good example of the Integrated Assessment Model approach to projections can be found in Paltsev et al (2005). Some of the difficulties of untangling the impacts of income and technology on emissions growth are tackled in Huntington (2005), among others.

The World Energy Council (2000) is a good source of information on availability of fossil fuels. The IEA (1995) have also produced an excellent report on this. The extraction costs of fossil fuels are also considered by Rogner (1997). The issues posed by exhaustible fossil fuels in the context of climate change are analysed in papers referenced in Chapter 2.

An, F. and A. Sauer, (2004): 'Comparison of passenger vehicle fuel economy and GHG emission standards around the world', Prepared for the Pew Centre on Global Climate Change, Virginia: Pew Centre.

Asadoorian, M.O., R.S. Eckaus and C.A. Schlosser (2006): 'Modeling climate feedbacks to energy demand: The case of China' MIT Global Change Joint Program Report 135, June, Cambridge, MA: MIT.

Ayres, R. U., Ayres, L. W. and Pokrovsky, V. (2005): 'On the efficiency of US electricity usage since 1900', Energy, 30(7): 1092–1145

Ayres and Warr (2005): 'Accounting for growth: the role of physical work', Structural Change and Economic Dynamics, 16(2): 181–209

Balassa, B. (1964): 'The purchasing power parity doctrine: a reappraisal', Journal of Political Economy, 72: 584–596

Bosworth, B.P. and S.M. Collins (2003): 'The empirics of growth: an update', Brookings Papers on Economic Activity, No. 2, 2003.

Castles, I. and D. Henderson (2003a): 'The IPCC emissions scenarios: an economic-statistical critique', Energy and Environment,14(2–3) May: 159–185

Castles, I. and D. Henderson (2003b): 'Economics, emissions scenarios and the work of the IPCC', Energy and Environment, 14 (4) July: 415–435

Chinese Academy of Social Sciences (2005): 'Understanding China's Energy Policy', Background Paper Prepared for Stern Review on the Economics of Climate Change by the Research Centre for Sustainable Development

EPA (forthcoming): 'Global anthropogenic non-CO_2 greenhouse-gas emissions: 1990–2020', US Environmental Protection Agency, Washington DC. Figures quoted from draft December 2005 version.

Grubb, M. (2001): 'Who's afraid of atmospheric stabilisation? Making the link between energy resources and climate change', Energy Policy, 29: 837–845

Harbaugh, B., A. Levinson and D. Wilson (2002): 'Reexamining the empirical evidence for an environmental Kuznets Curve,' Review of Economics and Statistics, 84 (3) August.

Holtsmark, B.J. (2006): 'Are global per capita CO_2 emissions likely to remain stable?', Energy & Environment, 17 (2) March.

Holtsmark, B.J. and K.H. Alfsen (2005): 'PPP correction of the IPCC emissions scenarios – does it matter?', Climatic Change, 68 (1–2) January: 11–19

Holtz-Eakin, D. and T.M. Selden (1995): 'Stoking the fires? CO_2 emissions and economic growth', Journal of Public Economics, 57, May: 85–101

Houghton, R.A. (2005): 'Tropical deforestation as a source of greenhouse-gas emissions', Tropical Deforestation and Climate Change, (eds.) P. Moutinho and S. Schwartzman, Belém: IPAM – Instituto de Pesquisa Ambiental da Amazônia ; Washington DC – USA : Environmental Defense. See: www.environmentaldefense.org/documents/4930_TropicalDeforestation_and_ClimateChange.pdf

Hunt, L.C., G. Judge, and Y. Ninomiya (2003): 'Modelling underlying energy demand trends', Chapter 9 in L.C. Hunt (ed), 'Energy in a competitive market: essays in honour of Colin Robinson', Cheltenham: Edward Elgar.

Huntington, H.G. (2005): 'US carbon emissions, technological progress and economic growth since 1870', Int. J. Global Energy Issues, 23 (4): 292–306

International Energy Agency (2005): 'Resources to reserves: oil and gas technologies for the energy markets of the future', Paris: OECD/IEA.

International Energy Agency (2006): 'Energy technology perspectives – scenarios & strategies to 2050', Paris: OECD/IEA.

International Energy Agency (in press): World Energy Outlook 2006, Paris: OECD/IEA.

Intergovernmental Panel on Climate change (2000): 'Emissions Scenarios', Special Report, N. Nakicenovic and R. Swart (eds.), Cambridge: Cambridge University Press.

Kaya, Y. (1990): 'Impact of carbon dioxide emission control on GNP growth: interpretation of pro-posed scenarios', paper presented to IPCC Energy and Industry Sub-Group, Response Strategies Working Group.

Lanne, M. and M. Liski (2004): 'Trends and breaks in per-capita carbon dioxide emissions, 1870–2028', The Energy Journal, 25 (4): 41–65

McKibbin, W., D. Pearce, and A. Stegman (2004): 'Long-run projections for climate-change scenarios', Brookings Discussion Papers in International Economics No 160, April, Washington, DC: The Brookings Institution.

McKibbin, W. and A. Stegman (2005): 'Convergence and per capita carbon emissions', Working Paper in International Economics No 4.05, May, Sydney: Lowy Institute fore International Policy.

McKitrick, R. and M.C. Strazicich (2005). 'Stationarity of global per capita carbon dioxide emissions: implications for global warming scenarios', University of Guelph Department of Economics Discussion Paper 2005–03.

Nakicenovic, N., A. Grübler, S. Gaffin et al. (2003): 'IPCC SRES revisited: a response', Energy & Environment, 14 (2–3) May: 187–214

Neumayer, E. (2004): 'National carbon dioxide emissions: geography matters', Area, 36 (1): 33–40

Paltsev, S., J.M. Reilly, H.D. Jacoby, et al. (2005): 'The MIT emissions prediction and policy analysis (EPPA) model: version 4', MIT Global Change Joint Program Report No. 125, August, Cambridge, MA: MIT.

Pindyck, R.S. (1999): 'The long-run evolution of energy prices', The Energy Journal, 20 (2): 1–27

Population Division of the Department of Economic and Social Affairs of the United Nations Secretariat (2005): 'World population prospects: the 2004 revision highlights' New York: United Nations.

Pritchett, L. (1997): 'Divergence, big time', Journal of Economic Perspectives, 11 (3): 3–17

Quah, D. (1996): 'The invisible hand and the weightless economy', London: London School of Economics Centre for Economic Performance.

Rogner, H.H. (1997): 'An assessment of world hydrocarbon resources', Annual Reviews of Energy and the Environment (22): 217–262

Schmalensee, R., T.M. Stoker and R.A. Judson (1998): 'World carbon dioxide emissions: 1950–2050', Review of Economics and Statistics, 80: 15–27

Seldon, T., and D. Song (1994): 'Environmental quality and development: is there a Kuznets curve for air pollution emissions?', Journal of Environmental Economics and Management, 27:7–162

Sue Wing, I. and R.S. Eckaus (2004): 'Explaining long-run changes in the energy intensity of the US economy', MIT Global Change Joint Program, Report No 116, September, Cambridge, MA: MIT.

World Energy Council (2000): World Energy Assessment: Energy and the challenge of sustainability, New York: United Nations Development Programme.

World Resources Institute (2005): 'Navigating the numbers', World Resources Institute, Washington DC.

World Resources Institute (2006): Climate Analysis Indicators Tool (CAIT) on-line database version 3.0., Washington, DC: World Resources Institute, available at: http://cait.wri.org

7A Climate Change and the Environmental Kuznets Curve

Some evidence indicates that, for local pollutants like oxides of nitrogen, sulphur dioxide and heavy metals, there is an inverted-U shaped relationship between income per head and emissions per head: the so-called 'environmental Kuznets curve', illustrated in Figure 7.A.1[52]. The usual rationale for such a curve is that the demand for environmental improvements is income elastic, although explanations based on structural changes in the economy have also been put forward. So the question arises, is there such a relationship for CO_2? If so, economic development would ultimately lead to falls in global emissions (although that would be highly unlikely before GHG concentrations had risen to destructive levels).

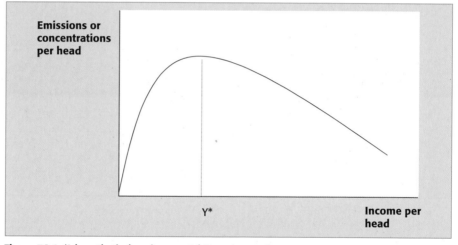

Figure 7A.1 'A hypothetical environmental Kuznets curve'

In the case of greenhouse gases, this argument is not very convincing. As societies become richer, they may want to improve their own environment, but they can do little about climate change by reducing their own CO_2 emissions alone. With CO_2, the global nature of the externality means that people in any particular high-income country cannot by themselves significantly affect global emissions and hence their own climate. This contrasts with the situation for the local pollutants for which environmental Kuznets curves have been estimated. It is easier than with greenhouse gases for the people affected to set up abatement incentives and appropriate political and regulatory mechanisms. Second, CO_2 had not been identified as a pollutant until around 20 years ago, so an explanation of past data based on the demand for environmental improvements does not convince.

Nevertheless, patterns like the one in Figure 7.4 suggest that further empirical investigation of the relationship between income and emissions is warranted.

[52] See Seldon and Song (1994) and Harbaugh et al (2002)

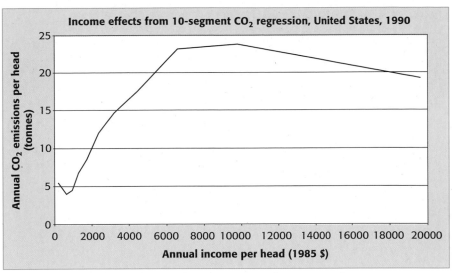

Figure 7A. 2 'Income effects from 10-segment CO_2 regression, USA, 1990'

Source: Schmalensee et al (1998)

The relationship could reflect changes in the structure of production as countries become better off, as well as or instead of changes in the pattern of demand for environmental improvements. Several empirical studies[53] have found that a relationship looking something like the first half of an environmental Kuznets curve exists for CO_2 (after allowing for some other explanatory factors in some, but not all, cases). Figure 7.A.2 illustrates this, using Schmalensee et al's estimates for the United States.

Even if this finding *were* robust, however, it does not imply that the global relationship between GDP per head and CO_2 emissions per head is likely to disappear soon. The estimated turning points at which CO_2 emissions start to fall are at very high incomes (for example, between $55.000 and $90,000 in Neumayer's cross-country study, in which the maximum income level observed in the data was $41,354). Poor and middle-income countries will have to grow for a long time before they get anywhere near these levels. Schmalensee et al found that, using their estimates – *with* an implied inverted-U shape – as the basis for a projection of future emissions, emissions growth was likely to be positive up to their forecast horizon of 2050; indeed, they forecast more rapid growth than in nearly all the 1992 IPCC scenarios, using the same assumptions as the IPCC for future population and income growth.

In any case, it is not clear that the link between emissions and income does disappear at high incomes. First, the apparent turning points in some of the studies may simply be statistical artefacts, reflecting the particular functional forms for the relationship assumed by the researchers[54]. Second, the apparent weakening of the link may result from ignoring the implications of past changes in energy technology; after controlling for the adoption of new technologies that, incidentally, were less carbon-intensive, the link may reappear, as argued by Huntington (2005).

---

[53] See, inter alia, Neumayer, op. cit., Holtz-Eakin and Selden (1995) and Schmalensee et al, op. cit.
[54] This is not the case with the 'piecewise segments' approach of Schmalensee et al.

8 The Challenge of Stabilisation

KEY MESSAGES

The world is already irrevocably committed to further climate changes, which will lead to adverse impacts in many areas. Global temperatures, and therefore the severity of impacts, will continue to rise unless the stock of greenhouse gases is stabilised. Urgent action is now required to prevent temperatures rising to even higher levels, lowering the risks of impacts that could otherwise seriously threaten lives and livelihoods worldwide.

Stabilisation – at whatever level – requires that annual emissions be brought down to the level that balances the Earth's natural capacity to remove greenhouse gases from the atmosphere. In the long term, global emissions will need to be reduced to less than 5 $GtCO_2e$, over 80% below current annual emissions, to maintain stabilisation. The longer emissions remain above the level of natural absorption, the higher the final stabilisation level will be.

Stabilisation cannot be achieved without global action to reduce emissions. Early action to stabilise this stock at a relatively low level will avoid the risk and cost of bigger cuts later. The longer action is delayed, the harder it will become.

Stabilising at or below 550 ppm CO_2e (around 440–500 ppm CO_2 only) would require global emissions to peak in the next 10–20 years, and then fall at a rate of at least 1–3% per year. By 2050, global emissions would need to be around 25% below current levels. These cuts will have to be made in the context of a world economy in 2050 that may be three to four times larger than today – so emissions per unit of GDP would need to be just one quarter of current levels by 2050.

Delaying the peak in global emissions from 2020 to 2030 would almost double the rate of reduction needed to stabilise at 550 ppm CO_2e. A further ten-year delay could make stabilisation at 550 ppm CO_2e impractical, unless early actions were taken to dramatically slow the growth in emissions prior to the peak.

To stabilise at 450 ppm CO_2e, without overshooting, global emissions would need to peak in the next 10 years and then fall at more than 5% per year, reaching 70% below current levels by 2050. This is likely to be unachievable with current and foreseeable technologies.

If carbon absorption were to weaken, future emissions would need to be cut even more rapidly to hit any given stabilisation target for atmospheric concentration.

Overshooting paths involve greater risks to the climate than if the stabilisation level were approached from below, as the world would experience at least a century of temperatures, and therefore impacts, close to those expected for the peak level of emissions. Some of these impacts might be irreversible. In addition, overshooting paths require that emissions be reduced to extremely low levels, below the level of natural absorption, which may not be feasible.

Energy systems are subject to very significant inertia. It is important to avoid getting 'locked into' long-lived high carbon technologies, and to invest early in low carbon alternatives.

8.1   Introduction

The stock of greenhouse gases in the atmosphere is already at 430 ppm CO_2e and currently rising at roughly 2.5 ppm every year. The previous chapter presented clear evidence that greenhouse gas emissions will continue to increase over the coming decades, forcing the stock of greenhouse gases upwards at an accelerating pace. Parts I and II demonstrated that, if emissions continue unabated, the world is likely to experience a radical transformation of its climate, with profound implications for our way of life.

Global mean temperatures will continue to rise unless the stock of greenhouse gases in the atmosphere is stabilised. This chapter considers the pace, scale and composition of emissions paths associated with stabilisation. This is a crucial foundation for examining the costs of stabilisation; which are discussed in the following two chapters.

The first section of this chapter looks at what different stabilisation levels mean for global temperature rises and presents the science of how to stabilise greenhouse gas levels. The following two sections go on to consider stabilisation of carbon dioxide and other gases in detail. Sections 8.4 and 8.5 use preliminary results from a simple model to examine the emissions cuts required to stabilise the stock of greenhouse gases in the range 450–550 ppm CO_2e, and the implications of delaying emissions cuts. The final section gives a more general discussion of the scale of the challenge of achieving stabilisation.

The focus on the range 450–550 ppm CO_2e is based on analyses presented in chapter 13, which conclude that stabilisation at levels below 450 ppm CO_2e would require immediate, substantial and rapid cuts in emissions that are likely to be extremely costly, whereas stabilisation above 550 ppm CO_2e would imply climatic risks that are very large and likely to be generally viewed as unacceptable.

8.2   Stabilising the stock of greenhouse gases

The higher the stabilisation level, the higher the ultimate average global temperature increase will be.

The relationship between stabilisation levels and temperature rise is not known precisely (chapter 1). Box 8.1 summarises recent studies that have tried to establish probability distributions for the ultimate temperature increase associated with given greenhouse gas levels. It shows the warming that is expected when the climate comes into equilibrium with the new level of greenhouse gases; it can be understood as the warming committed to in the long run. In most cases, this would be higher than the temperature change expected in 2100.

BOX 8.1 Likelihood of exceeding a temperature increase at equilibrium

This table provides an indicative range of likelihoods of exceeding a certain temperature change (at equilibrium) for a given stabilisation level (measured in CO_2 equivalent). For example, for a stock of greenhouse gases stabilised at 550 ppm CO_2e, recent studies suggest a 63–99% chance of exceeding a warming of 2°C relative to the pre-industrial.

The data shown is based on the analyses presented in Meinshausen (2006), which brings together climate sensitivity distributions from eleven recent studies (chapter 1). Here, the 'maximum' and 'minimum' columns give the maximum and minimum chance of exceeding a level of temperature increase across all eleven recent studies. The 'Hadley Centre' and 'IPCC TAR 2001' columns are based on Murphy *et al.* (2004) and Wigley and Raper (2001), respectively. These results lie close to the centre of the range of studies (Box 1.2). The 'IPCC TAR 2001' results reflect climate sensitivities of the seven coupled ocean-atmosphere climate models used in the IPCC TAR. The individual values should be treated as approximate.

The red shading indicates a 60 per cent chance of exceeding the temperature level; the amber shading a 40 per cent chance; yellow shading a 10 per cent chance; and the green shading a less than a 10 per cent chance.

Stabilisation Level (CO_2e)	Maximum	Hadley Centre Ensemble	IPCC TAR 2001 Ensemble	Minimum
Probability of exceeding 2°C (relative to pre-industrial levels)				
400	57%	33%	13%	8%
450	78%	78%	38%	26%
500	96%	96%	61%	48%
550	99%	99%	77%	63%
650	100%	100%	92%	82%
750	100%	100%	97%	90%
Probability of exceeding 3°C (relative to pre-industrial levels)				
400	34%	3%	1%	1%
450	50%	18%	6%	4%
500	61%	44%	18%	11%
550	69%	69%	32%	21%
650	94%	94%	57%	44%
750	99%	99%	74%	60%
Probability of exceeding 4°C (relative to pre-industrial levels)				
400	17%	1%	0%	0%
450	34%	3%	1%	0%
500	45%	11%	4%	2%
550	53%	24%	9%	6%
650	66%	58%	25%	16%
750	82%	82%	41%	29%
Probability of exceeding 5°C (relative to pre-industrial levels)				
400	3%	0%	0%	0%
450	21%	1%	0%	0%
500	32%	3%	1%	0%
550	41%	7%	2%	1%
650	53%	24%	9%	5%
750	62%	47%	19%	11%

Box 8.1 shows, for example, that stabilisation at 450 ppm CO_2e would lead to an around 5–20% chance of global mean temperatures ultimately exceeding 3°C above pre-industrial (from probabilities based on the IPCC Third Assessment Report (TAR) and recent Hadley Centre work). An increase of more than 3°C would entail very damaging physical, social and economic impacts, and heightened risks of catastrophic changes (chapter 3). For stabilisation at 550 ppm CO_2e, the chance of exceeding 3°C rises to 30–70%. At 650 ppm CO_2e, the chance rises further to 60–95%.

Stabilisation – at whatever level – requires that annual emissions be brought down to the level that balances the Earth's natural capacity to remove greenhouse gases from the atmosphere.

To stabilise greenhouse gas concentrations, emissions must be reduced to a level where they are equal to the rate of absorption/removal by natural processes. This level is different for different greenhouse gases. The longer global emissions remain above this level, the higher the stabilisation level will be. It is the *cumulative* emissions of greenhouse gases, less their cumulative removal from the atmosphere, for example by chemical processes or through absorption by the Earth's natural systems, that defines their concentration at stabilisation. The following section examines the stabilisation of carbon dioxide concentrations. The stabilisation of other gases in discussed separately in section 8.4.

8.3   Stabilising carbon dioxide concentrations

Carbon dioxide concentrations have risen by over one third, from 280 ppm pre-industrial to 380 ppm in 2005. The current concentration of carbon dioxide in the atmosphere accounts for around 70% of the total warming effect (the 'radiative forcing') of all Kyoto greenhouse gases[1].

Over the past two centuries, around 2000 $GtCO_2$ have been released into the atmosphere through human activities (mainly from burning fossil fuels and land-use changes)[2]. The Earth's soils, vegetation and oceans have absorbed an estimated 60% of these emissions, leaving 800 $GtCO_2$ to accumulate in the atmosphere. This corresponds to an increase in the concentration of carbon dioxide in the atmosphere of 100 parts per million (ppm), thus an *accumulation* of around 8 $GtCO_2$ corresponds to a 1 ppm rise in concentration.

Accordingly, a carbon dioxide concentration of 450 ppm, around 70 ppm more than today, would correspond to a further *accumulation* of around 550 $GtCO_2$ in the atmosphere. However, the cumulative *emissions* that would be expected to lead to this concentration level would be larger, as natural processes should continue to remove a substantial portion of future carbon dioxide emissions from the atmosphere.

---

[1] The conversion to radiative forcing is given in IPCC (2001).
[2] Extrapolating to 2005 from Prentice *et al.* (2001), which gives 1800 $GtCO_2$ total emissions in 2000 and a 90 ppm increase in atmospheric carbon dioxide concentration. The extrapolation assumes 2000 emissions to 2005.

Note that, a carbon dioxide concentration of 450 ppm would be equivalent to a total stock of greenhouse gases of at least 500 ppm CO_2e (depending on emissions of non-CO_2 gases).

Today, for every 15–20 GtCO_2 *emitted*, the concentration of carbon dioxide rises by a further 1 ppm, with natural processes removing the equivalent of roughly half of all emissions. But, the future strength of natural carbon absorption is uncertain. It will depend on a number of factors, including:

- The sensitivity of carbon absorbing systems, such as forests, to future climate changes.
- Direct human influences, such as clearing forests for agriculture.
- The sensitivity of natural processes to the rate of increase and level of carbon dioxide in the atmosphere. For example, higher levels of carbon dioxide can stimulate a higher rate of absorption by vegetation (the carbon fertilisation effect – chapter 3).

Assuming that climate does not affect carbon absorption, a recent study projects that stabilising carbon dioxide concentrations at 450 ppm would allow cumulative *emissions* of close to 2100 GtCO_2 between 2000 and 2100 (Figure 8.1)[3] (equivalent to roughly 60 years of emissions at today's rate). This means that approximately

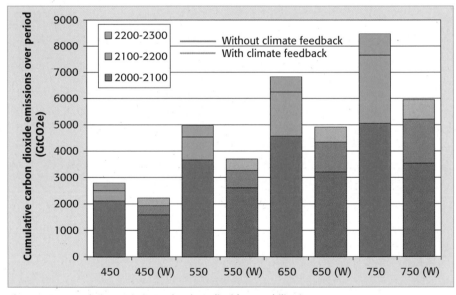

Figure 8.1 Cumulative emissions of carbon dioxide at stabilisation

This figure gives illustrative results from one study that shows the level of cumulative emissions between 2000 and 2300 for a range of stabilisation levels (carbon dioxide only). For the green bars, natural carbon absorption is not affected by the climate. The grey bars include the feedbacks between the climate and the carbon cycle (stabilisation levels labelled as (W)). Comparison of these sets of bars shows that if natural carbon absorption weakens (as predicted by the model used) then the level of cumulative emissions associated with a stabilisation goal reduces. The intervals on the bars show emissions to 2100 and 2200.

Source: based on Jones et al. (2006)

---

[3] Based on Jones *et al.* 2006, assuming no climate-carbon feedback.

75% of emissions would have been absorbed. Stabilising at 550 ppm CO_2 would allow roughly 3700 $GtCO_2$.

Land use management, such as afforestation and reforestation, can be used to enhance natural absorption, slowing the accumulation of greenhouse gases in the atmosphere and increasing the permissible cumulative level of human emissions at stabilisation. However, this can only be one part of a mitigation strategy; substantial emissions reduction will be required from many sectors to stabilise carbon dioxide concentrations (discussed further in chapter 9).

There is now strong evidence that natural carbon absorption will weaken as the world warms (chapter 1). This would make stabilisation more difficult to achieve.

A recent Hadley Centre study shows that if feedbacks between the climate and carbon cycle are included in a climate model, the resulting weakening of natural carbon absorption means that the cumulative emissions at stabilisation are dramatically reduced. Figure 8.1 shows that to stabilise carbon dioxide concentrations at 450–750 ppm, cumulative emissions must be 20–30% lower than previously estimated. For example, the cumulative emissions allowable to stabilise at 450 ppm CO_2 are reduced by 500 $GtCO_2$, or around fifteen years of global emissions at the current rate. This means that emissions would need to peak at a lower level, or be cut more rapidly, to achieve a desired stabilisation goal. The effects are particularly severe at higher stabilisation levels.

The uncertainties over future carbon absorption make a powerful argument for taking an approach that allows for the possibility that levels of effort may have to increase later to reach a given goal.

Not taking into account the uncertainty in future carbon absorption, including the risk of weakening carbon absorption, could lead the world to overshoot a stabilisation goal. As the scientific understanding of this effect strengthens, adjustments will need to be made to the estimates of trajectories consistent with different levels of stabilisation.

To stabilise concentrations of carbon dioxide in the long run, emissions will need to be cut by more than 80% from 2000 levels.

To achieve stabilisation, annual carbon dioxide emissions must be brought down to a level where they equal the rate of natural absorption. After stabilisation, the level of natural absorption will gradually fall as the vegetation sink is exhausted. This means that to maintain stabilisation, emissions would need to fall to the level of ocean uptake alone over a few centuries. This level is not well quantified, but recent work suggests that emissions may need to fall to roughly 5 $GtCO_2$e per year (more than 80% below current levels) by the second half of the next century[4]. On a timescale of a few hundred years, this could be considered a 'sustainable' rate of emissions[5]. However, in the long term, the rate of ocean uptake will also weaken, meaning that emissions may eventually need to fall below 1 $GtCO_2$e per year to maintain stabilisation.

---

[4] The two carbon cycle models used in the IPCC Third Assessment Report project emissions falling to around 3–9 $GtCO_2$ per year by around 2150–2300 (longer for higher stabilisation levels) (Prentice et al. (2001), Figure 3.13).

[5] See Jacobs (1991) for discussion of operationalising the concept of sustainability for complex issues.

Reducing annual emissions below the rate of natural absorption would lead to a fall in concentrations. However, such a recovery would be a very slow process; even if very low emissions were achieved, concentrations would only fall by a few parts per million (ppm) per year[6]. This rate would be further reduced if carbon absorption were to weaken as projected.

8.4   Stabilising concentrations of non-CO_2 gases

Non-CO_2 gases account for one quarter of the total 'global warming potential' of emissions and therefore, must play an important role in future mitigation strategies.

Global warming potentials (GWP) provide a way to compare greenhouse gases, which takes into account both the warming affect and lifetime[7] of different gases. The 100-year GWP is most commonly used; this is equal to *the ratio of the warming affect (radiative forcing) from 1 kg of a greenhouse gas to 1 kg of carbon dioxide over 100 years.* Over a hundred year time horizon, methane has a GWP twenty-three times that of carbon dioxide, nitrous oxide nearly 300 times and some fluorinated gases are thousands of times greater (Table 8.1).

This leads to a measure, also known as CO_2 equivalent (CO_2e), which weights emissions by their global warming potential. This measure is used as an exchange metric to compare the long-term impact of different emissions. Table 8.1 shows the portion of 2000 emissions made up by the different Kyoto greenhouse gases in

Table 8.1   Characteristics of Kyoto greenhouse gases

Despite the higher GWP of other greenhouse gases over a 100-year time horizon, carbon dioxide constitutes around three-quarters of the total GWP of emissions. This is because the vast majority of emissions, by weight, are carbon dioxide. HFCs and PFCs include many individual gases; the data shown are approximate ranges across these gases.

	Lifetime in the atmosphere (years)	100-year Global Warming (GWP) Potential	Percentage of 2000 emissions in CO_2e
Carbon dioxide	5–200	1	77%
Methane	10	23	14%
Nitrous Oxide	115	296	8%
Hydrofluorocarbons (HFCs)	1–250	10–12,000	0.5%
Perfluorocarbons (PFCs)	>2500	>5,500	0.2%
Sulphur Hexafluoride (SF$_6$)	3,200	22,200	1%

Source: Ramaswamy et al. (2001)[8] and emissions data from the WRI CAIT database[9].

[6]   For example, O'Neill and Oppenheimer (2005).
[7]   The lifetime of a gas is a measure of the average length of time that a molecule of gas remains in the atmosphere before it is removed by chemical or physical processes.
[8]   These estimates are from the Third Assessment Report of the IPCC (Ramaswamy et al. (2001)). The UNFCCC uses slightly different GWPs based on the Second Assessment Report (http://ghg.unfccc.int/gwp.html).
[9]   The World Resources Institute (WRI) Climate Analysis Indicators Tool (CAIT): http://cait.wri.org/

terms of CO_2e. Note that, in this Review, CO_2 equivalent emissions are defined differently to CO_2 equivalent concentrations, which consider the *instantaneous* warming effect of the gas in the atmosphere. For example, non-CO_2 Kyoto gases make up around one quarter of total emissions in terms of their long term warming potential in 2000 (Table 3.1). However, they account for around 30% of the total warming effect (the radiative forcing) of non-CO_2 gases in the atmosphere today.

As methane is removed from the atmosphere much more rapidly than carbon dioxide, its short term effect is even greater than is suggested by its 100-year GWP. However, over-reliance on abatement of gases with strong warming effects but short lifetimes could lock in long term impacts from the build up of carbon dioxide. Some gases, like HFCs, PFCs and SF_6, have both a stronger warming effect and longer lifetime than CO_2, therefore abating their emissions is very important in the long run.

The stock of different greenhouse gases at stabilisation will depend on the exact stabilisation strategy adopted. In the examples used in this chapter, stabilising the stock of all Kyoto greenhouse gases at 450–550 ppm CO_2e would mean stabilising carbon dioxide concentrations at around 400–490 ppm. More intensive carbon dioxide mitigation, relative to other gases, might lead to a lower fraction of carbon dioxide at stabilisation, and vice versa. Two recent cost optimising mitigation studies find that, at stabilisation, non-CO_2 Kyoto gases contribute around 10–20% of the total warming effect expressed in CO_2e[10]. Therefore, a stabilisation range of 450–550 ppm CO_2e, could mean carbon dioxide concentrations of 360–500 ppm. The cost implications of multi-gas strategies are discussed further in chapter 10.

It is the total warming effect (or radiative forcing), expressed as the stock in terms of CO_2 equivalent, which is critical in determining the impacts of climate change. For this reason, this Review discusses stabilisation in terms of the total stock of greenhouse gases.

8.5   Pathways to stabilisation

As discussed above, stabilisation at any level ultimately requires a cut in emissions down to less than 20% of current levels. The question then becomes one of how quickly stabilisation can be achieved. If action is slow and emissions stay high for a long time, the ultimate level of stabilisation will be higher than if early and ambitious action is taken.

The rate of emissions cuts required to meet a stabilisation goal is very sensitive to both the timing of the peak in global emissions, and its height. Delaying action now means more drastic emissions reductions over the coming decades.

There are a number of possible emissions trajectories that can achieve any given stabilisation goal. For example, emissions can peak early and decline gradually, or peak later and decline more rapidly. This is demonstrated in Figure 8.2, which shows illustrative pathways to stabilisation at 550 ppm CO_2e.

The height of the peak is also crucial. If early action is taken to substantially slow the growth in emissions prior to the peak, this will significantly reduce the

[10] For example, Meinshausen (2006) and US CCSP (2006)

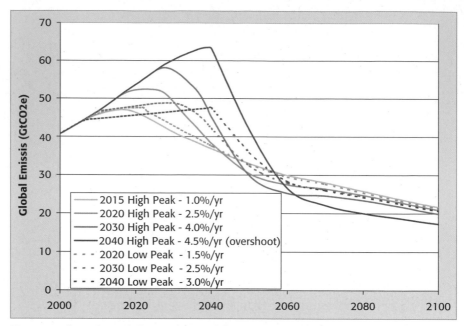

Figure 8.2 Illustrative emissions paths to stabilise at 550 ppm CO_2e.

The figure above shows six illustrative paths to stabilisation at 550 ppm CO_2e. The rates of emissions cuts are given in the legend and are the *maximum* 10-year average rate (see Table 8.2). The figure shows that delaying emissions cuts (shifting the peak to the right) means that emissions must be reduced more rapidly to achieve the same stabilisation goal. The rate of emissions cuts is also very sensitive to the height of the peak. For example, if emissions peak at 48 $GtCO_2$ rather than 52 $GtCO_2$ in 2020, the rate of cuts is reduced from 2.5%/yr to 1.5%/yr.

Source: Generated with the SiMCaP EQW model (Meinshausen et al. 2006)

required rate of reductions following the peak. For example, in Figure 8.2, if action is taken to ensure that emissions peak at only 7% higher than current levels, rather than 15% higher in 2020 to achieve stabilisation at 550 ppm CO_2e, the rate of reductions required after 2020 is almost halved.

If the required rate of emissions cuts is not achieved, the stock of greenhouse gases will overshoot the target level. Depending on the size of the overshoot, it could take at least a century to reduce concentrations back to a target level (discussed later in Box 8.2).

Table 8.2 gives examples of implied reduction rates for stabilisation levels between 550 ppm and 450 ppm CO_2e. A higher stabilisation level would require weaker cuts. For example, to stabilise at 650 ppm CO_2e, emissions could be around 20% above current levels by 2050, and 35% below current levels by 2100. As described in section 8.2, this higher stabilisation level would mean a much greater chance of exceeding high levels of warming and therefore, a higher risk of more adverse and unacceptable outcomes. The paths shown in Table 8.2 are based on one model and should be treated as indicative. Despite this, they provide a crucial illustration of the scale of the challenge. Further research is required to explore the uncertainties and inform more detailed strategies on future emissions paths.

Table 8.2 Illustrative emissions paths to stabilisation

The table below explores the sensitivity of rates of emissions reductions to the stabilisation level and timing and size of the peak in global emissions. These results were generated using the SiMCaP EQW model, as used in Meinshausen *et al.* (2006), and should be treated as indicative of the scale of emissions reductions required.

The table covers three stabilisation levels and a range of peak emissions dates from 2010 to 2040. The centre column shows the implied rate of global emissions reductions. The value shown is the *maximum* 10-year average rate. As shown in Figure 8.2, the rate of emissions reductions accelerates after the peak and then slows in the second half of the century. The *maximum* 10-year average rate is typically required in the 5–10 years following the peak in global emissions. The range of rates shown in each cell is important: the lower bound illustrates the rate for a low peak in global emissions (that is, action is taken to slow the rate of emissions growth prior to the peak) – in this example, these trajectories peak at not more than 10% above current levels; the upper bound assumes no substantial action prior to the peak (note that emissions in this case are still below IEA projections – see Figure 8.4).

The paths use the assumption of a maximum 10%/yr reduction rate. A symbol "-" indicates that stabilisation is not possible given this assumption. Italic figures indicate overshooting. The overshoots are numbered in brackets '[ ]' and details given below the table.

Stabilisation Level (CO_2e)	Date of peak global emissions	Global emissions reduction rate (% per year)	Percentage reduction in emissions below 2005* values	
			2050	2100
450 ppm	2010	7.0	70	75
	2020	–	–	–
500 ppm (falling to 450 ppm in 2150)	2010	3.0	50	75
	2020	4.0–6.0	60–70	75
	2030	*5.0[1]–5.5 [2]*	*50–60*	*75–80*
	2040	–	–	–
550 ppm	2015	1.0	25	50
	2020	1.5–2.5	25–30	50–55
	2030	2.5–4.0	25–30	50–55
	2040	*3.0–4.5 [3]*	*5–15*	*50–60*

Notes: overshoots: [1] to 520 ppm, [2] to 550 ppm, [3] to 600 ppm. 2005 emissions taken as 45 $GtCO_2e$/yr.
Source: Generated with the SiMCaP EQW model and averaged over multiple scenarios (Meinshausen et al. 2006)

To stabilise at 550 ppm CO_2e, global emissions would need to peak in the next 10–20 years and then fall by around 1–3% per year. Depending on the exact trajectory taken, global emissions would need to be around 25% lower than current levels by 2050, or around 30–35 $GtCO_2$.

If global emissions peak by 2015, then a reduction rate of 1% per year should be sufficient to achieve stabilisation at 550 ppm CO_2e (Table 8.2). This would mean immediate, substantial and global action to prepare for this transition. Given the current trajectory of emissions and inertia in the global economy, such an early peak in emissions looks very difficult. But the longer the peak is delayed, the faster emissions will have to fall afterwards. For a delay of 15 years in the peak, the rate of reduction must more than double, from 1% to between 2.5% and 4.0% per year, where the lower value assumes a lower peak in emissions (see Figure 8.2). Given that it is likely to be difficult to reduce emissions faster than around 3% per year (discussed in the following section), this emphasises the importance of urgent action now to slow the growth of global emissions, and therefore lower the peak.

A further 10-year delay would mean a reduction rate of at least 3% per year, assuming that action is taken to substantially slow emissions growth; if emissions growth is not slowed significantly, stabilisation at 550 ppm CO_2e may become unattainable without overshooting.

Stabilising at 450 ppm CO_2e or below, without overshooting, is likely to be very costly because it would require around 7% per year emission reductions.

Table 8.2 illustrates that even if emissions peaked in 2010, they would have to fall by around 7% per year to stabilise at 450 ppm CO_2e without overshooting[11]. This would take annual emissions to 70% below current levels, or around 13 $GtCO_2$ by 2050. This is an extremely rapid rate, which is likely to be very costly. For example, 13 $GtCO_2$ is roughly equivalent to the annual emissions from agriculture and transport alone today.

Achieving this could mean, for example, a rapid and complete decarbonisation of non-transport energy emissions, halting deforestation and substantial intensi-fication of sequestration activities. The achievability of stabilisation levels is dis-cussed in more detail in the following sections and in chapter 9.

Allowing the stock to peak at 500 ppm CO_2e before stabilising at 450 ppm (an 'overshooting' path to stabilisation, Box 8.2) would decrease the required annual reduction rate from around 7% to 3%, if emissions were to peak in 2010. However, overshooting paths, in general, involve greater risks.

An overshooting path to any stabilisation level would lead to greater impacts, as the world would experience a century or more of temperatures close to those expected for the peak level (discussed later in Figure 8.3). Given the large number of unknowns in the climate system, for example, threshold points and irreversible changes, overshooting is potentially high risk. In addition, if natural carbon absorption were to weaken as projected, it might be impossible to reduce con-centrations on timescales less than a few centuries.

Given the extreme rates of emissions cuts required to stabilise at 450 ppm CO_2e, in this case overshooting may be unavoidable. The risks involved in over-shooting can be reduced through minimising the size of the overshoot by taking substantial, early action to cut emissions.

8.6   Timing of Emissions Reductions

Pathways involving a late peak in emissions may effectively rule out lower stabil-isation trajectories and give less margin for error, making the world more vulner-able to unforeseen changes in the Earth's system.

Early abatement paths offer the option to switch to a lower emissions path if at a later date the world decides this is desirable. This might occur for example, if

[11] An atmospheric greenhouse gas level of 450 ppm is less than 10 years away, given that concen-trations are rising at 2.5 ppm per year (chapter 3). However, in the scenarios outlined in Table 8.1, aerosol cooling temporarily offsets some of the increase in greenhouse gases, giving more time to stabilise. This effect is illustrated in Box 8.2.

BOX 8.2 Overshooting paths to stabilisation

The figure below illustrates an overshooting path to stabilisation at 450 ppm CO_2e (or 400 ppm CO_2 only) – this is characterised by greenhouse gas levels peaking above the stabilisation goal and then reducing over a period of at least a century.

The light blue line shows the level of all Kyoto greenhouse gases in CO_2e (the Review definition) and the red line shows the level of carbon dioxide alone. The dark blue line shows a third measure of greenhouse gas level that includes aerosols and tropospheric ozone. This is the measure used in the Meinshausen *et al.* trajectories shown in this chapter. The gap between the two blue lines in the early period is mainly due to the cooling effect of aerosols. Critically, by 2050 the lines converge as it is assumed that aerosol emissions diminish.

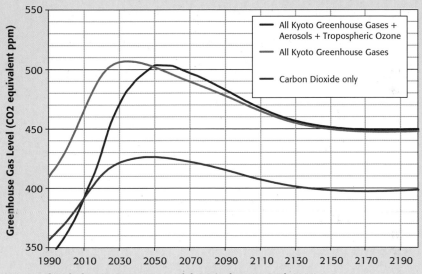

Source: Generated with the SiMCaP EQW model (Meinshausen et al. 2006)

natural carbon absorption weakened considerably (section 8.3) or the damages associated with a stabilisation goal were found to be greater than originally thought. Similarly, aiming for a lower stabilisation trajectory may be a sensible hedging strategy, as it is easier to adjust upwards to a higher trajectory than downwards to a lower one.

Late abatement trajectories carry higher risks in terms of climate impacts; overshooting stabilisation paths incur particularly high risks.

The impacts of climate change are not only dependent on the final stabilisation level, but also the path to stabilisation. Figure 8.3 shows that if emissions are accumulated more rapidly, this will lead to a more rapid rise in global temperatures. Figure 8.3 demonstrates the point made in the last section, that overshooting paths lead to particularly high risks, as temperatures rise more rapidly and to a higher level than if the target were approached from below.

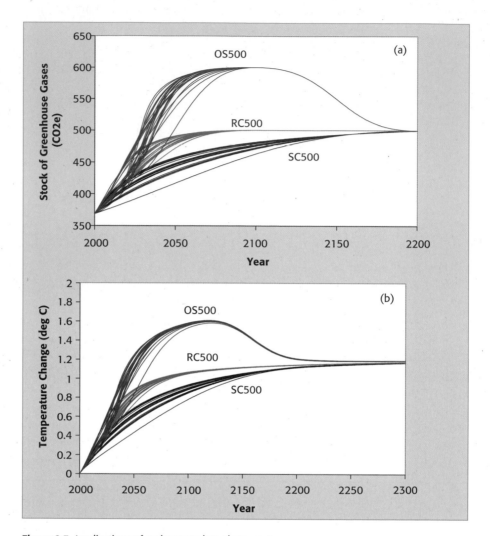

Figure 8.3 Implications of early versus late abatement

The figure below is an illustrative example of the rate of change in (a) the stock of greenhouse gases and (b) global mean temperatures, for a set of slow (SC, black), rapid (RC, blue) and overshooting (OS, red) paths to stabilisation at 500 ppm CO_2e.

On the slow paths, emissions cuts begin early and progress at a gradual pace, leading to a gradual increase in greenhouse gas concentrations and therefore, temperatures. On the rapid paths, reductions are delayed, requiring stronger emissions cuts later on. This leads to a more rapid increase in temperature as emissions are accumulated more rapidly early on. The overshooting path has even later action, causing concentrations and temperatures to rise rapidly, as well as peaking at a higher level before falling to the stabilisation level.

The higher rate of temperature rise associated with the delayed action paths (RC and OS) would increase the risk of more severe impacts. Temperatures associated with the overshooting path rise at more than twice the rate of the slow path (more than 0.2°C/decade) for around 80 years and rise to a level around 0.5°C higher. Many systems are sensitive to the rate of temperature increase, most notably ecosystems, which may be unable to adapt to such high rates of temperature change.

Source: redrawn from O'Neill and Oppenheimer (2004). The temperature calculations assume a climate sensitivity of 2.5°C (see chapter 1), giving an eventual warming of 2.1°C relative to pre-industrial.

Early abatement may imply lower long-term costs through limiting the accumulation of carbon-intensive capital stock in the short term.

Delaying action risks getting 'locked into' long-lived high carbon technologies. It is crucial to invest early in low carbon technologies. Technology policies are discussed in chapter 15.

Paths requiring very rapid emissions cuts are unlikely to be economically viable.

To meet any given stabilisation level, a late peak in emissions implies relatively rapid cuts in annual emissions over a sustained period thereafter. However, there is likely to be a maximum practical rate at which global emissions can be reduced. At the national level, there are examples of sustained emissions cuts of up to 1% per year associated with structural change in energy systems (Box 8.3).

BOX 8.3 Historical reductions in national emissions

Experience suggests it is difficult to secure emission cuts faster than about 1% per year except in instances of recession. Even when countries have adopted significant emission saving measures, national emissions often rose over the same period.

- **Nuclear power in France**: In the late 1970s, France invested heavily in nuclear power. Nuclear generation capacity increased 40-fold between 1977 and 2003 and emissions from the electricity and heat sector fell by 6% per year, against a background 125% increase in electricity demand. The reduction in total fossil fuel related emissions over the same period was less significant (0.6% per year) because of growth in other sectors.

- **Brazil's biofuels**: Brazil scaled up the share of biofuels in total road transport fuel from 1% to 25% from 1975 to 2002. This had the effect of slowing, but not reversing, the growth of road transport emissions, which rose by 2.8% per year with biofuels, but would otherwise have risen at around 3.6% per year. Total fossil fuel related emissions from Brazil rose by 3.1% pa over the same period.

- **Forest restoration in China**: China embarked on a series of measures to reduce deforestation and increase reforestation from the 1980s, with the aim of restoring forests and the environmental benefits they entail. Between 1990 and 2000 forested land increased by 18 m hectares from 16% to 18% of total land area[12]. Despite cuts in land use emissions of 29% per year between 1990 and 2000[13], total GHG emissions rose by 2.2% over the same period.

- **UK 'Dash for Gas'**: An increase in coal prices in the 1990s relative to gas encouraged a switch away from coal towards gas in power generation. Total GHG emissions fell by an average of 1% per year between 1990 and 2000.

- **Recession in Former USSR:** The economic transition and the associated downturn during the period 1989 to 1998 saw fossil fuel related emissions fall by an average of 5.2% per year.

Source for emission figures: WRI (2006) and IEA (2006).

---

[12] Zhu, Taylor, Feng (2004)
[13] Chapter 25 notes that some of this gain was offset by increased timber imports from outside China.

One is the UK 'dash for gas'; a second is France, which, by switching to a nuclear power-based economy, saw energy-related emissions fall by almost 1% per year between 1977 and 2003, whilst maintaining strong economic growth.

However, cuts in emissions greater than this have historically been associated only with economic recession or upheaval, for example, the emissions reduction of 5.2% per year for a decade associated with the economic transition and strong reduction in output in the former Soviet Union. These magnitudes of cuts suggest it is likely to be very challenging to reduce emissions by more than a few percent per year while maintaining strong economic growth.

The key reason for the difficulty in sustaining a rapid rate of annual emissions cuts is inertia in the economy. This has three main sources:

- First, capital stock lasts a number of years and for the duration it is in place, it locks the economy into a particular emissions pathway, as early capital stock retirement is likely to be costly. The extent and impact of this is illustrated in Box 8.3.
- Second, developing new lower emissions technology tends to be a slow process, because it takes time to learn about and develop new technologies. This is discussed in more detail in Chapter 9.

BOX 8.4 The implications for mitigation policy of long-lived capital stock

Power generation infrastructure typically has a very long lifespan, as does much energy-using capital stock. Examples are given below.

Infrastructure	Expected lifetime (years)
Hydro station	75++
Building	45+++
Coal station	45+
Nuclear station	30–60
Gas turbine	25
Aircraft	25–35
Motor vehicle	12–20

This means that once an investment is made, it can last for decades. A high-carbon or low-efficiency piece of capital stock will tend to lock the economy into a high emissions pathway. The only options are then early retirement of capital stock, which is usually uneconomic; or "retrofitting" cleaner technologies, which is invariably more expensive than building them in from the start. This highlights the need for policy to recognise the importance of capital stock replacement cycles, particularly at key moments, such as the next two decades when a large volume of the world's energy generation infrastructure is being built or replaced. Missing these opportunities will make future mitigation efforts much more difficult and expensive.

Source: World Business Council for Sustainable Development (2004) and IPCC (1999).

- Third, it takes time to change habits, preferences and institutional structures in favour of low-carbon alternatives. Chapter 15 discusses the importance of policy in shifting these.

These limits to the economically feasible speed of adjustment constrain the range of feasible stabilisation trajectories.

8.7 The Scale of the Challenge

Stabilisation at 550 ppm CO_2e requires emissions to peak in the next 10–20 years, and to decline at a substantial rate thereafter. Stabilisation at 450 ppm CO_2e requires even more urgent and strong action. But global emissions are currently on a rapidly rising trajectory, and under "business as usual" (BAU) will continue to rise for decades to come. The "mitigation gap" describes the difference between these divergent pathways.

To achieve stabilisation between 450 and 550 ppm CO_2e, the mitigation gap between BAU and the emissions path ranges from around 50–70 GtCO_2e per year by 2050.

Figure 8.4 plots expected trends in BAU emissions[14] against emission pathways for stabilisation levels in the range 450 to 550 ppm CO_2e. The exact size of the

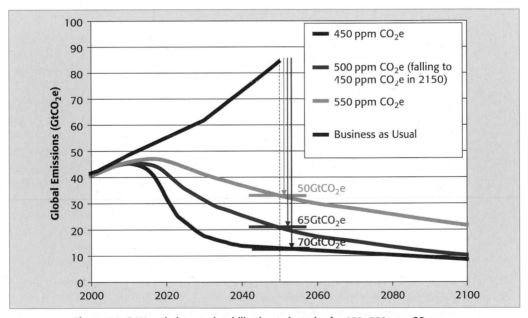

Figure 8.4 BAU emissions and stabilisation trajectories for 450–550 ppm CO_2e

The figure above shows illustrative pathways to stabilise greenhouse gas levels between 450 ppm and 550 ppm CO_2e. The blue line shows a business as usual (BAU) trajectory. The size of the mitigation gap is demonstrated for 2050. To stabilise at 450 ppm CO_2e (without overshooting) emissions must be more than 85% below BAU by 2050. Stabilisation at 550 ppm CO_2e would require emissions to be reduced by 60–65% below BAU. Table 8.2 gives the reductions relative to 2005 levels.

[14] Business as usual (BAU) used in this chapter is described in chapter 7.

mitigation gap depends on assumptions on BAU trajectories, and the stabilisation level chosen. In this example, it ranges from around 50 to 70 $GtCO_2e$ in 2050 to stabilise at 450–550 ppm CO_2e. For comparison, total global emissions are currently around 45 $GtCO_2e$ per year.

Another way to express the scale of the challenge is to look at how the relationship needs to change between emissions and the GDP and population (two of the key drivers of emissions). To meet a 550 ppm CO_2e stabilisation pathway, global average emissions per capita need to fall to half of current levels, and emissions per unit of GDP need to fall to one quarter of current levels by 2050. These are structural shifts on a major scale.

Stabilising greenhouse gas concentrations in the range 450–550 ppm CO_2e will require substantial action from both developed and developing regions.

Even if emissions from developed regions (defined in terms of Annex I countries[15]) could be reduced to zero in 2050, the rest of the world would still need to cut emissions by 40% from BAU to stabilise at 550 ppm CO_2e. For 450 ppm CO_2e, this rises to almost 80%. Emissions reductions in developed and developing countries are discussed further in Part VI.

Stabilisation at 550 ppm CO_2e or below is achievable, even with currently available technological options, and is consistent with economic growth.

An illustration of the extent and nature of technological change needed to make the transition to a low-carbon economy is provided by Socolow and Pacala (2004). They identify a 'menu' of options, each of which can deliver a distinct 'wedge' of savings of 3.7 $GtCO_2e$ (1 GtC) in 2055, or a cumulative saving of just over 90 $GtCO_2e$ (25 GtC) between 2005 and 2055. Each option involves technologies already commercially deployed somewhere in the world and no major technological breakthroughs are required. Some technologies are capable of delivering several wedges.

In their analysis, Socolow and Pacala only consider what effort is required to maintain carbon dioxide levels below 550 ppm (roughly equivalent to 610–690 ppm CO_2e when other gases are included) by implementing seven of their wedges. This is demonstrated in Figure 8.5.

While the Socolow and Pacala analysis does not explicitly explore how to stabilise at between 450 and 550 ppm CO_2e, it does provide a powerful illustration of the scale of action that would be required. It demonstrates that substantial emissions savings are achievable with currently available technologies and the importance of utilising a mix of options across several sectors. These conclusions are supported by many other studies undertaken by industry, governments and the scientific and engineering research community.

To meet a stabilisation level of 550 ppm CO_2e or below, a broad portfolio of measures would be required, with non-energy emissions being a very important part of the story.

---

[15] Annex I includes OECD, Russian Federation and Eastern European countries. This is discussed further in Part IV.

Fossil fuel related emissions from the energy sector in total would need to be reduced to below the current 26 $GtCO_2$ level, implying a very large cut from the BAU trajectory, which sees emissions more than doubling. This implies:

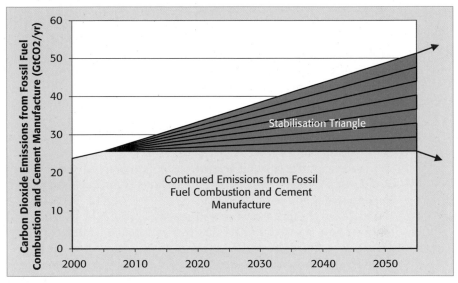

Figure 8.5 Socolow and Pacala's "wedges"

Socolow and Pacala compare a simple mitigation path for fossil fuel emissions with a projected BAU path. In the BAU path, fossil fuel CO_2 emissions grow to around 50 $GtCO_2$e in 2055. In the mitigation path, fossil fuel CO_2 emissions remain constant at 25 $GtCO_2$ until 2055. This mitigation trajectory should maintain carbon dioxide concentrations at around 550 ppm. The difference between BAU and the stabilisation trajectory is the *stabilisation triangle*. To demonstrate how these emissions savings can be achieved, this triangle is split into 7 equal wedges, each of which delivers 3.7 $GtCO_2$e (1 GtC) saving in 2055. Socolow and Pacala give a menu of fifteen measures that could achieve one wedge using currently available technologies. However, some wedges cannot be used together as they would double count emission savings. The text below gives four of these suggested measures.

Four abatement measures that could each deliver one 'wedge' (3.7 $GtCO_2$e) in 2055.

1. Replace coal power with an extra 2 million 1-MW-peak windmills (50 times the current capacity) occupying $30*10^6$ ha, on land or off shore.
2. Increase fuel economy for all cars from 30 to 60 mpg in 2055.
3. Cut carbon emissions by one-fourth in buildings and appliances in 2055.
4. Replace coal power with 700 GW of nuclear (twice the current capacity).

Source: Pacala and Socolow (2004)

- A reduction in demand for emissions-intensive goods and services, with both net reductions in demand, and efficiency improvements in key sectors including transport, industry, buildings, fossil fuel power generation.
- The electricity sector would have to be largely decarbonised by 2050, through a mixture of renewables, CCS and nuclear.
- The transport sector is still likely to be largely oil based by 2050, but efficiency gains will be needed to keep down growth; biofuels, and possibly some hydrogen or electric vehicles could have some impact. Aviation is unlikely to see technology breakthroughs, but there is potential for efficiency savings.

A portfolio of technologies will be required to achieve this. Different studies make different assumptions on what the mix might be. This is discussed further in chapter 9.

Emissions from deforestation are large, but are expected to fall gradually over the next fifty years as forest resources are exhausted (Annex 7.F). With the right policies and enforcement mechanisms in place, the rate of deforestation could be reduced and substantial emissions cuts achieved. Together with policies on afforestation and reforestation, net emissions from land-use changes could be reduced to less than zero – that is, land-use change could strengthen natural carbon dioxide absorption.

Emissions from agriculture will rise due to rising population and income, and by 2020 could be almost one third higher than their current levels of 5.7 GtCO2e. The implementation of measures to reduce agricultural emissions is difficult, but there is potential to slow the growth in emissions.

In practice the policy choices involved are complex; some actions are much more expensive than others, and there are also associated environmental and social impacts and constraints.

The following chapters discuss how to achieve cost-effective emissions cuts over the next few decades. These activities must be continued and intensified to maintain stabilisation in the long run. Over the next few centuries, section 8.3 showed that emissions would need to be brought down to approximately the level of agriculture alone today. Given that preliminary analyses indicate that it would be difficult to cut agricultural emissions (Chapter 9 and Annex 7.F), this means that, in the long term, net emissions (which includes sequestration from activities such as planting forests) from all other sectors would need to fall to zero.

8.8   Conclusions

Stabilising the stock of greenhouse gases in the range 450–550 ppm CO_2e requires urgent, substantial action to reduce emissions, firstly to ensure that emissions peak in the next few decades and secondly, to make the rate of decline in emissions as low as possible. If insufficient action is taken now to reduce emissions, stabilisation will become more difficult in the longer term, in terms of the speed of the transition required and the consequent costs of mitigation.

Stabilising greenhouse gas emissions is achievable through utilising a portfolio options, both technological and otherwise, across multiple sectors. The cost-effectiveness of these measures is discussed in detail in the following chapters.

References

The analyses of emissions trajectories presented in this chapter are based on those presented in den Elzen and Meinshausen (2005), using the same model (Meinshausen et al. 2006). These papers provide a clear and concise overview of the key issues associated with stabilisation paths. Pacala and Socolow (2004) discuss ways of filling the 'carbon gap' with currently available technologies.

Den Elzen, M. and M. Meinshausen (2005): 'Multi-gas emission pathways for meeting the EU 2°C climate target', in Schellnhuber et al. (eds.), Avoiding Dangerous Climate Change, Cambridge: Cambridge University Press.

Intergovernmental Panel on Climate Change (1999): 'Aviation and the global atmosphere', A Special Report of IPCC Working Groups I and III in collaboration with the Scientific Assessment Panel to the Montreal Protocol on Substances that Deplete the Ozone Layer [J.E. Penner, D.H. Lister, D.J. Griggs et al. (eds.)], Cambridge: Cambridge University Press.

International Energy Agency (2006): 'CO2 emissions from fossil fuel combustion on-line database', version 2005/06, Paris: International Energy Agency, available from http://data.iea.org/stats/eng/main.html

Jacobs, M. (1991): 'The green economy: Environment, sustainable development and the politics of the future, London: Pluto Press.

Jones, C.D., P.M. Cox and C. Huntingford (2006): 'Impact of climate-carbon feedbacks on emissions scenarios to achieve stabilisation', in Avoiding Dangerous Climate Change, Schellnhuber et al. (eds.), Cambridge: Cambridge University Press.

Meinshausen, M. (2006): 'What does a 2°C target mean for greenhouse gas concentrations? A brief analysis based on multi-gas emission pathways and several climate sensitivity uncertainty estimates', pp.265–280 in Avoiding dangerous climate change, H.J. Schellnhuber et al. (eds.), Cambridge: Cambridge University Press.

Meinshausen, M., D. van Vuuren, T.M.L. Wigley, et al. (2006): 'Multi-gas emission pathways to meet climate targets', Climatic Change, 75: 151–194

Murphy, J.M., D.M.H. Sexton, D.N. Barnett, et al. (2004): 'Quantification of modelling uncertainties in a large ensemble of climate change simulations', Nature 430: 768–772

O'Neill, B. and M. Oppenheimer (2004): 'Climate change impacts are sensitive to the concentration stabilization path'. Proceedings of the National Academy of Sciences 101(47): 16411–16416

Pacala, S. and R. Socolow (2004): 'Stabilization wedges: Solving the climate problem for the next 50 years with current technologies', Science, 305: 968–972

Prentice, I.C., G.D. Farquhar, M.J.R. Fasham, et al. (2001): 'The carbon cycle and atmospheric carbon dioxide', in Climate Change 2001: The Scientific Basis. Contribution of Working Group I to the Third Assessment Report of the Intergovernmental Panel on Climate Change [J. Houghton et al. (eds.)]. Cambridge: Cambridge University Press.

Ramaswamy, V. et al. (2001): 'Radiative forcing of climate change', in Climate Change 2001: The Scientific Basis. Contribution of Working Group I to the Third Assessment Report of the Intergovernmental Panel on Climate Change, [J. Houghton et al. (eds.)], Cambridge: Cambridge University Press.

US Climate Change Science Program (2006): 'Scenarios of greenhouse gas emissions and atmospheric concentrations and review of integrated scenario development and application', Public Review Draft of Synthesis and Assessment Product 2.1

Wigley, T.M.L. and S.C.B. Raper (2001): 'Interpretation of high projections for global-mean warming', Science 293: 451–454

World Business Council for Sustainable Development (2004): 'Facts and trends to 2050 – energy and climate change', World Business Council for Sustainable Development, Stevenage: Earthprint Ltd., available from www.wbcsd.org

World Resources Institute (2006): Climate Analysis Indicators Tool (CAIT) on-line database version 3.0, Washington, DC: World Resources Institute, 2006, available from http://cait.wri.org

Zhu, C., R. Taylor and G. Feng (2004): 'China's wood market, trade and the environment', available from: http://assets.panda.org/downloads/chinawoodmarkettradeenvironment.pdf

9 Identifying the Costs of Mitigation

KEY MESSAGES

Slowly reducing emissions of greenhouse gasses that cause climate change is likely to entail some costs. Costs include the expense of developing and deploying low-emission and high-efficiency technologies and the cost to consumers of switching spending from emissions-intensive to low-emission goods and services.

Fossil fuel emissions can be cut in several ways: reducing demand for carbon-intensive products, increasing energy efficiency, and switching to low-carbon technologies. **Non-fossil fuel emissions are also an important source of emission savings.** Costs will differ considerably depending on which methods and techniques are used where.

- **Reducing demand for emissions-intensive goods and services is part of the solution.** If prices start to reflect the full costs of production, including the greenhouse gas externality, consumers and firms will react by shifting to relatively cheaper low-carbon products. Increasing awareness of climate change is also likely to influence demand. But demand-side factors alone are unlikely to achieve all the emissions reductions required.

- **Efficiency gains offer opportunities both to save money and to reduce emissions,** but require the removal of barriers to the uptake of more efficient technologies and methods.

- **A range of low-carbon technologies is already available, although many are currently more expensive than fossil-fuel equivalents.** Cleaner and more efficient power, heat and transport technologies are needed to make radical emission cuts in the medium to long term. Their future costs are uncertain, but experience with other technologies has helped to develop an understanding of the key risks. The evidence indicates that efficiency is likely to increase and average costs to fall with scale and experience.

- **Reducing non-fossil fuel emissions** will also yield important emission savings. The cost of reducing emissions from deforestation, in particular, may be relatively low, if appropriate institutional and incentive structures are put in place and the countries facing this challenge receive adequate assistance. Emissions cuts will be more challenging to achieve in agriculture, the other main non-energy source.

A portfolio of technologies will be needed. Greenhouse gases are produced by a wide range of activities in many sectors, so it is highly unlikely that any single technology will deliver all the necessary emission savings. It is also uncertain which technologies will turn out to be cheapest, so a portfolio will be required for low-cost abatement.

An estimate of resource costs suggests that the annual cost of cutting total GHG to about three quarters of current levels by 2050, consistent with a 550 ppm CO_2e stabilisation level, will be in the range −1.0 to +3.5% of GDP, with an average estimate of approximately 1%. This depends on steady reductions in the cost of low-carbon technologies, relative to the cost of the technologies currently deployed, and improvements in energy efficiency. The range is wide because of the uncertainties as to future rates of innovation and fossil-fuel extraction costs. The better the policy, the lower the cost.

Mitigation costs will vary according to how and when emissions are cut. Without early, well-planned action, the costs of mitigating emissions will be greater.

9.1   Introduction

Vigorous action is urgently needed to slow down, halt and reverse the growth in greenhouse-gas (GHG) emissions, as the previous chapters have shown. This chapter considers the types of action necessary and the costs that are likely to be incurred.

This chapter outlines a conceptual framework for understanding the costs of reducing GHG emissions, and presents some upper estimates of costs to the global economy of reducing total emissions to three quarters of today's levels by 2050 (consistent with a 550 ppm CO_2e stabilisation trajectory, described in Chapter 8). The costs are worked out by looking at costs of individual emission saving technologies and measures. Chapter 10 looks at what macroeconomic models can say about how much it would cost to reduce emissions by a similar extent, and reaches similar conclusions. Chapter 10 also shows why a 450 ppm CO_2e target is likely to be unobtainable at reasonable cost.

Section 9.2 explains the nature of the costs involved in reducing emissions. Estimating the resource cost of achieving given reductions by adopting new de-carbonising technologies alone provides a good first approximation of the true cost. The costs of achieving reductions can be brought down, however, by sensible policies that encourage the use of a range of methods, including demand-switching and greater energy efficiency, so this approach to estimation is likely to exaggerate the true costs of mitigation.

Section 9.3 sets out the range of costs associated with different technologies and methods. The following four sections look at the potential and cost of tackling non-fossil fuel emissions (mainly from land-use change) and cutting fossil fuel related emissions (either by reducing demand, raising energy efficiency, or employing low-carbon technologies).

The overall costs to the global economy are estimated in Sections 9.7 and 9.8, using the resource-cost method. They are found to be in the region of −1.0 to 3.5% of GDP, with a central estimate of approximately 1% for mitigation consistent with a 550 ppm CO_2e stabilisation level. Different modelling approaches to calculating the cost of abatement generate estimates that span a wide range, as Chapter 10 will show. But they do not obscure the central conclusion that climate-change mitigation is technically and economically feasible at a cost of around 1% of GDP.

While these costs are not small, they are also not high enough seriously to compromise the world's future standard of living. A 1% cost increase is like a one-off 1% increase in the price index with nominal income unaffected (see Chapter 10). While that is not insignificant, most would regard it as manageable, and it is consistent with the ambitions of both developed and developing countries for economic growth. On the other hand, climate change, if left unchecked, could pose much greater threats to growth, as demonstrated by Part II of this Review.

9.2   Calculating the costs of cutting GHG emissions

Any costs to the economy of cutting GHG emissions, like other costs, will ultimately be borne by households.

Emission-intensive products will either become more expensive or impossible to buy. The costs of adjusting industrial structures will be reflected in pay and profits – with opportunities for new activities and challenges for old. The costs of adjusting industrial structures will be reflected in pay and profits – with opportunities for new activities and challenges for old. More resources will be used, at least for a while, in making currently emissions-intensive products in new ways, so fewer will be available for creating other goods and services. In considering how much mitigation to undertake, these costs should be compared with the future benefits of a better climate, together with the potential co-benefits of mitigation policies, such as greater energy efficiency and less local pollution, discussed in Chapter 12. The comparison is taken further in Chapter 13, where the costs of adaptation and mitigation are weighed up.

A simple first approximation to the cost of reducing emissions can be obtained by considering the probable cost of a simple set of technological and output changes that are likely to achieve those reductions.

One can measure the extra resources required to meet projected energy demand with known low-carbon technologies and assess a measures of the opportunity costs, for example, from forgone agricultural output in reducing deforestation. This is the approach taken below in Sections 9.7 and 9.8. If the costs were less than the benefits that the emissions reductions bring, it would be better to take the set of mitigation measures considered than do nothing. But there may be still better measures available[1].

The formal economics of marginal policy changes or reforms has been studied in a general equilibrium framework that includes market imperfections[2]. A reform, such as reducing GHG emissions by using extra resources, can be assessed in terms of the direct benefits of a marginal reform on consumers (the emission reduction and the reduced spending on fossil fuels), less the cost at shadow prices[3] of the extra resources.

---

[1]   A full comparison of the cost estimates used in the Review is given in Annex 9A on www.stern-review.org.uk.

[2]   See Drèze and Stern (1987 and 1990), Ahmad and Stern (1991) and Atkinson and Stern (1974).

[3]   Expressed informally, shadow prices are opportunity costs: they can often be determined by 'correcting' market prices for market imperfections. For a formal definition, see Drèze and Stern (1987 and 1990). In the models used there, the extra resources for emissions reductions represent a tightening of the general equilibrium constraint and the shadow prices times the quantities involved represent a summary of the overall general equilibrium repercussions.

The formal economics draws attention to two issues that are important in the case of climate-change policies. First, the policies need to bring about a large, or non-marginal, change. The marginal abatement cost (MAC) – the cost of reducing emissions by one unit – is an appropriate measuring device only in the case of small changes. For big changes, the marginal cost may change substantially with increased scale. Using the MAC that initially applies, when new technologies are first being deployed, would lead to an under-estimate of costs where marginal costs rise rapidly with the scale of emissions. This could happen, for example, if initially cheap supplies of raw materials start to run short. But it may over-estimate costs where abatement leads to reductions in marginal costs – for example, through induced technological improvements[4]. These issues will be discussed in more detail below, in the context of empirical estimates, where average and total costs of mitigation are examined as well as marginal costs.

It is important to keep the distinction between marginal and average costs in mind throughout, because they are likely to diverge over time. On the one hand, the marginal abatement cost should rise over time to remain equal to the social cost of carbon, which itself rises with the stock of greenhouse gases in the atmosphere (see Chapter 13). On the other hand, the average cost of abatement will be influenced not only by the increasing size of emissions reductions, but also by the pace at which technological progress brings down the total costs of any given level of abatement (see Box 9.6).

Second, as formal economics has shown, shadow prices and the market prices faced by producers are equal in a fairly broad range of circumstances, so market prices can generally be used in the calculations in this chapter. But an important example where they diverge is in the case of fossil fuels. Hydrocarbons are exhaustible natural resources, the supply of which is also affected by the market power of some of their owners, such as OPEC. As a result, the market prices of fossil fuels reflect not only the marginal costs of extracting the fuels from the ground but also elements of scarcity and monopoly rents, which are income transfers, not resource costs to the world as a whole. When calculating the offset to the global costs of climate-change policy from lower spending on fossil fuels, these rents should not be included[5].

If there are cheaper ways of reducing carbon emissions than the illustrative set of measures examined in this chapter, and there generally will be cheaper methods than any one particular set chosen by assumption, then the illustration gives an upper bound to total costs.

An illustration of how emissions can be reduced, and at what cost, by one particular simple set of actions should provide an over-estimate of the costs that will actually be involved in reducing emissions – as long as policies set the right incentives for the most cost-effective methods of mitigation to be used. Policy-makers cannot predict in detail the cheapest ways to achieve emission reductions, but they

---

[4] Similar issues to those arising for marginal changes arise in assessing instruments for reducing GHG emission although in the non-marginal changes, the distributions of costs and benefits can be of special importance.

[5] Of course, if the objective is to calculate the costs of climate-change mitigation to energy users rather than to the world as a whole, the rents can be included.

can encourage individual households and firms to find them. Thus the costs of mitigation will depend on the effectiveness of the policy tools chosen to deliver a reduction in GHG emissions. Possible tools include emission taxes, carbon taxation and tradable carbon quotas. Carbon pricing by means of any of these methods is likely to persuade consumers to reduce their spending on currently emissions-intensive products, a helpful channel of climate-change policy that is ignored in simple technology-based cost illustrations. Induced changes in the pattern of demand can help to bring down the total costs of mitigation, but consumers still suffer some loss of real income. Regulations requiring the use of certain technologies and/or imposing physical limits on emissions constitute another possible tool.

In assessing the impact of possible instruments, key issues include the structure of taxes and associated deadweight losses[6], the distribution of costs and benefits and whether or not they disrupt or enhance competitive processes. Some of these issues are tackled in simple ways by the model-based approaches to estimating costs of mitigation considered in Chapter 10. Chapter 14 considers the merits and demerits of different methods in further detail. That discussion also examines the notion of a 'double dividend' from raising taxes on 'public bads'. Chapter 11 uses UK input-output data to illustrate how extra costs proportional to carbon emissions would be distributed through the economy. If, for example, extra costs amounted to around $30/tCO_2 (£70/tC), it would result in an overall increase in UK consumer prices of around 1%. The analysis shows how this additional cost would be distributed in different ways across different sectors.

In examining whether mitigation by any particular method should be increased at the margin, and whether policies are cost-effective, the concept of marginal abatement cost (MAC) is central. There are many possible ways to reduce emissions, and many policy tools that could be used to do so. The costs of reductions will depend on the method chosen. One key test of the cost effectiveness of a possible plan of action is whether the MAC for each method is the same, as it should be if total costs are to be kept to a minimum. Otherwise, a saving could be made by switching at the margin from an option with a higher MAC to one with a lower MAC. This principle should be borne in mind in the discussion of different abatement opportunities below.

9.3   The range of abatement opportunities

The previous section set out a conceptual framework for thinking about the costs of reducing GHG emissions. The following sections look in more detail at estimates of the costs of different methods of achieving reductions.

This section sets out four main ways in which greenhouse-gas emissions can be reduced. The first is concerned with abating non-fossil-fuel emissions, and the

---

[6]   The deadweight loss to a tax on a good that raises $1 of revenue arises as follows. Suppose the government has raised $1 in tax revenue, and the consumer has paid this $1 in tax. But, in addition, the individual has reduced consumption in response to changes in prices and the firms producing the goods have lost profits. In the jargon of economics, the sum of the loss of consumer surplus and the loss of producer surplus exceeds the tax revenue.

latter three are about cutting fossil-fuel (energy-related) emissions. These are:

- To reduce non-fossil fuel emissions, particularly land use, agriculture and fugitive emissions
- To reduce demand for emission-intensive goods and services
- To improve energy efficiency, by getting the same outputs from fewer inputs
- To switch to technologies which produce fewer emissions and lower the carbon intensity of production

Annexes 7.B to 7.G[7] include some more detail on which technologies can be used to cut emissions in each sector, and the associated costs.

The array of abatement opportunities can be assessed in terms of their cost per unit of GHG reduction ($/tCO$_2$e), both at present and through time. In theory, abatement opportunities can be ranked along a continuum of the kind shown in Figure 9.1. This shows that some measures (such as improving energy efficiency and reducing deforestation) can be very cheap, and may even save money. Other measures, such as introducing hydrogen vehicles, may be a very expensive way to achieve emission reductions in the near term, until experience brings costs down.

The precise ranking of measures differs by country and sector. It may also change over time (represented in Figure 9.1 by arrows going from right to left), for example, research and development of hydrogen technology may bring the costs down in future (illustrated by the downward shift in the abatement curve over time).

For any single technology, marginal costs are likely to increase with the extent of abatement in the short term, as the types of land, labour and capital most suitable for the specific technology become scarcer. The rate of increase is likely to differ across regions, according to the constraints faced locally.

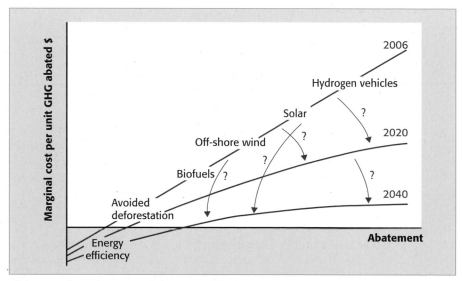

Figure 9.1 Illustrative marginal abatement option cost curve

[7] See www.sternreview.org.uk

For these reasons, flexibility in the type, timing and location of emissions reduction is crucial in keeping costs down. The implications for total costs of restricting this flexibility are discussed in more detail in Chapter 10. A test of whether there is enough flexibility is to consider whether the marginal costs of abatement are broadly the same in all sectors and countries; if not, the same amount of reductions could be made at lower cost by doing more where the marginal cost is low, and less where it is high.

9.4   Cutting non-fossil-fuel related emissions

Two-fifths of global emissions are from non-fossil fuel sources; there are opportunities here for low-cost emissions reductions, particularly in avoiding deforestation.

Non-fossil fuel emissions account for 40% of current global greenhouse-gas emissions, and are an important area of potential emissions savings. Emissions are mainly from non-energy sources, such as land use, agriculture and waste. Chapter 7 contains a full analysis of emission sources.

Almost 20% (8 $GtCO_2$/year) of total greenhouse-gas emissions are currently from deforestation. A study commissioned by the Review looking at 8 countries responsible for 70% of emissions found that, based upon the opportunity costs of the use of the land which would no longer be available for agriculture if deforestation were avoided, emission savings from avoided deforestation could yield reductions in CO_2 emissions for under $5/tCO_2$, possibly for as little as $1/tCO_2$ (see Box 9.1). In addition, large-scale reductions would require spending on administration and enforcement, as well as institutional and social changes. The transition would need to be carefully managed if it is to be effective.

Planting new forests (afforestation and reforestation) could save at least an additional 1 $GtCO_2$/yr, at a cost estimated at around $5/tCO_2$–$15/tCO_2$[8]. The full technical potential of forestry related measures would go beyond this. An IPCC report in 2000 estimated a technical potential of 4–6 $GtCO_2$/year from the planting of new forests alone between 1995 and 2050, 70% of which would come from tropical countries[9]. Revised estimates are expected from the Fourth Assessment Report of IPCC.

Changes to agricultural land management, such as changes to tilling practices[10], could save a further 1 $GtCO_2$/year at a cost of around $27/tCO_2e$ in 2020[11]. More recent analysis suggested savings could be as much as 1.8 $GtCO_2$ at $20/tCO_2$ in 2030[12]. The production of bioenergy crops would add further savings. In this chapter, this is discussed in the context of its application to emissions savings in other sectors (see Box 9.5). Biogas from animal wastes could also yield further savings.

---

[8] Benitez et al. (2005), using a land-cover database, together with econometric modelling and Sathaye et al. (2005).

[9] IPCC (2000) Chapter 3.

[10] Conservation tillage describes tillage methods that leave sufficient crop residue in place to reduce exposure of soil carbon to microbial activity and hence, conserve soil carbon stocks (IPCC (2001)).

[11] IPCC (2001). Revised estimates are expected from the Fourth Assessment Report of IPCC.

[12] Smith et al (2006, forthcoming).

BOX 9.1 The costs of reducing emissions by avoiding further deforestation

A substantial body of evidence suggests that action to prevent further deforestation would be relatively cheap compared with other types of mitigation.

Three types of costs arise from curbing deforestation. These are the opportunity cost foregone from preserving forest, the cost of administering and enforcing effective action, and the cost of managing the transition.

The opportunity cost to those who use the land directly can be estimated from the potential revenue per hectare of alternative land uses. These potential returns vary between uses. Oil palm and soya produce much higher returns than pastoral use, with net present values of up to $2000 per hectare compared with as little as $2 per hectare[13]. Timber is often harvested, particularly in South East Asia, where there is easy access to nearby markets and timber yields higher prices. Timber sales can offset the cost of clearing and converting land.

A study carried out for this Review[14] estimated opportunity costs on this basis for eight countries[15] that collectively are responsible for 70% of land-use emissions (responsible for 4.9 $GtCO_2$ today and 3.5 $GtCO_2$ in 2050 under BAU conditions). If all deforestation in these countries were to cease, the opportunity cost would amount to around $5–10 billion annually (approximately $1–2/$tCO_2$ on average). On the one hand, the opportunity cost in terms of national GDP could be higher than this, as the country would also forego added value from related activities, including processing agricultural products and timber. The size of the opportunity cost would then depend on how easily factors of production could be re-allocated to other activities. On the other hand, these estimates may overstate the true opportunity cost, as sustainable forest management could also yield timber and corresponding revenues. Furthermore, reducing emissions arising from accidental fires or unintended damage from logging may be lower than the opportunity costs suggest.

Other studies have estimated the cost of action using different methods, such as land-value studies assuming that the price of a piece of land approximates to the market expectation of the net present value of income from it, and econometric studies that estimate an assumed supply curve. In econometric studies[16], marginal costs have been projected as high as $30 t/$CO_2$ to eliminate all deforestation. High marginal values for the last pieces of forestland preserved are not inconsistent with a bottom-up approach based on average returns across large areas. These studies also suggest that costs are low for early action on a significant scale.

Action to address deforestation would also incur administrative, monitoring and enforcement costs for the government. But there would be significant economies of scale if action were to take place at a country level rather than on a project basis. Examination of such schemes suggests that the possible costs are likely to be small: perhaps $12 m to $93 m a year for these eight countries.

The policy challenges involved with avoiding further deforestation are discussed in Chapter 25.

---

[13] These figures are calculated from income over 30 years, using a discount rate of 10%, except for Indonesia, which uses 20%.
[14] See Grieg-Gran report prepared for the Stern Review (2006)
[15] Cameroon, Democratic Republic of Congo, Ghana, Bolivia, Brazil, Papua New Guinea, Indonesia, Malaysia.
[16] See for example Sohngen et al (2006)

The other main further sources of non-energy-related emissions, with estimates of economic potential for emissions reductions, are:

- Livestock, fertiliser and rice produce methane and nitrous oxide emissions. The IPCC (2001) suggested that around 1 $GtCO_2e$/year could be saved at a cost of up to \$27/tCO_2e[17] in 2020. However more recent analysis suggests that just 0.2 $GtCO_2e$/year might be saved at \$20/tCO_2e in 2030[18]. It is important to investigate ways of cutting this growing source of emissions.
- Wastage in the production of fossil fuels (so-called fugitive emissions) and other energy-related non-CO_2 emissions currently amount to around 2 $GtCO_2e$/year[19]. If fugitive emissions of non-CO_2 and CO_2 gases could be constrained to current levels, then savings could amount to 2.3 $GtCO_2e$/year and 0.2 $GtCO_2$/year respectively in 2050 on baseline levels[20].
- Waste is currently responsible for 1.4 $GtCO_2e$/year[21], of which over half is from landfill sites and most of the remainder from wastewater treatment. Reusing and recycling lead to less resources being required to produce new goods and a reduction in associated emissions. Technologies such as energy-recovering incinerators also help to reduce emissions. The IPCC estimate that 0.7 $GtCO_2e$/year could be saved in 2020, of which three quarters could be achieved at negative cost and one quarter at a cost of \$5/tCO_2e[22].
- Industrial processes used to make products such as adipic and nitric acid produce non-CO_2 emissions; the IPCC estimate that 0.4 $GtCO_2e$/year could be reduced from these sources in 2020 at a cost of less than \$3/tCO_2e[23]. The production of products such as aluminium and cement also involve a chemical process that release CO_2. Assuming that emissions from this source could be reduced by a similar proportion, savings could amount to 0.5 $GtCO_2e$ in 2050[24].

Table 9.1 summarises the possible cost-effective non-fossil fuel CO_2 emission savings for 2050 described above. These figures are very uncertain but the estimates for waste and industrial processes arguably represent a lower-end estimate because they come from IPCC studies looking at possible emission savings in 2020, and savings by 2050 could be higher. Some of these savings cost \$5/tCO_2e or less, and it is possible that more could be saved at a slightly higher cost, with the technical potential for land-use changes being particularly significant. Achieving these emission savings would mean non-fossil fuel emissions in 2050 would be almost 11 $GtCO_2e$ lower in 2050 than in the baseline case.

[17] IPCC (2001). Note this excludes savings from use of biomass and indirect emission reductions from fossil fuels via energy-efficiency measures.
[18] Smith et al (2006 forthcoming).
[19] EPA (forthcoming).
[20] Stern Review estimates. This is consistent with a mitigation scenario in which fossil-fuel use is limited to current levels or below by 2050, as in the work by Dennis Anderson described later in this chapter, and the IEA (2006) analysis discussed in Section 9.9.
[21] EPA (forthcoming).
[22] IPCC (2001)
[23] IPCC (2001)
[24] Stern Review estimate.

Table 9.1 Non-fossil-fuel emissions and savings, by sector

Sector	BAU emissions in 2050 $(GtCO_2e)$[25]	Savings in 2050 $(GtCO_2e)$	Abatement scenario emissions in 2050 $(GtCO_2e)$
Deforestation (CO_2)		3.5	
Afforestation & reforestation (CO_2)	5.0	1.0	−0.5
Land-management practices (CO_2)		1.0	
Agriculture (non-CO_2)		1.0	
Energy-related non-CO_2 emissions including fugitive emissions	18.8	2.3	14.3
Waste (non-CO_2)		0.7	
Industrial processes (non-CO_2)		0.4	
Industrial processes (CO_2)	2.1	0.5	1.6
Fugitive emissions (CO_2)	0.4	0.2	0.2
Total	**26.3**	**10.7**	**15.6**

9.5   Reducing the demand for carbon-intensive goods and services

One way of reducing emissions is to reduce the demand for greenhouse-gas-intensive goods and services like energy. Policies to reduce the amount of energy-intensive activity should include creating price signals that reflect the damage that the production of particular goods and services does to the atmosphere. These signals will encourage firms and households to switch their spending towards other, less emissions-intensive, goods and services.

Regulations, the provision of better information and changing consumer preferences can also help. If people's preferences evolve as a result of greater sensitivity to energy use, for instance to favour smaller, more fuel-efficient vehicles, they may perceive the burden from 'trading down' from a larger vehicle as small or even negative (see Chapter 17). Efforts to reduce the demand for emissions-intensive activities include reducing over-heating of buildings, reducing the use of energy-hungry appliances, and the development and use of more environmentally friendly forms of transport.

In some cases, there may be 'win-win' opportunities (for example, congestion charging may lead to a reduction in GHG emissions and also reduce journey time for motorists and bus users). But some demand-reduction measures may conflict with other policy objectives. For example, raising the cost of private transport could lead to social exclusion, especially in rural areas. Chapter 12 discusses in more detail how climate change policy may fit with other policy objectives. Part IV of the Review includes discussion of how policy can be designed to ensure that the climate change damage associated with emission-intensive goods and services is better reflected in their prices.

---

[25] For explanation of how BAU emissions were calculated, see Chapter 7.

9.6   Improving energy efficiency

Improving efficiency and avoiding waste offer opportunities to save both emissions and resources, though there may be obstacles to the adoption of these opportunities.

Energy efficiency refers to the proportion of energy within a fuel that is converted into a given final output. Improving efficiency means, for example, using less electricity to heat buildings to a given temperature, or using less petrol to drive a kilometre. The opportunities for reducing carbon emissions through the uptake of low-carbon energy sources, 'fuel switching', are not considered in this section.

The technical potential for efficiency improvements to reduce emissions and costs is substantial. Over the past century, efficiency in energy supply improved ten-fold or more in the industrial countries. Hannah's historical study[26] of the UK electricity industry, for example, reports that the consumption of coal was 10–25 lbs/kWh in 1891, 5 lbs/kWh in the first decade of the 20th century and 1.5 lbs/kWh by 1947; today it is about 0.7 lbs/kWh[27], a roughly 10-fold increase over the century in the efficiency of power generation alone.

There have also been impressive gains in the efficiency with which energy is utilised for heating, lighting, refrigeration and motive power for industry and transport, with the invention of the fluorescent light bulb, the substitution of gas for coal for heat, the invention of double glazing, the use of 'natural' systems for lighting, heating and cooling, the development of heat pumps, the use of loft and cavity-wall insulation, and many other innovations.

Furthermore, the possibilities for further gains are far from being exhausted, and are now much sought after by industry and commerce, particularly those engaged in energy-intensive processes. Many of these opportunities are yet to be incorporated fully into the capital stock. For example, the full hybrid car (which may also pave a path for electric and fuel-cell vehicles) offers the prospect of a step change in the fuel efficiency of vehicles, while new diode-based technologies have the potential to deliver marked reductions in the intensity of lighting.

However, the rate of uptake of efficiency measures is often slow, largely because of the existence of market barriers and failures. These include hidden and transaction costs such as the cost of the time needed to plan new investments; a lack of information about the available options; capital constraints; misaligned incentives; together with behavioural and organisational factors affecting economic rationality in decision-making. These are discussed in more detail in Chapter 17.

There is much debate about how big a reduction in emissions efficiency measures could in practice yield. The IEA studies summarised in Section 9.9 find that efficiency in the use of fossil fuels is likely to be the single largest source of fossil fuel-related emission savings in 2050, capable of reducing carbon emissions by up to 16 GtCO$_2$e per year by 2050. While estimates vary between studies, there is general agreement that the possibilities for further gains in efficiency are appreciable at each stage of energy conversion, across all sectors, end uses and economies.

Figure 9.2 provides a graphical representation of the estimated costs and abatement potential by 2020 for a selected sample of energy efficiency technologies across different sectors.

---

[26] See Hannah (1979)

[27] Assuming 40% thermal efficiency and a c.v. of coal of 8,000 kWh/tonne. Pounds (lbs) are a unit of weight: 1 lbs = 0.454 kg.

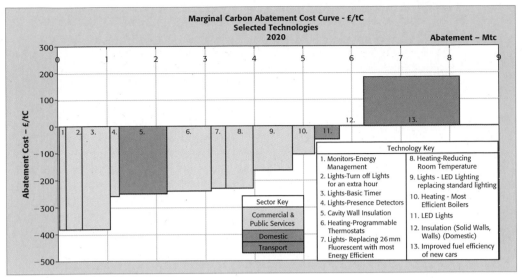

Figure 9.2 Aggregate carbon abatement cost curve for the UK – annual carbon savings by 2020[28]

Source: See notes

9.7   Low-carbon technologies

Options for low-emission energy technologies are developing rapidly, though many remain more expensive than conventional technologies.

This section examines the options for emissions reductions in the energy sector, their costs and how they are likely to move over time. The next section illustrates the costs of a set of policies in electricity and transport that could reduce emissions to levels consistent with a stabilisation path at 550 ppm CO_2e. A range of options is currently available for decarbonising energy use in electricity generation, transport and industry, all of which are amenable to significant further development. These include:-

- On and offshore wind.
- Wave and tidal energy projects.
- Solar energy (thermal and photovoltaic).
- Carbon capture and storage for electricity generation (provided the risk of leakage is minimised) – Box 9.2 sets out the state of this relatively new technology, and what is known about costs.
- The production of hydrogen for heat and transport fuels.
- Nuclear power, if the waste disposal and proliferation issues are dealt with. A new generation of reactors is being built in India, Russia and East Asia.

---

[28] This is intended to provide an indicative representation of average technology costs only (costs of individual technologies will, or course, vary). It draws together work on recent sectoral estimates undertaken by Enviros as part of the Energy Efficiency and Innovation Review (see www.defra. gov.uk/environment/energy/eeir/pdf/enviros-report.pdf) and drawing on data from the BRE and Enusim databases on the service sectors respectively, as well as Defra internal estimates for the domestic sector. The cost information presented here is based on a 3.5% social discount rate.

BOX 9.2 Carbon capture and storage (CCS)

No single technology or process will deliver the emission reductions needed to keep climate change within the targeted limits. But much attention is focused on the potential of Carbon Capture and Storage (CCS). This is the process of removing and storing carbon emissions from the exhaust gases of power stations and other large-scale emitters. If it proved effective, CCS could help reduce emissions from the flood of new coal-fired power stations planned over the next decades, especially in India and China[29].

CCS technologies have the significant advantage that their large-scale deployment could reconcile the continued use of fossil fuels over the medium to long term with the need for deep cuts in emissions. Nearly 70% of energy production will still come from fossil fuels by 2050 in the IEA's ACT MAP scenario[30]. In their base case, energy production doubles by 2050 with fossil fuels accounting for 85% of energy. The growth of coal use in OECD countries, India and China is a particular issue – the IEA forecast that without action a third of energy emissions will come from coal in 2030. Even with strong action to encourage the uptake of renewables and other low-carbon technologies, fossil fuels may still make up to half of all energy supply by 2050. Successfully stabilising emissions without CCS technology would require dramatic growth in other low-carbon technologies.

Once captured, the exhaust gases can be either processed and compressed into liquefied CO_2 or chemically changed into solid, inorganic carbonates. Captured CO_2 can be transported either through pipelines or by ship. The liquid or solid CO_2 can be stored in various ways. As a pressurised liquid, CO_2 can also be injected into oil fields to raise well pressure and increase flow rates from depleted wells. Norway's Statoil, for example, captures emissions from on-shore power stations and re-injects the captured CO_2 for such 'enhanced oil recovery' from its off-shore Sleipner oil field.

In most cases, the captured gas will be injected and stored in suitable, non-porous underground rock foundations such as depleted oil and gas wells, deep saline formations and old coalmines. Other theoretically possible but as yet largely untested ways of storing the CO_2 are to dissolve it deep within the ocean, store as an inorganic carbonate or use the CO_2 to produce hydrogen or various carbon-rich chemicals. Careful site evaluation is needed to ensure safe, long-term storage. Estimates of the potential geological storage capacity range from 1,700 to 11,100 $GtCO_2$ equivalent[31], or from to 70 to 450 years of the 2003 level of fossil-fuel-related emissions (24.5 $GtCO_2$[32]/year).

It is technically possible to capture emissions from virtually any source, but the economics of CCS favours capturing emissions from large sources producing concentrated CO_2 emissions (such as power stations, cement and petrochemical plants), to capture scale economies, and where it is possible to store the CO_2 close to the emission and capture point, to reduce transportation costs.

There are several obstacles to the deployment of CCS, including technological and cost barriers, particularly the need to improve energy efficiency in power stations adopting CCS. Others include regulatory and legal[33] barriers, such as the legal issues around the ownership of the CO_2 over long periods of time, the lack of safety standards and emission-recording

---

[29] Read (2006) discusses how if CCS technologies were to capture emissions from the use of biofuels this could create negative emissions, that is, sequestering carbon dioxide from the atmosphere.

[30] IEA (2006) – ACT MAP is a scenario that includes CCS and where emissions are constrained to near-current levels in 2050 following a technology 'push' for low-carbon technologies.

[31] IPCC (2005)

[32] Page 93 IEA (2005)

[33] At present sub-sea storage of CO_2 without enhanced oil recovery would be illegal.

guidelines. There are also environmental concerns that the CO_2 might leak or that building the necessary infrastructure might damage the local environment. Public opinion needs to be won over.

Employing CCS technology adds to the overall costs of power generation. But there is a wide range of estimates, partly reflecting the relatively untried nature of the technology and variety of possible methods and emission sources. The IPCC quotes a full range from zero to \$270 per tonne of CO_2. A range of central estimates from the IPCC and other sources[34] show the costs of coal-based CCS employment ranging from \$19 to \$49 per tonne of CO_2, with a range from \$22 to \$40 per tonne if lower-carbon gas is used. Some studies provide current estimates and some medium-term costs. A range of technologies is also considered, with and without CCS, and some with more basic generation technologies as the baseline[35]. The assumptions set have an important impact on cost estimates. The range of cost estimates will narrow when CCS technologies have been demonstrated but, until this occurs, the estimates remain speculative.

The IPCC special report on CCS suggested that it could provide between 15% and 55% of the cumulative mitigation effort until 2100. The IEA's Energy Technology Perspectives uses a scenario that keeps emissions to near current levels by 2050, with 14–16.2% of electricity generated from coal-fired power stations using CCS. This would deliver from 24.7–27.6% of emission reductions[36]. Sachs and Lackner[37] calculate that, if all projected fossil-fuel plants were CCS, it could save 17 $GtCO_2$ annually at a cost of 0.1% to 0.3% of GDP[38], and reduce global emissions by 2050 from their 554 ppm BAU to 508 ppm CO_2.

IEA modelling shows that, without CCS, marginal abatement costs would rise from \$25 to \$43 per tonne in Europe, and from \$25 to \$40 per tonne in China, while global emissions are 10% to 14% higher. This highlights the crucial role CCS is expected to play[39]. For more on international action and policies to encourage the demonstration and adoption of CCS technologies, see Section 24.3 and Box 24.8.

Reactors have either been commissioned or are close to being commissioned in France, Finland and the USA.

- Hydroelectric power, though environmental issues need to be considered and new sites will become increasingly scarce. The power output/storage ratio will also need to increase, to reduce the typical area inundated and increase the capacity of schemes to meet peak loads.
- Expansion of bioenergy for use in the power, transport, buildings and industry sectors from afforestation, crops, and organic wastes.
- Decentralised power generation, including micro-generation, combined heat and power (dCHP) using natural gas or biomass in the first instance, and hydrogen derived from low-carbon sources in the long term.
- Fuel cells with hydrogen as a fuel for transport (with hydrogen produced by a low-carbon method).
- Hybrid- and electric-vehicle technology (with electricity generated by a low-carbon method).

[34] Sources include MIT, SPRU, UK CCS, IPCC, UK Energy Review, Sachs and Lackner.
[35] Some compare CCGT, IGCC and supercritical/basic pulverised coal with and without CCS while others compare IGCC with CCS to pulverised coal without or an alternative fossil-fuel mix.
[36] At a cost of \$0.9 trillion around \$23 per tonne.
[37] Sachs and Lackner, 2005
[38] \$280 to \$840 billion at \$19–\$49/tCO_2.
[39] Page 61 IEA, 2006

Most low-carbon technologies are currently more expensive than using fossil fuels.

Estimates of the costs per unit of energy of substituting low-carbon-emitting energy sources for fossil fuels over the next 10–20 years are presented in Box 9.3; the

---

BOX 9.3 Costs of low-carbon technologies relative to fossil-fuel technologies replaced

This figure shows estimates by Anderson[40] of costs of technologies in 2015, 2025 and 2050 used to constrain fossil fuel emissions in 2050 at today's levels[41]. For most technologies, the unit cost as a proportion of the fossil-fuel alternative is expected to fall over time, largely because of learning effects (discussed below). But, as a technology comes up against increasing constraints and extends beyond its minimum efficient scale of production, the fall in unit costs may begin to reverse. The ranges quoted reflect judgements about the likely probability distribution for unit costs and allow for the variability of fossil-fuel prices (see text below and Section 9.8 for a further discussion of the treatment of uncertainties). The 0% line indicates that costs are the same as the corresponding fossil-fuel option.

Unit costs of energy technologies expressed as a percentage of the fossil-fuel alternative (in 2015, 2025, 2050)

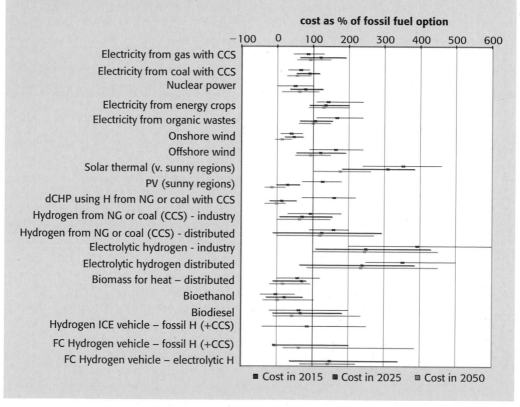

---

40 Paper by Dennis Anderson, "Costs and Finance of Carbon Abatement in the Energy Sector", published on the Stern Review web site.

41 For central electricity generation, the cost ratios reflect the generation costs (including the capital costs of generation capacity), but exclude transmission and distribution. The costs of the latter are, however, included in the estimates for decentralised generation. The average costs of energy from the fossil-fuel technologies are 2.5 p/kWh for central generation, 8 p/kWh for decentralised generation, £4/GJ for industrial gas, $6/GJ for domestic gas, and 30p/litre (exclusive of excise taxes) for vehicle fuels; all are subject to the range of uncertainties noted in the text.

technologies shown cover electricity supply, the gas markets (mainly for heat) and transport. The costs are expressed as a central estimate, with a range.

Even in the near to medium term, the uncertainties are very large. The costs of technologies vary with their stage of development, and on specific regional situations and resource endowments, including the costs and availability of specific types of fossil fuels, the availability of land for bioenergy or sites for wind and nuclear power. Other factors include climatic suitability in the case of solar 'insolation' (incident solar energy) and concentrated emission sources (in the case of CCS). In recent years, oil prices have swung over a range of more than $50 per barrel and industrial gas from $4 to $9/GJ; such swings alone can shift the relative costs of the alternatives to fossil fuels by factors or two or three or more. In principle, estimates of global costs should be based on the extraction costs of fossil fuels, not their market prices, which include a significant but uncertain proportion of rents (see Section 9.2).

The cost of technologies tends to fall over time, because of learning and economies of scale.

Historical experience shows that technological development does not stand still in the energy or other sectors. There have been major advances in the efficiency of fossil-fuel use; similar progress can also be expected for low-carbon technologies as the state of knowledge progresses.

Box 9.4 shows cost trends for selected low-carbon technologies. Economists have fitted 'learning curves' to such data to estimate how much costs might decline with investment and operating experience, as measured by cumulative investment. 'Learning' is of course an important contributor to cost reductions, but should be seen as one aspect of several factors at work. These include:

- The development of new generations of materials and design concepts through R&D and the insights gained from investment and operating experience—for example, from current efforts to develop thin-film and organic solar cells, or in new materials and catalysts for fuel cells and hydrogen production and use;
- Opportunities for batch production arising from the modularity of some emerging technologies, such as solar PV. This leads to scale economies in production; to associated technical developments in manufacture; to the reduction of lead times for investments, often to a few months, as compared with three to six years or longer for conventional plant; and to the more rapid feedback of experience;
- R&D to seek further improvements and solve problems encountered with investments in place;
- Opportunities for scale economies in the provision of supporting services in installation and use of new technologies, the costs of which are appreciable when markets are small. For example, if specialised barges are required to install and service off-shore wind turbines, the equipment is much more efficiently utilised in a farm of 100 turbines than in one with just ten, and of course if there are many offshore wind farms in the project pipeline.

The effects of the likely fall in costs with R&D and investment are reflected in the estimates for medium-term costs shown in Box 9.3. There is a general shift down in the expected costs of the alternatives to fossil fuels, in some cases to the point

BOX 9.4 Evidence on learning rates in energy technologies

A number of key energy technologies in use today have experienced cost reductions consistent with the theories of learning and scale economies. The diagram below shows historical learning rates for a number of technologies. The number in brackets gives an indication of the speed of learning: 97%, for instance, means that unit costs are 97% of their previous level after each doubling of installed capacity (3% cheaper).

Cost evolution and learning rates for selected technologies

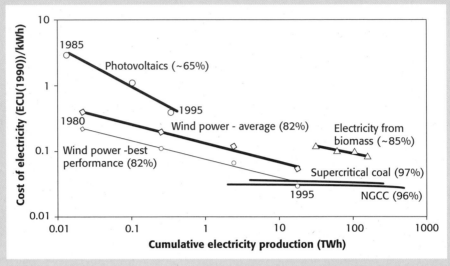

Source: IEA (2000) pp21

After early applications in manufacturing and production (1930s) and business management, strategy and organisation studies, the past decade has seen the application of learning curves as an analytical tool for energy technologies (see IEA, 2000). The majority of published learning-rate estimates relevant to climate change relate to electricity-generation technologies. In Figure 9.5 above, estimates of learning rates from different technologies[42] span a wide range, from around 3% to over 35% cost reductions associated with a doubling of output capacity.

   Using evidence on learning to project likely technology-cost changes suffers from selection bias, as technologies that fail to experience cost reductions drop out of the market and are then not included in studies. In order to correct for this, the learning and experience curves used to guide the cost exercise in this chapter take account of the high risks associated with new technologies. Moreover, the projected cost reductions are based on a far broader range of factors than just 'learning', as discussed in the main text.

where they overlap under combinations of higher fossil-fuel prices and higher rates of technical progress.

   In addition, the rankings of the technologies change, with some that are currently more expensive becoming cheaper with investment and innovation.

[42] Note different time periods for different technologies.

Examples are solar energy in sunny regions and decentralised sources of combined heat and power (see Chapter 25). Nevertheless, most unit energy costs seem likely to remain higher than fossil fuels, and policies over the next 25 years should be based on this assumption. These are, of course, in the main costs borne in the first place by the private sector, although the public power sector is large in many countries. It will be the role of policy to shift the distribution of relative costs faced by investors in the low-carbon options downward relative to those of higher carbon options (see Part IV).

Costs, constraints and energy systems in the longer term

Moving to the longer term highlights the dangers of thinking in terms of individual technologies instead of energy systems. Most technologies can be expected to progress further and see unit costs reduced. But all will run into limitations that can be addressed only by developments elsewhere in the energy system. For example:

- *Energy Storage.* With the exception of biofuels, and hydrogen and batteries using low carbon energy sources, all the low carbon technologies are concerned with the instantaneous generation of electricity or heat. A major R&D effort on energy storage and storage systems will be crucial for the achievement of a low-carbon energy system. This is important for progress in transport, and for expanding the use of low-carbon technologies, for reasons discussed below.
- *Decarbonising transport.* The transport sector is still likely to remain oil-based for several decades, and efficiency gains will be important for keeping emissions down. Increasing use of biofuels will also be important. In the long term, decarbonising transport will also depend on progress in decarbonising electricity generation and on developments in hydrogen production. The main technological options currently being considered for decarbonising transport (other than the contributions of biofuels and efficiency) are hydrogen and battery-electric vehicles. Much will depend on transport systems too, including road pricing, intelligent infrastructure, public transport and urban design.
- *Nuclear power and base-load electricity generation.* A nuclear power plant is cheapest to operate continuously as base-load generation is expensive to shut down. There are possibilities of 'load following' from nuclear power, but this will reduce capacity utilisation and raise costs. Most of the load following (where output of the power plant is varied to meet the changes in the load) will be provided by fossil-fuel plant in the absence of investments in energy-storage systems. In addition, of course, there are issues of waste disposal and proliferation to be addressed.
- *Intermittent renewables.* Renewables such as solar power and wind power only generate electricity when the natural resource is available. This leads to unpredictable and intermittent supply, creating a need for back-up generation. The cost estimates presented here allow for investment in and the fuel used in doing this, but, for high levels of market penetration, more efficient storage systems will be needed.

- *Bioenergy from crops.* Biomass can yield carbon savings in the transport, power generation, industry and building sectors. However exploitation of conventional biomass on a large scale could lead to problems of competition with agriculture for land and water resources, depending on crop practices and policies. This is discussed in Box 9.5.

BOX 9.5 Biomass: emission saving potential and costs

Biomass, the use of crops to produce energy for use in the power generation, transport, industry and buildings sectors, could yield significant emission savings in the transport, power and industry sectors. When biomass is grown, it absorbs carbon from the atmosphere during the photosynthesis process; when the crop is burnt, the carbon is released again. Biomass is not a zero carbon technology because of the emissions from agriculture and the energy used in conversion. For example, when used in transport, emissions savings from biofuel vary from 10–90% compared to petrol depending on the source of biofuel and production technique used.

Biomass crops include starch and sugar crops such as maize and sugar cane, and oil crops such as sunflower, rapeseed and palm oil. These biocrops are often referred to as first generation biomass because the technologies for converting them into energy are well developed. The highest yielding biocrops tend to be water-intensive and require good quality land, but some other biocrops can be grown on lower quality land with little water.

Research is now focusing on finding ways of converting lignocellulosic materials (such as trees, grasses and waste materials) into energy (so-called second generation technology).

The technical potential of biomass could be very substantial. On optimistic assumptions, the total primary bioenergy potential could reach 4,800–12,000 Mtoe by 2050[43] (compared with anticipated energy demand under BAU conditions of 22,000 Mtoe in 2050). Half of the primary biomass would come from dedicated cropland and half would be lignocellulosic biomass (residues and waste converted into energy). 125–150 million ha would be required for biomass crops (10% of all arable land worldwide, roughly the size of France and Spain together). However this analysis does not take into account the potentially significant impacts on local environment, water and land resources, discussed in Section 12.6. The extent to which biomass can be produced sustainably and cost effectively will depend on developments in lignocellulosic technology and to what extent marginal and low-quality land is used for growing crops.

The economically viable potential for biomass is somewhat smaller, and has been estimated at up to 2,600 Mtoe, almost a tripling of current biomass use. According to the IEA, this would result in an emission reduction of 2 to 3 $GtCO_2e$/year on baseline levels by 2050 at $25/t$CO_2$ (though the actual estimate can vary widely around this depending on oil prices). If it is assumed that one-third of biomass were used for transport fuels by 2050, for example, it could meet 10% of road transport fuel demand, compared with 1% now. This could grow to 20% under more optimistic assumptions. Biomass costs vary both by crop and by country; current production costs are lowest in parts of Southern and Central Africa and Latin America.

This analysis excludes the possible emission savings from biogas (methane and CO_2 collected from decomposing manure). This technology is discussed in Box 17.7.

---

[43] All the emission saving and cost estimates in this box come from IEA analysis. IEA (2006) and IEA (in press).

- *The availability and long-term integrity of sites for carbon capture and storage.* This may set limits to the long-term contribution of CCS to a low-carbon economy, depending on whether alternative ways of storing carbon are discovered in time. It nevertheless remains an important option given the continued use of cheap fossil fuels, particularly coal, over the coming decades.

- *Electricity and gas infrastructure.* Infrastructure services and their management would also change fundamentally with the emergence of small-scale decentralised generation and CHP, and with hydrogen as an energy-carrying and storage medium for the transport and heat markets. There will also be new opportunities for demand management through new metering and information and control technologies.

These limitations mean that all technologies will run into increasing marginal cost as their uptake expands, which will offset to some extent the likely reductions in cost as developments in the technology occur. Some of the constraints might be removed – research is ongoing, for example, on storing carbon in solid form (see Box 9.2). On the other hand, economies of scale and induced innovation will serve to bring down costs. Overall, a phased use of technologies across the board is likely to limit the cost burden of mitigating and sequestering GHGs.

In the current and next generation of investments over the next 20 years, the costs of climate change mitigation will probably be low, as some of the more familiar and easier options are exploited first. But as the scale of mitigation activities expands, at some point the problems posed by storage and the need to develop new systems and infrastructures must be overcome, particularly to meet the needs of transport. This is expected to raise costs (see below).

When looking forward over a period of several decades, however, there is also significant scope for surprises and breakthroughs in technology. This is one of the reasons why it is recommended that R&D and demonstration efforts are increased, both nationally and internationally (see discussion in Chapters 16 and 24). Such surprises may take the form of discoveries and innovations not currently factored into mainstream engineering analysis of energy futures[44].

The conclusion to be drawn from the analysis of the costs and risks associated with developing the various technologies, from the uncertainties as to their rates of development, and from the known limitations of each, is that no single technology, or even a small subset of technologies, can shoulder the task of climate-change mitigation alone. If carbon emissions are to be reduced on the scale shown to be necessary for stabilisation in Chapter 8, then policies must encourage the development of a portfolio of options; this will act both to reduce risks and improve the chances of success. Chapter 16 of this Review discusses how this can be done.

---

[44] Examples might be polymer-based PVs, with prospects for 'reel-to-reel' or batch processing; the generation of hydrogen directly from the action of sunlight on water in the presence of a catalyst (photo-electrolysis); novel methods and materials for hydrogen storage; small and large-scale energy storage devices more generally, including one known as the regenerable fuel cell; nuclear fusion; and new technologies and practices for improving energy efficiency. In addition, the technologies currently under development will also offer scope for 'learning-by-doing' and scale economies in manufacture and use.

9.8   A technology-based approach to costing mitigation of fossil fuel emissions

This section presents the results of calculations undertaken for this review by Dennis Anderson[45]. It illustrates how fossil-fuel (energy) emissions could be cut from 24 $GtCO_2e$/year in 2002 to 18 $GtCO_2e$/year in 2050 and how much this would cost. Together with the non-fossil fuel savings outlined in Table 9.1, this would be consistent with a 550 ppm CO_2e stabilisation trajectory in 2050 (outlined in Chapter 8).

A key advantage of this exercise is that it is data-driven, transparent, and easy to understand. It builds on the analysis of options in the preceding section. It illustrates one approach and establishes a benchmark. This will lead to an upward bias in the estimated costs, as there are many options, some of which will appear along the way with appropriate R&D, which will be cheaper. Like any such exercise, however, it depends on its assumptions. An independent technology-based study has recently been carried out by the IEA (see Section 9.9), which comes up with rather lower cost estimates. The next chapter reviews studies based on an economy-wide approach that attempt to incorporate some economic responses to policy instruments. These are broadly consistent with the results presented here.

The exercise here assumes that energy-related emissions at first rise and are then reduced to 18 $GtCO_2$/year through a combination of improvements in energy efficiency and switching to less emission-intensive technologies. This calculation looks only at fossil fuel related CO_2 emissions, and excludes possible knock-on effects on non-fossil fuel emissions. The precise approach used and assumptions made are detailed in the full paper[46].

Figure 9.3 presents the estimated BAU[47] energy-related CO_2 emissions over the period to 2075 and the abatement trajectory associated with reducing emissions to reach current levels by 2050. The abatement trajectory demonstrates a peak in emissions at 29 $GtCO_2$/year in 2025 before falling back to 18 $GtCO_2$/year in 2050, and falling further to reach 7 $GtCO_2$/year in 2075.

A combination of technologies, together with advances in efficiency, are needed to meet the stabilisation path.

For each technology, assumptions are made on plausible rates of uptake over time[48]. It is assumed, for the purposes of simplification, that as the rate of uptake of individual technologies is modest, they will not run into significant problems of increasing marginal cost (as discussed above in Section 9.7). Assumptions are also made on the potential for energy-efficiency improvements. These assumptions can be used to calculate an average cost of abatement. Estimates of the

---

[45] Dennis Anderson is Emeritus Professor of Energy and Environmental Studies at Imperial College London, and was formerly the Senior Energy Adviser and an economist at the World Bank, Chief Economist of Shell and an engineer in the electricity supply industry.

[46] Paper by Dennis Anderson, published on the Stern Review web site, "Costs and Finance of Carbon Abatement in the Energy Sector."

[47] This analysis assumes that fossil fuels emissions reach 61 $GtCO_2$/year in 2050 under BAU conditions. Note this is slightly greater than the BAU projection of fossil fuel emissions used in Chapter 8 and parts of Chapter 7 (of 58 $GtCO_2$/year in 2050).

[48] More detail on the assumptions made can be found in Anderson (2006).

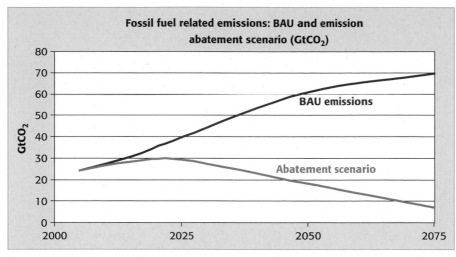

Figure 9.3 Emissions scenarios

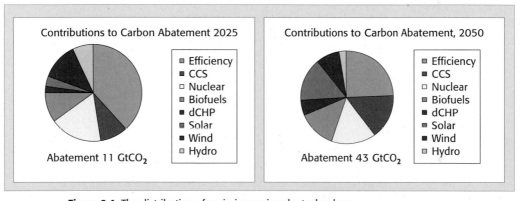

Figure 9.4 The distribution of emission savings by technology

additional contribution of energy efficiency and technological inputs to abatement are shown in Figure 9.4. The implications for sources of electricity and composition of road transport vehicle fleet are illustrated in the full paper.

An average cost of abatement per tonne of carbon can be constructed by calculating the cost of each technology (as in Box 9.3) weighted by the assumed take-up, and comparing this with the emissions reductions achieved by these technologies against fossil-fuel alternatives. This is shown in Figure 9.5, where upper and lower bounds represent best estimates of 90% confidence intervals.

The costs of carbon abatement are expected to decline by half over the next 20 years, because of the factors discussed above, and then by a further third by 2050. But the longer-term estimates of shifting to a low-carbon energy system span a very broad range, as indicated in the figure, and may even be broader than indicated here. This reflects the inescapable uncertainties inherent in forecasting over a long time period, as discussed above. It should be noted that, although average costs may fall, marginal costs are likely to be on a rising trajectory through time, in line with the social cost of carbon; this is explained in Box 9.6.

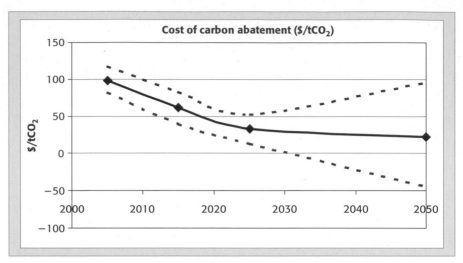

Figure 9.5 Average cost of reducing fossil fuel emissions to 18 $GtCO_2$ in 2050*

*The red lines give uncertainty bounds around the central estimate. These have been calculated using Monte Carlo analysis. For each technology, the full range of possible costs (typically ±30% for new technologies, ±20% for established ones) is specified. Similarly, future oil prices are specified as probability distributions ranging from $20 to over $80 per barrel, as are gas prices (£2–6/GJ), coal prices and future energy demands (to allow for the uncertain rate of uptake of energy efficiency). This produces a probability distribution that is the basis for the ranges given.

The global cost of reducing total GHG emissions to three quarters of current levels (consistent with 550 ppm CO_2e stabilisation trajectory) is estimated at around $1 trillion in 2050 or 1% of GDP in that year, with a range of −1.0% to 3.5% depending on the assumptions made.

Anderson's central case estimate of the total cost of reducing fossil fuel emissions to around 18 $GtCO_2$e/year (compared to 24 $GtCO_2$/year in 2002) is estimated at $930 bn, or less than 1% of GDP in 2050 (see table 9.2). In the analysis by Anderson, this is associated with a saving of 43 $GtCO_2$ of fossil fuel emissions relative to baseline, at an average abatement cost of $22/$tCO_2$/year in 2050. However these costs vary according to the underlying assumptions, so these are explored below.

The sensitivity of the cost estimates to different assumptions is presented in Table 9.3[49]; costs are shown as a percentage of world product. Over the next 20 years, it is virtually certain that the costs of providing energy will rise with the transition to low-carbon fuels, barring shocks in oil and gas supplies. Over the longer term, the estimates are less precise and, as one would expect, are sensitive to the future prices of fossil fuels, to assumptions as to energy efficiency, and indeed to the prices of the low-carbon technologies, such as carbon capture and storage.

Overall, the estimates range from −1.0% (a positive contribution to growth) to around 3.5% of world product by 2050, and are within the range of a large number of other studies discussed below in the next chapter. The estimates fan out in precisely the same way as those for the costs per tonne of carbon abatement shown in Figure 9.5, and for precisely the same reasons[50].

[49] A full specification of the different cases are set out in the full paper.
[50] Rows (ii) and (iii) provide a rough estimate of the confidence intervals associated with the estimates in row (i).

BOX 9.6 The relationship between marginal and average costs over time

It is important not to confuse average costs with marginal costs or the prevailing carbon price. The carbon price should reflect the social cost of carbon and be rising with time, because of increased additional damages per unit of GHG at higher concentrations of gases in the atmosphere (see Chapter 13). Rising prices should encourage abatement projects with successively higher marginal costs. This does not necessarily mean that the average costs will rise. Indeed, in this analysis, average costs are assumed to fall, quickly at first and then tending to level off (Figure 9.5). At any time, marginal costs will tend to be above average costs as the most costly projects are undertaken last.

At the same time, however, innovation, learning and experience – driven through innovation policy – will lower the cost of producing any given level of output using any specific technology. This is shown in the figure below, which traces the costs of a specific technology through time.

Despite more extensive use of the technology and rising costs on the margin through time (reflecting the rising carbon price), the average cost of the technology may continue to fall. The key point to note is that marginal costs might be rising even where average costs are falling (or at least rising more slowly), as a growing range of technologies are used more and more intensively.

Illustrative cost per unit of GHG abated for a specific technology

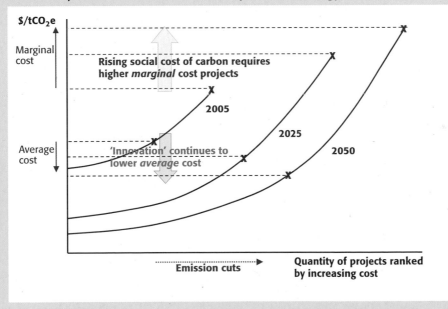

Table 9.2 Annual total costs of reducing fossil fuel emissions to 18 GtCO$_2$ in 2050

	2015	2025	2050
Average cost of abatement, \$/t CO$_2$	61	33	22
Emissions Abated GtCO$_2$ (relative to emissions in BAU)	2.2	10.7	42.6
Total cost of abatement, \$ billion per year:	134	349	930

Table 9.3 Sensitivity analysis of global costs of cutting fossil fuel emissions to 18 GtCO₂ in 2050 (costs expressed as % of world GDP)[a]

Case	2015	2025	2050
(i) Central case	0.3	0.7	1.0
(ii) Pessimistic technology case	0.4	0.9	3.3
(iii) Optimistic technology case	0.2	0.2	−1.0
(iv) Low future oil and gas prices	0.4	1.1	2.4
(v) High future oil and gas prices	0.2	0.5	0.2
(vi) High costs of carbon capture and storage	0.3	0.8	1.9
(vii) A lower rate of growth of energy demand	0.3	0.5	0.7
(viii) A higher rate of growth of energy demand	0.3	0.6	1.0
(ix) Including incremental vehicle costs[b]			
• Means	0.4	0.8	1.4
• Ranges	0.3–0.5	0.5–1.1	−0.6–3.5

[a] The world product in 2005 was approximately $35 trillion (£22 trillion at the PPP rate of $1.6/£). It is assumed to rise to $110 trillion (£70 trillion) by 2050, a growth rate of 2.5% per year, or 1 ½–2% in the OECD countries and 4–4$$% in the developing countries.
[b] Assuming the incremental costs of a hydrogen fuelled vehicle using an internal combustion engine are £2,300 in 2025 and $1400 in 2050, and for a hydrogen fuelled fuel cell vehicle £5000 in 2025 declining to £1700 by 2050. (Ranges of ~ ±30% are taken about these averages for the fuel cell vehicle.)

Assumptions as to future oil and gas prices and rates of innovation clearly make a large difference to the estimates. Combinations of a return to low oil and gas prices and low rates of innovation lead to higher costs, while higher oil and gas prices and rates of innovation point to possibly beneficial effects on growth (even ignoring the benefits of climate change mitigation). Another cost, which requires attention, is the incremental cost of hydrogen vehicles (case ix). Costly investment in hydrogen cars would significantly increase the costs associated with this element of mitigation. However, in so far as such costs might induce a switch out of mitigation in the transport sector towards alternatives with lower MACs, these estimates are likely to overstate the true cost impact on the whole economy.

The fossil fuel emission abatement costs outlined in table 9.2 together with the non-fossil fuel emission savings presented in Table 9.1 would be sufficient to bring global GHG emissions to around 34 GtCO₂e in 2050, which is consistent with a 550 ppm CO₂e stabilisation trajectory. The cost of this is estimated at under $1 trillion in 2050 (or 1% of GDP in that year).

In absolute terms, the costs are high, but are within the capacity of policies and industry to generate the required financial resources. For the economy as a whole, a 1% extra cost would be like a one-off increase in the price index by one percentage point (with unchanged nominal income profiles), although the impact will be significantly more for energy-intensive sectors (see Chapter 11). Economies have in the past dealt with much more rapid changes in relative prices and shocks from exchange-rate changes of much larger magnitude.

9.9   Other technology-based studies on cost

Other modellers have also taken a technology-based approach to looking at emissions reductions and costs. The IEA, in particular, have done detailed work based on their global energy models on the technological and economic feasibility of cutting emissions below business as usual, while also meeting other energy-policy goals.

The recent Energy Technology Perspectives report (2006) looks at a number of scenarios for reducing energy-related emissions from baseline levels by 2050. Scenarios vary in their assumptions about factors such as rates of efficiency improvements in various technologies. Box 9.7 sets out the scenarios in the report,

BOX 9.7 Sources of fossil fuel related emission savings in 2050

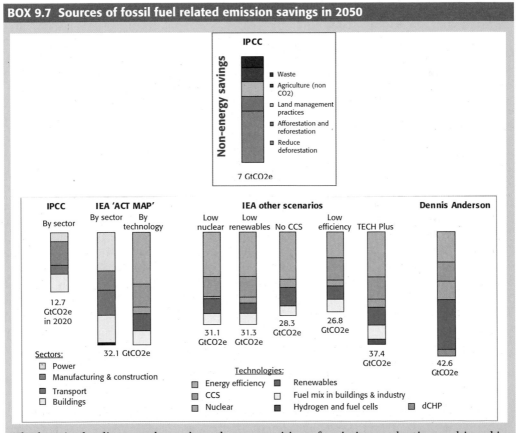

The bars in the diagram above show the composition of emissions reductions achieved in different models. The IPCC work relates to emissions savings in 2020, while the others relate to emissions savings in 2050. Separately, the IPCC have also estimated plausible emissions savings from non-energy sectors (discussed in Section 9.4).

The IPCC reviewed studies on the extent to which emissions could be cut in the power, manufacturing and construction, transport and buildings sectors. They find that for a cost of less than $25/tCO_2e$, emissions could be cut by 10.8–14.7 $GtCO_2e$ in 2020. The savings presented in the diagram are around the mid-point of this range.

The IEA Energy Technology Perspectives report sets out a range of scenarios for reducing energy-related CO_2 emissions by 2050, based on a marginal abatement cost of $25/t$CO_2$ in 2050, and investment in research and development of new technologies. The 'ACT MAP' scenario is the central scenario; the others make different assumptions on, for instance, the success of CCS technology and the ability to improve energy efficiency. Total emission savings range from 27 to 37 $GtCO_2$/year. In all scenarios, the IEA find that the CO_2 intensity of power generation is half current levels by 2050. However there is much less progress in the transport sector in all scenarios apart from TECH PLUS because further abatement from transport is too expensive. To achieve further emission cuts beyond 2050, transport would have to be decarbonised.

and compares this with work by the IPCC, as well as the technology-based estimates by Anderson set out in this chapter.

These studies make different assumptions about the quantity of abatement achieved, and the exact mix of technologies and efficiency measures used to achieve this. But all agree on some basic points. These are that energy efficiency will make up a very significant proportion of the total; that a portfolio of low-carbon technologies will be needed; and that CCS will be particularly important, given the continued use in fossil fuels.

The report also looks at the additional costs for the power-generation sector of achieving emissions cuts. It finds that in the main alternative policy scenario ('ACT MAP'), which brings energy-related emissions down to near current levels by 2050, additional investments of $7.9 trillion would be needed over the next 45 years in low-carbon power technologies, compared with the baseline scenario. However, there would be $4.5 trillion less spent on fossil-fuel power plants, in part because of lower electricity demand due to energy-efficiency improvements. In addition, there would be significant savings in transmission and distribution costs, and fuel costs; taking these into account brings the total net cost to only $100 bn over 45 years.

The forthcoming World Energy Outlook (2006) depicts an Alternative Policy Scenario that shows how the global energy market could evolve if countries were to adopt all of the policies they are currently considering related to energy security and energy-related CO_2 emissions. This Alternative Policy Scenario cuts fossil fuel emissions by more than 6 $GtCO_2$/year against the Reference Scenario by 2030, and finds that there is little difference in the investment requirements[51]. The World Energy Outlook (2006) also looks at a more radical path that would bring energy-related CO_2 emissions back to current levels by 2030, through more aggressive action on energy efficiency and transport and energy technologies, including the use of second generation biofuels and carbon capture and storage.

9.10   Conclusion

The technology-based analysis discussed in this chapter identifies one set of ways in which total GHG emissions could be reduced to three-quarters of current levels by 2050 (consistent with a 550 ppm CO_2e stabilisation trajectory). The costs

---

[51] The alternative policy scenario entails more investment in energy efficient infrastructure, but less investment in energy production and distribution. These effects broadly cancel one another out so investment requirements are about the same as in the reference case.

of doing so amount to under $1 trillion in 2050, which is relatively modest in relation to the level and expansion of economic output over the next 50 years, which in any scenario of economic success is likely to be over one hundred times this amount. They equate to around $1 \pm 2\frac{1}{2}\%$ of annual GDP – with the IEA analysis suggesting that the costs could be close to zero. As discussed in the next chapter, this finding is broadly consistent with macroeconomic modelling exercises. Chapter 10 also looks at the possible cost implications of aiming for more restrictive stabilisation targets such as 450 ppm CO_2e.

This resource-cost analysis suggests that a globally rational world should be able to tackle climate change at low cost. However, the more imperfect, less rational, and less global policy is, the more expensive it will be. This will also be examined further in the next chapter.

References

Relatively little work has been done looking cost effective emission savings possible from non-fossil fuel sources. The IPCC Working Group III Third Assessment Report (TAR, published in 2001) is the best source of non-fossil fuel emission savings, while work commissioned for the Stern Review by Grieg-Gran covers the latest analysis on tacking deforestation. IPCC has also produced estimates of fossil fuel related emission savings (2001). IPCC emission saving estimates are expected to be updated in the Fourth Assessment Report (to be published 2007). The International Energy Agency has produced a series of publications on how to cut fossil fuel emissions cost effectively; their most up to date estimates of aggregate sector-wide results are presented in the Energy Technology Perspectives (2006) and World Energy Outlook 2006 (in press). Dennis Anderson produced a simple analysis of how fossil fuel emissions can be reduced for the Stern Review, looking forward to 2075 (full paper published on Stern Review web site).

Ahmad, E. and N.H. Stern (1991): 'The theory and practice of tax reform in developing countries', Cambridge: Cambridge University Press.

Anderson, D. (2006) 'Costs and finance of carbon abatement in the energy sector'. Paper for the Stern Review ' available from www.sternreview.org.uk

Atkinson, A.B. and N.H. Stern (1974) 'Pigou, taxation and public goods', Review of Economic Studies, 41(1): 119–128

Benitez, P.C., I. McCallum, M. Obersteiner and Y. Yamagata (2005): 'Global potential for carbon sequestration: Geographical distribution, country risk and policy implications'. Ecological Economics.

Dreze, J. and N.H. Stern (1987): 'The theory of cost-benefit analysis', in Chapter 14, A.J. Auerbach and M. Feldstein (eds.), Handbook of Public Economics volume 2.

Dreze, J. and N.H. Stern (1990): 'Policy reform, shadow prices, and market prices', Journal of Public Economics, Oxford: Elsevier, 42(1): 1–45.

EPA (forthcoming): 'Global anthropogenic Non-CO_2 greenhouse-gas emissions: 1990–2020', US Environmental Protection Agency, Washington DC. Figures quoted from draft December 2005 version.

Grieg-Gran, M. (2006): 'The Cost of Avoiding Deforestation', Report prepared for the Stern Review, International Institute for Environment and Development.

Hannah, L. (1979): 'Electricity before nationalisation: A study of the development of the electricity supply industry in Britain to 1948', Baltimore and London: The Johns Hopkins University Press.

International Energy Agency (2000): 'Experience curves for energy technology policy', Paris: OECD/IEA.

International Energy Agency (2005): 'World Energy Outlook 2005', Paris: OECD/IEA.

International Energy Agency (2006): 'Energy Technology Perspectives: Scenarios and Strategies to 2050' Paris: OECD/IEA.

International Energy Agency (in press): 'World Energy Outlook 2006' Paris: OECD.

Intergovernmental Panel on Climate Change (2000): 'Land-use, land-use change and forestry', Special Report of the Intergovernmental Panel on Climate Change [Watson RT, Noble IR, Bolin B et al. (eds.)], Cambridge: Cambridge University Press.

Intergovernmental Panel on Climate Change (2001): Climate Change 2001: 'Mitigation'. Contribution of Working Group III to the Third Assessment Report of the Intergovernmental Panel on Climate Change [Metz B, Davidson O, Swart R and Pan J (eds.)], Cambridge: Cambridge University Press.

Intergovernmental Panel on Climate Change (2005): " 'IPCC Special Report on Carbon Capture and Storage", Cambridge: Cambridge University Press, November, available from http://www.ipcc.ch/activity/ccsspm.pdf

Read, P. (2006): 'Carbon Cycle Management with Biotic Fixation and Long-term Sinks', in Avoiding Dangerous Climate Change, H.J. Schellnhuber et al. (eds.), Cambridge: Cambridge University Press, pp. 373–378.

Sachs, J. and K. Lackner (2005): 'A robust strategy for sustainable energy,' Brookings Papers on Economic Activity, Issue 2, Washington, D.C: The Brookings Institution.

Sathaye, J., W. Makundi, L. Dale and P. Chan (2005 in press): 'GHG mitigation potential, costs and benefits in global forests: a dynamic partial equilibrium approach'. Energy Journal.

Smith, P., D. Martino, Z. Cai, et al. (2006 in press): 'Greenhouse-gas mitigation in agriculture', Philosophical Transactions of the Royal Society, B.

Sohngen B and R. Mendelsohn (2003): 'An optimal control model of forest carbon sequestration', American Journal of Agricultural Economics, 85 (2): 448–457, doi: 10.1111/1467-8276.00133

10 Macroeconomic Models of Costs

KEY MESSAGES

Broader behavioural modelling exercises suggest a wide range of costs of climate-change mitigation and abatement, mostly lying in the range −2 to +5% of annual GDP by 2050 for a variety of stabilisation paths. These capture a range of factors, including the shift away from carbon-intensive goods and services throughout economies as carbon prices rise, but differ widely in their assumptions about technologies and costs.

Overall, the expected annual cost of achieving emissions reductions, consistent with an emissions trajectory leading to stabilisation at around 500–550 ppm CO_2e, is likely to be around 1% of GDP by 2050, with a range of +/−3%, reflecting uncertainties over the scale of mitigation required, the pace of technological innovation and the degree of policy flexibility.

Costs are likely to rise significantly as mitigation efforts become more ambitious or sudden, suggesting that efforts to reduce emissions rapidly are likely to be very costly.

The models arriving at the higher cost estimates for a given stabilisation path make assumptions about technological progress that are pessimistic by historical standards and improbable given the cost reductions in low-emissions technologies likely to take place as their use is scaled up.

Flexibility over the sector, technology, location, timing and type of emissions reductions is important in keeping costs down. By focusing mainly on energy and mainly on CO_2, many of the model exercises overlook some low-cost abatement opportunities and are likely to over-estimate costs. Spreading the mitigation effort widely across sectors and countries will help to ensure that emissions are reduced where is it cheapest to do so, making policy cost-effective.

While cost estimates in these ranges are not trivial, they are also not high enough seriously to compromise the world's future standard of living – unlike climate change itself, which, if left unchecked, could pose much greater threats to growth (see Chapter 6). An annual cost rising to 1% of GDP by 2050 poses little threat to standards of living, given that economic output in the OECD countries is likely to rise in real terms by over 200% by then, and in developing regions as a whole by 400% or more.

How far costs are kept down will depend on the design and application of policy regimes in allowing for 'what', 'where' and 'when' flexibility in seeking low-cost approaches. Action will be required to bring forward low-GHG technologies, while giving the private sector a clear signal of the long-term policy environment (see Part IV).

Well-formulated policies with global reach and flexibility across sectors will allow strong economic growth to be sustained in both developed and developing countries, while making deep cuts in emissions.

10.1   Introduction

The previous chapter calculated the price impact of increasing fossil-fuel costs on the economy and then developed a detailed technology-based estimation approach, in which the costs of a full range of low GHG technologies were compared with fossil fuels for a path with strong carbon emissions abatement. A low-carbon economy with manageable costs is possible, but will require a portfolio of technologies to be developed. Overall, the economy-wide costs were found to be around 1% of GDP, though there remains a wide range reflecting uncertainty over future innovation rates and future fossil-fuel extraction costs and prices.

The focus of this chapter is a comparison of more detailed behavioural modelling exercises, drawing on a comparative analysis of international modelling studies. Different models have been tailored to tackle a range of different questions in estimating the total global costs of moving to a low-GHG economy. Section 10.2 highlights the results from these key models. The models impose a variety of assumptions, which are identified in section 10.3 and reflect uncertainty about the real world and differences of view about the appropriate model structure and, in turn, yield a range of costs estimates. The section investigates the degree to which specific model structures and characteristics affect cost estimates, in order to draw conclusions about which estimates are the most plausible and what factors in the real world are likely to influence them. Section 10.4 puts these estimated costs into a global perspective. There are also important questions about how these costs will be distributed, winners and losers, and the implications of countries moving at different speeds. These are examined further in Chapter 11.

The inter-model comparison reaffirms the conclusion that climate-change mitigation is technically and economically feasible with mid-century costs most likely to be around 1% of GDP, +/−3%.

Nevertheless, the full range of cost estimates in the broader studies is even wider. This reflects the greater number of uncertainties in the more detailed studies, not only over future costs and the treatment of innovation, but also over the behaviour of producers and consumers and the degree of policy flexibility across the globe. Any models that attempt to replicate consumer and producer behaviours over decades must be highly speculative. Particular aspects can drive particular results especially if they are 'run forward' into the distant future. Such are the difficulties of analysing issues that affect millions of people over long time horizons. However, such modelling exercises are essential, and the presence of such a broad and growing range of studies makes it possible to draw judgements on what are the key assumptions.

10.2   Costs of emissions-saving measures: results from other models

A broader assessment of mitigation costs requires a thorough modelling of consumer and producer behaviour, as well as the cost and choice of low-GHG technologies.

There have been a number of modelling exercises that attempt to determine equilibrium allocations of energy and non-energy emissions, costs and prices (including carbon prices), consistent with changing behaviour by firms and households. The cost estimates that emerge from these models depend on the assumptions that drive key relationships, such as the assumed ease with which consumers and producers can substitute into low-GHG activities, the degree of foresight in making investment decisions and the role of technology in the evolution of costs.

To estimate how costs can be kept as low as possible, models should cover a broad range of sectors and gases, as mitigation can take many forms, including land-use and industrial-process emissions.

Most models, however, are restricted to estimating the cost of altered fossil-fuel combustion applied mostly to carbon, as this reduces model complexity. Although fossil-fuel combustion accounts for more than three-quarters of developed economies' carbon emissions, this simplifying assumption will tend to over-estimate costs, as many low-cost mitigation opportunities in other sectors are left out (for example, energy efficiency, non-CO_2 emissions mitigation in general, and reduced emissions from deforestation; see Chapter 9). Some of the most up-to-date and extensive comparisons surveyed in this section include:

- Stanford University's Energy Modelling Forum (EMF);
- the meta-analysis study by Fischer and Morgenstern (Resources For the Future (2005));
- the International Energy Agency accelerated technology scenarios;
- the IPCC survey of modelling results;
- the Innovation Modelling Comparison Project (IMCP);
- the Meta-Analysis of IMCP model projections by Barker et al (2006);
- the draft US CCSP Synthesis and Assessment of "Scenarios of Greenhouse-Gas Emissions and Atmospheric Concentrations and Review of Integrated Scenario Development and Application" (June 2006).

The wide range of model results reflects the design of the models and their choice of assumptions, which itself reflects the uncertainties and differing approaches inherent in projecting the future.

Figure 10.1 uses Barker's combined three-model dataset to show the reduction in annual CO_2 emissions from the baseline and the associated changes in world GDP. Although most of the model estimates for 2050 are clustered in the −2 to 5% of GDP loss in the final-year cost range, these costs depend on a range of assumptions. The full range of estimates drawn from a variety of stabilisation paths and years extends from −4% of GDP (that is, net gains) to +15% of GDP costs. A notable feature, examined in more detail below, is the greater-than-proportionate increase in costs to any rise in the amount of mitigation.

This variation in cost estimates is driven by a diversity of characteristics in individual models. To take two examples, the AIM model shows a marked rise in costs towards 2100, reflecting the use of only one option – energy conservation – being induced by climate policy, so that costs rise substantially as this option becomes exhausted. At the opposite extreme, the E3MG global econometric model assumes market failures due to increasing returns and unemployed resources in the base case. This means that additional energy-sector investment, and associated innovation driven by stabilisation constraints, act to *increase* world GDP. The fact that

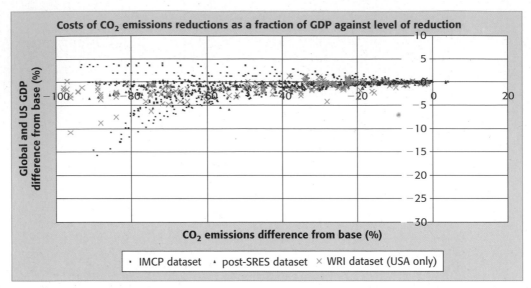

Figure 10.1 Scatter plot of model cost projections

Source: Barker et al. (2006)

there is such a broad range of studies and assumptions is welcome, making it possible to use meta-analysis[1] to determine what factors drive the results.

Model comparison exercises help to identify the reasons why the results vary.

To make sense of the growing range of estimates generated, model comparison exercises have attempted to synthesise the main findings of these models. This has helped to make more transparent the differences between the assumptions in different models. A meta-analysis of leading model simulations, undertaken for the Stern Review by Terry Barker[2], shows that some of the higher cost estimates come from models with limited substitution opportunities, little technological learning, and limited flexibility about when and where to cut emissions[3].

The meta-analysis work essentially treats the output of each model as data, and then quantifies the importance of parameters and assumptions common to the various models in generating results. The analysis generates an overarching model, based on estimates of the impacts of individual model characteristics. This can be used to predict costs as a percentage of world GDP in any year, for any given mitigation strategy. Table 10.1 shows estimated costs in 2030 for stabilisation at 450 ppm CO_2. This corresponds with approximately 500–550 ppm CO_2e, assuming adjustments in the emissions of other gases such that, at stabilisation, 10–20% of total CO_2e will be composed of non-CO_2 gases (see Chapter 8).

[1] In statistics, a meta-analysis combines the results of several studies that tackle a set of related research hypotheses. In order to overcome the problem of reduced statistical power in individual studies with small sample sizes, analysing the results from a group of studies can allow more accurate data analysis.

[2] Terry Barker is the Director of the Cambridge Centre for Climate Change Mitigation Research (4CMR), Department of Land Economy, University of Cambridge, Leader of the Tyndall Centre's research programme on Integrated Assessment Modelling and Chairman of Cambridge Econometrics. He is a Coordinating Lead Author in the IPCC's Fourth Assessment Report, due 2007, for the chapter covering mitigation from a cross-sectoral perspective.

[3] Barker et al. (2006) but see also Barker et al. (2004) and Barker (2006)

Table 10.1 Meta-analysis estimates, contributions to cost reductions

	Percentage point GDP
Worst case assumptions	−3.4
Active revenue recycling[4]	1.9
CGE model	1.5
Induced technology	1.3
Non-climate benefit	1.0
International mechanisms	0.7
'Backstop' technology	0.6
Climate benefit	0.2
Total extra assumptions	7.3
Best-case assumptions	3.9

Source: Barker et al. 2006.
Average impact of model assumptions on world GDP in 2030 for stabilisation at 450 ppm CO_2 (approximately 500–550 ppm CO_2e)
(% point levels difference from base model run)

A feature of the model is that it can effectively switch on or off the factors identified as being statistically and economically significant in cutting costs. For example, the 'worst case' assumption assumes that all the identified cost-cutting factors are switched off – in this case, costs total 3.4% of GDP. At the other extreme, the 'best case' projection assumes all the identified cost-cutting factors are active, in which case mitigation yields net benefits to the world economy to the tune of 3.9% of GDP. (Table 10.1 lists the individual estimated contributions to costs from the identified assumptions – a positive percentage point contribution represents the average reduction in costs when the parameter is 'switched on').

It is immediately obvious that no model includes all of these assumptions to the extent suggested here. This is because in practice, not all the cost-cutting factors are likely to apply to the extent indicated here, and the impact of each assumption is likely to be exaggerated (for example the active recycling parameter is based on the data from only one model[2]).

Nevertheless, the exercise suggests that the inclusion in individual models of induced technology, averted non-climate-change damages (such as air pollution) and international emission-trading mechanisms (such as carbon trading and CDM flows), can limit costs substantially.

The time paths of costs also depend crucially on assumptions contained within the modelling exercises. A number of models show costs rising as a proportion of output through to the end of the century, as the rising social cost of carbon

---

[4] The parameter can be interpreted as switched 'off' for models where no account is taken of revenues (effectively only the changes in relative prices are modelled) and 'on' for models where the revenues are recycled in some way. Unfortunately, the data underpinning this parameter are thin: among the IMCP models, only E3MG models the use of revenues at all.

requires ever more costly mitigation options to be utilised. Other models show a peak in costs around mid-century, after which point costs fall as a proportion of GDP, reflecting cost reductions resulting from increased innovation (see Section 10.3). In addition, greater disaggregation of regions, sectors and fuel types allow more opportunities for substitution and hence tend to lower the overall costs of GHG mitigation, as does the presence of a 'backstop' technology[5].

10.3   Key assumptions affecting cost estimates

Other model-comparison exercises, including studies broadening the scope to include non-carbon emissions, draw similar conclusions to the Barker study. A number of key factors emerge that have a strong influence in determining cost estimates. These explain not only the different estimates generated by the models, but also some of the uncertainties surrounding potential costs in the real world. These considerations are central, not only to generating realistic and plausible cost estimates, but also to formulating policies that might keep costs low for any given mitigation scenario. The overarching conclusion of the model studies is that costs can be moderated significantly if many options are pursued in parallel and new technologies are phased in gradually, and if policies designed to induce new technologies start sooner rather than later. The details will be quantified bellow, but the following key features are central to determining cost estimates.

Assumed baseline emissions determine the level of ambition.

The cost of stabilising GHG emissions depends on the amount of additional mitigation required. This is given by the 'mitigation gap' between the emissions goal and the 'business as usual' (BAU) emissions profile projected in the absence of climate-change policies. Scenarios with larger emissions in the BAU scenario will require greater reductions to reach specific targets, and will tend to be more costly. Large differences in baseline scenarios reflect genuine uncertainty about BAU trends, and different projected paths of global economic development.

The 2004 EMF study found a marked divergence in baseline Annex 1 (rich) country emissions projections from around 2040. Rich-country emissions begin at around 26 GtCO$_2$ at the start of the century and then rise to a range of 40–50 GtCO$_2$ by mid-century. By 2100, the range of BAU projections fans out dramatically. Some baseline scenarios show emissions dropping back towards levels at the start of the century while others show emissions rising towards 95 GtCO$_2$; there is an even spread between these extremes. These different paths encompass a variety of assumptions about energy efficiency, GHG intensity and output growth, as well as about exogenous technological progress and land-use policies.

---

[5] Under the assumption of a 'backstop' technology, energy becomes elastic in supply and the price of energy is determined independently of the level of demand. Thus, 'backstop' technologies imply lower abatement costs with the introduction of carbon taxes. The 'backstop' price may vary through technical change. For example, wind, solar, tidal and geothermal resources may serve as 'backstop' technologies, whereas nuclear fission is generally not, because of its reliance on a potentially limited supply of uranium. In practice, very few technologies will be entirely elastic in supply: even wind farms may run out of sites, and the best spots for catching and transporting electricity from the sun may be exhausted quickly.

Technological change will determine costs through time.

Costs vary substantially between studies, depending on the assumed rate of technological learning, the number of learning technologies included in the analysis and the time frame considered[6]. Many of the higher cost estimates tend to originate from models without a detailed specification of alternative technological options. The Barker study found that the inclusion of induced technical change could lower the estimated costs of stabilisation by one or two percentage points of GDP by 2030 (see table 10.1). All the main studies found that the availability of a non-GHG 'backstop' (see above) lowered predicted costs if the option came into play. Chapter 16 shows that climate policies are necessary to provide the incentive for low-GHG technologies. Without a 'loud, legal and long' carbon price signal, in addition to direct support for R&D, the technologies will not emerge with sufficient impact (see Part IV).

How far costs are kept down will depend on the design and application of policy regimes in allowing for 'what', 'where' and 'when' flexibility in seeking low-cost approaches. Action will be required to bring forward low-GHG technologies, while giving the private sector a clear signal of the long-term policy environment (see Part IV).

Abatement costs are lower when there is 'what' flexibility: flexibility over how emission savings are achieved, with a wide choice of sectors and technologies and the inclusion of non-CO$_2$ emissions.

Flexibility between sectors. It will be cheaper, per tonne of GHG, to cut emissions from some sectors rather than others because there will be a larger selection of better-developed technologies in some. For example, the range of emission-saving technologies in the power generation sector is currently better developed than in the transport sector. However, this does not mean that the sectors with a lack of technology options do nothing in the meantime. Indeed, innovation policies will be crucial in bringing forward clean technologies so that they are ready for introduction in the long term. The potential for cost-effective emission saving is also likely to be less in those sectors in which low-cost mitigation options have already been undertaken. Similarly, flexibility at emissions from a range of consumption options and economic sectors is also likely to reduce modelled costs. Models that are restricted to a narrow range of sectors with inelastic demand, for example, parts of the transport sector, will tend to estimate very high costs for a given amount of mitigation (see Section 10.2).

Flexibility between technologies. Using a portfolio of technologies is cheaper because individual technologies are prone to increasing marginal costs of abatement, making it cheaper to switch to an alternative technology or measure to secure further savings. There is also a lot of uncertainty about which technologies will turn out to be cheapest so it is best to keep a range of technology options open. It is impossible to predict accurately which technologies will experience breakthroughs that cause costs to fall and which will not.

---

[6] Grubb et al. (2006). See also Grubler et al. (1999), Nakićenović (2000), Jaffe et al. (2003) and Köhler (2006)

Flexibility between gases. Broadening the scope of mitigation in the cost-modelling exercises to include non-CO_2 gases has the potential to lower the costs by opening up additional low-cost abatement opportunities. A model comparison by the Energy Modelling Forum[7] has shown that including non-carbon greenhouse gases (NCGGs) in mitigation analysis can achieve the same climate goal at considerably lower costs than a CO_2-only strategy. The study found that model estimates of costs to attain a given mitigation path fell by about 30–40% relative to a CO_2-only approach, with the largest benefits occurring in the first decades of the scenario period, with abatement costs on the margin falling by as much as 80%. It is notable that the impacts on costs are very substantial in comparison to the much smaller contribution of NCGGs to overall emissions, reflecting the low-cost mitigation options and the increase in flexibility of abatement options from incorporating a multi-gas approach[8,9].

However, given that climate change is a product of the stock of greenhouse gases in the atmosphere, the lifetime of gases in the atmosphere also has to be taken into account (see Chapter 8). Strategies that focus too much on some of the shorter-lived gases risk locking in to high future stocks of the longer-lived gases, particularly CO_2.

Some countries can cut emissions more cheaply than other countries, so 'where' flexibility is important.

Flexibility over the distribution of emission-saving efforts across the globe will also help to lower abatement costs, because some countries have cheaper abatement options than others[10].

- **The natural resource endowments of some countries will make some forms of emissions abatement cheaper than in other countries.** For example, emission reduction from deforestation will only be possible where there are substantial deforestation emissions. Brazil is well suited to growing sugar, which can be used to produce biofuel cheaply, although, to the extent that biofuels can be transported, other countries are also likely to benefit. Brazil, like many other developing countries, also has a very good wind resource. In addition, the solar resources of developing countries are immense, the incident solar energy per m^2 being 2–2.5 times greater than in most of Europe, and it is better distributed throughout the year (see Chapter 9).
- **Countries that have already largely decarbonised their energy sector are likely to find further savings there expensive.** They will tend to focus on the

---

[7] EMF-21; see Weyant et al. (2004), van Vuuren et al. (2006)

[8] The EMF found that as much as half of agriculture, waste and other non-CO_2 emissions could be cut at relatively low cost. The study looked at how the world might meet a stabilisation objective if it selected the least-cost abatement among energy-related CO_2 emissions and non-CO_2 emissions (but not land use). Two stabilisation scenarios were compared (aimed at stabilising emissions to 650 ppm CO_2e): one in which only energy-related CO_2 emissions could be cut; and another in which energy-related CO_2 emissions and non-CO_2 gases could be reduced. In the 'energy-related CO_2 emissions only' scenario, CO_2 emissions fall by 75% on baseline levels in 2100. Some non-CO_2 gases also fall as an indirect consequence. In the multi-gas scenario, CO_2 emissions fall by a lesser extent (67% by 2100) and there are significant cuts in the non-CO_2 gases (CH4 falling by 52%, N_2O by 38%, F-gases by 73%). CO_2 remains the major contributor to emission savings, because it represents the biggest share in GHG emissions.

[9] Babiker et al. (2004)

[10] Discussion of which countries should pay for this abatement effort is a separate question. Part IV looks at how policy should be designed to achieve emissions reductions, while Chapter 11 examines the possible impacts on national competitiveness.

scope for emissions cuts elsewhere. Energy-efficiency measures are typically among the cheapest abatement options, and energy efficiency varies hugely by country. For example, unit energy and carbon intensity are particularly low in Switzerland (1.2 toe/$GDP and 59 tc/$GDP respectively in 2002), reflecting the compositional structure of output and the use of low-carbon energy production. By contrast, Russia and Uzbekistan remain very energy- and carbon-intensive (12.5 toe/$GDP and 840 tc/$GDP respectively for Uzbekistan in 2002), partly reflecting aging capital stock and price subsidies in the energy market (see, for example, Box 12.3 on gas flaring in Russia).

- **It will also be cheaper to pursue emission cuts in countries that are in the process of making big capital investments.** The timing of emission savings will also differ by country, according to when capital stock is retired and when savings from longer-term investments such as innovation programmes come to fruition. Countries such as India and China are expected to increase their capital infrastructure substantially over coming decades, with China alone accounting for around 15% of total global energy investment. If they use low-emission technologies, emission savings can be 'locked in' for the lifetime of the asset. It is much cheaper to build a new piece of capital equipment using low-emission technology than to retro-fit dirty capital stock.

The Barker study also found that the presence of international mechanisms under the Kyoto Protocol (which include international emissions trading, joint implementation and the Clean Development Mechanism) allow for greater flexibility about where cuts are made across the globe. This has the potential to reduce costs of stabilising atmospheric GHG concentrations at approximately 500–550 ppm CO_2e by almost a full percentage point of world GDP[11,12]. Similarly, Babiker et al. (2001) concluded that limits on 'where' flexibility, through the restriction of trading between sectors of the US economy, can substantially increase costs, by up to 80% by 2030.

Changes in consumer and producer behaviour through time are uncertain, so 'when' flexibility is desirable.

The timing of emission cuts can influence total abatement cost and the policy implications. It makes good economic sense to reduce emissions at the time at which it is cheapest to do so. Thus, to the extent that future abatement costs are expected to be lower, the total cost of abatement can be reduced by delaying emission cuts. However, as Chapter 8 set out, limits on the ability to cut emissions rapidly, due to the inertia in the global economy, mean that delays to action can imply very high costs later.

---

[11] Richels et al. (1998) found that international co-operation through trade in emission rights is essential to reduce mitigation costs of the Kyoto protocol. The magnitude of the savings would depend on several factors including the number of participating countries and the shape of each country's marginal abatement cost curve. Weyant and Hill (1999) assessed the importance of emissions permits and found that they had the potential to reduce OECD costs by 0.1ppt to 0.9ppt by as early as 2010.

[12] For example, Reilly et al. (2004) compare the effectiveness of two GHG abatement regimes: a global regime of non-CO_2 gas abatement, and a regime that is globally less comprehensive and mimics the present ratification of the Kyoto Protocol. The study found that, by 2100, the abatement programme that is globally comprehensive, but has limited coverage of gases (non-CO_2 only), might be as much as twice as effective at limiting global mean temperature increases and less expensive than the Kyoto framework.

Also, as discussed above, the evolution of energy technologies to date strongly suggests that there is a relationship between policy effort on innovation and technology cost. Early policy action on mitigation can reduce the costs of emission-saving technologies (as discussed in Chapter 15).

Cost-effective planning and substituting activities across time require policy stability, as well as accurate information and well-functioning capital markets. Models that allow for perfect foresight together with endogenous investment possibilities tend to show much reduced costs. Perfect foresight is not an assertion to be taken literally, but it does show the importance of policy being transparent and predictable, so that people can plan ahead efficiently.

The ambition of policy has an impact on estimates of costs.

A common feature of the model projections was the presence of increasing marginal costs to mitigation. This applies not just to the total mitigation achieved, but also the speed at which it is brought about. This means that each additional unit reduction of GHG becomes more expensive as abatement increases in ambition and also in speed. Chapter 13 discusses findings from model comparisons and shows a non-linear acceleration of costs as more ambitious stabilisation paths are pursued. The relative absence of energy model results for stabilisation concentrations below 500 ppm CO_2e is explained by the fact that carbon-energy models found very significant costs associated with moving below 450 ppm, as the number of affordable mitigation options was quickly exhausted. Some models were unable to converge on a solution at such low stabilisation levels, reflecting the absence of mitigation options and inflexibilities in the diffusion of 'backstop' technologies.

In general, model comparisons find that the cost of stabilising emissions at 500–550 ppm CO_2e would be around a third of doing so at 450–500 ppm CO_2e.

The lesson here is to avoid doing too much, too fast, and to pace the flow of mitigation appropriately. For example great uncertainty remains as to the costs of very deep reductions. Digging down to emissions reductions of 60–80% or more relative to baseline will require progress in reducing emissions from industrial processes, aviation, and a number of areas where it is presently hard to envisage cost-effective approaches. Thus a great deal depends on assumptions about technological advance (see Chapters 9, 16 and 24). The IMCP studies of cost impacts to 2050 of aiming for around 500–550 ppm CO_2e were below 1% of GDP for all but one model (IMACLIM), but they diverged afterwards. By 2100, some fell while others rose sharply, reflecting the greater uncertainty about the costs of seeking out successive new mitigation sources.

Consequently, the average expected cost is likely to remain around 1% of GDP from mid-century, but the range of uncertainty is likely to grow through time.

Potential co-benefits need to be considered.

The range of possible co-benefits is discussed in detail in Chapter 12. The Barker meta-analysis found that including co-benefits could reduce estimated mitigation costs by 1% of GDP. Such models estimate, for example, the monetary value of improved health due to reduced pollution and the offsetting of allocative efficiency losses through reductions in distortionary taxation. Pearce (1996) highlighted

BOX 10.1 The relationship between marginal and average carbon cost estimates

It is important to distinguish marginal from average carbon costs. In general, the marginal cost of carbon mitigation will rise as mitigation becomes more expensive, as low-cost options are exhausted and diminishing returns to scale are encountered. But the impact on overall costs to the economy is measured by the average cost of mitigation, which will be lower than those on the margin.

In some cases, for example, where energy efficiency increases or where induced technology reduces the costs of mitigation, average costs might not rise and could be zero or negative, even where costs on the margin are positive and rising. The correlation from plotting carbon tax rates against losses in GDP from the IMCP study is only 0.37; a survey for the US Congress by Lasky (2003) showed a similar low correlation from model results on the US costs of Kyoto (2003, p.92).

Changes in the marginal carbon cost are related, but do not correspond one-for-one, to the average cost of mitigation. The social cost of carbon will tend to rise as the stock of atmospheric GHGs, and associated damages, rises. The marginal abatement cost will also rise, reflecting this, but average abatement costs may fall (see Chapter 9). This explains why some of the models with a high social cost of carbon, and corresponding high carbon price, show very low average costs. The high carbon price is assumed to be necessary to induce benefits from energy efficiency, technological innovation and other co-benefits such as lower pollution. In some cases, these result in a reduction in average costs that raise GDP above the baseline when a stabilisation goal is imposed. This also explains why the work by Anderson (Chapter 9) shows a falling average cost of carbon through time consistent with rising costs on the margin.

Most models represent incentives to change emissions trajectories in terms of the marginal carbon price required. This not only changes specific investments according to carbon content, but also triggers technical change through the various mechanisms considered in the models, including through various forms of knowledge investment. The IMCP project (Grubb et al. 2006) charts the evolution of carbon prices required to achieve stabilisation and shows that they span a wide range, with a variety of profiles through time. For stabilisation at 450 ppm (around 500–550 ppm CO_2e), most models show carbon prices start off low and rise to US\$360/tCO_2 +/− 150% by 2030, and are in the range US\$180-900/tCO_2 by 2050, as the social cost of carbon increases and more expensive mitigation options need to be encouraged on the margin in order to meet an abatement goal.

After that, they diverge significantly: some increase sharply as the social cost of carbon continues to rise. Others level off as the carbon stock and corresponding social cost of carbon stabilise and a breadth of mitigation options and technologies serve to meet the stabilisation objective. Once again, rising marginal carbon prices need not mean that GDP impacts grow proportionately, as new technologies and improved energy efficiency will reduce the economy's dependence on carbon, narrowing the economic base subject to the higher carbon taxation.

studies from the UK and Norway showing benefits of reduced air pollution that offset the costs of carbon dioxide abatement costs by between 30% and 100%. A more recent review of the literature[13] came to similar conclusions, noting that developing countries would tend to have higher ancillary benefits from GHG mitigation

---

[13] OECD et al. 2000

compared with developed countries, since, in general, they currently incur greater costs from air pollution.

Analyses carried out under the Clean Air for Europe programme suggest cost savings as high as 40% of GHG mitigation costs are possible from the co-ordination of climate and air pollution policies[14]. Mitigation through land-use reform has implications for social welfare (including enhanced food security and improved clean-water access), better environmental services (such as higher water quality and better soil retention), and greater economic welfare through the impact on output prices and production[15]. These factors are difficult to measure with accuracy, but are potentially important and are discussed further in Chapter 12.

10.4   Understanding the scale of total global costs

Overall, the model simulations demonstrate that costs depend on the design and application of policy, the degree of global policy flexibility, and, whether or not governments send the right signals to markets and get the most efficient mix of investment. If mitigation policy is timed poorly, or if cheap global mitigation options are overlooked, the costs can be high.

To put these costs into perspective, the estimated effects of even ambitious climate change policies on economic output are estimated to be small – around 1% or less of national and world product, averaged across the next 50 to 100 years – provided policy instruments are applied efficiently and flexibly across a range of options around the globe. This will require early action to retard growth in the stock of GHGs, identify low-cost opportunities and prevent locking-in to high GHG infrastructure. The numbers involved in stabilising emissions are potentially large in absolute terms – maybe hundreds of billions of dollars annually (1% of current world GDP equates to approximately $350–400 billion) – but are small in relation to the level and growth of output.

For example, if mitigation costs 1% of world GDP by 2100, relative to the hypothetical 'no climate change' baseline, this is equivalent to the average growth rate of annual GDP over the period dropping from 2.5% to 2.49%. GDP in 2100 would still be approximately 940% higher than today, as opposed to 950% higher if there were no climate-change to tackle. Alternatively, one can think of the price level being 1% higher through time, with the same GDP level and growth rate. The same level of real output is reached around four or five months later than would be the case in the absence of mitigation costs[16].

The illustration of costs above assumes no change in the baseline growth rate relative to the various mitigation scenarios, that is, it takes no account of climate-change damages. In practice, by 2100, the impacts of climate change make it likely that the 'business as usual' level of world GDP will be lower than the post-mitigation profile (see Chapters 6 and 13). Hence stabilising at levels around 500–550 ppm CO_2e need not cost more than a year's deferral of economic growth over the century

---

[14] Syri et al. 2001

[15] A difficulty in evaluating the exact benefits of climate polices to air pollution is the different spatial and temporal scales of the two issues being considered. GHGs are long-lived and hence global in their impact while air pollutants are shorter-lived and tend to be more regional or local in their impacts.

[16] See, for example, Azar (2002)

with broad-based, sensible and comprehensive policies. Once damages are accounted for, mitigation clearly protects growth, while failing to mitigate does not.

The mitigation costs modelled in this chapter are unlikely to make the same kind of material difference to household lifestyles and global welfare as those which would arise with the probable impact of dangerous climate change, in the absence of mitigation (see section II). The importance of weighing together the costs, benefits and uncertainties through time is emphasised in Chapter 13.

10.5 Conclusion

This chapter draws on a range of model estimates with a variety of assumptions. A detailed analysis of the key drivers of costs suggests the estimated effects of ambitious policies to stabilise atmospheric GHGs on economic output can be kept small, rising to around 1% of national and world product averaged over the next fifty years.

By 2050, models suggest a plausible range of costs from −2% (net gains) to +4% of GDP, with this range growing towards the end of the century, because of the uncertainties about the required amount of mitigation, the pace of technological innovation and the efficiency with which policy is applied across the globe. Critically, these costs rise sharply as mitigation becomes more ambitious or sudden.

Whether or not the costs are actually minimised will depend on the design and application of policy regimes in allowing for 'what', 'where' and 'when' flexibility, and taking action to bring forward low-GHG technologies while giving the private sector a clear signal of the long-term policy environment.

These costs, however, will not be evenly distributed. Issues around the likely distribution of costs are explored in the next chapter. Possible opportunities and benefits arising from climate-change policy also need to be taken into account in any serious consideration of what the true costs will be, and of the implications of moving at different speeds. These are examined further in Chapter 12.

References

Volume 2 of Jorgenson's book "Growth" and also Ricci (2003) provide a rigorous and thorough basis for understanding the theoretical framework against which to assess the costs of environmental regulation and GHG mitigation. The special edition of Energy Economics 2004 is also recommended and includes a crystal-clear introduction to modelling issues by John Weyant. The study by Fischer and Morgenstern (2005) offers a comprehensive introduction to the key modelling issues, explaining divergent modelling results in terms of modelling assumptions, while highlighting the importance of 'what', 'where', 'when' flexibility. Van Vuuren et al. (2006) are among those who take this a step further by allowing for multi-gas flexibility in modelling scenarios.

Edenhofer et al. (2006) review the results of ten IMCP energy modelling exercises examining the costs associated with different stabilisation paths, the dynamics of carbon prices and the importance of key assumptions, in particular, induced innovation. Barker et al. (2006) use a more a quantitative approach to synthesise the results of different model projections and examine the importance of induced technological innovation. Using a meta-analysis estimation technique, they attempt to quantify how important various modelling assumptions are in determining cost estimates for different mitigation scenarios.

Azar C. and S.H. Schneider (2002): 'Are the economic costs of stabilising the atmosphere prohibitive?' Ecological Economics **42**:(2002) 73–80

Babiker M.H., J. Reilly, J.M. Mayer, et al. (2001): 'The emissions prediction and policy analysis (EPPA) model: revisions, sensitivities and comparisons of results', Cambridge, MA: MIT Press.

Babiker M.H., J. Reilly, J.M. Mayer, et al. (2004): 'Modelling non-CO_2 Greenhouse-Gas Abatement', Netherlands Springer pp175–186, October.

Barker T., J. Köhler and M. Villena (2002): 'The costs of greenhouse-gas abatement: A Meta-Analysis of post-SRES mitigation scenarios.' Environmental Economics and Policy Studies 5(2): 135–166

Barker, T., M.S. Qureshi and J. Köhler (2006): 'The costs of greenhouse-gas mitigation with induced technological change: A Meta-Analysis of estimates in the literature', 4CMR, Cambridge Centre for Climate Change Mitigation Research, Cambridge: University of Cambridge.

Barker, T. (2004): 'Economic theory and the transition to sustainability: a comparison of general-equilibrium and space-time-economics approaches', Working Paper 62, November, Norwich: Tyndall Centre for Climate Change Research.

Edenhofer, O., K. Lessmann and M. Grubb (2005): 'Bringing down the costs of climate protection – Lessons from a modelling comparison exercise', Submission to Stern Review on the Economics of Climate Change, December.

Edenhofer, O., M. Grubb, K. C. Kemfert et al. (2006): 'Induced technological change: exploring its implications for the economics of atmospheric stabilization: Synthesis Report from the Innovation Modelling Comparison Project', The Energy Journal, Endogenous Technological Change and the Economics of Atmospheric Stabilisation Special Issue, April, 2006.

Fischer, C. and R.D. Morgenstern (2005): 'carbon abatement costs: why the wide range of estimates?' Resources for the Future, September, Discussion Paper 03–42 November.

Grubb, M., C. Carraro and J. Schellnhuber (2006): 'Technological change for atmospheric stabilisation: introductory overview to the Innovation Modelling Comparison Project' in The Energy Journal, Endogenous Technological Change and the Economics of Atmospheric Stabilisation Special Issue, April, 2006.

Grubb, M., J. Hourcade, O. Edenhofer and N. Nakicenovic (2005): 'Framing the Economics of Climate Change: an international perspective', Submission to the Stern Review on the Economics of Climate Change, December.

Grübler, A., N. Nakićenović and D.G. Viictor (1999): 'Modelling technological change: implications for the global environment', Annual Review Energy Environ 24: 545–69

Jaffe A.B., R.G. Newell and R.N. Stavins, (2003): 'Technological change and the environment' Handbook of Environmental Economics, Maler KG and Vincent JR Ed., 1: 461–516

Jorgenson, D.W. and P.J. Wilcoxen (1990): 'Environmental regulation and U.S. economic growth' Rand Journal of Economics, 21, No. 2, (Summer): 314–340

Jorgenson, D.W. (1998): 'Growth' Vol 2, Energy, the Environment, and Economic Growth. Cambridge, MA: MIT Press.

Köhler, J., M. Grubb, D. Popp and O. Edenhofer (2006): 'The transition to endogenous technical change in climate-economy models: a technical overview to the Innovation Modelling Comparison Project'. Energy Journal, 27: 17–55

Lasky, M. (2003): 'The economic costs of reducing emissions of greenhouse gases: a survey of economic models'. CBO Technical Paper No. 2003-03 (May), Washington, DC: CBO.

Nakićenović, N. and A. Gritsevskyi. (2000): 'Modeling uncertainty of induced technological change', Laxenburg: International Institute for Applied Systems Analysis.

Organization for Economic Co-operation and Development, Resources for the Future, Intergovernmental Panel on Climate Change, and World Resources Institute, (2000): 'Ancillary benefits and costs of greenhouse-gas mitigation. Workshop on assessing the ancillary benefits and costs of greenhouse-gas mitigation strategies'. Washington DC, OECD.

Pearce, D.W., W.R. Cline, A. Achanta, et al. (1996): 'The social costs of climate change: greenhouse damage and the benefits of control', in Climate change 1995: economic and social dimensions of climate change, Cambridge: Cambridge University Press, pp. 183–224.

Reilly, J.M., M.C. Sarofim, S. Paltsev and R.G. Prinn (2004): 'The role of non-CO_2 greenhouse-gases in climate policy: analysis using the MIT IGSM', MIT Joint Program on the Science and Policy of Global Change, No. 114 (Aug) available from http://hdl.handle.net/1721.1/5543

Ricci, F. (2003): 'Channels of transmission of environmental policy to economic growth: a survey of the theory', Thema Universit'e de Cergy-Pontoise, July.

Richels, R.G. and A.S. Manne (1998): 'The Kyoto Protocol: A cost-effective strategy for meeting environmental objectives?' California: EPRI, Stanford University.

Sijm J.P.M. (2004): 'Induced technological change and spillovers in climate policy modeling: an assessment', ECN, December, Petten: ECN Beleids studies.

Syri, S., M. Amann, P. Capros, et al. (2001): Low-CO_2 energy pathways and regional air pollution in Europe. Energy Policy, 29(11): 871–884

Weyant, J.P. and J. Hill (1999): 'The costs of the Kyoto protocol: a multi-model evaluation', Energy Journal, Introduction and Overview Pages vii–xliv

Weyant, J.P. (ed.). (2004): 'EMF introduction and overview. Alternative technology strategies for climate change policy', Energy Economics Special Issue, **26**: 501–755

van Vuuren, D.P., J. Cofala, H.E. Eerens, et al. (2006): 'Exploring the ancillary benefits of the Kyoto Protocol for air pollution in Europe', Energy Policy, **34**: 444–460

van Vuuren, D.P., J.P. Weyant and F. de la Chesnaye (2006): 'Multi-gas scenarios to stabilize radiative forcing', Energy Economics, **28**(1): 102–120, January.

11 Structural Change and Competitiveness

> **KEY MESSAGES**
>
> **The costs of mitigation will not be felt uniformly across countries and sectors.** Greenhouse-gas-intensive sectors, and countries, will require the most structural adjustment, and the timing of action by different countries will affect the balance of costs and benefits.
>
> **If some countries move more quickly than others in implementing carbon reduction policies, there are concerns that carbon-intensive industries will locate in countries without such policies in place.** A relatively small number of carbon-intensive industries could suffer significant impacts as an inevitable consequence of properly pricing the cost of greenhouse-gas (GHG) emissions.
>
> **The empirical evidence on trade and location decisions, however, suggests that only a small number of the worst affected sectors have internationally mobile plant and processes.** Moreover, to the extent that these firms are open to competition this tends to come predominately from countries within regional trading blocs. This suggests that action at this regional level will contain the competitiveness impact.
>
> **Trade diversion and relocation are less likely, the stronger the expectation of eventual global action** as firms take long-term decisions when investing in plant and equipment that will produce for decades.
>
> **International sectoral agreements for GHG-intensive industries could play an important role** in promoting international action for keeping down competitiveness impacts for individual countries.
>
> **Even where industries are internationally mobile, environmental policies are only one determinant of plant and production location decisions.** Other factors such as the quality of the capital stock and workforce, access to technologies, infrastructure and proximity to markets are usually more important determinants of industrial location and trade than pollution restrictions.

11.1 Introduction

All economies undergo continuous structural change through time. Indeed, the most successful economies are those that have the flexibility and dynamism to cope with and embrace change. Action to address climate change will require policies that deter greenhouse-gas emitting activities, and stimulate a further phase of structural change.

One concern is that under different speeds of action, policies might be dispro-portionately costly to countries or companies that act faster, as they might lose energy-intensive production and exports to those who act more slowly. This could lead to relocation that simply transfers, rather than reduces, global emis-sions, making the costs borne by more active countries self-defeating.

Even where action is taken on a more uniform collective basis, concern remains that different countries will be affected differently. Some countries have developed comparative advantages in GHG-intensive sectors and would be hit hardest by attempts to rein-in emissions and shift activity away from such production.

The "competitiveness" of a firm or country is defined in terms of relative per-formance. An uncompetitive firm risks losing market share and going out of busi-ness. On the other hand, a country cannot "close", but low competitiveness means the economy is likely to grow more slowly with lower real wage growth and enjoy fewer opportunities than more competitive economies. At the national level, pro-moting competitiveness means applying policies and re-vamping institutions to enable the economy to adapt more flexibly to new markets and opportunities, and facilitate the changes needed to raise productivity. Carefully designed, flexible poli-cies to encourage GHG mitigation and stimulate innovation need not be inconsis-tent with enhancing national competitiveness. On the contrary, the innovation associated with tackling climate change could trigger a new wave of growth and creativity in the global economy. It is up to individuals, countries, governments and companies to tailor their policies and actions to seize the opportunities.

Section 11.2 looks at the likely distribution of carbon costs across industrial sec-tors and assesses their exposure to international competition. Section 11.3 examines evidence behind firms' location decisions and the degree to which environmental regulations influence trade patterns. Climate change policies may also help meet other goals, such as enhanced energy security, reduced local pollution and energy market reform and these issues are addressed in detail in the next chapter.

11.2   Distribution of costs and implications for competitiveness

To assess the likely impact of carbon costing, a disaggregated assessment of fossil fuel inputs into various production processes is required. For many countries, this can be by analysing whole economy disaggregated Input-Output tables. Using the UK as a detailed case study, direct and indirect carbon costs can be applied to various fossil fuel inputs, and traced through the production process, to final goods prices, see Box 11.1. This reveals the carbon intensity of produc-tion. It also gives a crude estimate of the final impact on total consumer prices, and so reflects the reduction in consumer purchasing power.[1]

The impacts of action to tackle climate change are unevenly distributed between sectors

---

[1] This assumes no behavioural response and no substitution opportunities and 100% pass through of costs. It is in theory possible to use older full supply-use Input-Output tables and the inverse Leontief matrix to gauge the rough magnitude of higher order indirect impacts. The study has not done this, but extending the analysis to include more multipliers shows the num-bers converging to zero pretty quickly suggesting this analysis offers a close approximation.

Input-Output tables can be used to look at the distribution of carbon costs across sectors of the economy. For illustrative purposes, the UK, with energy intensity close to the OECD average, is used as a case study of disaggregated cost impacts. However, the lessons drawn for the UK need not be applicable to all countries, even within the OECD.

An illustrative carbon price of £70/tC ($30/tCO$_2$)[2] can be traced through the economy's disaggregated production process, to final consumer prices. Adding the carbon price raises the cost of fossil fuel energy in proportion to carbon intensity of each fossil fuel input (oil, gas and coal) see Box 11.1.

The overall impact is to raise consumer prices by just over one per cent on the assumption of a full cost pass-through. However, the impact on costs and prices in the most carbon-intensive industries, either directly or indirectly through say, their consumption of electricity, is considerably higher. In the UK, six industries out of 123 would face an increase in variable costs of 5% or more as a result of the impact of carbon pricing on higher energy costs (see table 11A.1 at end). In these industries prices would have to rise by the following amounts for profits to remain unchanged:

- gas supply and distribution (25%);
- refined petroleum (24%);
- electricity production and distribution (16%);
- cement (9%);
- fertilisers (5%);
- fishing (5%)

Although this analysis is restricted to the UK, it is these same industries, together with metals, chemicals, paper/pulp, and transport that dominate global carbon emissions from fossil fuels the world over. The competitiveness impacts in these sectors will be reduced to the extent that they are not highly traded. In the UK, combined export and import intensity for these sectors is below 50% (see Box 11.3)[3].

Electricity and gas distribution for example are almost entirely domestic, and to the extent energy intensive industries do trade, this is mostly within the EU. Trade intensity falls by a factor of two to seven for the key energy-intensive industries when measured in terms of non-EU trade only (see Annex table 11A.1 for details of trade intensity among carbon-intensive activities). Nevertheless:

- The magnitude of the impact on a small number of sectors is such that it could provide incentives for import substitution and incentives to relocate to countries with more relaxed mitigation regimes, even though these sectors are not currently characterised by high trade intensity. Further, many industries suffering smaller price increases are more open to trade; these include

---

[2] This figure is illustrative, but the impact on prices is linear so the results can be appropriately factored up/down drawn for different carbon costs. Ideally this figure should correspond with the social cost of carbon (see Chapter 13), which to put it into context, is slightly above prices quoted in the European Emissions Trading scheme – ETS – over the much of the past year. It is important to distinguish tonnes of carbon from carbon dioxide as the two measures are used interchangeably. £1/tC = £0.273/tCO$_2$ so £70/tC = £19/tCO$_2$. Exchange rates are calculated at 2003 purchasing power parities.

[3] Trade intensity defined as total and exports of goods and services as a percentage of total supply of goods and services, plus imports of goods and services as a percentage of total demand for goods and services. Output is defined as gross, so the maximum value attainable is 200.

Box 11.1 Potential costs to firms and consumers; UK Input-Output study

The primary users of fossil fuels (oil, gas and coal) as direct inputs include refined petrol, electricity, gas distribution, the fossil fuel extraction industries and fertiliser production. Figure A shows the share of oil, gas and coal in variable cost for these primary users.

Input-Output analysis can trace the impact of carbon pricing on secondary users of oil, gas and coal – defined as those industries that use inputs from the primary oil, gas and coal users such as electricity. Outputs from these sectors are then fed in as inputs to other sectors, and so on. For illustrative purposes, Figure B shows the impact of a carbon price of £70/tC, but the effects are linear with respect to price and so different impacts for different prices can be assessed using the appropriate multiple. Chapter 9 showed that although the average abatement cost may fall as new technologies arise, the marginal abatement cost is likely to rise with time, reflecting the rising social cost of carbon as the atmospheric carbon stock increases. As industry becomes decarbonised, the whole-economy impact is likely to begin to fall. But going the other way will be the rising social cost of carbon and the corresponding marginal abatement cost (this is illustrated in Box 9.6). This will have an increasing impact on costs in remaining carbon-intensive sectors.

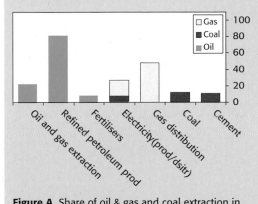

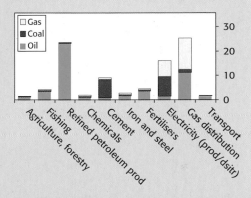

Figure A Share of oil & gas and coal extraction in variable costs, percent

Figure B Product price increases from £70/tC pricing (full pass-through), percent

The largest users of petroleum-products include agriculture, forestry and fishing, chemicals and the transportation sectors. The main users of coal are electricity and cement. The main users of electricity include the electricity sector itself, a number of manufacturing industries and the utilities supplying gas and water.

Total fossil fuel energy costs account for 3% of variable costs in UK production. When the illustrative carbon price of £70/tC ($30/tCO$_2$) is applied, whole economy production costs might be expected to rise by just over 1%. Only 19 out of 123 sectors, accounting for less than 5% of total UK output, would see variable costs increase of more than 2% and only six would undergo an increase of 5% or more[4].

---

[4] Full industry listings for all 123 Standard Industrial Classification (SIC) sectors are given in annex table 11A.1.

Mapping costs through to final consumer goods prices, the aggregate impact on consumer prices of a £70/tC would be of the order of a 1.0% one-off increase in costs, with oil's contribution accounting for just under half and the remainder split between gas and coal[5].

oil and gas extraction or air transport. The competitiveness impacts will be reduced if climate change action is coordinated globally.

- It is likely that some sectors (for example steel and cement or even electricity for a more inter-connected country) may be more vulnerable in countries bordering more relaxed mitigation regimes. Such countries should conduct similar Input-Output exercises to assess the vulnerability of their tradable sectors.
- In addition, there is a problem of aggregation. Aluminium smelting for example is among the most heavily energy-intensive industrial processes. Yet the upstream process is classed under broader 'non-ferrous metals' (of which aluminium accounts for around half). Hence although it is correct to conclude that overall value-added is not at much as risk, to infer that aluminium production is not at risk would be wrong. In general, upstream metal production tends to be both the most energy-intensive and tradable component, something that analysis at a broad level of aggregation may not reveal.

The forgoing analysis offers an indication of the distribution of static costs among various sectors from pricing-in the cost of GHG emissions. However, there is a risk that action to reduce GHG emissions could generate dynamic costs, for example, scrapping capital prematurely and de-skilling workers might retard the economy's ability to grow. Before assessing these costs, it is important to re-emphasise that under 'business as usual' policies, dynamic costs relating to early capital scrapping and adjustment are liable to be even larger in the medium term. Timely investment will reduce the impact of climate change. Chapter 8 showed that a smooth transition to a low GHG environment with early action to reduce emissions is likely to limit adjustment costs.

The dynamic impacts from a transition to a low-GHG economy should be small. The change in relative prices that is likely to result from adopting the social cost of carbon into production activities is well within the 'normal' range of variation in prices experienced in an open economy. Input cost variations from recent fluctuations in the exchange rate and the world oil price, for example, are likely to far exceed the short-run primary energy cost increases from a carbon tax required to reflect the damage from emissions (see Box 11.2).

The economic literature investigating the impact of energy cost changes focuses disproportionately on resource, capital and energy-intensive sectors and firms. While this is understandable from a policy perspective, since regulation is likely to disproportionately affect these sectors, it also indicates a significant gap in data on other sectors in particular services, which constitute up to three quarters of some developed economies output.

[5] It is in theory possible to use older full supply-use Input-Output tables and the inverse Leontief matrix to gauge the rough magnitude of this higher order indirect impact. Because data disaggregated to a level commodity output per unit of domestically met final demand has not been published in the UK since 1993, the study has not adopted this approach and has not been able to follow the impact through the entire supply-chain. However, extending the analysis to include more multipliers seems to make little difference to the results, suggesting the numbers presented here are a close approximation to the price impacts that would be derived using an up-to-date inverse Leontief.

Box 11.2 Vulnerability to energy shocks: lessons from oil and gas prices

Past energy price movements can be used to illustrate the likely economic impact of carbon pricing. Energy costs constitute a small part of total gross output costs, in most developed economies under 5%, in contrast to, say, labour costs, which account for up to a third of total gross output costs. Nevertheless, past movements in energy costs can offer a guide to the potential impact of carbon proving.

UK I-O tables show that oil and gas together account for more than ninety percent by final value of UK fossil fuel energy consumption, but only three-quarters of fossil fuel emissions, as coal is more carbon-intensive. The I-O data reveal that a £10/tC ($4/tCO$_2$) carbon price would have a similar impact on producer prices as a $1.6/bl rise in oil prices *with a proportionate gas price increase.*

To put this in context, the sterling oil price has risen 240% in real terms from its level over most of the period 1986–1997($18/bl) to around $69/bl (as of May 2006), and by 150% in real terms since 2003 (average), when the price of Brent crude hovered at around $26/bl for most of the year. On this basis, the change in the real oil price since 2003, assuming a proportionate changes in gas prices, is likely to have had a similar impact on the economy as unchanged oil and gas prices and the imposition of a £260/tC ($132/tCO$_2$) carbon price[6]. Or, alternatively, a £70/tC ($30/tCO$_2$) carbon resource cost is likely to have a similar impact as a $11/bl real oil price increase (at 2003 prices), according to I-O tables.

Gross estimate of impact on UK consumer prices and GDP*

Brent spot price $ per barrel (real)		Equivalent Carbon cost		Consumer prices, % change	GDP % change (producer prices)
		£/T carbon	$/T CO2		
2003 average,	26.3	0	0	0.0	0.0
	38	70	30	0.9	−1.2
	40	84	37	1.1	−1.5
	60	206	90	2.6	−3.6
	80	329	143	4.2	−5.7
	100	451	196	5.8	−7.9

*Uses 2003 prices and Input-output tables; assumes no substitution in producer processes or consumption patterns and assumes all revenues are lost to economy.
Source: Stern using 2003 UK Input-Output tables, Carbon Trust carbon intensity and UK DTI energy price statistics.

In practice, the overall impact on GDP from oil and gas price rises is likely to have been far smaller than suggested here at the national and global level. This is because the rise in the oil price in part reflects a transfer of rent to low marginal cost oil exporters, who in turn will spend more on imported goods and services from oil-importers. The presence of rent in the oil price means the impact on GDP is likely to be over-estimated even for oil importers. Furthermore, to the extent that carbon taxes generate transfers within the economy, the

---

[6] The exercise assumes that gas prices change in full proportion with oil prices, but that coal prices remain unchanged. In reality oil and gas prices tend to co-move as they are partial substitutes within a fossil fuel energy market and are linked contractually.

impact on GDP will also be exaggerated. Finally, the use of fixed Input-Output tables assume consumer and producer behaviour is static.

In practice, costs will be lowered as firms and consumers switch out of more expensive carbon-intensive activities. Consequently, the total impact of both carbon pricing and oil price changes on GDP will be lower than the numbers presented here, which should be regarded as an illustrative upper-end estimate of the costs of mitigation in the energy sector for applying any given carbon price.

The recent rise in the brent spot price, US $ per barrel (2003 prices)

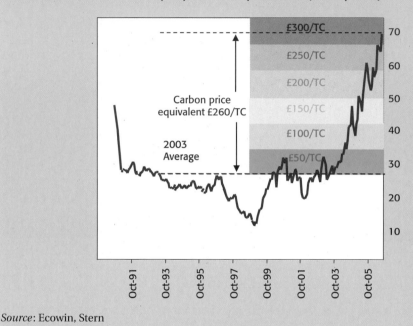

Source: Ecowin, Stern

The analysis also assumes that carbon costs are fully passed through to final prices. In practice this need not be the case, especially for tradable sectors that face sensitive demand and are likely to "price-to-markets" to avoid a loss of market share. In addition, the presence of competing inputs, and the opportunity to change processes and reduce emissions, also serve to limit the impact on both profits and prices. However, this analysis still gives an indication of which sectors are most vulnerable to a profit squeeze if carbon pricing is applied to emissions.

The nature of the policy instrument and the framework under which it is applied will also lead to sectoral distributions of costs. For example:

- Who bears the costs/gains from emissions trading depends on whether the allowances are auctioned or given out for free.
- The scope of trading schemes also matters. The EU ETS, for example, extends to primary carbon-intensive sectors, but does not allocate permits to secondary users, such as the aluminium sector, which relies heavily on electricity inputs.[7]

---

[7]   For analysis of the structure and impact of the EU ETS see: Frontier Economics (2006); Carbon Trust (2004); Grubb (2004); Neuhoff (2006); Sijm et al. (2005) OXERA (2004) and Reinaud (2004).

- The structure of the electricity market also helps determine outcomes. In highly regulated or nationalised electricity markets, for example, carbon costs are not necessarily passed through, in which case the impact would be felt through the public finances. With regulation limiting cost pass-through in a private sector industry, there will be a squeeze on profits with impacts felt by shareholders. Different impacts will be felt across the globe, but the analysis here gives an indication of the sectors likely to be directly affected.

International sectoral agreements for such industries could play an important role in both promoting international action and keeping down competitiveness impacts for individual countries. Chapter 22 shows how emissions intensities within sectors often vary greatly across the world, so a focus on transferring and deploying technology through sectoral approaches could reduce intensities relatively quickly. Global coverage of particular sectors that are internationally exposed to competition and produce relatively homogenous products can reduce the impact of mitigation policy on competitiveness. A sectoral approach may also make it easier to fund the gap between technologies in developed and developing countries.

Countries most reliant on energy-intensive goods and services may be hardest hit.

The question of the distribution of additional costs applies to countries also. Some small agricultural or commodity-based economies rely heavily on long-distance transport to deliver products to markets while some newly-industrialising countries are particularly energy-intensive. Primary energy consumption as a percent of GDP is generally three or four times higher in the developing world than in the OECD,[8] though in rapidly growing sectors and countries such as China and India, primary energy consumption per unit output has fallen sharply as new, efficient infrastructure is installed (see Section 7.3). Some of these countries may benefit from energy efficiency improvements and energy market reforms that could lower real costs, but the distribution of costs raises issues relating to design of policies and different speeds of action required to help with the transition in certain countries and sectors (see Part VI).

The impact on oil and fossil fuel producers will depend on the future energy market and the rate of economic diversification in the relevant economies during the transition, which will open up new opportunities for exploiting and exporting renewable energy and new technologies such as carbon capture and storage. Producers of less carbon-intensive fossil fuels, such as gas, will tend to benefit relative to coal or lignite producers.

Where transfers are involved, the extra burden on rich countries need not be significant given the disparities in global income. For illustration, assume GHG stabilisation requires a commitment of 1% of world GDP annually to tackle climate change. If, in the initial decades, the richest 20% of the world's population, which produce 80% of the world's output and income, agreed to pay 20% more – or 1.2% of GDP, this would allow the poorer 80% of the world's population to shoulder costs equivalent to only 0.2% of GDP.[9] Similarly, transfers to compensate

---

[8]   International Energy Agency (2005).

[9]   OECD economies account for 15% of the world's population and just over 75% of world output in terms of GDP at current prices using World Bank Statistics (2004). Use of market prices overstates the real value of output in rich countries relative to poorer countries because equivalent non-tradable output in general tends to be cheaper in poorer countries. However, in terms of ability to transfer income globally at market exchange rates, market prices are the appropriate measure.

countries facing disproportionately large and costly adjustments to the structure of their economies could also be borne at relatively small cost, if distributed evenly at a global level. Questions of how the costs of mitigation should be borne internationally are discussed in Part VI of this report.

11.3   Carbon mitigation policies and industrial location

The impact on industrial location if countries move at different speeds is likely to be limited

The transitional costs associated with implementing GHG reduction policies faster in one country than in another were outlined in the previous section. In the long run, however (when by definition, resources are fully employed and the impact for any single country is limited to the relocation of production and employment between industries), openness to trade allows for cheap imports to substitute domestic production in polluting sectors subject to GHG pricing. This is likely to reduce the long-run costs of GHG mitigation to consumers, while some domestic GHG-intensive firms that are relatively open to trade lose market share.

A reduction in GHG-intensive activities is the ultimate goal of policies designed to reduce emissions. However, this aim is most efficiently achieved in an environment of global collective action (see Part VI). This is because if some countries move faster than others, the possible relocation of firms to areas with weaker GHG policies could reduce output in countries implementing active climate change policies by more than the desired amount (that is, the amount that would prevail in the case where *all* countries adopted efficient GHG policies). At the same time, global emissions would fall by less than the desired amount if polluters simply re-locate to jurisdictions with less active climate change policies.[10]

This risk should not be exaggerated. To the extent that energy-intensive industry is open to trade, the bulk of this tends to be limited to within regional trading blocks. UK Input-Output tables, for example, suggest trade diversion is likely to be reduced where action is taken at an EU level (see Box 11.3). However, several

---

Box 11.3   The risk of trade diversion and firm relocation – a UK Input–Output case-study

By changing relative prices, GHG abatement will reduce demand for GHG-intensive products. Sectors open to competition from countries not enforcing abatement policies will not be able to pass on costs to consumers without risking market share. The short-run response to such elastic demand is likely to be lower profits. In the long run, with capital being mobile, firms are likely to make location decisions on the basis of changing comparative advantages.

I-O analysis helps identify which industries are likely to suffer trade diversion and consider relocation: in general the list is short. Continuing with the £70/tC ($30/tCO$_2$) carbon price example, the figure below maps likely output price changes against exposure to foreign trade.[11] With the exception of refined petrol and coal, fuel costs are not particularly exposed

---

[10] The 'desired amount' refers to the amount consistent with relative comparative advantages in an 'ideal' world with collective action, where gains form trade are maximised.
[11] This is defined as exports of goods and services as a percentage of total supply of goods and services, plus imports of goods and services as a percentage of total demand for goods and services. Output is defined as gross, so the maximum value attainable is 200.

to foreign trade. Under carbon pricing, the price of electricity and gas distribution is set to rise by more than 15%, but output is destined almost exclusively for domestic markets. In all other cases, price increases are limited to below – mostly well below – 10%.

Vulnerable industries: price sensitivity and trade exposure, percent

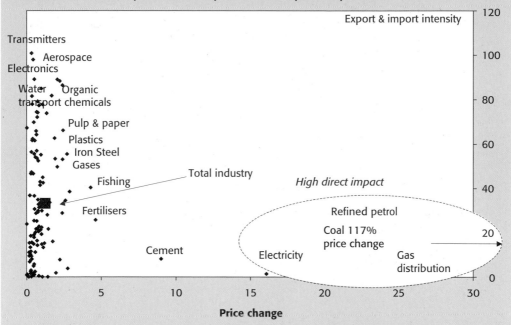

The bulk of the economy is not vulnerable to foreign competition as a result of energy price rises. However, a few sectors are. Apart from refined petrol, these include fishing, coal, paper and pulp, iron and steel, fertilisers, air and water transport, chemicals, plastics, fibres and non-ferrous metals, of which aluminium accounts for approximately half of value added. In addition, the level of aggregation used in I-O analysis masks the likelihood that certain processes and facilities within sectors will be both highly energy-intensive and exposed to global competition.

The impact on competitiveness will depend not only on the strength of international competition in the markets concerned, but also the geographical origin of that competition. Many of the proposed carbon abatement measures (such as the EU ETS) are likely to take place at an EU level and energy-intensive sectors tend to trade very little outside the EU.

Trade intensity falls seven-fold in the cement industry when restricted to non-EU countries, as cement is bulky and hard to transport over long distances. Trade in fresh agricultural produce drops by a factor of 5 when restricted only to non-EU countries. The next largest drop in trade occurs in pulp and paper, plastics and fibres. Here trade intensity is quartered at the non-EU level. Trade intensity in plastics and iron and steel and land-transport as well as fishing and fertilisers drop by two-thirds. Trade intensity for air transport and refinery products halves in line with the average for all sectors (complete non-EU trade intensities are listed in Annex table 11A.1). All of these sectors are fossil fuel-intensive; suggesting that restrictions applied at the EU level would greatly diminish the competitiveness impact of carbon restrictions.

sectors are open to trade outside the EU. To the extent that variations in the climate change policy regime between countries result in trade diversion in these sectors the impact on GHG emissions will be reduced.

Trade diversion and relocation are also less likely, the stronger the expectation of eventual global action. Firms need to take long-term decisions when investing in plant and equipment intended for decades of production. One illustration of this effect is the growing aluminium sector in Iceland. Iceland has attracted aluminium producers from Europe and the US partly because a far greater reliance on renewable electricity generation has reduced its exposure to prices increases, as a result of the move to GHG regulations (see Box 11.4).

Box 11.4 Aluminium production in Iceland

Over the last six years, Iceland has become the largest producer of primary aluminium in the world on a per capita basis. The growth in aluminium production is the result both of expansion of an existing smelter originally built in 1969 and construction of a new greenfield smelter owned by an American concern and operated since 1998. The near-future looks set to see a continuing sharp increase in aluminium production in Iceland. Both existing plants have plans for large expansions in the near future. These projects are forecast to boost aluminium production in Iceland to about one million tonnes a year, making Iceland the largest aluminium producer in western Europe.

Power-intensive operations like aluminium smelters are run by large and relatively footloose international companies. Iceland has access to both the European and US aluminium market, but its main advantage is the availability of water and emission-free, renewable energy. Emissions of CO_2 from electricity production per capita in Iceland is the lowest in the OECD: 70% of its primary energy consumption is met by domestic, sustainable energy resources. Iceland is also taking action to reduce emissions of fluorinated compounds associated with aluminium smelting. Expectations of future globalisation action to mitigate GHG emissions is already acting as a key driver in attracting investment of energy-intensive sectors away from high GHG energy suppliers and towards countries with renewable energy sources.

The impact on location and trade is likely to be more substantial for mitigating countries bordering large trade-partners with more relaxed regimes, such as Canada which borders the US, and Spain which is close to North Africa. For example, Canada's most important trading partner, the United States, has not signed the Kyoto Protocol, raising concerns of a negative competitive impact on Canada's energy-intensive industry.[12] However, even for open markets such as Canada and the US, or states within the EU, firms tend to be reluctant to relocate or trade across borders, when they have markets in the home nation. This so-called 'home-bias' effect is surprisingly powerful and the consequent necessity for firms to locate within borders to access local markets limits the degree to which they are footloose in their ability to relocate when faced with carbon pricing.[13]

[12] For an interesting discussions see the Canadian Government's *Industry Canada* (2002) report, as well as the representations of the Canadian Plastic Industry Association.

[13] This was the finding of McCallum's seminal 1995 paper, further reinforced by subsequent discussions such as Helliwell's assessment of Canadian-US economic relations, and Berger and Nitsch's (2005) gravity model of intra-EU trade, both of which found significant evidence of home-bias where borders inhibit trade despite open markets and short distances.

Theory suggests that country-specific factors, such as the size and quality of the capital stock and workforce, access to technologies and infrastructure, proximity to large consumer markets and trading partners, and other factor endowments are likely to be the most important determinants of location and trade. In addition, the business tax and regulatory environment, agglomeration economies, employment law and sunk capital costs are also key determinants. These factors are unlikely to be much affected by GHG mitigation policies. Overall, empirical evidence supports the theory, and suggests environmental policies do affect pollution-intensive trade and production on the margin, but there is little evidence of major relocations.[14,15]

Environmental policies are only one determinant of plant and production location decisions. Costs imposed by tighter pollution regulation are not a major determinant of trade and location patterns, even for those sectors most likely to be affected by such regulation.

The bulk of the world's polluting industries remain located in OECD countries despite tighter emissions standards.[16] By the same token, 2003 UK Input-Output tables show that around 75% of UK trade in the output of carbon-intensive industries is with EU countries with broadly similar environmental standards, with little tendency for such products to be imported from less stringent environmental jurisdictions.

One way of assessing the impact of environmental regulations is to see if greater trade openness has led to a relocation of polluting industries to poorer countries, which have not tightened environmental standards. Antweiler, Copeland and Taylor (2001) calculated country-specific elasticities of pollution concentrations with respect to an increase in openness over the latter part of the twentieth century (Figure 11.1). A positive value for a country implies that trade liberalisation shifts pollution-intensive production towards that country, in effect signalling that it has a comparative advantage in such production.

Perhaps surprisingly, the study found that rich countries have tended to have unexploited comparative advantages in pollution-intensive production and tend to have positive values for the elasticity while poorer countries tend to have negative values. This indicates that opening up trade will on average shift polluting production towards richer countries. The authors offer this as support for the view that factor endowments such as capital intensity, availability of technology and skilled labour, and access to markets and technologies are the key determinants of

---

[14] See Copeland and Taylor (2004) for one of the most thorough-going theoretical and empirical investigation into environmental regulations and location decisions. See also Levinson et al (2003), Smita et al. (2004) Greenstone (2002), Cole et al. (2003), Ederington et al (2000, 2003), Jeppesen (2002), Xing et al. (2002), UNDP (2005).

[15] The analysis by Smita et al. (2004) confirms that other factors are likely to be more significant determinants of international location and direct investment decisions – factors such as the availability of infrastructure, agglomeration economies and access to large consumer markets. The study of the influence of air pollution regulations carried out by AEA Metroeconomica found that "it is extremely difficult to assess the impact of air pollution on relocation from the other factors that determine location decisions."

[16] Low and Yeats (1992) reported that over 90% of all 'dirty-good' production in 1988 was in OECD countries. This fact alone suggests the location of dirty-good production across the globe reflects much more than weak environmental regulations. See also Trefler (1993) and Mani et al (1997).

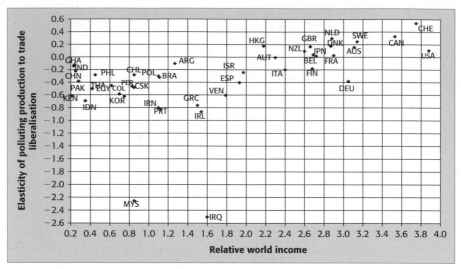

Figure 11.1 Trade liberalisation reveals 'comparative advantages' in pollution

Source: Anweiler et al.

environmentally sensitive firms' location decisions. Such factors outweigh rich countries' tendency to apply tighter environmental restrictions in determining firms' location decisions.

11.4   Conclusion

The competitiveness threat arising if some countries move quicker than others in mitigating GHGs is, for most countries, not a macro-economic one, but certain processes and facilities could be exposed in the transition to a low-emissions environment, with new plant diverted to countries or regions with less active climate change policies.

However, if early and gradual action is taken regionally, such impacts are likely to apply to only a very narrow subset of production in a few states with little impact on the economy as a whole. There is likely to be a differentiation in a country's attractiveness as an investment location towards less carbon-intensive activities, but with well-designed policies and flexible institutions there will also be new opportunities in innovative sectors.

Environmental policies are only one determinant of plant and production location decisions. Even for those sectors most likely to be affected by such regulation, factors such as the quality of the capital stock and workforce, access to technologies and infrastructure and the efficiency of the tax and regulation system are more significant. Proximity to markets and suppliers is another important determinant of location and trade. These fundamental factors will always be the key drivers of overall national competitiveness and dynamic economic performance.

Focusing on the costs of mitigation is not the whole story: there are a number of non-climate change related benefits that countries which take action to mitigate GHGs will benefit from; these are outlined in the next chapter.

References

General discussions defining competitiveness are few and far between, reflecting the fact that the definition varies depending on the context. An entertaining account of the problems associated with defining "competitiveness" and the limitations to the notion when applied at a national level can be found in Krugman (1994) and at a more applied level in Azar (2005). There are a number of very thorough and well-researched sectoral analyses of the competitiveness impact of climate change policies; particularly informative are Demailly and Quirion's (2006) study of competitiveness in the European cement industry and Berman and Bui's account of location decisions in the fossil fuel price sensitive refinery sector. There are also a host of in-depth studies of specific regional policies, in particular the competitiveness impact of the EU ETS. Among the many notable reports listed below are: Frontier Economics (2006); Oxera (2006); Grubb Neuhoff (2006), and Reinaud (2004).

Perhaps the most authoritative and comprehensive account of the evolving literature on firms' location decisions in the presence of differential national environmental policies can be found in Copeland and Taylor (2004). Smita et al. (2004) and Lowe and Yeats also undertake in-depth analyses of the degree to which environmental regulations influence trade patters. McCallum (1995) and Nitsch and Berger (2005) provide illustrations of the impact of country borders in containing trade, even where borders are open and goods are highly tradable.

Antweiler, W., B.R. Copeland and M.S. Taylor, (2001): 'Is free trade good for the environment?' American Economic Review **91**(4): 877–908

Azar, C. (2005): 'Post-Kyoto climate policy targets: costs and competitiveness implications', Climate Policy journal 309–328

Berman, E. and L. Bui (1999): 'Environmental regulation and productivity: evidence from oil refineries' Review of Economics and Statistics 2001 **83**(3): 498–510, available from http://www.mitpressjournals.org/doi/abs/10.1162/00346530152480144

Brunnermeier, S.B. and Cohen M. (2003): 'Determinants of environmental innovation in U.S. manufacturing industries', Journal of Environmental Economics & Management, **45**(2): 278–293

Burtraw, D. (1996) "Trading emissions to clean the air: exchanges few but savings many," *Resources*, **122** (Winter): 3–6

The Carbon Trust (2004): 'The European Emissions Trading Scheme: implications for industrial competitiveness', London: The Carbon Trust.

The Carbon Trust (2005): 'The UK climate change programme: potential evolution of business and the public sector', London: The Carbon Trust.

Cole, M.A. and R.J.R. Elliott (2003): 'Do environmental regulations influence trade patterns? Testing old and new trade theories' The World Economy **26** (8):1163–1186

Copeland, B.R. and S. Taylor (2004): 'Trade, growth and the environment', Journal of Economic Literature, **XLII**: 7–71

Department for Environment, Food and Rural Affairs (2006): 'Synthesis of climate change policy evaluations' April. London: Defra.

Demailly, D. and P. Quirion (2006): 'CO_2 abatement, competitiveness and leakage in the European cement industry under the EU ETS: grandfathering versus output-based allocation' April, CESifo, Venice: Venice International University.

Ederington, J. and J. Minier (2000): 'Environmental regulation and trade flows', Journal of Economic Literature, Jan 19 2001, available from http://fmwww.bc.edu/RePEc/es2000/1507.pdf

Ederington, J. and J. Minier (2003): 'Is environmental policy a secondary trade barrier? An empirical analysis'. Canadian Journal of Economics, **36**(1): 137–154

Frontier Economics (2006): 'Competitiveness impacts of the EU ETS'.

Greenstone, M. (2002): 'The Impacts of environmental regulations on industrial activity: evidence from the 1970 and 1977 Clean Air Act Amendments and the Census of Manufactures', Journal of Political Economy, **110**: 1175–1219

Grubb, M. (2004): Implications of EU ETS for competitiveness of energy-intensive industry. Presentation at IEA/IETA/EPRI workshop on Greenhouse-Gas Emissions Trading, Paris, 4–5 October 2004.

Grubb, M., and K. Neuhoff (2006): 'Allocation and competitiveness in the EU emissions trading scheme: policy overview', Climate Policy **6**: 7–30

Henderson, D.J. and D.L. Millimet (2001): 'Pollution abatement costs and foreign direct investment inflows to U.S. states: a nonparametric reassessment' Review of Economics and Statistics Vol. **84**(4): 691–703

Helliwell, J.F. (1998): 'How much do national borders matter?' Washington, DC: Brookings Institution Press.

IEA (2005) 'CO$_2$ Emissions from fuel combustion 1971–2003', OECD/IEA, Paris

Jeppesen, T., J.A. List and H. Folmer (2002): 'Environmental regulations and new plant location decisions: evidence from a Meta-Analysis' Journal of Regional Science 42:19 – February 2002

Klepper, G. and S. Peterson (2004): 'The EU emissions trading scheme allowance prices, trade flows and competitiveness effects' European Environment, 14(4), 2004. Pages 201–218 Kiel Institute for World Economics, Germany.

Krugman, P. (1994): 'Competitiveness: a dangerous obsession', Foreign Affairs, March.

Levinson, A. and S. Taylor (2003): 'Trade and the environment: unmasking the pollution haven hypothesis' Georgetown University Working Paper Washington, DC: Georgetown University.

Low, P. and A. Yeats (1992): 'Do 'dirty' industries migrate?', in P. Low (ed.), International trade and the environment, Washington, DC: World Bank Discussion Paper 159 pp. 89–104

Mani M, and D. Wheeler (1997): 'In search of pollution havens? Dirty industry migration in the world economy', World Bank working paper #16. Washington, DC: World Bank.

McCallum, J. (1995): 'National borders matter: Canada-U.S. regional trade patterns', American Economic Review, American Economic Association, vol. 85(3): 615–23, June.

Newbery, D.M. (2003): 'Sectoral dimensions of sustainable development: energy and transport', available from http://www.unece.org/ead/sem/sem2003/papers/newbery.pdf

Neuhoff ,K., J. Barquin, M. Boots, et al. (2004): 'Network constrained models of liberalized electricity markets: the devil is in the details. Internal report', Energy research Centre of the Netherlands (ECN), Amsterdam: Petten.

Neuhoff, K. (2006): 'The EU ETS: allocation, competitiveness and longer term design options', submission to Stern Review.

Nitsch, V. and H. Berger (2005): 'Zooming out: the trade effect of the euro in historical perspective'. CESifo Working Paper Series No. 1435.

Oxera (2004): 'CO$_2$ Emissions Trading: How will it affect UK Industry?' Report prepared for The Carbon Trust, Oxford: Oxera.

Oxera (2006): 'Modelling the competitiveness impact of policy proposals arising from the Energy Review?' Report prepared for the Department of Trade and Industry, Oxford: Oxera.

Oates, W., Palmer and Portney (1995): 'Tightening environmental standards: the benefit-cost or the no-cost paradigm?' Journal of Economic Perspectives 9(4), Autumn: 119–132

Reinaud, J. (2004): 'Industrial competitiveness under the European Union Emissions Trading Scheme', Paris: IEA.

Sijm, J.P.M., S.J.A. Bakker, Y.H.W. Chen Harmsen and W. Lise (2005): 'CO$_2$ price dynamics: The implications of EU emissions trading for the price of electricity', Energy research Centre of the Netherlands (ECN), Amsterdam: Petten.

Smita, B., S.B. Brunnermeier and A. Levinson (2004): 'Examining the evidence on environmental regulations and industry' Location Journal of Environment & Development, 13(1): 6–41 March

Trefler, D. (1993): 'Trade liberalization and the theory of endogenous protection: an econometric study of US import policy', Journal of Political Economy. 101(1): 138–60

Williams, E., K. Macdonald and V. Kind (2002): 'Unravelling the competitiveness debate', European Environment 12(5): 284–290 in Special Issue: Environmental Policy in Europe: Assessing the Costs of Compliance Williams E (ed.).

United Nations Development Programme (2005): 'International co-operation behind national borders: an inventory of domestic policy measures aimed at internalising cross-border spillovers adversely affecting the global environment', UNDP Office of Development Studies, New York June 28, available from http://www.thenewpublicfinance.org/background/domestic_spillovers.pdf

Xing, Y. and C.D. Kolstad (2002): 'Do lax environmental regulations attract foreign investment?' Environmental & Resource Economics, 21(1): 1–2

11A

Key Statistics for 123 UK Production Sectors

Annex Table 11A.1 Key statistics for 123 UK production sectors (ranked by carbon intensity).[18]

	Carbon intensity (ppt change at £70/tC)	Energy % total costs	Export and import intensity* Non-EU	Export and import intensity*	Percent total UK output
Metal ores extraction	0.00	0.00	67.17	62.86	0.00
Private households with employed persons	0.00	0.00	0.78	0.33	0.50
Financial intermediation services indirectly measured	0.00	0.00	23.82	10.75	−4.68
Letting of dwellings	0.03	0.07	1.10	0.47	7.90
Owning and dealing in real estate	0.08	0.23	0.35	0.20	1.89
Estate agent activities	0.11	0.29	0.11	0.06	0.50
Membership organizations nec	0.14	0.37	0.00	0.00	0.59
Legal activities	0.16	0.43	11.04	6.58	1.39
Market research, management consultancy	0.17	0.46	9.44	5.58	1.15
Architectural activities and technical consultancy	0.17	0.47	15.31	8.98	1.95
Accountancy services	0.20	0.53	6.77	3.96	0.99
Other business services	0.20	0.55	36.76	21.98	3.53
Computer services	0.23	0.60	13.32	5.76	2.93
Insurance and pension funds	0.24	0.67	10.15	8.10	2.36
Other service activities	0.25	0.68	2.28	1.16	0.64
Recreational services	0.26	0.64	18.47	10.64	2.87
Health and veterinary services	0.26	0.59	1.49	0.63	4.99
Advertising	0.27	0.72	11.49	6.53	0.67
Footwear	0.27	0.60	46.59	21.14	0.03
Banking and finance	0.27	0.78	7.66	4.56	4.05
Education	0.28	0.68	2.88	1.57	6.01

[18] by 123 industry Standard Industrial Classification (SIC) level

Annex Table 11A.1 (*Cont.*)

	Carbon intensity (ppt change at £70/tC)	Energy % total costs	Export and import intensity* Non-EU	Export and import intensity*	Percent total UK output
Auxiliary financial services	0.30	0.73	56.36	35.31	0.88
Transmitters for TV, radio and phone	0.30	0.64	100.70	24.66	0.14
Telecommunications	0.31	0.82	9.28	4.27	2.29
Receivers for TV and radio	0.31	0.63	47.26	24.36	0.08
Social work activities	0.31	0.84	0.03	0.02	1.80
Construction	0.32	0.77	0.23	0.09	6.20
Office machinery & computers	0.33	0.69	81.43	31.86	0.24
Tobacco products	0.33	0.84	15.53	8.03	0.12
Ancillary transport services	0.33	0.97	8.03	3.94	1.81
Medical and precision instruments	0.35	0.80	61.6	33.79	0.56
Pharmaceuticals	0.36	0.77	77.84	31.7	0.64
Leather goods	0.38	0.82	62.28	34.31	0.02
Aircraft and spacecraft	0.41	0.90	97.8	64.35	0.54
Researchy and development	0.42	1.10	46.57	27.48	0.42
Motor vehicle distribution and repair, automotive fuel retail	0.43	1.22	1.04	0.48	2.24
Renting of machinery etc	0.45	1.25	4.87	2.48	1.07
Printing and publishing	0.45	0.90	14.87	7.02	1.64
Jewellery and related products	0.45	0.89	69.7	54.02	0.04
Retail distribution	0.47	1.26	1.68	0.70	5.73
Confectionery	0.47	0.80	17.80	4.48	0.22
Other transport equipment	0.47	1.10	25.34	12.58	0.10
Hotels, catering, pubs etc	0.48	1.26	19.02	8.38	3.32
Postal and courier services	0.48	1.37	5.69	2.71	0.86
Electronic components	0.49	0.89	88.97	40.31	0.13
Electrical equipment nec	0.49	1.10	55.50	24.19	0.21
Wearing apparel and fur products	0.49	1.02	36.55	22.00	0.17
Public administration and defence	0.49	1.31	0.96	0.58	5.12

Annex Table 11A.1 (*Cont.*)

	Carbon intensity (ppt change at £70/tC)	Energy % total costs	Export and import intensity* Non-EU	Export and import intensity*	Percent total UK output
Soap and toilet preparations	0.51	1.15	30.60	8.91	0.20
Motor vehicles	0.52	1.10	61.50	14.54	0.85
Sewage and sanitary services	0.54	1.47	2.33	1.15	0.67
Railway transport	0.56	1.40	11.11	4.67	0.29
Made-up textiles	0.56	1.30	20.02	12.84	0.07
Cutlery, tools etc	0.56	1.27	54	22.75	0.15
Other food products	0.61	1.47	28.70	7.94	0.26
Electric motors and generators etc	0.61	1.42	65.78	32.83	0.23
Furniture	0.62	1.48	21.64	8.29	0.37
Agricultural machinery	0.63	1.48	64.12	19.21	0.05
Machine tools	0.64	1.40	74.32	33.24	0.07
General purpose machinery	0.65	1.56	56.89	22.56	0.40
Weapons and ammunition	0.65	1.31	25.19	14.51	0.06
Insulated wire and cable	0.67	1.37	53.54	24.58	0.04
Soft drinks and mineral waters	0.67	1.44	16.32	3.93	0.10
Special purpose machinery	0.68	1.59	72.01	35.36	0.27
Meat processing	0.70	1.80	21.72	4.83	0.34
Bread, biscuits etc	0.70	1.60	14.22	2.72	0.32
Mechanical power equipment	0.71	1.51	79.07	41.72	0.26
Knitted goods	0.72	1.48	74.07	40.57	0.04
Domestic appliances nec	0.73	1.76	34.84	13.75	0.11
Alcoholic beverages	0.73	1.71	29.24	13.36	0.29
Paints, varnishes, printing ink etc	0.74	1.67	29.78	8.75	0.12
Rubber products	0.76	1.70	52.40	17.45	0.16
Wood and wood products	0.77	1.95	32.75	10.07	0.28
Sports goods and toys	0.78	1.94	20.48	12.46	0.05
Water supply	0.80	1.56	0.42	0.21	0.30
Pesticides	0.80	1.83	77.22	30.00	0.05
Grain milling and starch	0.81	2.01	22.74	5.38	0.10
Metal boilers and radiators	0.81	1.78	31.36	7.21	0.07

Annex Table 11A.1 (*Cont.*)

	Carbon intensity (ppt change at £70/tC)	Energy % total costs	Export and import intensity* Non-EU	Export and import intensity*	Percent total UK output
Wholesale distribution	0.82	2.48	–	–	4.41
Textile fibres	0.87	1.68	41.41	18.12	0.03
Other metal products	0.88	2.03	42.92	18.03	0.24
Plastic products	0.90	1.99	33.69	11.10	0.63
Dairy products	0.91	2.56	21.26	3.66	0.14
Other textiles	0.93	1.85	55.12	19.46	0.05
Other chemical products	0.96	2.22	84.83	34.01	0.17
Carpets and rugs	0.97	2.23	19.26	4.09	0.03
Miscellaneous manufacturing nec & recycling	0.97	2.39	22.33	13.03	0.20
Animal feed	0.99	2.34	14.74	3.35	0.07
Fish and fruit processing	0.99	2.56	29.87	12.38	0.20
Metal forging, pressing, etc	1.03	2.46	0.00	0.00	0.46
Textile weaving	1.04	1.78	77.76	36.85	0.03
Shipbuilding and repair	1.05	2.36	44.94	28.82	0.10
Ceramic goods	1.08	2.42	42.51	18.75	0.08
Structural metal products	1.09	2.47	13.27	4.56	0.30
Paper and paperboard products	1.17	2.02	15.19	3.99	0.28
Coal extraction	1.22	7.24	33.24	24.76	0.05
Non-ferrous metals	1.32	2.36	73.75	39.90	0.10
Agriculture	1.37	3.96	27.99	11.34	0.96
Metal castings	1.40	2.84	0.00	0.00	0.07
Forestry	1.44	4.18	21.64	6.90	0.03
Glass and glass products	1.53	3.44	33.62	9.55	0.14
Water transport	1.65	5.26	81.65	28.76	0.24
Articles of concrete, stone etc	1.73	2.96	15.97	4.67	0.25
Plastics & synthetic resins etc	1.85	4.57	62.56	15.31	0.12
Oil and gas extraction	1.89	5.73	53.30	30.28	2.06
Textile finishing	1.95	3.34	1.76	0.80	0.03
Other mining and quarrying	2.03	4.64	88.90	61.53	0.16
Industrial gases and dyes	2.03	4.31	49.69	20.32	0.09
Man-made fibres	2.21	4.60	88.19	24.96	0.02

Annex Table 11A.1 (*Cont.*)

	Carbon intensity (ppt change at £70/tC)	Energy % total costs	Export and import intensity* Non-EU	Export and import intensity*	Percent total UK output
Other land transport	2.21	7.04	7.74	2.33	1.94
Sugar	2.37	3.20	28.83	22.36	0.04
Organic chemicals	2.38	6.27	86.31	31.19	0.17
Air transport	2.39	7.64	53.03	23.82	0.55
Pulp, paper and paperboard	2.42	4.23	66.07	16.52	0.10
Inorganic chemicals	2.58	5.64	34.51	11.75	0.06
Iron and steel	2.69	7.02	55.40	18.32	0.12
Structural clay products	2.73	6.61	3.78	0.63	0.04
Oils and fats	2.86	5.87	38.48	14.49	0.02
Fishing	4.28	12.78	40.35	14.74	0.04
Fertilisers	4.61	13.31	25.69	9.54	0.02
Cement, lime and plaster	9.00	5.00	8.11	1.20	0.05
Electricity production and distribution	16.07	26.70	1.35	0.11	1.08
Refined petroleum	23.44	72.83	25.66	11.75	0.27
Gas distribution	25.36	42.90	0.32	0.18	0.36

* Trade intensity defined as total and non-EU exports of goods and services as a percentage of total supply of goods and services, plus total and non-EU imports of goods and services as a percentage of total demand for goods and services. Output is defined as gross, so the maximum value attainable is 200.

12 Opportunities and Wider Benefits from Climate Policies

KEY MESSAGES

The transition to a low-emissions global economy will open many new opportunities across a wide range of industries and services. Markets for low carbon energy products are likely to be worth at least $500 bn per year by 2050, and perhaps much more. Individual companies and countries should position themselves to take advantage of these opportunities.

Financial markets also face big opportunities to develop new trading and financial instruments across a broad range including carbon trading, financing clean energy, greater energy efficiency, and insurance.

Climate change policy can help to root out existing inefficiencies. At the company level, implementing climate policies can draw attention to money-saving opportunities. At the economy-wide level, climate change policy can be a lever for reforming inefficient energy systems and removing distorting energy subsidies on which governments spend around $250 bn a year.

Policies on climate change can also help to achieve other objectives, including enhanced energy security and environmental protection. These co-benefits can significantly reduce the overall cost to the economy of reducing greenhouse gas emissions. There may be tensions between climate change mitigation and other objectives, which need to be handled carefully, but as long as policies are well designed, the co-benefits will be more significant than the conflicts.

12.1 Introduction

Climate change policies will lead to structural shifts in energy production and use, and in other emissions-intensive activities. Whilst the previous chapters focused on the resource costs and competitiveness implications of this change, this chapter considers the opportunities that this shift will create. This is discussed in Section 12.2.

In addition, climate change policies may have wider benefits, which narrow cost estimates will often fail to take into account. Section 12.3 looks at the ways in which climate change policies have wider benefits through helping to root out existing inefficiencies at the company or country level.

Section 12.4 considers how climate policies can contribute to other energy policy goals, such as enhanced energy security and lower air pollution. Conversely, policies aimed at other objectives can be tailored to help to make climate change policies more effective. Energy market reform aimed at eliminating energy subsidies and other distortions is an important example, and is considered in Section 12.5.

In other areas, there may be tensions. The use of coal in certain major energy-using countries, for instance, presents challenges for climate change mitigation – although the use of carbon capture and storage can sustain opportunities for coal. Climate change mitigation policies also have important overlaps with broader environmental protection policies, which are discussed in Section 12.6.

Thinking about these issues in an integrated way is important in understanding the costs and benefits of action on climate change. Policymakers can then design policy in a way that avoids conflicts, and takes full advantage of the significant co-benefits that are available.

12.2   Opportunities from growing markets

Markets for low-carbon energy sources are growing rapidly

Whilst some carbon-intensive activities will be challenged by the shift to a low-carbon economy, others will gain. Enormous investment will be required in alternative technologies and processes. Supplying these will create fast-growing new markets, which are potential sources of growth for companies, sectors and countries.

The current size of the market for renewable energy generation products alone is estimated at $38 billion, providing employment opportunities for around 1.7 million people. It is a rapidly growing market, driven by a combination of high fossil fuel prices, and strong government policies on climate change and renewable energy. Growth of the sector in 2005 was 25%[1].

Within this overall total, some markets are growing at an even more rapid rate. The total global installed capacity of solar PV rose by 55% in 2005, driven by strong policy incentives in Germany, Japan and elsewhere[2], and the market for wind power by nearly 50%[3]. The market capitalisation of solar companies grew thirty-eightfold to $27 billion in the 12 months to August 2006, according to Credit Suisse[4]. Growth in biofuels uptake was not quite as rapid, but there was still a 15% rise to 2005, making the total market over $15 billion.

Growth rates in these markets will continue to be strong, creating opportunities for business and for employment opportunities.

Looking forward, whilst some of these very rapid rates may not be sustained, policies to tackle climate change will be a driver for a prolonged period of strong growth in the markets for low-carbon energy technology, equipment and construction. The fact that governments in many countries are also promoting these new industries for energy security purposes (Section 12.5) will only strengthen this effect.

---

[1]   REN21 (2006).
[2]   Renewables Global Status Report, 2006 update: REN21.
[3]   Clean Edge (2006).
[4]   Quoted in Business Week, "Wall Street's New Love Affair", August 14 2006.

One estimate of the future market for low-carbon energy technologies can be derived from the IEA's Energy Technology Perspectives report. This estimates the total investment required in low-carbon power generation technologies in a scenario where total energy emissions are brought back down to today's levels by 2050[5]. It finds that cumulative investment in these technologies by 2050 would be over $13 trillion, accounting for over 60% of all power generation by this date. The annual market for low-carbon technologies would then be over $500 bn per year. Other estimates are still higher: recent research commissioned by Shell Springboard suggests that the global market for emissions reductions could be worth $1 trillion cumulatively over the next five years, and over $2 trillion per year by 2050.[6]

The massive shift towards low-carbon technologies will be accompanied by a shift in employment patterns. If it is assumed that jobs rise from the current level of 1.7 million in line with the scale of investment, over 25 million people will be working in these sectors worldwide by 2050.

Climate change also presents opportunities for financial markets

Capital markets, banks and other financial institutions will have a vital role in raising and allocating the trillions of dollars needed to finance investment in low-carbon technology and the companies producing the new technologies. The power companies will also require access to large, long-term funds to finance the adoption of new technology and methods, both to conform to new low-carbon legislation and to satisfy rising global power demand from growing populations enjoying higher living standards.

The new industries will create new opportunities for start-up, small and medium enterprises[7] as well as large multinationals. Linked to this, specialist funds focusing on clean energy start-ups and other specialist engineering, research and marketing companies are emerging. Clean technology investment has already moved from being a niche investment activity into the mainstream; clean technology was the third largest category of venture capital investment in the US in the second quarter of 2006[8].

The insurance sector will face both higher risks and broader opportunities, but will require much greater access to long-term capital funding to be able to underwrite the increased risks and costs of extreme weather events[9]. Higher risks will demand higher premiums and will require insurance companies to look hard at their pricing; of what is expected to become a wider range of weather and climate-related insurance products[10].

The development of carbon trading markets also presents an important opportunity to the financial sector. Trading on global carbon markets is now worth over $10 bn annually with the EU ETS accounting for over $8 bn of this[11]. Expansions of the EU ETS to new sectors, and the likely establishment of trading schemes in

---

[5] This investment excludes the transport sector, but includes nuclear, hydropower, and carbon capture and storage.

[6] Shell Springboard (2006). This is an estimate of total expenditure on carbon abatement, and so would include all emission reduction sources. Figures are based on a central scenario.

[7] See, for instance, Shell Springboard (2006).

[8] Cleantech Venture Network (2006).

[9] Salmon and Weston (2006).

[10] See Ceres (2006).

[11] World Bank (2006a).

other countries and regions is expected to lead to a big growth in this market. Calculations by the Stern Review as a hypothetical exercise show that if developed countries all had carbon markets covering all fossil fuels, the overall market size would grow 200%, and if markets were established in all the top 20 emitting countries, it would grow 400% (the analysis behind these numbers can be found in Chapter 22).

This large and growing market will need intermediaries. Some key players are set out in Box 12.1. The City of London, as one of the world's leading financial centres, is well positioned to take advantage of the opportunities; the most actively traded emissions exchange, ECX, is located and cleared in London, dealing in more than twice the volume of its nearest competitor[12].

Companies and countries should position themselves now to take advantage of these opportunities

---

BOX 12.1 Financial intermediaries and climate change

The transition involved in moving to a low-carbon economy creates opportunities and new markets for financial intermediaries. Emissions trading schemes in particular require a number of key financial, legal, technical and professional intermediaries to underpin and facilitate a liquid trading market. These include:

Corporate and project finance: trillions of dollars will be required over the coming decades to finance investments in developing and installing new technologies. Creative new financing methods will be needed to finance emission reduction projects in the developing world. And emissions trading will require the development of services needed to manage compliance and spread best practice.

MRV services (monitoring, reporting and verification): these are the key features for measuring and auditing emissions. MRV services are required to ensure that one tonne of carbon emitted or reduced in one place is equivalent to one tonne of carbon emitted or reduced elsewhere.

Brokers: are needed to facilitate trading between individual firms or groups within a scheme, as well as offering services to firms not covered by the scheme who can sell emission reductions from their projects.

Carbon asset management and strategy: reducing carbon can imply complex and inter-related processes and ways of working at a company level. New opportunities will arise for consultancy services to help companies manage these processes.

Registry services: these are needed to manage access to and use of the registry accounts that hold allowances necessary for surrender to the regulator.

Legal services: these will be needed to manage the contractual relationships involved in trading and other schemes.

Trading services: the transition to a low carbon economy offers growing opportunities for trading activities of all kinds, including futures trading and the development of new derivates markets.

---

[12] CEAC (2006).

There are numerous examples of forward-looking companies which are now positioning themselves to take advantage of these growth markets, ranging from innovative high-technology start-up firms to some of the world's largest companies.

Likewise, governments can seek to position their economy to take advantage of the opportunities. Countries with sound macroeconomic management, flexible markets, and attractive conditions for inward investment can hope to win strong shares of the growing clean energy market. But particular countries may also find that for historical or geographical reasons, or because of their endowment of scientific or technical expertise, they have advantages in the development of particular technologies. There may be grounds for government intervention to support their development, particularly if promising technologies are far from market and needs to be scaled up to realise their full potential – Chapter 16 discusses how market failures and uncertainties over future policy justify action in this area.

Implementing ambitious climate change goals and policies may also help to create a fertile climate for clean energy companies. Hanemann et. al. (2006) analysed the economic impact of California taking the lead in adopting policies to reduce GHG emissions. They concluded that, if it acts now, California can gain a competitive advantage, by becoming a leader in the new technologies and industries that will develop globally as international action to curb GHG emissions strengthens. They estimate that this could increase gross state product by $60 billion, and create 20,000 new jobs, by 2020.

12.3   Climate change policy as a spur to efficiency and productivity

Climate change policies can be a general spur to greater efficiency, cost reduction and innovation for the private sector

Predictions of the costs of environmental regulations often turn out to be overestimates. Hodges (1997) compared all cases of emission reduction regulations for which successive cost estimates were available, a dozen in total. He found that in all cases except one (CFCs where costs were only 30% below expectations due to the accelerated timetable for phase-out of the chemical), the early estimates were at least double the later ones, and often much greater.

One example is the elimination of CFCs in car air conditioners. Early industry estimates suggested this would increase the price of a new car by between $650 and $1200. By 1997, the cost was $40 to $400[13].

When such numbers come to light, companies are often accused of inflating initial cost estimates to support their lobbying efforts. But there is a more positive side to the story. The dramatic reduction in costs is often a result of the process of innovation, particularly when a regulatory change results in a significant increase in the scale of production.

And the process of complying with new policies may reveal hidden inefficiencies which firms can root out, saving money in the process (Box 12.2).

---

[13] American Prospect, "Polluted Data", November 1997.

BOX 12.2 Reducing business costs through tackling climate change

An increasing number of private and public sector organisations are discovering the potential to reduce the cost of goods and services they supply to the market. A study of 74 companies drawn from 18 sectors in 11 countries including North America, Europe, Asia, and Australasia revealed gross savings of $11.6 billion, including[14]:

- BASF, the multi-national conglomerate and chemical producer, has reduced GHG emissions by 38% between 1990 and 2002 through a series of process changes and efficiency measures which cut annual costs by 500 million euros at one site alone;

- BP established a target to reduce GHG emissions by 10% on 1990 levels by 2010, which it achieved nine years ahead of schedule, while delivering around $650 million in net present value savings through increased operational efficiency and improved energy management. Between 2001 and 2004, the organisation contributed a further 4 MtC of emission reductions through energy and flare reduction projects. $350 million investment in energy efficiency is planned over 5 years from 2004.

- Kodak began tracking its greenhouse gas emissions in the 1990s, and set five-year goals for emissions reductions. To help to achieve this, the company performed short, focused energy assessments – "Energy Kaizens" – across different areas of its business, aimed at reducing waste. Between 1999 and 2003, this and other initiatives resulted in overall savings of $10 million.

Tackling climate change may also have more far-reaching effects on the efficiency and productivity of economies. Schumpeter (1942)[15] developed the concept of "creative destruction" to describe how breakthrough innovations could sweep aside the established economic status quo, and unleash a burst of creativity, investment and economic growth which ushers in a new socio-economic era. Historical examples of this include the introduction of the railways, the invention of electricity, and more recently, the IT revolution. Dealing with climate change will also involve fundamental changes worldwide, particularly to energy systems.

In particular, the shift to low-carbon energy technologies will result in a transformation of energy systems; the implications of this are explored in the following sections.

12.4 The links between climate change policy and other energy policy goals

Climate change policies cannot be disconnected from policies in other areas, particularly energy policy. Where such synergies can be found, they can reduce the effective cost of emissions reductions considerably. There may also be tensions in some areas, if climate change policies undermine other policy goals. But as long as policies are well designed, the co-benefits should outweigh the conflicts.

---

[14] The Climate Group (2005).
[15] See also Aghion and Howitt (1999).

Climate change and energy security drivers will often work in the same direction, although there are important exceptions

Energy security is a key policy goal for many developed and developing countries alike. Although often understood as referring mainly to the geopolitical risks of physical interruption of supply, a broader definition would encompass other risks to secure, reliable and competitive energy, including problems with domestic energy infrastructure.

Energy efficiency is one way to meet climate change and energy security objectives at the same time. Policies to promote efficiency have an immediate impact on emissions. More efficient use of energy reduces energy demand and puts less pressure on generation and distribution networks and lowers the need to import energy or fuels. For developing countries in particular, who often have relatively low energy efficiency, this is an attractive option. Indirectly, they also help with local air pollution, by limiting the growth in generation.

Improving efficiency within the power sector itself has similar effects. Box 12.3 gives an example of the scale of the potential to reduce emissions from making fossil fuel production processes more efficient.

BOX 12.3 Economic opportunities from reducing gas-flaring in Russia

In total, leaks from the fossil fuel extraction and distribution account for around 4% of global greenhouse gas emissions. Within this, gas flaring – the burning of waste gas from oil fields, refineries and industrial plants – accounts for 0.4% of global emissions. Increasingly, there has been a move to capture these gases, driven by economic as much as environmental reasons. This is by no means universal, and in some countries the potential for emissions savings in this area remains significant.

The post-Soviet collapse of Russia's energy-intensive economy cut carbon emissions and left it with a surplus of transferable emission quotas under the Kyoto protocol. Decades of under-investment, however, mean that current 6–7 per cent GDP growth, spurred by higher energy and commodity prices, is both raising emissions and putting pressure on the infrastructure. Sustaining growth requires very large energy and related infrastructure investment. In June 2006 the government approved a $90 bn investment programme to replace ageing coal and nuclear generating plants, increase generating capacity and strengthen the grid system.

A recent IEA report[16] on Russian gas flaring, however, indicates that without accompanying price and structural reforms, especially in the gas sector, investment alone is unlikely to deliver the full potential for efficiency gains or reductions in GHGs.

The report indicates that low prices for domestic gas, coupled with Gazprom's monopoly over access to both domestic and export gas pipelines and the high levels of waste and inefficient technology, restrict its ability to satisfy rising export and domestic demand, and to reduce both gas losses and GHG emissions.

In 2004 Gazprom lost nearly 70 billion cubic metres (bcm) of the nearly 700 bcm of natural gas which flowed through its network because of leaks and high wastage from inefficient compressors. Gas related emissions amounted to nearly 300 MtCO2e of GHG, including 43 MtCO2e from the 15 bcm of gas flared off, mainly by oil companies unable to

---

[16] IEA (2006).

gain access to Gazprom's pipes. On this basis, Russia accounted for around ten per cent of natural gas flared off globally every year. However, an independent study conducted by the IEA and the US National Oceanic and Atmospheric administration, calibrated from satellite images of flares in the main west Siberian oilfields, indicated however that up to 60 bcm of gas may be lost through flaring – over a third of the estimated global total[17].

Gas flaring represents a clear illustration of the potential efficiency gains from new technology linked to more rational pricing policies and other structural reforms. These would also yield significant climate change mitigation benefits.

A more diverse energy mix can be an effective hedge against problems in the supply of any single fuel. As climate change policy tends to encourage a more diverse energy mix, it is generally good for energy security. And conversely, policies carried out for energy security reasons may have benefits for climate change. The expansion of a range of sources of renewable power and, where appropriate, of nuclear energy can reduce the exposure of economies to fluctuations in fossil fuel prices, as well as reducing import dependence.

Coal is an important exception to this rule. Coal is much more carbon intensive than other fossil fuels: coal combustion emits almost twice as much carbon dioxide per unit of energy as does the combustion of natural gas (the amount from crude oil combustion falls between coal and natural gas[18]). Many major energy-using countries have abundant domestic coal supplies, and hence see coal as having an important role in enhancing energy security. China, in particular, is already the world's largest coal producer; its consumption of coal is likely to double over the 20 years between 2000 and 2020[19].

As well as using coal directly for energy production, coal-producing countries including the US, Australia, China and South Africa are investing in coal-to-liquids technology, which would allow them to reduce their dependence on imported oil and use domestic coal to meet some of the demand for transport fuel. But it has been estimated that "well-to-wheel" (full lifecycle) emissions from the production and use of coal-to-liquids in road transport are almost double those from using crude oil[20].

However, extensive deployment of carbon capture and storage (as discussed in Chapter 9), can reconcile the use of coal with the emissions reductions necessary for stabilising greenhouse gases in the atmosphere.

Supporting sufficient investment in generation and distribution capacity also requires a sound framework capable of bringing forward required investment. Clear, long-term credible signals about climate policy are a critical part of this. If there is uncertainty about the future direction of climate change policy, energy companies may delay investment, with serious consequences for security of supply. This is discussed in more detail in Chapter 15.

---

[17] IEA (2006).
[18] Energy Information Administration (1993).
[19] Chinese Academy of Social Sciences (2006).
[20] Well-to-wheels emissions from fuels such as gasoline are around 27.5 pounds of CO2 per gallon of fuel. This compares with 49.5 pounds per gallon from coal-to-liquids, assuming the CO2 from the refining process is released into the atmosphere. See Natural Resources Defence Council (2006).

Access to energy is a priority for economic development

There are currently 1.6 billion people in the world without access to modern energy services[21]. This restricts both their quality of life, and their ability to be economically productive. Providing poor people with access to energy is a very high priority for many developing countries, and can have significant co-benefits in reducing local pollution, as the next section discusses.

Increasing the number of energy consumers, by providing access to energy, would tend to push emissions upwards. But well-designed policies present opportunities for meeting several objectives at once. New renewable technologies, developed with climate change objectives in mind, can help to overcome barriers to access to energy. Microgeneration technologies (see Box 17.3 in Chapter 17) such as small-scale solar and hydropower, in particular, remove the need to be connected to the grid, and so help raise availability and reduce the cost of electrification in rural areas. And as discussed below, the replacement of low-quality biomass energy with modern energy can cut emissions and pollution.

As well as access, affordability is a key issue in both developed and developing countries. Poverty is determined by people's capacity to earn in relation to prices. Energy prices are one significant aspect, along with food and other essentials.

But it is inappropriate to deal with poverty by distorting the price of energy. Addressing income distribution issues directly is more effective. There are a number of ways to achieve this. One is indexing social transfers to a price index, taking account of different consumption patterns of poorer groups in the relevant price index for those groups. Other more direct means include making special transfers to those with special energy needs such as the elderly, and the use of "lifeline tariffs", whereby people using a minimal amount of power pay a sharply reduced tariff for a fixed maximum number of units.

Climate change policies can help to reduce local air pollution, with important benefits for health

Measures to reduce energy use, and to reduce the carbon intensity of energy generation, can have benefits for local air quality. Most obviously, switching from fossil fuels to renewables, or from coal to gas, can significantly the levels of air pollution resulting from fossil fuel burning.

A recent study by the European Environment Agency[22] showed that the additional benefits of an emissions scenario aimed at limiting global mean temperature increase to 2°C would lead to savings on the implementation of existing air pollution control measures of €10 billion per year in Europe, and additional avoided health costs of between €16–46 billion per year.

Local air pollution has a serious impact on public health and the quality of life. These impacts are particularly severe in developing countries, where only malnutrition, unsafe sex and lack of clean water and adequate sanitation are greater health threats than indoor air pollution[23]. In China, a recent study[24] showed that

[21] World Bank, "1.6 billion people still lack access to electricity today", press release, 18 September 2006.
[22] Air Quality and ancillary benefits of climate change, EEA, Copenhagen, 2006
[23] WHO (2006).
[24] Aunan et al (2006)

for CO_2 reductions up to 10–20%, air pollution and other benefits more than off-set the costs of action.

Forthcoming analysis from the IEA (Box 12.4) shows that combustion of trad-itional biomass for cooking and heating in developing countries is associated with high GHG emissions and adverse indoor air quality and health impacts, which switching to a cleaner fuel could reduce.

Sometimes climate change objectives will conflict with local air quality aims. This is a particular issue in transport. In road transport, switching from petrol to diesel reduces CO_2 emissions, but increases local air pollution (PM10 and NOx emissions). High blends of biodiesel can also emit slightly more NOx than conventional diesel. The US and EU are in the process of implementing stronger policies to reduce CO_2 emissions from diesel vehicles, although this will take time to have an effect.

In the case of aviation, there are multiple links between objectives[25]. One of the ways of achieving CO_2 improvements in aircraft is to increase combustion

---

BOX 12.4 Use of traditional biomass in developing countries

In developing countries, 2.5 bn people depend on traditional biomass such as fuel wood and charcoal as their primary fuel for cooking and heating because it is a cheap source of fuel. The emissions associated with this biomass are relatively high because it is not com-busted completely or efficiently. Aside from the climate change impact, combustion of biomass is associated with a range of detrimental effects on health, poverty and local environment including:-

- Smoke from biomass from cooking and heating was estimated to cause 1.3 m prema-ture deaths in 2002. Women and children are most severely affected because they spend most time in the home doing domestic tasks. More than half the deaths are children because their immune systems are poorly equipped to deal with the local air pollution.

- Time spent collecting the biomass is time that could otherwise be spent by women or children in education or other productive work. The collection of biomass may also involve hard physical labour that deteriorates the health of the women and children doing it.

- Collection of biomass causes localised deforestation and land degradation. If animal dung is used as a fuel rather than a fertiliser then soil fertility suffers. The widespread use of fuel wood and charcoal can mean local resources getting used up so people have to travel further to collect it.

Switching away from traditional biomass towards modern, cleaner cooking fuels can save GHG emissions and reduce the health, poverty and local environment concerns outlined above. The UN Millennium Project has adopted a target of reducing by 50% the number of households using traditional biomass as their primary fuel by 2015; this means giving an extra 1.3 bn people access to clean fuels by this date. If this were achieved by switching these users to liquid petroleum gas, it would cost $1.5 bn per year for new stoves and can-isters, increase global demand for oil by just 0.7% in 2015, and result in a small reduction in GHG emissions.

Source: IEA (in press).

---

[25] See European Commission (2005).

temperatures in engines. However, this increases levels of NOx, an important local air pollutant. Other measures to improve fuel efficiency and CO_2 performance, such as reducing aircraft weight, have benefits for local air pollution. And there are complex relationships between gases emitted at altitude – there are suggestions, for instance, that more modern engines have a greater tendency to produce condensation trails, which intensify warming effects (see Box 15.6, Chapter 15). Further technological advances in aircraft construction will be important in meeting both climate change and air pollution objectives simultaneously.

Policies to meet air pollution and climate change goals are not always compatible. But if governments wish to meet both objectives together, then there can be considerable cost savings compared to pursuing both separately.

12.5   The role of pricing and regulatory reforms in the energy markets

Pricing and regulatory reforms in the energy markets are important both for effective climate change policy, and for long-term productivity and efficiency

Many countries have a long history of subsidising particular fuels: coal, oil, nuclear power, electricity for rural areas, and more recently renewable energy. With the important exceptions of support mechanisms for R&D and innovation (see Chapter 16), these are a source of economic distortion and loss. Furthermore there has been a strong historical bias toward the more polluting fuels. The liberalisation of energy markets that began to take place in many countries in the late 1980s and early 1990s was seen as a means of reducing these subsidies, which in some cases had reached extraordinary proportions. By 1998 they had declined worldwide, but still amounted to nearly $250 billion per year, of which over $80 billion were in the OECD countries and over $160 billion in developing countries (see Table 12.1). These transfers are on broadly the same scale as the average incremental costs of an investment programme required for the world to embark on a substantial policy of climate change mitigation over the next twenty years (see Chapter 9). The IEA estimate that world energy subsidies were still $250 billion in 2005, of which subsidies to oil products amounted to $90 billion[26].

Applied in the form of tax credits and incentives for innovation, subsidies can and do serve an economic purpose. However, the prevailing subsidies are for the most part not applied to this end. The inefficiencies associated with subsidies have been reviewed by economists many times over the past decades, and can be simply stated:

- subsidies stimulate unnecessary consumption and waste, and more generally are a source of economic inefficiency in that the low price is associated with low benefits on the margin relative to the cost of production;
- tend to benefit the middle and higher income groups, so impacting income distribution in a negative way, particularly in developing countries where poor people lack access to energy;
- by undermining the capacity of the industry to earn returns directly on the basis of cost-reflecting prices, subsidies undermine the managerial (or 'X') efficiency of the industry, and also its capacity to finance its expansion;

[26] IEA (in press).

Table 12.1 Energy Subsidies by Source $ billion (data for 1995–1998 period)

	OECD Countries	Countries not in OECD	Total
Coal	30	23	53
Oil	19	33	52
Gas	8	38	46
All fossil fuels	**57**	**94**	**151**
Electricity	–	48	48
Nuclear	16	?	16
Renewables and energy efficiency	9	?	9
Cost of bankruptsy bail-out	0	20	20
Total	**82**	**162**	**244**

Source: de Moor (2001) and van Beers and de Moor (2001). Another perspective on subsidies is provided by Myers, N. and J. Kent (1998) 'Perverse Subsidies: Tax $s Undercutting our Economies and Environment Alike', Winnipeg, IISD.

- lead to wasteful lobbying and rent-seeking by groups trying to maintain or increase subsidies;
- when applied to fossil fuels, subsidies discourage the development of and investment in low carbon alternatives, including investment in carbon capture and storage.

To the extent that climate change policy triggers wider energy reform, it would have great supplementary benefits, as long as the transition is well managed. And for carbon price signals to work well, it is essential that the energy market also works well.

An example of the costs of energy market inefficiencies, and the way in which reforms can deliver environmental and other goals, is given in Box 12.5 for India.

BOX 12.5 Fuelling India's growth and development

India's economic growth is constrained by an inadequate power supply that results in frequent blackouts and poor reliability. Subsidised tariffs to residential and agricultural consumers,[27] low investment in transmission and distribution systems, inadequate maintenance, and high levels of distribution losses, theft and uncollected bills place the State Electricity Boards (SEBs, which form the basis of India's power system) under severe financial difficulties.[28] These losses and subsidies are a significant drain on budgets and can result in public spending on vital areas such as health and education being crowded out. Annual power sector losses associated with inefficiencies and theft are estimated at over $5 billion – more than it would cost to support India's primary health care system.[29]

---

[27] The tariff structure, for example, violates the fundamental principle of economics whereby tariffs should reflect the actual cost of service. In practice, industry is charged the highest tariff despite having the least costs of supply, whilst agriculture has the lowest tariff and the highest cost of service.

[28] World Bank (2001).

[29] World Bank (2006b).

The demand shortages facing India – 56% of Indian households have no electricity supply – create incentives for getting generation plants on line as rapidly as possible. These priorities in turn favour reliable, conventional, coal-fired units.[30] The use of coal for the bulk of electricity generation presents particular challenges. Coal mining is dangerous, and its transportation creates environmental problems of its own. Coal also produces pollutants such as sulphur dioxide that damage local air quality, causing further problems for human health and the environment. These issues are exacerbated by the low energy efficiency of India's coal-fired power plants, combined with India's policies of high import tariffs on high-quality coal and subsidies on low-quality domestic coal. The use of CCS technology will be an important way to reconcile the cost and convenience advantages of coal, with environmental goals.

The Government of India has set out an energy policy to help address these constraints and concerns. The broad objective of this policy is to reliably meet the demand for energy services of all sectors at competitive prices, through "safe, clean and convenient forms of energy at the least-cost in a technically efficient, economically viable and environmentally sustainable manner".[31] With sufficient effort made in improving energy efficiency and conservation, for example, the Government of India has stated that it would be possible to reduce the country's energy intensity by up to 25% from current levels.[32] Progress in achieving the goals and objectives of their energy policy, ranging from improving energy efficiency to promoting the use of renewables, will also make a significant contribution to reducing future GHG emissions from India.

12.6   Climate change mitigation and environmental protection

This section looks at the links between climate change and broader environmental protection goals. One area where these links are particularly strong is deforestation. Policies that prevent deforestation can have significant benefits for communities dependant on forests, for water management and biodiversity. Some of these are set out in Box 12.6.

BOX 12.6 Co-benefits of ending deforestation

Protection/Preservation of biodiversity: Tropical forests house 70% of the Earth's plants and animals. Without forest conservation, many of the world's plant and animal species face extinction this century. Essential natural resources are found in frontier forests that cannot be recreated.

Research and development: Frontier forests in Brazil, Colombia and Indonesia are home to the greatest plant biodiversity in the world. Destroying these forests destroys the source of essential pharmaceutical ingredients; 40–50% of drugs in the market have an origin in natural products[33], with 42% of the sales of the top 25 selling drugs worldwide either biologicals, natural products, or derived from natural products[34].

[30] World Bank (2006b).
[31] Government of India (2006: xiii).
[32] Government of India (2006).
[33] www.fic.nih.gov/programs/research_grants/icbg/index.htm
[34] CBI (2005).

Indigenous peoples and sustainability: About 50 million people are believed to be living in tropical forests, with the Amazonian forests home to around 1 million people of 400 different indigenous groups. Forest conservation affects people beyond those who inhabit them. Over 90% of the 1.2 billion people living in extreme poverty depend on forests for some part of their livelihoods[35].

Tourism: Forests provide opportunities for recreation for an increasingly wealthy and urbanised population. Brazil had a five-fold increase in tourists between 1991 and 1999, with 3.5 m people visiting Brazil's 150 Conservation Areas.

Consequences for vulnerability to extreme weather events: Forests systems can play an important role in watersheds, and their loss can lead to an increase in flooding. In November 2005 a flash flood occurred in Langkat, Indonesia that killed 103 people with hundreds more missing. The Mount Leuser National Park had lost up to 22% of its forest cover due to logging and, combined with high rainfall, had caused a landslide to occur[36].

In 2004, 3000 people died in Haiti after a tropical storm, while only 18 people across the border in the Dominican Republic died. The difference has been linked to extensive deforestation in Haiti where political turmoil and poverty have lead to the destruction of 98% of original forest cover[37]. Mangrove forests, depleted by 35% (see Millennium Ecosystem Assessment 2005) play an important role in coastal defence, as well as providing important nursery grounds for fish stocks. Areas with healthy mangrove or tree cover were significantly less likely to have experienced major damage in the 2004 tsunami[38].

Reducing GHG emissions from agriculture could also have benefits for local environment and health. For example, in China, nitrous oxide emissions associated with overuse of fertiliser contributes to acid rain, causes severe eutrophication of the China Sea and damage to health through contamination of drinking water. Cutting these emissions could help to reduce these effects[39].

However, climate change mitigation may, if poorly implemented, undermine sustainable development. Chapter 9 discussed the technical potential of biomass to save emissions in the power, transport, industry and buildings sectors. But if the crops are grown at very large scale through intensive, large-scale monoculture, then this has the potential to cause serious environmental impacts. These may include the increased use of pesticides; a loss of biodiversity and natural habitats[40]; and social problems and displacement of indigenous peoples.

Mitigation policies can also sometimes be designed in a way that helps countries cope with existing climate variability and adapt to future climate change. Better design of building stock, for instance, can both reduce the demand for space heating and cooling and provide greater resilience to a changing climate.

While there are important links between mitigation and development, it is important to assess policy development against the full range of opportunities to meet

---

[35] World Bank (2006): 'Forests and Forestry' available from http://web.worldbank.org/WBSITE/
       EXTERNAL/TOPICS/EXTARD/EXTFORESTS/0,,menuPK:985797~pagePK:149018~piPK:1490
       93~theSitePK:985785,00.html
[36] Jakarta Post (2003): Rampant deforestation blamed for Langkat flash flood. 05/11/2003.
[37] Secretariat of the Convention on Biological Diversity (2006).
[38] Secretariat of the Convention on Biological Diversity (2006).
[39] Norse (2006).
[40] See, for instance, European Environmental Bureau (2006).

climate goals and the full range of options to achieve the Millennium Development Goals (see Michaelowa 2005). As with other co-benefits, the key is that well designed policy can realise the synergies between different goals, as well as the limits to this. For example, to improve education levels in developing countries, schools could be supplied with low emission energy supplies, or more trained teachers. Both interventions will be associated with a wide range of different costs and benefits, which should be weighed up when considering which option is preferred.

12.7   Conclusion

Whilst climate change presents clear challenges and costs to the global economy, it also presents opportunities. Markets for clean energy technologies are set for a prolonged period of rapid growth, and will be worth hundreds of billions of dollars a year in a few decades' time. Companies and countries should position themselves to take advantage of these growth markets.

It is also important to consider the wider impacts of climate change policy. As well as helping to root out existing inefficiencies, climate change policy can also help to achieve other policies and goals, particularly around energy policy and sustainable development.

A full understanding of these interlinkeages is key to designing policy in a way that minimises the areas of conflict between goals, and to reap the benefits of the opportunities and synergies that exist.

References

Aghion, P. and P. Howitt (1999): 'On the macroeconomic consequences of major technological change', in General Purpose Technologies and Economic Growth, ed. E. Helpman, Cambridge, MA: MIT Press.

Aunan et al. (2006): 'Benefits and costs to China of a climate policy', Environment and Development Economics, accepted for publication

Confederation of British Industry (2005): 'EU market survey natural ingredients for pharmaceuticals' Rotterdam: CBI.

CEAC (2006): "Emissions trading and the City of London", report to the City of London, September 2006.

Ceres (2006): 'From Risk to Opportunity: How insurers can proactively and profitably manage climate change', Ceres, August 2006

Chinese Academy of Social Sciences (2006): available from http://www.hm-treasury.gov.uk/media/5FB/FE/Climate_Change_CASS_final_report.pdf

Clean Edge (2006): 'Clean Energy Trends', San Francisco, Clean Edge Inc., available from http://www.cleanedge.com/reports/trends2006.pdf

Cleantech Venture Network (2006): 'Cleantech Becomes Third Largest Venture Capital Investment Category with $843 Million Invested in Q2 2006', Press release August 10 2006

Energy Information Administration (1993): 'Emissions of greenhouse gases in the United States 1985–1990', DOE/EIA-0573, Washington, DC: EIA, p. 16.

European Commission (2005): 'Giving Wings to Emissions Trading: Inclusion of aviation into the European Emissions Trading System – design and impacts', European Commission, report reference ENV.C.2/ETU/2004/0074r

European Environmental Bureau (2006): 'Fuelling extinction? Unsustainable biofuels threaten the environment', Brussels: EEB, BirdLife International and European Federation for Transport and the Environment.

Government of India, Planning Commission (2006): 'Integrated Energy Policy: Report of the Expert Committee'. New Delhi: Government of India, Planning Commission, August 2006.

Hanemann M.W. and A.E. Farrell (2006): 'Managing greenhouse gas emissions in California', California: The California Climate Change Center at UC Berkeley.

Hodges, H. (1997): 'Falling prices, cost of complying with environmental regulations almost always less than advertised.' Washington, DC: Economic Policy Institute, available from http://www.epinet.org/briefingpapers/bp69.pdf

International Energy Agency (2006): 'Optimising Russian natural gas', Paris: OECD/IEA

International Energy Agency (in press): 'World Energy Outlook 2006', Paris: OECD/IEA.

Millennium Ecosystem Assessment (2005): 'Ecosystems and Human Well-being: Synthesis'. Washington, DC: Island Press.

Natural Resources Defence Council (2006): Testimony of David G. Hawkins to the Committee on Energy and Natural Resources, 24 April 2006.

Norse D. (2006): 'Key trends in emissions from agriculture and use of policy instruments', available from www.sternreview.org.uk

REN21 (2006): 'Renewables Global Status Report 2006 Update', Washington, DC: REN21 Secretariat and Washington DC: Worldwatch Institute.

Salmon, M. and Weston, S. (2006): 'Open letter to the Stern Review on the economics of climate change', available from www.sternreview.org.uk

Schumpeter, J. (1942): 'Capitalism, Socialism and Democracy', New York: Harper.

Secretariat of the Convention on Biological Diversity (2006): 'Global Biodiversity Outlook 2'. Montreal: CBD available from http://www.biodiv.org

Shell Springboard (2006): 'The Business Opportunities for SMEs in tackling the causes of climate change', report by Vivid Economics for Shell Springboard, October 2006.

Swart, R., M. Amann, F. Raes and W. Tuinstra (2004): 'A good climate for clean air: linkages between climate change and air pollution, an editorial essay', Climatic Change Journal, **66**(3): 263–269

The Climate Group (2005), 'Carbon down profits up – Second Edition 2005'.

World Bank (2006a): 'State and Trends of the Carbon Market', Washington, DC: World Bank.

World Bank (2006b): Clean energy and development: towards an investment framework. Washington, DC: World Bank.

World Bank (2001): Fuelling India's Growth and development: World Bank support for India's Energy Sector. Washington, DC: World Bank.

World Health Organisation (2006): 'Fuel for Life', Geneva: WHO.

World Resources Institute (2005): 'Growing in the greenhouse: protecting the climate by putting development first' [R. Bradley and K.A. Baumert], Washington, DC: WRI.

13 Towards a Goal for Climate-Change Policy

KEY MESSAGES

Reducing the expected adverse impacts of climate change is both highly desirable and feasible. The need for strong action can be demonstrated in three ways: by comparing disaggregated estimates of the damages from climate change with the costs of specific mitigation strategies, by using models that take some account of interactions in the climate system and the global economy, and by comparing the marginal costs of abatement with the social cost of carbon.

The science and economics both suggest that a shared international understanding of the desired goals of climate-change policy would be a valuable foundation for action. Among these goals, aiming for a particular target range for the ultimate concentration of greenhouse gases (GHGs) in the atmosphere would provide an understandable and useful guide to policy-makers. It would also help policy-makers and interested parties at all levels to monitor the effectiveness of action and, crucially, anchor a global price for carbon. Any long-term goal would need to be kept under review and adjusted as scientific and economic understanding developed.

However, the first key decision, to be taken as soon as possible, is that strong action is indeed necessary and urgent. This does not require immediate agreement on a precise stabilisation goal. But it does require agreement on the importance of starting to take steps in the right direction while the shared understanding is being developed.

Measuring and comparing the expected benefits and costs over time of different potential policy goals can provide guidance to help decide how much to do and how quickly. Given the nature of current uncertainties explored in this Review, and the ethical issues involved, analysis can only suggest a range for action.

The current evidence suggests aiming for stabilisation somewhere within the range 450–550 ppm CO_2e. Anything higher would substantially increase risks of very harmful impacts but would only reduce the expected costs of mitigation by comparatively little. Anything lower would impose very high adjustment costs in the near term for relatively small gains and might not even be feasible, not least because of past delays in taking strong action.

For similar reasons, weak action over the next 20 to 30 years, by which time GHG concentrations could already be around 500 ppm CO_2e, would make it very costly or even impossible to stabilise at 550 ppm CO_2e. **There is a high price to delay.** Delay in taking action on climate change would lead both to more climate change and, ultimately, higher mitigation costs.

Uncertainty is an argument for a more, not less, demanding goal, because of the size of the adverse climate-change impacts in the worse-case scenarios.

Policy should be more ambitious, the more societies dislike bearing risks, the more they are concerned about climate-change impacts hitting poorer people harder, the more optimistic they are about technology opportunities, and the less they discount future generations' welfare purely because they live later. The choice of objective will also depend on judgements about political feasibility. These are decisions with such globally significant implications that they will rightly be the subject of a broad public debate at a national and international level.

The ultimate concentration of greenhouse gases anchors the trajectory for the social cost of carbon. **The social cost of carbon is likely to increase steadily over time, in line with the expected rising costs of climate-change-induced damage. Policy should therefore ensure that abatement efforts at the margin also intensify over time. But policy-makers should also spur on the development of technology that can drive down the average costs of abatement.** The social cost of carbon will be lower at any given time with sensible climate-change policies and efficient low-carbon technologies than under 'business as usual'.

Even if all emissions stopped tomorrow, the accumulated momentum behind climate change would ensure that global mean temperatures would still continue to rise over the next 30 to 50 years. Thus **adaptation is the only means to reduce the now-unavoidable costs of climate change over the next few decades. But adaptation also entails costs, and cannot cancel out all the effects of climate change.** Adaptation must go hand in hand with mitigation because, otherwise, the pace and scale of climate change will pose insurmountable barriers to the effectiveness of adaptation.

13.1 Introduction

It is important to use both science and economics to inform policies aimed at slowing and eventually bringing a stop to human-induced climate change.

Science reveals the nature of the dangers and provides the foundations for the technologies that can enable the world to avoid them. Economics offers a framework that can help policy-makers decide how much action to take, and with what policy instruments. It can also help people understand the issues and form views about both appropriate behaviour and policies. The scientific and economic framework provides a structure for the discussions necessary to get to grips with the global challenge and guidance in setting rational and consistent national and international policies.

Reducing the expected adverse impacts of climate change is both desirable and feasible.

Previous chapters argued that, without mitigation efforts, future economic activity would generate rising greenhouse gas emissions that would impose unacceptably high economic and social costs across the entire world. Fortunately, technology and innovation can help rein back emissions over time to bring human-induced climate change to a halt. This chapter first makes the case for strong action now, and then discusses how a shared understanding around the world of the nature of the challenge can guide that action on two fronts: mitigation and adaptation.

13.2   The need for strong and urgent action

The case for strong action can be examined in three ways: a 'bottom-up' approach, comparing estimates of the damages from unrestrained climate change with the costs of specific mitigation strategies; a 'model-based' approach taking account of interactions in the climate system and the global economy; and a 'price-based' approach, comparing the marginal costs of abatement with the social cost of carbon.

The 'bottom-up' approach was adopted in Chapters 3, 4 and 5 of this Review for the heterogeneous impacts of climate change, and in Chapters 8 and 9 for the scale and costs of possible mitigation strategies. If global temperatures continue to rise, there will be mounting risks of serious harm to economies, societies and ecosystems, mediated through many and varied changes to local climates. The impacts will be inequitable. It is not necessary to add these up formally into a single monetary aggregate to come to a judgement that human-induced climate change could ultimately be extremely costly. Chapter 7 showed that, without action, greenhouse-gas emissions will continue to grow, so these risks must be taken seriously. But Chapter 9 showed that it is possible to identify technological options for stabilising greenhouse gas concentrations in the atmosphere that would cost of around 1% of world gross world product – moderate in comparison with the high cost of potential impacts. The options considered there are not the only ways of tackling the problem, nor necessarily the best. But they do demonstrate that the problem can be tackled. And there will be valuable co-benefits, such as reductions in local air pollution.

The 'model-based' approach was illustrated in Chapter 6 for the impacts, and Chapter 10 for the costs, of mitigation. Models make it easier to consider the quantitative implications of different degrees of action and can build in some behavioural responses, both to climate change and the policy instruments used to combat it. But they do so at the cost of considerable simplification. They also require explicit decisions about the ethical framework appropriate for aggregating costs and benefits of action. The model results surveyed in this Review point in the same direction as the 'bottom up' evidence: the benefits of strong action clearly outweigh the costs.

In broad brush terms, spending somewhere in the region of 1% of gross world product on average forever could prevent the world losing the equivalent of 5–20% of gross world product for ever, using the approach to discounting explained in Chapters 2 and 6.

This can be thought of as akin to an investment. Putting together estimates of benefits and costs of mitigation through time, as in Figures 13.1 and 13.2, shows how incurring relatively modest net costs this century (peaking around 2050) can earn a big return later on, because of the size of the damages averted. These charts are quantitative analogues to the schematic diagram in Figure 2.4 comparing a 'business as usual' trajectory with a mitigation path. They are drawn

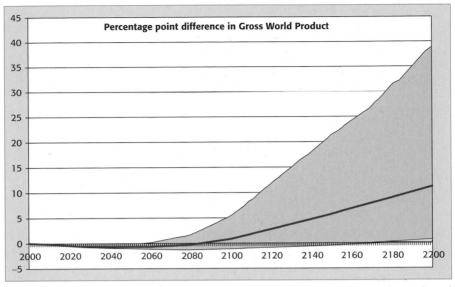

Figure 13.1 'Output gap' between the '550 ppm CO_2e and 1% GWP mitigation cost' scenario and BAU scenario, mean and 5th–95th percentile range

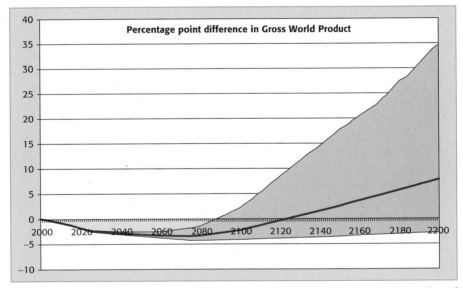

Figure 13.2 'Output gap' between the '550 ppm CO_2e and 4% GWP mitigation cost' scenario and BAU scenario, mean and 5th–95th percentile range

assuming mitigation costs to be a constant 1% (Figure 13.1) and 4% (Figure 13.2) of gross world product and taking a 'business as usual' scenario with baseline climate sensitivity, some risk of catastrophes and a rough-and-ready estimate of non-market impacts. As explained in Chapter 6, this is now likely to underestimate the sensitivity of the climate to greenhouse-gas emissions. Also, the charts focus on impacts measured in terms of how they might affect output, not wellbeing; in other words, they do not reflect the more appropriate approach to dealing with risk, as advocated in Chapter 2. But the range between the 5th and 95th percentiles of the distribution of possible impacts under the specific scenario is shown.

The 'price-based' approach compares the marginal cost of abatement of emissions with the 'social cost' of greenhouse gases. Consider, for example, the social cost of carbon – that is, the impact of emitting an extra unit of carbon at any particular time on the present value (at that time) of expected wellbeing or utility[1]. The extra emission adds to the stock of carbon in the atmosphere for the lifetime of the relevant gas, and hence increases radiative forcing for a long time. The size of the impact depends not only on the lifetime of the gas, but also on the size of the stock of greenhouse gases while it is in the atmosphere, and how uncertain climate-change impacts in the future are valued and discounted. The social cost of carbon has to be expressed in terms of a numeraire, such as current consumption, and is a relative price. If this price is higher than the cost, at that time, of stopping the emission of the extra unit of carbon – the marginal abatement cost – then it is worth undertaking the extra abatement, as it will generate a net benefit. In other words, if the marginal cost of abatement is lower than the marginal cost of the long-lasting damage caused by climate change, it is profitable to invest in abatement.

The 'price-based' approach points out that estimates of the social cost of carbon along 'business as usual' trajectories are much higher than the marginal abatement cost today. The academic literature provides a wide range of estimates of the social cost of carbon, spanning three orders of magnitude, from less than £0/tC (in year 2000 prices) to over £1000/tC (see Box 13.1), or equivalently from less than \$0/tCO_2 to over \$400/tCO_2. This is obviously an extremely broad range and as such makes a policy driven by pricing based on an estimate of the social cost of carbon difficult to apply. The mean value of the estimates in the studies surveyed by Tol was around \$29/tCO_2 (2000 US\$), although he draws attention to many studies with a much lower figure than this.

The modelling approach that was illustrated in Chapter 6 of this Review also indicates the sensitivities of estimates of the social cost of carbon to assumptions about discounting, equity weighting and other aspects of its calculation, as described by Tol, Downing and others. Preliminary analysis of the model used in Chapter 6 points to a number around \$85/tCO_2 (year 2000 prices) for the central 'business as usual' case, using the PAGE2002 valuation of non-market impacts. It should be remembered that this model is different from its predecessors, in that it incorporates both explicit modelling of the role of risk, using standard approaches to the economics of risk, and makes some allowance for catastrophe risk and non-market costs, albeit in an oversimplified way. In our view, these are very important aspects of the social cost of carbon, which should indeed be included in its calculation even though they are very difficult to assess. We would

---

[1]   The social cost of carbon and carbon price discussed here are convenient shorthand for the social cost (and corresponding price) for each individual greenhouse gas. Their relative social costs, or 'exchange rate', depend on their relative global warming potential (GWP) over a given period and when that warming potential is effective, as the latter determines the economic valuation of the damage done. Suppose there were a gas with a life in the atmosphere one tenth that of CO_2 but with ten times the GWP while it is there. The social cost of that gas today would be less than the social cost of CO_2, because it would have its effect on the world while the total stock of greenhouse gases was lower on average, so that its marginal impact would be less in economic terms.

BOX 13.1 Estimates of the social cost of carbon

Downing et al (2005), in a study for DEFRA, drew the following conclusions from the review of the range of estimates of the social cost of carbon:

- The estimates span at least three orders of magnitude, from 0 to over £1000/tC (2000 £), reflecting uncertainties in climate and impacts, coverage of sectors and extremes, and choices of decision variables

- A lower benchmark of £35/tC is reasonable for a global decision context committed to reducing the threat of dangerous climate change. It includes a modest level of aversion to extreme risks, relatively low discount rates and equity weighting

- An upper benchmark for global policy contexts is more difficult to deduce from the present state of the art, but the risk of higher values for the social cost of carbon is significant.

The Downing study draws on Tol (2005), who gathered 103 estimates from 28 published studies. Tol notes that the range of estimates is strongly right-skewed: the mode was $2/tC (1995 US$), the median was $14/tC, the mean $93/tC and the 95th percentile $350/tC. He also finds that studies that used a lower discount rate, and those that used equity weighting across regions with different average incomes per head generated higher estimates and larger uncertainties. The studies did not use a standard reference scenario, but in general considered 'business as usual' trajectories. (See also Watkiss et al (2005) on the use of the social cost of carbon in policy-making and Clarkson and Deyes (2002) for earlier work on the social cost of carbon in a UK context.)

NB conversion rates:

£100/tC (2000 prices) = $116/tC (1995 prices) = $35.70/tCO$_2$ (2000 prices)

---

therefore point to numbers for the 'business as usual' social cost of carbon well above (perhaps a factor of three times) the Tol mean of $29/tCO$_2$ and the 'lower central' estimate of around $13/tCO$_2$ in the recent study for DEFRA (Watkiss et al. (2005)). But they are well below the upper end of the range in the literature (by a factor of four or five). Nevertheless, we are keenly aware of the sensitivity of estimates to the assumptions that are made. Closer examination of this issue – and a narrowing of the range of estimates, if possible – is a high priority for research.

The case for strong action from the perspective of comparing the 'business as usual' social cost of carbon and the marginal abatement cost is powerful, even if one takes Tol's mean or the Watkiss lower benchmark as the value of the former, when one compares it with the opportunities for low-cost reductions in emissions and, indeed, for those that make money (see Chapter 9). It is still more powerful if one takes higher numbers for the social cost of carbon, as we would suggest is appropriate, and also recognises that the SCC will increase over time, because of the current and prospective increases in the stock of greenhouse gases in the atmosphere.

All three of these approaches would lead to exactly the same estimate of the net benefits of climate-change policies and the same extent of action if models were perfect and policy-makers had full information about the world. In practice, these conditions do not hold, so the three perspectives can be used to cross-check the broad conclusions from adopting any one of them.

13.3   Setting objectives for action

Having made the case for strong action, there remains the challenge of formulating more specific objectives, so that human-induced climate change is slowed and brought to a halt without unnecessary costs. The science and economics both suggest that a shared international understanding of what the objectives of climate-change policy should be would be a valuable foundation for policy.

The problem is global. Policy-makers in different countries cannot choose their own global climate. If they differ about what they think the world needs to achieve, not only will many of them be disappointed, the distribution of efforts to reduce emissions will be inefficient and inequitable. The benefits of a shared understanding include creating consensus on the scale of the problem and a common appreciation of the size of the challenge for both mitigation and adaptation. It would provide a foundation for discussion of mutual responsibilities in tackling the challenge. At a national and individual level, it would reduce uncertainty about future policy, facilitating long-term planning and making it more likely that both adaptation and mitigation would be appropriate and cost-effective.

The ultimate objective of stopping human-induced climate change can be translated into a variety of possible long-term global goals to give guidance about the strength of measures necessary.

Table 13.1 below summarises five types of goal, each defining key stages along the causal chain from emissions to atmospheric concentrations, to global temperature changes and finally to impacts.

These different types of goal are not necessarily inconsistent, and some are more suited to particular roles than others. Public concern focuses on impacts to be avoided, and this is indeed the language of the UNFCCC, which defines the ultimate objective of the Convention as "...*to achieve...stabilisation of greenhouse gas concentrations in the atmosphere at a level that would prevent dangerous anthropogenic interference with the climate system. Such a level should be achieved within a time-frame sufficient to allow ecosystems to adapt naturally to climate change, to ensure that food production is not threatened and to enable economic development to proceed in a sustainable manner."* However, this does not provide a quantitative guide to policy-makers on the action required. The EU has defined a temperature threshold – limiting the global average temperature change to less than 2°C above pre-industrial. This goal allows policy-makers and the public to debate the level of tolerable impacts in relation to one simple index, but it does not provide a transparent link to the level of mitigation action that must be undertaken.

The analysis presented in Chapter 8, linking cumulative emissions first to long-run concentrations in the atmosphere, and then to the probabilities of different ultimate temperature outcomes, provides an alternative basis for long-term goals. It is one that allows the level of and uncertainty about both impacts and the

Table 13.1 Five types of goal

	Advantages	Disadvantages
Maximum tolerable level of impacts (e.g. no more than a doubling of the current population under water stress)	– Linked directly to the consequences to avoid.	– Scientific, economic and ethical difficulties in defining which impacts are important and what level of change can be tolerated. – Uncertainties in linking avoidance of a specific impact to human action. – Success not measurable until too late to take further action.
Global mean warming (above a baseline)	– Can be linked to impacts (with a degree of uncertainty). – One quantifiable variable.	– Uncertainties in linking goal with specific human actions. – Lags in time between temperature changes and human influence, so difficult to measure success of human actions in moving towards the goal.
Concentration(s) of greenhouse gases (or radiative forcing)	– One quantifiable variable. – Can be linked to human actions (with a degree of uncertainty). – Success in moving towards the goal is measurable quickly.	– Uncertainties about the magnitude of the avoided impacts.
Cumulative emissions of greenhouse gases (over a given time period)	– One quantifiable variable. – Directly linked to human actions. – Success in moving towards the goal is measurable quickly.	– Uncertainties about the magnitude of the avoided impacts.
Reduction in annual emissions by a specific date	– One quantifiable variable. – Success in moving towards the goal is measurable quickly.	– Uncertainties about the magnitude of the avoided impacts. – Does not address the problem that impacts are a function of stocks not flows. – May limit 'what, where, when' flexibility and so push up costs

costs of mitigation to be debated together. Once a shared understanding of what the broad objectives of policy should be has been established, it is useful to go further and translate it into terms that can guide the levels at which the instruments of policy should be set.

Any operational goal should be closely related to the ultimate impacts on well-being that policy seeks to avoid. But, if it is to guide policy-makers in adjusting policy sensibly over time, progress towards it must also be easy to monitor. The goal therefore should be clear, simple and specific; it must be possible to use new information regularly to assess whether recent observations of the variable targeted are consistent with hitting the goal. Policy-makers must also have some means of adjusting policy settings to alter the trajectory of the variable targeted. Seeing policy-makers adjust policy settings in this way to keep their aim on the goal would also build the credibility of climate-change policies. This is very important, if private individuals and firms are to play their full part in bringing about the necessary changes in behaviour.

A goal for atmospheric concentrations would allow policy-makers to monitor progress in a timely fashion and, if the world were going off course, adjust policy instrument settings to correct the direction of travel.

The rest of this chapter focuses on the question of what concentration of greenhouse gases in the atmosphere, measured in CO_2 equivalent, to aim for. Policy instruments should be set to make the expected long-run outcome for concentration (on the basis of today's knowledge) equal to this level. Atmospheric concentration is closer than cumulative emissions in the causal chain to the impacts with which climate-change policy is ultimately concerned. And, compared with other possible formulations of policy aspirations such as global temperature change, observations of atmospheric concentration allow more rapid feedback to policy settings[2].

Such a goal is a device to help structure and calibrate climate-change policy. But it is only a means to an end – limiting climate change – and it is useful to keep that ultimate objective in mind. Other intermediate and local goals (for example, national limits for individual countries' annual emissions or effective carbon-tax rates) may also help to move economies towards the long-run objective and to monitor the success of policy, given the long time it will take to achieve stabilisation – as long as they are consistent with, and subsidiary to, the primary goal. They may also be necessary as stepping-stones towards the adoption of a more comprehensive and coherent global objective, given the time it is likely to take to reach a shared understanding of what needs to be done. The danger is that multiple objectives may reduce the efficiency with which the main one is pursued. Part VI of the Review considers some of the problems of turning an international objective into obligations for national governments. This chapter sidesteps those problems in order to focus on what economics suggests might be desirable characteristics of the set of local, national and supranational policies that emerge from the political process.

However, the key decision required now is that strong action is both urgent and necessary. That does not require immediate agreement on a precise stabilisation goal.

It is important to start taking steps in the right direction while the shared understanding is being developed.

13.4 The economics of choosing a goal for global action

Measuring and comparing the expected benefits and costs over time associated with different stabilisation levels can provide guidance to help decide how much to do and how quickly.

---

[2] Cumulative emissions are closer to the policy-induced emissions reductions that incur the *costs* of mitigating climate change. The choice between the two goals comes down to how the costs and benefits of missing the goal by some amount differ in the two cases, given uncertainty about the relationship between the two variables due to uncertainty about the functioning of carbon 'sinks', etc. This is related to the issue of whether setting greenhouse-gas prices or quotas is preferable in the face of uncertainty (see Chapter 14); the arguments there imply that, for the long run, a concentration goal is to be preferred).

Estimates need to take account of the great uncertainties about climate-change damages and mitigation costs that remain even when a specific stabilisation goal is being considered. The time dimension is also important. A different stabilisation goal entails a different trajectory of emissions through time, so analysis should not simply compare the costs and benefits of extra emission reductions this year. Instead, one needs to compare incremental changes in the present values of current and future costs and benefits.

The marginal benefits of a lower stabilisation level reflect the expected impact on people's wellbeing of achieving a lower expected ultimate temperature change and a reduced risk of extreme outcomes. Risk will increase along the path towards stabilisation and cannot be accounted for simply by comparing ultimate stabilisation levels. As Chapter 2 showed, this requires judgements about how wellbeing is affected by risk, uncertainty and the distribution of the impacts of climate change across individuals and societies. Subjective assessments have to be made where objective evidence about risks is limited, particularly those associated with more extreme climate change. These assessments should adopt a consistent approach towards risk and uncertainty, reflecting the degree of risk aversion people decide is appropriate in this setting.

The marginal costs of aiming for a lower stabilisation level reflect the need to speed up the introduction of mitigation measures, such as development of low-carbon technologies and switching demand away from carbon-intensive goods and services. Stabilisation, however, requires emissions to be cut to below 5 $GtCO_2e$ eventually, to the Earth's natural annual absorption limit, whatever the specific GHG stock level chosen (Chapter 8).

Figure 13.3 illustrates the approach sketched here. The figure shows in schematic fashion how the incremental or marginal benefits and costs of a programme of action change through time (in terms of present values) as successively lower goals are considered. As explained in Chapter 2, the benefits (and the

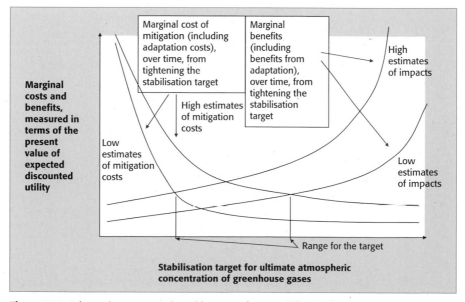

Figure 13.3 Schematic representation of how to select a stabilisation level

costs) of action should be thought of in terms of the expected impacts on wellbeing over time, appropriately discounted, not simply monetary amounts. That allows for risk weighting, risk aversion and considerations of fairness across individuals and generations to be incorporated in the analysis. For simplicity, two 'marginal benefits' curves are drawn to remind the reader of the huge uncertainties. In practice, people differ about the weights they attach to different sorts of climate-change impacts. There is scope for legitimate debate about how they should be aggregated to compare them with the costs of mitigation.

The costs of mitigation, too, should be thought of in terms of their impact on broad measures of wellbeing. It matters on whom the costs fall, when they are incurred and what the uncertainties about them are. Figure 13.2 shows two curves, for high and low estimates of the incremental costs of tougher action to curb emissions. They are drawn with the costs rising more sharply as the stabilisation level considered becomes lower and lower. The ideal objective is where the marginal benefits of tougher action equal the marginal costs. Given the uncertainty about both sides of the ledger, this approach cannot pin down a precise number but can, as the chart indicates, suggest a range in which it should lie. The range excludes levels where either the incremental costs of mitigation or the incremental climate-change impacts are rising very rapidly.

Uncertainty is an argument for setting a more demanding long-term policy, not less, because of the asymmetry between unexpectedly fortunate outcomes and unexpectedly bad ones.

Suppose there is a probability distribution for the scale of physical impacts associated with a given increase in atmospheric concentrations of greenhouse gases. As one moves up the probability distribution, the consequences for global wellbeing become worse. But, more than that, the consequences are likely to get worse at an accelerating rate, for two reasons. First, the higher the temperature, the more rapidly adverse impacts are likely to increase. Second, the worse the outcome, the lower will be the incomes of people affected by them, so any monetary impact will have a bigger impact on wellbeing[3].

There is a second line of reasoning linking uncertainty with stronger action. There is an asymmetry due to the very great difficulty of reducing the atmospheric concentration of greenhouse gases. Increases are irreversible in the short to medium run (and very difficult even in the ultra-long run, on our current understanding). If new information is collected that implies that climate-change impacts are likely to be *worse* than we now think, we cannot go back to the concentration level that would have been desirable had we had the new information earlier. But if the improvement in knowledge implies that a *less* demanding goal is appropriate, it is easy to allow the concentration level to rise faster. In other words, there is an option value to choosing a lower goal than would be picked if no improvements in our understanding of the science and economics were anticipated. The 'option value' argument is not, however, clear-cut[4]. There is also an option value associated with delaying investment in long-lived structures, plant

---

[3] More formally, we take impacts to be convex in atmospheric concentration and note that the expected utility of a range of outcomes is lower than the utility of the expected outcome, if marginal utility declines with income. This is discussed further in Chapter 2.
[4] See, for example, Kolstad (1996), Pindyck (2000) and Ingham and Ulph (2005)

and equipment for greenhouse gas abatement. Investments in physical capital, like cumulative emissions, are largely irreversible, so there is an option value to deferring them. That argues for a higher level of annual emissions than otherwise desirable.

Some of the parameters that modellers have treated as uncertain, such as discount factors and equity weights, reflect societies' preferences. In the process of agreeing an international stabilisation objective, or at least narrowing its range, discussions have to resolve, or at least reduce disagreement over, the issues of social choice lying behind these uncertainties.

As explained in Chapter 2 and its appendix, this Review argues for using a low rate of pure time preference and assuming a declining marginal utility of consumption as consumption increases across time, people and states of nature. However, the magnitude of the risks described in Part II of this Review suggests that a broad range of perspectives on these two issues indicates the need for strong action to mitigate emissions.

Given this framework, the evidence on the costs and benefits of mitigation reviewed in the chapters above can give a good indication of upper and lower limits that might be set for the extent of action, as argued below. The policy debate should seek some indication of where within these limits international collective action should aim[5]. But it is vital that, while a shared understanding permitting agreement on a common goal is being developed, initial actions to reduce emissions are not delayed.

There is room for debate about precisely how fast emissions need to be brought down, but not about the direction in which the world now has to move.

13.5   Climate-change impacts and the stabilisation level

Expected climate-change impacts rise with the atmospheric concentration of greenhouse gases, because the probability distributions for the long-run global temperature move upwards. The evidence strongly suggests that 550 ppm CO_2e would be a dangerous place to be, with substantial risks of very unpleasant outcomes.

Figure 13.4 illustrates how the risk of various impacts occurring is associated with different stabilisation levels[6] (see also Box 8.1 for frequency distributions of the range of temperature increases associated with various stabilisation levels in a selection of climate models). The top section shows the 5–95% probability ranges

---

[5]   If policy-makers adopt a zone rather than a single number as a goal, recognising that no policy is able to ensure that a point goal can be hit precisely, it should be within these upper and lower limits. It would also be desirable if the zone were considerably narrower than the span of those limits, so as not to weaken substantially the discipline on policy-makers to adjust policy settings if it looks as if the goal is not going to be met. Too wide a target zone also increases the risk of different policy-makers around the world choosing policy settings that are inconsistent with each other.

[6]   Where the risk is defined using subjective probabilities based on current knowledge of climate sensitivity – the relationship between greenhouse-gas concentration and temperatures.

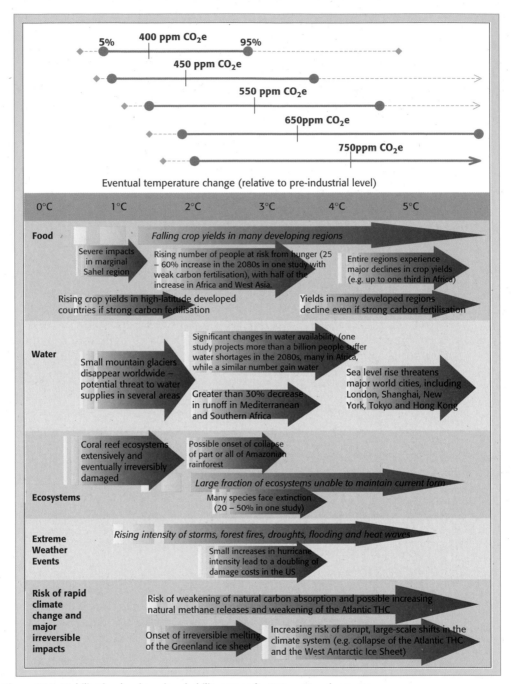

Figure 13.4 Stabilisation levels and probability ranges for temperature increases

The figure above illustrates the types of impacts that could be experienced as the world comes into equilibrium with higher greenhouse gas levels. The top panel shows the range of temperatures projected at stabilisation levels between 400 ppm and 750 ppm CO_2e at equilibrium. The solid horizontal lines indicate the 5–95% range based on climate sensitivity estimates from the IPCC TAR 2001 (Wigley and Raper (2001)) and a recent Hadley Centre ensemble study (Murphy et al. (2004)). The vertical line indicates the mean of the 50th percentile point. The dashed lines show the 5 – 95% range based on eleven recent studies (Meinshausen (2006)). The bottom panel illustrates the range of impacts expected at different levels of warming. The relationship between global average temperature changes and regional climate changes is very uncertain, especially with regard to changes in precipitation (see Box 3.2). This figure shows potential changes based on current scientific literature.

of temperature increases projected at different stabilisation levels; the central marker is the 50th percentile point. The bottom section shows the projected impacts. At some point, the risks of experiencing some extremely damaging phenomena begin to become significant. Such phenomena include:

- Irreversible losses of ecosystems and extinction of a significant fraction of species.
- Deaths of hundreds of millions of people (due to food and water shortages, disease or extreme weather events).
- Social upheaval, large-scale conflict and population movements, possibly triggered by severe declines in food production and water supplies (globally or over large vulnerable areas), massive coastal inundation (due to collapse of ice sheets) and extreme weather events.
- Major, irreversible changes to the Earth system, such as collapse of the Atlantic thermohaline circulation and acceleration of climate change due to carbon-cycle feedbacks (such as weakening carbon absorption and higher methane releases) – at high temperatures, stabilisation may prove more difficult, or impossible, because such feedbacks may take the world past irreversible tipping points (chapter 8).

The expected impacts of climate change on well-being in the broadest sense are likely to accelerate as the stock of greenhouse gases increases, as argued in Chapter 3. The expected benefits of extra mitigation will therefore increase with the stabilisation level[7]. In Figure 13.2, the marginal benefit curve is therefore drawn as rising increasingly steeply with the stabilisation level. There are four main reasons:

- As global mean temperatures increase, several specific climate impacts are likely to increase more and more rapidly: in other words, the relationship is convex. Examples include the relationship between windstorm wind-speed and the value of damage to buildings (IAG (2005)) and new estimates of the relationship between temperature and crop yields (Schlenker and Roberts (2006));
- Different elements of the climate system may interact in such a way that the combined impacts rise more and more rapidly with temperature;
- As global mean temperatures increase several degrees above pre-industrial levels, existing stresses would be more and more likely to trigger the most severe impacts of climate change that arise from interactions with societies, namely social upheaval, large-scale conflict and population movements;
- As global mean temperatures increase, so does the risk that positive feedbacks in the climate system, such as permafrost melting and weakening carbon sinks, kick in.

---

[7] There is, however, considerable uncertainty about how climate-change effects will evolve as temperatures rise, as many of the hypothesised effects are expected to take place or intensify outside the temperature range experienced by humankind, and so cannot be verified by empirical observation. One characteristic of the climate physics works in the opposite direction: the expected rise in temperature is a function of the *proportional* increase in the stock of greenhouse gases, not its *absolute* increase. As a result, some integrated assessment models, for example Nordhaus' DICE model, have S-shaped functions to represent the costs of climate-change impacts.

The uncertainties about impacts make it impossible to quantify exactly where the marginal impacts of climate change will rise more sharply. However, across the current body of evidence, two approximate global turning points appear to exist, at around 2–3°C and 4–5°C above pre-industrial levels:

- At roughly 2–3°C above pre-industrial, a significant fraction of species would exceed their adaptive capacity and, therefore, rates of extinction would rise. This level is associated with a sharp decline in crop yields in developing counties (and possibly developed counties) and some of the first major changes in natural systems, such as some tropical forests becoming unsustainable, irreversible melting of the Greenland ice sheet and significant changes to the global carbon cycle (accelerating the accumulation of greenhouse gases).
- At around 4–5°C above pre-industrial, the risk of major abrupt changes in the climate system would increase markedly. At this level, global food production would be likely to fall significantly (even under optimistic assumptions), as crop yields fell in developed countries.

Few studies have examined explicitly the benefits of choosing a lower stabilisation level. Generally, those that have done so show that the benefits vary across sectors. For example, in reducing the stabilisation temperature from 3.5°C to 2.5°C, significant benefits to ecosystems and in the number of people exposed to water stress have been estimated[8]. However, such evidence is strongly model-dependent and, therefore, subject to significant uncertainties.

Recent integrated assessment models (discussed in Chapter 6) have attempted to capture some of these uncertainties by representing damage functions stochastically. These cover several dimensions, including the risk of major abrupt changes in the climate systems (they do not, however, generally include estimates of the potential costs of social disruption). They also take account of adaptation to climate change to varying extents. Chapter 6 notes that such models show a steep increase in marginal costs with rising temperature. The PAGE2002 model, used in chapter 6, has the advantage of allowing for the uncertainty in the literature about several dimensions of impacts. It permits a comparison of the probability distribution of projected gross world product net of the cost of climate change with the hypothetical gross world product without climate change, for a given increase in global mean temperature, thus providing an estimate of climate-change costs (see Table 13.2, where estimates include some measure of 'non-market' impacts). The costs of climate change as a proportion of gross world product are modelled as an uncertain function of the increase in temperature, among other factors.

Thus, for example, according to PAGE2002, if the temperature increase rises from 2°C to 3°C, the mean damage estimate increases from 0.6% to 1.4% of gross world product; but the 'worst case' – the 95th percentile of the probability distribution – goes from 4.0% to 9.1%. These costs fall disproportionately on low-latitude, low-income regions, but there are significant net costs in higher-latitude regions, too.

The estimates of the costs of impacts suggest that the mean expected damages rise significantly if the global temperature change rises from 3°C to 4°C and even

---

[8]   Arnell et al. (2004)

Table 13.2 Estimates of the costs of climate change by temperature increase, as a proportion of gross world product, from PAGE2002

	Mean expected cost	5th percentile	95th percentile
2°C	0.6%	0.2%	4.0%
3°C	1.4%	0.3%	9.1%
4°C	2.6%	0.4%	15.5%
5°C	4.5%	0.6%	23.3%

Source: Hope (2003)

more from 4°C to 5°C. But the damages associated with a 'worst case' scenario – the 95th percentile of the distribution – rise more rapidly still.

On the basis of current scientific understanding, it is no longer possible to prevent all risk of dangerous climate change.

Box 8.1 showed how the risk of exceeding these temperature thresholds rises at stabilisation levels of 450, 550, 650, and 750 ppm CO_2e. This box implies:

- Even if the world were able to stabilise at current concentrations, it is already possible that the ultimate global average temperature increase will exceed 2°C
- At 450 ppm CO_2e, there is already a 18% chance of exceeding 3°C, according to the Hadley ensemble reported in the table, but a very high chance of staying below 4°C
- By 550 ppm CO_2e, there is a 24% chance that temperatures will exceed 4°C, but less than a 10% chance that temperatures will exceed 5°C.

It can be seen that a move above 550 ppm CO_2e would entail considerable additional costs of climate change, taking into account the further increases in the risks of extreme outcomes.

Our work with the PAGE model suggests that, allowing for uncertainty, if the world stabilises at 550 ppm CO_2e, climate change impacts could have an effect equivalent to reducing consumption today and forever by about 1.1%[9]. As Chapter 6 showed, this compares with around 11% in the corresponding 'business as usual' case – ten times as high. With stabilisation at 450 ppm CO_2e, the percentage loss would be reduced to 0.6%, so choosing the tougher goal 'buys' about 0.5% of consumption now and forever. Choosing 550 ppm instead of 650 ppm CO_2e 'buys' about 0.6%. As with all models, these numbers reflect heroic assumptions about the valuation of potential impacts, although, as Chapter 6 explains, they reflect an attempt to ensure the model calibration reflects the nature of the problem faced. They also entail explicit judgements about some of the ethical issues involved. In addition, the PAGE2002 model is not ideal for

[9] These figures are based on the 'broad impacts, standard climate sensitivity' case among the scenarios considered in Chapter 6. As such, they do not allow for equity weighting; if they did, the estimates in the text would be higher. They would also be higher if higher estimates of climate sensitivity, incorporating more amplifying feedback mechanisms, were used. The valuation of non-market impacts is particularly difficult and dependent on ethical judgements, as explained in Chapter 6.

analysing stabilisation trajectories. Nevertheless, all integrated assessment models are sensitive to the assumptions and they should be taken as only indicative of the quantitative impacts, given those assumptions. It should be noted that the results quoted from Chapter 6 leave out much that is important, and the other models referred to there leave out more.

13.6   The costs of mitigation and the stabilisation level

The lower the stabilisation level chosen, the faster the technological changes necessary to bring about a low-carbon society will have to be implemented.

Stabilising close to the current level of greenhouse-gas concentration would require implausibly rapid reductions in emissions, because the technologies currently available to achieve such reductions are still very expensive[10] and the appropriate structures, plant and equipment are not yet in place. Hitting 450 ppm CO_2e, for example, appears very difficult to achieve with the current and foreseeable technologies, as suggested in Chapter 8. It would require an early peak in emissions, very rapid emission cuts (more than 5% per year), and reductions by 2030 of around 70%. Even with such cuts, the stock of greenhouse gases covered by the Kyoto Protocol would initially overshoot, their effect temporarily masked by aerosols (so that there would be only a very small overshoot in radiative forcing)[11]. Costs would start to rise very rapidly if emissions had to be reduced sharply before the existing capital stock in emissions-producing industries would otherwise be replaced and at a speed that made structural adjustments in economies very abrupt and hence expensive. Abrupt changes to economies can themselves trigger wider impacts, such as social instability, that are not covered in economic models of the costs of mitigation.

Technological change eventually has to get annual emissions down to their long-run sustainable levels without having to accelerate sharply the retirement of the existing capital stock, if costs are to be contained. Model-based estimates of the present value of the costs of setting a tougher stabilisation objective are not widely available in the literature. That reflects, among other factors, the unavoidable uncertainties about the pace and costs of future innovation. In principle, such estimates ought to reflect the incidence of the mitigation costs, which ultimately fall on the consumers of currently GHG-intensive goods and services, as well as their monetary value (just as the incidence of climate-change impacts matters as well as their level), but there has been little investigation of this aspect of the problem.

However, there are some estimates to help as a guide. Chapter 9 in effect argued that the extra mitigation costs incurred by stabilising at around 550 ppm

---

[10] Costs of delivering any particular level of abatement are likely to decline with investment and experience; see Chapters 9 and 16.

[11] The world is already at around 430 ppm CO_2e if only the greenhouse gases covered by the Kyoto Protocol are included; but aerosols reduce current radiative forcing. The projection reported in the text assumes that the aerosol affect diminishes over time, but for a period counteracts a temporary rise in Kyoto greenhouse gases above 450 ppm CO_2e. As the concentration of greenhouse gases is rising at around 2.5 ppm CO_2e per year, and annual emissions are increasing, 450 ppm CO_2e could be reached in less than ten years.

CO_2e instead of allowing business to continue as usual would probably be of the order of 1% of gross world product. Choosing a lower goal would cost more, a higher goal less. Some studies of costs give more of an indication of their sensitivity to the stabilisation objective. For example, the study by Edenhofer et al (2006), averaging over five models, provides the following estimates of cost increases from choosing a lower stabilisation goal:

Table 13.3 Some model-based estimates of the increase in mitigation costs from reducing a stabilisation goal (discounted percentage of gross world output), by discount rate used

	5% pa	'Green Book'	2% pa	1% pa	0% pa
Moving from 500 ppm to 450 ppm CO_2	0.25%	0.39%	0.43%	0.51%	0.58%
Moving from 550 ppm to 500 ppm CO_2	0.06%	0.11%	0.12%	0.14%	0.18%

Source: adapted from Edenhofer et al. (2006); 'Green Book' is a declining discount rate over time, as in HM Treasury Green Book project-appraisal guidance.

It is important to note that these results are tentative, and that there is still much debate about the role of induced technological progress, the focus of the study. Nevertheless, the bottom line in Table 13.3 suggests that the extra mitigation costs from choosing a goal of around 500 ppm instead of 550 ppm CO_2 would be small, ranging from 0.06% to 0.18% of gross world output, depending on how much future costs are discounted. In terms of a CO_2e goal, this is similar to going from 600–700 ppm to 550–650 ppm, depending on what happens to non-CO_2 greenhouse gases (see Chapter 8). The extra costs of choosing a goal of 450 ppm CO_2 instead of 500 ppm CO_2 would be higher, ranging from 0.25% to 0.58%; this is similar to going from 550–650 ppm CO_2e to 500–550 ppm CO_2e. None of the discount schemes used are the same as the one used in Chapter 6 of this Review, as the discount rates are not path-dependent. However, as stabilisation reduces the chances of very bad outcomes compared with 'business as usual', the discounting issue is less important than when evaluating potential impacts without mitigation. It is important to note that the studies concerned take the year 2000 as a baseline. Given the probable cumulative emissions since then, the goals would now be more difficult and expensive to hit.

The recent US Climate Change Science Program draft report on scenarios of greenhouse gas emissions and atmospheric concentrations also provides useful estimates, reporting for various points in time the percentage change in gross world product expected due to adopting policies to meet four different stabilisation goals[12]. Again, the studies covered take 2000 as the base year. The implications for incremental costs (as a fraction of gross world output) of adopting successively tougher goals are summarised in Table 13.4 below. These studies were not designed with the objective of this chapter in mind, of course, and the draft is subject to revision, so the estimates should be regarded as suggestive of magnitudes, not definitive.

[12] US CCSP Synthesis and Assessment Product 2.1, Part A: 'Scenarios of Greenhouse Gas Emissions and Atmospheric Concentrations', Draft for public comment, June 26, 2006.

Table 13.4 Some model-based estimates of the incremental savings in mitigation costs from relaxing a stabilisation goal (% of gross world output in the relevant year)

Incremental change	Model	2020	2040	2060	2080	2100
Moving from around 550 ppm to around 450 ppm CO_2 (670 ppm to 525 ppm CO_2e)	IGSM	1.6%	2.9%	4.4%	6.2%	9.3%
	MERGE	0.7%	1.3%	1.5%	1.2%	0.7%
	MiniCAM	0.2%	0.6%	1.0%	0.8%	0.6%
Moving from around 650 ppm to around 550 ppm CO_2 (820 ppm to 670 ppm CO_2e)	IGSM	0.3%	0.8%	1.4%	2.1%	3.7%
	MERGE	0.0%	0.1%	0.3%	0.4%	0.5%
	MiniCAM	0.0%	0.1%	0.3%	0.4%	0.3%
Moving from around 750 ppm to 650 ppm CO_2 (970 ppm to 820 ppm CO_2e)	IGSM	0.1%	0.2%	0.5%	0.9%	1.4%
	MERGE	0.0%	0.0%	0.1%	0.1%	0.1%
	MiniCAM	0.0%	0.0%	0.0%	0.1%	0.3%

Source: Adapted from US CCSP Synthesis and Assessment Product 2.1, Part A: 'Scenarios of Greenhouse Gas Emissions and Atmospheric Concentrations', Draft for public comment, June 26, 2006[13]

Table 13.4 shows in the bottom panel that the extra costs incurred by adopting an objective of around 820 ppm instead of 970 ppm CO_2e are very small, and, for two of the three models (MERGE and MiniCAM in the middle panel), aiming for around 670 ppm instead of 820 ppm CO_2e also costs little. According to the same two models, choosing 525 ppm instead of 670 ppm CO_2e increases costs by around 1% of gross world product, the amount varying somewhat over time. The most pessimistic model here generates considerably higher estimates for the total yearly costs of mitigation, reflecting its relatively high trajectory for 'business as usual' emissions and relatively pessimistic assumptions about the likely pace of innovation in low-carbon technologies. The studies suggest that mitigation costs start to rise sharply towards the bottom of the ranges of stabilisation levels considered.

Delay will make it more difficult and more expensive to stabilise at or below 550 ppm CO_2e.

All of these studies take as a starting point the year 2000. If it takes 20 years or so before strong policies are put in place globally, it is likely that the world would already be at somewhere around 500 ppm CO_2e, making it very difficult and expensive then to take action to stabilise at around 550 ppm.

13.7   A range for the stabilisation objective

Integrated assessment models have been used in a number of studies to compare the marginal costs and marginal benefits of climate-change policy over time. But many of the estimates in the literature do not take into account the latest science or treat risk and uncertainty appropriately. Doing so would bring down the stabilisation level desired.

---

[13] The ranges in terms of CO_2e are derived from the long-run constraints on total radiative forcing in the modelling exercise.

In some cases, the models have been used to estimate the 'optimal' amount of mitigation that maximises benefits less costs. These studies recommend that greenhouse gas emissions be reduced below business-as-usual forecasts, but the reductions suggested have been modest. For example, on the basis of the climate sensitivities and assessments available at the time the studies were undertaken,

- Nordhaus and Boyer (1999) found that the optimal global mitigation effort reduces atmospheric concentrations of carbon dioxide from 557 ppm in 2100 (business-as-usual) to 538 ppm. This reduces the global mean temperature from an estimated 2.42°C above 1900 levels to 2.33°C;
- Tol (1997) found that the optimal mitigation effort reduces the global mean temperature in 2100 from around 4°C above 1990 levels to between around 3.6°C and 3.9°C, depending on whether countries cooperate and on the costs of mitigation;
- Manne et al. (1995) did not use their model to find the optimal reduction in emissions, but the policy option they explored that delivers the highest net benefits reduces atmospheric concentrations of carbon dioxide from around 800 ppm in 2100 to around 750 ppm, reducing global mean temperature from around 3.25°C above 1990 levels to around 3°C.

However, the optimal amount of mitigation may in fact be greater than these studies have suggested. Above all, they carry out cost-benefit analysis appropriate for the appraisal of small projects, but we have argued in Chapter 2 that this method is not suitable for the appraisal of global climate change policy, because of the very large uncertainties faced. As a result, these studies underestimate the risks associated with large amounts of warming. Neither does any of these studies place much weight on benefits and costs accruing to future generations, as a consequence of their ethical choices about how to discount future consumption. Manne et al. apply a much higher discount rate to utility than do we in Chapter 6. Nordhaus and Boyer assume relatively low and slowing economic growth in the future, which reduces future warming. Tol estimates relatively modest costs of climate change, even at global mean temperatures 5-6°C above pre-industrial levels. Recent scientific developments have placed more emphasis on the dangers of amplifying feedbacks of global temperature increases and the risks of crossing irreversible tipping points than these models have embodied.

Given the paucity of estimates of the appropriate stabilisation level and the disadvantages of the ones that exist, this chapter does not propose a specific numerical goal. Instead, it explores how economic analysis can at least help suggest upper and lower limits to the range for an atmospheric concentration goal. Allowing for the current uncertainties, the evidence suggests that the upper limit to the stabilisation range should not be above 550 ppm CO_2e.

Putting together our results on the valuation of climate-change impacts with the mitigation-cost studies suggests that the benefits of choosing a lower stabilisation goal clearly outweigh the costs until one reaches 550–600 ppm CO_2e. But around this level the cost-benefit calculus starts to get less clear-cut. The incremental mitigation costs of choosing 500–550 ppm instead of 550–600 ppm CO_2e are three to four times as much as the incremental costs of choosing 550– 600 ppm instead of 600–650 ppm CO_2e, according to the numbers in Edenhofer et al. The higher mitigation costs incurred if 500–550 ppm is chosen instead of 550–600 ppm CO_2e might be of similar size to the incremental benefits. They would be bigger if induced technological change were inadequate or 'business as usual' emissions were at the higher end of projections, as in the IGSM projections reported in Table 13.4.

As far as the climate-change impacts are concerned, the incremental benefits might be bigger than these calculations allow – for example, if policy-makers are more risk-averse than the PAGE calculations assumed or attach more weight to non-market impacts. Nevertheless, in choosing an upper limit to the stabilisation range, one needs to consider what is appropriate if climate-change impacts turn out to be towards the low end of their probability distribution (for a given atmospheric concentration) and mitigation costs towards the high end of their distribution. Following broadly this approach, but assuming mitigation costs are brought down over time by induced technological change, we suggest an upper limit of 550 ppm CO_2e.

The lower limit to the stabilisation range is determined by the level at which further tightening of the goal becomes prohibitively expensive. On the basis of current evidence, stabilisation at 450 ppm CO_2e or below is likely to be very difficult and costly.

Cost estimates derived from modelling exercises suggest that costs as a share of gross world product would increase sharply if a very ambitious goal were adopted (see Chapter 10). It is instructive that cost modelling exercises rarely consider stabilisation below 500 ppm CO_2e. Edenhofer et al point out that some of the models in their study simply cannot find a way of achieving 450 ppm CO_2e. Even stabilising at 550 ppm CO_2e would require complete transformation of the power sector. 450 ppm CO_2e would in addition require very large and early reductions of emissions from transport, for which technologies are further away from deployment. Given that atmospheric greenhouse gas levels are now at 430 ppm CO_2e, increasing at around 2.5 ppm/yr, the feasibility of hitting 450 ppm CO_2e without overshooting is very much in doubt. And it would be unwise to assume that any overshoot could be clawed back.

The evidence on the benefits and costs of mitigation at different atmospheric concentrations in our view suggests that the stabilisation goal should lie within the range 450–550 ppm CO_2e.

The longer action is delayed, the higher will be the lowest stabilisation level achievable. The suggested range reflects in particular the judgements that:

- Any assessment of the costs of climate change must take into account uncertainty about impacts and allow for risk aversion. Because of the risk of very adverse impacts, extreme events and amplifying feedbacks, this implies adopting a tougher goal than if uncertainty were ignored
- Proper weight should be given to the interests of future generations. Future individuals should be given the same weight in ethical calculations as those currently alive, if it is certain that they will exist. But, as there is uncertainty about the existence of future generations, it is appropriate to apply some rate of discounting over time. That points to the use of a positive, but small, rate of pure time preference (see Chapter 2 and its appendix)
- Proper attention should be paid to the distribution of climate-change impacts, in particular to the disproportionate impact on poor people

- Productivity growth in low-greenhouse-gas activities will speed up if there is more output from and investment in these activities
- The speed of decarbonisation is constrained by the current state of technology and the availability of resources for investment in low-carbon structures, plant, equipment and processes.

It is clear that studies of climate-change impacts and of mitigation costs do not yet establish a narrow range for the level at which the atmospheric concentrations of greenhouse gases should be stabilised. More research is needed to narrow the range further. There will always be disagreements about the size of the risks being run, the appropriate policy stance towards risk, and the valuation of social, economic and ecological impacts into the far future. But the range suggested here provides room for negotiation and debate about these. And we would argue that agreement on the range stated does not require signing up to all of the judgements specified above. In presenting the arguments, for example, we have omitted a number of important factors that are likely to point to still higher costs of climate change and thus still higher benefits of lower emissions and a lower stabilisation goal.

In any case, agreement requires discussion and negotiation about the ethical issues involved. Chapter 6 demonstrates that taking proper account of the non-marginal nature of the risks from climate change leads to a higher estimate of risk-adjusted losses of wellbeing than if the larger risks are ignored or submerged in simple averages. Those who weigh more heavily the potential costs of the climate change possible at any given stabilisation level will argue for a goal towards the lower end of the range. Greater risk aversion and more concern for equity across regions and generations will push in the same direction. But those who are pessimistic about the direction and pace of technological developments or who believe emissions under 'business as usual' will grow more rapidly than generally expected will tend to advocate a goal towards the upper limit, other things being equal.

The EU has adopted an objective, endorsed by a large number of NGOs and policy think-tanks, to limit global average temperature change to less than 2°C relative to pre-industrial levels. This goal is based on a precautionary approach. A peak temperature increase of less than 2°C would strongly reduce the risks of climate-change impacts, and might be sufficient to avoid certain thresholds for major irreversible change – including the melting of ice-sheets, the loss of major rainforests, and the point at which the natural vegetation becomes a source of emissions rather than a sink. Some would argue that the implications of exceeding the 2°C limit are sufficiently severe to justify action at any cost. Others have criticised the 2°C limit as arbitrary, and have raised questions about the feasibility of the action that is required to maintain a high degree of confidence of staying below this level. Recent research on the uncertainties surrounding temperature projections suggests that at 450 ppm CO_2e there would already be a more-than-evens chance of exceeding 2°C (see Chapter 8). This highlights the need for urgent action and the importance of keeping quantitative objectives under review, so that they can be updated to reflect the latest scientific and economic analysis.

Some of the uncertainties will be resolved by continuing progress in the science of climate change, but ethics and social values will always have a crucial part to play in decision-making. The precise choice of policy objective will depend on values, attitudes to risk and judgements about the political feasibility of the objective. It is a decision with significant implications that will rightly be the subject of a broad public and international debate.

13.8   Implications for emissions reductions and atmospheric concentrations

Stabilisation of atmospheric concentrations implies that annual greenhouse-gas emissions must peak and then fall, eventually reaching the level that the Earth system can absorb annually, which is likely to be below 5 $GtCO_2e$.

At the moment, annual emissions are over 40 $GtCO_2e$. Chapter 8 showed how, for the range of stabilisation levels considered here, annual emissions should start falling within the next 20 years, if implausibly high reduction rates are to be avoided later on. Global emissions will have to be between 25% and 75% lower than current levels by 2050. That illustrates the fact that, even at the high end of the stabilisation range, major changes in energy systems and land use are required within the next 50 years.

While annual emissions are likely to rise first and then fall, atmospheric concentrations are likely to continue to rise until the long-term objective is reached.

For any given stabilisation level, overshooting entails increased risks of climate change, by increasing the chances of triggering extreme events associated with higher concentration levels than the goal, and amplifying feedbacks on concentration levels. The expected impacts on wellbeing associated with any stabilisation level are thus likely to be smaller if overshooting is avoided. As reducing emissions in agriculture appears relatively difficult, and that sector accounts for more than 5 $GtCO_2e$ per year by itself already, stabilisation is likely ultimately (well beyond 2050) to require complete decarbonisation of all other activities and some net sequestration of carbon from the atmosphere (e.g. by growing and burning biofuels, and capturing and storing the resultant carbon emissions, or by afforestation). Overshooting and return require that annual emissions can at some stage be reduced for a period below the level consistent with a stable level of the stock of greenhouse gases. On the basis of the current economic and technological outlook, that is likely to be very difficult.

Setting up a long-run stabilisation goal does not, however, preclude future revisions to make it more ambitious, if either technological progress is more far-reaching than anticipated or the expected impacts of rises in concentration levels rise. But, equally, unexpected difficulties in driving technical progress or a downward revision in expected impacts of climate change would warrant a less challenging goal. Given the pervasive uncertainties about both costs and benefits of climate-change policies, it is essential that any policy regime incorporate from the outset mechanisms to update the long-run goal in a transparent fashion in response to new developments in the science or economics.

The precise trajectory of annual emissions will depend on, among other factors, how climate-change policy is implemented, the pace of economic growth and the extent of innovation, particularly in the energy sector. Chapter 9 demonstrated that mitigation is more likely to be carried out cost effectively if policy encourages 'what, where and when' flexibility, so setting a precise trajectory as a firm intermediate objective is likely to be unnecessarily costly. Trajectories can nevertheless give a guide as to whether emissions are on course to reach the long-term goal.

13.9   The social cost of carbon

Calculations of the social cost of carbon have commonly been used to show the price that the world has to pay, if no action is taken on climate change, for each tonne of gas emitted – as in Section 13.2. But the concept can also be used to evaluate the damages along a stabilisation trajectory[14].

Choosing a concentration level to aim for also anchors a trajectory for the social cost of carbon. Without having a specific stabilisation goal in mind, it is difficult to calibrate what the carbon price should be – or, more generally, how strong action should be. The social cost of carbon will be lower at any given time with sensible climate-change policies than under 'business as usual'.

The social cost of carbon will be lower, the lower the ultimate stabilisation level. The social cost of carbon depends on the overall strategy for mitigating climate change and can help support that strategy, for instance by helping to evaluate abatement proposals. But it should not be seen as the driver of strategy. If the ultimate stabilisation goal has been chosen sensibly, the social cost of carbon along the stabilisation trajectory should be a good guide to the carbon price needed to help persuade firms to make the carbon-saving investments and undertake the research and development that would help deliver the necessary changes and entice consumers to buy fewer GHG-intensive goods and services. However, as Part IV of this Review argues, carbon pricing is only part of what needs to be done to bring down emissions.

If the concentration of carbon in the atmosphere rises steadily towards its long-run stabilisation level (so there is no overshooting), and expected climate-change damages accelerate with concentrations, the social cost of carbon will rise steadily over time, too[15]. An extra unit of carbon will do more damage at the margin the later it is emitted, because it will be around in the atmosphere while concentrations are higher, and higher concentrations mean larger climate-change impacts at the margin[16].

---

[14] The social cost of carbon is well defined along any specific emissions trajectory, not only stabilisation trajectories, as the usual calculations of 'business as usual' SCCs illustrate.

[15] This requires that the convexity of the relationship between expected damages (in terms of broad measures of wellbeing) and global mean temperature increases outweighs the declining marginal impact of increases in concentration on temperature as concentration rises.

[16] The social cost of carbon can also be thought of as the shadow price of carbon if there are no other distortions in the economy, apart from the greenhouse-gas externality, affected by emissions. The shadow-price path over time will depend on the precise dynamics of expected growth, climate-change impacts, the rate of removal of CO_2 from the atmosphere, discount rates and the marginal utility of income. The social cost of carbon is likely to rise faster, the higher is expected economic growth, the higher the rate at which total impacts rise with concentrations, the higher the decay rate of the greenhouse gases, and the higher the pure rate of time preference.

The social cost of carbon will be lower at any given time with sensible climate-change policies than under 'business as usual', because concentrations will be lower at all points in time. Hence, *for given assumptions about discounting and the other relevant factors*, the social cost of carbon associated with sensible emissions strategies is likely to be considerably lower than estimates reviewed in the recent DEFRA study, which were based on various 'business as usual' scenarios[17].

The social cost of carbon will also be lower if the efficiency of emissions-abatement methods improves rapidly and new low-carbon technologies prove to be cheap and easy to spread around the world. In that case, it would be worthwhile undertaking more mitigation and a lower stabilisation level would be appropriate. The lower stabilisation level and path drive down the SCC – better technology is a means to that end. Policy nevertheless has to be strong enough to bring about the changes in technology and energy demand necessary to stabilise at the chosen level.

Compared with the assumptions lying behind the estimates of the social cost of carbon reported in the DEFRA study, there are a number of aspects of this Review's framework of analysis that tend to push up the implied social cost of carbon. These include:

- The adoption of a full 'expected utility' approach to valuation of impacts, allowing risk aversion to give more weight to the possibility of bad outcomes
- Greater weight given to 'non-market' outcomes, especially life chances in poor countries[18]
- The use of a low pure rate of time preference, reflecting the view that this rate should be based largely on the probability that future generations exist, rather than their having some more lowly ethical status[19]
- Equity weighting
- The weight given to recent work on uncertainty about climate sensitivity
- The weight given to recent work on amplifying-feedback risks within the climate system to global temperatures and the risks of extreme events

Policy should ensure that abatement efforts intensify over time. Emissions reductions should be driven to the point where their marginal costs keep pace with the rising social cost of carbon.

Firms and individuals are likely to undertake abatement activities up to the point where the marginal costs of reducing carbon emissions are equal to the carbon price, given by the social cost of carbon associated with the desired trajectory. Anticipated improvements in the overall efficiency of emissions reductions should be reflected in quantity adjustments – lower emissions – not a fall in the price of carbon. The rising SCC is driven by the rising atmospheric concentration of greenhouse gases and the marginal abatement costs are brought into equality with the SCC by firms' and households' reactions to the carbon price. This is illustrated in Box 13.2.

---

[17] Watkiss et al. (2005)

[18] While we have counselled against excessively formal monetary approaches to the value of life, losses of life from climate change nevertheless should weigh heavily in any assessment of damages from climate change.

[19] Note that this is not the same as a low discount rate. The higher the growth rate, the higher the discount rate (see Chapter 2 and its appendix).

BOX 13.2   The relationship between the social cost of carbon and emissions reductions

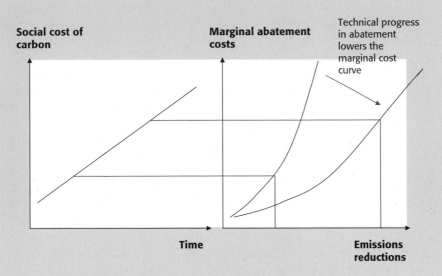

Up to the long-run stabilisation goal, the social cost of carbon will rise over time because marginal damage costs do so. This is because atmospheric concentrations are expected to rise and damage costs are expected to be convex in temperature (i.e. there is increasing marginal damage); these effects are assumed to outweigh the declining marginal impact of the stock of gases on global temperature at higher temperatures.

The price of carbon should reflect the social cost of carbon. In any given year, abatement will then occur up to this price, as set out in the right-hand panel of the diagram above. Over time, technical progress will reduce the total cost of any particular level of abatement, so that at any given price there will be more emission reductions.

The diagram reflects a world of certainty. In practice, neither climate-change damages nor abatement costs can be known with certainty in advance. If the abatement-cost curve illustrated in the right-hand panel were to fall persistently faster than expected, that would warrant revising the stabilisation goal downwards, so that the path for the social cost of carbon in the left-hand panel would shift downwards.

Marginal abatement costs are a measure of effort. If in any region or sector they fall below the estimated social cost of carbon, not enough is being done – unless emissions have ceased. Over time, it may become much easier to reduce emissions in some sectors. Some models suggest an eventual fall in marginal abatement costs in the energy sector, for example, as a result of technological progress. If that does happen, the sector can become completely decarbonised. But elsewhere, where complete decarbonisation will not have taken place – for example, transport – efforts should increase over time and the marginal abatement cost should continue to rise. But policy-makers should foster the development of technology that can drive down the *average* costs of abatement over time.

Delay in taking action on climate change will increase total costs and raise the whole trajectory for the social cost of carbon. The difference between the social cost of carbon on the 'business as usual' trajectory and on stabilisation trajectories reflects the fact that a tonne of greenhouse gas emitted is more harmful and more costly, the higher concentration levels are allowed to go. Delay allows excessive accumulation of greenhouse gases, giving decision-makers a worse starting position for implementing policies.

BOX 13.3   The social cost of carbon and stabilisation

Pearce (2005)[20] reports a range of estimates of the social cost of carbon on 'optimal' paths towards stabilisation goals. The approach of Nordhaus and Boyer (2000) is perhaps closest in spirit to ours. They derive an estimate of only $2.48/tCO_2$ (converted to CO_2, year 2000 prices) for 2001–2010. But they have a low 'business as usual' scenario, do not apply equity weighting and use a discount rate of 3%, which is a little higher than our approach would usually imply.

Further work on what social cost of carbon corresponds to potential stabilisation levels is needed. Current studies disagree about the values and use different methods to tie down the trajectory through time. The US CCSP review reports values of $20/tCO_2$, $2/tCO_2$ and $5/tCO_2$ in 2020 for a stabilisation level of 550 ppm CO_2e in the three studies covered. Edenhofer et al. report estimates of the social cost of carbon ranging from 0 to around $12/tCO_2$ in 2010 for the same stabilisation level (year 2000 prices). Most of the models reviewed envisage the social cost of carbon rising over time, with the level and rate of growth sufficient to pull through the required technologies and reductions in demand for carbon-intensive goods and services.

Preliminary calculations with the model used in Chapter 6 suggest that the current social cost of carbon with business as usual might be around $85/tCO_2$ (year 2000 prices), taking the baseline climate sensitivity assumption used there, if some account is taken of non-market impacts and the risk of catastrophes, subject to all the important caveats discussed in Chapter 6. But along a trajectory towards 550 ppm CO_2e, the social cost of carbon would be around $30/tCO_2$ and along a trajectory to 450 ppm CO_2e around $25/tCO_2$e. These numbers indicate roughly where the range for the policy-induced price of emissions should be if the ethical judgements and assumptions about impacts and uncertainty underlying the exercise in Chapter 6 are accepted.

It would only make sense to have chosen a 550 ppm CO_2e target in the first place if a carbon-price path starting at $30/tCO_2$ had been judged likely to be sufficient (together with other policies) to pull through over time the deployment of the technological innovations required. Similarly, it would only make sense to have chosen a 450 ppm CO_2e target if a price path starting at $25/tCO_2$e had been judged sufficient to bring through the technology needed.

The social cost of carbon[21] can be used to calculate an estimate of the benefits of climate-change policy. The gross benefits of policy for a particular year can be approximated by

$$(SCC_H \times E_H) - (SCC_S \times E_S)$$

[20] Pearce (2005)

[21] The social cost of carbon has to be expressed in terms of some numeraire. Typically the change in consumption that brings about the same impact on the present value of expected utility is used. But that depends on the level of consumption one starts with, so the numeraire differs when comparing significantly different paths. Hence these calculations are strictly valid only if consumption along one or other of the two paths (or some weighted average) is used as numeraire for the calculation of both SCCs.

where SCC denotes the social cost of carbon, E the annual level of emissions, the subscript H the high 'business as usual' trajectory and the subscript S the stabilisation trajectory[22]. This is the net present value of the flow of damages from emissions on the high path less the net present value of the flow of damages on the lower path. With sensible policies ensuring that marginal abatement costs equal the social cost of carbon along the stabilisation trajectory, and assuming for simplicity's sake that marginal abatement cost is equal to average abatement cost[23], the annual costs of abatement can be approximated by

$$SCC_S \times (E_H - E_S)$$

Hence benefits less costs are equal to

$$(SCC_H \times E_H) - (SCC_S \times E_S) - (SCC_S \times (E_H - E_S)) = (SCC_H - SCC_S) \times E_H$$

Thus an approximation of the net present value of the benefits of climate-change policy in any given year can be obtained by multiplying 'business as usual' emissions by the difference between the social costs of carbon on the two trajectories. Calculations for this Review suggest that the social cost of carbon on a reasonable stabilisation trajectory may be around one-third the level on the 'business as usual' trajectory, implying that the net present value of applying an appropriate climate-change policy this year might be of the order of $2.3 – 2.5 trillion. This is not an estimate of costs and benefits falling in this year, but of the costs and benefits through time that could flow from decisions this year; many of these costs and benefits will be in the medium- and long-term future. It is very important, however, to stress that such estimates reflect a large number of underlying assumptions, many of which are very tentative or specific to the ethical perspectives adopted.

13.10   The role of adaptation

Adaptation as well as mitigation can reduce the negative impacts of future climate change.

Adaptation reduces the damage costs of climate change that does occur (and allows beneficial opportunities to be taken), but does nothing direct to prevent climate change and is in itself part of the cost of climate change. Mitigation prevents climate change and the damage costs that follow. Stabilisation at lower levels would entail less spending on adaptation, because the change in climate would be smaller. That needs to be taken into account when considering how total costs change with changes in the ultimate stabilisation level. Similarly, for lower stabilisation levels, a given increase in spending on adaptation is likely to have a bigger effect in lowering the costs of climate change than the same increase at higher concentration levels (because of declining returns to scale for adaptation activities)[24].

---

[22] Because the social cost of carbon is a function of the stock of greenhouse gases, not the flow of emissions, it is insensitive to the variation of emissions in a single year.

[23] This is equivalent to assuming constant returns to scale in abatement over time. In fact, we would expect the average abatement cost to be lower than the marginal abatement cost, with dynamic returns to scale reducing them over time, so this simplification gives an underestimate of the benefits of climate-change policy.

[24] Part V considers adaptation in detail. The key point here is that adaptation is likely to become more expensive and less effective as global temperatures rise further.

There are important differences between adaptation and mitigation that differentiate their roles in policy.

First, while those paying the costs will often capture the benefits of adaptation at the local level, the benefits of mitigation are global and are experienced over the long run. Second, because of inertia in the climate system, past emissions of greenhouse gases will drive increases in global mean temperature for another several decades. Thus mitigation will have a negligible effect in reducing the cost of climate change over the next 30–50 years: adaptation is the only means to do so.

Adaptation can efficiently reduce the costs of climate change while atmospheric concentrations of greenhouse gases are being stabilised.

A stabilisation goal facilitates adaptation by allowing a better understanding to develop of what ultimately societies will have to adapt to. Work using Integrated Assessment Models (IAMs, discussed in Chapter 6) has identified significant opportunities to reduce damage costs through adaptation. There are many reasons other than assumptions about adaptation why the predictions of one model differ from another[25]. It is nevertheless intuitive that those models with the most comprehensive adaptation processes estimate the lowest damage costs and highest adaptation benefits[26]. Studies at a more local level of the costs and benefits of adaptation usually point to net benefits, so some is likely to take place, although policy measures are often required to overcome barriers (see Part V). Adaptation will have a particular role to play in low-income regions, where vulnerability to climate change is higher. In such regions, there are strong complementarities between development policies in general and adaptation actions in particular.

There are further examples of complementarities:

- Mitigation reduces the likelihood of dangerous climate change, which makes adaptation either infeasible or very costly;
- Mitigation reduces uncertainty about the range of possible climate outcomes requiring adaptation decisions. Uncertainty is a clear impediment to successful adaptation.

In the longer run, both adaptation and mitigation will be required to reduce climate-change damage in cost-effective and sustainable ways.

They should not be regarded as alternatives. Part II outlined why the damage costs of climate change are likely to increase more rapidly as global mean temperatures increase. As Part V explains in more detail, attempts at adaptation would not be an adequate response to the pace and magnitude of climate change at high global mean temperatures compared with pre-industrial levels. Ecosystems, for instance, cannot physically keep pace with the shifts in climatic conditions implied. The adaptation that remains viable is likely to be very costly. Without mitigation, little can reduce the underlying acceleration in climate-change impacts as temperatures rise. This is why promoting development in developing economies, while vital in its own right and helpful in building the capacity to adapt, is not an adequate response by itself. Mitigation is the key to reducing the probability of

---

[25] Hanemann (2000).
[26] In particular, Mendelsohn et al. (2000).

dangerous climate change, given the scale of the challenge. A strategy of mitigation plus adaptation is superior to 'business as usual' plus adaptation, and
requires less spending on adaptation.

13.11   Conclusions

This chapter has considered in broad terms what climate-change policy should
aim to achieve, given the evidence about the risks of serious damages from climate change and the costs of cutting greenhouse-gas emissions. The first priority
is to strengthen global action to slow and stop human-induced climate change
and to start undertaking the necessary adaptation to the change that will happen
before stability is established. The benefits of doing more clearly outweigh the
costs. Delay would entail more climate change and eventually higher costs of
tackling the problem. The nature of the uncertainties in the science and economics warrants more action not less.

Once the case for stronger global action is accepted, the question arises, how
much? We have argued the merits of organising the discussion of this problem
around the idea of a goal for the ultimate concentration of greenhouse gases
in the atmosphere. Choosing a specific level or range for such a goal should help
to make policies around the world more consistent, coherent and cost-effective.
In particular, choosing a goal helps to define and anchor a path for the carbon
price, a key tool for implementing climate-change policy. The next part of this
Review examines in more detail the types of policy instruments that need to be
used to reduce greenhouse-gas emissions cost-effectively and on the scale
required.

References

The issues involved in choosing an optimal level of atmospheric concentrations of greenhouse
gases are explored comprehensively in the context of a particular model in Nordhaus and Boyer
(1999). Some of the challenges posed by the great uncertainties surrounding climate change are
ably surveyed in Ingham and Ulph (2005). The social cost of carbon, in principle and practice is
discussed thoroughly in Downing et al (2005) and Watkiss et al (2005).

Arnell, N.W., M.J.L. Livermore, S. Kovats et al. (2004): 'Climate and socio-economic scenarios for
    global-scale climate change impacts assessments: characterising the SRES storylines', Global
    Environmental Change 14:3–20
Clarkson, R., and K. Deyes (2002): 'Estimating the social cost of carbon emissions', GES Working
    Paper 140, London: HM Treasury.
Downing, T.E., D. Anthoff, R. Butterfield et al. (2005): 'Social cost of carbon: a closer look at uncer-
    tainty'. London: Department of Environment, Food, and Rural Affairs (DEFRA), available from
    http://www.DEFRA.gov.uk/ENVIRONMENT/climatechange/carboncost/index.htm
Edenhofer, O., K. Lessmann, C. Kemfert, et al. (2006): 'Induced technological change: exploring its
    implications for the economics of atmospheric stabilization: synthesis report from the innova-
    tion modeling comparison project', The Energy Journal, special issue, April: 57–108
Hanemann, W.M. (2000): 'Adaptation and its measurement' Climatic Change 45 (3–4): 571–581
Hope, C. (2003): 'The marginal impacts of CO2, CH4 and SF6 emissions,' Judge Institute of
    Management Research Paper No.2003/10, Cambridge, UK, University of Cambridge, Judge
    Institute of Management.
Ingham, A, and Ulph, A (2005): 'Uncertainty and climate-change policy' in Helm, D (2005):
    Climate-change policy, Oxford: Oxford University Press.

IAG (2005): 'Evidence to the Stern Review on the economics of climate change', Melbourne: Insurance Australia Group, available from http://www.sternreview.org.uk

Kolstad, C. (1996): 'Fundamental irreversibilities in stock externalities', Journal of Public Economics, **60**: 221–233

Mendelsohn, R.O., W.N. Morrison, M.E. Schlesinger and N.G. Andronova (1998): 'Country-specific market impacts of climate change', Climatic Change 45(3-4): 553–569. (change the citation to Mendelsohn et al. (1998).

Nordhaus, W., and J.G. Boyer (1999): 'Roll the DICE Again: Economic Models of Global Warming', Cambridge, MA: MIT Press.

Pindyck, R. (2000): 'Irreversibilities and the timing of environmental policy', Resource and Energy Economics, **22**: 233–259

Manne, A. and R. Richels, 1995: 'The greenhouse debate: economic efficiency, burden sharing and hedging strategies'. The Energy Journal, **16(4)**, 1–37.

Pearce, D. (2005): 'The social cost of carbon' in Helm, D (2005): 'Climate-change policy', Oxford: Oxford University Press.

Schlenker W. and M.J. Roberts (2006): 'Nonlinear effects of weather on corn yields', Review of Agricultural Economics, **28**: in press.

Tol, R.S.J. (1997): 'On the optimal control of carbon dioxide emissions: an application of FUND' Environmental Modelling and Assessment, **2**, 151–163.

Tol, R.S.J. (2005): 'The marginal damage costs of carbon dioxide emissions: an assessment of the uncertainties', Energy Policy, **33**: 2064–2074.

US CCSP Synthesis and Assessment Product 2.1, Part A: 'Scenarios of greenhouse gas emissions and atmospheric concentrations', Draft for public comment, June 26, 2006.

Watkiss, P. et al. (2005): 'The social cost of carbon', London: DEFRA, December.

IV Policy Responses for Mitigation

The first half of this Review has considered the evidence on the economic impacts of climate change itself, and the economics of stabilising greenhouses in the atmosphere. Parts IV, V and VI now look at the policy response.

The first essential element of climate change policy is carbon pricing. Greenhouse gases are, in economic terms, an externality: those who produce greenhouse gas do not face the full consequences of the costs of their actions themselves. Putting an appropriate price on carbon, through taxes, trading or regulation, means that people pay the full social cost of their actions. This will lead individuals and businesses to switch away from high-carbon goods and services, and to invest in low-carbon alternatives.

But the presence of a range of other market failures and barriers mean that carbon pricing alone is not sufficient. Technology policy, the second element of a climate change strategy, is vital to bring forward the range of low-carbon and high-efficiency technologies that will be needed to make deep emissions cuts. Research and development, demonstration, and market support policies can all help to drive innovation, and motivate a response by the private sector.

Policies to remove the barriers to behavioural change are a third critical element. Opportunities for cost-effective mitigation options are not always taken up, because of a lack of information, the complexity of the choices available, or the upfront cost. Policies on regulation, information and financing are therefore important. And a shared understanding of the nature of climate change and its consequences should be fostered through evidence, education, persuasion and discussion.

The credibility of policies is key; this will need to be built over time. In the transitional period, it is important for governments to consider how to avoid the risks that long-lived investments may be made in high-carbon infrastructure.

Part IV is structured as follows:

- **Chapter 14** looks at the principles of carbon pricing policies, focusing particularly on the difference between taxation and trading approaches.
- **Chapter 15** considers the practical application of carbon pricing, including the importance of credibility and good policy design, and the applicability of policies to different sectors.
- **Chapter 16** discusses the motivation for, and design of, technology policies.
- **Chapter 17** looks at policies aimed at removing barriers to action, particularly in relation to the take-up of opportunities for energy efficiency, and at how policies can help to change preferences and behaviour.

14 Harnessing Markets for Mitigation – The Role of Taxation and Trading

KEY MESSAGES

- **Agreeing a quantitative global stabilisation target range for the stock of greenhouse gases (GHGs) in the atmosphere is an important and useful foundation for overall policy.** It is an efficient way to control the **risk** of catastrophic climate change in the long term. Short term policies to achieve emissions reductions will need to be consistent with this long-term stabilisation goal.

- **In the short term, using price-driven instruments (through tax or trading) will allow flexibility in how, where and when emission reductions are made, providing opportunities and incentives to keep down the cost of mitigation.** The price signal should reflect the marginal damage caused by emissions, and rise over time to reflect the increasing damages as the stock of GHGs grows. For efficiency, it should be common across sectors and countries.

- **In theory, taxes or tradable quotas could establish this common price signal across countries and sectors.** There can also be a role for regulation in setting an implicit price where market-based mechanisms alone prove ineffective. In practice, tradable quota systems – such as the EU's emissions-trading scheme – may be the most straightforward way of establishing a common price signal across countries. To promote cost-effectiveness, they also need flexibility in the timing of emissions reductions.

- **Both taxes and tradable quotas have the potential to raise public revenues.** In the case of tradable quotas, this will occur only if some firms pay for allowances (through an auction or sale). Over time, there are good economic reasons for moving towards greater use of auctioning, though the transition must be carefully managed to ensure a robust revenue base.

- **The global distributional impact of climate-change policy is also critical.** Issues of equity are likely to be central to securing agreement on the way forward. Under the existing Kyoto protocol, participating developed countries have agreed binding commitments to reduce emissions. Within such a system, company-level trading schemes such as the EU ETS, which allow emission reductions to be made in the most cost-effective **location** – either within the EU, or elsewhere – can then drive financial flows between countries and promote, in an equitable way, accelerated mitigation in developing countries.

- At the **national – or regional** – level, governments will want to choose a policy framework that is suited to their specific circumstances. Tax policy, tradable quotas and regulation can all play a role. In practice, some administrations are likely to place greater emphasis on trading, others on taxation and possibly some on regulation.

14.1   Introduction

This chapter focuses on the first and key element of a mitigation strategy – how best to ensure GHG emissions are priced to reflect the damage they cause.

This chapter focuses on the principles of policy and, in particular, on the efficiency, equity and public finance implications of tax and tradable quotas. Chapter 15 follows with a detailed discussion of the practical issues associated with the implementation of tax and trading schemes.

Section 14.2 begins by setting out the basic theory of externalities as this applies to climate change. Based on this, Section 14.3 sets out two overarching principles for reducing GHG emissions efficiently. First, abatement should occur just up to the point where the costs of going any further would outweigh the extra benefits. Second, a common price signal is needed across countries and sectors to ensure that emission reductions are delivered in the most cost-effective way.

Section 14.4 explores the policy implications of the significant risks and uncertainties surrounding both the impacts of climate change, and the costs of abatement. It concludes that a long-term quantity ceiling – or stabilisation target – should be used to limit the total stock of GHGs in the atmosphere. In the short term, to keep down the costs of mitigation, the amount of abatement should be driven by a common price signal across countries and sectors, and there should be flexibility in how, where and when reductions are made. Over time, the price signal should trend upwards, as the social cost of carbon is likely to increase as concentrations rise towards the long-term stabilisation goal.

These sections conclude that both taxes and tradable quotas have the potential to deliver emission reductions efficiently. The other key dimensions of climate change policy – tackling market failures that limit the development low carbon technologies, and removing barriers to behavioural change are discussed in Chapter 16 and Chapter 17 respectively.

The penultimate section of the chapter considers the public-finance aspects of taxes and tradable quotas. Finally, Section 14.8 briefly considers the international dimension of carbon-pricing policy. These international issues are treated in greater depth in Part VI of this Review – in particular, the challenge of how national action can be co-ordinated and linked at the international level to support the achievement of a long-run stabilisation goal is considered in Chapter 22.

14.2   Designing policy to reduce the impact of the greenhouse-gas externality

As described in Chapter 2, the climate change problem is an international and intergenerational issue.

Climate change is a far more complicated negative externality than, for example, pollution (such as smog) or congestion (such as traffic jams). Key features of the greenhouse-gas externality are:

- it is a global externality, as the damage from emissions is broadly the same regardless of where they are emitted, but the impacts are likely to fall very unevenly around the world;

- its impacts are not immediately tangible, but are likely to be felt some way into the future. There are significant differences in the short-run and long-run implications of greenhouse-gas emissions. It is the stock of carbon in the atmosphere that drives climate change, rather than the annual flow of emissions. Once released, carbon dioxide remains in the atmosphere for up to 100 years;
- there is uncertainty around the scale and timing of the impacts and about when irreversible damage from emission concentrations will occur;
- the effects are potentially on a massive scale.

These characteristics have implications for the most appropriate policy response to climate change. In the standard theory of externalities[1], there are four ways in which negative externalities can be approached:

- a tax can be introduced so that emitters face the full social cost of their emissions[2] ie. a carbon price can be established that reflects the damage caused by emissions;
- quantity restrictions can limit the volume of emissions, using a 'command and control' approach;
- a full set of property rights can be allocated among those causing the externality and/or those affected (in this case including future generations), which can underpin bargaining or trading[3];
- a single organisation can be created which brings those causing the externality together with all those affected[4].

In practice, cap-and-trade systems tend to combine aspects of the second and third approach above. They control the overall quantity of emissions, by establishing binding emissions commitments. Within this quantity ceiling, entities covered by the scheme – such as firms, countries or individuals – are then free to choose how best – and where – to deliver emission reductions within the scheme. The largest example of a cap-and-trade scheme for GHG emissions is the EU's Emissions Trading Scheme, and there are a range of other national or regional emissions trading schemes, including the US Regional Greenhouse Gas Initiative and the Chicago Climate Exchange.

The Kyoto Protocol established intergovernmental emissions trading for those countries that took quantified commitments to reduce GHG emissions, as well as other mechanisms to increase the flexibility of trading across all Parties to the Protocol. The Kyoto Protocol and its flexible mechanisms are discussed in detail in Chapter 22.

Whatever approach is taken, the key aim of climate-change policy should be to ensure that those generating GHGs, wherever they may be, face a marginal cost of emissions that reflects the damage they cause. This encourages emitters to invest in alternative, low-carbon technologies, and consumers of GHG-intensive goods and services to change their spending patterns in response to the increase in relative prices.

[1] Developed mainly in the first half of the last century.
[2] Pigou (1920) showed how taxes can establish a marginal cost to polluters equal to the marginal damage caused by their pollution.
[3] Coase (1960)
[4] Meade (1951). This is not discussed further, as it is clearly not a practical option in relation to climate change.

14.3   Delivering carbon reductions efficiently

Where markets are well functioning, two conditions must hold to reduce GHG emissions efficiently[5]:

- Abatement should take place up to the point where the benefits of further emission reductions are just balanced by the costs. Or – put another way – abatement should occur up to the point where the marginal social cost of carbon is equal to the marginal cost of abatement. This is a necessary condition for choosing the appropriate level of emissions, and hence setting a long-term stabilisation target (and is explained fully in Chapter 13).
- To deliver reductions at least cost, a common price signal is required across countries and different sectors of their economies at a given point in time. For example, if the marginal cost of reduction is lower in country A than in country B, then abatement costs could be reduced by doing a little more reduction in country A, and a little less in country B.

In ideal conditions – perfectly competitive markets, perfect information and certainty, and no transaction costs – both taxes and quantity controls, if correctly designed, can meet these criteria, and be used to establish a common price signal across countries and sectors. Taxes can set the global price of greenhouse gases, and emitters can then choose how much to emit. Alternatively, a total quota (or ceiling) for global emissions can be set and tradable quotas can then determine market prices.[6]

Without market imperfections and uncertainty, and with an appropriate specification of taxes and quotas (entailing an allocation of property rights), both approaches would produce the same price level and quantity of emissions[7]. The remainder of this chapter, and Chapter 15, consider how the considerable uncertainties and imperfections that exist in the real world affect the choice and design of policy.

14.4   Efficiency under uncertainty – the implications for climate-change policy

Substantial uncertainty exists around the timing and scale of impacts, as well as the costs of abatement. In such circumstances, prices and quantity controls are no longer equivalent and policy instruments will need to be chosen with care to reduce GHG emissions efficiently.

Weitzman (1974) examined how price (here tax) and quota or quantity-control instruments compare where there is uncertainty about the costs and benefits of action, and how this affects the comparative efficiency of the two instruments[8]. A price instrument sets a price for a required service or good and lets markets

---

[5]   These conditions abstract from uncertainty and market imperfections.

[6]   Continuous trading is necessary to ensure a common price between auctions/ allocations.

[7]   But it is worth noting that even if these ideal conditions were to hold, the nature of the climate-change problem means there are limitations to the applicability of some of the policy options set out above. In particular, a **full set** of property rights cannot be allocated, because many of those affected by the impacts of climate change are yet to be born. It is not possible for them to bargain with the current emitters for the impacts that they will have to endure.

[8]   Weitzman (1974)

determine its supply. In contrast, a quota instrument specifies a particular level of supply. Applying the Weitzman analysis to pollution:

- Prices are preferable where the benefits of making further reductions in pollution change less with the level of pollution than do the costs of delivering these reductions i.e. when the marginal damage curve – or the marginal social cost of carbon - is relatively flat, compared with the marginal abatement cost curve, as pollution rises.
- Quantity controls are preferable where the benefits of further reductions increase more with the level of pollution than do the costs of delivering these reductions i.e. there are potentially large and sharply rising costs associated with exceeding a given level of pollution.

Box 14.1 sets out these economic arguments in detail[9].

BOX 14.1 Prices versus quantities in the short term and long term.

Figure (A) illustrates how Weitzman's analysis is applied in the climate-change case. If the emissions reductions are measured over a short period, say a year, the expected marginal benefits of abatement are flat or gently decreasing as the quantity of emission reduction increases (from left to right). This reflects the fact that variations in emissions in any single year are unlikely to have a significant effect on the ultimate stock of greenhouse gases. The expected marginal costs of abatement (MAC_E), however, are steeply increasing as abatement activity intensifies; firms find it progressively more difficult to reduce emissions, unless they can adjust their capital stock and choice of technology (assumed by definition to be impossible in the short term).

If it were known with certainty that the marginal costs of abatement were given by the schedule MAC_E, the policy-maker should set the rate of the emission tax to equal T_E, given by the intersection of the schedule with the marginal benefits of abatement, also assumed to be known. The optimal quantity of emission quotas or allowances allocated (Q_E) would also be given by this intersection, giving rise to an equilibrium price in a perfectly competitive allowance market of P_E. The choice of quota or tax would not matter in this case.

However, following the exposition in Hepburn (2006), suppose that the real marginal costs of abatement in the period are not known with certainty in advance and turn out to be higher at every point, as represented by the curve MAC_{REAL}, and that the policy-maker cannot adjust the policy instrument in anticipation. In this case, the optimal quantity of allowances to be allocated would in fact turn out to have been Q_{REAL}. In Figure 14.1, the efficiency loss caused by issuing Q_E instead of Q_{REAL} allowances is given by the large blue triangle. If instead a tax had been set at T_E, the efficiency loss resulting from having set a slightly lower tax rate than turns out to have been warranted, is given by the small red triangle. Thus it is often argued that a tax is superior to a quota as an instrument of climate-change policy[10] in the short run. As Chapter 2 explains, however, diagrams like that in Figure (A) need to be interpreted with great care, as the positions of both the curves may depend on policy settings in earlier and later periods.

---

[9] This box draws on the exposition in Hepburn (2006).

[10] The direct allocation of non-tradable allowances requires information about relative costs across firms, as well as total costs, and so is likely to be even less efficient, given the uncertainties in the real world, than promoting perfect competition in the market for allowances.

(A) The efficiency of taxes and tradable allowances in climate-change mitigation in the short term.

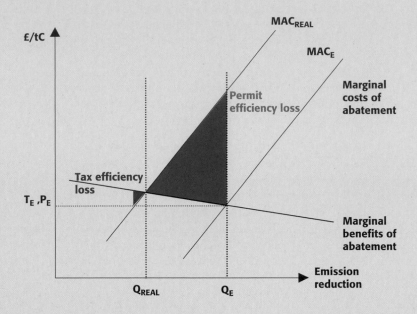

Figure (B) illustrates the situation in the long term, with the cumulative emissions reductions required to reach the ultimate stabilisation target on the x-axis now, instead of annual emissions reductions as in Figure (A). The curve representing the marginal benefits of abatement is steeply decreasing, as more and more abatement effort is put in (put another way, the costs of the impacts of climate change increase steeply as cumulative emissions increase). But the marginal costs of abatement are only gently increasing as a function of abatement effort, since in the long run there is more flexibility. In the certainty case with MAC_E as the true cost of abatement curve, Q_E is the appropriate cumulative quota, while T_E is the equivalent tax[11]. But if MAC_E represents the expected costs of abatement and MAC_{REAL} the higher ex post actual costs, the efficiency loss implied by setting the tax at T_E (the blue triangle) is now much larger than that implied by setting the quantity of tradable allowances at Q_E. Of course, if the policy-maker is able to revise the tax or quota schedule as information comes in about the marginal abatement costs function, s/he can do better than keeping either schedule fixed.

(B) The efficiency of taxes and tradable allowances in climate-change mitigation in the long term.

See figure in next page

   This contrast between short-term and long-term marginal cost and marginal benefit curves gives rise to the problem of how to combine a tax-like regime in the short term with

---

[11] Strictly, there is an intertemporal tax schedule that generates cumulative emissions reductions Q_E

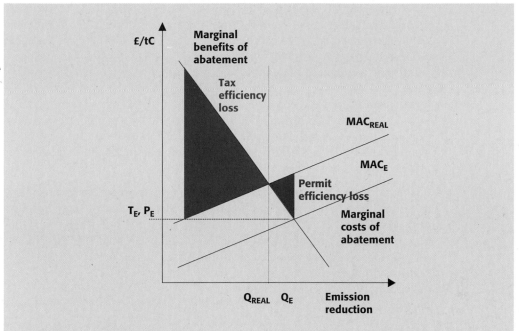

£/tC

Marginal benefits of abatement

Tax efficiency loss

MAC_{REAL}

MAC_E

Permit efficiency loss

T_E, P_E

Marginal costs of abatement

Q_{REAL} Q_E Emission reduction

a quantitative constraint in the long term. A rule is needed for updating the tax in the light of new information about costs over the long term and the ex post quantity of emissions.

In the case of climate change, these arguments indicate that the most efficient instrument – over a particular time horizon – will depend on:

- how the total costs of abatement change with the level of emissions;
- how the total benefits of abatement change with the level of emissions;
- the degree of uncertainty about both costs and benefits of abatement.

Chapter 8 explains that it is the total stock of GHGs in the atmosphere that drives the damage from climate change. In economic terms, this means that the marginal damage associated with emitting one more unit of carbon is likely to be more or less constant over short periods of time. Thus, in the short-term, the marginal damage curve is likely to be fairly flat. But over the long term, as the stock of GHGs grows, marginal damages are likely to rise and – as the stock reaches critical levels – marginal damages may rise sharply. In other words, the damage function is likely to be strongly convex (as discussed in Part Two and Chapter 13)[12].

On the other side of the equation, many uncertainties remain about the marginal costs of abatement. Many new technologies that could be used to reduce carbon

[12] To the extent that damages may relate to the *rate* of climate change, the relationship is more complex, but it remains true that the damage curve is likely to respond most to cumulative emissions over several years or even decades.

emissions are not yet in widespread use. Trying to abate rapidly in the short term – when the capital in industries emitting greenhouse gases is fixed and technologies are given – can quickly become costly for firms, as the marginal cost of abatement is likely to rise sharply[13]. In particular, if the costs of abatement prove to be unexpectedly high, then setting a fixed quantity target in the short term could prove unexpectedly costly. Over the long term – as the capital stock is replaced and new lower-carbon technologies become available – the marginal costs of abating in the long term are likely to be broadly flat, or, put another way, bounded relative to incomes. The implications are explained more fully in Box 14.1.

These characteristics of the costs and benefits of abatement and damage from emissions suggest three things:

- Policy instruments should distinguish between the short term and long term, ensuring that short-term policy outcomes are consistent with achieving long-term goals[14];
- The policy-maker should have a clear long-term goal for stabilising concentrations of greenhouse gases in the atmosphere. This reflects, first, the likelihood that marginal damages (relative to incomes) will accelerate as cumulative emissions rise and, second, that the marginal costs of abatement (relative to incomes) are likely to be relatively flat in the long term once new technologies are available.
- In the short term, the policy-maker will want to choose a flexible approach[15] to achieving this long-term goal, reflecting the likelihood that marginal damages will be more or less constant, and there will be risks of sharply rising costs from forcing abatement too rapidly.

In practical terms, this means that a long-term stabilisation target should be used to establish a quantity ceiling to limit the total stock of carbon over time. Short-term policies (based on tax, trading or in some circumstances regulation) will then need to be consistent with this long-term stabilisation goal. In the short term, the amount of abatement should be driven by a common price signal across countries and sectors, and should not be rigidly fixed[16].

This common price signal could – in principle – be delivered through taxation or tradable quotas. A country can levy taxes without consultation with another, but harmonisation requires agreement. In practice, therefore, it may prove difficult to use taxes to deliver a common price signal in the absence of political commitment to move towards a harmonised carbon tax across different countries. In contrast, to the extent that a tradable quota scheme embraces both different countries and sectors, it may be an effective way of delivering a consistent price signal across a wide area – though this, of course, requires agreement on the mechanics of the scheme. International co-ordination issues are fully discussed in Chapter 22 – here it is sufficient to note that building consensus on the best way forward will be critical to achieving a long-run stabilisation goal.

[13] For a discussion of the relative abatement costs and marginal benefits of climate change see, for example, Lydon (2002) and Pizer (2002). Both conclude that the marginal damage curve is relatively flat – at least in the short term – and, as such, there are strong arguments for flexibility in the quantity of abatement in the short term, subject to a fixed carbon price.

[14] The short term is defined as the period during which the capital stock is essentially fixed. This will vary from sector to sector.

[15] With respect to the size of emission reductions.

[16] One option is to combine price controls within a quota trading system in the short term. This is discussed more in Chapter 15.

14.5   Setting short term policies to meet the long term goal

The key question that arises from the previous section is how to combine a price instrument that allows flexibility about where, when and what emissions are reduced in the short term, with a long-term quantity constraint. In particular, the challenge is how to ensure that the short-term policy framework remains on track to deliver the long-term stabilisation goal.

There are two important aspects to this:

- having established the long-term stabilisation goal, the price of carbon is likely to rise over time, because the damage caused by further emissions at the margin-the social cost of carbon- is likely to increase as concentrations rise towards this agreed long-term quantity constraint;
- short-term tax or trading policies will then need to be consistent with delivering this long-term quantitative goal.

In the short-term, applying these principles to tax and trading, this means that:

- In a tax-based regime, the tax should be set to reflect the marginal damage caused by emissions. Abatement should then occur up to the point where the marginal cost of abatement is equal to this tax. See Box 14.2.
- In a tradable-quota scheme, the parameters of the scheme – notably the total quota allocation – should be set with a view to generating a market price that is consistent with the social cost of carbon (SCC). In practice – and within the time period between allocations in a tradable-quota system – the market price may be higher or lower than the SCC. This is because the actual market price will reflect both the quota-driven demand for carbon reductions and the marginal cost of delivering reductions in the most cost-effective location. Ex-post, the trading period will therefore deliver abatement up to where the marginal abatement cost equals the actual market price.

BOX 14.2 The social cost of carbon and the carbon price

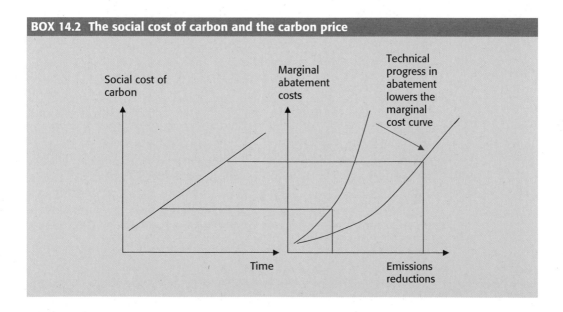

Up to the long-term stabilisation goal, the social cost of carbon will rise over time, because marginal damage costs also rise. This is because atmospheric concentrations are expected to rise, so that temperatures are likely to rise; marginal damage costs are expected to rise with temperature. These effects are assumed to outweigh the declining marginal impact of the stock of gases on global temperature at higher temperatures.

As GHG concentrations move towards the stabilisation goal, the price of carbon should reflect the social cost of carbon. In any given year, abatement should then occur up to where the marginal cost is equal to this price, as set out in the right-hand part of the diagram above. If, over time, technical progress reduces the marginal cost of abatement, then at any given price level there should be more emission reductions.

In the case of either tax or trading, clear revision rules are therefore necessary to ensure that short-term policies remain on track to meet the long-term stabilisation goal. In particular, the short-term policy framework should be able to take systematic account of the latest scientific information on climate change, as well as improved understanding of abatement costs.

The framework within which any principles for revisions apply must be clear, credible, predictable and set over long time horizons, say 20 years, with regular points, say every five years, to review new evidence, analysis and information[17]. Chapter 22 discusses the challenge of achieving this at an international level.

Revision rules for climate-change policies can be compared to setting interest rates within a well-specified inflation-targeting regime[18]. The stabilisation target is analogous to the inflation target. In the UK, the Monetary Policy Committee each month sets a short-term policy instrument, the interest rate on central-bank money, until their next meeting, in order to keep inflation on track to hit its target. The analogy with climate-change policy would be the setting of a tax rate or an emissions trading quota for, say, a five year period, with firms and households making their own decisions about emissions reductions subject to that carbon-price path and their expectations about policy-makers' commitment to the long-term stabilisation goal.

The analogy is not, however, exact. First, there is widespread agreement about the appropriate long-term goal for monetary policy – price stability, which corresponds to a small positive measured inflation rate. In the climate-change case, there is not yet agreement about the stabilisation level at which that stability should be achieved. Second, the stabilisation objective is likely to have to be revised intermittently – possibly by a large amount – to reflect improved scientific and economic understanding of the climate-change problem, whereas the definition of price stability in terms of a specific inflation measure is less problematic. And third, the locus of decision-making in monetary policy clearly lies with the monetary authority of the country for which the inflation rate is measured, whereas climate change requires international collective action.

Nevertheless, the comparison with an inflation-targeting regime draws attention to the importance of building the credibility of policy-makers. This requires

---

[17] Newell et al (2005)
[18] This analogy has been explored by Helm et al (2005).

clarity about the ultimate objective of policy and giving policy-makers control over an instrument that can change private-sector behaviour. It also means announcing the principles governing changes in the policy instrument in advance, giving policy-makers incentives to keep aiming at the ultimate target, and holding policy-makers accountable for their actions.

14.6   The interaction between carbon pricing and fossil fuel markets

Imperfections in the markets for exhaustible resources and energy could have important interactions with carbon-pricing policy that should also be considered.

Carbon emissions come from energy production and use across various sectors (see Chapter 7). Much of this energy is generated using exhaustible resources such as oil. In the face of climate change policy, the owners of the natural resource may be willing to reduce producer prices substantially in order to sell off the commodity before it becomes obsolete or of a much lower value. Thus any carbon-pricing policy would need to be carefully designed to ensure it does not accelerate the pace with which carbon-intensive exhaustible resources are used up. The policy implications of this – as well as market imperfections more generally – are explored in Box 14.3.

BOX 14.3 Efficiency market structure and exhaustible resources

Energy and related markets have pervasive market imperfections that will affect the efficiency of a given policy instrument[19]. For example, the collusive behaviour of the OPEC cartel can make it difficult to predict what the final impact on market prices will be from either a tax or a quota-driven carbon price. Thus, on the one hand, OPEC might respond to a carbon tax by further restricting supply, pushing up producer prices and retaining most of their rents. On the other hand, they may choose to retain market share and extract a lower rent[20] with little change in carbon emissions[21].

Where the input prices concerned relate to fossil fuels, the policy must also take account of the fact that such fuels are exhaustible natural resources. Prices to consumers will reflect both the marginal costs of extraction and a scarcity rent (which reflects the stock of the natural resource relative to the expected demand schedule over time). In these circumstances, attempts to reduce carbon emissions through tax measures (imposing the social cost on polluters) may simply lead to a fall in producer prices, with little change in consumption and therefore carbon emissions. In some models, the incidence of the tax would fall wholly on the resource owner's rent. For the same reason, the introduction of new renewable-energy technologies may simply accelerate the use of carbon-intensive energy sources[22] – as the owners of the natural resource try to sell them off before they

---

[19] See Blyth and Hamilton (2006) for background discussion on the nature of electricity markets, interaction with fossil fuel markets and issues to consider for policy approaches to introducing climate policy to electricity systems.

[20] This would shift rents from OPEC to Kyoto countries.

[21] Hepburn (2006)

[22] The economic theory of exhaustible natural resources is exposited in Hotelling (1931) and Dasgupta & Heal (1979).

become obsolete or fall sharply in value. In these circumstances – for some market structures, and in the absence of carbon capture and storage – optimal tax theory can suggest that a declining ad valorem[23] tax rate over time may eventually be desirable, to delay fossil-fuel consumption and push back in time the impacts of climate change[24]. In this case, the tax rate through time reflects more than the social cost of carbon, as it is also takes account of these other market dynamics. The key point here is that there are many complexities that should be considered.[25]

Under a tradable quota system, the price associated with an emissions quota may be much higher than expected if exhaustible-resource pricing is ignored. In effect, rent may be transferred from the owners of fossil fuels to the owners of the allowances (or issuers, if allowances are auctioned). More generally, if trading creates rents, it may undermine the acceptability of policy and lead to gaming, wasted resources in rent-seeking, and possibly corruption. Where incumbent firms enjoy rents, they may also discourage competition and new entry.

14.7   Public finance issues

Both taxes and tradable quotas can be used to raise public funds. Carbon taxes automatically raise public revenues, but tradable-quota systems only have the potential to raise public revenue if firms have to purchase the quotas from government through a sale or auction.

Carbon taxes automatically transfer funds from emitting industries to the public revenue. This transfer may be used to:

- enhance the revenue base[26];
- limit the overall tax burden on the industry affected through revenue recycling[27];
- reduce taxes elsewhere in the economy;

Revenue recycling to the industry can encourage emitters to reduce GHG emissions, without increasing their overall tax burden relative to other parts of the economy[28]. The advantage of this approach is that it can ease the initial impact of the scheme for those industries facing the greatest increase in costs, and therefore ease the transition where carbon taxes are introduced. As the introduction of carbon pricing through taxation is a change to the rules of the game (which will

---

[23] Ad valorem taxes are based on the value or price of a good or service. The alternative to ad-valorem taxation is a fixed-rate tax, where the tax base is the quantity of something, regardless of its price.

[24] There is a debate about whether the tax rate should first rise and then fall. See Ulph & Ulph (1994) and Sinclair (1994).

[25] For a more detailed discussion, see Newbery (2005).

[26] In practice, the overall impact on the revenue base may be limited, if taxes are reduced elsewhere in the economy.

[27] The ultimate incidence of the tax is on the industries' customers and – in the absence of perfect competition – shareholders.

[28] Although, as already noted, in a competitive industry the tax will ultimately fall on the consumer.

affect shareholders in the short run), there is a case for some transitional arrangements. Over time, however, recycling may discourage or slow the necessary exit of firms from the polluting sectors. Monitoring and protecting the position of incumbents in this way could also reduce competition.

Alternatively, revenue from carbon taxes can be used to reduce taxes elsewhere in the economy. In such circumstances, the revenue from the carbon tax is sometimes argued to generate a so-called 'double dividend' by allowing other distortionary taxes to be reduced.

But this argument needs some care. There is no doubt that environmental taxes have the special virtue of reducing 'public bads', at the same time as they generate revenue. Reducing the 'bad' is indeed central to any assessment of this type of tax. But arguments invoking the so-called 'double dividend' as sometimes advanced in general terms (i.e. that there is always a double dividend), can be incorrect. Putting the reduced public bad to one side for a moment, there is a 'dead-weight' loss to the economy from raising any tax on the margin. Whether it is greater or less with goods associated with carbon (compared with other goods or services) is unclear and depends on the circumstances. For example, where energy is subsidised, reducing the subsidy (equivalent to raising the tax) will probably be a gain in terms of reducing deadweight losses. Note, however, that where other taxes have been optimally set – and abstracting from the externality – then the dead-weight loss on the margin from increasing any one tax will be exactly the same as the loss on another and there will clearly be no 'double dividend' in this context.

This is not an argument against raising revenue through pricing GHG emissions. On the contrary: there are strong benefits from ensuring that GHG emissions are properly priced to reflect the damage they cause. Thus GHG taxes have the clear additional benefit relative to other ways of raising revenue of reducing a 'bad'. Where that benefit has not been adequately recognised, they will be under-used relative to other forms of taxation.

In contrast, a quota-based system will not automatically raise revenue unless firms must initially purchase some or all quotas from the government in either an auction or a direct sale. In constrast, if quotas are allocated for free, then the asset is passed to the private sector and the benefits ultimately accrue to the owners and shareholders of the firms involved[29]. In the short term, there may be reasons for introducing auctioning slowly – to ease the transition to a new policy environment. Equally, finance ministries will want to ensure that the overall tax revenue base is reliable and predictable: revenues from auctioning may be less predictable than those from taxation. In the long term, however, there is little economic justification for such transfers from the public sector to individual firms and their shareholders[30].

Free allocation of quotas to business also has a number of other potential drawbacks. These are discussed in more detail in the next chapter, which focuses on practical issues associated with the implementation of tax and trading schemes.

---

[29] To the extent that firms are able to pass on to consumers the increase in marginal production costs, a system with free quotas may be regressive (because shareholders tend to be wealthier than the general population).

[30] Where the ultimate incidence of the tax falls on customers, they pay a price of carbon, but there is no benefit to the wider revenue base.

In summary, a tax-based approach will automatically generate public revenues, whereas a tradable-quota approach will only generate revenues if quotas are sold. Requiring firms to pay for the right to pollute is consistent with a move to raise revenue via the taxation of 'bads' rather than 'goods'[31]. In the case of climate change, where understanding of the potential damage caused by emissions continues to improve, there is a strong argument for shifting the balance of taxation. In the case of tradable quotas, there are good economic reasons for moving towards greater use of auctioning over time, though the transition will need to be carefully managed – in particular, to ensure a robust revenue base.

14.8   Co-ordinating action across countries

The mitigation of climate change requires co-ordinated action across different countries. In thinking about the differences between tax and tradable quotas, it is therefore important to recognise the different implications they have for market-driven financial flows between countries.

Chapter 22 will explore the challenges in building up broadly similar price signals for carbon around the world. Issues of equity – as discussed in Chapter 2 – are likely to be central to creating frameworks that support this goal. It is therefore important to consider how taxes and tradable quota systems may differ in the relative ease with which they can drive financial flows between countries.

In theory, either a tax or a tradable quota system could drive financial flows from the developed to developing countries. Under a tax-based system, revenues raised will in the first instance flow to national governments. An additional mechanism would need to be put in place to transfer resources to developing countries.

Under a tradable-quota system, there are a number of ways that governments in rich countries can drive flows, either through direct purchase of quotas allocated to developing countries or through the creation of company-level trading where companies have access to credits for emissions reductions created in developing countries. In this case, financial flows between sectors and/or countries can occur automatically as carbon emitters search for the most cost-effective way of reducing emissions. The opportunities and challenges in these areas are discussed in detail in Chapters 22 and 23.

In summary, financial flows from developed to developing countries can occur under either a tax or tradable-quota system. However, market-driven financial flows will only occur automatically under the latter route, and only at sufficient scale if national quotas are set appropriately.

14.9   The performance of taxation and trading against principles of efficiency, equity and public finance considerations

In terms of the criteria discussed above – efficiency, equity and public finance – carbon taxes perform well against the efficiency and public finance criteria, as they:

- can contribute to establishing a consistent price signal across regions and sectors. However, this may prove difficult if a country perceives that it is acting in

---

[31] Were auctioning to substitute in whole or in part for taxation, it would be important to manage the revenue base to underpin the sustainability of the public finances.

isolation, and – as discussed in chapter 22 – there are many reasons why achieving a common price signal through harmonising taxes across countries is likely to be difficult to achieve;
- raise public revenues;
- can be kept stable, and thus do not risk fluctuations in the marginal costs that could increase the total costs of mitigation policy.

However,

- they do not automatically generate financial flows to developing countries in search of the most efficient carbon reductions.

In terms of the criteria discussed above – efficiency, equity and the impact on public finances – the strengths of a tradable quota scheme are:

- to the extent that the scheme embraces different sectors and countries, it will establish a common price signal and therefore have the potential to drive carbon reductions efficiently;
- to the extent that inter-country trading is allowed, it will ensure carbon reductions are made in the most cost-effective location, and automatically drive private-sector financial flows between regions;
- if allowances are sold or auctioned, then the scheme also has the potential to generate public revenues.

Some countries may make substantial use of tax measures to reduce GHG emissions. Others may place greater emphasis on participation in emissions trading schemes or, indeed, regulation. Some countries may choose a mix of all three depending on the sector, other policies, market structures, and political and constitutional opportunities and constraints.

The effectiveness of any tax or emissions trading scheme depends on its credibility and on good design. Investors need a credible and predictable policy framework on which to base their investment decisions; and good design is important to ensure effectiveness and efficiency. This is discussed in detail in the next chapter.

Carbon-pricing policy is only one element of a policy response to climate change. There are a range of other market failures and barriers to action which must be tackled. For this reason, carbon pricing policy should sit alongside technology policies, and policies to remove the behavioural barriers to action. These two further objectives are discussed in Chapter 16 and Chapter 17 respectively.

14.10   Conclusion – building policies for the future

A shared understanding of the long-term goals for stabilisation is a crucial guide to climate change policy-making: it narrows down strongly the range of acceptable emissions paths, and establishes a long-term goal for policy. But, from year to year, flexibility in when, where and how reductions are made will reduce the costs of meeting these goals. Policies should adapt to changing circumstances as the costs and benefits of climate change become clearer over time. This means that short-term policy may be revised periodically to take account of information, as and when it comes, so as to keep on track towards meeting a long-term goal.

This need for both a long-term goal, and consistent short-term policy to meet this, should guide action at the international and national level to price carbon.

At the international level, this means that the key policy objectives for tackling climate change should include:

- Choosing a policy regime that:
  i. in the long term, will stabilise the concentration of greenhouse gases in the atmosphere, and establish a long-term quantity goal to limit the risk of catastrophic damage;
  ii. in the short term, uses a price signal (tax or trading) to drive emission reductions, thus avoiding unexpectedly high abatement costs by setting short-term quantity constraints that are too rigid.
- Establishing a consistent price signal across countries and sectors to reduce GHG emissions. This price signal should reflect the damage caused by carbon emissions.

In theory, either taxes or tradable quotas – and in specific circumstances regulation – can play a role in establishing a common price signal. Chapter 22 discusses the potential difficulties of co-ordinating national policies to achieve this.

Both taxes and tradable quotas can contribute to raising public revenues. Under a tradable quota scheme, this depends on using a degree of auctioning and, over time, there are sound economic reasons for doing so. However, this would need to be well managed, understanding fully the implications for governments' revenue flows, and ensuring that these remain predictable and reliable.

Taxes and tradable quotas can both support the financing of carbon reductions across different countries. However, only a tradable-quota system will do this automatically, provided there is an appropriate initial distribution of quotas and structure of rules.

At the national – or regional level – governments will want to tailor a package of measures that suits their specific circumstances, including the existing tax and governance system, participation in regional initiatives to reduce emissions (eg. via trading schemes), and the structure of the economy and characteristics of specific sectors.

Some may choose to focus on regional trading initiatives, others on taxation and others may make greater use of regulation. The factors influencing this choice are discussed in the following chapter.

References

Useful background reading that summarises the debate on the use of price or quantity instruments are included in the list below. The seminal article by Weitzman (1974) is a technical exposition of the arguments. Pizer (2002) is good outline of the debate in terms of international climate change policy. Hepburn (2006) outlines a clear application of the Weitzman analysis to climate change policy, including the trade-off between credible commitments and flexibility.

Blyth, W. & K. Hamilton (2006): 'Aligning Climate and Energy Policy: Creating incentives to invest in low carbon technologies in the context of linked markets for fossil fuel, electricity and carbon', Energy, Environment and Development Programme, Royal Institute of International Affairs, April 21 2006, London

Coase, R.H. (1960): 'The problem of social cost', Journal of Law & Economics, 3 (1): 1–44

Dasgupta, P. and G. Heal (1979): 'Economic theory and exhaustible resources', Cambridge: Cambridge University Press

Helm, D., Hepburn, C., and Mash, R. (2005), 'Credible carbon policy', in Helm, D. (ed), Climate Change policy, Oxford, UK: Oxford University Press, chapter 14.

(Also available as Helm, D., Hepburn, C., & Mash, R. (2003), 'Credible carbon policy', Oxford Review of Economic Policy).

Hepburn, C. (2006): 'Regulating by prices, quantities or both: an update and an overview', Oxford Review of Economic Policy, 22(2): 226–247

Hotelling, H (1931): 'The economics of exhaustible resources', Journal of Political Economy, 39: 137–175

Lydon, P., (2002): 'Greenhouse warming and efficient climate protection policy, with discussion of regulation by "price" or by "quantity"', Working Paper 2002-5, Berkeley, CA: Institute of Governmental Studies.

Meade, J.E. (1951): 'External economies and diseconomies in a competitive situation', Economic Journal, Vol. 62, No. 245, 54–67

Newberry, D. (2005): 'Why tax energy? Towards a more rational policy', Energy Journal, 26(3): 1–40

Newell, R., Pizer, W. and Zhang, J (2005): 'Managing permit markets to stabilise prices', Journal of Environmental and Resource Economics, 31(2): 133–157

Pigou, A.C. (1920): 'The economics of welfare', Macmillan, London.

Pizer, W.A (2002): 'Combining price and quantity controls to mitigate global climate change', Journal of Public Economics, 85: 409–534

Sinclair, P.J.N (1994): 'On the optimum trend of fossil fuel taxation', Oxford Economic Papers, 46, 869–877

Ulph, A and Ulph, D (1994): 'The optimal time path of a carbon tax', Oxford Economic Papers, 46, 857–868

Weitzman, M.L (1974): 'Prices versus quantities', Review of Economic Studies, 41 (4):477–491

15 Carbon Pricing and Emissions Markets in Practice

KEY MESSAGES

Both tax and trading can be used to create an explicit **price for carbon;** and regulation can create an implicit price.

For all these instruments, **credibility, flexibility and predictability are vital to effective policy design.**

A lack of credible policy may undermine the effectiveness of carbon pricing, as well as creating uncertainties for firms considering large, long-term investments.

To establish the credibility of carbon pricing globally will take time. During the transition period, governments should consider how to deal with investments in long-lived assets which risk locking economies into a high-carbon trajectory.

To reap the benefits of emissions trading, deep and liquid markets and well designed rules are important. Broadening the scope of schemes will tend to lower costs and reduce volatility. Increasing the use of auctioning is likely to have benefits for efficiency, distribution and potentially the public finances.

Decisions made now on the third phase of the EU Emissions Trading Scheme pose an opportunity for the scheme to influence, and be the nucleus of, future global carbon markets.

The establishment of common incentives across different sectors is important for efficiency. The overall structure of incentives, however, will reflect other market failures and complexities within the sectors concerned, as well as the climate change externality.

The characteristics of different sectors will influence the design and choice of policy tool. Transaction costs of a trading scheme, for instance, will tend to be higher in sectors where there are many emission sources. The existing framework of national policies in these sectors will be an important influence on policy choice.

15.1 Introduction

This chapter considers how markets for emission reductions can be built on the principles considered in Chapter 14. The application of these principles requires careful analysis of the context of specific economies and institutional structures– at the national, international, regional or sectoral levels.

Section 15.2 discusses the importance of designing policies in a way which creates confidence in the future existence of a robust carbon price, so that businesses and individuals can plan their investment decisions accordingly. The current use of emissions trading schemes is discussed in Section 15.3, and 15.4 focuses particularly on the issues around creating a credible carbon price in emissions trading schemes.

The choice and design of such policy instruments also depends on the specific sectoral context. Policies which work for one sector may be inappropriate for another, although a common price is still needed across sectors for efficiency in the costs of mitigation. The relationship between climate change policy and other objectives, such as energy security and local air pollution, is also important. These issues are discussed in 15.5.

Carbon pricing is only one part of a strategy to tackle climate change. It must be complemented by measures to support the development of technologies, and to remove the barriers to behavioural change, particularly around take-up of energy efficiency. These two elements are discussed in Chapters 16 and 17.

15.2 Carbon pricing and investment decisions

Investors need a predictable carbon policy

Businesses always have to take uncertainties into account when making investment decisions. Factors such as the future oil price, changes in consumer demand, and even the weather can affect the future profitability of an investment. Business decision-makers make judgements on how these factors are likely to evolve over time.

But unlike many other uncertainties that firms face, climate change policy is created solely by governments. To be successful, a carbon pricing policy must therefore be based on a framework that enables investors to have confidence that carbon policy will be maintained over sequential periods into the future.

Serious doubt over the future viability of a policy, or its stringency, risks imposing costs without having a significant impact on behaviour, so increasing the cost of mitigation. Creating an expectation that a policy is very likely to be sustained over a long period is critical to its effectiveness.

Credibility, flexibility and predictability are key to effective policy

Three essential elements for an effective policy framework are credibility (belief that the policy will endure, and be enforced); flexibility (the ability to change the policy in response to new information and changing circumstances); and predictability (setting out the circumstances and procedures under which the policy will change). These apply to any type of policy, including the technology and regulatory measures set out in the following chapters, but are particularly pertinent to carbon pricing.

A key issue for credibility is whether the policy commands support from a range of interest groups. Public opinion is particularly important: sustained pressure from the public for action on climate change gives politicians the confidence to take measures which they might otherwise deem too risky or unpopular. It must also make sense within an international context: if there are good prospects for a robust international framework, this will greatly enhance the credibility of national goals for emissions reductions.

As Chapter 14 has discussed, the flexibility to adjust policy in the short term is an important principle for efficient pricing under conditions of uncertainty.

Policy must be robust to changing circumstances and changing knowledge. If policy is seen to be excessively rigid, its credibility may suffer, as people perceive a risk that it will be dropped altogether if circumstances change.

Building in predictable and transparent revision rules from the start is the best way to maintain confidence in the policy, whilst also allowing flexibility in its application.

Issues of credibility are particularly important for investments in long-lived capital stock

Taking a long-term view on the carbon price is particularly important for businesses investing in long-lived assets[1]. Assets such as power stations, industrial plant and buildings last for many decades, and businesses making investment decisions on these assets often have longer time horizons than many governments.

If businesses believe that carbon prices will rise in the long run to match the damage costs of emissions over time, this should lead them to invest in low-carbon rather than high-carbon assets. But in the transitional period, where the credibility of carbon pricing is being established worldwide, there is a risk that future carbon prices are not properly factored into business decision-making, and investments may be made in long-lived, high-carbon assets.

This could lock economies into a high-carbon trajectory, making future mitigation efforts more expensive. Governments should take careful account of this: as well as providing as much clarity as possible about future carbon pricing policies, they should also consider whether any additional measures may be justified to reduce the risks[2].

Uncertainty about the long-term future framework for carbon pricing is also a reason why additional measures to encourage the development of low-carbon technologies are important. This is discussed in Chapter 16.

Policy uncertainty not only undermines climate change policy – it can also undermine security of supply, by creating an incentive to delay investment decisions.

Uncertainty about the future existence or overall direction of policy creates difficulties for how businesses respond. There is a risk that businesses will adopt a 'wait and see' attitude, delaying their investment decisions until the policy direction becomes clearer.

Blyth and Yang (2006) look at the incentives for a company faced with a decision on whether to invest in high-carbon or low-carbon infrastructure. If a decision is expected at some point in the future about whether or not a new climate change policy will be introduced, a company which makes its investment decision now, risks a loss later if it makes the wrong call on policy. If it waits until the policy is agreed, it can make a more informed choice. Given this uncertainty, a much higher expected profit level would be required to trigger the investment now[3].

In the energy sector, such delays in investment could create serious problems for a country's security of supply. Modelling work by Blyth and Yang (2006) indicates that an increase in the period of relative carbon price stability from 5 to

---

[1]   See Helm et al (2005) which argues that credibility problems in recent UK energy and carbon policy have costs for meeting objectives on energy and climate change. The irreversibility of energy investments and the risk of governments reneging on commitments to carbon commitments imply a need for a more consistent policy framework.

[2]   Grubb et al (1995), Lecocq et al (1998).

[3]   See Blyth and Sullivan (2006)

10 years (which could equate to increasing the length of an allocation period in a trading scheme) could reduce the size of the investment thresholds arising from uncertainty by a factor of 2 or more[4].

Credibility may also vary between policy instruments

Credibility may vary between different types of policy instrument. For instance, taxation provides governments with a revenue stream, and there tends to be an expectation that it will not be in a government's interests to abolish it. Regulation may be more effective in countries with a culture of using command and control methods, or where there are political or administrative problems with raising taxes or with tax collection. Specific national circumstances, including constitutional structures, the stability of political institutions and the quality of legal infrastructures and enforcement, play a key role in determining what credible policy is.

Another important element is the level at which policy takes place. Regulation or trading schemes which are agreed at the EU level, for instance, are difficult to reverse, and hence may be seen as more credible than some national policies.

The issues surrounding credibility in trading schemes are discussed in detail in the following section.

15.3   Experience in emissions trading

As outlined in Chapter 14, emissions trading has several benefits. Emissions trading schemes can deliver least-cost emission reductions by allowing reductions to occur wherever they are cheapest. A key corollary benefit to this is that it generates automatic transfers between countries, while delivering the least-cost reductions. In many instances, introducing trading schemes is also an easier mechanism through which to achieve a common carbon price across countries than attempts to harmonise taxes. As such, trading schemes can be used to introduce carbon pricing, without risking carbon leakage and competitiveness implications between participating countries. Emissions trading is therefore a very powerful tool in the framework for addressing climate change at an international level.

Emissions trading is not new to environmental policy. Trading in emissions has been used to reduce sulphur dioxide and nitrous oxide emissions that cause acid rain in the US since 1995[5]. The experience of this scheme increased interest in the potential use of emissions trading to tackle climate change – particularly due to its potential cost effectiveness compared to the use of regulation. Burtaw (1996) estimated that emissions trading under the US Acid Rain Program saved 50% of the costs compared to command and control.

The use of carbon trading schemes is expanding

During the 1990s, as experience of emissions trading for air pollution grew in the US, the EU began to consider the potential of using trading to help meet its Kyoto target emission reduction obligations. The European Commission presented a 'Green Paper' in 2000 that proposed the use of emissions trading. It showed that

---

[4] See Blyth and Yang (2006)
[5] See www.epa.gov/airmarkets/arp/index.html for more detail on the US Acid Rain Program.

a comprehensive trading scheme could reduce compliance costs of meeting Kyoto by a third, compared to a scenario with no trading instrument[6].

The EU has since gone on to implement a trading scheme in major energy intensive and energy generation sectors, and in so doing, established the world's largest greenhouse gas emissions market. Launched in January 2005, the EU emissions trading scheme (EU ETS) is still in its infancy. The scheme will enter a second, longer phase in 2008, with a major review on the scheme's design from 2013 to be launched in 2007. Box 15.1 describes how the EU ETS works, and discusses the experience of the scheme to date.

BOX 15.1 The European Union Emissions Trading Scheme (EU ETS)

The EU ETS is the first international emissions trading scheme. It established a uniform price of carbon for greenhouse gas emissions from specific heavy industry activities in the 25 EU member states. Phase One of the scheme was launched on 1 Jaunary 2005 and runs to the end of 2007. Phase Two runs from 2008-12, and the scheme will continue with further phases beyond 2012. Participation is mandatory for emissions from industrial sectors specified in the scheme. These currently include energy generation, metal production, cement, bricks, and pulp and paper[7].

Member states decide, through their National Allocation Plans (NAPs), on the quota or total allocation of allowances for each phase within their country, and on how these are distributed between companies. The plans are subject to approval by the European Commission. They must demonstrate that allocation levels will not exceed expected emission levels in sectors, and are in line with broader plans to make reductions to meet Kyoto targets[8]. Allowances are then issued to all firms on the basis of the NAP. Firms in the scheme must provide an annual report on their emissions, which is audited by a third party.

In Phase One, the scheme covers less than 40% of all EU25 GHG emissions[9], with the permit market over the three-year period worth around US $115 billion[9a]. The majority of permits are currently allocated for free to installations included in the scheme (only 0.2% of all allowances will be auctioned in Phase One[10]), and most member states have prevented the banking of allowances between the two phases. An allowance market has developed through trade exchanges and brokers, with the City of London emerging as an important location for trading. Traded volumes have grown steadily (see below). The price of allowances has been in the range of €10 to €25 per tonne of CO_2 for most of the period, with a steep price drop in April 2006.

---

[6] The 2000 Green Paper estimated the cost of meeting Kyoto as €9 billion euros without trading, €7.2 billion with trading amongst energy producers only, €6.9 billion with trading among energy producers and energy intensive industry and €6 billion with trading among all sectors. See EC (2000).

[7] The scheme covers emissions from heat and energy use from installations of a particular size in these sectors. See EC (2003) for more detail on the scope of the EU ETS

[8] Articles 9 to 11 and Annex III of EU (2003) outline the criteria for allocation in the NAP

[9] Based on emission estimates for €4, 25 countries in WRI (2006)

[9a] This assumes around 2 billion tonnes of allowances are allocated each year for three years, and that the average allowance price is €19 (€15)

[10] Schleich and Betz (2005)

The market for EU allowances (EUAs) –prices and volumes

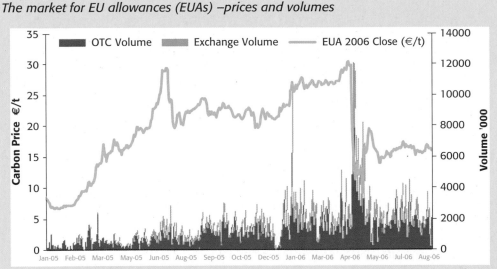

Source: Data taken from Point Carbon, www.pointcarbon.com

Early experience in the scheme has highlighted a number of important issues:

- **The potential for emissions trading schemes to generate demand for emissions reductions in developing countries**: the Linking Directive has enabled EU-based industry to purchase carbon reductions from the cheapest source, including projects and programmes being implemented in the developing world through the use of the Clean Development Mechanism[11]. This has driven growing interest of EU firms in the CDM market, particularly as CDM credits can be used in either phase of the scheme. The CDM market volume grew threefold between 2005 and 2006, to 374 million tonnes (CO_2e)[12], much of this driven by demand from the EU ETS[12].

- **The importance of long term confidence in the future of the scheme:** the EU ETS will continue with a third phase beyond 2012. But companies would like greater clarity over what the EU ETS will look like in Phase III and beyond in order to help judge the impact on their investment decisions. A survey to discover the issues that need to be considered in the review of the EU ETS put the need for certainty on future design issues in the scheme as a top priority[13]. The majority of those surveyed also stated they would prefer allocation decisions to be made a few years in advance of trading periods, and trading periods be lengthened to around 10 years.

- **The impact of imperfect information on prices:** at the start of trading in January 2005, traders had limited information on supply and demand for emission allowances. In particular, the NAPs did not contain clear data on the assumptions lying behind the

---

[11] The Clean Development Mechanism is one of the flexible mechanisms under the Kyoto Protocol. Its operation is discussed in detail in Chapter 23.

[12] Capoor and Ambrosi (2006) state that European and Japanese private entities dominated the buy-side of the CDM market in 2005 and 2006, taking up almost 90% of transacted project emissions credits.

[13] See McKinsey et al (2005) for details of the survey of governments, companies and NGO views on issues for the Review of the EU ETS. For UK companies, see also UKBCSE and The Climate Group (2006)

projections of emissions used as the basis for allocations. The release of the first data on actual emissions from the scheme's participants in April 2006 led to a sharp downward correction in prices (see figure above), as the data showed that the initial NAP allocations exceeded emissions in most sectors of the scheme[14]. The volatility that this caused demonstrates the importance of transparency in initial allocation plans.

- **The difficulties of ensuring scarcity in the market:** overall allocation in the EU ETS market is not set centrally. Rather, it is the sum of 25 individual member state decisions, subject to approval by the Commission. As such, total EU allocation is an outcome of many decisions at various levels, with a risk of gaming on allocation levels between member states if they make their decisions expecting allocation levels will be higher elsewhere in Europe. It has therefore been difficult to ensure scarcity in the EU ETS market. As a result, the total EU wide allocation in Phase One is estimated to be only 1% below projected "business as usual" emissions[15,16]. This underlines the need for stringent criteria on allocation levels for member states, and robust decisions by the European Commission on NAPs to ensure scarcity in the scheme.

- **The need for robust administrative systems:** the methods used to determine allocations placed considerable demands on companies to collect, verify and submit historical data on emissions.In addition, to ensure confidence in compliance standards across the EU on measuring emissions[17], companies had to set up monitoring, reporting and verification systems in line with EU guidelines[18]. Costs were high for small firms that had low annual emissions included in the EU ETS; requests to reconsider the minimum size of plants included in the scheme have subsequently been made by both member states and business.[19]

The growing importance of the use of emissions trading markets to price carbon is also illustrated by the scope of trading schemes planned or already operating across the world. Norway introduced emissions trading in January 2005 for major energy plants and heavy industry. New South Wales (Australia) already operates a mandatory baseline-and-credit scheme for electricity retailers. Japan and South Korea are also running pilot programmes for a limited number of companies.

Elsewhere, the biggest plans for new emissions trading markets are in the USA, through the Regional Greenhouse Gas Initiative (RGGI) from January 2009[20], and California's plans for using a cap and trade scheme from 2008[21]. Switzerland and Canada also plan to implement trading schemes as part of their programmes to meet Kyoto commitments.The voluntary market for carbon reductions is also growing, driven by demand from both companies and individuals looking to

---

[14] Grubb et al (2006)
[15] Grubb et al (2006)
[16] EC (2005)
[17] See Kruger and Egenhofer (2005). Also, some countries such as the UK went further asking firms to provide verification of data submitted by firms on historic emissions which were baselines for initial allocations.
[18] See EC (2004) for details of these guidelines.
[19] See Egenhofer and Fujiwara (2005)
[20] RGGI covers Connecticut, Delaware, Maine, New Hampshire, New Jersey, New York and Vermont. See www.rggi.org for more details.
[21] See announcements by the Governor of the State of California, www.climatechange.ca.gov

reduce or offset their emissions[22]. The CCX (Chicago Climate Exchange) is an example of a voluntary carbon market. Since December 2003, US based companies that take on voluntary targets to reduce GHG emissions have used this market to achieve their targets.

The following section outlines the design issues that impact on trading scheme efficiency and market effectiveness.

15.4 Designing efficient and well-functioning emissions trading schemes

To reap the benefits of emissions trading, deep and liquid markets and well-designed rules are important

Emissions trading schemes will, necessarily, deliver carbon prices that vary over time. But a degree of price stability through the emergence of a predictable average price within the emissions trading mechanism is important, particularly for businesses planning long-term investments. And the efficient operation of the scheme, including its impact on incentives, is important to achieve least-cost reductions.

One option to limit the bounds of price movements is to supplement the market instrument itself with price controls, such as formal price caps and price floors[23]. Although this approach has some attractions in principle, there are significant problems with its practical implementation and effectiveness, including the implications for the feasibility of linking with other schemes. These are set out in Box 15.2.

Fundamentally, to ensure confidence in a stable long-term carbon price, and to realise the full efficiency benefits of any trading scheme, the creation of deep, liquid and efficient markets is essential. Several factors can facilitate this:

- **Broadening the scope** of the scheme, to include more gases, more countries, and international credits;
- Ensuring appropriate **scarcity** in the system;
- **Lengthening the trading periods**, to provide longer-term confidence;
- Designing appropriate **allocation schemes**; and
- Promoting **transparency**.

The following sections discuss these in more detail.

Broadening the scope of the scheme will tend to lower costs and reduce volatility

In general, the deeper and more liquid a market, the harder it is for any individual trade to affect the overall price level, and hence the less volatile the market will tend to be. Introducing different economic sectors or countries to a market can also reduce the impact of a shock in any one sector on the scheme as a whole. In addition, the greater the degree of flexibility about what type of emissions reductions are made and where they are made, the lower the cost will be.

There are a number of ways to widen the scope of trading schemes. One is to widen the number of sectors and activities covered by an individual

---

[22] See Butzengeiger (2005) and Taiyab (2006) for more on markets for voluntary carbon offsets
[23] See, for instance, Pizer (2002) and Pizer (2005)

BOX 15.2 Price caps and floors in emission trading schemes

As explained in Hepburn et al (2006), a hybrid instrument can in principle be tailored to ensure that in the long term, an overall quantity ceiling is achieved, but that in the short term there is sufficient flexibility to avoid temporarily very high marginal abatement costs. This would help to achieve the balance of long-term certainty and short-term flexibility discussed in Chapter 14.

Price caps (or 'safety valves')[24] supply allowances on demand if the agreed ceiling price is hit, and would eliminate the risk of price spikes. Price floors would stop the carbon price from falling below a minimum level. They can be implemented in a number of ways, including through a levy that only becomes operational once the floor is breached, or by guaranteeing a minimum future quota price to emitters, by entering a contract to buy permits (which the government can then sell back to the market)[25] – although the risks to the public finances from this latter route should be taken seriously.

However, people would still have to believe that the caps and floors themselves will not be changed. There are also risks that the imposition of a cap alone would damage incentives for investing in low carbon technologies as it sets an upper limit on the future expected price, lowering potential returns to low carbon technology[26].

Importantly, the use of different price caps and floors in different schemes would compromise the efficiency of regional trading schemes- there are risks of carbon leakage and unintended transfers across jurisdictions with different carbon price ranges. As such, to operate efficiently, price caps and floors would need to be the same across all participating countries. Agreeing a common price cap or floor across countries is likely to suffer from the same difficulties as any attempt to harmonise carbon taxes more generally. Even if countries within a single scheme could agree a cap or floor, this would present an obstacle to linking to other schemes with different rules. This is a drawback to the practical applicability of these methods.

scheme. Some of the practical issues associated with this are discussed in Section 15.5 below.

Another is to offer access to flexible mechanisms such as Joint Implementation (JI) or the Clean Development Mechanism (CDM)[27]. This expands the options for generating credits for emissions reductions to most parts of the world, maximising the opportunities for efficiency. The environmental benefits of using these credits will depend on the credits representing a real reduction on what emission levels would otherwise have been (the 'business as usual' level of emissions). Countries that can generate CDM credits do not have binding caps on emissions, and are often fast changing economies; as such, establishing a credible estimate of what a business as usual baseline is, and whether reductions would have taken place in the absence of the CDM project, can be complex[28]. Chapter 23 examines this in more detail.

[24] See Jacoby and Ellerman (2004)
[25] Helm and Hepburn (2005)
[26] Blyth and Yang (2006) modelling shows that in principle, price caps and floors would reduce uncertainty on future prices, but as people need to believe that caps will stay, the impact is limited. Stronger effects on reducing uncertainty come from lengthening the period of price stability from 5 to 10 years as discussed above.
[27] These mechanisms are discussed fully in Chapter 23.
[28] The CDM Executive Board approves methodologies for baseline setting in CDM projects. See Chapter 23.

Linking different national or regional cap and trade schemes is also desirable on efficiency grounds, but, to reap the efficiency benefits, the schemes should be broadly similar in design. The practical issues of linking are discussed in Chapter 22.

The introduction of new sectors, and linking to new regions, can cause some short-term price instability, as there is uncertainty over the net impacts of newly included sectors and their response to the scheme. But the impact on long-term stability should still be positive.

As well as bringing extra depth and liquidity into markets, commonality or linking of schemes avoids the leakage, confusion and inefficiency of parallel schemes with different carbon prices. In any one area or country, a single or unified scheme is better than a proliferation of schemes.

The degree of scarcity in the market is important in determining prices

To facilitate more stable carbon markets, allocation levels should be consistent with overall national, regional or multilateral emissions reductions targets, and be clearly below expected 'business as usual' (BAU) emissions. This is complicated by the uncertainties in predicting future emissions over an entire trading period.

The first phase of the EU ETS illustrates this. Allocation decisions were based on projections of BAU emissions of the sectors in the scheme, many of which appear to have been overestimated, meaning that total EU allocation was just 1% under projections of BAU of the whole EU ETS. In contrast, earlier emissions trading schemes such as the US Sulphur Dioxide trading programme, had allocation levels at around 50% below baseline emissions[29].

The degree of scarcity in a scheme depends not just on the cap which is set for the scheme itself, but also on whether or not companies are permitted to use credits for emission reductions that are generated in areas without a cap, such as those from the CDM. As long as these credits represent real emission reductions, there is little reason to limit their use, as cost-efficiency demands that emissions reductions are made wherever this is cheapest.

If allowing the use of mechanisms such as the CDM turns out to deliver large quantities of low-cost reductions into a trading scheme, then, at the time when allocations for subsequent periods for the scheme are set, the cap may need to be tightened to ensure that the carbon price continues to reflect the social cost of carbon, and is consistent with the achievement of the long-term goal for stabilisation. The impact of CDM credits on the price should be considered alongside other emerging information on the costs and benefits, as part of the revision process for allocations.

Greater certainty on the evolution of prices over future trading periods, and banking and borrowing between periods, can help to smooth compliance over time and investment cycles

Longer trading periods in trading schemes can help to smooth compliance over time and investment cycles, as they allow the private sector to have greater control over the timing of the response to carbon policy. They also reduce policy risk to the extent that they suggest a deeper commitment to carbon policy. However,

---

[29] See Grubb and Neuhoff (2006) for a discussion of the use of projections and price volatility in the EU ETS.

excessively long commitment periods limit policymakers' flexibility in responding to changing information and circumstances. As the previous chapter discussed, this is important in order to keep down the overall costs of carbon pricing to the economy[30], and to readjust targets as more information on climate change itself is gathered.

The key issues for investor confidence are a commitment to the long-term future of the scheme and predictability in its overall shape and rules. This predictabilty can be achieved through establishing revision rules for future allocation periods. For instance, governments may announce that future allocations will be contingent on factors such as the price of permits in the preceding period. They could also announce a target range for prices[31] (which should be in line with the expected trajectory for the social cost of carbon – see Chapter 13). Setting out expectations on issues such as expansion to new sectors, or the use of CDM, could also be important. These principles could be set over a very long time period of perhaps 10 to 20 years, with allocations made at more regular intervals.

Within this framework, banking, and possibly borrowing, can be used to create links between different phases of a trading scheme. Banking is the ability to carry over unused quotas from one period to another, and borrowing the ability to use or purchase quotas from a future period in the current period. This allows trading to take place across commitment periods, as well as across sectors and countries. This can improve flexibility, as well as reducing the risk of price spikes or crashes at the end of trading periods discussed above.

Some existing emission trading schemes already allow banking. Banking should help to encourage early emission reductions where this is more cost effective[32]. For example, the heavy use of banking in the US Acid Rain Program has been seen by some as a success in terms of delivering early reductions and improving efficiency. Ellerman and Pontero (2005) found that 30% of allowances were banked between 1995-99 (Phase One of the programme). Firms made efficient decisions to make earlier reductions and bank allowances forward, due to the expectation of tighter caps in future phases. As a result, in total, emissions reduced in Phase One were twice that required to the meet the cap.

In contrast, very few existing emissions trading schemes have made use of borrowing. The main reason why borrowing has been restricted in existing trading schemes is credibility and compliance, including the risk of borrowing simply being offset by compensating increases in allocations in future periods. In theory, unrestricted borrowing could delay emissions reductions indefinitely, thus raising the risk of 'overshooting' a long run quantity ceiling. A credible enforcement strategy, and long-term principles for allocation, are therefore essential to ensure that reductions borrowed from the future are real and delivered.

Where there are longer periods within which compliance is possible, and a clearer view of the longer term direction of carbon policies, liquid futures markets in carbon are more likely to emerge, and hedging instruments will be developed that allow firms to manage price uncertainty more systematically.

---

[30] Helm and Hepburn (2006)
[31] See Newell et al (2005) for an example of how such revision rules could work.
[32] However, unrestricted banking can also allow emissions to be concentrated in time (Tietenberg,1998) – and such hoards of emissions could have high associated damage costs compared to dispersed emissions.

The choice and design of allocation methodology is an important determinant of both efficiency and distributional impact

Permits in an emissions trading scheme can be allocated for free, or sold (usually, though not necessarily, through auction[33]). It is possible to combine these – for instance, the EU ETS allowed for up to 5% of permits to be auctioned in Phase One, and 10% in Phase Two.

In principle and assuming perfect competition, free allocation and auctioning should both be equally efficient. In both cases, businesses face the same marginal costs arising from the emission of an extra tonne of carbon dioxide, and should therefore make the same decision on whether or not to emit in either case.

But this argument is static, ignores the structure of markets and takes no account of distributional or public finance issues. In reality the methods differ in two important respects. First, free allocation methodologies can dampen incentives to incorporate the cost of carbon into decision making consistently, and distort competition. Thus they slow adjustment and potentially raise the overall cost of compliance.

Second, they differ in their distributional impact. Free allocations give companies lump sum transfers in the form of carbon allowances; depending on market structure and demand.Such transfers may result in windfall profits. Not surprisingly, free permits are generally favoured by existing players in an industry. Auctioning leads to financial transfers to governments, which may have benefits for the public finances, depending on whether this is a new revenue flow or a substitute for other sources of finance.

These issues were raised in the preceding chapter, and are explored in the next two sections.

Free allocations can significantly distort incentives

There are a number of reasons why emissions trading schemes based on free allocation may distort incentives for emissions reductions:

- If there is an expectation that the baseline year upon which free allocations are based will be updated, participants have incentives to invest in dirty infrastructure and emit more now to get more free allowances in the future[34]. A one-off allocation based on past emissions (or grandfathering) over all trading periods is one way of avoiding this. However, as a trading scheme matures, the relevance of past emission levels may become a less and less relevant basis for the likely emissions of each plant, say ten or more years later.
- Free allocations can act as a disincentive to new entry to a market, restricting competition and reducing efficiency. If incumbents receive free allowances, but new plants must purchase allowances, free allocations directly create barriers to entry, meaning that the provision of free allocations for new plants may be required[35]. In turn, the rules for free allocations to new plants may indirectly

---

[33] The discussion in this section assumes that the sale of permits to industry would happen through auctioning. Other methods are also possible, such as direct sales; these are not discussed fully here, but would be subject to some of the same arguments.

[34] Neuhoff et al (2006) also find that in an international emissions trading scheme, if updating is used in one country but not others, it equates to free riding by the country that uses updating.

[35] In an international trading scheme, if one country has free allowances for new plants, there are compeitiveness implications if other countries do not. This logic drove all 25 EU member states chose to set aside of allowances for new entrant plants that total around 5% of all EU allowances.

distort incentives: if allocations are given in proportion to the expected emissions from the new plant, they may reward higher-carbon technologies[36].

- There may also be disincentives to exit from markets. The existence of 'use it or lose it' closure rules, which mean a plant must be open in order to receive free allowances, may prevent the closure of inefficient plants.This would mean emission levels are higher than if plants could keep allowances if they shut down, or had no free allowances to begin with[37].

- Under auctioning, with no lump sum of free allowances, businesses will face upfront costs in buying permits to cover their emissions. This will tend to bring management attention to the importance of making efficient decisions that fully account for the cost of carbon. Free allocations may not have the same behavioural impact, delaying adjustments to making effective decisions on carbon compliance[38].

Free allocation methodologies can therefore seriously reduce the dynamic efficiency of a trading scheme, making the cost of reductions higher in the longer term than would otherwise be the case.

Benchmarking the emissions needed for efficient low carbon technologies for both existing and new plants is an alternative basis for issuing free allocations. It offers the opportunity to more clearly 'reward' clean technologies, and penalise carbon intensive technology by developing an average 'rate' of emissions for particular fuels, technologies or plant sizes. The more standardised a benchmark is, the more effective benchmarking is likely to be[39]. Benchmarking can also be used specifically for new entrants, by allocating on the basis of the most efficient technologies available[40].

Auctioning can avoid many of the incentive problems associated with free allocation, although good design is necessary to avoid introducing new inefficiencies. Small, frequent auctions may be more effective in limiting any market power that may exist in the permit market[41]. In principle, to ensure an efficient outcome, the auction method should promote competition and participation for small as well as larger emitters. While one auction at the beginning of the permit period may minimise administration costs, it may also carry a risk of larger players buying the majority of permits and extracting oligopoly rents in the secondary permit market. More frequent auctions also allow for all players to adjust bids and learn from experience of early auctions, and may be helpful in promoting price stability[42]. Given the administrative costs of the data required for free allocation methodologies, auctioning may also offer lower administrative costs to both firms and governments.

---

[36] Modelling of the UK electricity sector in Neuhoff et al (2006), demonstrates that free allowances for new plant using high carbon technologies, can increase overall emissions. The existence of a 'use it or lose it' closure rule for EU ETS allocations will reduce plant retirement rates and reduce investment in new plants, causing higher emission levels.

[37] In the EU ETS, most member states had 'use it or lose it' closure rules, mainly due to the rules for free allocation to new plants. In Germany, a 'transfer rule' allowed allowances from old plants to be retained if a new plant was built. This still risks new plants receiving higher allocation levels than needed.

[38] Hepburn et al (2006a)

[39] Neuhoff et al (2006) show that for generation plants in the EU ETS, benchmarks based on plant capacity as opposed to fuel and technology specific benchmarks are the least distorting.

[40] The use of benchmarking on the basis of low carbon technology emission rates is an option and has been used in the EU ETS NAPs of some member states. See DTI (2005) for an example of the use of benchmarks for 'new entrant' plants in the UK

[41] Hepburn et al (2006a) considers auction design in the EU ETS

[42] Hepburn et al (2006a)

Using free allocation has benefits for managing the transition to emissions trading, but risks creating substantial windfall profits

Free allocations and auctioning have very different distributional impacts. This has led to a debate over whether allocation methods will affect the profitability of firms, as well as the implications for competitiveness. Carbon pricing will most affect the operating costs of energy intensive industries that compete in international markets, such as non-ferrous metals and some chemicals sectors (see Chapter 11). In the first instance, as auctioning and free allocation both impose the same marginal cost on emissions (as the carbon price is the same), the profit maximising quantity and price for any company should be the same in each case, and there should be no impact on the fundamental risks to competitiveness from the choice of allocation method.

There is, however, an important difference in terms of the impact on companies' balance sheet, which may have competitiveness implications[43]. A firm with free allocations that competes against other firms who face the cost of carbon but do not have free allowances, would be in an advantageous direct position in the sense that it receives a subsidy. It could for example, use this to capture market share by a period of low prices. However, if a firm competes against other firms who do not face a cost of carbon, the 'subsidy' of free allowances may be used to maintain its competitiveness, rather than gain competitive advantage over other firms.

This subsidy effect means that free allocations may have an important role to play in managing the transition to carbon pricing. Full auctioning imposes an immediate hit on companies' balance sheets equivalent to the full cost of all their emissions, whereas free allocation means that companies only have to pay for the cost of any additional permits they need to purchase. This difference in upfront costs may be important, particularly for firms that have significant sunk costs in existing assets and need to invest in lower-carbon assets in response.

In terms of the impact on firms' profits, free or purchased allowances are one factor influencing whether firms face profit or losses from the introduction of a trading scheme. Emissions trading increases the marginal costs of production, but the extent to which firms have to internalise these costs and therefore suffer reduced profits, will depend on:

- whether they can pass on costs to consumers (which depends on market structure and the shape of the demand curve for the good);
- whether they have ways of reducing emissions themselves which are cheaper than buying allowances (cost effective abatement); and
- whether they have some free allowances that can compensate for increased marginal costs

A firm that receives free allowances equal to its existing emissions can make the same profits as before from unchanged production activities, provided the market price for its output is unchanged – or do still better by responding to the new price for carbon. What happens to the market price for its product will depend on industrial structure.

---

[43] Smale et al (2006) show that marginal cost increases from the EU ETS most affects the competitiveness of the aluminium sector as it competes in a very global market, and does not get free allowances to compensate-the aluminium sector is currently not directly covered by the scheme, but still faces higher electricity prices.

If firms are in perfectly competitive markets, the increase in marginal costs from emissions trading will be fully reflected in prices to consumers, and (in the absence of abatement) profits will stay the same as before the scheme's introduction. Any free allowances they receive equate to windfall profits[44]. But where firms operate in markets where there is international competition and/or very elastic demand and so are unable to pass on costs, free allowances can act to maintain profitability by compensating for the increasing operating costs and reduced revenue that may be necessary to maintain market share[45].

Nevertheless, whatever the market structure, it is important that free allocations are only temporary. They may be necessary to manage a transition, but if permanently used, they would distort competition and emission reductions will be below their efficient levels.

The creation of robust institutions, and the collection and provision of reliable information, are important for efficiency

Price stability can also be encouraged by the provision of robust information. In particular, transparent and regular information on actual emissions of scheme participants, as well as on the intial allocations, will help to reveal the basis of market demand and supply.

The importance of information of this kind is illustrated by the experience of the EU ETS when the first verified emissions data of installations included in the scheme were published in March 2006. As Box 15.2 showed, prices dropped sharply in response, as it was clear that, for many firms, actual emissions were well below the number of allowances given to them at the start of the scheme. Revealing information on actual emissions more regularly through the trading period would help limit this volatility. Such requirements for more frequent information releases would, however, impose additional costs on emitters, implying that these requests may need to be limited to the largest emitters.

The quality of monitoring, reporting and verification standards is integral to confidence in a trading scheme. A transparent and well enforced system of measuring and reporting emissions is crucial for securing the environmental credibility of a scheme as well as free trade across plants. Monitoring, reporting and verification (MRV) rules ensure that a tonne of carbon emitted or reduced in one plant is equal to a tonne of carbon emitted or reduced in a different plant[46].

Just as these issues are important in national and regional emissions trading schemes, the emergence of a liquid and efficient global carbon market has similar requirements. Indeed, to facilitate such a market, the EU and others wanting to develop global emissions trading will need to build on existing institutions to develop trading infrastructure. The World Bank emphasises that this includes ensuring strong legal bases to enforce compliance in the jurisdictions of participating

---

[44] Sijm et al (2006) show that in the EU ETS, free allocation to electricity generation companies has created substantial windfall profits while consumers have faced increased electricity prices to reflect allowance costs.

[45] To maintain profits, commentators state various levels of free allocation as necessary, they need not be 100%. See, for instance, work by Bovenberg and Goulder (2001), Smale et al (2006), Vollebergh et al (1997), Quirion (2003) on allocation and profitability. Also Hepburn et al (2006b) provide a generalised theoretical framework, including an analysis of asymmetric market structure and apply this to four EU ETS sectors.

[46] Kruger and Egenhofer (2005)

firms and agreeing on minimum standards for monitoring, reporting and verification of emissions. Institutions that can deliver predictable and transparent information for emissions markets will also be vital, as will general oversight on the transparency of financial services that support trading such as securities, derivative products or hedge funds[47].

Drawing out implications for the future of the EU emissions trading scheme

The EU ETS will continue beyond 2012 with a third phase. The details of Phase III have yet to be determined, and will be considered in the European Commission's review of the EU ETS in 2007. The review will propose developments in the scheme, drawing on the experience of the EU ETS to date. In particular, it will consider the expansion of the scheme to other sectors (including transport) and links to other trading schemes.

Decisions made now on the third phase of the scheme that will run post 2012, pose an opportunity for the EU ETS – the most important emissions trading market – to influence other emerging markets, as well as to be the nucleus of future global carbon markets. Based on the analysis in this section, there are certain key principles to consider in taking the EU ETS scheme forward. These are set out in Box 15.3.

BOX 15.3 Principles for the future design of the EU ETS

A credible signal

- Setting out a **credible long-term vision** for the overall scheme over the next few decades could boost investor's confidence that carbon pricing will exist in the EU going forward

- The overall EU limit on emissions should be set at a level that **ensures scarcity** in the allowance market. Stringent criteria for allocation volumes across all EU sectors are necessary.

- To realise efficiency in the scheme, and minimise perverse incentives, there should be a move to **greater use of auctioning** in the longer term, although some free allocation may be important to manage short-term transitional issues[48].

- Where free allocation is necessary, standardised **benchmarking** is a better alternative to grandfathering and updating.

A deep and liquid market

- **Clear and frequent information on emissions** during the trading period would improve the efficient operation of the market, reducing the risks of unnecessary price spikes.

- Clear and predictable **revision rules** for future trading periods, with the possibility of **banking** between periods, would help smooth prices over time, and improve credibility

- **Broadening participation** to other major industrial sectors, and to sectors such as aviation, would help deepen the market[49].

- Enabling the EU ETS to **link with other emerging trading schemes** (including in the USA and Japan) could improve liquidity as well as establish the ETS scheme as the nucleus of a global carbon market.

---

[47] Capoor and Ambrosi (2006)
[48] See Neuhoff et al (2006) for more on free allocation and perverse incentives in the EU ETS
[49] See Environment Agency (2006) for more detail on expansion options in the EU ETS.

- **Allowing use of emission reductions from the developing world** (such as the CDM or its successor) can continue to benefit both the efficiency of the EU scheme as well as the transfer of low carbon technology to the developing world

15.5   Carbon pricing across sectors of the economy

Abatement costs are minimised when the carbon price is equalised across sectors

As discussed in Chapter 9, sectors vary widely in terms of the current availability and average cost of abatement options. The cost of avoiding deforestation, for instance, appears to be relatively low compared with the cost of many low-carbon power generation options; by contrast, in aviation, although there are some opportunities for efficiency gains, options for technology switching are currently very limited.

As discussed in the previous chapter, to minimise the total cost of abatement, the carbon price (whether explicit via a tax or trading instrument, or implicit via regulation) should be equalised across sectors. When the carbon price is applied to sectors with cheap abatement options, initially, emissions will tend to decline more; when applied to sectors with more expensive abatement options, the degree of abatement will be less than in cheaper abatement sectors. At the same time, the price increase for the output of the latter sectors will be, and should be, greater.

This means that from an efficiency perspective, sectors with expensive abatement options should not be excluded from carbon pricing; but neither should they be subject to a different higher carbon price in that sector in order to achieve abatement.

As well as carbon pricing, governments should also look at the use of technology policies and efficiency policies across sectors – these are considered in the following two chapters. It is also important to consider climate change policy within the context of meeting other policy objectives within sectors, including its interaction with the treatment of externalities such as local air pollution and congestion.

The overall structure and scale of policy incentives will therefore reflect other market failures and complexities within the sectors concerned, as well as the climate change externality. As economies make the transition to full carbon pricing, they may in practice use a mix of instruments.

How the characteristics of different sectors affect choice and design of instrument

The characteristics of sectors may influence the choice and design of the carbon pricing instrument. The underlying economic structures in which the emitters operate in sectors will differ, with implications for the attractiveness of using tax, trade or regulation instruments.

Some of the relevant features of different sectors include:

- Transaction costs: this may be affected by the number and dispersion of emitters, and the institutional arrangements for monitoring and pricing.

- Carbon leakage: this is the risk that emissions-intensive activity moves to an area not subject to a carbon constraint. The choice and design of an instrument may have implications for carbon leakage and competitiveness.
- Distributional impacts: depending on the market structure of the sector, the choice of policy instrument may have different implications for who bears the cost.
- Existing frameworks: policy choices will be influenced by existing national policy frameworks and regulatory structures.

It is also important to consider where in the value chain to price carbon. If "upstream" emissions are priced (for instance, at the power station or oil refinery), it is not necessary to price "downstream" emissions as well (for instance, in domestic buildings or individual vehicles). However, Chapter 17 focuses particularly on policies to enable investments in energy efficiency by the end-user, which are not discussed separately here.

The following sections analyse how these factors influence policy choice in power and heavy industry, road transport and aviation, and agriculture.

Power and heavy industry

At a global level, power and heavy industry (such as iron and steel, cement, aluminium, paper industries and chemical and petrochemicals) are large emitters. Because of their high carbon intensity, these sectors are likely to be very sensitive to carbon pricing. They typically invest in very long-lived capital infrastructure such as power plant or heavy machinery, so a clear indication of the future direction of carbon pricing policy is particularly important to them.

Power markets in particular are characterised by imperfect market structures, including state monopolies, regulatory constraints, and often large-scale subsidies. The interaction of carbon pricing with these imperfections is complex. Other industries such as paper and chemicals are more decentralised and deregulated. But overall, sources of emissions are concentrated amongst a relatively few, large, stationary installations, where emissions can be effectively measured and monitored.

The concentrated nature of emissions from these sources make them, in principle, well suited to emissions trading. As already discussed, the first and second phases of the EU ETS cover emissions from these sectors. Other trading schemes have a similar focus – the Regional Greenhouse Gas Initiative in the north-east of the USA, for instance, will cover only the power sector.

However, trading is not the only option. Tax could also be an effective mechanism, and would have the advantage of providing greater price predictability. Examples of countries using taxation to meet climate change goals in these sectors include the UK, which has used the Climate Change Levy, a revenue-neutral mechanism which encourages emissions reductions across sectors including industry; and Norway, which introduced a carbon tax in the early 1990s, covering much of its heavy industry as well as the transport sector (Box 15.4).

Heavy industries compete in international markets, and as Chapter 11 illustrated, there are some risks to competitiveness and of carbon leakage from the use of carbon policy in such sectors. In terms of tax and trading instruments, there may be a difference in impact if taxes cannot be harmonised globally. This is because an international trading scheme imposes a uniform carbon price across

BOX 15.4 A carbon tax in practice: Norway[50]

Like other Scandinavian countries, Norway introduced a carbon tax in the early 1990s. The tax was to form part of substantial shift in fiscal policy as Norway aimed to use the revenue generated by environmental taxes to help reduce distorting labour taxes.

The Norwegian carbon tax initially covered 60 percent of all Norwegian energy related CO_2 emissions. There are several sectors that were exempted from the tax, including cement, foreign shipping, and fisheries. Natural gas and electricity production are also exempt, although virtually all Norway's electricity production is from carbon-free hydro-electric power. Partial exemptions apply to sectors including domestic aviation and ship-ping, and pulp and paper.

The tax generates substantial revenues; in 1993 the tax represented 0.7 percent of total revenue, which by 2001 had increased to 1.7 percent. The tax is estimated to have reduced CO_2 emissions by approximately 2.3% between 1990 and 1999[51]. Overall in Norway, between 1990-1999 GDP grew by approximately 23 percent, yet emissions only grew by roughly 4 percent over the same period, indicating a decoupling of emissions growth from economic growth.

There is also some evidence that the tax helped to provide incentives for technological innovation. The Sleipner gas field is one of the largest gas producers in the Norwegian sec-tor of the North Sea. The gas it produces contains a higher CO_2 content than is needed for the gas to burn properly. With the imposition of a carbon tax the implied annual tax bill to Statoil, the state oil company, was approximately $50m for releasing the excess CO_2. This induced Statoil researchers to investigate the storing of excess carbon dioxide in a nearby geological formation. After several years of study, a commercial plant was installed on the Sleipner platform in time for the start of production in 1996. Experience with this plant has made an important contribution to the understanding of carbon capture and storage technology.

However, there have been some difficulties in the implementation of the tax:

● The impact of the tax on industry was weakened because of numerous exemptions put in place because of competitiveness concerns. This created a complex scheme, and blunted the incentive for industry to modify or upgrade existing plants.

● The carbon tax did not reflect the actual level of carbon emitted from fuels. For instance, low and high-emission diesel fuels are taxed at the same level, despite causing different levels of environmental damage.

● Although Norway, Sweden, Finland and Denmark all put carbon taxes in place in the early 1990s, they have not been able to harmonise their approaches – demonstrating the difficulties of co-ordinating tax policy internationally, even amongst a relatively small group of countries.

countries, minimising competitiveness implications for countries within the scheme, whereas taxes may impose different costs in different countries.

Regulatory measures have not played a major role in these sectors, although these have been used for other pollutants in the power sector, the EU's Large Combustion Plants Directive being one example. The concentrated number of

[50] This draws on Ekins and Barker (2001)
[51] Bruvoll and Larsen (2002)

companies and sources of emissions may make formal or informal sectoral agreements on best practice an effective complement to carbon pricing – this is discussed in Chapter 22.

Road transport

Although the production of fuel for road transport is centralised at oil refineries, most of the emissions from road transport come from a very large number of individual cars and other vehicles. Demand for transport tends to rise with income. There is considerable scope to improve efficiency in the sector, although the responsiveness of demand to price is low, and breakthrough technologies such as hydrogen are still some years away.

Many countries currently levy a road transport fuel tax. Fuel taxes are a close proxy for a carbon tax because fuel consumption closely reflects emissions. They are frequently aimed at other externalities at the same time (discussed further below), and have the advantage of providing a steady revenue stream to the government. Another example is taxes on purchase or annual car taxes, which can be calibrated by the efficiency of the vehicle.

However, it is also possible to use emissions trading in the road transport sector (see Box 15.5). A possible risk of including road transport in an emissions trading scheme is that permit prices and oil prices might move in tandem, thus exacerbating the extent of oil price fluctuations facing the motorist (in contrast to taxes,

---

BOX 15.5 Ways to include road transport in an emissions trading scheme

There are three main ways in which emissions from road transport could be included in an emissions trading scheme; they differ according to whom the permits are allocated to.

- Motorists. Individual motorists would have to surrender permits whenever they purchased fuel. Quantity instruments might be better than prices at encouraging motorists to reduce their consumption of fuel. However, there would probably be high transaction costs associated with this approach.

- Refineries. Refineries located in the region of the scheme, would have to buy permits to cover the emissions generated when the fuel that they produce is used in vehicles. It would probably be necessary to couple this approach with border adjustments to the price of imported fuel to avoid carbon leakage. Border adjustments are discussed in detail in Chapter 22.

- Manufacturers. Vehicle manufacturers would be faced with a target for fuel efficiency of the average vehicle sold and, to the extent that they exceeded this target, they would have to buy permits to cover the excess expected lifetime carbon emissions from fuel inefficient vehicles. However, future emissions from these vehicles would be uncertain, making this hard to reconcile with trading schemes based on actual emissions.

The European Commission is currently reviewing the operation of the EU ETS, including whether it should be extended to include other sectors such as road transport.

The inclusion of aviation, road, rail and maritime could increase the size of the EU ETS by up to 50% (such that the EU ETS would cover around 55% of total EU 25 greenhouse emissions, and a larger proportion of total CO_2 emissions), with benefits for liquidity[52].

---

[52] Estimates based on emission estimates for EU 25 in 2000 from WRI (2006).

which are levied as a fixed amount rather than a percentage of fuel price charged, meaning that the fuel price is prone to less variation).

Regulatory measures play an important role in the transport sectors in many countries. Vehicle standards – which may be mandatory or voluntary – can put an implicit value on carbon, by restricting the availability of less efficient vehicles. These measures are discussed in more detail in Chapter 17.

In practice, a combination of policies may be justified. Existing policy frameworks and institutional structures in countries will be an important determinant of policy choice. Countries with a history of high fuel taxes, for instance, would need to think very carefully about the public finance implications of switching to trading with free allocations; voluntary standards might be very effective in countries with a strong tradition of co-operation between government and business, but much less so in countries with a different culture.

As in other sectors, climate change is not the only market failure in the transport sector and there are important interactions with other policy goals. Congestion, for instance, imposes external costs on other motorists by increasing their journey time. Congestion pricing and carbon pricing are very similar approaches from an economic point of view - they both price for an externality. Congestion charging could have a positive or negative impact on carbon emissions from transport, depending on how the instrument is designed and level at which the charge is set.

Aviation

Aviation faces some difficult challenges. Whilst there is potential for incremental improvements in efficiency to continue, more radical options for emissions cuts are very limited. The international nature of aviation also makes the choice of carbon pricing instrument complex. Internationally coordinated taxes are difficult to implement, since it is contrary to International Civil Aviation Organisation (ICAO) rules to levy fuel tax on fuel carried on international services[53]. The majority of the many bilateral air service agreements that regulate international air services also forbid taxation of fuel taken on board. Partly for this reason, levels of taxation in the aviation sector globally are currently low relative to road transport fuel taxes. This contributes to congestion and capacity limits at airports – a form of rationing, which is an inefficient way of regulating demand.

While either tax or trading would, in principle, be effective ways to price emissions from this sector, the choice of tax, trading or other instruments is likely to be driven as much by political viability as by the economics. Chapter 22 will discuss further the issues of international co-ordination of policy in this area (as well as in shipping, which faces similar issues). A lack of international co-ordination could lead to serious carbon leakage issues, as aircraft would have incentives to fuel up in countries without a carbon price in place.

The level of the carbon price faced by aviation should reflect the full contribution of emissions from aviation to climate change. As outlined in Box 15.6, the impact of aviation on the global warming (radiative forcing) effect is expected to be two to four times higher than the impact of the CO_2 emissions alone by 2050. This should be taken into account, either through the design of a tax or trading scheme, through both in tandem, or by using additional complementary measures.

---

[53] Article 24 of Chicago Convention exempts fuel for international services from fuel duty. See ICAO (2006).

BOX 15.6 The impact of aviation on climate change

Aviation CO_2 emissions currently account for 0.7 Gt CO_2[54] (1.6% of global GHG emissions). However the impact of aviation on climate change is greater than these figures suggest because of other gases released by aircraft and their effects at high altitude. For example, water vapour emitted at high altitude often triggers the formation of condensation trails, which tend to warm the earth's surface. There is also a highly uncertain global warming effect from cirrus clouds (clouds of ice crystals) that can be created by aircraft.

In 2050 under 'business as usual' projections, CO_2 emissions from aviation would represent 2.5% of global GHG emissions[55]. However taking into account the non-CO_2 effects of aviation would mean that it would account for around 5% of the total warming effect (radiative forcing) in 2050[56].

The uncertainties over the overall impact of aviation on climate change mean that there is currently no internationally recognised method of converting CO2 emissions into the full CO2 equivalent quantity.

Agriculture and land use

Agricultural emissions come from a large number of small emitters (farms), over three quarters of which are in developing and transition economies. Emissions from agriculture depend on the specific farming practices employed and the local environment conditions. Since the sources tend to be distributed, there would be high transaction costs associated with actual measurement of GHG at the point of emission.

An alternative approach in this sector would be to focus on pricing GHG emission 'proxies'. For example, excessive use of fertiliser or high nutrient livestock feeds is associated with high emissions, but by appropriate pricing, emissions can be reduced. However in practice, in many developing countries fertiliser is actually subsidised, largely to support the incomes of farmers. In many countries it is a somewhat regressive subsidy, as it is the richer farmers or agribusinesses who gain most.

Difficulties associated with measuring emissions are also the reason why it is difficult to incorporate GHG emissions from agriculture into a trading scheme. However there are examples of projects that have overcome these problems and enabled farmers who adopt sustainable agriculture practices, to sell their emission savings on to others via voluntary schemes; this issue is discussed further in Chapter 25.

Inadequate water pricing can intensify the problems of weak fertiliser pricing, since water and fertiliser are complementary inputs – additional fertiliser works much better with stronger irrigation.

---

[54] WRI (2005).

[55] Aviation BAU CO2 emissions in 2050 estimated at 2.3 GtCO2, from WBCSD (2004). Total GHG emissions in 2050 estimated at 84 GtCO2e (for discussion of how calculated, see Chapter 7).

[56] IPCC (1999). This assumes that the warming effect (radiative forcing) of aviation is 2 to 4 times greater than the effect of the CO2 emissions alone. This could be an overestimate because recent research by Sausen et al (2005) suggests the warming ratio is closer to 2. It could be an underestimate because both estimates exclude the highly uncertain possible warming effects of cirrus clouds.

Many countries have adopted regulation of agricultural practices. For example, regulations for the use of water in growing rice, the quantity and type of fertiliser used in crop production, or the treatment of manure. Regulations are often location specific, because local conditions influence best practice. However, in developing countries, enforcement of regulations can be difficult because they may not have the institutional structures or resources to allocate to this task. Better pricing of inputs is generally a preferable route: income support to poor farmers or agricultural workers can be organised in much better ways than subsidised inputs.

There are complex challenges involved with the inclusion of deforestation, the major cause of land use emissions, in carbon trading schemes. These are discussed in detail in Chapter 25.

15.6   Conclusions

Chapter 14 discussed how, at the global level, policymakers need both a shared understanding of a long-run stabilisation goal, and the flexibility to revise short-run policies over time.

At the national – or regional level – policy makers will want to achieve these goals in a way that builds on existing policies, and creates confidence in the future existence of a carbon price. In particular, they will want to assess how carbon pricing (through either taxation, tradable quotas or regulation) will interact with existing market structures, and existing policies (for instance, to encourage the development of renewable energy or petrol taxes).

Governments will want to tailor a package of measures that suits their specific circumstances. Some may choose to focus on regional trading initiatives, others on taxation and others may make greater use of regulation. The key goal of policy should be to establish common incentives across different sectors, using the most appropriate mechanism for a particular sector. With market failures elsewhere, other objectives, and the costs of adjustment associated with long-lived capital, it will be important to look at both the simple price or tax options, as well as quotas and regulation to see what incentives in particular sectors really work.

Carbon pricing is only one element of a policy approach to climate change. The following two chapters discuss the role of technology policy, and policies to influence attitudes and behaviours, particularly in regard to energy efficiency. All three elements are important to achieve lowest cost emissions reductions.

References

A number of useful background readings on issues covered in this chapter are worth noting.

A general approach to uncertainty and investment is covered in 'Investment under uncertainty', by Avinash K. Dixit, A and Robert S. Pindyck, Princeton University Press, 1994.

A broad discussion of the technical and economic issues of emissions trading and existing schemes is in 'Act Locally, Trade Globally:Emissions Trading for Climate Policy', by Richard Baron and Cedric Philibert, IEA, 2005. Detailed analysis of the EU Emissions Trading Scheme is covered in a special issue on EU ETS in Climate Policy, Volume 6, No.1, 2006.

A useful summary of issues for decisions on including economic sectors in an emissions trading scheme is illustrated by presentations at a Stern Review seminar on 'Taxes versus trade in the transport sector', June 2006 (publication forthcoming at www.sternreview.org.uk).

Blyth, W., and M. Yang (2006): 'The effect of price controls on investment incentives', presentation to the Sixth Annual Workshop on Greenhouse Gas Emission Trading, Paris: IEA/IETA /EPRI, September 2006, available from
http://www.iea.org/Textbase/work/2006/ghget/Blyth.pdf
Blyth, W. and M. Yang (forthcoming), 'Impact of climate change policy uncertainty on power generation investments', Paris: IEA.
Blyth, W. and R. Sullivan (2006): 'Climate change policy uncertainty and the electricity industry:implications and unintended consequences', Energy, Environment and Development Programme, Briefing Paper 06/02, London: Royal Institute of International Affairs.
Bovenberg, A.L. and L.H. Goulder (2000): 'Neutralizing the adverse impacts of CO_2 abatement policies: what does it cost?', Discussion Paper, 00-27, Washington DC: Resources for the Future.
Bruvoll, A. and B. M. Larsen (2002): 'Greenhouse gas emissions in Norway – Do carbon taxes work?', Discussion Paper 337, Norway: Research Department of Statistics.
Burtraw, D. (1996): 'Cost savings sans allowance trades: evaluating the SO_2 emissions trading program to date", Washington, DC: Resources for the Future.
Butzengeiger, S., (2005): 'Voluntary compensation of greenhouse gas emissions: selection criteria and implications for the international climate policy system', HWWI Policy Paper, Hamburg: Hamburg Institute of International Economics.
California State Government (2006): 'Governor Schwarzenegger signs landmark legislation to reduce greenhouse gas emissions', Press Release 09/27/2006, GAAS:684:06, , Los Angeles: Office of the Governor of the State of California.
Capoor, K. and P. Ambrosi (2006): 'State and Trends of the Carbon Market 2006', Washington, DC: World Bank.
Department of Trade and Industry (2005): 'EU Emissions Trading Scheme: calculating the free allocation for new entrants', Report produced for the Department of Trade and Industry, AEAT, London: DTI.
EC (2000): 'Green Paper on greenhouse gas emissions trading with the European Union (presented by the Commission), Brussels: EC.
EC (2003): 'Directive 2003/87/EC of the European Parliament and of the Council of 13 October 2003 establishing a scheme for greenhouse gas emission allowance trading within the Community and amending Council Directive 96/61/EC', Official Journal L 275, 25/10/2003 P. 0032–0046
EC (2004): 'Commission Decision of 29 January 2004 establishing guidelines for the monitoring and reporting of greenhouse gas emissions pursuant to Directive 2003/87/EC of the European Parliament and of the Council', European Commission, Brussels: EC.
EC (2005): 'Emissions trading: Commission approves last allocation plan ending NAP marathon', Press Release IP/05/762, 20/06/2005, European Commission, Brussels: EC, available from http://europa.eu/rapid/pressReleasesAction.do?reference=IP/05/762&format=HTML&aged=0&language=EN&guiLanguage=en
Egenhofer, C. and N. Fujiwara (2005): 'Reviewing the EU Emissions Trading Scheme- Priorities for short-term implementation of the second round of allocation: Part 1',
Report of a CEPS Task Force, Brussels, Centre for European Policy Studies.
Ekins, P. and T. Barker (2001): 'Carbon taxes and carbon emissions trading', Journal of Economic Surveys, 2001, **15**(3)
Ellerman, A. and J. Pontero (2005): 'The Efficiency and Robustness of Allowance Banking in the US Acid Rain Programme', Working Paper 0505, Centre for Energy and Environmental Policy Research, Massachusetts Institute of Technology, Massachusetts
Environment Agency (2006): LETS Update: decision makers summary report, Bristol: Environment Agency.
Grubb, M. (2006): 'Climate change impacts, energy and development: annual bank conference on development economics', Tokyo, 2006.
Grubb, M., T. Chauis and M. Ha-Duong (1995) : 'The economics of changing course: implications of adaptability and inertia for optimal climate policy', Energy Policy **23**(4): 1–14
Grubb, M. and K. Neuhoff (2006): 'Allocation and competitiveness in the EU emissions trading scheme: policy overview', Climate Policy **6** (2006): 7–30
Helm, D. and C. Hepburn (2005): 'Carbon contracts and energy policy: an outline proposal', Oxford: Oxford Economic Papers.

Helm, D., C. Hepburn, and R. Mash (2005): 'Credible carbon policy', in Helm, D. (ed), Climate Change policy, Oxford: Oxford University Press, Chapter 14. (Also available as Helm, D., Hepburn, C., and Mash, R. (2003), 'Credible carbon policy', Oxford Review of Economic Policy) .

Hepburn, C., M. Grubb, K. Neuhoff (2006a): 'Auctioning of EUETS Phase II allowances: how and why?', Climate Policy **6** (2006): 135-158

Hepburn, C., J. Quah and R. Ritz (fortcoming 2006b): 'Emissions trading and profit-neutral grandfathering', Oxford University Economics Department, Discussion Paper, Oxford: Oxford University.

International Civil Aviation Authority (2006): 'Convention on International Civil Aviation', Doc 7300/9, Ninth Edition, Montreal: ICAO, 2006.

Intergovernmental Panel on Climate Change (1999): 'Aviation and the global atmosphere - summary for policy makers', available at www.ipcc.ch/pub/av(e).pdf .

Jacoby, H.D. and A.D Ellerman (2004): 'The safety valve and climate policy', Energy Policy, **32** (4) 2004: 481-491

Kruger, J and C. Egenhofer (2005): 'Confidence through compliance in emissions trading markets', Paper prepared for the International Network for Environmental Compliance and Enforcement (INECE) workshop, November 15-18 2005, Washington, DC: INECE.

Lecocq F. , J-C.Hourcade, and M.Haduong (1998): 'Decision-making under uncertainty and inertia constraints: implications of the when flexibility', Energy Economics, **20** (1998): 539-555.

McKinsey and Ecofys (2005): 'Review of the EU Emissions Trading Scheme: survey highlights', DG Environment, Brussels: European Commission.

Neuhoff, K., K. Keats, and M. Sato (2006): 'Allocation, incentives and distortions: the impact of the EU ETS emissions allowance allocations to the electricity sector', Climate Policy **6** (2006): 71-89

Newell, R., W. Pizer, and J. Zhang, (2005): 'Managing permit markets to stabilize prices', Environmental and Resource Economics, **31**: 133-157

Pizer, W.A (2002): 'Combining price and quantity controls to mitigate global climate change', Journal of Public Economics, 85: 409- 534.

Pizer, W.A (2005): 'Climate policy design under uncertainty', Washington, DC: Resources for the Future,

Quirion, P., (2003): 'Allocation of CO_2 allowances and competitiveness: A case study on the European iron and steel industry', mimeo

Hepburn, C., Quah, J., & Ritz, R. (forthcoming 2006): 'Emissions trading and profit neutral grandfathering', Oxford University, Economics Department, Discussion Paper, Oxford: Oxford University.

Sausen, R., I. Isaksen, V. Grewe et al. (2005): 'Aviation radiative forcing in 2000: an update on IPCC (1999)', Meteorologische Zeitschrift, **14**(4): 555-561

Schleich, J. and R. Betz (2005): 'Incentives for Energy Efficiency and Innovation in the European Emissions Trading System', Stockholm: European Council for an Energy Efficient Economy.

Sijm, J., K. Neuhoff, Y. Chen (2006): 'CO_2 cost pass-through and windfall profits in the power sector', Climate Policy **6** (2006): 49–72

Smale, R., M. Hartley, C. Hepburn (2006) : 'The impact of CO_2 emissions trading on firm profits and market prices', Climate Policy **6** (2006): 31-48

Taiyab, N., (2006): 'Exploring the market for voluntary carbon offsets', Markets for Environmental Services Series, Number 8, London: IIED.

Tietenberg, T. (1998): 'Tradable Permits and the Control of Air Pollution in the United States' Colby College, Department of Economics, Working Paper.

UK Business Council for Sustainable Energy and The Climate Group (2006): 'Business views on International Climate and Energy Policy, April 2006, London: UKBCSE.

Vollebergh, H. DeVries & Koutstaal, P., (1997): 'Hybrid carbon incentive mechanisms and political acceptability', Environmental and Resource Economics, **9**: 43-63

World Business Council for Sustainable Development (2004): 'Mobility 2030: Meeting the Challenges to Sustainability', Geneva: WBCSD.

World Resources Institute (2005): 'Navigating the Numbers', Washington DC: WRI.

World Resources Institute (2006): Climate Indicators Tool (CAIT) on-line database, available from the World Resources Institute: http://cait.wri.org

16 Accelerating Technological Innovation

KEY MESSAGES

Effective action on the scale required to tackle climate change requires a widespread shift to new or improved technology in key sectors such as power generation, transport and energy use. Technological progress can also help reduce emissions from agriculture and other sources and improve adaptation capacity.

The private sector plays the major role in R&D and technology diffusion. But closer collaboration between government and industry will further stimulate the development of a broad portfolio of low carbon technologies and reduce costs. Co-operation can also help overcome longer-term problems, such as the need for energy storage systems, for both stationary applications and transport, to enable the market shares of low-carbon supply technologies to be increased substantially.

Carbon pricing alone will not be sufficient to reduce emissions on the scale and pace required as:

- Future pricing policies of governments and international agreements should be made as credible as possible but cannot be 100% credible.

- The uncertainties and risks both of climate change, and the development and deployment of the technologies to address it, are of such scale and urgency that the economics of risk points to policies to support the development and use of a portfolio of low-carbon technology options.

- The positive externalities of efforts to develop them will be appreciable, and the time periods and uncertainties are such that there can be major difficulties in financing through capital markets.

Governments can help foster change in industry and the research community through a range of instruments:

- **Carbon pricing**, through carbon taxes, tradable carbon permits, carbon contracts and/or implicitly through regulation will itself directly support the research for new ways to reduce emissions;

- **Raising the level of support for R&D** and demonstration projects, both in public research institutions and the private sector;

- **Support for early stage commercialisation investments in some sectors.**

Such policies should be complemented by tackling institutional and other non-market barriers to the deployment of new technologies.

These issues will vary across sectors with some, such as electricity generation and transport, requiring more attention than others.

Governments are already using a combination of market-based incentives, regulations and standards to develop new technologies. These efforts should increase in the coming decades.

Our modelling suggests that, in addition to a carbon price, **deployment incentives for low-emission technologies should increase two to five times globally** from current levels of around $33 billion.

Global public energy R&D funding should double, to around $20 billion, for the development of a diverse portfolio of technologies.

16.1   Introduction

Stabilisation of greenhouse gases in the atmosphere will require the deployment of low-carbon and high-efficiency technologies on a large scale. A range of technologies is already available, but most have higher costs than existing fossil-fuel-based options. Others are yet to be developed. Bringing forward a range of technologies that are competitive enough, with a carbon price, for firms to adopt is an urgent priority.

In the absence of any other market failures, introducing a fully credible carbon price path for applying over the whole time horizon relevant for investment would theoretically be enough to encourage suitable technologies to develop. Profit-maximising firms would respond to the creation of the path of carbon prices by adjusting their research and development efforts in order to reap returns in the future. This chapter sets out why this is unlikely to be sufficient in practice, why other supporting measures will be required, and what form they could take.

This chapter starts by examining the process of innovation and how it relates to the challenge of climate change mitigation, exploring how market failures may lead to innovation being under-delivered in the economy as a whole. Section 16.3 looks more closely at the drivers for technology development in key sectors related to climate change. It finds that clean energy technologies face particularly strong barriers – which, combined with the urgency of the challenge, supports the case for governments to set a strong technology policy framework that drives action by the private sector.

Section 16.4 outlines the policy framework required to encourage climate related technologies. Section 16.5 discusses one element of this framework – policies to encourage research, development and demonstration. Such policies are often funded directly by government, but it is critical that they leverage in private sector expertise and funding.

Investment in Research and Development (R&D) should be complemented by policies to create markets and drive deployment, which is discussed in Section 16.6. A wide range of policies already exist in this area; this section draws together evidence on what works best in delivering a response from business.

A range of complementary policies, including patenting, regulatory measures and network issues are also important; these issues are examined in Section 16.7.

Regulation is discussed in the context of mitigation more generally, and in particular in relation to energy efficiency in Chapter 17.

Overall, an ambitious and sustained increase in the global scale of effort on technology development is required if technologies are to be delivered within the timescales required. The decline in global public and private sector R&D spending should be reversed. And deployment incentives will have to increase two to five-fold worldwide in order to support the scale of uptake required to drive cost reductions in technologies and, with the carbon price, make them competitive with existing fossil fuel options. In Chapter 24, we return to the issue of technological development, considering what forms of international co-operation can help to reduce the costs and accelerate the process of innovation.

16.2   The innovation process

Innovation is crucial in reducing costs of technologies. A better understanding of this complex process is required to work out what policies may be required to encourage firms to deliver the low-emission technologies of the future.

Defining innovation

Innovation is the successful exploitation of new ideas[1]. Freeman identified four types of innovation in relation to technological change[2]:

- Incremental innovations represent the continuous improvements of existing products through improved quality, design and performance, as has occurred with car engines;
- Radical innovations are new inventions that lead to a significant departure from previous production methods, such as hybrid cars;
- Changes in the technological systems occur at the system level when a cluster of radical innovations impact on several branches of the economy, as would take place in a shift to a low-emission economy;
- Changes of techno-economic paradigm occur when technology change impacts on every other branch of the economy, the internet is an example.

Many of the incentives and barriers to progress for these different types of technological change are very different from each other.

Innovation is about much more than invention: it is a process over time

Joseph Schumpeter identified three stages of the innovation process: invention as the first practical demonstration of an idea; innovation as the first commercial application; and diffusion as the spreading of the technology or process throughout the market. The traditional representation of the diffusion process is by an S-shaped curve, in which the take-up of the new technology begins slowly, then 'takes off' and achieves a period of rapid diffusion, before gradually slowing down as saturation levels are reached. He proposed the idea of 'creative destruction' to

[1]   DTI (2003)
[2]   Freeman (1992)

describe the process of replacement of old firms and old products by innovative new firms and products.

There is an opportunity for significant profits for firms as the new product takes off and this drives investment in the earlier stages. High profits, coupled with the risk of being left behind, can drive several other firms to invest through a competitive process of keeping up. As incumbent firms have an incentive to innovate in order to gain a competitive advantage, and recognising that innovation is typically a cumulative process that builds on existing progress, market competition can stimulate innovation[3]. As competition increases, and more firms move closer to the existing technological frontier of incumbents, the expected future profits of the incumbents are diminished unless they innovate further. Such models imply a hump-shaped relationship between the degree of product market competition and innovation, as originally suggested by Schumpeter.

An expanded version of this 'stages' model of innovation that broadens the invention stage into basic R&D, applied R&D and demonstration is shown in the subsequent figure. In this chapter the term R&D will be used but this will also cover the demonstration stage[4]. The commercialisation and market accumulation phases represent early deployment in the market place, where high initial cost or other factors may mean quite low levels of uptake.

This model is useful for characterising stages of development, but it fails to capture many complexities of the innovation process, so it should be recognised as a useful simplification. A more detailed characterisation of innovation in each market can be applied to particular markets using a systems approach[6]. The transition

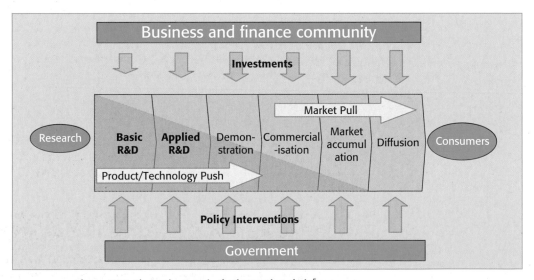

Figure 16.1 The main steps in the innovation chain[5]

[3] Aghion et al (2002): Monopolists do not have competitive pressures to innovate while intense competition means firms may lack the resource or extra profit for the innovator may be competed away too quickly to be worthwhile.
[4] R,D&D (Research, Development and Demonstration) can be used for this but it can lead to confusion over the final D as some of the literature uses deployment or diffusion in the same acronym.
[5] Grubb (2004)
[6] For an excellent overview of innovation theory see Foxon (2003)

between the stages is not automatic; many products fail at each stage of development. There are also further linkages between stages, with further progress in basic and applied R&D affecting products already in the market and learning also having an impact on R&D.

Experience curves can lead to lock-in to existing technologies

As outlined in Section 9.7 dynamic increasing returns, such as economies of scale and learning effects, can arise during production and lead to costs falling as production increases. These vary by sector with some, such as pharmaceuticals, experiencing minimal cost reductions while others fall by several orders of magnitude. These benefits lead to experience curves as shown in Box 9.4.

Experience curves illustrate that new technologies may not become cost effective until significant investment has been made and experience developed. Significant learning effects may reduce the incentive to invest in innovation, if companies wait until the innovator has already proven a market for a new cost effective technology. This is an industry version of a collective action problem with its associated free-rider issues.

Dynamic increasing returns can also lead to path dependency and 'lock-in' of established technologies. In Figure 16.2, the market dominant technology (turquoise line) has already been through a process of learning. The red line represents a new technology, which has the potential to compete. As production increases the cost of the new technology falls because of dynamic increasing returns, shown by the red line below. In this case, the price of the new technology does ultimately fall below the level of the dominant technology. Some technological progress can also be expected for incumbent dominant technologies but existing deployment will have realised much of the learning[7].

The learning cost of the new technology is how much more the new technology costs than the existing technology; shown by the dotted area where the red line is above the blue. During this period, the incumbent technology remains cheaper, and

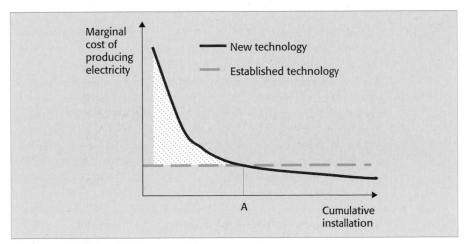

Figure 16.2 Illustrative experience curve for a new technology

---

[7]   The learning rate is the cost reduction for a doubling of production and this requires much more deployment after significant levels of investment.

the company either has to sell at a loss, or find consumers willing to pay a premium price for its new product. So, for products such as new consumer electronics, niche markets of "early adopters" exist. These consumers are willing to pay the higher price as they place a high value on the function or image of the product.

The learning cost must be borne upfront; the benefits are uncertain, because of uncertainty about future product prices and technological development, and come only after point A when, in this case, the technology becomes cheaper than the old alternative. If, as is the case in some sectors, the time before the technology becomes competitive might span decades and the learning costs are high, private sector firms and capital markets may be unwilling to take the risk and the technology will not be developed, especially if there is a potential free-rider problem.

Innovation produces benefits above and beyond those enjoyed by the individual firm ('knowledge spillovers'); this means that it will be undersupplied

Information is a public good. Once new information has been created, it is virtually costless to pass on. This means that an individual company may be unable to capture the full economic benefit of its investment in innovation. These knowledge externalities (or spillovers) from technological development will tend to limit innovation.

There are two types of policy response to spillovers. The first is the enforcement of private property rights through patenting and other forms of protection for the innovator. This is likely to be more useful for individual products than for breakthroughs in processes or know-how, or in basic science. The disadvantage of rigid patent protection is that it may slow the process of innovation, by preventing competing firms from building on each others' progress. Designing intellectual property systems becomes especially difficult in fields where the research process is cumulative, as in information technology[8]. Innovation often builds on a number of existing ideas. Strong protection for the innovators of first generation products can easily be counterproductive if it limits access to necessary knowledge or research tools for follow-on innovators, or allows patenting to be used as a strategic barrier to potential competitors. Transaction costs, the equity implications of giving firms monopoly rights (and profits) and further barriers such as regulation may prevent the use of property rights as the sole incentive to innovate. Also much of value may be in tacit knowledge ('know-how' and 'gardeners' craft') rather than patentable ideas and techniques.

Another broad category of support is direct government funding of innovation, particularly at the level of basic science. This can take many forms, such as funding university research, tax breaks and ensuring a supply of trained scientists.

Significant cross-border spillovers and a globalised market for most technologies offer an incentive for countries to free-ride on others who incur the learning cost and then simply import the technology at a later date[9]. The basic scientific and technical knowledge created by a public R&D programme in one country can spillover to other countries with the capacity to utilise this progress. While some of the leaning by doing will be captured in local skills and within local firms, this may not be enough to justify the learning costs incurred nationally.

---

[8] Scotchmer (1991)
[9] Barreto and Klaassen (2004)

International patent arrangements, such as the Trade Related International Property Rights agreement (TRIPs[10]), provides some protection, but Intellectual Property Right (IPR) can be hard to enforce internationally. Knowledge is cheap to copy if not embodied in human capital, physical capital or networks, so R&D spillovers are potentially large. A country that introduces a deployment support mechanism and successfully reduces the cost of that technology also delivers benefits to other countries. Intellectual property right issues are discussed in more detail in Section 23.4.

International co-operation can also help to address this by supporting formal or informal reciprocity between RD&D programmes. This is explored in Chapter 24.

Where there are long-term social returns from innovation, it may also be undersupplied

Government intervention is justified when there is a departure between social and private cost, for example, when private firms do not consider an environmental externality in their investment decisions, or when the benefits are very long-term (as with climate change mitigation) and outside the planning horizons of private investments. Private firms focus on private costs and benefits and private discount rates to satisfy their shareholders. But this can lead to a greater emphasis on short-term profit and reduce the emphasis on innovations and other low-carbon investments that would lead to long-term environmental improvements.

16.3   Innovation for low-emission technologies

The factors described above are common to innovation in any sector of the economy. The key question is whether there are reasons to expect the barriers to innovation in low-emission technologies to be higher than other sectors, justifying more active policies. This section discusses factors specific to environmental innovation and in particular two key climate change sectors – power generation and transport.

Lack of certainty over the future pricing of the carbon externality will reduce the incentive to innovate

Environmental innovation can be defined[11] as innovation that occurs in environmental technologies or processes that either control pollutant emissions or alter the production processes to reduce or prevent emissions. These technologies are distinguished by their vital role in maintaining the 'public good' of a clean environment. Failure to take account of an environmental externality ensures that there will be under-provision or slower innovation[12].

In the case of climate change, a robust expectation of a carbon price in the long term is required to encourage investments in developing low-carbon technologies. As the preceding two chapters have discussed, carbon pricing is only in its

---

[10] The agreement on Trade Related Intellectual Property Rights (TRIPs) is an international treaty administered by the World Trade Organization which sets down minimum standards for most forms of intellectual property regulation within all WTO member countries.
[11] Taylor, Rubin and Nemet (2006)
[12] Anderson et al (2001); Jaffe, Newell and Stavins (2004) and (2003)

infancy, and even where implemented, uncertainties remain over the durability of the signal over the long term. The next chapter outlines instances in which regulation may be an appropriate response to lack of certainty. This means there will tend to be under-investment in low-carbon technologies. The urgency of the problem (as outlined in Chapter 13) means that technology development may not be able to wait for robust global carbon pricing. Without appropriate incentives private firms and capital markets are less likely to invest in developing low-emission technologies.

There are additional market failures and barriers to innovation in the power generation sector

Innovation in the power generation sector is key to decarbonising the global economy. As shown in Chapter 10, the power sector will need to be at least 60% decarbonised by 2050[13] to keep on track for greenhouse gas stabilisation trajectories at or below 550ppm CO_2e.

For reasons that this section will explore the sector is characterised by low levels of research and development expenditure by firms. In the USA, the R&D intensity (R&D as a share of total turnover) of the power sector was 0.5% compared to 3.3% in the car industry, 8% in the electronics industry and 15% in the pharmaceutical sector[14]. OECD figures for 2002 found an R&D intensity of 0.33% compared to 2.65% for the overall manufacturing sector[15]. Unlike in many other sectors, public R&D represents a significant proportion, around two thirds of the total R&D investment[16].

The available data[17] on energy R&D expenditure show a downward trend in both the public and private sector, despite the increased prominence of energy security and climate change. Public support for energy R&D has declined despite a rising trend in total public R&D. In the early 1980s, energy R&D budgets were, in real terms, twice as high as now, largely in response to the oil crises of the 1970s.

Private energy R&D has followed a similar trend and remains below the level of public R&D. The declines in public and private R&D have been attributed to three factors. *First*, energy R&D budgets had been expanded greatly in the 1970s in response to the oil price shocks in the period, and there was a search for alternatives to imported oil. With the oil price collapse in the 1980s and the generally low energy prices in the 1990s, concerns about energy security diminished, and were mirrored in a relaxation of the R&D effort. Recent rises in oil prices have not, yet, led to a significant increase in energy R&D. *Second*, following the liberalisation of energy markets in the 1990s, competitive forces shifted the focus from long-term investments such as R&D towards the utilisation of existing plant and deploying well-developed technologies and resources - particularly of natural gas for power and heat, themselves the product of R&D and investment over the previous three decades. *Third*, there were huge declines in R&D expenditures on nuclear power following the experiences of many countries with cost over-runs, construction delays, and the growth of

---

[13] This is consistent with the ACT scenarios p86 IEA, 2006 which would also require eliminating land use change emissions to put us on a path to stabilising at 550ppm CO_2e
[14] Alic, Mowery and Rubin (2003)
[15] Page 35: OECD, (2006)
[16] There are doubts as to the accuracy of the data and the IEA's general view is that private energy R&D is considerably higher than public energy R&D (though this still represents a significant share).
[17] Page 33–37: OECD (2006)

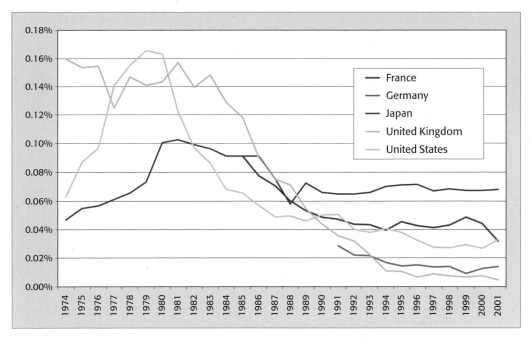

Figure 16.3 Public energy R&D investments as a share of GDP[18]

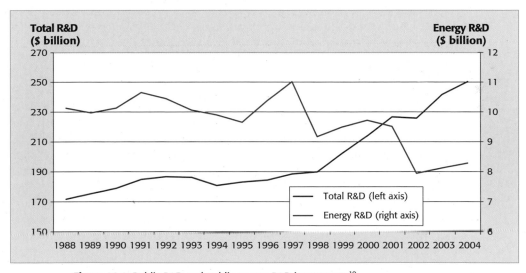

Figure 16.4 Public R&D and public energy R&D investments[19]

public concerns about reactor safety, nuclear proliferation and nuclear waste disposal. In 1974, electricity from nuclear fission and fusion accounted for 79% of the public energy R&D budget; it still accounts for 40%. Apart from nuclear technologies, energy R&D budgets decreased across the board (Figure 16.8).

[18] Source: IEA R&D database http://www.iea.org/Textbase/stats/rd.asp Categories covered broken down in IEA total Figure 16.8
[19] OECD countries Page 32: OECD (2006)

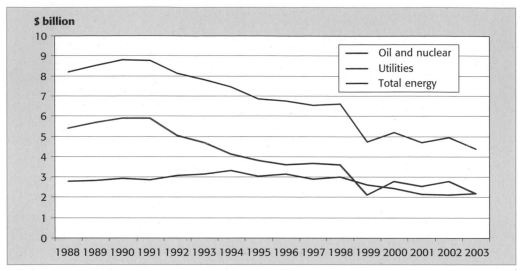

Figure 16.5 Trends in private sector energy R&D[20]

The sector's characteristics explain the low levels of R&D

There are a number of ways to interpret these statistics, but they suggest that private returns to R&D are relatively low in the sector. There are four distinct factors which help explain this.

The first factor is the nature of the learning process. Evidence from historical development of energy-related technologies shows that the learning process is particularly important for new power generation technologies, and that it typically takes several decades before they become commercially viable. Box 9.4 shows historical learning curves for energy technologies.

If early-stage technologies could be sold at a high price, companies could recover this learning cost. In some markets, such as IT, there are a significant number of 'early adopters' willing to pay a high price for a new product. These 'niche markets' allow innovating companies to sell new and higher-cost products at an early profit. Later, when economies of scale and learning bring down the cost, the product can be sold to the mass market. Mobile phones are a classic example. The earliest phones cost significantly more but there were people willing to pay this price.

In the absence of niche markets the innovating firm is forced to pay the learning cost, as a new product can be sold only at a price that is competitive with the incumbent. This may mean that firms would initially have to sell their new product at a loss, in the hope that as they scale up, costs will reduce and they can make a profit. If this loss-making period lasts too long, the firm will not survive.

In the power sector, niche markets are very limited in the absence of government policy, because of the homogeneous nature of the end-product (electricity). Only a very small number of consumers have proved willing to pay extra for carbon-free electricity. As cost reductions typically take several decades this leaves a significant financing gap which capital markets are unable to fill. Compounding this, the power generation sector also operates in a highly regulated environment and tends to be risk averse and wary of taking on technologies that may prove

---

[20] Source Page 35 OECD (2006); For US evidence see Kammen and Nemet (2005)

costlier or less reliable. Together, these factors mean that energy generation technologies can fall into a 'valley of death', where despite a concept being shown to work and have long-term profit potential they fail to find a market.

For energy technologies, R&D is only the beginning of the story. There is continual feedback between learning from experience in the market, and further R&D activity. There is a dependence on tacit knowledge and a series of incremental innovations in which spillovers play an important role and reduce the potential benefits of intellectual property rights. This is in strong contrast with pharmaceutical sector. For a new drug, the major expense is R&D. Once a drug has been invented and proven, comparatively little further research is required and limited economies of scale and learning effects can be expected.

The second factor is infrastructure. National grids are usually tailored towards the operation of centralised power plants and thus favour their performance. Technologies that do not easily fit into these networks may struggle to enter the market, even if the technology itself is commercially viable. This applies to distributed generation as most grids are not suited to receive electricity from many small sources. Large-scale renewables may also encounter problems if they are sited in areas far from existing grids. Carbon capture and storage also faces a network issue, though a different one; the transport of large quantities of CO_2, which will require major new pipeline infrastructures, with significant costs.

The third factor is the presence of significant existing market distortions. In a liberalised energy market, investors, operators and consumers should face the full cost of their decisions. But this is not the case in many economies or energy sectors. Many policies distort the market in favour of existing fossil fuel technologies[21], despite the greenhouse gas and other externalities. Direct and indirect subsidies are the most obvious. As discussed in Section 12.5 the estimated subsidy for fossil fuels is between $20-30 billion for OECD countries in 2002 and $150-250 billion per year globally[22]. The IEA estimate that world energy subsidies were $250 billion in 2005 of which subsidies to oil products amounted to $90 billion[23]. Such subsidies compound any failure to internalise the environmental externality of greenhouse gases, and affect the incentive to innovate by reducing the expectations of innovators that their products will be able to compete with existing choices.

Finally, the nature of competition within the market may not be conducive to innovation. A limited number of firms, sometimes only one, generally dominate electricity markets, while electricity distribution is a 'natural' monopoly. Both factors will generally lead to low levels of competition, which, as outlined in Section 16.1, will generally lead to less innovation as there is less pressure to stay ahead of competitors. The market is also usually regulated by the government, which reduces the incentive to invest in innovation if there is a risk that the regulator may prevent firms from reaping the full benefits of successful innovative investments.

[21] Neuhoff (2005).
[22] Source: REN21 (2005) which cites; UNEP & IEA. (2002). Reforming Energy Subsidies. Paris. www.uneptie.org/energy/publications/pdfs/En-SubsidiesReform.pdf Also Johansson, T. & Turkenburg, W. state in (2004). Policies for renewable energy in the European Union and its member states: an overview. *Energy for Sustainable Development* 8(1): 5-24.that "at present, subsidies to conventional energy are on the order of $250 billion per year" and $244 billion per annum between 1995 and 1998 (34% OECD) in Pershing, J. and Mackenzie (2004) Removing Subsidies.Leveling the Playing Field for Renewable Energy Technologies. Thematic Background Paper. International Conference for Renewable Energies, Bonn (2004)
[23] WEO, (in press)

These barriers will also affect the deployment of existing technologies

The nature of competition, existing infrastructure and existing distortions affect not only the process of developing new technologies; these sector-specific factors can also reduce the effectiveness of policies to internalise the carbon externality. They inhibit the power of the market to encourage a shift to low-carbon technologies, even when they are already cost-effective and especially if they are not. The generation sector usually favours more traditional (high-carbon) energy systems because of human, technical and institutional capacity. Historically driven by economies of scale, the electricity system becomes easily locked into a technological trajectory that demonstrates momentum and is thereby resistant to the technical change that will be necessary in a shift to a low-carbon economy[24].

Despite advances in the transport sector, radical change may not be delivered by the markets

Transport currently represents 14% of global emissions, and has been the fastest growing source of emissions because of continued growth of car transport and rapid expansion of air transport. Innovation has been dominated by incremental improvements to existing technologies, which depend on oil. These, however, have been more than offset by the growth in demand and shift towards more powerful and heavier vehicles. The increase in weight is partly due to increased size and partly to additional safety measures. The improvements in the internal combustion engine from a century of learning by doing, the efficiency of fossil fuel as an energy source and the existence of a petrol distribution network lead to some 'lock-in' to existing technologies. Behavioural inertia compounds this 'lock-in' as consumers are also accustomed to existing technologies.

Certain features of road transport suggest further innovative activity could be delivered through market forces. Although there is no explicit carbon price for road fuel, high and stable fuel taxes[25] in most developed countries provide an incentive for the development of more efficient vehicles. Niche markets also exist which help innovative products in transport markets to attract a premium. These factors together help to explain how hybrid vehicles have been developed and are now starting to penetrate markets, with only very limited government support: some consumers are content to pay a premium for what can be a cleaner and more fuel-efficient product. There is also a small number of large global firms in this sector, each of which have the resources to make significant innovation investments and progress. They can also be less concerned about international spillovers as they operate in several markets.

Incremental energy efficiency improvements are expected to continue in the transport sector. These will be stimulated both by fuel savings and, as they have been in the past, by government regulation. Both the hybrid car, and later, the fuel cell vehicle, are capable of doubling the fuel efficiency of road vehicles, whilst behavioural changes - perhaps encouraged, for example, by congestion pricing or intelligent infrastructure[26] - could lead to further improvements.

---

[24] Amin (2000)

[25] There are exceptions in the case of biofuels with many countries offering incentives through tax incentives.

[26] Intelligent infrastructure uses information to encourage efficient use of transport systems. http://www.foresight.gov.uk/Intelligent_Infrastructure_Systems/Index.htm

Markets alone, however, may struggle to deliver more radical changes to transport technologies such as plug-in hybrids or other electrical vehicles. Alternative fuels (such as biofuel blends beyond 5-10%, electricity or hydrogen) may require new networks, the cost of which is unlikely to be met without incentives provided by public policy. The environmental benefit of alternative transport fuels will depend on how they are produced. For example, the benefit of electric and hydrogen cars is limited if the electricity and hydrogen is produced from high emission sources. Obstacles to the commercial deployment of hydrogen cell vehicles, such as the cost of hydrogen vehicles and low-carbon hydrogen production, and the requirement to develop hydrogen storage further, ensure it is unlikely that such vehicles will be widely available commercially for at least another 15 to 20 years.

In Brazil policies to encourage biofuels over the past 30 years through regulation, duty incentives and production subsidies have led to biofuels now accounting for 13% of total road fuel consumption, compared with a 3% worldwide average in 2004. Other countries are now introducing policies to increase the level of biofuels in their fuel mix. Box 16.1 shows how some governments are already acting to create conditions for hydrogen technologies to be used. Making hydrogen fuel cell cars commercial is likely to require further breakthroughs in fundamental science, which may be too large to be delivered by a single company, and are likely to be subject to knowledge spillovers.

BOX 16.1 Hydrogen for transport

Hydrogen could potentially offer complete diversification away from oil and provide very low carbon transport. Hydrogen would be best suited to road vehicles. The main ways of producing hydrogen are by electrolysis of water, or by reforming hydrocarbons. Once produced, hydrogen can be stored as a liquid, a compressed gas, or chemically (bonded within the chemical structure of advanced materials). Hydrogen could release its energy content for use in powering road vehicles by combustion in a hydrogen internal combustion engine or a fuel cell. Fuel cells convert hydrogen and oxygen into water in a process that generates electricity. They are almost silent in operation, highly efficient, and produce only water as a by-product. Hydrogen can produce as little as 5% of the emissions of conventional fuel if produced by low-emission technologies.[27]

There are several hydrogen projects around the world including:

- Norway: plans for a 580 km hydrogen corridor between Oslo and Stavanger in a joint project between the private sector, local government and non-government organisations. The first hydrogen station opened in August 2006

- Denmark and Sweden: interested in extending the Norwegian hydrogen corridor

- Iceland: home to the first hydrogen fuelling station in April 2003 and it is proposed that Iceland could be a hydrogen economy by 2030

- EU: trial of hydrogen buses

- China: hydrogen buses to be used at the Beijing Olympics in 2008

- California: plans to introduce hydrogen in 21 interstate highway filling stations

[27] E4tech, (2006)

The development of alternative technologies in the road transport sector will be important for reducing emissions from other transport sectors such as the aviation, rail and maritime sectors. The local nature of bus usage allows the use of a centralised fuel source and this has led to early demonstration use of hydrogen in buses (see Box 16.1). In other sectors, such as aviation where weight and safety are prominent concerns, early commercial development is unlikely to take place and will be dependent on development in other areas first. The capital stock in the aviation, maritime and rail sectors (ships, planes and trains) lasts several times longer than road vehicles so this may result in a slower rate of take-up of alternative technologies. The emissions associated with rail transport can be reduced through decarbonising the fuel mix through biofuels or low carbon electricity generation. In the aviation sector improved air traffic management and reduced weight, through the use of alternative and advanced materials, can add to continued improvements in the efficiency of existing technologies.

Innovation will also play a role by addressing emissions in other sectors, reducing demand and enabling adaptation to climate change.

Innovation has enabled energy efficiency savings, for example, through compact fluorescent and diode based lights and automated control systems. Furthermore, innovation is likely to continue to increase the potential for energy efficiency savings. Energy efficiency innovation has often been in the form of incremental improvements but there is also a role for more radical progress that may require support. Some markets (such as the cement industry in some developing countries including China and building refurbishment in most countries) are made up of small local firms not large multinationals, which are less likely to undertake

---

BOX 16.2 The scope for innovation to reduce emissions from agriculture

Research into fertilisers and crop varieties associated with lower GHG emissions could help fight climate change[28]. In some instances it may be possible to develop crops that both reduce emissions and have higher yields in a world with more climate change (see Box 26.3).

Another important research area in agriculture will be how to enhance carbon storage in soils, complementing the need to understand emissions from soils (see Section 25.4). The economic potential for enhanced storage is estimated at 1 GtCO2e in 2020, but the technical potential is much greater (see Section 9.6).

Research into sustainable farming practices (such as agroforestry) suitable to local conditions could lead to a reduction in GHG emissions and may also improve crop yields. It could reduce GHG emissions directly by reducing the need to use fertilisers, and indirectly by reducing the emissions from industry and transport sectors to produce the fertiliser[29].

Research into livestock feeds, breeds and feeding practices could also help reduce methane emissions from livestock.

In addition to using biomass energy (see Box 9.5), agriculture, and associated manufacturing industries, have the potential to displace fossil-based inputs for sectors such as chemicals, pharmaceuticals, manufacturing and buildings using a wide range of products made from renewable sources.

---

[28] Norse (2006).
[29] Box 25.4 provides further examples of sustainable farming practices.

research since their resources and potential rewards are smaller. In addition, R&D, for example, in building technologies and urban planning could have a profound impact on the emissions attributed to buildings and increase climate resilience. Chapter 17 discussed energy efficiency in more detail.

Direct emissions from industrial sectors such as cement, chemical and iron and steel can also benefit from further innovation, whether it is in these sectors or in other lower-carbon products that can be substitutes. Innovation in the agricultural sector, discussed in a mitigation context in Box 16.2 above, can also help improve the capacity to adapt to the impacts of climate change. New crop varieties can improve yield resilience to climate change[30]. The Consultative Group on International Agricultural Research (CGIAR) will have a role to play in responding to the climate challenge through innovation in the agricultural sector (see Box 24.4). The development and dissemination of other adaptation technologies is examined in Chapter 19.

16.4   Policy implications for climate change technologies

Policy should be aimed at bringing a portfolio of low-emission technology options to commercial viability

Innovation is, by its nature, unpredictable. Some technologies will succeed and others will fail. The uncertainty and risks inherent in developing low-emission technologies are ideally suited to a portfolio approach. Experience from other areas of investment decisions under uncertainty[31] clearly suggests that the most effective response to the uncertainty of returns is to develop a portfolio. While markets will tend to deliver the least-cost short-term option, it is possible they may ignore technologies that could ultimately deliver huge cost savings in the long term.

As Part III set out, a portfolio of technologies will also be needed to reduce emissions in key sectors, because of the constraints acting on individual technologies. These constraints and energy security issues mean that a portfolio will be required to achieve reductions at the scale required. There is an option value to developing alternatives as it enables greater and potentially less costly abatement in the future. The introduction of new options makes the marginal abatement cost curve (see Section 9.3) more elastic. Early development of economically viable alternatives also avoids the problem of 'locking in' high-carbon capital stock for decades, which would also increase future marginal abatement costs. Policies to encourage low-emission technologies can be seen as a hedge against the risk of high abatement costs.

There are costs associated with developing a portfolio. Developing options involves paying the learning cost for more technologies. But policymakers should also bear in mind links to other policy objectives. A greater diversity in sources of energy, for instance, will tend to provide benefits to security of supply, as well as climate change. There is thus a type of externality from creating a new option in terms of risk reduction as well as potential cost reduction. Firms by themselves do not have the same perspective and weight on these criteria as broader society. The next section looks at how the development of a suitable portfolio can be encouraged.

---

[30] IRRI (2006).
[31] Pindyck and Dixit (1994)

Developing a portfolio requires a combination of government interventions including carbon pricing, R&D support and, in some sectors, technology-specific early stage deployment support. These should be complemented by policies to address non-market barriers.

Alongside carbon pricing and the further factors identified in Chapter 17, supporting the development of low-emission technologies can be seen as an important element of climate policy. The further from market the product, given some reasonable probability of success, the greater the prima facie case for policy intervention. In the area of pure research, spillovers can be very significant and direct funding by government support is often warranted. Closer to the market, the required financing flows are larger, and the private returns to individual companies are potentially greater. The government's role here is to provide a credible and clear policy framework to drive private-sector investment.

The area in the innovation process between pure research and technologies ready for commercialisation is more complex. Different sectors may justify different types of intervention. In the electricity market, in particular, deployment policies are likely to be required to bring technologies up to scale. How this support is delivered is important and raises issues about how technology neutral policy should be, which will be discussed later in this chapter in Section 16.6.

This diagram summarises the links between two of the elements of climate policy. The introduction of the carbon price reduces the learning cost since the new technology, for example a renewable, in this illustrative figure becomes cost effective at point B rather than point A, reducing the size of the learning cost represented by the dotted area. Earlier in the learning curve, deployment support is required to reduce the costs of the technology to the point where the market will adopt the technology. It is the earlier stages of innovation, research, development and demonstration which develop the technology to the point that deployment can begin.

Across the whole process, non-market barriers need to be identified and, where appropriate, overcome. Without policy incentives when required, support will be

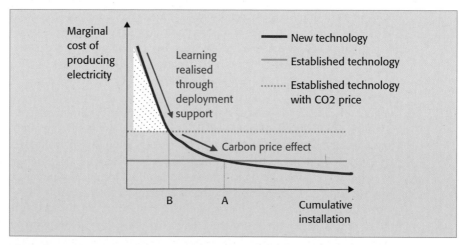

Figure 16.6 Interaction between carbon pricing and deployment support[32]

---

[32] In this figure the policy encourage learning but firms may be prepared to undertake investments in anticipation of technological progress or carbon price incentives.

unbalanced, and bottlenecks are likely to appear in the innovation process[33]. This would reduce the cost effectiveness at each other stage of support, by increasing the cost of the technology and delaying or preventing its adoption.

Uncertainties, both with respect to climate change and technology development, argue *for* investment in technology development. Uncertainties in irreversible investments argue for postponing policies until the uncertainties are reduced. However, uncertainties, especially with respect to technology development, will not be reduced exogenously with the 'passage of time' but endogenously through investment and the feedback and experience it provides.

Most of the development and deployment of new technologies will be undertaken by the private sector; the role of governments is to provide a stable framework of incentives

Deployment support is generally funded through passing on increased prices to the consumers. But it should still be viewed, alongside public R&D support, as a subsidy and should thus be subject to close scrutiny and, if possible, time limited. The private sector will be the main driver for these new technologies. Deployment support provides a market to encourage firms to invest and relies on market competition to provide the stimulus for cost reductions. Both public R&D and deployment support are expected to have a positive impact on private R&D.

In some sectors the benefits from innovation can be captured by firms without direct support for deployment, other than bringing down institutional barriers and via setting standards. This is particularly so in sectors that rely on incremental innovations to improve efficiency, rather than a step change in technology, since the cost gap is unlikely to be so large. In these sectors firms may be comfortable to invest in the learning cost of developing low-emission technologies.

Firms with products that are associated with greenhouse gas emissions are increasingly seeking to diversify in order to ensure their long-run profitability. Oil firms are increasingly investing in low-emission energy sources. General Electric's Ecomagination initiative has seen the sale of energy efficient and environmentally advanced products and services rise to $10.1 billion in 2005, up from $6.2 billion in 2004 - with orders nearly doubling to $17 billion. GE's R&D in cleaner technologies was $700m in 2005 and expected to rise to $1.5 billion per annum by 2010.[34] Indeed in a number of countries the private sector is running ahead of government policy and taking a view on where such policy is likely to go in the future which is in advance of what the current government is doing.

R&D and deployment support have been effective in encouraging the development of generation technologies in the past

Determining the benefits of both R&D and deployment is not easy. Studies have often successfully identified a benefit from R&D but without sufficient accuracy to determine what the appropriate level of R&D should be. Estimating the appropriate level is made more difficult by the broad range of activities that can be

---

[33] Weak demand-side policies risk wasting R&D investments see Norberg-Bohm and Loiter (1999) and Deutch (2005)

[34] Source GE press release May 2006: http://home.businesswire.com/portal/site/ge/index.jsp?ndmViewId=news_view&newsId=20060517005223&newsLang=en&ndmConfigId=1001109&vnsId=681

classed as R&D. Ultimately the benefits of developing technologies will depend on the amount of abatement that is achieved (and thus the avoided impacts) and the long-term marginal costs of abating across all the other sectors within the economy (linked to the carbon price), both of which are uncertain.

However, some evidence provides indications of the effectiveness of policy in promoting the development of technologies:

- **Estimates of R&D benefits**. Private returns from economy-wide R&D have been estimated at 20-30% whilst the estimated social rate of return was around 50%[35] While it is private-sector not public-sector R&D that has been positively linked with growth, the public-sector R&D can play a vital role in stimulating private spending up to the potential point of crowding out[36]. It also plays an important role in preserving the 'public good' nature of major scientific advances. Examples of valuable breakthroughs stimulated by public R&D must be weighed up alongside examples of wasteful projects.
- **Historical evidence**. Examining the history of existing energy technologies and the prominent role that public R&D and initial deployment have played in their development illustrates the potential effectiveness of technology policy. Extensive and prolonged public support and private markets were both instrumental in the development of all generating technologies. Military R&D, the US space programme and learning from other markets have also been crucial to the process of innovation in the energy sector. This highlights the spillovers that occur between sectors and the need to avoid too narrow an R&D focus. This experience has been mirrored in other sectors such as civil aviation and digital technologies where the source has also been military. Perhaps this is related to the fact that US public defence R&D was eight times greater than that for energy R&D in 2006 (US Federal Budget Authority). Historical R&D and deployment support has delivered the technological choices of the present with many R&D investments that may have seemed wasteful in the 1980s, such as investments in renewable energy and synfuels, now bearing fruit. The technological choices of the coming decades are likely to develop from current R&D.

BOX 16.3 Development of existing technology options[37]

Nuclear: From the early stages of the Cold War, the Atomic Energy Commission in the US, created primarily to oversee the development of nuclear weapons, also promoted civilian nuclear power. Alic et al[38] argue that by exploiting the 'peaceful atom' Washington hoped to demonstrate US technological prowess and perhaps regain moral high ground after the atomic devastation of 1945. The focus on weapons left the non-defence R&D disorganised and starved of funds and failed to address the practical issues and uncertainties of commercial reactor design. The government's monopoly of nuclear information, necessary to prevent the spreading of sensitive information, meant state R&D was crucial to development.

---

[35] Kammen and Margolis (1999)
[36] When public expenditure limits private expenditure by starving it of potential resources such as scientists OECD (2005)
[37] Alic, Mowery and Rubin (2003)
[38] Alic, Mowery and Rubin (2003)

Gas: The basic R&D for gas turbine technology was carried out for military jet engines during World War II. Since then developments in material sciences and turbine design have been crucial to the technological innovation that has made gas turbines the most popular technology for electricity generation in recent years. Cooling technology from the drilling industry and space exploration played an important role. In the 1980s improvements came from untapped innovations in jet engine technology from decades of experience in civil aviation. Competitive costs have also been helped by low capital costs, reliability, modularity and lower pollution levels.

Wind: The first electric windmills were developed in 1888 and reliable wind energy has been available since the 1920s. Stand-alone turbines were popular in the Midwestern USA prior to centrally generated power in the 1940s. Little progress was made until the oil shocks led to further investment and deployment, particularly in Denmark (where a 30% capital tax break (1979-1989) mandated electricity prices (85% of retail) and a 10% target in 1981 led to considerable deployment) and California where public support led to extensive deployment in the 1980s. Recent renewable support programmes and technological progress have encouraged an average annual growth rate of over 28% over the past ten years[39].

Photovoltaics: The first PV cells were designed for the space programme in the late 1950s. They were very expensive and converted less than 2% of the solar energy to electricity. Four decades of steady development, in the early phases stimulated by the space programme, have seen efficiency rise to nearly 25% of the solar energy in laboratories, and costs of commercial cells have fallen by orders of magnitude. The need for storage or ancillary power sources have held the technology back but there have been some niche markets in remote locations and, opportunities to reduce peak demand in locations where solar peaks and demand peaks coincide.

Public support has been important. A study by Norberg-Bohm[40] found that, of 20 key innovations in the past 30 years, only one of the 14 they could source was funded entirely by the private sector and nine were totally public. Recent deployment support led the PV market to grow by 34% in 2005. Nemet[41] explored in more detail how the innovation process occurred. He found that, of recent cost reductions, 43% were due to economies of scale, 30% to efficiency gains from R&D and learning-by-doing, 12% due to reduced silicon costs (a spillover from the IT industry).

- **Learning curve analysis.** Learning curves, as shown in Box 9.4 and in other studies[42], show that increased deployment is linked with cost reductions suggesting that further deployment will reduce the cost of low-emission technologies. There is a question of causation since cost reductions may lead to greater deployment; so attempts to force the reverse may lead to disappointing learning rates. The data shows technologies starting from different points and achieving

[39] Global Wind Energy Council 13 http://www.gwec.net/index.php?id=13
[40] Norberg-Bohm (2000)
[41] Source: Nemet, in press
[42] For an example Taylor, Rubin and Nemet (2006)

very different learning rates. The increasing returns from scale shown in these curves can be used to justify deployment support, but the potential of the technologies must be evaluated and compared with the costs of development.

16.5   Research, development and demonstration policies

Government has an important role in directly funding skills and basic knowledge creation for science and technology

At the pure science end of the spectrum, the knowledge created has less direct commercial application and exhibits the characteristics of a 'public good'. At the applied end of R&D, there is likely to be a greater emphasis on private research, though there still may be a role for some public funding.

Governments also fund the education and training of scientists and engineers. Modelling for this review suggests that the output of low-carbon technologies in the energy sector will need to expand nearly 20-fold over the next 40-50 years to stabilise emissions, requiring new generations of engineers and scientists to work on energy-technology development and use. The prominent role of the challenge of climate change may act as an inspiration to a new generation of scientists and spur a wider interest in science.

R&D funding should avoid volatility to enable the research base to thrive. Funding cycles in some countries have exhibited 'roller-coaster' variations between years, which have made it harder for laboratories to attract, develop, and maintain human capital. Such volatility can also reduce investors' confidence in the likely returns of private R&D. Kammen[43] found levels changed by more than 30% in half the observed years. Similarly it may be difficult to expand research capacity very quickly as the skilled researchers may not be available. Governments should seek to avoid such variability, especially in response to short-term fuel price fluctuations. The allocation of public R&D funds should continue to rely on the valuable peer review process and this should include post-project evaluations and review to maximise the learning from the research. Research with clear objectives but without over-commitment to narrow specifications or performance criteria can eliminate wasteful expenditures[44] and allow researchers more time to apply to their research interests and be creative.

Governments should seek to ensure that, in broad terms, the priorities of publicly funded institutions reflect those of society. The expertise of the researchers creates an information asymmetry with policymakers facing a challenge in selecting suitable projects. Arms-length organisations and expert panels such as research-funding bodies may be best placed to direct funding to individual projects.

Three types of funding are required for university research funding.

- Basic research time and resources for academic staff to pursue research that interests them.
- Research programme funding (such as research councils) that directs funding towards important areas.

---

[43] Kammen (2004)
[44] Newell and Chow (2004)

- Funding to encourage the transfer of knowledge outside the institution. The dissemination of information encourages progress to be applied and built on by other researchers and industry and ensures that it not be unnecessarily duplicated elsewhere.

Research should cover a broad base and not just focus on what are currently considered key technologies, including basic science and some funding to research the more innovative ideas[45] to address climate change. Historical examples of technological progress when the research was not directed towards specific economic applications (such as developments in nanotechnology, lasers and the transistor) highlight the importance of open-ended problem specification. There must be an appropriate balance between basic science and applied research projects[46]. Increases in energy R&D (as discussed in the final section of this chapter) can be complemented by increased funding for science generally. The potential scale of increase in basic science will vary by country depending on their current level and research capabilities[47].

There may also be a case for demonstration funding to prove viability and reduce risk. An example of this is the UK DTI's 'Wave and Tidal Stream Energy Demonstration Scheme' that will support demonstration projects undertaken by private firms. This has many features to encourage the projects and maximise learning through provision of test site and facilities and systematic comparison of competing alternatives. Governments can help such projects through providing infrastructure. Demonstration projects are best conducted or at least managed by the private sector.[48]

Energy storage is worthy of particular attention

Inherent uncertainty on fruitful areas of research ensures that governments should be cautious against picking winners. However, some areas of research suggest significant potential through a combination of the probability of success, lead-times and global reward for success. Priorities for scientific progress in the energy sector should include PV (silicon and non-silicon based), biofuel conversion technologies, fusion, and material science.

As markets expand, all the key low carbon primary energy sources will run into constraints. Nuclear power will be confined to base-load electricity generation unless energy storage is available to enable its energy to follow loads and contribute to the markets for transport fuels. Intermittent renewable energy forms with backup generation will face the same problem. Electricity generation from fossil fuels with carbon capture and storage will likewise be unable to enter the transport markets unless improved and lower cost forms of hydrogen storage or new battery technology are developed. Solar energy can in theory meet the world's energy needs

---

[45] For some examples, see Gibbs (2006)

[46] Newell and Chow (2004)

[47] In 2004 the UK Government published a ten-year Science and Innovation Investment Framework, which set a challenging ambition for public and private investment in R&D to rise from 1.9% to 2.5% of UK GDP, in partnership with business; as well as the policies to underpin this. An additional £1 billion will be invested in science and innovation between 2005-2008, equivalent to real annual growth of 5.8% and to continue to increase investment in the public science base at least in line with economic growth. http://www.dti.gov.uk/science/science-funding/framework/page9306.html

[48] Newell and Chow (2004)

BOX 16.4 Second generation biofuels

Cellulosic ethanol is a not-yet-commercialized fuel derived from woody biomass. In his 2006 State of the Union address, Bush praised the fuel's potential to curb the nation's "addiction to foreign oil". A joint study by the Departments of Agriculture and Energy[49] concludes that U.S. biomass feedstocks could produce enough ethanol to displace 30 percent of the nation's gasoline consumption by 2030.

In May 2006, Goldman Sachs & Co became the first major Wall Street firm to invest in the technology. Goldman Sachs & Co invested more than $26 million in Iogen Corp., an Ottawa-based company that operates the world's first and only demonstration facility that converts straw, corn stalks, switchgrass and other agricultural materials to ethanol. Iogen hopes to begin construction on North America's first commercial cellulosic ethanol plant next year.

In September 2006 Richard Branson announced plans to invest $3 billion in mitigating climate change. Some of this will be invested in Virgin Fuels, which will develop biofuels including cellulosic ethanol.

many times over, but will, like energy from wind, waves and tides, eventually depend on the storage problem being solved.

The analysis of the costs of climate change mitigation in Chapter 9 provides further confirmation of the need for an expansion of RD&D activities in energy storage technologies. A failure to develop such technologies will inevitably increase the costs of mitigation once low-emission options for electricity generation are exploited. In contrast, success in this area will allow low-emission sources to provide energy in other sectors, such as transport. Current R&D and demonstration efforts on hydrogen production and storage along with other promising options for storing energy (such as advanced battery concepts) should be increased. This should include research on devices that convert the stored energy, such as the fuel cell.

In the case of applied energy research, partnership between the public and private sectors is key

It is important that public R&D leverages private R&D and encourages commercialisation. Ultimately the products will be brought into the market by private firms who have a better knowledge of markets, and, so it is important that public R&D maintains the flow of knowledge by ensuring public R&D complements the efforts of the private sector.

The growth and direction of private R&D efforts will be a product of the incentives for low-emission investments provided by the structure of markets and public policies. Public R&D should aim to complement, not compete, with private R&D, generally by concentrating on more fundamental, longer-term possibilities, and by sharing in the risks of some larger-scale projects such as CCS. In many areas the private sector will make research investments without public support, as has been the case recently on advanced biofuels (see Box 16.4).

The OECD[50] found that economic growth was closely linked to general private R&D, not public R&D, but that public R&D plays a vital role in stimulating private spending. There is evidence[51] from the energy sector that patents do track public

---

[49] US Departments of Agriculture and Energy (2005)
[50] OECD (2005)
[51] Kammen and Nemet (2005)

R&D closely, which suggests that they successfully spur innovation and private sector innovation. R&D collaboration between the public and private-sector is one way of reducing the cost and risks of R&D.

The public sector could fund private sector research through competitive research funding, with private sector companies bidding for public funds as public organisations currently do from research councils. Prizes to reward innovation can be used to encourage breakthroughs. Historically they have proved very successful but defining a suitable prize can be problematic[52]. An alternative approach, as suggested for the pharmaceutical sector, is to commit to purchase new products to reward those that successfully innovate.[53]

BOX 16.5 Public-private research models - UK Energy Technologies Institute[54]

In 2006, the UK launched the Energy Technologies Institute (ETI). It will be funded on a 50:50 basis between private companies and the public sector with the government prepared to provide £500 million, creating the potential for a £1 billion institute over a minimum lifetime of ten years.

The institute will aim to accelerate the pace and volume of research directed towards the eventual deployment of the most promising research results. ETI will work to existing UK energy policy goals including a 60% reduction in emissions by 2050.

The ETI will select, commission, fund, manage and, where appropriate, undertake research programmes. Most investment will focus on a small number of key technology areas that have greatest promise for deployment and contributing to low-emission secure energy supplies.

16.6 Deployment policy

A wide range of policies to encourage deployment are already in use.

In addition to direct emissions pricing through taxes and trading and R&D support, there are strong arguments in favour of supporting deployment in some sectors when spillovers, lock-in to existing technologies, or capital market failures prevent the development of potentially low-cost alternatives. Without support the market may never select those technologies that are further from the market but may nevertheless eventually prove cheapest. Policies to support deployment exist throughout the world including many non-OECD countries[55]. China and India have both encouraged large-scale renewable deployment in recent years and now have respectively the largest and fifth largest renewable energy capacity worldwide[56].

There is some deployment support for clean technologies in most developed countries. The mechanism of support takes many forms though the costs are generally passed onto the consumer. The presence of a carbon price reduces the

---

[52] Newell and Wilson (2005)
[53] Kremer and Glennerster (2004)
[54] http://www.dti.gov.uk/science/science-funding/eti/page34027.html
[55] Page 20 REN 21 Renewables global status report 2005 - See page 20 REN 21 (2005)
[56] Figures from 2005 - excluding large scale hydropower. Page 6 REN 21 (2006)

cost and requirement for deployment support. Deployment support is generally a small component of price when spread across all consumption (see Box 16.7) but does add to the impact of carbon pricing on electricity prices. Policymakers should consider the impact of deployment support on energy prices over time. Consumers will be paying for the development of technologies that benefit consumers in the future.

The deployment mechanisms described in Box 16.6 can be characterised as price or quantity support, with some tradable approaches containing elements of both. The costs of these policies are generally passed directly on to consumers though some are financed from general taxation. When quantity deployment instruments are not tradable, the policymaker should consider whether there are sufficient incentives to strive for cost reductions and whether the supplier can profit from passing an excessive cost burden onto the consumer. If the level of a price deployment instrument is too low no deployment will occur, while if it is too high large volumes of deployment will occur with financial rewards for participants

---

BOX 16.6 Examples of existing deployment incentives

- **Fiscal incentives**: including reduced taxes on biofuels in the UK and the US; investment tax credits.

- **Capital grants** for demonstrator projects and programmes: clean coal programmes in the US; PV 'rooftop' programmes in the US, Germany and Japan; investments in marine renewables in the UK and Portugal; and numerous other technologies in their demonstration phase.

- **Feed-in tariffs** are a fixed price support mechanism that is usually combined with a regulatory incentive to purchase output: examples include wind and PVs in Germany; biofuels and wind in Austria; wind and solar schemes in Spain, supplemented by 'bonus prices'; wind in Holland.

- **Quota based schemes**: the Renewable Portfolio Standards in twenty three US States; the vehicle fleet efficiency standards in California

- **Tradable quotas**: the Renewables Obligation and Renewable Transport Fuels Obligation in the UK.

- **Tenders for tranches of output** (the former UK Non Fossil Fuel Obligation) with increased output prices subsidised out of the revenues from a general levy on electricity tariffs.

- **Subsidy** of the infrastructure costs of connecting new technologies to networks.

- **Procurement policies of public monopolies**: This was the approach historically of the public monopolies in electricity for purchase of nuclear power throughout the OECD; it is currently the approach in China. It is often combined with regulatory agreements to permit recovery of costs, soft loans by governments, and, in the case of nuclear waste, government assumption of liabilities.

- **Procurement policies of national and local governments**: these include demonstrator projects on public buildings; use of fuel cells and solar technologies by defence and aerospace industries; hydrogen fuel cell buses and taxis in cities; energy efficiency in buildings.

which are essentially government created rents. With tradable quantity instruments, the market is left to determine the price, usually with tradable certificates between firms. This does lead to price uncertainty. If the quantity is too high, bottlenecks may lead to a high cost. If the quantity is too low, there may not be sufficient economies of scale to reduce the cost.

Both sets of instruments have proved effective but existing experience favours price-based support mechanisms. Comparisons between deployment support through tradable quotas and feed-in tariff price support suggest that feed-in mechanisms achieve larger deployment at lower costs[57]. Central to this is the assurance of long-term price guarantees. The German scheme, as described in Box 16.7 below, provides legally guaranteed revenue streams for up to twenty years if the technology remains functional. Whilst recognising the importance of

---

BOX 16.7   Deployment support in Germany

Feed-in tariffs have been introduced in Germany to encourage the deployment of onshore and offshore wind, biomass, hydropower, geothermal and solar PV[58]. The aim is to meet Germany's renewable energy goals of 12.5% of gross electricity consumption in 2010 and 20% in 2020. The policy also aims to encourage the development of renewable technologies, reduce external costs and increase the security of supply.

Each generation technology is eligible for a different rate. Within technologies the rate varies depending on the size and type. Solar energy receives between €0.457 to 0.624 per kWh while wind receives €0.055 to 0.091 per kWh. Once the technology is built the rate is guaranteed for 20 years. The level of support for deployment in subsequent years declines over time by 1% to 6.5% each year with the rate of decline derived from estimated learning curves[59].

In 2005 10.2% of electricity came from renewables (70% supported with feed-in tariffs) the Federal Environment Ministry (BMU) estimate that the current act will save 52 million tonnes on CO_2 in 2010. The average level of feed-in tariff was €0.0953 per kWh in 2005 (compared to an average cost of displaced energy of €0.047 kWh). The total level of subsidy was €2.4 billion Euro at a cost shared all consumers of €0.0056 per kWh (3% of household electricity costs)[60]. There are an estimated 170,000 people working in the renewable sector with an industry turnover of €8.7 billion.[61]

The 43.7 TWh of electricity covered by the feed in tariffs was split mostly between wind (61%), biomass (19%) and hydropower (18%). It has succeeded in supporting several technologies. Solar accounted for 2% (0.2% of total electricity) with an average growth rate of over 90% over the last four years. Despite photovoltaic's low share Germany has a significant proportion of the global market with 58% of the capacity installed globally in 2005 (39% of the total installed capacity) and 23% of global production.[62]

---

[57] Butler and Neuhoff (2005); EC (2005); Ragwitz, and Huber (2005); Fouquet et al (2005)
[58] Originally introduced in 1991 with the Electricity Feed Act this was replaced in 2000 with the broader Act on Granting Priority to Renewable Energy Sources (Renewable Energy Sources Act) and amended in 2004 http://www.ipf-renewables2004.de/en/dokumente/RES-Act-Germany_2004.pdf
[59] Small hydropower does not decline and is guaranteed for 30 years and large hydropower only 15 years.
[60] BMU (2006a)
[61] BMU (2006b)
[62] http://www.iea-pvps.org/isr/index.htm

planning regimes for both PV and wind, the levels of deployment are much greater in the German scheme and the prices are lower than comparable tradable support mechanisms (though greater deployment increases the total cost in terms of the premium paid by consumers). Contrary to criticisms of the feed-in tariff, analysis suggests that competition is greater than in the UK Renewable Obligation Certificate scheme. These benefits are logical as the technologies are already prone to considerable price uncertainties and the price uncertainty of tradable deployment support mechanisms amplifies this uncertainty. Uncertainty discourages investment and increases the cost of capital as the risks associated with the uncertain rewards require greater rewards.

Regulation can also be used to encourage deployment, for example by reducing uncertainty and accelerating spillover effects, and may be preferable in certain markets (see Chapter 17 for details). Performance standards encourage uptake and innovation in efficient technologies by establishing efficiency requirements for particular goods, in particular encouraging incremental innovation Alternatively, technology specific design standards can be targeted directly at the cleanest technologies by mandating their application or banning alternatives.

There are already considerable sums of money spent on supporting technology deployment. It is estimated that $10 billion[63] was spent in 2004 on renewable deployment, around $16 billion is spent each year supporting existing nuclear energy and around $6.4 billion[64] is spent each year supporting biofuels. The total support for these low-carbon energy sources is thus $33 billion each year. Such sums are dwarfed by the existing subsidies for fossil fuels worldwide that are estimated at $150 billion to 250 billion each year. All these costs are generally paid by the consumer.

Technology-neutral incentives should be complemented by focused incentives to bring forward a portfolio of technologies

Policy frameworks can be designed to treat support to all low-carbon technologies in a 'technology-neutral' way. The dangers of public officials 'picking winners' should point to this as the starting point in most sectors. Markets and profit orientated decisions, where the decision maker is forced to look carefully at cost and risk are better at finding the likely commercial successes. However, the externalities, uncertainties and capital market problems in some sectors combine with the urgency of results and specificity of some of the technological problems that need to be solved when tackling climate change, all point to the necessity to examine the issues around particular technologies and ensure that a portfolio develops.

The policy framework of deployment support could differentiate between technologies, offering greater support to those further from commercialisation, or having particular strategic or national importance. This differentiation can be achieved several ways, including technology-specific quotas, or increased levels of price support for certain technologies. Policies to correct the carbon externality (taxes/trading) are, and should continue to be, technology neutral. Technology neutrality is also desirable for deployment support if the aim is to deliver least

---

[63] Deployment share of figure page 16 REN 21, 2005 grossed up to global figure based on IEA deployment figures. Nuclear figure from same source.

[64] Based on global production of 40 billion litres and on an average support of £0.1 per litre and a PPP exchange rate of $1.6 to £1

cost reductions to meet short-term targets, since the market will deliver the least-cost technology.

However, as has already been discussed, the process of learning means that longer-established technologies will tend to have a price advantage over newer technologies, and untargeted support will favour these more developed technologies and bring them still further down the learning curve. This effect can be seen in markets using technology-neutral instruments: in the USA, onshore wind accounts for 92% of new capacity in green power markets[65].

This concentration on near-to-market technologies will tend to work to the exclusion of other promising technologies, which means that only a very narrow portfolio of technologies will be supported, rather than the broad range which Part III of this report shows are required. This means technology neutrality may be cost efficient in the short term, but not over time.

Most deployment support in the electricity generation sector has been targeted towards renewable and nuclear technologies. However, significant reductions are also expected from other sources. As highlighted in Box 9.2 carbon capture and storage (CCS) is a technology expected to deliver a significant portion of the emission reductions. The forecast growth in emissions from coal, especially in China and India, means CCS technology has particular importance. Failure to develop viable CCS technology, while traditional fossil fuel generation is deployed across the globe, risks locking-in a high emissions trajectory. The demonstration and deployment of CCS is discussed in more detail in Chapter 24. Stabilising emissions below 550ppm CO_2e will require reducing emissions from electricity generation by about 60%[66]. Without CCS that would require a dramatic shift away from existing fossil-fuel technologies.[67]

Policies should have a clear review process and exit strategies, and governments must accept that some technologies will fail.

Uncertainty over the economies of scale and learning-by-doing means that some technological failures are inevitable. Technological failures can still create valuable knowledge, and the closing of technological avenues narrows the investment options and increases confidence in other technologies (as they face less alternatives). The Arrow-Lind theorem[68] states that governments are generally large enough to be risk neutral as they are large enough to spread the risk and thus have a role to play in undertaking riskier investments. It is not a mistake per se to buy insurance or a hedge that later is not needed and that is in many ways a suitable analogy for fostering a wider portfolio of viable technologies than the market would do by itself[69].

Credibility is also important to policy design. Policies benefit from providing clear, bankable, signals to business. There is a role for monitoring and for a clear exit strategy to prevent excessive costs and signal the ultimate goal of these policies: competition on a level playing field. A good example has been the Japanese

---

[65] Bird and Swezey (2005)
[66] This is consistent with the IEA ACT scenarios see Box 9.7
[67] For more on CCS see Boxes 9.2 and 24.8 and Section 24.3
[68] Arrow and Lind (1970)
[69] Deutch (2005)

rebates in the 'Solar Roofs' programme, which have declined gradually over time, from 50% of installed cost in 1994 to 12% in 2002 when the scheme ended.

Alternative approaches can also help spur the deployment of new innovations. For example, extension services, the application of scientific research and new knowledge to agricultural practices through farmer education, had a significant impact on the deployment of new crop varieties during the Green Revolution. Also, organisations such as the Carbon Trust in the UK, Sustainable Development Technologies Canada, established by governments but independent of them to allow the application of business acumen, have proved successful in encouraging investment in the development and demonstration of clean technologies. They can play an important role at each stage of the technology process, from R&D to ensuring their widespread deployment once they have become cost effective. They have proved especially successful in acting as a "stamp of approval" that spurs further venture capital investment. Finding niche markets and building these into large-scale commercialisation opportunities is a key challenge for companies with promising low carbon technologies. These organisations are at the forefront of identifying niche markets for commercialisation of new technologies and promoting public-private investment in deployment.

16.7   Other supporting policies

Other policies have an important impact on the viability of technologies.

There are many other policy options available to governments that can affect technology deployment and adoption. Governments set policies such as the planning regime and building standards. How these are set can have an important impact on the adoption of new technologies. They can constrain deployment either directly or indirectly by increasing costs. Regulations can stifle innovation, but if well designed they can drive innovation. Depending how these are set, they can act as a subsidy to low-emission alternative technologies or to traditional fossil fuels. Setting the balance is difficult, since their impacts are hard to value. But they must be considered since they can have an important effect on the outcome.

- The intellectual property regime can act as an incentive to the innovator, but the granting of the property right can also slow the dissemination of technological progress and prohibit others from building on this innovation. Managing this balance is an important challenge for policymakers.
- Planning and licensing regulations have proven a significant factor for nuclear, wind and micro-generation technologies. Planning can significantly increase costs or, in many cases, prevent investments taking place. Local considerations must be set against wider national or global concerns.
- It is important how governments treat risks and liabilities such as waste, safety or decommissioning costs for nuclear power or liabilities for CO_2 leakage from CCS schemes. Governments can bear some of these costs but, unless suppliers and ultimately consumers are charged for this insurance, it will be a subsidy.
- Network issues are particularly important for energy and transport technologies. The existing transport network and infrastructure, especially fuel stations, is tailored to fossil fuel technologies.

- Intermittent technologies such as wind and solar may be charged a premium if they require back-up sources. How this is treated can directly affect economic viability, depending on the extent of the back-up generation required and the premium charged.

- Micro-generation technologies can sell electricity back to the grid and do not incur the same distribution costs and transmission losses as traditional much larger sources. The terms under which such issues are resolved has an important impact on the economics of these technologies. Commercially proven low-carbon technologies require regulatory frameworks that recognise their value, in terms of flexibility and modularity[70], within a distributed energy system. Regulators should innovate in response to the challenge of integrating these technologies to exploit their potential, and unlock the resultant opportunities that arise from shifting the generation mix away from centralised sources.

- Capacity constraints may arise because of a shortage in a required resource. For example, there may be a shortage of skilled labour to install a new technology.

- There are other institutional and even cultural barriers that can be overcome. Public acceptability has proven an issue for both wind and nuclear and this may also be the case for hydrogen vehicles. Consumers may have problems in finding and installing new technologies. Providing information of the risks and justification of particular technologies can help overcome these barriers.

16.8   The scale of action required

Extending and expanding existing deployment incentives will be key

Deployment policies encourage the private sector to develop and deploy low-carbon technologies. The resulting cost reductions will help reduce the cost of mitigation in the future (as explained in Chapter 10). Consumers generally pay the cost of deployment support in the form of higher prices. Deployment support represents only a proportion of the cost of the technology as it leverages private funds that pay for the market price element of the final cost.

It is estimated that existing deployment support for renewables, biofuels and nuclear energy is $33 billion each year (see Section 16.6). The IEA's Energy Technology Perspectives[71] looks at the impact of policies to increase the rate of technological development. It assumes that $720 billion of investment in deployment support occurs over the next two to three decades. This estimate is on top of an assumed carbon price (whether through tax, trading or implicitly in regulation) of $25 per tonne of CO_2. If the IEA figure is assumed to be additional to the existing effort, it suggests an increase of deployment incentives of between 73% and 109%, depending on whether this increase is spread over two or three decades.

The calculations shown in Section 9.8 include estimates of the level of deployment incentives required to encourage sufficient deployment of new technologies (consistent with a 550ppm CO_2e stabilisation level). The central estimates from this work are that the level of support required will have to increase deployment

[70] Small-scale permits incremental additions in capacity unlike large technologies such as nuclear generation.
[71] Page 58, IEA (2006)

incentives by 176% in 2015 and 393% in 2025[72]. These estimates are additional to an assumed a carbon price at a level of $25 per tonne of CO_2.

At this price the abatement options are forecast to become cost effective by 2075 so the level of support tails off to zero by this time. If policies lead to a price much higher than this before the technologies are cost effective then less support will be required. Conversely if no carbon price exists the level of support required will have to increase (by a limited amount initially but by much larger amounts in the longer term). While most of this cost is expected to be passed on to consumers, firms may be prepared to incur a proportion of this learning cost in order to gain a competitive advantage.

Such levels of support do represent significant sums but are modest when compared with overall levels of investment in energy supply infrastructure ($20 trillion up to 2030[73]) or even estimates of current levels of fossil-fuel subsidy as shown in the graph below.[74]

The level of support required to develop abatement technologies depends on the carbon price and the rate of technological progress, which are both uncertain. It is clear from these numbers that the level of support should increase in the decades to come, especially in the absence of carbon pricing. Based on the numbers above, an increase of 2-5 times current levels over the next 20 years should help encourage the requisite levels of deployment though this level should be evaluated as these uncertainties are resolved.

The scale is, however, not the only issue. It is important that this support is well structured to encourage innovation at low cost. A diverse portfolio of investments

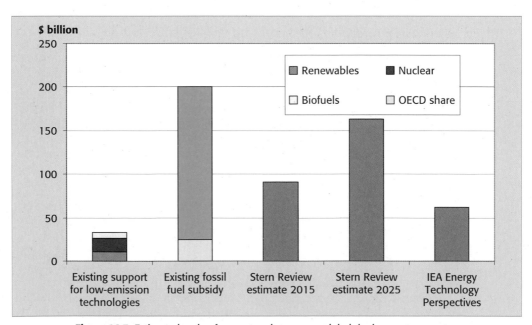

Figure 16.7 Estimated scale of current and necessary global deployment support

[72] See papers by Dennis Anderson available at www.sternreview.org.uk
[73] IEA (in press)
[74] In this graph mid points in the fossil fuel subsidy range is used in and the IEA increase made over a 20 year period.

is required as it is uncertain which technologies will prove cheapest and constraints on individual technologies will ensure that a mix is necessary. Those technologies that are likely to be the cheapest warrant more investment and these may not be those that are the currently the lowest cost. This requires a reorientation of public support towards technologies that are further from widespread diffusion.

Some countries are already offering significant support for new technologies but globally this support is patchy. Issues on coordinating deployment support internationally to achieve the required diversity and scale are examined in Chapter 24.

Global energy R&D funding is at a low level and should rise

Though benefits of R&D are difficult to evaluate accurately a diverse range of indicators illustrate the benefits of R&D investments. Global public energy R&D support has declined significantly since the 1980s and this trend should reverse to encourage cost reductions in existing low-carbon technologies and the development of new low-carbon technological options. The IEA R&D database shows a decline of 50% in low-emission R&D[75] between 1980 and 2004. This decline has occurred while overall government R&D has increased significantly[76]. A recent IEA publication on RD&D priorities[77] strongly recommends that governments consider restoring their energy RD&D budgets at least to the levels seen, in the early 1980s. This would involve doubling the budget from the current level of around $10 billion[78]. This is an appropriate first step that would equate to global levels of public energy R&D around **$20 billion** each year.

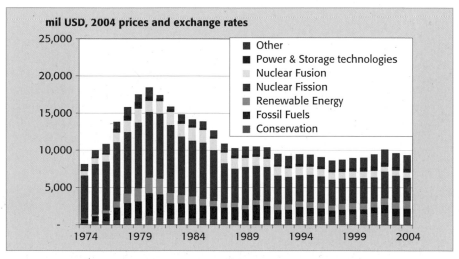

Figure 16.8 Public energy R&D in IEA countries[79]

[75] For countries available includes renewables, conservation and nuclear. The decline is 36% excluding nuclear.
[76] OECD R&D database shows total public R&D increasing by nearly 50% between 1988 and 2004 whilst public energy R&D declined by nearly 20% over the same period.
[77] Page 19 OECD (2006)
[78] 2005 figure Source: IEA R&D database http://www.iea.org/Textbase/stats/rd.asp
[79] Source: IEA Energy R&D Statistics

The directions of the effort should also change. A generation ago, the focus was on nuclear power and fossil fuels, including synthetic oil fuels from gas and coal, with comparatively few resources expended on conservation and renewable energy. Now the R&D efforts going into carbon capture and storage, conservation, the full range of renewable energy technologies, hydrogen production and use, fuel cells, and energy storage technologies and systems should all be much larger.

A phased increase in funding, within established frameworks for research priorities, would allow for the expansion in institutional capacity and increased expertise required to use the funding effectively. A proportion of this public money should target be designed to encourage private funds, as is proposed for the UK's Energy Technology Institute (see Box 16.5).

Private R&D should rise in response to market signals. Private energy R&D in OECD countries fell in recent times from around $8.5bn at the end of the 1980s to around $4.5bn in 2003[80]. Significant increases in public energy R&D and deployment support combined with carbon pricing should all help reverse this trend and encourage an upswing in private R&D levels.

This is not just about the total level of support. How this money is spent is crucial. It is important that the funding is spread across a wide range of ideas. It is also important that it is structured to provide stability to researchers while still providing healthy competition. There should be rigorous assessment of these expenditures to ensure that they maintained at an appropriate level. Approaches to encourage international co-operation to achieve these goals are explored in Chapter 24.

16.9   Conclusions

This chapter explores the process of innovation and discovers that externality from the environmental impact of greenhouse gas emissions exacerbates existing market imperfections, limiting the incentive to develop low-carbon technologies. This provides a strong case for supporting the development of new and existing low-carbon technologies, particularly in a number of key climate change sectors. The power of market forces is the key driver of innovation and technical change but this role should be supplemented with direct public support for R&D and, in some sectors, policies designed to create new markets. Such policies are required to deliver an effective portfolio of low-carbon technologies in the future.

References

For an excellent overview of innovation theory and its application in the case of climate change see Foxon, T J (2003), "Inducing innovation for a low-carbon future: drivers, barriers and policies" (full reference and link below). A good exploration of the barriers within the electricity generation sector can be found in Neuhoff, K. (2005), "Large-Scale Deployment of Renewables for Electricity Generation".

A good overview of innovation in the climate change context and technology policy can be found in Grubb (2004), "Technology Innovation and Climate Change Policy: an overview of issues and options" and a thorough exploration of the market failures and evidence on policy instruments can be found in both Jaffe, Newell and Stavins publications listed below. An excellent review of

[80]   Page 35, OECD (2006)

existing US based R&D policies and their effectiveness is Alic, Mowery and Rubin (2003), "U.S. Technology and Innovation Policies: Lessons for Climate Change". Newell and Chow also provide some valuable insights reviewing the literature on the effectiveness of energy R&D policy.

Aghion, P., N. Bloom, R. Blundell, R. Griffith and P. Howitt (2002). 'Competition and innovation: an inverted U relationship', NBER Working Paper No. 9269, London: UCL.

Alic, J., D. Mowery, and E. Rubin (2003): 'U.S. technology and innovation policies: Lessons for climate change', Virginia: Pew Center on Global Climate Change, available from http://www.pewclimate.org/docUploads/US%20Technology%20%26%20Innovation%20Policies%20%28pdf%29%2Epdf

Amin, A-L. (2000): 'The power of networks – renewable electricity in India and South Africa', unpublished thesis, SPRU, University of Sussex.

Anderson, D., C. Clark, T. Foxon, et al. (2001): 'Innovation and the environment: challenges and policy options for the UK'. A report for the UK Economic and Social Science Research Council. Imperial College London, Centre for Energy Policy and Technology (ICEPT) and the Fabian Society, London: Imperial College.

Arrow, K. and R. Lind (1970): 'Uncertainty and the evaluation of public investment decisions', American Economic Review, **60**: 364–78

Barreto, L. and G. Klaassen (2004): 'Emissions trading and the role of learning-by-doing spillovers in the "bottom-up" energy-systems ERIS model'. International Journal of Energy Technology and Policy (Special Issue on Experience Curves).

Bird, L. and B. Swezey (2005): 'Estimates of new renewable energy capacity serving U.S. green power markets' available from http://www.eere.energy.gov/greenpower/resources/tables/new_gp_cap.shtml

BMU (Germany's Federal Ministry for the Environment, Nature Conservation and Nuclear Safety). (2006a): 'What electricity from renewable energies costs', available from www.bmu.de/files/pdfs/allgemein/application/pdf/electricity_costs.pdf

BMU (2006b): Renewable energy sources in figures – national and international development available from http://www.bmu.de/files/english/renewable_energy/downloads/application/pdf/broschuere_ee_zahlen_en.pdf

Butler, L. and K. Neuhoff (2005): 'Comparison of feed in tariff, quota and auction mechanisms to support wind power development', Cambridge Working Papers in Economics 0503, Faculty of Economics (formerly DAE), Cambridge: University of Cambridge.

Chow, J. and R. Newell (2004): 'A retrospective review of the performance of energy R&D RFF discussion paper' [in draft] available from: http://www.energycommission.org/files/finalReport/VI.1.e%20-%20Retrospective%20of%20Energy%20R&D%20Perf.pdf

Deutch, J. (2005): 'What should the Government do to encourage technical change in the energy sector?', MIT Joint Program on the Science and Policy of Global Change Report No.120. available from http://mit.edu/globalchange/www/MITJPSPGC_Rpt120.pdf

Department of Trade and Industry 2003): 'Innovation report - competing in the global economy: the innovation challenge' available from http://www.dti.gov.uk/files/file12093.pdf

E4tech. (2006): 'UK carbon reduction potential from technologies in the transport sector', prepared for the UK Department for Transport and the Energy Review team, available from http://www.dti.gov.uk/files/file31647.pdf

European Community (2005): 'The support of electricity from renewable energy sources', Communication from the Commission available from http://ec.europa.eu/energy/res/biomass_action_plan/doc/2005_12_07_comm_biomass_electricity_en.pdf

Fouquet D., C. Grotz, J. Sawin and N. Vassilakos (2005): 'Reflections on a possible unified eu financial support scheme for renewable energy systems (res): a comparison of minimum-price and quota systems and an analysis of market conditions'. Brussels and Washington, DC: European Renewable Energies Federation and Worldwatch Institute.

Foxon, T. J. (2003):. 'Inducing innovation for a low-carbon future: drivers, barriers and policies', London: The Carbon Trust, available from http://www.thecarbontrust.co.uk/Publications/publicationdetail.htm?productid=CT-2003-07

Freeman, C. (1992): 'The economics of hope', New York, Pinter Publishers.

Gibbs, W. (2006): 'Plan B for energy', Scientific American, September 2006 issue, New York: Scientific American.

Grubb, M. (2004): 'Technology innovation and climate change policy: an overview of issues and options', Keio Economic Studies. Vol.XLI. no.2. available from http://www.econ.cam.ac.uk/faculty/grubb/publications/J38.pdf

International Energy Agency (2006): '255" |Energy technology perspectives - scenarios & strategies to 2050', Paris: OECD/IEA.

International Energy Agency (in press): 'World Energy Outlook 2006', Paris: OECD/IEA.

International Rice Research Institute (2006): 'Climate change and rice cropping systems: Potential adaptation and mitigation strategies', available from www.sternreview.org.uk

Jaffe, A., R. Newell and R. Stavins (2004): 'A tale of two market failures', RFF Discussion paper, available from http://www.rff.org/Documents/RFF-DP-04-38.pdf

Jaffe, A., R. Newell and R. Stavins. (2003): 'Technological change and the environment', Handbook of Environmental Economics K. Mäler and J. Vincent (eds.), Amsterdam: North-Holland/Elsevier Science pp. 461–516

Kammen, D.M. and R. Margolis (1999): 'Evidence of under-investment in energy R&D in the United States and the impact of federal policy', Energy Policy, Oxford: Elsevier.

Kammen, D.M. (2004): 'Renewable energy options for the emerging economy: advances, opportunities and obstacles. Background paper for 'The 10-50 Solution: Technologies and Policies for a Low-Carbon Future' Pew Center & NCEP Conference, Washington, D.C., March 25-26, 2004, available from http://rael.berkeley.edu/old-site/kammen.pew.pdf

Kammen, D.M. and G.F. Nemet. (2005): 'Real numbers: reversing the incredible shrinking energy R&D budget', Issues in Science Technology, 22: 84–88

Kremer, M. and R. Glennerster (2004): 'Strong medicine: creating incentives for pharmaceutical research on neglected diseases', Princeton: Princeton University Press.

Nemet, G. F. (in press): 'Beyond the learning curve: factors influencing cost reductions in photovoltaics'. Energy Policy, Oxford: Elsevier.

Neuhoff, K. (2005): 'Large-scale deployment of renewables for electricity generation', Oxford Review of Economic Policy', Oxford University Press, 21 (1): 88–110, Spring.

Newell R. C. and N.E Wilson (2005): 'Technology prizes for climate change mitigation, resources for the future discussion paper #05-33', available from http://www.rff.org/rff/Documents/RFF-DP-05-33.pdf

Norberg-Bohm, V. and J. Loiter (1999): 'Public roles in the development of new energy technologies: The case of wind turbines'. Energy Policy, Oxford: Elsevier.

Norberg-Bohm, V. (2000): 'Creating incentives for environmentally enhancing technological change: lessons from 30 years of U.S. energy technology policy'. Technological Forecasting and Social Change 65: 125–148, available from http://bcsia.ksg.harvard.edu/publication.cfm?program=CORE&ctype=article&item_id=222

Norse D. (2006): 'Key trends in emissions from agriculture and use of policy instruments', available from www.sternreview.org.uk

OECD (2005): 'Innovation in the business sector working paper 459' Paris: OECD, available from http://www.olis.oecd.org/olis/2005doc.nsf/43bb6130e5e86e5fc12569fa005d004c/5d1216660b7d83dcc12570ce00322178/$FILE/JT00195405.PDF

OECD (2006): 'Do we have the right R&D priorities and programmes to support energy technologies of the future'. 18th Round Table on Sustainable Development background paper Paris: OECD, available from http://www.oecd.org/dataoecd/47/9/37047380.pdf

Pindyck, A.K. and R.S. Dixit (1994): 'Investment under uncertainty' Princeton, NJ: University Press.

Ragwitz, M. and C. Huber (2005): 'Feed-in systems in Germany and Spain and a comparison', Germany: Fraunhofer Institut für Systemtechnik und Innovationsforschung.

REN21 Renewable Energy Policy Network, (2005): 'Renewables 2005 Global Status Report'. Washington, DC: Worldwatch, available from http://www.ren21.net/globalstatusreport/RE2005_Global_Status_Report.pdf

REN 21. (2006): 'Renewables Global Status Report: 2006 update', Washington, DC: Worldwatch, available from http://www.ren21.net/globalstatusreport/download/RE_GSR_2006_Update.pdf

Scotchmer, S. (1991):' Standing on the shoulders of giants: cumulative research and the patent law', Journal of Economic Perspectives 5 (1): 29-41, Winter. Reprinted in The Economics of Technical Change (1993): Mansfield, E. and E. Mansfield, (eds.), Cheltenham: Edward Elgar Publishing.

Taylor, M., E. Rubin and G. Nemet (2006): 'The role of technological innovation in meeting California's greenhouse gas emissions targets', available from http://calclimate.berkeley.edu/3_Innovation_and_Policy.pdf

US Departments of Agriculture and Energy (2005): 'Biomass as feedstock for a bioenergy and bioproducts industry: the technical feasibility of a billion-ton annual supply', available from http://feedstockreview.ornl.gov/pdf/billion_ton_vision.pdf

17 Beyond Carbon Markets and Technology

KEY MESSAGES

Policies to price greenhouse gases, and support technology development, are fundamental to tackling climate change. However, **even if these measures are taken, barriers and market imperfections may still inhibit action, particularly on energy efficiency.**

These barriers and failures include hidden and **transaction costs** such as the cost of the time needed to plan new investments; **lack of information** about available options; capital constraints; misaligned incentives; as well as **behavioural and organisational factors** affecting economic rationality in decision-making.

These market imperfections result in **significant obstacles to the uptake** of cost-effective mitigation, and weakened drivers for innovation, particularly in markets for **energy efficiency** measures.

Policy responses which can help to overcome these barriers in markets affecting demand for energy include:

- **Regulation**: Regulation has an important role, for example in product and building markets by: communicating policy intentions to global audiences; reducing uncertainty, complexity and transaction costs; inducing technological innovation; and avoiding technology lock-in, for example where the credibility of carbon markets is still being established.

- **Information**: Policies to promote: performance labels, certificates and endorsements; more informative energy bills; wider adoption of energy use displays and meters; the dissemination of best practice; or wider carbon disclosure help consumers and firms make sounder decisions and stimulate more competitive markets for more energy efficient goods and services.

- **Financing**: Private investment is key to raising energy efficiency. Generally, policy should seek to tax negative externalities rather than subsidise preferable outcomes, and address the source of market failures and barriers. Investment in public sector energy conservation can reduce emissions, improve public services, fostering innovation and change across the supply chain and set an example to wider society.

Careful **appraisal, design, implementation and management** helps minimise the cost and increase the effectiveness of regulatory, information and financing measures. Energy contracting can reduce the costs of raising efficiency through economies of scale and specialisation.

Fostering a shared understanding of the nature and consequences of climate change and its solutions is critical both in shaping behaviour and preferences, particularly in relation to their housing, transport and food consumption decisions, and in underpinning national and international political action and commitment.

Governments cannot force this understanding, but can be a catalyst for dialogue through evidence, education, persuasion and discussion. And governments, businesses and individuals can all help to promote action through **demonstrating leadership**.

17.1   Introduction

Chapters 14, 15 and 16 have outlined the arguments, and appropriate policies, for establishing well-functioning carbon markets and encouraging technological research, development and diffusion. These are necessary to provide incentives and enable mitigation responses by households and firms. However, alone, they are not sufficient to elicit the necessary scale of investment and behavioural responses from households and firms due to the presence of failures and barriers in many relevant markets.

These obstacles are outlined in Section 17.2, in particular in relation to actions and investments for energy saving (although the framework is broadly applicable to other aspects of mitigation such as fuel switching). The significant untapped energy efficiency potential which exists, for example, in the buildings, transport, industry, agriculture and power sectors provides evidence of the impact of these failures and barriers.

Sections 17.3 to 17.5 outline the role of regulation, information and financing policies in responding to obstacles to energy efficiency:

- *Regulation:* such as forward-looking standards stimulate innovation by reducing uncertainty for innovators; encourage investment by increasing the costs and commercial risks of inaction for firms; and reduce technology costs by facilitating scale economies. In some respects regulation involves the creation of an implicit carbon price;
- *Information:* encourages efficient consumption and production decisions by raising awareness of the full energy costs and climate impacts; evidence and guidance on how to assess options and reduce energy bills can explicitly shape the direction and priorities for innovation;
- *Financing:* can accelerate the uptake of energy efficiency in both private and public sector.

Section 17.6 outlines issues relating to policy delivery. Section 17.7 discusses the role of public policy, information, education and discussion in influencing the perceptions and attitudes of individuals, firms and communities towards both adopting environmentally responsible behaviour and co-operating to reduce the impacts of climate change.

17.2   Market failures and responses to incentives

Behaviour is driven by a number of factors, not just financial costs and benefits.

For the most part, investment decisions in energy-using technologies rest on the balance of financial costs and benefits facing an individual or firm: for example, how much additional investment is required, what is the (opportunity) cost of capital and, in comparison, how much energy is the investment expected to save?

However, consumers and firms frequently do not make energy efficiency investments that appear cost-effective.[1] The IEA estimate that unexploited energy efficiency potential offers the single largest opportunity for emissions reductions, with major potential across all major end uses and in all economies. For example, energy efficiency accounts for between 31% and 53% of CO_2 emissions reductions by 2050 under the accelerated technology scenario (see Chapter 9 for a discussion of sources and costs of mitigation).[2]

It is difficult to explain low take up of energy efficiency as purely a rational response to investment under uncertainty.[3] This implies the existence of one or more of a potentially wider set of costs, market failures, or 'barriers'[4] to 'rational' behaviour and motivation. These fall into three main groups:[5]

- Financial and 'hidden' costs and benefits;
- Multiple objectives, conflicting signals, or, information and other market failures;
- Behavioural and motivational factors.

These are illustrated in the Figure 17.1 below. Standard economic theory of rational decision-making under uncertainty is important in understanding each. However, moving down this list, systems and behavioural theories of decision-making are progressively more relevant.

An assessment of the case for action has to take into account the existence of "hidden" costs and benefits.

---

[1] Individuals and firms should invest until the expected savings are equal to the opportunity cost of borrowing or saving (assuming risk neutrality). Studies suggest that individuals and firms appear to place a low value on future energy savings. Their decisions expressed in terms of standard methods of appraisal would imply average discount rates of the order of 30% or more. See, for example, analysis of consumer behaviour in markets for room air conditioners and home insulation in the US during the 1970's and 1980's by Hausman (1979) and Hartman and Doane (1986)). Also see Train (1985).

[2] IEA (2006)

[3] For example, Metcalf (1994) applies portfolio theory to show that investors should observe *lower* discount rates relative to the opportunity cost of capital, because reduced exposure to energy costs hedges against other risks. Dixit and Pindyck (1994) use 'option value' theory to explain relatively *higher* discount rates however Sanstad et al. (1995) show empirically, that these are not sufficient to explain the low take up of energy efficiency investment.

[4] See for example Blumstein et al. (1980), Grubb (1990). Also, see Mills (2002) for analysis of impacts of barriers on energy demand for lighting.

[5] Adapted from the Carbon Trust, *The UK Climate Change Programme: Potential Evolution for Business and the Public Sector*. London: The Carbon Trust. This framework was originally designed to evaluate markets for energy conservation in the business and public sector. However, it can be applied more broadly to other sectors and to other areas of mitigation such as fuel switching.

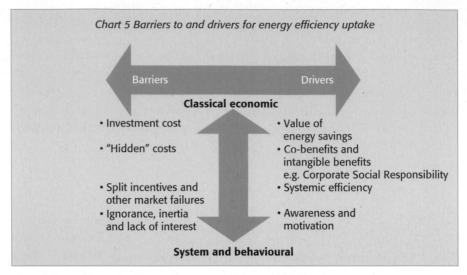

Figure 17.1 Barriers to and drivers for energy efficiency uptake[6]

Source: The Carbon Trust.

The primary driver of much investment in energy-using technologies is the balance of financial costs and benefits facing an individual or firm. However, accounting for "hidden" costs, such as those associated with researching different options, taking time off work to wait in for tradesmen, or the opportunity cost of devoting managerial time to efficiency projects is required for an assessment of the full range of costs and benefits.[7] These hidden costs may be counter-balanced by wider benefits such as reduced risk exposure to energy price volatility, or reputational benefits from demonstrating environmental responsibility.

Hidden or transaction costs are difficult to measure. One study found search and information costs of energy efficiency measures of between 3% and 8% of total investment costs.[8] Box 17.1 below summarises research highlighting the likelihood of significant transaction costs associated with energy efficiency measures. In general, these wider costs are expected to have most significant impact among small and medium-level energy users such as households, non-energy intensive and particularly small firms, as well as the public sector.

Individuals and firms are not always aware of the full costs and benefits of energy conservation, are capital constrained, or do not have sufficient incentives to invest.

Reliable, accessible and easily understandable information is important in making consumers and firms aware of the full lifetime costs and benefits of an economic decision, and hence supporting good decision-making. Whilst there are information difficulties in many or most markets, they may be particularly powerful in relation to energy efficiency measures.

---

[6] Framework designed in relation to energy efficiency markets but applicable more generally to mitigation (including fuel switching).

[7] Much of this argument relates to issues of transaction costs, see for example Williamson (1981, 1985).

[8] Hein and Blok (1994)

BOX 17.1 Estimating the costs of energy savings

Joskow and Marron (1992) undertook a study of the costs of information and particularly investment programs undertaken by energy suppliers designed to reduce demand among residential, commercial and industrial customers in the US. The authors identified a tendency for studies to underestimate the costs of actions to save energy,[9] in particular:

- *Supplier transaction costs*: full accounting for all administrative costs was likely to increase the cost per kWh saved by 10% to 20%. Supplier administration costs were likely to exceed 30% of the total for commercial and industrial programs;

- *Customer transaction costs and 'free riding'*: customer transaction costs varied from close to zero to close to 100% of the direct investment costs across the programmes sampled. 'Free riding'[10] was considered a significant risk particularly among the heaviest energy users within any target group. It was estimated that full accounting of these factors was likely to increase costs of demand side management programmes by about 25% to 50%;

- *Energy saving measurement issues*: The study identified significant methodological issues estimating energy savings given diverse, dynamic patterns of customer demand and limited availability of baseline information. In addition, they identified a tendency for widely used ex post engineering based forecasts to significantly overstate economic savings. Overall, accurate measurement of energy savings was considered likely to increase estimated costs by about 50%.

Capital and/or asset market failures also inhibit action. For example, a lack of available capital prevents people investing in more energy efficient processes which typically have higher upfront costs (but are cheaper overall when evaluated over a longer period). Restricted access to capital is especially common among poor households and small firms, particularly in developing countries.

Incentive failures restrict the effectiveness of price instruments. An example in the buildings sector is the 'landlord-tenant' problem in which landlords do not invest in the energy efficiency of their asset, because tenants benefit from lower energy bills, and more efficient capital typically does not command sufficiently higher rents.

Individuals and firms are not always able to make effective decisions involving complex and uncertain outcomes. Social and institutional norms and expectations strongly influence decision-making, although these norms are not immutable.

Some economists have suggested that people use simple decision rules when faced with complexity, uncertainty or risk.[11] For example, many people are unable to calculate the long-run value of energy savings, or have difficulties determining appropriate responses to risks and uncertainties around future energy costs or the

---

[9] Study compared costs against results of research by the Electric Power Research Institute and Rocky Mountain Institute (Lovins)

[10] An individuals or firm that takes advantage of financial support for a particular energy efficiency measure who would have invested without the additional incentive is a free rider in this context. This differs from the use of the term in the context of international agreements on climate change where non-signatories enjoy the benefits of mitigation but do not incur the costs, see Chapter 21.

[11] Kahneman & Tversky (1979, 1986, 1992) developed the idea of 'prospect theory' in which people determine the value of an outcome based on a reference point.

potential impacts of climate change. As a result, individuals and firms commonly make decisions which simply meet their needs, rather than undertaking complex analysis to determine the best possible decision.[12]

Shared social and institutional norms are important determinants of behaviour.[13] Individuals and firms behave habitually and in response to social customs and expectations. This leads to 'path dependency", which limits their responses to policies designed to raise efficiency (or encourage fuel switching). However, these norms change over time in response to a whole range of factors, including the influence of the media and action by governments. Developing and encouraging a shared concept of what responsible behaviour is, and of the consequences of irresponsible actions, is therefore an important aspect of policy (see Section 17.7).

17.3   Policy responses: regulation and standards, direct controls

Regulatory measures are less efficient and flexible than market mechanisms in the context of perfect markets, but can be an efficient response to the challenge of irremovable or unavoidable imperfections.

This section discusses the economic rationale for different types of regulatory policy instruments. As Chapter 14 discussed, regulatory measures are generally less efficient than market mechanisms when applied to perfect markets. However, the existence of market failures and barriers outlined in the previous section mean that there are circumstances in which standards and regulations have an important role to play.

Regulatory measures may be appropriate either instead of, or complementary to, tax or trading instruments, and can be more effective and efficient in a number of important circumstances, in particular to:

- Reduce the complexity faced by consumers or firms, by restricting or removing the availability of inefficient (or polluting) technologies, for example through banning of Chloroflourocarbons (or CFC's) in cooling systems;
- Cut the transaction costs associated with investments, through measures, for example by simplifying planning rules relating to the installation of micro-generation technologies;
- Overcome barriers to the transmission of incentives throughout the supply chain, for example, agreements with cable and satellite television providers have resulted in significant improvements in the efficiency of licensed 'set top' boxes;
- Stimulate competition and innovation, by signaling policy intentions, reducing uncertainty and increasing scale in markets for outputs of technological innovation;
- Promote efficiency through strategic coordination of key markets, for example by reducing long-run transport demand through integrated land-use planning and infrastructure development;

---

[12] See Simon, H.A. (1959) for concept of 'satisficing'. See also transcript of 2005 Bowman Lecture: Energy Demand – Rethinking from Basics, Professor David Fisk submitted to Stern Review Call for Evidence http://www.hm-treasury.gov.uk/media/F7E/46/climatechange-fisk_1.pdf

[13] This is commonly known as 'evolutionary' or 'procedural' rationality. See, for example, Goldstein, D. (2002), Decanio (1998)

- Overcome practical constraints on policymakers to imposing the appropriate explicit carbon price,[14] for example where this may be politically difficult to achieve or administratively expensive to implement directly through markets;
- Avoid capital stock 'lock in', particularly in markets which are subject to lengthy capital replacement cycles, for example buildings and power sectors.[15] This may be important where the credibility of carbon markets is still being established (issues discussed in Chapter 15).

Regulatory approaches, in contrast with market mechanisms, place a value on reducing greenhouse gas emissions implicitly rather than explicitly and can help reduce obstacles associated with information or other market failures. This value can be calculated by dividing the cost of the measure (to firms, consumers and regulators) by the estimated savings in greenhouse gas emissions. From the point of view of maximizing efficiency losses, it is important that the implied value of carbon, at the margin, is broadly the same whether market mechanisms or regulatory measures are used.

Performance standards help to limit energy demand by removing inefficient products from the market, and promoting mass diffusion of more efficient alternatives.

Performance standards establish requirements to achieve particular levels of energy efficiency or carbon intensity without prescribing how they are delivered. This can take the form of a minimum standard for a particular type of good, or a requirement on their average performance (commonly known as a 'fleet averages').[16]

Standards encourage the removal of poorly performing equipment from the market completely, or improve availability and uptake of more efficient alternatives. In addition, by projecting the future levels of performance which will be required, standards have the potential to encourage innovation towards the production of more efficient products: for example, US federal energy efficiency standards on room air conditioner and gas water heaters are estimated to have elicited energy efficiency improvements of approximately 2% per annum.[17]

The overall costs of regulation depend on the precise policy context. It is likely that performance standards induce the creation and adoption of new technologies although at some real opportunity cost.[18] Nevertheless, there are opportunities to promote efficiency at very low, or even negative cost, for example in certain product markets. Box 17.2 shows examples of effective performance based regulations. Section 17.6 outlines issues relating to design and implementation of performance standards.

---

[14] Equal to the expected marginal environmental cost.
[15] Note that, in some circumstances, poorly designed and managed regulation can cause technology lock in.
[16] Fleet averages, such as Corporate Average Fuel Economy vehicle standards, place average performance requirements on a particular type of good, thereby not mandating the removal of the poorest quality but rather incentivising patterns in the overall distribution of the efficiencies of products sold.
[17] Newell et al. (1999) using a model of induced product characteristics. Greening et al (1997) estimated the impacts of 1990 and 1993 national efficiency standards on the refrigerators and freezer units, using hedonic price functions, and found that the quality-adjusted price fell after implementation of standards. See also Magat (1979). However, in other instances, studies found no clear evidence of performance standards impacting on technological innovation. See For example, see Bellas (1998), Jaffe and Stavins (1995).
[18] See, for example, Palmer et al. (1995)

BOX 17.2   Successful performance standards programmes

Buildings: Building codes have been applied in many different countries.[19] In California, they are estimated to have saved approximately 10,000 GWh of electricity roughly equal to 4% of annual electricity use in 2003.[20] Studies of codes applied in Massachusetts and Colorado in have also demonstrated their potential to deliver energy saving.[21] In the UK, building regulations are expected to yield a cumulative saving of 1.4 MtC02 per year in 2010.[22] The EU Commission established a framework to realize an estimated cost-effective savings potential of around 22% of present consumption in buildings across the EU by 2010 as part of the European Energy Performance of Buildings Directive. In China, regulations are estimated to apply to buildings with a floor space of approximately 500 million square meters (among a total of approximately 40 billion nationwide) and have saved 36 MtCO2.[23]

Appliances: Since the introduction of federal standards by the US Department of Energy in 1978, total government programme expenditure is equivalent to US$2 per household. This is estimated to have delivered US$1,270 per household of net-present-value savings to the U.S. economy during the lifetimes of the products affected. Projected annual residential carbon reductions in 2020 due to these appliance standards are approximately 37 MtC02, an amount roughly equal to 9% of projected US residential carbon emissions in 2020.[24]

China first introduced appliance standards in 1989 and expanded their application rapidly during the 1990's to include, for example: refrigerators, fluorescent ballasts and lamps, and room air-conditioners. By 2010, energy savings are estimated to reach 33.5 TWh, or about 9% of China's residential electricity. This is equivalent to a CO_2 emission reduction of 11.3MtC02.[25] A more recent study highlighted the potential for significant energy savings in the longer term from more stringent performance standards on three major residential end uses: household refrigeration, air-conditioning, and water heating.[26]

Transport: Japan's Top Runner scheme, a leading programme of fleet averages in which future average performance requirements are based on current best available technologies, applies to a range of energy using products.[27] It is estimated to have delivered energy savings

[19] An OECD study: *Environmentally Sustainable Buildings – Challenges and Policies* found that 19 out of 20 countries surveyed had legislated mandatory building: http://www1.oecd.org/publications/e-book/9703011E.PDF#search=%22OECD%20study%3A%20Environmentally%20Sustainable%20Buildings%20-%20Challenges%20and%20Policies%20%22

[20] California Energy Commission (2005): http://www.energy.ca.gov/2005publications/CEC-400-2005-043/CEC-400-2005-043.PDF

[21] Evaluation of New Home Energy Efficiency: An assessment of the 1996 Fort Collins residential energy code and benchmark study of design, construction and performance for homes built between 1994 and 1999. Summary report June 2002 : http://www.estar.com/publications/Evaluation_of_New_Home_Energy_Efficiency.pdf XENERGY, 2001: Impact analysis of the Massachusetts 1998 residential energy Code revisions: http://www.energycodes.gov/implement/pdfs/Massachusetts_rpt.pdf

[22] Regulatory Impact Assessment, 2006 amendment to part L building regulation http://communities.gov.uk/pub/308/RegulatoryImpactAssessmentPartLandApprovedDocumentF2006_id1164308.pdf

[23] New Era of China Building Energy Saving, Speech by Mr. Zhang Qingfeng, Chairman of China Council of Construction Technology, April 10th

[24] Meyers (2002). Savings evaluated by comparing against base case estimated without policy intervention

[25] China Markets Group, Lawrence Berkeley Laboratories: http://china.lbl.gov/china_buildings-asl-standards.html

[26] Lin (2006)

[27] 'Top Runner' fleet average requirements are agreed on a voluntary basis between the Japanese government and industry. They apply to approximately 18 different groups of energy using technologies in a range of markets including appliances, heaters and vehicles.

on diesel passenger vehicles of 15% between 1995 and 2005 (and 7% on diesel freight vehicles). By 2010, it is expected to deliver energy savings on gasoline passenger vehicles of 23% (and 13% on passenger freight vehicles).[28]

In response to the introduction of Corporate Average Fuel Economy (CAFE) standards in the USA in 1975, the average fuel economy of new cars almost doubled and that of light trucks increased by 55% from 1975 to 1988.[29] Without these efficiency improvements it is estimated that the US car and light truck fleet would have consumed an additional 2.8 million barrels of gasoline per day in the year 2000 (about 14% of 2002 consumption levels).[30] However, the average rated fuel economy of new cars and light trucks combined declined from a high of 25.9 miles per gallon in 1987 to 23.9 miles per gallon in 2002, partly because of the shift from cars towards less efficient sport utility vehicles, pick-up trucks and minivans (which were classified as cargo transport under CAFE standards).

Design standards are inflexible, but can create scale economies for strategically important technologies.

Design standards mandate, or prohibit, the use of a particular technology. For example, CFC gases were prohibited in refrigerators in favour of alternative coolants, following the Montreal Protocol in 1987 and the establishment of a strong causal link with ozone depletion. Design standards and prohibitions are inflexible measures and, as such, risk being inefficient relative to performance standards or market mechanisms.

However, their application may be appropriate where a particular technological solution is highly preferable (or undesirable in the case of prohibitions) in the short term, where it is considered imperative to accelerate 'pull through' and create scale economies for a particular technology in the medium or longer term, or where alternative measures have proved unsuccessful. The need for medium term 'pull' through, for example, is likely to apply in the context of certain carbon capture and storage technologies since coal is a particularly damaging source of GHG's while it is likely to be widely used in power markets in a number of countries on grounds of cost and energy security (see Chapters 16 and 24 for details).

Urban design and land use planning regulations have the potential to facilitate a less energy intensive society, while balancing a range of wider economic and social objectives.

Planning rules and regulations balance a complex range of economic, social, and environmental objectives. However, their design and implementation can have important implications for mitigating climate change and also has the potential to influence the resilience to the impacts of climate change, for example, in the management of flood risks or water scarcity (these issues are examined in Part 5 of the report).

Achieving planning permission is often an important transaction cost when installing renewable energy technologies, such as wind turbines or solar panels, or energy conservation measures such as solar water heaters. This applies to both large-scale commercial as well as microgeneration installations (see Box 17.3 below).

---

[28] Top Runner Programme: Developing the World's Best Energy Efficient Appliance, Energy Conservation Centre Japan (2005): http://www.eccj.or.jp/top_runner/index.html

[29] Geller & Nadel (1994)

[30] National Academy of Sciences (2002) http://newton.nap.edu/books/0309076013/html/ 111.html

BOX 17.3 Microgeneration technologies

Microgeneration technologies produce thermal and/or electrical energy. Examples include small-scale wind, solar, hydro or combined heat and power installations, as well as heat pumps and solar water heaters. According to the Energy Saving Trust, micro-generation could supply 30–40% of UK electricity demand by 2050.[31]

Deployment of microgeneration capacity has the potential to reduce the carbon intensity of industrial, commercial, public as well as residential buildings and developments. In addition, it can reduce energy wastage compared to centralised systems.[32] Greater uptake could be driven by: consumers, energy suppliers and firms selling energy services, and the implementation of private wire networks by planners and developers (see Box 17.9 on Woking).

However, many of the technologies are currently expensive relative to the delivered price of conventional energy sources. Enabling investors to sell excess electricity at the real-time market price, and subject to distribution or other charges reflecting limited demand on low voltage networks, is key to their cost effectiveness: the use of smart meters in microgeneration installations is an important enabler.[33] Appropriate regulatory frameworks for energy markets and distribution networks are also important to achieving a level playing field.

Incentives to consumers and energy suppliers could accelerate the reduction of technology costs and promote diffusion. Finally, relaxation of planning rules also has the potential to reduce transaction costs and promote network effects through heightened awareness of these technologies.

Spacial and strategic planning can affect patterns of energy consumption. Higher-density urban environments, for example, typically consume less energy for transport and in buildings. In addition, land use controls such as restrictions on the availability and pricing of parking spaces, the use of pedestrian zones and parks, and land use zonal strategies (including congestion charging), have the potential to support integrated public transport to reduce the use of private motor vehicles.

Higher energy prices and rising congestion require central and municipal planners to develop mass transit systems to cope with inner city and suburban traffic such as: bus rapid transit, urban trams and relatively cheap light railway systems, in addition to subways for larger, higher density metropolitan centres. Such systems lead to large gains in energy efficiency and reduced emissions as passengers transfer from private cars to public transport.

The development of Dongtan in China provides an important example of the potential for sustainable urban development across the rapidly urbanising transition and developing economies of the world (see Box 17.4).

---

[31] Energy Savings Trust, Potential for Microgeneration Study and Analysis (2005) http://www.dti.gov.uk/files/file27558.pdf

[32] For example, an estimated 20% of the UK's CO_2 emissions result from energy wasted in the combustion, transmission and distribution of energy from centralised fossil fuel power plants. Greenpeace, Decentralising power: an energy revolution for the 21st century generation, transmission and distribution http://www.greenpeace.org.uk/MultimediaFiles/Live/FullReport/7154.pdf#search=%22greenpeace%20%2B%20microgeneration%22

[33] Unlocking the power house: policy and system change for domestic micro-generation in the UK. http://www.sussex.ac.uk/spru/documents/unlocking_the_power_house_report.pdf

BOX 17.4 Dongtan, Eco-City, Shanghai

Dongtan is situated on Chongming Island off the coast of Shanghai. This rural area is undergoing a rapid economic transformation into an 'eco city', facilitated by the construction of the Shanghai Yangtze River Tunnel bridge, which began in 2004, linking this region directly to the Shanghai conurbation.

Project engineers at Arup are working with Shanghai Industrial Investment Company to develop and construct Dongtan, an 86-square kilometer project, into a prosperous city which achieves a stable balance between economy, society and the environment. The city is being developed in phases but is expected to have a population of 25,000 by 2010 and around 80,000 after 2020, growing to a total of several hundred thousands in the longer term.

Dongtan will have highly energy efficient buildings powered by renewable energy sources including wind, solar and biofuels. Its energy intensity will be reduced through the use of passive energy systems: for example by making full use of natural sunlight to light public and private spaces or by varying the heights of buildings to reduce heating and cooling arising from adverse weather conditions. In addition, its waste will be recycled and composted.

Chinese policy makers and planners have been impressive in scaling up best practice to help achieve their objective to reduce the ratio of energy demand to output by 20% over 5 years. In the case of Dongtan, a high-speed rail link to Shanghai is planned, while the city itself is being designed in a compact, inter-linked way, supported by mixed patterns of land use, and a network of pedestrian and cycle routes, in order to reduce the demand for private motorised transport (and associated infrastructure costs).[34]

17.4 Policy responses: information policy

Information policies can achieve a number of objectives.

Well-designed information policies can:

- Provide people with a fuller picture of the economic and environmental consequences of their actions;
- Stimulate and provide the framework for market innovation and competition in environmentally friendly goods and services, for example through performance indicators and labels;
- Reduce the transaction costs associated with investments, by providing information on the energy use characteristics of different products or processes;
- Prompt people to take responsible action, by informing them about the wider implications of their choices and by highlighting public policy priorities.

Information policies take a number of forms. This section discusses a few generic types and their potential market applications including: labeling and certification, billing and metering, and policies to disseminate best practice.

Labels, certificates and endorsements raise the visibility of energy costs in investment decisions, promote innovation in product markets, and support procurement initiatives.

[34] Further information is available in the publication: Shanghai Dongtan: An Eco City, published by SIIC Dongtan Investment & Development (Holdings) Co., Ltd. Arup

The energy use, costs and environmental consequences of purchasing decisions commonly have low visibility, particularly when compared to the purchase price of a good.[35] Where such labels do exist, they can have a significant impact on consumer behaviour: organic certification and the FAIRTRADE mark are two examples (see Section 17.7 discussion of preferences for environmentally and socially responsible production and consumption).

In the field of energy efficiency, labels, certificates and endorsements support more rational purchasing decisions, by allowing people to make comparisons between competing goods on the basis of their operating cost and environmental impact. They also make it cheaper and easier for firms or the public sector to implement sustainable procurement policies.

Box 17.5 highlights a number of successful schemes. These vary in design, and include labels giving comparative information on energy use, and endorsements which state that a product meets a particular standard.

There are considerable opportunities for broader or more stringent application of performance and endorsement labels in key product areas such as: domestic lighting, consumer electronics, white goods, electric motors, boilers, air conditioning units, and office equipment.[36] Biogas is an example of an agricultural product that could have value as a renewable substitute for fossil fuels; establishing product standards supported by labeling can allow consumer demand to help to create this market.

The cost and regulatory burden of such measures should be taken into account when designing them; Section 17.6 outlines key principles for effective design and management. Such measures may be much more powerful if they are applied at an international level. The issues involved in this are discussed in Chapter 24.

BOX 17.5 Successful labels, certificates and endorsements in the US and EU

USA: The US *Energy Star* one of the best-known information and endorsement programmes, applying to over 30 products. It is estimated to have delivered annual savings of US$4.9 billion savings in 2002 (an increase of almost 30% over 2001). This is targeted to rise to US$55 billion in 2010 and US$140 billion in 2020.[37]

EU: The introduction of an EU labelling scheme on refrigerators is estimated to have delivered one-third of the 29% improvement in the energy efficiency of refrigeration products between 1992 and late 1999.[38] The figure below shows a clear and strong evolution of the market toward higher-efficiency products since the introduction of the EU label (contrasting favourably with the predominantly flat efficiency trends immediately prior to its announcement).

---

[35] Hassett and Metcalf (1995), for example, showed that consumers were much more responsive to changes in installation cost than change in energy prices. This is also inferred by the findings of Jaffe and Stavins (1995) which showed that consumers were about three times as sensitive to changes in technology costs than changes in energy prices.

[36] See for example IEA (2003), Lin (2006)

[37] Webber et al. (2004). Figures discounted at 4%. Potential savings of US$160 in 2010 and $US390 in 2020 are projected if 100% of products within particular classes are energy star compliant.

[38] Bertoldi (2000)

Impact of the EU refrigerator energy label: sales of refrigerators in the EU by energy label class 1992–2003.

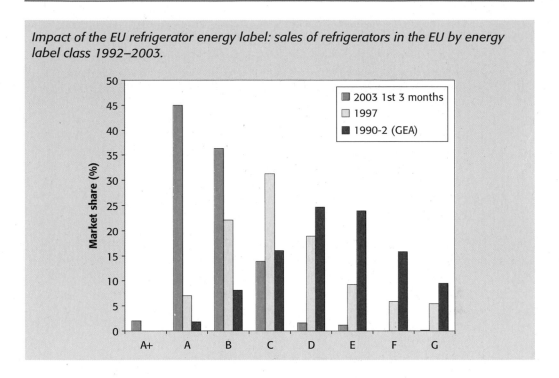

Regular and accurate energy billing, as well as displays and smart meters have the potential to promote conservation among energy users and reduce the operating costs of utilities.

Giving individuals and firms accurate and timely information on their energy use can act as a spur to investment in energy efficiency and the adoption of energy saving behaviours. New technologies are now available which have the potential to make this a much more powerful tool.

- **Energy bills** are most effective when they are regular, accurate, and informative. Bills which reveal historical patterns of energy consumption, and/or details on how consumption levels compare with a similar household or firm, are potentially effective in encouraging a response;[39] However, many people receive irregular bills, which are often based on estimated levels of consumption.[40] This problem is most prevalent among those consuming small and moderate quantities of energy such as households, small firms and those in non-energy intensive, service, or public sectors;
- **Real time electricity displays** inform consumers on energy consumption levels (and associated costs) directly and in real time. Estimated to cost in the region of £2–6 annually over 5 years,[41] they have been successful in encouraging energy

---

[39] Darby S. (2000) Wilhite, Hoivik and Olsen (1999) Eide and Kempton (2000) A recent survey for Ofgem suggested that consumers in the UK preferred bar charts highlighting consumption levels compared to relevant historical periods. http://www.ofgem.gov.uk/temp/ofgem/cache/cmsattach/8401_consumer_fdbak_pref.pdf

[40] For example, the UK Energy Review (2006) estimated that between 25 and 50% of all energy bills from UK energy suppliers were based on estimates.

[41] DTI Energy Review Report (2006) http://www.dti.gov.uk/files/file31890.pdf

conservation behaviours among households resulting in average reductions of 6.5% (net of technology costs).[42] Further development of a comparable display technology for metered gas supplies might extend these opportunities;

- **Smart meters** provide customers with sophisticated energy price and cost information. Those with "time of use" functionality enable flexible energy pricing. This allows suppliers to impose a higher price for peak-time energy, resulting in load shifting and consequently reducing base load capacity needs. Trials in California, for example, indicated reductions in peak period energy use by residential customers of between 8% and 17%;[43]

Smart meters with an 'export facility' encourage the diffusion of micro-generation capacity by enabling people to be paid at a different rate for the supply of their electricity into the local distribution network – which is critical to the cost effectiveness of these technologies in the medium term. Purchase and installation of smart meters are estimated to cost between £40 and £180 depending on function.[44] In addition to savings enjoyed by customers able to reduce peak level demand, Californian utilities recovered over 90% of the initial technology cost through savings made in metering, billing and systems.[45]

Sharing best practice encourages and enables individuals and firms to increase energy efficiency.

The energy efficiency of individuals and firms often varies widely within the same market. In transport, for example, particular styles of driving are more efficient than others. An in-car technology known as gear shift indicators which informs motorists when they should change gear in order to maximise fuel efficiency for any given engine speed could improve fuel economy by up to 5%.[46] In addition, methodologies for identifying best practice, for example through benchmarking, also have the potential to support wider policies on mitigation (see Box 17.6).

In the buildings sector, for example, large numbers of poor quality and inefficient buildings are constructed despite the existence of a range of cost effective technologies and design techniques. Training architects, designers and construction technicians on the principles and application of 'sustainable' design and efficient technologies, and on relevant policy frameworks develops market capacity to supply efficient buildings. However, coordinating different elements of the construction industries is a key barrier.[47]

The long term cost effective energy efficiency potential of a building is heavily determined by decisions made at the design phase (although there are widespread opportunities to retro fit technologies especially given the lengthy capital replacement cycle of buildings and often low performance of existing stock). As such, polices which target this window of opportunity may have significant potential to reduce emissions from buildings, especially in fast growing construction markets.

---

[42] A summary of the various studies can be found in: Darby S. (2006)
[43] California Energy Commission (2005) IEA (2006) identifies potential energy savings of 5–15% from 'smart' meters.
[44] DTI Energy Review Report (2006)
[45] California Energy Commission (2005)
[46] Presentation by Toyota as Stern Review Transport Seminar 12 January 2006 http://www.hm-treasury.gov.uk/media/B70/64/stern_transportseminar_toyota.pdf
[47] Lovins (1992), Golove and Eto (1996)

BOX 17.6 Benchmarking: driving conservation and facilitating mitigation policy

Benchmarking enables sharing of best practice and helps identify and encourage energy conservation opportunities. For example, the G8 communiqué from Gleneagles 2005 called on the IEA to benchmark the most efficient coal fired power stations and to identify ways of sharing best practice globally.[49] As previously outlined, benchmarking consumption patterns on energy bills has the potential to drive conservation among consumers and firms.

In addition, benchmarking methodologies facilitate the formulation and delivery of mitigation policies. For example, the UK used benchmarking to determine the allocations for new installations in the first phase of the EU ETS, and extended the methodology to incumbent large electricity producers in phase II. Under this approach, plants received emissions rights based on their capacity, output, and the carbon intensity of the particular generating technology. Individual emission rights were then reduced by a common factor calculated to meet the sector-wide cap. This provides an alternative approach to the allocations based on either the historic or projected emissions from individual installations (see Chapter 15 for issues on trading schemes and allocations).

In addition, benchmarking can be instrumental in determining a baseline upon which to formulate voluntary agreements (see Box 23.6 on the 1000 enterprises scheme in China), or establish an accreditation process under any technology based application of the CDM (see Box 23.5).

In the UK, the Carbon Trust, an independent but largely publicly funded company provides a range of advisory services to business of all sizes as well as the public sector. In 2005/06, the organisation helped its customers save between 1.1 and 1.6 MtC02 and identify potential savings of 3.9 MtC02 annually at an average lifetime programme administration cost of £5-7/tC02.[48]

Information provision, in conjunction with policies to deliver appropriate energy pricing, has strong potential to elicit energy savings. However, realising this requires effective intervention targeted across a broad range of sectors and economic activities.

17.5 Policy responses: financing mitigation

Investment by the private sector in efficiency measures is central to raising efficiency; governments have a limited but important role in supporting this.

Private investment is key to transforming the efficiency of energy-using markets. Generally speaking, if energy efficiency measures have a positive net present value there is little case for governments to intervene directly in their financing. For example, it should be a decision for energy supply companies whether to invest in facilitating demand reductions among customers or additional generating capacity depending on assessments of relative cost effectiveness.

---

[48] Caron Trust Annual Report 2005/6: www.carbontrust.co.uk Readers should also note active support for energy efficiency by the Energy Savings Trust. Information available at http://www.est.org.uk/

[49] http://www.fco.gov.uk/Files/kfile/PostG8_Gleneagles_Communique,0.pdf

In general, it is preferable to tax negative externalities rather than subsidise preferable outcomes.[50] Where possible, it is desirable to foster solutions to barriers or market imperfections, such as capital or technology market failures, at source for example, through markets for insurance or microcredit.[51] However, where such options are not available, carefully targeted provision of direct financial incentives such as loans, subsidies, and tax rebates are appropriate, in particular where:

- *Capital market failure*: Households or firms face a shortage or lack of access to capital. This may be particularly relevant to poorer households and to firms in developing countries (see Chapter 23 in relation to financing international energy efficiency). Alternatively, larger scale private investment, for example in major infrastructure projects, may be limited due to long return periods or a lack of credibility in carbon markets;
- *Technology market failure*: Support may significantly reduce long run technology costs. For example, direct support for next generation lighting technologies or micorgeneration technologies may increase the overall emissions reduction potential of the buildings sector by promoting economies scale markets and encouraging innovation for these technologies;
- *Delivery of wider policy objectives*: Financial support can create opportunities to deliver wider climate-related or social policy objectives. For example, in providing financial incentives, for example on building insulation, it may also be possible deliver information on a wider range of technologies such as advanced window glazing or lighting control systems. Alternatively, revenue from energy taxation or trading schemes may be used to overcome distributional and other perverse effects of policy.

There are examples in which incentives such as loans, subsidies, and tax rebates by public bodies, non-governmental organisation or energy suppliers have delivered significant energy savings: US demand side management programmes (of which the majority are financial incentives), for instance, saved approximately $1.78 billion of energy in 2000. This is at a cost equivalent to 3.4 cents kWh (less than half of the cost of end use consumption).[52] The Carbon Trust offers interest-free loans to small and medium sized firms in the UK to purchase energy efficient equipment. These realised 25 kT of CO_2 reductions in 2005/6 at a lifetime programme cost of £9 t/C02.[53] Box 17.7 outlines an example in which information provision and financing support can help overcome barriers to reducing emissions from agriculture.

Specialist management by energy service companies has the potential to reduce the cost of conserving energy among both private and public sector organisations (compared to a direct delivery mechanism). This is set out in Box 17.8 below.

---

[50] The costs of subsidies, for example, may be increased by the tendency for households or firms to take advantage of financial support for a particular energy efficiency measure who would have invested without the additional incentive: see Box 17.1.

[51] Microcredit is a form or finance designed to target poor people without sufficient collateral to have access to affordable private capital. See Yunus, M., Banker to the Poor: Micro-Lending and the Battle Against World Poverty

[52] Gillingham, Newell and Palmer (2004). Statistic assumes all energy saved is electricity and includes utility costs only.

[53] Caron Trust Annual Report 2005/6: www.carbontrust.co.uk

BOX 17.7 Support for deployment of anaerobic digesters in US agriculture

Anaerobic digesters store manure and allow it to decay in the absence of oxygen, producing biogas (a mixture of methane and CO2) which can be captured and combusted as an alternative to fossil fuels. Furthermore, heat generated in the process can be used, for example, to warm water or livestock units. The digestion process may also increase the value of the manure as a fertiliser.

Barriers to the uptake of this technology include upfront investment costs (estimated to be $500–600/cow)[54]; lack of information about the technology; high transaction costs associated with using the biogas as a power source; and planning regulations on the building of anaerobic digestors.

In the US, the AgSTAR programme encouraged the adoption of this technology by providing information to farmers.[55] State and federal funding was also made available in the form of interest subsidy payments, tax exemptions and loans.[56] In the last two years, the number of digesters in the US has more than doubled, reducing emissions by 0.6 MtCO2e annually and generating 120 million kWh of energy.[57]

BOX 17.8 Energy service contracting

Energy service contracting is a form of financial market transformation in which responsibility for designing, managing, or financing energy-using processes is outsourced to a third party (commonly known as an energy service company). In return, the company receives direct payment or a share of the financial benefits of delivered energy savings.

Energy service contracting can reduce energy costs by employing economies of scale and specialisation to overcome failures and barriers both within, and external to, industrial, commercial, public sector clients and, occasionally, households. Individual contracts vary widely but service companies may undertake audits, invest, install and/or manage energy systems.

Energy service markets are well established in countries such as the US, Germany and Austria. They are difficult to define but it is estimated that the US energy services industry has brought $8–15billion in net benefits.[58] In London, energy service contracting is at the heart of urban planners strategy to deliver low carbon energy solutions.[59]

Policy makers create the conditions for these markets to develop by: encouraging efficient energy and carbon markets, enabling service companies to access markets in public sector efficiency and by acting to facilitate local availability of capital (see Chapter 23 in relation to financing international efficiency).

---

[54] Minnesota Project (2002) Final report: Haubenschild Farms Anaerobic Digester: http://www.mnproject.org/pdf/Haubyrptupdated.pdf

[55] EPA AgSTAR Program, www.epa.gov/agstar

[56] EPA AgSTAR Funding on-farm biogas recovery systems: a guide to federal and state resources: http://www.epa.gov/agstar/pdf/ag_fund_doc.pdf

[57] EPA "AgSTAR digest winter 2006" http://www.epa.gov/agstar/pdf/2006digest.pdf

[58] Goldman et al (2005). Figure dependent on choice of discount rate.

[59] The London Climate Change Agency recently established the London ESCO, a public/private joint venture energy service company, with EDF Energy to deliver a range of planned mitigation projects, including the zero carbon development project recently announced by the Mayor. See: http://www.london.gov.uk/mayor/environment/energy/climate-change/edf-energy.jsp and *http://www.lcca.co.uk*.

Public sector investment in energy conservation has the potential to both reduce emissions and save public money

Public authorities are commonly the largest energy consumers in an economy, typically 10–20% of gross domestic product in both industrial and developing countries and a similar share of building floor space, energy use, and greenhouse gas emissions.[60]

There is widespread potential for cost-effective energy conservation across government buildings and state owned industrial facilities. For example, the public sector emits approximately 11% of the UK's total carbon emissions, and it is estimated that over 13% of this could be saved in a cost effective way.[61]

Raising energy efficiency in the public sector can both save public money and reduce emissions. In addition, there may be indirect benefits through fostering innovation and change across the supply chain, and demonstrating the desirability of, and potential for, action to wider society. Woking is an example of how effective this can be Box 17.9).

BOX 17.9 Woking Borough Council

Woking Borough Council is at the forefront of local authority efforts to tackle climate change in the UK.[62] In 2002, the Council adopted a comprehensive Climate Change Strategy designed to reduce greenhouse gas emissions, adapt to climate change, and promote sustainable development.

Between 1991 and March 2005, the Council's policies reduced energy consumption by almost 51% and carbon dioxide emissions by 79% across its own buildings. Between, March 2004 and March 2005, the Council purchased 82% of its electrical and thermal energy requirements from sustainable sources.

In 1999, the Council established an energy services company, Thameswey Energy Ltd., in conjunction with a commercial business partner, to finance sustainable and renewable energy projects. It has been instrumental, for example, in enabling the Council to install the town centre Combined Heat and Power station, which provides electricity, heat and power to the Civic Offices, the Holiday Inn Hotel and a number of other town centre customers. The Council also has a number of PV projects, accounting for approximately 10% of the UK's total installed capacity.

Woking Council is taking a leading role in promoting energy conservation and reducing carbon intensity across the municipality. It sponsors an energy efficiency advice centre, which provides free energy saving advice to residents. Furthermore the Council is currently investigating, in conjunction with Thameswey Ltd., the potential to deliver a number of wind turbines installations together with 1,000 low carbon homes with embedded micro generation across the Borough.

---

[60] Harris *et al.*, (2005, 2004, 2003)
[61] Carbon Trust (2005). Figures valid for 2002 based on a discount rate of 15% which is higher than the appropriate discount rates currently identified in the 'Green Book'.
[62] See the Councils climate change strategy for further information. http://www.woking.gov.uk/ environment/climatechangestrategy/climatechange.pdf

However, many of the barriers outlined in the earlier part of this chapter apply to the public sector, including capital constraints, information failures, landlord-tenant incentive failures, as well as institutional and behavioural barriers. Key issues in raising public sector efficiency include:

- *Allocating resources and overcoming capital constraints:* Short-term budgeting processes in the public sector may hinder the delivery of energy efficiency. Private sector energy contracting may also be useful in leveraging private investment in the public sector (see Box 17.9 for examples of such partnerships in London and Woking);
- *Establishing targets on energy efficiency:* As in the private sector, high-level targets can overcome behavioural and institutional barriers by focusing management attention and establishing accountability for delivery. Grading and comparisons between government departments and public organizations can further promote this competitive dimension;
- *Driving efficiency through public sector reform:* Reform of public services and state-owned enterprises, including the closure of inefficient facilities or their merging under more effective management, can directly drive energy efficiency. Examples include industrial restructuring and consolidation in China's iron and steel industry, and the power sector reforms discussed in Chapter 12;
- *Coordinated investment and planning of infrastructure and energy systems:* Coordinating systems such as water, waste, transport, and power can achieve energy savings. For example, planners in London are introducing cooling systems onto the underground network using absorption chilling technologies which convert waste heat from the buildings above;
- *Driving efficiency through procurement:* Governments are major procurers of energy using products (the US federal government alone accounts for 10% of the total market for energy using products).[63] Purchasing life-cycle cost-effective products reduces future public expenditure, as well as fostering innovation and driving the wider market in energy efficient products (see Box 17.10).

BOX 17.10   Driving efficiency through procurement

Since 1999, US guidelines have been in place requiring federal agencies to purchase Energy Star products over alternatives and, in product categories not covered by the endorsement scheme, only those products in the upper 25% of the distribution of efficiencies in the product class. It is estimated that this commitment will save between $160 and $620 million (or between 3% and 12% of total energy use in federal buildings) by 2010.[64] The size of the federal market delivers high participation rates among manufacturers: an estimated 95% of monitors, 90% of computers and almost 100% of printers sold are Energy Star compliant.[65]

Several US state and municipal governments have helped fuel market changes by adopting the federal efficiency criteria for their own purchases. If agencies at all levels of government adopt these same criteria, estimated electricity savings in the US would be

[63] Gillingham, Newell and Palmer (2004)
[64] Harris and Johnson (2000) Harris et al (2005)
[65] Webber et al. (2004)

18 TWh/year, allowing government agencies (and taxpayers) to save at least US$1 billion/year on their energy bills.[66]

The PROST study concluded that, for the EU as a whole, public sector investments of about € 80 million/year in program management and incremental purchase costs for buying energy-efficient products could reduce annual government energy costs by up to €12 billion/year.[67]

17.6 Policy delivery

Effective policy appraisal, design, implementation and management is essential in keeping down the costs and maximizing the effectiveness of policies to promote energy efficiency to firms, consumers and governments.

This section outlines general principles of policy delivery which help to reduce the costs to consumers, firms and governments and raise the effectiveness of polices to promote energy efficiency. In particular, it focuses on issues relating to the delivery of energy efficiency labelling, certification and endorsements as well as performance standards. Key principles are:

- *Effective policy signalling:* Paradoxically, the mark of a low-cost policy action is often the absence of an observable step-change in market behaviour, where planning, investment and market delivery mechanisms are allowed to respond, within normal economic cycles and in advance of the enforcement date. Good policy communication is essential to this process. Evidence of pre-commitment, perhaps in the form of voluntary agreements, throughout the supply chain indicates market preparedness. For example, transparent USA/ EU negotiations to revise Energy Star specifications for information and communication technologies (ICT), supported by a well informed dialogue with industry and experts on the technical potential, is expected to result in a very high level of compliance (with minimal impact on the price of new equipment) in advance of the new standards coming into force in Summer 2007;

- *Policy appraisal and prioritisation:* Thorough engineering, market and economic assessments of the likely costs and benefits of individual policy approaches enable strategic decisions on policy priorities.[68] Many product markets, such as those for appliances or ICT, are extremely dynamic, requiring regular re-appraisal of policy priorities. For example, the EU market for mobile phones has grown from hundreds of thousands to tens of millions in just a few years. Policy makers will need to respond to the challenge of rapid

---

[66] Harris and Johnson, (2000)

[67] Harnessing the Power of the Public Purse: Final report from the European PROST study on energy efficiency in the public sector http://195.178.164.205/library_links/downloads/procurement/PROST/PROST-fullreport.pdf

[68] Understanding this balance requires consideration of the risk of perverse incentives. For example, regulations which become stricter over time may delay the retirement of inefficient plant by making new installations relatively more expensive. See for example, Maloney and Brady (1988), Nelson et al. (1993), Stewart (1981), Gollop and Roberts (1983), McCubbins et al (1989). However, such secondary barriers may be correctable by, for example, suitable fiscal instruments.

project growth in demand for products such as: ICT technologies, power supplies, and digital television reception platforms ('set top boxes');

- *Monitoring and flexibility:* Careful and regular evaluation helps sustain a positive balance of costs and benefits throughout the lifecycle of a policy. As set out in Chapter 15, a degree of flexibility is required at the design stage to allow for a response to changing circumstances; for example, as a result of the success of the EU labelling scheme on refrigerators outlined in Box 17.5, the market is now saturated with 'A' performance graded products requiring the introduction of A+, A++ performance classifications;

- *Verification and reporting:* Well-defined testing protocols and procedures are particularly important foundations for the implementation of labels, endorsements and standards. Sound verification processes are essential to maintain policy credibility among producers, intermediaries, consumers and governments. For example, poor compliance is commonly cited as the key barrier increasing energy savings from building regulations, particularly in the developing world and transition economies where supporting institutional frameworks are typically weaker.

Policies can be mandatory, the subject of a voluntary agreement between public authorities and industry, or industry led. None of these approaches is universally preferable or appropriate. Regulatory policies may depend on the tacit agreement of industry and end-users. Voluntary strategies typically depend on implicit of explicit policy commitments to support the desired market transition, for example by regulatory underpinning or other sanctions. The choice of implementing strategy depends on:

- *Political culture of the implementing country:* public authorities often prefer to mandate policy to increase certainty around policy delivery. However, countries such as Japan have a strong culture of implementing policy based on voluntary consensus, which has been successful in ensuring high compliance with its Top Runner programme (see Box 17.2);

- *Market structure:* Voluntary agreements may be more readily achievable where capacity is concentrated among relatively few producers or retailers (and where there is some form of recognition of that commitment by government in its broader policy). For example, an EU voluntary agreement on set top boxes[69] has been successful in raising energy efficiency of satellite and cable platforms following support from major service providers. However less complete coverage of the more disparate market for freeview platforms, coupled with tough price competition, has resulted in relatively weaker improvements in standby and operating performance;

- *Implementation cost:* Regulatory approaches may be expensive to implement in some sectors. In agriculture, for instance, enforcement of regulations could be costly because sources of emissions are diffuse. Developing countries, in particular, may not have resources to establish or strengthen the required institutional structures or allocate appropriate resources more generally. However, the long run costs of inaction are often higher;

- *Timing:* Voluntary or industry led agreements may be quicker to implement, which may be useful where product markets are growing quickly or unexpectedly.

---

[69] The EU Code of Conduct for Digital Television Systems 2003

Regional or international action may take longer to organize than national action, but may be more powerful. Government objectives may be delivered faster and more efficiently by participating in and influencing established co-operative structures (for instance EU adoption of certain Energy Star protocols – see Box 17.4 for an outline of Energy Star and Chapter 24 for details on international policy management);

- *Delivery risk:* Information asymmetries between firms and governments on the costs and potential for innovation mean that voluntary and industry led measures may not achieve the full cost effective energy savings potential.[70] Investment in data collection help support more ambitious, cost-effective policy.[71]

The IEA publication on 'Labels and Standards' (2000) provides a useful outline of key principles and steps for developing policy while its report entitled 'Cool Appliances: Policy Strategies for Energy-Efficient Homes' (2003) is an excellent guide to consumer product markets. International aspects of the design, implementation and monitoring of tests and standards are outlined in Chapter 24).

17.7   Building a shared concept of responsible behaviour

Individual preferences play a key role, both in shaping behaviour, and in underpinning political action.

Most of economics assumes that individuals have fixed preferences and systems of valuations. It then examines policy largely in terms of 'sticks' and 'carrots', with the objective to increase welfare relative to this given set of preferences. This theory is powerful and central to most of the analysis of this Review, however it does not reflect the whole story.

Much of public policy is actually about changing attitudes. In particular, there are two broad areas where policy makers may focus in the context of climate change: seeking to change notions of responsible behaviour, and promoting the willingness to co-operate. Examples of the former in other areas include policies towards pensions, smoking and recycling while those of the latter include neighbourhood watch schemes on crime and community services more generally.

In the case of climate change, individual preferences play a particularly important role. Dangerous climate change cannot be avoided solely through high level international agreements; it will take behavioural change by individuals and communities, particularly in relation to their housing, transport and food consumption decisions.[72] There is clear evidence of shift towards environmentally and socially responsible consumption and production. For example, global sales of Fairtrade products increased by 37% to 1.1 billion Euro in 2005.[73]

---

[70] Cadot and Sinclair-Desgagne (1996) developed a game theoretic model solution for setting performance targets given asymmetric information regarding cost of technological advance.

[71] IEA/OECD (2003) Estimated data collection costs of approximately $1million to support revision of performance standards per product class.

[72] See 'I will if you will: towards sustainable consumption', a report by the Sustainable Development Commission. http://www.sd-commission.org.uk/publications/downloads/I_Will_If_You_Will.pdf

[73] Fairtrade Organisation Annual Report 2005: http://www.fairtrade.net/fileadmin/user_upload/content/FLO_Annual_Report_05.pdf

The actions and attitudes of individuals also matter when it comes to international collective action by governments. The most important force that will generate and sustain this action is domestic political demand in the key countries or regions (see Chapter 21 for discussion of collective action issues). Policies should therefore aim to create a shared understanding of the key issues. This is again an area where "policy" cannot be confined to the sticks/carrots and structural analysis standard in economics, although to emphasise once more that these approaches are absolutely crucial and, indeed, underlie most of the policy analysis of this report.

Refusing to move the argument beyond one of 'sticks' and 'carrots' would miss much that is important to policy formation on climate change. Alongside the influence of preferences in the community, leadership by governments, businesses and individuals is important in demonstrating how change is possible.

Governments can help shape preferences and behaviour through education, persuasion and discussion.

Crude attempts by government to "tell people what's best for them" tend to fail, and in any case raise ethical problems (see Chapter 2). The acceptability of "persuasion" requires public debate.[74] This dialogue may involve a range of actors, including the public sector, communities and individuals, NGOs, the media, and business. The public authorities can play a key role in helping to bring these elements together. For "government by discussion" as advocated by John Stuart Mill to work well, evidence and balanced argument which cuts through the complexity are crucial.

Polices designed to change preferences raise issues around the moral authority for action. There are examples of unacceptable public actions, such as deliberate misinformation in propaganda campaigns. However, most would view action to promote the understanding of climate change as appropriate – and, in fact, would view a failure to do so as irresponsible. This requires bringing to public attention the interests of those who might be ignored, such as future generations and those in poorer countries, and thinking through consequences of actions, as opposed to advancing the interests of narrow groups or excluding sections of the population.

The way in which issues and responses are communicated is critical. However, evidence suggests that people often see climate change discourse as confusing, contradictory and chaotic:[75] some approaches are alarmist, emphasising the scale of the problem (often rightly) but failing to acknowledge the potential for real action in response; others cast doubt on the human causes of climate change or optimistically assume that no response is necessary (Box 21.6 outlines public attitudes to climate change internationally.

Effective climate change discourse creates the conditions for positive behaviours by:

- Clear exposition of the existence and causes of the problem;
- Emphasising the potential for action using simple, positive messages. In particular, by tackling the disparity between the scale of the problem and the potential actions of households and firms so that the necessity of individual responses is broadly understood;

[74] See John Stewart Mill, 'On Liberty', where he advocated an approach to democracy based on government by discussion.

[75] See report commissioned by the Institute of Public Policy Research entitled, 'Warm Words: How are we telling the climate story and can we tell it better?' http://www.ippr.org.uk/publicationsandreports/publication.asp?id=485

- Targeting groups which share values (rather than demographics), working with individuals and community leaders to disseminate key messages, and using both evidential and moral arguments to engage people.

Ultimately, climate friendly behaviour will have to become well understood and highly valued (not simply the subject of campaign issues) in order for it to become a mass phenomenon.

Schools have an especially important role. Educating people from an early age about how our actions influence the environment is a vital element in promoting responsible behaviour. Creative and practical ways can be found to help pupils translate the study of climate change into actions in their everyday lives. For instance, practical examples of sustainability, such as installing wind turbines in school grounds, can help to provide pupils both with an understanding of the consequences of their actions and a tangible example of how behaviour, incentives and technologies can provide solutions.

Responsible behaviour can be encouraged through leadership.

Building a shared understanding of the problem, and of what responsible action means, is a key element in action. Leadership by the public sector, business, investors, communities and individuals can provide reassurance not only that action is possible, but also that it often has wider financial and other benefits.

Actions by central, regional and local governments and cities can have important demonstration effects that can be influence wider action, both by other governments and by the general public. Box 17.11 outlines California as an example

---

BOX 17.11   California: treating energy efficiency as a resource

California is the sixth largest economy in the world and has a long history of successful energy efficiency and conservation programs including building and appliance standards, and demand side reduction by the state's investor-owned and publicly owned utilities. This has resulted in lower energy intensity compared with other states or the country as a whole. Many of California's policies have been forerunners to federal government interventions establishing, for example, the nation's first standards for residential and non-residential buildings in 1978.

As of 2004, the state's Building and Appliance Standards and energy efficiency incentive and education programs have cumulatively saved more than 40,000 GWh of electricity and 12,000 MW of peak electricity, equivalent to 24 500 MW power plants. This has also increased fuel security, improved the competitiveness of its businesses, and saved consumers money.

In 2004, the California authorities adopted a set of aggressive energy conservation goals designed to help save the equivalent of 30,000 GWh between 2004 and 2013. If achieved, this would meet up to 59% of the investor-owned utilities' additional electricity requirements, and increase natural gas savings by 116% over the period.

To help support the delivery of these goals, the authorities have significantly increased allocations of public funding for cost effective energy efficiency programs to reduce peak electricity demand and increase natural gas efficiency. In addition, new appliance and building standards were introduced in 2005.[76]

---

[76] Californian Energy Commission (2005)

of an administration which has deliberately positioned itself as a leader, both in order to gain economic advantage through efficiency gains and technology development, and to inspire action both by its citizens and elsewhere.

A rapidly growing number of businesses are taking action on climate change policy. As discussed in Chapter 12, many are motivated by the desire to combine environmental responsibility and business profitability by increasing the energy efficiency of their business operations, or entering fast-growing environmental technology markets. The Carbon Disclosure Project provides evidence of a growth in the desire of businesses to report carbon footprints to investors.[77]

Many are also deliberately positioning themselves as leaders in this area. This may be driven by a desire to demonstrate responsible behaviour to the public and investors and use their leadership position to influence both government policy the conditions in which other businesses operate. For example, the Corporate Leaders Group on Climate Change recently called upon the UK Prime Minister to take bold steps to reduce climate change.[78]

Investors can also be a powerful voice for responsible action by businesses. The Socially Responsible Investment (SRI) movement grew out of a desire from individuals and organisations such as churches to invest their money in a way compatible with their own beliefs about what responsible behaviour means. Funds managed using some element of SRI principles have grown rapidly, with US assets under management totaling $2.29 trillion, almost 10% of assets under management in that country.[79]

More recently, concerns about how businesses treat social, ethical and environmental issues have become a more mainstream issue for investors, with a growing appreciation that failing to take account of these risks can directly threaten a company's financial health and reputation, for example, California state administration recently filed a law suit against 6 major vehicle manufacturers for alleged contributions to climate change. Organisations such as the Investor Network on Climate Risk in the US, and the Institutional Investor Network on Climate Change, have brought together concerned investors to have a dialogue with businesses on how they are responding to the challenge of climate change, and to encourage those who have neglected the issue so far to give it their active consideration.

17.8   Conclusion

Widespread failures and barriers in many relevant markets result in significant untapped energy efficiency potential in the buildings, transport, industry, agriculture and power sectors. These obstacles mean it is necessary to go beyond policies to establish carbon markets and encourage technological research, development and diffusion.

---

[77] Complete responses of GHG emissions from the world's largest 500 companies were up from 59% in 2005 to 71% in 2006. Carbon Disclosure Report 2006: http://www.cdproject.net/download.asp?file=cdp4_ft500_report.pdf

[78] http://www.cpi.cam.ac.uk/bep/clgcc/downloads/pressrelease_2006.pdf

[79] Social Investment Forum, January 2006: http://www.socialinvest.org/areas/news/2005 Trends. htm. This figure includes funds which involve at least one of the following elements: screening, shareholder engagement, and community investment.

Regulation can stimulate innovation by reducing uncertainty for innovators; encourage investment by increasing the costs and commercial risks of inaction for firms; reduce technology costs by facilitating scale economies, and influence more efficient outcomes in markets such as buildings, transport and energy using products. Policies to promote information, for example through labels, education programmes or technologies such as smart meters and real time displays, can encourage and develop capacity among households and firms to change their behaviour or make investments in energy savings.

Private investment is key to transforming the efficiency of energy-using markets. Generally, policy should seek to tax negative externalities rather than subsidise preferable outcomes, and address the source of market failures and barriers wherever possible (although there are cases for limited direct financial support to firms and individuals). Investment in public sector energy conservation can reduce emissions, improve public services, foster innovation and change across the supply chain and set an example to wider society.

Individual preferences play a key role, both in shaping behaviour and demand for goods and services affecting the environment, as well as in underpinning political action. Public policy on climate change should seek to change notions of what responsible behaviour means, and promote the willingness to co-operate. Education and promotion of clear discourse on the potential risks, costs and benefits together with leadership by the governments, businesses, investors, communities and individuals on the potential for action is critical.

References

The general reader seeking an overview of markets for energy efficiency should refer to the IEA's *Energy Technology Perspectives 2006*, which provides extensive information about failures and barriers, technological solutions, and policy options in sectors such as buildings, transport and industry. The Carbon Trust's publication for the UK Climate Change Programme, *The UK Climate Change Programme: Potential Evolution for Business and the Public Sector*, also provides a useful framework for understanding energy efficiency in different markets which can be applied more broadly. Chapter 6 of Michael Hanneman's M*anaging Greenhouse Gas Emissions in California*, informs the reader on a range of issues relating to energy efficiency including the debate between economists advocating market failures versus market barriers as a basis for policy intervention. The IEA's publication, the *experience of energy saving policies and programmes in IEA countries: learning from the critics*, highlights many of the criticisms commonly leveled at policies to promote energy efficiency and provides a useful introduction to more policy focused literature.

Bellas, A. S. (1998): 'Empirical evidence of advances in Scrubber Technology,' Resource and Energy Economics, 20:4 (December): 327–343
Bertoldi, P. (2000): 'European Union efforts to promote more efficient equipment', European Commission, Directorate General for Energy, Brussels: EC.
Blumstein, C., and B. Kreig, L. Schipper, C. York. (1980): 'Overcoming social and institutional barriers to energy efficiency', Energy 5 (4): 355–372
Cadot, O., and B. Sinclair-Desgagne (1996): 'Innovation under the threat of stricter environmental standards', in Environmental Policy and Market Structure, C. Carraro et al. (eds)., Dordrecht: Kluwer Academic Publishers, pp 131–141.
Carbon Trust (2005): 'The UK Climate Change Programme: potential evolution for business and the public sector;. London: The Carbon Trust.
DeCanio, Stephen J. (1998): 'The efficiency paradox: bureaucratic and organizational barriers to profitable energy-saving investments.' Energy Policy 26 (5), April: 441–454
Darby, S. (2000): 'Making it obvious: designing feedback into energy consumption'. Proceedings, 2nd International Conference on Energy Efficiency in Household Appliances and Lighting. Italian Association of Energy Economists/ EC-SAVE programme.

Darby S. (2006): 'The effectiveness of feedback on energy consumption'. A review for Defra of the literature on metering, billing and direct displays, London: Defra.

Dixit, A. K., and R. S. Pindyck (1994): 'investment under uncertainty'. Princeton, New Jersey: Princeton University Press.

Eide, A. and W. Kempton (2000): 'Comparative energy information in the US: lessons learned from a pilot innovative billing program', presented at AIEE 2nd International Conference on Energy Efficiency in Household Appliances and Lighting, Naples, Italy.

Geller, H. and S. Nadel (1994). 'Market transformation strategies to promote end-use efficiency.' Annual Review of Energy and the Environment 19: 301–46

Gillingham, K., R. Newell and K. Palmer (2004): Retrospective examination of demand-side energy efficiency, Washington, DC: Resources for the Future.

Goldman, C., N. Hopper, J. Osborn, and T. Singer (2005): 'Review of U.S. ESCO industry market trends: An Empirical Analysis of Project Data', LBNL-52320. January 2005, Berkeley, CA: Lawrence Berkeley National Laboratory.

Goldstein, D. (2002): 'Theoretical perspectives on strategic environmental management'. Journal of Evolutionary Economics 12: 495–524

Gollop, F.M., and M.J. Roberts (1983): 'Environmental regulations and productivity growth: The case of fossil-fueled electric power generation', Journal of Political Economy 91: 654–674

Golove, W.H. and J.H. Eto. (1996): 'Market barriers to energy efficiency: a critical reappraisal of the rationale for public policies to promote energy efficiency'. LBL-38059. Berkeley, CA: Lawrence Berkeley National Laboratory.

Greening, L.A., A.H. Sanstad and J.E. McMahon. (1997): 'Effects of appliance standards on product price and attributes: an hedonic pricing model'. Journal of Regulatory Economics 11: 181–194

Grubb, M.(1990): 'Energy policies and the Greenhouse Effect, vol.1 Policy Appraisal', Dartmouth: Chatham House.

Hanneman (2005): 'Managing greenhouse gas emissions in California', Berkeley, CA: California Climate Change Center.

Harris, J., et al. (2005): 'Public Sector Leadership: transforming the market for efficient products and services', in Proceedings of the 2005 ECEEE Summer Study: Energy Savings: What Works & Who Delivers? May 30 – 4 June, Mandelieu: ECEEE.

Harris, J., and F. Johnson (2000): 'Potential energy, cost, and C02 savings from energy-efficient government purchasing', in Proceedings of the ACEEE Summer Study on Energy-efficient Buildings, Asilomar, CA: ECEEE.

Harris, J., et al. (2004): 'Energy-efficient purchasing by state and local government: triggering a landslide down the slippery slope to market transformation', in Proceedings of the 2004 ACEEE Summer Study on Energy Efficiency in Buildings, Asilomar, CA: ECEEE.

Harris, J., et al. (2003): Using government purchasing power to reduce equipment standby power. In: Proceedings of the 2003 ECEEE Summer Study, Energy Intelligent Solutions for Climate, Security and Sustainable Development, June 2–7, 2003, St. Raphaël: ECEEE.

Hartman, R.S. and M.J. Doane (1986): 'Household discount rate revisited.' The Energy Journal 7(1): 139–148

Hassett, K.A., and G.E. Metcalf (1995): 'Energy tax credits and residential conservation investment: Evidence form panel data', Journal of Public Economics 57:201–217

Hausman, J. A. (1979): 'Individual discount rates and the purchase and utilization of energy-using durables', The Bell Journal of Economics 10(1): 33–54

Hein, L. and K. Blok (1994): 'Transaction costs of energy efficiency improvement.' Proceedings. European Council for an Energy-Efficient Economy.

International Energy Agency (2000): 'Labels and standards', Paris: OECD/IEA.

International Energy Agency (2002): 'Reducing standby power waste to less than 1 watt: a relevant global strategy that delivers', Paris: OECD/IEA.

International Energy Agency (2003): 'Cool Appliances: Policy Strategies for Energy-Efficient Homes', Paris: OECD/IEA.

International Energy Agency (2005): 'The experience of energy saving policies and programmes in IEA countries: learning from the critics', Paris: OECD/IEA.

International Energy Agency (2006): 'Energy technology perspectives', Paris: OECD/IEA.

Jaffe, A.B. and R.N. Stavins (1995): 'Dynamic incentives of environmental regulations: The effects of alternative policy instruments on technology diffusion', Journal of Environmental Economics & Management 29: S43–S63

Joskow, P., and D. Marron (1992) 'What does a negawatt really cost? Evidence from utility conservation programs.' The Energy Journal 13 (4): 41–74

Kahneman, D. and A. Tversky (1992): ' Advances in prospect theory: cumulative representation of uncertainty,' Journal of Risk and Uncertainty, Springer, 5(4): 297–323, October.

Kahneman, D. and A. Tversky (1986) 'Rational choice and the framing of decisions,' Journal of Business, University of Chicago Press, 59(4): S251–78, October

Kahneman, D. and A. Tversky (1979) 'Prospect Theory: An analysis of decision under risk,' Econometrica, Econometric Society, 47(2): 263–91, March.

Lovins, A. (1992): 'Energy-efficient buildings: institutional barriers and opportunities'. Boulder, CO: ESource, Inc.

Lin, J. (2006): 'Mitigating carbon emissions: the potential of improving efficiency of household appliances in China', Berkeley, CA: Lawrence Berkeley National Laboratory.

S. Meyers, J. McMahon, M. McNeil and X. Liu (2002): 'Realized and prospective impacts of U.S energy efficiency standards for residential appliances'. p. 42, June.

Magat, W.A. (1979): 'The effects of environmental regulation on innovation', Law and Contemporary Problems 43:3–25

Maloney, M.T., and G.L. Brady (1988): 'Capital turnover and marketable pollution rights', Journal of Law and Economics 31:203–226

McCubbins, M.D., R.G. Noll and B.R. Weingast (1989): 'Structure and process, politics and policy: Administrative arrangements and the political control of agencies', Virginia Law Review 75:431–482

Metcalf, G. E. 1994. 'Economics and rational conservation policy.' Energy Policy 22 (10), 81

Mills, E. (2002), 'Why we're here: the $230-billion global lighting energy bill', Proceedings from the 5th European Conference on Energy Efficient Lighting, held in Nice, France, 29–31 May, pp. 369–395

Nelson, R., T. Tietenberg and M. Donihue (1993): 'Differential environmental regulation: effects on electric utility capital turnover and emissions', Review of Economics and Statistics 75:368–373

Newell R., A.B. Jaffe and R. N. Stavins (1999): 'The induced innovation hypothesis and energy-saving technological change', The Quarterly Journal of Economics, 114(3): 941–975

Palmer, K., W.E. Oates and P.R. Portney (1995): 'Tightening environmental standards: The benefit-cost or the no-cost paradigm?' Journal of Economic Perspectives 9:119–132

Sanstad, A. H. and C. Blumstein, S.E. Stoft (1995): 'How high energy-efficiency investments?' Energy Policy 23 (9): 739–743

Simon, H.A. (1959): 'Theories of decision-making in economics and behavioural science'. American Economic Review XLIX: 253–283

Stewart, R.B. (1981): 'Regulation, innovation, and administrative law: A conceptual framework', California Law Review 69:1256–1270

Train, K. (1985) 'Discount rates in consumers', Energy-Related Decisions: A Review of the Literature.' Energy 10 (12): 1243–1253

Webber, C.A., R. E. Brown, and M.McWhinney (2004): '2003 status report savings estimates for the energy star(R) voluntary labeling program' (November 9, 2004). Berkeley, CA: Lawrence Berkeley National Laboratory.

Wilhite H, A., Hoivik and J-G. Olsen (1999): Advances in the use of consumption feedback information in energy billing: the experiences of a Norwegian energy utility. Proceedings, European Council for an Energy-Efficient Economy, 1999.Panel III, 02, Brussels: EC.

Williamson, O. E. (1981) 'The economics of organization: the transaction cost approach.' American Journal of Sociology 87 (3), November: 548–577

Williamson, O. (1985): 'The economic institutions of capitalism'. New York: The Free Press

Yunus, M., (1999): 'Banker to the poor: micro-lending and the battle against world poverty' New York: Public Affairs.

V   Policy Responses for Adaptation

Part V of the Review analyses adaptation as a response to climate change.

Climate is a pervasive factor in social and economic development – one so universally present and so deeply ingrained that it is barely noticed until things go wrong. People are adapted to the distinct climate of the place where they live. This is most obvious in productive sectors such as agriculture, where the choice of crops and the mode of cultivation have been finely tailored over decades, even centuries, to the prevailing climate. But the same is true for other economic sectors that are obviously weather-dependent, such as forestry, water resources, and recreation. It is also evident in how people live their daily lives, for instance in working practices.

Adaptation will be crucial in reducing vulnerability to climate change and is the only way to cope with the impacts that are inevitable over the next few decades. In regions that may benefit from small amounts of warming, adaptation will help to reap the rewards. It provides an impetus to adjust economic activity in vulnerable sectors and to support sustainable development, especially in developing countries. But it is not an easy option, and it can only reduce, not remove, the impacts. There will be some residual cost – either the impacts themselves or the cost of adapting. Without early and strong mitigation, the costs of adaptation rise sharply.

Part V is structured as follows.

- **Chapter 18** outlines key adaptation concepts and sets out an economic framework for adaptation.
- **Chapter 19** examines the barriers and constraints to adaptation identified in this chapter. It sets out how governments in the developed world can promote adaptation by providing information and a policy framework for individuals to respond to market signals.
- **Chapter 20** explores the particular issue of how developing countries can adapt to climate change. Developing countries lack the infrastructure, financial means, and access to public services that would otherwise help them adapt. The chapter shows the importance of support from the international community, and the need for investment in global public goods such as the development of resistant crops.

18 Understanding the Economics of Adaptation

KEY MESSAGES

Adaptation is crucial to deal with the unavoidable impacts of climate change to which the world is already committed. It will be especially important in developing countries that will be hit hardest and soonest by climate change.

Adaptation can mute the impacts, but cannot by itself solve the problem of climate change. Adaptation will be important to limit the negative impacts of climate change. However, even with adaptation there will be residual costs. For example, if farmers switch to more climate resistant but lower yielding crops.

There are limits to what adaptation can achieve. As the magnitude and speed of unabated climate change increase, the relative effectiveness of adaptation will diminish. In natural systems, there are clear limits to the speed with which species and ecosystems can migrate or adjust. For human societies, there are also limits – for example, if sea level rise leaves some nation states uninhabitable.

Without strong and early mitigation, the physical limits to – and costs of – adaptation will grow rapidly. This will be especially so in developing countries, and underlines the need to press ahead with mitigation.

Adaptation will in most cases provide local benefits, realised without long lag times, in contrast to mitigation. Therefore some adaptation will occur autonomously, as individuals respond to market or environmental changes. Much will take place at the local level. Autonomous adaptation may also prove very costly for the poorest in society.

But adaptation is complex and many constraints have to be overcome. Governments have a role to play in making adaptation happen, starting now, providing both policy guidelines and economic and institutional support to the private sector and civil society. Other aspects of adaptation, such as major infrastructure decisions, will require greater foresight and planning, while some, such as knowledge and technology, will be of global benefit.

Studies in climate-sensitive sectors point to many adaptation options that will provide benefits in excess of cost. But quantitative information on the costs and benefits of economy-wide adaptation is currently limited.

Adaptation will be a key response to reduce vulnerability to climate change. Part II highlighted the significant impacts of climate change around the world. The Earth has already warmed by 0.7° C since around 1900. Even if all emissions stopped tomorrow, the Earth will warm by a further 0.5 - 1° C over coming decades due to the considerable inertia in the climate system. On current trends, global temperatures could rise by 2 - 3° C within the next fifty years or so, with several degrees more warming by the end of the century if emissions continue to grow.

But adaptation is not an easy or cost-free option. This Chapter outlines key adaptation concepts and sets out an economic framework for adaptation. It highlights that adaptation is unlikely to reduce the net costs of climate change to zero – namely there will be limits. There will often be residual damages from climate change and adaptation itself will bring costs. The final part of the chapter outlines why policies may be required to overcome barriers and constraints to adaptation in anticipation of future impacts. These policy responses are outlined in more detail in Chapters 19 and 20 for developed and developing countries, respectively.

But even with a policy framework in place, there will be limits to or sharply rising costs of adaptation – for the most vulnerable at moderate levels of warming (e.g. ecosystems, the poorest regions), and for all parts of the world with higher amounts of climate change (4 or 5° C of warming). Developing countries are especially vulnerable to the negative effects of climate change. They are geographically vulnerable, located where climate change is likely to have often damaging impacts, and – as explained in Chapter 20 – are likely to have the least capacity to adapt. Chapter 26 in Part VI picks up this story and outlines how the international community can help developing countries deal with these impacts.

18.1   Role of adaptation

Adaptation is a vital part of a response to the challenge of climate change. It is the only way to deal with the unavoidable impacts of climate change to which the world is already committed, and additionally offers an opportunity to adjust economic activity in vulnerable sectors and support sustainable development.

A broad definition of adaptation, following the IPCC, is any adjustment in natural or human systems in response to actual or expected climatic stimuli or their effects, which moderates harm or exploits beneficial opportunities.[1] The objective of **adaptation** is to reduce vulnerability to climatic change and variability, thereby reducing their negative impacts (Figure 18.1). It should also enhance the capability to capture any benefits of climate change. Hence adaptation, together with mitigation, is an important response strategy. Without early and strong mitigation, the costs of adaptation will rise, and countries' and individuals' ability to adapt effectively will be constrained.

Adaptation can operate at two broad levels:[2]

- **Building adaptive capacity** – creating the information and conditions (regulatory, institutional, managerial) that are needed to support adaptation. Measures to build adaptive capacity range from understanding the potential impacts of climate change, and the options for adaptation (i.e. undertaking

---

[1]   Intergovernmental Panel on Climate Change (IPCC) (2001), Chapter 18
[2]   UKCIP (2005) Measuring progress, Chapter 4

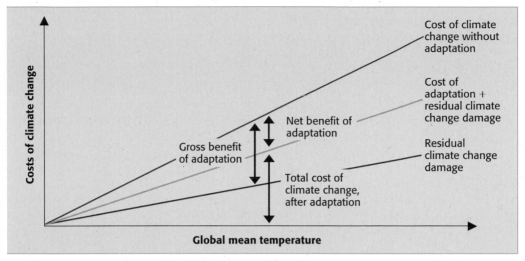

Figure 18.1 The role of adaptation in reducing climate change damages

Adaptation will reduce the negative impacts of climate change (and increase the positive impacts), but there will almost always be residual damage, often very large. The gross benefit of adaptation is the damage avoided. The net benefit of adaptation is the damage avoided, less the cost of adaptation.

The residual cost of climate damage plus the cost of adaptation is the cost of climate change, after adaptation.

For the sake of simplicity, the relationships between rising temperatures and the different costs of climate change/adaptation are shown as linear. In reality, Part II and Chapter 13 demonstrated that the costs of climate change are likely to accelerate with increasing temperature, while the net benefit of adaptation is likely to fall relative to the cost of climate change.

impact studies and identifying vulnerabilities), to piloting specific actions and accumulating the resources necessary to implement actions.

- **Delivering adaptation actions** – taking steps that will help to reduce vulnerability to climate risks or to exploit opportunities. Examples include: planting different crops and altering the timing of crop planting; and investing in physical infrastructure to protect against specific climate risks, such as flood defences or new reservoirs.

18.2 Adaptation perspectives

Some adaptation will occur autonomously, as individuals respond to changes in the physical, market or other circumstances in which they find themselves. Other aspects will require greater foresight and planning, e.g. major infrastructure decisions.

Adaptation is different from mitigation because: (i) it will in most cases provide local benefits, and (ii) these benefits will typically be realised without long lag times. As such, many actions will be taken 'naturally' by private actors such as individuals, households and businesses in response to actual or expected climate change, without the active intervention of policy. This is known as 'autonomous' adaptation.

Table 18.1 Examples of adaptation in practice

Type of response to climate change	Autonomous	Policy-driven
Short-run	• Making short-run adjustments, e.g. changing crop planting dates • Spreading the loss, e.g. pooling risk through insurance	• Developing greater understanding of climate risks, e.g. researching risks and carrying out a vulnerability assessment • Improving emergency response, e.g. early-warning systems
Long-run	• Investing in climate resilience if future effects relatively well understood and benefits easy to capture fully, e.g. localised irrigation on farms	• Investing to create or modify major infrastructure, e.g. larger reservoir storage, increased drainage capacity, higher sea-walls • Avoiding the impacts, e.g. land use planning to restrict development in floodplains or in areas of increasing aridity.

In contrast, policy-driven adaptation can be defined as the result of a deliberate policy decision.[3] Autonomous adaptation is undertaken in the main by the private sector (and in unmanaged natural ecosystems), while policy-driven adaptation is associated with public agencies (Table 18.1) - either in that they set policies to encourage and inform adaptation or they take direct action themselves, such as public investment. There are likely to be exceptions to this broad-brush rule, but it is useful in identifying the role of policy. *The extent to which society can rely on autonomous adaptation to reduce the costs of climate change essentially defines the need for further policy.* Costs may be lower in some cases if action is planned and coordinated, such as a single water-harvesting reservoir for a whole river catchment rather than only relying on individual household water harvesting. The primary barriers to autonomous adaptation will be discussed in Section 18.5.

The distinction between short-run and long-run adaptation is linked to the appropriate pace and flexibility of adaptation options (Box 18.1). In the short run, the decision maker's response to climate change and variability is constrained by a fixed capital stock (e.g. physical infrastructure), so that the principal options available are restricted to variable inputs to production. For example, a farmer can switch crops and postpone or bring forward planting dates in response to forecasts about the forthcoming growing season. On the other hand, major investments in irrigation infrastructure cannot be made reactively on such a timescale. Evaluating such investments requires expectations to be formed on costs and benefits over several decades, which places a challenging requirement on climate and weather forecasting. If the climate changes faster than expected, infrastructure could become obsolete before its planned design-life or require a costly retrofit to increase resilience.

---

[3] This is sometimes referred to as "planned adaptation" in the literature.

BOX 18.1   Adaptation actions with fixed and variable capital stock

The difference between short-run and long-run adaptation decisions can be explained using the following illustrative diagram. In any given year, the output of an economy is generated using three types of inputs, capital K, variable inputs to production and environmental quality E.

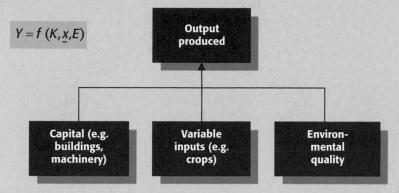

$$Y = f(K, \underset{\sim}{x}, E)$$

In the short run – and given a change in E due to climatic change and variability over a short period of time (e.g. one year) – the decision maker, who seeks to maximise the net profits of production, can respond only by changing variable inputs (shown as red in the diagram below).

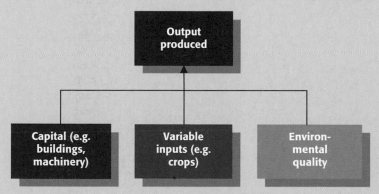

In contrast, in the long run (e.g. over 30 years), the decision maker can respond by changing both variable inputs and the capital stock to maximise profits given a change.

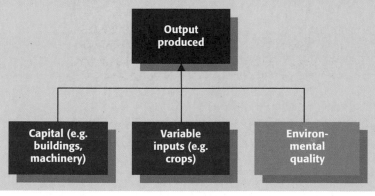

Adaptation will occur in practice in response to particular climate events and in the context of other socio-economic changes.

Responding to changed climate and weather (for example the appearance of stronger and more frequent floods or storms) is often an important first step for adaptation. Enhancing these responses to prepare for future impacts is the second step – for example, by using drought-resistant crops or improving flood defences. Many decisions to adapt will be made autonomously, within existing communities, markets and regulatory frameworks. This has important consequences for the way economists understand and appraise adaptation policy.

First, much adaptation will be triggered by the way climate change is experienced. Climate variability and in particular extreme weather, such as summer heat waves or storms, are likely to constitute important signals, alongside the dissemination of knowledge and information. Since adaptive capacity is related to income and capabilities, the most vulnerable in society will experience the same negative climate impacts more acutely.

Second, many adaptation decisions involve a measure of habit and custom, especially smaller decisions made by, for example, individuals, households and small businesses on short time-scales and with small amounts of resources. This effect may limit the extent to which such adaptations will be orientated towards maximising net benefits in an economic and social sense, since 'custom' may have been based on responding to past climate patterns.

Decisions about the timing and amount of adaptation require that costs and benefits are compared.

An appraisal of any particular method of adaptation should compare the benefits – which are the avoided damages of climate change – with the costs, appropriately discounted over time (see Chapter 2 and Annex 2A for a discussion of discounting). The adaptation route that is chosen should be the one that yields the highest net benefit, having taken account of the risks and uncertainties surrounding climate change (see Box 18.2 for a risk management framework).[4]

BOX 18.2 Adaptation costs and benefits.

The table below presents a simple framework for thinking about the costs and benefits of adaptation.[5] The columns reflect two climate scenarios, one with and one without climate change (T_0 and T_1 respectively). The two rows represent two adaptation options, one which is best to pursue without climate change and one which is best to pursue with climate change (A_0 and A_1 respectively).

The top left box represents the initial situation, where society is adapted to the current climate (T_0, A_0). The bottom right box represents a situation where society adapts (A_0 to A_1) to a change in climate from T_0 to T_1.

The top right box represents a situation where society fails to adapt to the change in climate. Finally, the bottom left box represents a counterfactual situation where society

---

[4]   Callaway and Hellmuth (2006); Willows and Connell (2003).
[5]   Drawing on a framework originally presented by Fankhauser (1997) and modified by Callaway (2004).

undertakes adaptation (A_0 to A_1), but the climate does not in the end change. This is an example of the type of situation that could arise if climate does not change in the anticipated way.

Adaptation costs and benefits

Adaptation Type	Existing Climate (T_0)	Altered Climate (T_1)
Adaptation to existing climate (A_0)	Existing climate. Society is adapted to existing climate: (T_0, A_0), or Base Case	Altered climate. Society is adapted to existing climate: (T_1, A_0).
Adaptation to altered climate (A_1)	Existing climate. Society is adapted to altered climate: (T_0, A_1).	Altered climate. Society is adapted to altered climate: (T_1, A_1).

The various costs and benefits of adapting to climate change follow from this, and can be thought of along the following lines:

- *Climate change damage* is the welfare loss associated with moving from the base climate (top left) to a changed climate without adaptation (top right): $W(T_1,A_0) - W(T_0,A_0)$.

- *Net benefits of adaptation* are the reduction in damage achieved by adapting to the changed climate (net of the costs of doing so), subtracting the top right box from the bottom right box: $W(T_1,A_1) - W(T_1,A_0)$.

- *Climate change damage after adaptation* is the difference between social welfare in the bottom right box and in the top left box: $W(T_1,A_1) - W(T_0,A_0)$.

Uncertainty over the nature of future climate change is implicit in this framework, and is one of the principal challenges facing climate policy. The second table below therefore modifies the framework to illustrate the trade-offs facing those planning adaptation under uncertainty.[6] The decision to implement an adaptation strategy should take account of the balance of risks and costs of planning for climate change that does not occur and vice versa.

Where the cost of planning for climate change is low, but the risks posed by climate change are high (top right box), there is a comparatively unambiguous case for adaptation. In contrast, where the costs of adaptation are high but the risks posed by climate change are low (bottom left box), the proposed adaptation responses may be disproportionate to the risks faced. Where the costs of planning for climate change and the risks of climate change are both low (top left box), there is little risk to the situation and the downsides are small, regardless of the choice made. In contrast, where the costs of both 'mistakes' are high, the stakes and risks are very high for the planner.

Cost of planning for climate change	Risks of climate change	
	Low	High
Low	Low risk	Plan for climate change
High	Don't plan for climate change	High risk

---

[6]   Callaway and Hellmuth (2006).

More quantitative information on the costs and benefits of economy-wide adaptation is required. For some specific sectors - such as coastal defences and agriculture – some studies indicate that efficient adaptation could reduce climate damages substantially.

As Chapter 6 explained, adaptation is an important component of integrated assessment models that estimate the economy-wide cost of climate change at the regional and global levels. However, these models are currently of limited use in quantifying the costs and benefits of adaptation, because the assumptions made about adaptation are largely implicit. Adaptation costs and benefits are rarely reported separately.[7]

However, for some sectors that are especially vulnerable to climate change, illustrative studies have been undertaken. As with the IAMs discussed in Chapter 6, many assumptions must be made to project costs and benefits over long periods of time. Assumptions about population and economic growth are especially important for evaluating the benefits of adaptation expressed in terms of avoided damage.

For coastal protection, the avoided damages of climate change can be calculated from the value of land, infrastructure, activities and so on protected by sea walls, while the cost of sea walls can be calculated by scaling up from engineering estimates of construction costs. Coastal protection should – in theory – occur up to the point where the cost of the next unit of protection is just equal to the benefit. In general, these studies suggest that high levels of protection may be economically efficient and reduce the costs of land loss substantially.[8] According to one recently analysis, the effectiveness of adaptation declines with higher amounts of sea level rise. This analysis found that for 0.5-m of sea level rise damage costs were reduced by 80 – 90% with enhanced coastal protection than without, while the costs were only reduced by 10 – 70% for 1-m of sea level rise.[9] For most countries, protection costs based on these calculations are likely to be below 0.1% of GDP, at least for rises up to 0.5-m. But for low-lying countries or regions, costs could reach almost 1% of GDP.[10] For 1-m of sea level rise, the costs could exceed several percent of GDP for the most vulnerable nations.[11]

In agriculture, adaptation responses could be even more diverse, ranging from low-cost farm-level actions – such as choice of crop variety, changes in the planting date, and local irrigation – to economy-wide adjustments – including availability of new cultivars, large-scale expansion of irrigation in areas previously

---

[7] Tol *et al.* (1998)

[8] Fankhauser (1995) assumes no population or GDP growth and finds that almost total protection of all coastal cities and harbours in OECD countries would be optimal (e.g. greater than 95% land area protected) and around 80% of open coastline. By allowing for population and GDP growth in line with IPCC scenarios, Nicholls and Tol (2006) find that protecting at least 70% of coastline in most parts of the world could be an optimal protection response.

[9] Anthoff *et al.* (2006) analysing data from Nicholls and Tol (2006) for the decade 2080 - 2089. Costs were calculated as net present value in US $ billion (1995 prices). Damage costs include value of dryland and wetland lost and costs of displaced people (assumed in this study to be three times average per capita income). The ranges represent results for different IPCC socio-economic scenarios with different population and per capita GDP growth trajectories over time.

[10] Analysis in Nicholls and Tol (2006)

[11] This analysis considers only protection costs required to manage loss of land from permanent inundation and not the costs of protection to deal with episodic flooding, which could cause damages an order of magnitude greater (Chapter 5).

Table 18.2 Benefits of adaptation in agriculture

Study	Climate scenario	Type of adaptation	Climate Impacts		Impact change of adaptation
			without adaptation	with adaptation	
Easterling *et al* (1993) Missouri, Iowa, Nebraska, Kansas (MINK)	1930s climate analogue; base year 1980s	Change in planting date tillage practices, change in crops, improved irrigation and crop drought resistance	**Yield change ($ bn)** −1.33 to −2.71	−0.53 to − 1.92	**% impact reduction** 29 – 60
Rosenzweig and Parry (1994)	2 × CO$_2$ base year 2060	Small shifts in planting date (<1 month), change in crops, additional irrigation ('level 1 adaptation')	**Change in cereal production %**		**% impact reduction**
Developed countries			−3.5 to 11.3	4.0 to 14.0	24 – > 100
Developing countries			−10.8 to −11.00	−9.0 to −12.0	9 – 17
World			−1.2 to −7.6	0.0 to −5.0	34 – 100
Adams et al. (1993) United States	2 × CO$_2$ base year 1990	As Rosenzweig and Parry (1994)	**Welfare change ($ bn)** 2.15 to −13.00	10.82 to − 9.03	**% impact reduction** >100
Reilly et al. (1994)[a]	2 × CO$_2$ base year 1989	As Rosenzweig and Parry (1994)	**Welfare change ($ bn)**		**% impact reduction**
Developing countries					
GDP/cap <$500			−2.07 to −19.83	−0.21 to −10.67	26–90
GDP/cap $500 – 2000			−1.80 to −15.01	−0.43 to −10.67	41–76
GDP/cap >$2000			−0.33 to − 0.82	−0.60 to − 1.02	20–46
E. Europe and former			1.89 to −10.96	2.42 to −4.88	29–56
USSR			2.67 to −15.10	5.82 to −6.47	57 −> 100
OECD			−0.13 to −61.23	7.00 to −37.62	39 −> 100
World					

[a] Based on Rosenzweig and Parry (1994) yield date
Source: Table reproduced from Tol et al. (1998)

only rain-fed, widespread fertiliser application, regional/national shifts in planting date. Some studies suggest that relatively simple and low-cost adaptive measures, such as change in planting date and increased irrigation, could reduce yield losses by at least 30 - 60% compared with no adaptation (Table 18.2).[12] But adaptation gains will be realised only by individuals or economies with the capacity to undertake such adjustments. The costs of implementing adaptation, particularly the transition and learning costs associated with changes in farming regime, have not been clearly evaluated.

18.3   Barriers and limits to adaptation

In many cases, market forces are unlikely to lead to efficient adaptation.

Broadly, there are three reasons for this:

- **Uncertainty and imperfect information;**
- **Missing and misaligned markets,** including public goods;
- **Financial constraints,** particularly those faced by the poor.

Policies can reduce these problems (see Chapters 19 and 20). But policy-makers themselves face imperfect information and have their own organisational challenges. Difficult policy choices may not always be tackled head-on.[13]

Uncertainty and imperfect information

Alongside an increase in global temperatures, climate change will bring increases in regional temperatures, changes in patterns of rainfall, rising sea levels, and increases in extreme events (heatwaves, droughts, floods, storms). High-quality information on future climate change at the regional scale is important for a market-based mechanism that drives successful adaptation responses. In particular, information is required for markets to operate efficiently. Without a robust and reliable understanding about the likely consequences of climate change, it is difficult for individuals – or firms – to weigh up the costs and benefits of investing in adaptation. Uncertainty in climate change projections could therefore act as a significant impediment to adaptation. The uncertainty will never be completely resolved, but should become more constrained as our understanding of the system improves.

As this understanding improves and develops, there may also be a role for markets in providing information to individuals. For example, better developed insurance markets would help to create clear price signals – for example through differentiated insurance premia - about the risks associated with climate change. Thus premia associated with buildings in high flood risk areas might be expected to be higher than those on buildings in less vulnerable locations.

---

[12] Reviewed in Tol *et al.* (1998)

[13] Lonsdale *et al.* (2005) explored these challenges in the Atlantis Project, where key London decision-makers faced a collapse of the West Antarctic Ice Sheet beginning in 2030, and a 30% chance of a 5-metre rise in sea level by 2130. They found that a delay to approve construction of an outer barrier in the Thames by decision-makers meant that abandonment of parts of London became the only adaptation option.

Missing and misaligned markets, including public goods

Autonomous adaptation is more likely when the benefits will accrue solely – or predominantly – to those investing in adaptation. For sectors that are characterised by short planning horizons – and where there is less uncertainty about the potential impacts of climate change - successful adaptive responses may therefore be driven by autonomous decisions.

However, effective adaptation of long-term investment patterns (such as climate-proofing buildings and defensive infrastructure) could prove challenging for private markets, especially with uncertain information. Decisions that leave a long-lasting legacy require private agents to weigh the uncertain future benefits of adaptation against its more certain current costs (see also Box 18.2). Even if the benefits of adaptation can be realised over a relatively short time-horizon, unless those paying the costs can fully reap the benefits, then there will be a barrier to adaptation. For example, there will be little financial incentive for developers to increase resilience of new buildings unless property buyers discriminate between properties on the basis of vulnerability to future climate.

Evidence from the United States suggests that consumers often fail to adopt even low-cost protection against weather hazards. A report by the Wharton Center for Risk Management and Decision Processes cites major surveys of residents in hurricane- and earthquake-prone areas of the USA. It found that, in the majority of cases, no special efforts had been made to protect homes.[14] Willingness-to-pay research suggests that many property owners are reluctant to invest in cost-effective protection measures, because they do not make the implied trade-off between spending money on risk prevention measures now in return for potential benefits over time. Some may not be in a position to finance the investment. Some may expect the government to bail them out.[15] Others may believe that the benefits of investment will not be capitalised in the value of the home.

Some adaptive responses not only provide private benefits to those who have paid for them, they also provide benefits – or positive spillovers – to the wider economy. In such circumstances, the private sector is unlikely to invest in adaptation up to the socially desirable level because they are unable to capture the full benefits of the investment. In some cases, there may be little – or no – private adaptation because the necessary adaptive response is effectively a 'public good' in the technical economic sense[16]. Public goods occur where those who fail to pay for something cannot be excluded from enjoying its benefits, and where one person's consumption of a good does not diminish the amount available for others. In the case of climate change, relevant pubic goods include research to improve our understanding of climate change and its likely impacts, coastal protection and emergency disaster planning. These – and the appropriate policy response – are discussed more fully in chapters 19 and 20.

Financial constraints and distributional impacts

Upfront investment in adaptive capacity and adaptation actions will be financially constrained for those on low incomes. In many developing countries,

[14] Kleindorfer and Kunreuther (2000)
[15] Kydland and Prescott (1977)
[16] Samuelson (1954).

financial resources in general are already extremely limited, and poverty already limits the ability to cope with and recover from climate shocks – particularly when combined with other stresses (Chapter 20 discusses the particular challenge faced by developing countries).

Equally, across all countries, it will be the poorest in society that have the least capacity to adapt (Chapters 4 and 5 in Part II). Thus, the impacts of climate change could exacerbate existing inequalities by limiting the ability of poor people to afford insurance cover or to pay for defensive actions. Social safety nets that function in emergencies could be of great importance here: for example cash or food for work schemes, such as those involved in employment guarantee schemes in India, can play a very important role in droughts.

Even with an appropriate policy framework, adaptation will be constrained both by uncertainty and technical limits to adaptation.

An inherent difficulty for long-term adaptation decisions is uncertainty, due to limitations in our scientific knowledge of a highly complex climate system and the likely impacts of perturbing it. Even as scientific understanding improves, there will always remain some residual uncertainty, as the size of impacts also depend on global efforts to control greenhouse gas emissions. Effective adaptation will involve decisions that are robust to a range of plausible climate futures and are flexible so they can be modified relatively easily. But there will always be a cost to hedging bets in this way, compared to the expert 'optimal' adaptation strategy that is revealed only with the benefit of hindsight.

There are clear limits to adaptation in natural ecosystems. Even small changes in climate may be disruptive for some ecosystems (e.g. coral reefs, mangrove swamps) and will be exacerbated by existing stresses, such as pollution. Beyond certain thresholds, natural systems may be unable to adapt at all, such as mountainous habitats where the species have nowhere to migrate.

But even for human society, there are technical limits to the ability to adapt to abrupt and large-scale climate change, such as a rapid onset of monsoon failure in parts of South Asia. Sudden or severe impacts triggered by warming could test the adaptive limits of human systems. Very high temperatures alone could become lethal, while lack of water will undermine people's ability to survive in a particular area, such as regions that depend on glacier meltwater. Rising sea levels will severely challenge the survival of low-lying countries and regions such as the Maldives or the Pacific Islands, and could result in the abandonment of some highly populated coastal regions, including several European cities.[17]

18.4   Conclusions

There are many ways that people, governments and economic agents of all kinds can adapt to climate change. Indeed, adaptation has always occurred in response

---

[17] Tol *et al.* (2006) investigated possible responses of society to 5 – 6 m of sea level rise following collapse of the West Antarctic Ice Sheet. The scenarios were developed from case studies based on interviews with stakeholders and experts. In the Rhone delta, the most likely option would be retreat. In the Thames Estuary, there could be a mix of protection and retreat with parts of the city turned into a Venice-style canal city. In the Netherlands, the initial response would be protection, followed by retreat from areas of low economic value, with eventual abandonment of some large cities, like Amsterdam and Rotterdam.

to changes in the climate system. However, adaptation by private individuals will have to be bolstered by government support in a variety of ways, if countries and regions are to rise to the challenge of climate change this century and beyond.

Uncertainty and imperfect information, missing and misaligned markets, and financial and distributional constraints, especially on the poorest in society, will present barriers to adaptation to climate change. Chapters 19 and 20 discuss the role of both markets and government in helping to promote effective adaptation in developed and developing countries.

In all cases, however, it is important to recognise the limits to adaptation. Although it can mute the impacts of climate change, it cannot by itself solve the problems posed by high and rapidly increasing temperatures. Even for relatively low amounts of warming, there are natural and technical constraints to adaptation – as is made vividly clear in low-lying coastal regions. Equally, without strong and early mitigation, the physical limits to – and costs of – adaptation will grow rapidly.

References

Developing an economic framework for examining adaptation was one of the primary objectives of a workshop hosted by the Stern Review in London on 9 May 2006. Several valuable papers were presented at the workshop, all of which are on the Review website (http://www. sternreview. org.uk). Sam Fankhauser summarised his previous work on developing an economic framework for adaptation (see Fankhauser 1997), while Molly Hellmuth presented her work with Mac Callaway (building on Callaway 2004). Frans Berkhout provided a valuable complement to these papers, discussing how we should understand adaptation by private individuals and the role of the public sector in that light (drawing on Berkhout, 2005). On the measurement of adaptation costs and benefits, Tol *et al.* (1998) reviewed evidence from a range of sources. Their discussion of estimates generated by global integrated assessment models is valuable.

Anthoff, D., R. Nicholls, R.S.J. Tol, and A.T. Vafeidis (2006): 'Global and regional exposure to large rises in sea-level', Research report prepared for the Stern Review, Tyndall Centre Working Paper 96, Norwich: Tyndall Centre, available from http://www.tyndall.ac.uk/publications/working_papers/twp96.pdf

Berkhout, F. (2005): 'Rationales for adaptation in EU climate change policies', Climate Policy **5**: 377–391

Burton, I. and M. van Aalst (1999): Come hell or high water – integrating climate change vulnerability and adaptation into Bank work', World Bank Environment Department Paper No 72, Washington, DC: World Bank.

Callaway, J.M. (2004): 'The benefits and costs of adapting to climate variability and climate change', The Benefits of Climate Change Policies, Morlot J.C. and S. Agrawala (eds.), Paris: OECD, pp. 113–157.

Callaway, J.M. and M.E. Hellmuth (2006): 'Climate risk management for development: economic considerations', presented at Stern Review Workshop on "Economics of Adaptation", 15 May 2006, available from http://www.sternreview.org.uk

Fankhauser, S. (1995): 'Protection versus retreat: the economic costs of sea-level rise', Environment and Planning A **27**: 299–319

Fankhauser, S. (1997): 'The costs of adapting to climate change', Working Paper 13, Washington, DC: Global Environment Facility.

Fankhauser, S. (2006): 'The economics of adaptation', presented at Stern Review Workshop on "Economics of Adaptation", 15 May 2006, available from http://www.sternreview.org.uk

Goklany, I.M. (2005): 'Evidence for the Stern Review on the economics of climate change', available from http://www.sternreview.org.uk

Jorgenson, D.W., R.J. Goettle et al. (2004): 'US market consequences of global climate change', Washington, DC: Pew Center for Global Climate Change.

Kleindorfer, P.R. and H. Kunreuther (2000): 'Managing catastrophe risk', Regulation **23**: 26–31, available from http://www.cato.org/pubs/regulation/regv23n4/kleindorfer.pdf

Kydland, F.E. and C.E. Prescott (1977): 'Rules rather than discretion: the inconsistency of optimal plans', Journal of Political Economy **85**(3): 473–491

Lonsdale, K., T.E. Downing et al. (2005): 'Results from a dialogue on responses to an extreme sea level rise scenario in the Thames Region, England', Contribution to the Atlantis EC Project, available from http://www.uni-hamburg.de/Wiss/FB/15/Sustainability/annex13.pdf

Metroeconomica (2004): 'Costing the impacts of climate change in the UK', Oxford: UK Climate Impacts Programme, available from http://www.ukcip.org.uk

Nicholls, R.J. and R.S.J. Tol (2006): 'Impacts and responses to sea-level rise: a global analysis of the SRES scenarios over the 21st century', Philosophical Transactions of the Royal Society A 364: 1073–1095

Pearce, D.W. (2005): 'The social cost of carbon', in Helm, D. (ed.), Climate-change policy. Oxford: Oxford University Press.

Samuelson, P. (1954): 'The pure theory of public expenditure', Review of Economics and Statistics 36(4): 387–389

Schelling, T.C. (1992): 'Some economics of global warming', American Economic Review 82: 1–14

Smith, J.B., R.S.J. Tol, S. Ragland and S. Fankhauser (1998): 'Proactive Adaptation to Climate Change: Three Case Studies on Infrastructure Investments', Working Paper D-98/03, Institute for Environmental Studies, Amsterdam: Free University.

Tol, R.S.J., S. Fankhauser and J.B. Smith (1998): 'The scope for adaptation to climate change: what can we learn from the impact literature?' Global Environmental Change 8: 109–123

Tol, R.S.J., M. Bohn and T.E. Downing et al. (2006): 'Adaptation to five metres of sea level rise', Journal of Risk Research 9: 467–482

Willows, R.I. and R.K. Connell (eds.) (2003): 'Climate adaptation: risk, uncertainty and decision-making', UKCIP Technical Report, Oxford: UK Climate Impacts Programme.

World Bank (2006): 'Clean energy and development: towards an investment framework', Washington, DC: World Bank.

19 Adaptation in the Developed World

KEY MESSAGES

In developed countries, adaptation will be required to reduce the costs and disruption caused by climate change, particularly from extreme weather events like storms, floods and heatwaves. Adaptation will also help take advantage of any opportunities, such as development of new crops or increased tourism potential. But at higher temperatures, the costs of adaptation will rise sharply and the residual damages remain large. The additional costs of making new infrastructure and buildings more resilient to climate change in OECD countries could range from **$15 – 150 billion each year (0.05 – 0.5% of GDP)**, with higher costs possible with the prospect of higher temperatures in the future.

Markets that respond to climate information will stimulate adaptation amongst individuals and firms. Risk-based insurance schemes, for example, provide strong signals about the size of climate risks and encourage better risk management.

In developed countries, progress on adaptation is still at an early stage, even though market structures are well developed and the capacity to adapt is relatively high. Market forces alone are unlikely to deliver the full response necessary to deal with the serious risks from climate change.

Government has a role in providing a clear policy framework to guide effective adaptation by individuals and firms in the medium and longer term. There are four key areas:

- **High-quality climate information** will help drive efficient markets. Improved regional climate predictions will be critical, particularly for rainfall and storm patterns.

- **Land-use planning and performance standards** should encourage both private and public investment in buildings, long-lived capital and infrastructure to take account of climate change.

- **Government can contribute through long-term polices for climate-sensitive public goods**, such as natural resources protection, coastal protection, and emergency preparedness.

- **A financial safety net may be required to help the poorest in society** who are most vulnerable and least able to afford protection (including insurance).

19.1   Introduction

Adaptation will reduce the costs and disruption caused by climate change. Governments can promote adaptation by providing information and clear policy frameworks to encourage individuals and firms to respond to market signals.

While those in developing countries will be hit hardest by the impacts of climate change, developed countries will not be immune, particularly from extreme weather events (Part II).[1] Adaptation will be required to reduce the costs and disruption caused by climate change in the long term and take advantage of any future opportunities. Much adaptation will be a local response by private actors to a changing climate. Individuals and businesses will respond to climate change – both by reacting to specific climate events, such as floods, droughts, or heatwaves, and also in anticipation of future trends. But incomplete information and other market imperfections mean that long-term policies will be required to complement these individual responses (Chapter 18). Failing to do so could incur large costs, especially from the very serious risks associated with larger amounts of warming. This chapter sets out key economic principles to underpin a broad policy framework to promote sound adaptation in the public and private sectors, many of which also apply to developing countries (Chapter 20).

19.2   Adaptation costs and prospects in the developed world

At higher temperatures, the costs of adaptation will rise sharply and the residual damages will remain large. The additional costs of making new infrastructure and buildings resilient to climate change in OECD countries could range from $15 – 150 billion each year (0.05 – 0.5% of GDP), with higher costs reflecting the prospect of higher temperatures in future.

In the developed world, some sectors may experience benefits from climate change for moderate levels of warming up to 2 – 3°C, particularly in higher latitude regions. Here, adaptation may allow developed countries to enhance such benefits. Farmers could switch to crops more suitable for warmer climates, such as grapes for wine. And some regions may be able to develop their summer tourism industries, as traditional tourist areas in the Mediterranean, for example, suffer from extreme heat and increasing water shortages.

But the negative impacts will become increasingly serious with rising temperatures and a rising risk of abrupt and large-scale changes (Chapter 6). Growing water shortages in regions with an already dry Mediterranean-like climate (Southern Europe, California, Australia) will also require costly investment in reservoirs and other measures to manage water stress and shortages. The UK Environment Agency has estimated that 10 – 15% of increased reservoir capacity may be required to address potential water deficits could cost the UK $5.5 billion (£3 billion).[2]

---

[1] O'Brien *et al.* (2006)
[2] Environment Agency (2005) – cost at 2005 prices. This assumes some level of demand management.

Infrastructure is particularly vulnerable to heavier floods and storms, in part because OECD economies invest around 20% of GDP or roughly $5.5 trillion in fixed capital each year, of which just over one-quarter typically goes into construction ($1.5 trillion - mostly for infrastructure and buildings). The additional costs of adapting this investment to a higher-risk future could be $15 – 150 billion each year (0.05 – 0.5% of GDP), with one-third of the costs borne by the US and one-fifth in Japan.[3] This preliminary cost calculation assumes that adaptation requires extra investment of 1 – 10% to limit future damages from climate change. For temperature rises of 3 or 4°C, these calculations are likely to scale as a constant proportion of GDP, as GDP grows. But the costs will rise sharply if temperatures increase further to 5 or 6°C, as expected if emissions continue to grow and feedbacks amplify the initial warming effect.

Stronger flood defences to protect infrastructure from storm surge damage will form a significant part of the extra spending. In the UK, the Foresight study estimated that a cumulative increase in investment of $18 – 56 million (£10 – 30 million) each and every year for the next 80 years would be required to prevent the costs of flood damages escalating in the UK. Defending New Orleans alone from flooding during a Category-5 hurricane is expected to cost around $32 billion.[4]

Markets that respond to climate information will stimulate adaptation among individuals and firms.

Developed countries typically have well-established markets with individuals and firms modifying their behaviour in response to price signals. Markets that respond to changing climate risks will stimulate adaptation in the private sector ("autonomous" adaptation – see Chapter 18). Adaptation is likely to be most responsive to market signals in sectors dominated by traded goods, such as agriculture, timber and energy. Government action may be required to set up more effective pricing mechanisms to encourage more efficient use of goods such as water where property rights are often poorly defined (Section 19.4).[5]

Insurance provides another important mechanism through which market signals can drive adaptation. Insurance has a long history of driving risk management through pricing risk, providing incentives to reduce risk, and imposing risk-related terms on policies.[6] By accurately measuring and pricing today's climate risks, insurance can help incentivise the first steps towards adaptation. The extra cost of insurance can act as a disincentive to build on high flood risk areas. Market signals of this kind encourage individuals or firms to reduce their present-day risk to weather

---

[3]   The $15 – 150 billion range for OECD countries comes from assuming that additional costs of 1 – 10% of the total amount invested in construction each year ($1.5 trillion) are required to make new buildings and infrastructure more resilient to climate change. The original analysis was carried out by Simms *et al.* (2004) who assumed adaptation costs of 1 – 5% of construction from initial research by ERM (2000). Higher estimates, such as 10%, are possible, particularly with the prospect of higher temperatures in the future. A similar calculation by the World Bank (details in Chapter 19) assumes that additional costs of 10 – 20% of investment portfolios may be required for adaptation, with the result that the total adaptation costs in developing countries of $9 – 41 billion are of a similar magnitude despite lower levels of overall investment.

[4]   Hallegatte (2006)

[5]   Mendelsohn (2006) provides an interesting example of how autonomous adaptation may occur in the agriculture sector, and how a mixture of private and public adaptation may be required in the water sector where property rights are poorly defined in many parts of the world.

[6]   Kovacs (2006); Lloyd's of London (2006)

damage, because of the cost saving associated with taking steps to manage climate risks. Encouraging action that improves society's resilience to current climate today should improve robustness to climate change in the future. Over time additional adaptation may be required to deal with longer-term effects of climate change.

In developed countries, progress on adaptation is still at an early stage, even though market structures are well developed and the capacity to adapt is relatively high.

Market forces alone, however, are unlikely to deliver the full response required to deal with the challenge of climate change (Chapter 18). This does not mean that government should manage each individual response to climate change. Rather governments should put in place a set of policies that provide individuals and firms with better information and the appropriate regulatory framework to help markets stimulate adaptation.

Many developed countries have conducted detailed studies on projected climate change impacts and vulnerability in key sectors, but only a handful of governments are moving towards implementing adaptation initiatives.[7] Some governments are beginning to create policy frameworks for adaptation.[8] But even in the UK, where awareness on adaptation is relatively high, practical measures to prepare for climate change are limited and remain largely confined to the public sector.[9]

19.3   Providing information and tools

High quality information on climate change will drive efficient markets for adaptation. Improved regional climate predictions will be critical, particularly for rainfall and storm patterns.

To make rational and effective adaptation decisions, organisations require detailed information about the full economic impacts of climate change in space and time (more detail in Chapter 18). Clear information will help ensure that climate risks are properly priced in the market. For example, production of flood hazard maps will increase house-buyers' awareness of flood risk and what individuals can do. They will also potentially influence land and house prices. In the UK, there is some evidence that house prices have decreased in areas that have flooded recently, because of concerns about lack of insurance cover and greater understanding of the risks.[10]

---

[7] Gagnon-Lebrun and Agrawala (2005) – for example, in the Netherlands, the US and New Zealand

[8] For example, Adaptation Policy Frameworks in the UK http://www.defra.gov.uk/environment/climatechange/uk/adapt/policyframe.htm and in Finland http://www.ymparisto.fi/default.asp?contentid=165496&lan=en

[9] Tompkins *et al.* (2005) - a recent survey from the Tyndall Centre found evidence of relatively high levels of awareness of the need to adapt with UK stakeholders, particularly those in the public sector (e.g. government, local authorities, agencies), but very few, if any, specific adaptation actions that have been undertaken in response to expected climate change.

[10] Royal Institution of Chartered Surveyors (2004)

The scale and complexity of climate information make it unlikely that individual organisations will undertake basic research into future changes. Generic but high-quality information on climate change could be considered a public good (Section 19.5). Government-funded research programmes have advanced our understanding of climate change substantially. A central challenge for adaptation remains the uncertainty in climate predictions, particularly changing regional rainfall patterns (see Chapter 1), which are a key determinant of many likely adaptation requirements, for example the size and location of new sewers to cope with heavier downpours.[11] Improved regional climate predictions will help to integrate climate risk into long-term planning and provide a rationale for adaptation action.

High-quality climate information is an important starting point for adaptation, but effective communication to stakeholders will also be required. Information should not be too complex and should provide practical pointers without being excessively prescriptive, because local choice and flexibility are important.[12] The UK Climate Impacts Programme has developed an important tool for helping stakeholders deal with risk and uncertainty and incorporate climate change into project appraisal (Box 19.1). The programme overall has been instrumental in raising awareness of adaptation issues among a broad range of stakeholders in the UK and driving forward the first steps towards adaptation actions.

BOX 19.1 UKCIP adaptation wizard

The Government has established the UK Climate Impacts Programme (UKCIP) to provide individuals and organisations with the necessary tools and information on climate impacts to allow them to adapt successfully to the changing climate. The UKCIP (2005) Adaptation Wizard has been set up to help organisations move from a simple understanding of climate change to integration of climate change into decision-making. The Wizard draws heavily on Willows and Connell (2003) and provides web-based tools for four stages of adaptation:

- Scoping the impacts

- Quantifying risks

- Decision-making and action planning

- Adaptation strategy review.

One of the most valuable UKCIP tools is an up-to-date set of climate change scenarios that are available free of charge and used by a wide range of stakeholders, including local authorities, public agencies, and businesses. New scenarios will be published in 2008 that quantify risks and uncertainties in a more robust and quantitative manner to help stakeholders plan adaptation strategies. UKCIP has further tools on handling uncertainty and costing the impacts.

Source: UKCIP (2005)

---

[11] Heavy storms in London in August 2004 killed thousands of fish when more than 600,000 tonnes of untreated sewage was forced into the River Thames, because the sudden downpour overloaded the city's network of Victorian sewers.

[12] Chapter 17 considers the different ways that can be used to communicate climate change information to the public and how these link to regulation and standards.

19.4   Is there a role for regulation in overcoming market barriers to adaptation?

Land-use planning and performance standards could be used to encourage both private and public investment in buildings, long-lived capital and infrastructure to take account of climate change.

Infrastructure should be an important focus of adaptation efforts, because decisions taken today leave a long legacy for future generations when the impacts of climate change will be felt most sharply. OECD countries currently invest $1.5 trillion each year in construction of new infrastructure and buildings. Effective adaptation of long-term investments is unlikely to occur through market dynamics alone when there is limited incentive to invest today to avoid future losses for the next generation.[13] Given the uncertainty and imperfections in property markets, the investor may lack confidence that extra resilience will be fully reflected in resale value in future. Decisions that leave a long-lasting legacy for future generations require private agents to weigh the uncertain future benefits of adaptation against its more certain current costs (Chapter 18). Individuals and firms will require sufficient information to build long-term horizons and make adaptation decisions that fully reflect the risks and net benefits over the lifetime of the decision.

Some market intervention may be required in order to promote the proper pricing of risks of climate change in long-term investment decisions. Regulatory measures are often less efficient and flexible than market mechanisms, but may have an important role to play in avoiding unanticipated early obsolescence of capital stock (more detail on capital "lock in" in Chapter 17). Policies will be more efficient if they encourage private individuals and firms to take explicit account of the economic costs of climate change in their decision-making, rather than simply imposing prescriptive design standards. A developer will make a rational decision about whether to increase the long-term resilience of infrastructure or to design buildings with shorter lifespan if required to consider the impacts of climate change over the lifetime of the property.

Where the risks of climate change are clear and substantial, a planned approach that allows for changes in line with natural replacement cycles avoids costly retrofits or the abandonment of infrastructure before the end of its otherwise useful life. Where there is less certainty and the risks are moderate, no-regrets options may be most appropriate - namely those actions that offer net cost savings today regardless of the eventual amount of climate change, for example reducing vulnerability to current climate variability such as floods and storms.[14] In some cases, even relatively simple structural measures could yield both short- and long-term benefits to climate variability and change, such as bracing and securing roof trusses and walls using straps, clips or adhesives to reduce hurricane damages.[15] Property-owners in

---

[13] Mendelsohn (2000)

[14] Fankhauser *et al.* (1999); "no regrets" describe projects that have a positive net present value across a range of climate change outcomes.

[15] Kleindorfer and Kunreuther (2000) considered how simple hurricane protection measures could reduce the annual expected hurricane damage costs for a sample of the population in Miami by 25% ($9 million without measures, $6.8 million), with concurrent decreases in annual cost to homeowners of $1.5 million (10% decrease in cost), measured as sum of insurance premium, expected deductible losses and annual cost of prevention measures (7% discount rate, 20 year time horizon).

the US Gulf States who implemented all the recommended hurricane protection methods suffered only one-eighth of the damages from Hurricane Katrina than those that did not implement such methods. The result was that investment by property-owners of $2.5 million avoided damages of over $500 million.[16] This is a prime example of cost-effective adaptation.

Land-use decisions leave a substantial legacy. The costs for future generations may not be taken into account in market-based decisions today. There is also a moral hazard issue – private individuals may take greater risks if they think the government will bail them out because of political pressure.[17] Market signals alone, however improved, cannot carry the full weight of policy. The planning system will be a key tool for encouraging both private and public investment towards locations that are less vulnerable to climate risks today and in the future. Limiting construction of new developments in the floodplain may be an important element of a sustainable response to managing flood risk in the long term (Box 19.2).

In certain circumstances, performance standards that include headroom for climate change could reduce vulnerability to unpredictable weather, such as flash-flooding or storms. Whether and how such standards are introduced and implemented will depend on the size of the risk and the degree to which an individual's action affects others in the community. When there is a significant negative externality, the case for market intervention will be stronger. For example, individual decisions to pave over front-gardens in London have led to a loss of permeable drainage surface equivalent to 22 times the size of Hyde Park, increasing the city's vulnerability to flash-flooding substantially.[18] Each individual decision may be rational, but in aggregate this loss of permeable land will leave a legacy for future generations living in London.

BOX 19.2 Land-use planning and climate change: South East England housing case study

In February 2003, the UK Government set out its plans to provide 200,000 new homes above existing targets in the South East by 2016 to reduce the pressure on the country's housing stock. The Communities Plan identified four growth areas as the focus for the initial wave of additional housing in the South East – Thames Gateway, Ashford, the M11 corridor, and the South Midlands. These areas were chosen, in part, due to their high concentrations of brownfield sites close to existing urban centres, but face a growing risk of flooding associated with climate change.

Research by the Association of British Insurers (ABI) has shown that rigorous application of the Government's planning policy for floodplains[19] could be one of the most effective ways to control the risks from flooding and climate change.

- Moving properties off the floodplain and accommodating them in non-floodplain parts of development sites reduced flood risk by 89 - 96% for all growth areas except Thames Gateway.

- In Thames Gateway where more than 90% of the land targeted for development lies in the floodplain, a sequential approach that allocates housing to the lowest risks parts of the floodplain could reduce flood losses by 40 - 52% for the initial tranche of new housing.

- Overall, effective use of land-use planning could reduce annual flood losses from new housing by more than 50%.

---

[16] Mills and Lecomte (2006)
[17] Kydland and Prescott (1977)
[18] London Assembly Environment Committee (2005)
[19] Office of the Deputy Prime Minister (2005)

The alternatives to land-use planning were more costly – increased investment in flood defences to offset the uplift in national flood risk, and adding to construction costs through building in flood-resilience.

Source: Association of British Insurers (2005b)

In many countries, government plays a role in financing long-term infrastructure investment. Here, the nature of the arrangement between public and private sector in the provision of infrastructure will influence the form of any market intervention that may be required.

- Where infrastructure is provided through targeted public investment, resilience to climate change can be established through direct government action, for example (i) locating winter roads off ice and onto land in Manitoba, (ii) upgrading the Thames Barrier, which protects London from flooding (details in Box 19.4)
- Where the regulatory framework allows for infrastructure provision through the private sector, the operation of the arrangements should be flexible enough to allow for consideration of climate change. For example, in the UK, water companies are responsible for reservoir provision,[20] energy companies are responsible for power lines, transport providers are responsible for track maintenance, and private firms now manage some public construction projects.

Public procurement could be a useful vehicle for highlighting best practice in incorporating adaptation in investment decisions[21] – and may also drive forward demand for adaptation services to help guide private sector decisions.

19.5   Incorporating climate change into long-term policies for public and publicly provided goods

Government's own long-term polices for climate-sensitive public goods, such as natural resources protection, coastal protection, and emergency preparedness, should take account of climate change to control future costs (Box 19.3).

As well as providing a clear policy framework for investment decisions, government sets long-term policies for public and publicly provided goods that supply community services (Chapter 18). Examples of specific relevance to climate change include: flood and coastal protection (Box 19.3); public health and safety (Box 19.4); and natural resource protection. The risks of not taking action could leave a significant public liability – either because the private sector will no longer

---

[20] Water companies in the UK are able to examine the impact of climate change on future head-room allowances for water supply. However, even here, action on climate change remains limited to research and impact assessment, rather than specific adaptation measures (Arnell and Delaney 2006).

[21] Acclimatise (2005) identify that a changing climate could affect income, operating costs and financing costs for PFI projects, with potential knock-on effects for investor and market confidence.

BOX 19.3 Public sector adaptation examples

(a) Winter roads in Manitoba, Canada

The province of Manitoba uses winter roads constructed from snow and ice to transport essential goods (fuel, food, and building supplies) to its remote northern communities. The extent of this network is equivalent to building a road from Winnipeg to Vancouver every winter, a distance of approximately 2,000 km^2. After an extremely warm winter in 1997-98 when the roads could not be opened. 1 million kg of food had to be airlifted to communities at a cost of $50 million (Canadian), so Manitoba began the process of moving 600 km of roads from ice-based routes. Instead, Manitoba located routes on land, shifted the main access points further north, and installed permanent bridges over critical river crossings. *Source: Manitoba Transportation and Government Services (2006)*

(b) Managing flood risk in London

Climate change will put London at greater risk from flooding in future years. Many flood-plain areas are undergoing regeneration, putting more people, buildings and infrastructure at risk. Flooding would cause immense disruption to London's commercial activities, and could cause direct damage equivalent to around £50 billion (plus wider financial disruption). Climate change could increase the maintenance costs of flood defences in the Thames over 100 years from £3.8 billion without climate change (£1.1 billion, Green Book discounted) to £5.3 – £6.8 billion (£1.9 - £2.8 billion, Green Book discounted) with climate change. Following the 1953 East Coast floods the Thames Barrier and associated defences were planned and built over a 30-year period to protect London to a high standard from tidal flooding. The design of the Barrier allowed for sea level rise but did not make any specific allowance for changes in river flows or the height of North Sea storm surges. Although the defences offer a high level of protection from today's risks, they will only provide protection of 1-in-1000 years until 2030. After that, the risk increases, potentially reaching 1-in-50 years by the end of the century without any active intervention to upgrade capital defences. Slight modifications could extend the useful life of the defences by a few more years, but in the long term a more strategic approach is required. The Environment Agency has set up the Thames Estuary 2100 project to develop a flood risk management strategy for the next 100 years and explicitly factor in adaptation to climate change using a risk-based decision-testing framework. The project is developing decision pathways to retain flexibility over the timing and types of flood management measures as understanding about climate change increases. For example, introducing non-structural measures, such as flood storage, could delay more intrusive and expensive measures, such as construction of a new barrier, which could cost several billion. *Source: Environment Agency (2005)*

(c) Protecting Venice

Flood events in Venice have been increasing in frequency throughout the 20th century. At the beginning of this century, St Mark's Square flooded less than 10 times a year. By 1990 it was flooding around 40 times a year and in 1996 it flooded almost 100 times. Without further protection, sea level rise this century will lead to the flooding of St Mark's Square every day. In December 2001, the then Italian Prime Minister, Silvio Berlusconi, approved a $2.6 billion (€2.3 billion) scheme, known by the acronym of MOSE, to protect the city from the rising tides. The scheme consists of 78 metal gates placed across the three main inlets of the lagoon. These gates can be raised ahead of a storm surge to separate the city from the sea. The plans have been controversial. The current design is only able to cope with around 20 cm more of sea level rise, while many climate models predict around 50 cm by the end of the century. Environment campaigners have contested the design, arguing that the gates will disrupt the lagoon's delicately balanced ecosystem.

Source: Nosengo (2003)

BOX 19.4 Heatwave adaptations

With the recognition that heat is a growing mortality risk factor, many cities around the world are developing sophisticated heatwave warning systems. Climate change effects in cities are compounded by the urban heat island effect, which can maintain night temperatures several degrees above the surrounding rural area (chapter 3). Several international organisations are collaborating to promote good-practice in warning systems that deal with the impact of extreme heat on human health.

(a) France heatwave plan ("plan canicule")

Following the summer 2003 heatwave (the hottest three-month period recorded in France), which caused an estimated 15,000 extra deaths, the French Government prepared a national heatwave plan (plan canicule). The plan consists of four different levels of intervention.

1. Vigilance – Active every year from June to September to monitor action plans and keep the public informed.
2. Alert – Trigger public services at national and regional level when temperatures exceed critical levels.
3. Intervention – Medical and social intervention when the heatwave is already underway.
4. Requisition – Reinforce existing plans and apply exceptional measures when a heatwave is long lasting, for example through use of government transport and calling in the army.

The national plan is supported by a series of action plans that focus on particular vulnerabilities – (i) care homes for the elderly; (ii) medical emergency services; (iii) emergency alert system; and (iv) Paris.

Source: ONERC (2005)

(b) Philadelphia Heat Health Warning System

The system forecasts periods up to two days in advance when there is a high risk of a weather-system associated with heat-related mortality (more than four deaths expected). Once a warning is issued, the city of Philadelphia and its public agencies put in place a series of actions to minimise the dangers of the heatwave, including:

- TV, radio stations and newspapers are asked to publicised the upcoming conditions, along with information on how to avoid heat-related illnesses.
- Promotion of a "buddy" system – media announcements encourage friends, relatives and neighbours to visit elderly people during the hot weather and make sure they have sufficient water and proper ventilation to cope with the weather.
- Telephone "Heatline" to provide information and counselling to the public on avoidance of heat stress.
- Department of Public Health mobile field teams make home visits to vulnerable households.

- Nursing homes advised on how best to protect their residents, supported by visits from field teams.
- Emergency services increase staffing levels.
- Homeless agency increases outreach activities to assist those on the streets.
- Air-conditioned shelter facilities set up for high-risk individuals.

Source: Acclimatise (2006)

carry the risk, for example by refusing to offer flood insurance, or because of sharply rising costs of disaster recovery and public safety. However, adaptation policies will require careful cost-effectiveness analysis before implementation to prevent any wasteful expenditure on remote risks and inadequate expenditure on present-day risks.

Protecting natural systems could prove particularly challenging. The impacts of climate change on species and biodiversity are expected to be harmful for most levels of warming, because of the limited ability of plants and animals to migrate fast enough to new areas with suitable climate (Chapter 3). In addition, the effects of urbanisation, barriers to migration paths, and fragmentation of the landscape also severely limit species' ability to move. For those species that can move rapidly in line with the changing climate, finding new food and suitable living conditions could prove challenging. Climate change will require nature conservation efforts to extend out from the current approach of fixed protected areas. Conservation efforts will increasingly be required to operate at the landscape scale with larger contiguous tracts of land that can better accommodate species movement. Policies for nature protection should be sufficiently flexible to allow for species' movement across the landscape, through a variety of measures to reduce the fragmentation of the landscape and make the intervening countryside more permeable to wildlife, for example use of wildlife corridors or "biodiversity islands".

19.6   Spreading risk and protecting the vulnerable

Risk-based insurance schemes will encourage good risk management behaviours, but may require a financial safety net to protect those who are most vulnerable and cannot afford protection.

Many developed countries have mature insurance markets that provide additional adaptive capacity by spreading the risks of extreme weather events across a large pool of individuals or businesses. Without any insurance system or state-backed compensation at all, the costs of weather disasters will lead to crushing personal and business liabilities. However for rapidly escalating costs, even insurance capacity may not be sufficient to cover the costs, leading to restricted coverage or the use of alternative risk transfer mechanisms, such as weather derivatives or catastrophe bonds.

In a world of identical individuals where everyone faces the same risk, full risk pooling maximises overall welfare because average utility in a world of risk pooling is greater than an individual's expected utility where in some years they may have to pay the full cost of an extreme event.[22] In reality, individuals in a population face different risks. In this case, the nature of the insurance model used affects the outcome.[23]

- If everyone contributes equally to the pool, the costs of extreme events for those at greater risk are cross subsidised by those at lower risk.[24] This could act as a social safety net to protect those in society who are most vulnerable to the impacts of climate change. Government-backed insurance systems may cause such a subsidy effect, because the premiums are drawn implicitly from tax income and are unrelated to the risk of extreme events.[25] But, if no deductibles or limits are included in program design, this model creates moral hazard by offering no reward for those who take steps to reduce their vulnerability to climate change.

- If those at greatest risk contribute most to the pool and those who avoid risk pay least, the risks are pooled in proportion to their size. Private insurance markets may lead to such segmentation (risk based pricing), because competition between insurance providers drives firms to match individual premiums to the expected payout.[26] Risk-based pricing is efficient – it distributes the costs of weather amongst the insured on the basis of risk and encourages behaviours that reduce the risks. However, such a market-based approach could leave the most vulnerable financially excluded. From an equity perspective, government may wish to create a financial safety net to protect those who are most vulnerable to climate change and cannot afford protection.

But insurance systems will face challenges with operation of risk-sharing approaches if the risks reach very high levels.[27] The capital required to support a functioning insurance market will rise sharply in line with the rising costs of extreme weather (Chapter 5). At a global level, risk sharing works effectively where the risks are independent, but climate change will raise the frequency of very serious weather events in all the large insurance markets.[28] As a result, there will be a greater chance of several large events in one year and the insurance industry may struggle to cope. Finding alternative sources of capital to diversify the risk may help to some degree,[29] but ultimately the costs may become too large for the industry to bear.

---

[22] In other words, individuals perceive a greater damage from a loss of $10,000 than a benefit from a gain of $10,000, and would refuse a 50/50 gamble of that amount. This is because of the (assumed) concavity of the income-utility function.

[23] US GAO (2005) and Association of British Insurers (2005a) both provide a useful summary of insurance for natural catastrophes in different markets.

[24] However, those not directly affected can still be materially influenced indirectly, e.g. through community-wide curtailment of economic activity or loss of jobs due to business interruptions.

[25] For example, the NatCat model in France or the National Flood Insurance Program in the USA

[26] For example, in the UK, insurers have complete freedom over pricing and terms of cover. As insurers develop more sophisticated tools for quantifying risk (e.g. flood maps down to individual properties), prices increasingly reflect weather risks.

[27] Dlugolecki (2004); Lloyd's of London (2006)

[28] Association of British Insurers (2005a)

[29] Salmon and Weston (2006)

19.7   Conclusion

Adaptation could reduce the costs of climate change in developed countries, provided policies are put in place to overcome market barriers to private action. But at higher temperatures, the costs of adaptation will rise sharply and the residual damages remain large.

While some sectors of the developed world may experience benefits from climate change for moderate levels of warming (2 – 3°C), the costs will rise sharply with increasing temperatures. Adaptation can make an important difference to reducing some of these costs – but there will be limits, as the relative effectiveness diminishes. The residual damages after adaptation are likely to increase faster than the total costs, and adaptation itself will become more expensive. Preliminary estimates suggest that adapting infrastructure and buildings to climate change could increase costs by 1 – 10% taking the total for OECD countries to $15 – 150 billion each year.

These calculations assume 3 or 4°C of temperature rise, but the costs are likely to rise sharply if temperatures increase further to 5 or 6°C (as expected if emissions continue to grow and feedbacks amplify the initial warming effect). At this level, very serious risks of abrupt and large-scale change come into play. For human societies, absolute limits will be crossed once a region loses an essential but non-substitutable resource, such as glacier meltwater that supplies water to over a billion people during the dry season. Populations will then have little option but to migrate to another region of the world. At very high temperatures, the physical geography would change so strongly that the human and economic geography would be recast too. The full consequences of such effects are still uncertain, but they are likely to involve large movements of populations that would affect all countries of the world and present a new and very difficult dimension to adaptation.

References

Andrew Simms and colleagues from the new economics foundation produced one of the first assessments of the costs of adaptation for developed countries (Simms et al. 2004). The figures in the report are relatively (and necessarily) crude, but provide an indication of the scale of adaptation costs in the developed world. Few robust assessments of adaptation costs are available. More work has been undertaken to understand the relative roles of the public sector and private individuals in adapting to climate change in developed countries – for example carefully reasoned papers by the Australian Greenhouse Office (2005), Prof Frans Berkhout (2005) and most recently Prof Robert Mendelsohn (2006). This kind of analysis has been complemented by work to catalogue the extent of adaptation action in developed countries, including a recent paper on national government adaptation within the OECD (Gagnon-Lebrun and Agrawala 2005), a survey of public and private adaptation in the UK (Tompkins et al. 2005) and in Norway (O'Brien et al. 2006), and a review of adaptation in major world cities (Acclimatise 2006). The insurance industry has also produced several reports examining the role of insurance in promoting good risk management and protection against extreme weather through market signals and awareness raising (Dlugolecki 2004, Mills and Lecomte 2006, Lloyd's of London 2006).

Acclimatise (2005): 'Climate change risks for PFI/PPP projects', Briefing Note, Southwell: Acclimatise, available from http://www.acclimatise.uk.com/resources/briefing-notes

Acclimatise (2006): 'Review of adaptation to climate risks by cities', London: London Climate Change Partnership, available from http://www.london.gov.uk/climatechange-partnership/adapting-jul06.jsp

Arnell, N.W., and E.K. Delaney (2006): 'Adapting to climate change: public water supply in England and Wales', Climatic Change, DOI: 10.1007/s10584-006-9067-9

Association of British Insurers (2005a): 'Financial risks of climate change', London: Association of British Insurers, available from http://www.abi.org.uk/climatechange

Association of British Insurers (2005b): Making communities sustainable: managing flood risks in the Government's growth areas, London: Association of British Insurers, available from http://www.abi.org.uk/housing

Australian Greenhouse Office (2005): 'Climate change risk and vulnerability: promoting an efficient adaptation response in Australia', Canberra: Department of the Environment and Heritage, Australian Government

Berkhout, F. (2005): Rationales for adaptation in EU climate policies, Climate Policy 5: 377 – 391

Dlugolecki, A. (2004): 'A changing climate for insurance', London: Association of British Insurers, available from http://www.abi.org.uk/climatechange

Environment Agency (2005): 'Evidence to the Stern Review on the economics of climate change', available from http://www.sternreview.org.uk

Environmental Resources Management (2000): 'Potential UK adaptation strategies for climate change', London: Department for Environment, Transport and the Regions (DETR)

Fankhauser, S., J.B. Smith and R.S.J. Tol (1999): 'Weathering climate change: some simple rules to guide adaptation decisions', Ecological Economics 30: 67 - 78

Gagnon-Lebrun F. and S. Agrawala (2005): 'Progress on adaptation to climate change in developed countries: an analysis of broad trends', Working Party on Global and Structural Policies, Paris: OECD, available from http://www.oecd.org/dataoecd/49/18/37178873.pdf

Hallegatte, S. (2006): 'A cost-benefit analysis of the New Orleans flood protection system', (Regulatory Analysis 06-02), Washington, DC: AEI-Brookings Joint Center for Regulatory Studies, available from http://www.aei-brookings.org

Kleindorfer, P.R. and H. Kunreuther (2000): 'Managing catastrophe risk, Regulation' 23: 26 – 31, http://www.cato.org/pubs/regulation/regv23n4/kleindorfer.pdf

Kovacs, P. (2006): Hope for the best and prepare for the worst: how Canada's insurers stay a step ahead of climate change, Policy Options, Dec/Jan 2006: 53 – 56

Kydland, F.E.and E.C. Prescott (1977): 'Rules rather than discretion: the inconsistency of optimal plans', Journal of Political Economy, 85: 473–492

Lloyd's of London (2006): 'Climate change: adapt or bust?', London: Lloyd's of London, available from http://www.lloyds.com/360

London Assembly Environment Committee (2005): 'Crazy paving: the environmental importance of London's front gardens', London: Greater London Authority, available from http://www.london.gov.uk/assembly/reports/environment/frontgardens.pdf

Manitoba Transportation and Government Services (2006): 'Winter roads in Manitoba', Canada: Manitoba Government, available from http://www.gov.mb.ca/tgs/hwyinfo/winterroads

Mendelsohn, R. (2000): 'Efficient adaptation to climate change', Climatic Change 45: 583 – 600

Mendelsohn, R. (2006): 'The role of markets and governments in helping society adapt to a changing climate', Climatic Change 78: 203 - 215

Mills, E. and Lecomte, E. (2006): 'From risk to opportunity: how insurers can proactively and profitably manage climate change, Boston, MA: CERES, available from http://www.ceres.org/news/news_item.php?nid=221

National Audit Office (2004): 'Out of sight – not out of mind: Ofwat and the public sewer network in England and Wales', London: The Stationary Office, available from http://www.nao.org.uk/pn/03-04/0304161.htm

Nosengo, N. (2003): 'Save our city!', Nature 424: 608 - 609

O'Brien, K., S. Eriksen, L. Sygna, and L.O. Naess (2006): 'Questioning complacency: climate change impacts, vulnerability, and adaptation in Norway', Ambio 35: 50 -56

Office of the Deputy Prime Minister (2005): 'Consultation on Planning Policy Statement 25: development and flood risk', Wetherby: ODPM Publications, available from http://www.communities.gov.uk/index.asp?id=1162059

ONERC (2005): 'Un climat á la derive: comment s'adapter, Rapport de l'ONERC au Premier Ministre et au Parlement, Paris: Observatoire National sur les Effects du Rechauffement Climatique

Royal Institution of Chartered Surveyors (2004): 'The impact of flooding on residential property values', London: RICS Foundation, available from http://www.rics.org/NR/rdonlyres/ DF-DBBBEB-7F01-42FA-B338-2860945C4DAE/0/Effect_of_flooding_report.pdf

Salmon, M. and S. Weston (2006): 'Evidence to the Stern Review on the economics of climate change', available from http://www.sternreview.org.uk

Simms, A., D. Woodward and P. Kjell (2004): 'Cast adrift – how the rich are leaving the poor to sink in a warming world', London: new economics foundation, available from http://www.neweconomics.org/gen/z_sys_PublicationDetail.aspx?pid=200

Tompkins, E.L., E. Boyd and S.A. Nicholson-Cole (2005): 'Linking adaptation research and practice', Research Report, London: Defra, available from http://www.defra.gov.uk/science/project_data/DocumentLibrary/GA01077/GA01077_2664_FRP.pdf

UKCIP (2005): 'The Adaptation Wizard' (Prototype Version 1.0), Oxford: UK Climate Impacts Programme, available from http://www.ukcip.org.uk/resources/tools/adapt.asp

US Government Accountability Office (GAO) (2005): 'Catastrophe risk: U.S. and European approaches to insure natural catastrophe and terrorism risks', Report GAO-05-199, Washington, DC: US Government Accountability Office

Willows, R.I. and R.K. Connell (eds.) (2003): 'Climate adaptation: risk, uncertainty and decision-making', UKCIP Technical Report, Oxford: UK Climate Impacts Programme, available from http://www.ukcip.org.uk

20   Adaptation in the Developing World

KEY MESSAGES

Adaptation to mute the impact of climate change will be essential in the poorer parts of the world. The poorest countries will be especially hard hit by climate change, with millions potentially pushed deeper into poverty.

Development itself is key to adaptation. Much adaptation should be an extension of good development practice and reduce vulnerability by:

- Promoting growth and diversification of economic activity;

- Investing in health and education;

- Enhancing resilience to disasters and improving disaster management;

- Promoting risk-pooling, including social safety nets for the poorest.

Putting the right policy frameworks in place will encourage and facilitate effective adaptation by households, communities and firms. Poverty and development constraints will present obstacles to adaptation but focused development policies can reduce these obstacles.

Adaptation actions should be integrated into development policy and planning at every level. This will incur incremental adaptation costs relative to plans that ignore climate change. But ignoring climate change is not a viable option – inaction will be far more costly than adaptation.

Adaptation costs are hard to estimate, because of uncertainty about the precise impacts of climate change and its multiple effects. But they are likely to run into tens of billions of dollars. This makes it still more important for developed countries to honour their existing commitments to increase aid sharply and help the world's poorest countries adapt to climate change. More work is needed to determine the costs of adaptation.

Without global action to mitigate climate change, both the impacts and adaptation costs will be much larger, and so will be the need for richer countries to help the poorer and most exposed countries. The costs of climate change can be reduced through both adaptation and mitigation, but adaptation is the only way to cope with impacts of climate change over the next few decades.

20.1   Introduction

It is the countries with fewest resources which are most likely to bear the greatest burden of climate change in terms of loss of life, adverse effect on income and growth, and damage to living standards generally. Developing countries – and especially the low-income countries in tropical and sub-tropical regions – are expected to suffer most, and soonest, from climate change. They are especially vulnerable to the effects of climate change, because of their existing exposure to an already fragile environment and their economic and social sensitivity to climate change. And their poverty reduces their capacity to adapt (discussed in Chapter 4).

As in developed countries, much adaptation will be a local response by individuals to a changing climate. Households, communities, and firms respond autonomously to climate change and extreme variability in ways that help to reduce its harmful effects. Yet these autonomous responses will be likely to fall far short of what is necessary, given current vulnerabilities and the scale of future impacts. Sections 20.2 and 20.3 set out the essential role that governments will have to play in reducing this vulnerability through good development practice, including better disaster risk management and use of social safety nets to protect the most vulnerable.

Governments will also have a specific role in establishing the policy frameworks to encourage adaptation by private individuals and firms – in particular to address information uncertainties, ensure transparency of transactions, and tackle financial/non-financial constraints that will reduce the capacity for autonomous adaptation (as discussed in Chapters 18 and 19). Three aspects will be especially important for governments:

- Providing high quality information;
- Effective land-use planning and performance standards; and
- Ensuring that major planning and public sector investment decisions take account of climate change.

The application of these principles is context specific and will vary from country to country. Developing countries face additional constraints and obstacles that will require even greater effort by governments, as discussed in Section 20.4. Section 20.5 sets out a range of estimated costs of adaptation in developing countries. The chapter concludes with a preview of the necessity of international assistance for adaptation (discussed in detail in Chapter 26).

20.2   Adaptation prospects in the developing world

Individuals, firms and civil society will respond to a changing climate as far as their knowledge and resources allow. But they will require support from their government to overcome barriers and increase adaptive capacity.

Individuals, firms, and civil society will have a central role in responding to climate change. People and local companies will typically have better information than governments about their own specific situations, as well as stronger incentives to act. This is testified by examples of autonomous action taken in response to extreme weather events. For example:

- In parts of the Mahanadi floodplains in India's Orissa state, farmers usually cultivate a local variety of paddy (Champeswar) that is tolerant to water stagnation to reduce agricultural output loss;[1] and
- In the Sahel, a drought in the 1980s was greater than one in the 1970s but the losses associated with the later drought were far less as people effectively adapted and increased their resilience to the impacts of a hostile climate.[2]

However, many poor people face a plethora of constraints – linked mainly to low-income levels and poverty – which limit their ability to react autonomously to climate change, as set out in Chapter 4. Unless action is taken, these constraints will be compounded as developing countries are exposed to more frequent and intense extreme weather events. As in developed countries, individuals and firms will naturally react in response to market signals. For example, if climate-induced water scarcity results in higher prices, firms and households are likely to become more efficient in their use of water. But in many developing countries public water utilities do not provide water services to poor people but only to firms and better-off consumers – and at artificially low, subsidised prices. In such cases, existing structures and price systems limit autonomous adaptation and actually increase the burden on the poorest.

Governments have an important potential role in helping people to build their adaptive capacity through good development practice. In addition, developing-country governments – as with developed – have an essential role in supplying information and ensuring that markets provide appropriate signals. This is in addition to providing necessary infrastructure and public services. But already stretched local and national administrations face additional burdens with the need to adapt governmental activities to climate change, and ensure that both public and private sectors exploit whatever comparative advantages they have in adapting to the stresses of climate change. Box 20.1 sets out a summary of the various measures that governments should take to strengthen adaptation, discussed in more detail below.

BOX 20.1 Measures to strengthen adaptation

As discussed above, development itself is the most effective way to promote adaptation to climate change, because development increases resilience and reduces vulnerabilities. Beyond that broad development focus, fully integrating climate change will require ensuring that adaptation concerns are reflected across many aspects of government policy. Some of the required measures for strengthening adaptation include:

- **Ensuring access to high-quality information about the impacts of climate change and carrying out vulnerability assessments.** Early warning systems and information distribution systems help to anticipate and prevent disasters.

---

[1] Roy et al (2006). Champeswar can sustain almost seven days submergence
[2] Nkomo et al (2006) This resilience can be broadly attributed to a) reliance on local networks and groups, b) local savings schemes, many of them based on regular membership fees, c) a changing role of the state and linkages between countries and to global aid systems, and d) regional co-operation, such as the CILSS grouping (the Permanent Interstate Committee for the Fight against Drought in Sahelian countries founded in 1973 in the aftermath of the 1970s drought).

- **Increasing the resilience of livelihoods and infrastructure** using existing knowledge and coping strategies.

- **Improving governance,** including a transparent and accountable policy and decision-making process and an active civil society.

- **Empowering communities** so that they participate in assessments and feed their knowledge into the process at crucial points.

- **Integrating climate change impacts** in issues in all national, sub-national and sectoral planning processes and macro-economic projections. The national budget process is key here.

- **Encouraging a core ministry** with a broad mandate, such as finance, economics or planning, to be fully involved in mainstreaming adaptation.

Source: Adapted from Sperling (2003)

20.3 The foundations of the policy response: building on good development practice

Much of what governments should do in relation to adaptation is what they should be doing anyway – that is, implementing good development practice. This is key to reducing the vulnerability of developing countries to climate change and raising their capacity to adapt. Climate change concerns simply lend greater urgency to these core tasks of government and, as discussed in Chapter 26, the role of the international community in supporting adaptation in developing countries. This was noted in Chapter 4 where rapid growth, as being experienced in China and India for example, will equip these countries with the economic resources to invest in appropriate policies and tools to better manage the effects of climate change. In some circumstances, there may be additional costs which the international community will have a role in helping to finance (see Chapter 26), bearing in mind the differences in income and historical responsibility for the bulk of past emissions.

If individuals and communities are empowered by development and rendered less vulnerable overall, they will be better able to adapt to climate change.

By empowering individuals with the tools to shape and improve their own lives and livelihoods – in other words, by promoting development broadly – governments will also strengthen individuals' ability to respond autonomously to climate change. Economic diversification, for example, is typically a core feature of development *and* is one of the best defences against economic shocks. It typically reduces the dependence of households, and the economy more broadly, on climate-sensitive sectors such as agriculture. It also increases the flexibility of the economy and individuals to adjust to sudden or gradual changes in the climate. Broad development measures will improve the lives of millions today and reduce individuals' vulnerability to climate change. In some cases it will also reduce the risks of these impacts occurring in the first place. For example:

- The control of malaria benefits millions of people today *and* will reduce the extent to which climate change will expose people to greater risk of malaria infection in the future;

- Greater access to education and reproductive health care for women will improve their lives and opportunities today *and* help control the rapid rate of population growth in developing countries, reducing pressures on existing resources.[3]

Good development practice will also serve to better equip people through building and developing their resilience. This is demonstrated in the case of Bangladesh where vulnerability to extreme weather events has been reduced in part through good development (Box 20.2).

BOX 20.2 Reducing vulnerability in Bangladesh

Bangladesh has been identified as the "most disaster-prone" of all countries, having suffered 170 large-scale disasters between 1970 and 1998.[4] Substantial investments have been made in recent years to reduce vulnerability to extreme climate variability (with the recovery following the 1998 floods more rapid than predicted) including: a structural change in agriculture, with an increase in the planting of much lower risk dry season irrigated rice; better internal market integration; and increased private food imports. Bangladesh's dependence on agriculture has also been reduced by an increase in export-oriented garment manufacturing. These developments were aided by higher credit penetration, including micro credit, increased remittances from abroad, and increased donor assistance. General development support has contributed to reducing the economy's sensitivity to extreme climate variability.

Source: ODI (2005)

Key areas for development action that will help to reduce vulnerability to the effects of climate change include:

- Progress on achieving income and food security and on overcoming the structural causes of famine/insecurity;
- Building robust education and health systems, including eradication of malaria, cholera, and other diseases associated with water;
- Better urban planning and provision of public services and infrastructure; and
- Better gender equality.

Improving access to micro-finance to help create assets and income will also be important as much of the funding for autonomous adaptation in developing countries will have to come from domestic sources and much of the action will be by households and small firms. Access to insurance and reinsurance services, savings and credit facilities, and flows such as financing for disaster preparedness measures and remittances will also be important to help protect the most vulnerable from climate change (discussed below and in Chapter 26).

It is important to note, however, that not all development policies and practice will be beneficial from a climate change perspective, and in some cases will actually

---

[3] For example, evidence suggests that educated women are more likely to seek medical care, improve sanitation practices and choose to have fewer children. Econometric studies have found that an extra year of female schooling reduces female fertility by between 5 to 10%. World Bank (2001); Summers (1992).

[4] World Bank's recent Water Resources Assistance Strategy report for Bangladesh

increase vulnerability to the impacts of climate change. This is known as maladaptation.[5] Maladaptation is commonly caused by a lack of information on the potential external effects of policies and practices on other sectors, or a lack of consideration given to these effects. The destruction of coastal mangroves is a prime example. Mangroves provide a wide range of services, including protecting against floods, coastal erosion and storm surges. Despite their importance, mangroves are being cleared in countries such as Bangladesh and Fiji to make way for agriculture, urbanisation and tourism. Shrimp-farming, for example, took-off in Bangladesh as an export industry in the mid 1980s. While this provided incomes it also encouraged the deliberate inundation of land with brackish water during periods of low salinity to increase shrimp production. As these fragile ecosystems are destroyed, so vulnerability to climate change is increased. More integrated planning and management is widely recognised as an effective mechanism for strengthening sustainable development, as discussed in Section 20.4 below.

Improving disaster preparedness and management saves lives, but it also promotes early and cost-effective adaptation to climate-change risks.

Natural disasters exact far greater economic costs in developing countries than developed countries, in relative terms, and can cause setbacks to economic and social development in developing countries. As a result, private, public and international resources and assistance is increasingly being diverted through humanitarian responses and reconstruction needs to deal with natural disasters, as discussed in Chapter 4. For example, OECD estimates show that "emergency and distress assistance" from donors has risen from an average of 4.8% of total Official Development Assistance (ODA) in 1990 to 1994 to 7.2% in 1999 to 2003, reaching 7.8% of ODA (more than $6 billion) in 2003.[6]

It will typically be more effective – in terms of both lives saved and finances – to invest in better disaster preparedness and management. Macro-level assessments show that disaster risk reduction (DRR) measures have a high benefit-to-cost ratio. The US Geological Survey and the World Bank estimated that an investment of $40billion would have prevented losses of $280 billion in the 1990s.[7] And the savings are not just hypothetical:

- *China*: the $3.15 billion spent on flood control between 1960 and 2000 is estimated to have averted losses of some $12 billion;[8]
- *Brazil*: the Rio flood reconstruction and prevention project yielded an internal rate of return exceeding 50%;[9]
- *India*: disaster mitigation and preparedness programmes in Andhra Pradesh yielded a benefit/cost ratio of 13.38;[10]

---

[5] Burton (1996, 1997)
[6] OECD (2004) cited in ERM (2006). Note that a part of the increase in damages may be due to improved monitoring and reporting and increases in income.
[7] Cited in Environmental Resources Management (2006); Benson (1998). Figures are indicative as consistent methodologies were not used to prepare estimates.
[8] Benson (1998) cited in ERM (2006)
[9] ProVention (2005) cited in ERM (2006)
[10] Venton and Venton (2004), cited in ERM (2006)

- *Vietnam*: a mangrove-planting project aimed at protecting coastal populations from typhoons and storms yielded an estimated benefit/cost ratio of 52 over the period 1994 to 2001.[11]

Thus a focus on climate change reinforces an earlier development lesson: not only do disaster preparedness and emergency planning save lives and property, they are also highly cost-effective. DRR measures can also bring significant developmental benefits in normal times. Raised flood shelters in Bangladesh are used on a day-to-day basis as schools or clinics, for example, and boreholes drilled to protect against drought provide water that is cleaner and easier to access than alternative sources.[12]

At the margin, it will be important to ensure that disaster risk assessments take new climate-change risks into account. Otherwise, maladaptation can be the result (as discussed above and in Chapter 18). This was the case in Bangladesh where flood defences had been designed for lower levels of floods. Because those defences were poorly maintained and were in any event inadequate for the higher flood levels of recent years, they became counter-productive, eventually trapping and prolonging the floods of 1999.[13]

Governments should also promote risk-sharing approaches, through insurance and pooling of disaster risks.

Insurance is another area – closely related to disaster preparedness – in which climate-change considerations reinforce what governments should already be doing on developmental grounds.[14] Well-functioning insurance markets share risk across individuals, regions, and countries, reducing the welfare effects of negative shocks of all types, whether climate-related or not. Risk-based insurance schemes can also reduce the costs of climate change by encouraging good risk-management behaviours, as discussed in Chapter 19. For example, by providing the incentives to meet standards on building design and construction, they encourage action to reduce risk. In addition, insurance may also act as a catalyst for autonomous adaptation by providing information through its measurement and pricing of climate risks. With the expected increase in climate-related shocks, governments now have even more reason to promote well-functioning insurance markets, as described in Chapter 18 and 19. Low-cost micro-insurance options, particularly weather derivatives, could be a mechanism for sharing risk in the poorest countries. Promoting private-sector involvement and investment in disaster risk management in developing countries should be high on the agenda for governments.[15] Box 20.3 provides examples of innovative programmes in this area.

However, as noted in Chapter 4, these insurance markets will often fail to emerge autonomously in developing countries through poorly developed financial markets, low income levels with which to purchase the insurance and lack of robust information. While approximately a third of natural-disaster losses are

---

[11] IFRC (2002) cited in ERM (2006)
[12] ERM (2006)
[13] Bangladesh is now integrating long-term climate risks into disaster management.
[14] A lot of work has been done on this by UNEP Finance Initiative (http://www.unepfi.org/).
[15] Mechler et al (2006)

BOX 20.3 Pilot risk-sharing ventures in the developing world

A number of recent initiatives have pioneered micro-insurance and weather derivative instruments in the developing world:

- A weather insurance initiative was launched in 2003 in India by a group of companies called BASIX. It has already grown from 230 farmers in one state to 6,703 customers across six states for 2005, and it has generated much broader interest in weather-related insurance in India, with other insurance companies now beginning to offer the product;[16]

- The World Food Programme has a pilot drought insurance project in Ethiopia. The WFP secured contingency funding through a Paris-based reinsurer to set this up, and ensured data availability through capacity-building at the National Meteorological Agency. A drought index now tracks agricultural seasonal development through 26 weather stations reporting daily;[17]

- DFID is launching two pilot projects in Bangladesh to offer weather-based index insurance at the community level. These projects illustrate the possible convergence of micro-finance and complementary community-based programmes with more sophisticated market-based financial instruments;

- In Malawi, an index-based weather derivative is offered to groundnut farmers as part of a loan package organised by the National Smallholder Farmers Association (with technical assistance from the World Bank and Swiss development cooperation);

- In the coastal Andhra Pradesh region of India, micro-insurance services have been provided as part of the voluntary Disaster Preparedness Programme to groups of women with a minimum size of 250 members. The Oriental Insurance Company offers affordable cover to poor communities through cross-subsidy with the wider insurance market. In addition, Oxfam pays half the premium.

insured in high-income countries, less than 3% of such losses to households and businesses are insured in developing countries.[18] And only a small number of schemes offering weather derivatives or micro-insurance for disaster risks have been implemented in developing countries to date.[19] For insurance markets to work effectively, insurance companies need access to accurate forecasts of climate change effects and the damage it may cause. This is currently a major constraint in developing countries that will have to be addressed if insurance provision is to play an important role in disaster risk management. There is also a limit to the ability of insurance companies to spread risk as they will be unwilling or unable to insure against an event with a very high probability of occurring. In some cases the price of individual premiums will become unaffordable because of the high risks. At the same time, if risks increase in several insurance markets at once, then insurance companies may find it harder to spread risks and therefore be less willing to provide insurance at affordable rates.

[16] World Bank (2005a)
[17] WFP (2006)
[18] Munich Re (2005)
[19] The ProVention Consortium (http://www.proventionconsortium.org/) is actively pursuing the agenda.

Effective social safety nets will also be important to protect those who are most vulnerable and cannot afford protection. One example is the set of safety net programmes that were announced in Indonesia in response to the economic, natural and political crisis between 1997–98. The employment creation programmes – which relied on self-selection targeting – were found to be far more effective in reaching the most vulnerable households than programmes based on health subsidies and subsidised rice sales.[20] Equally, the Employment Guarantee Scheme (EPG) in the Indian state of Maharashtra has provided wage labour opportunities since the 1970s that help buffer households from the effects of poor harvests and other negative shocks.[21]

20.4   New policies focused on climate change

Investing in climate resilience has implications for each country's investments in natural, physical, human, technological, and social capital.

While many of the policies that promote adaptation will already form part of national governments' priorities, others may not. Beyond reducing vulnerability through a broad suite of development activities, effective adaptation may also require governments to address specific market failures and barriers that limit effective adaptation.[22] Box 20.4 highlights a range of issues to consider. However, as the impacts of climate change are difficult to predict accurately, any adaptation

BOX 20.4   Investing in adaptation

- **Encouraging technology transfer and supporting flows of knowledge:** governments can deliver better climate forecasts, and spread information about climate-resilient crop varieties and irrigation schemes. Regional Climate Outlook Forums for example, provide guidance on the probabilities of rainfall to farmers in Africa and South America;

- **Human capital:** investing in health and education raises the effectiveness of explaining to communities and individuals how their climate is changing, and why and how they should adapt in ways which effectively integrate climate risks into the development process;

- **Physical capital:** governments can make long-term infrastructure more climate resilient through building codes and regulations, land-use zoning, river management, and warning systems.[23] Some adaptation may require higher maintenance costs for basic infrastructure

---

[20] Pritchett et al (2002)

[21] A problem with scaling up an EPG scheme however is the lack of 'high value' works from local plans to provide employment. Sen (2004:11) discusses these challenges in the case of India and noted that "for any commitment of expenditure, the opportunity costs have to be scrutinized, and employment guarantee is no exception to this".

[22] Berkhout (2005)

[23] Climate norms including climate variability and extreme events should already be taken into account in infrastructure development. Climate is a factor in, for example, the design of domestic, industrial and commercial buildings, roads, bridges, drainage systems, water supply and sanitation systems, irrigation and hydroelectric power installations. Improving climate resilience will go a step beyond this. Burton and van Aalst (1999)

such as re-building or diverting dirt roads. Additional protective investments such as flood barriers and sea walls will also be required;

- **Social capital:** Supporting social networks, institutions and governance arrangements to strengthen safety nets for poor people in response to natural disasters. Many traditional risk-sharing mechanisms based on social capital, such as asset pooling and kinship networks, are less likely to be effective when climate change simultaneously damages families and households in an entire region. The same is true of traditional coping mechanisms like selling assets;[24]

- **Natural capital:** governments can help protect the resilience of natural systems to support the livelihoods of poor people, for example by planting mangrove belts to buffer the coastal erosion impacts of sea level rise.

Source: Sperling et al (2003)

strategy should be as flexible as possible, able to respond to new information, and robust enough to be cost-effective across a range of possible future scenarios.

Many of the interventions required in developed countries, as set out in Chapter 18 and 19, will also be needed by developing countries including:

- Better information on climate change and more accurate weather forecasting;
- Regulation to overcome market barriers to adaptation; and
- Incorporating climate change into long-term policies.

Whilst the principles may be similar, their application may be different. Development constraints – including high levels of illiteracy and poor governance – will present obstacles to the effectiveness of these policies. The significance of these constraints will vary from country to country, with low-income countries likely to face the greatest challenge. And additional adaptation measures will be required by developing countries that face acute threats of rising sea levels or desertification. Developed countries can play a strong leadership role on adaptation but these wide-ranging constraints have to be addressed to ensure early and effective adaptation in developing countries.

Governments have an important role to play in raising awareness. But there are barriers that should be addressed.

Individuals, firms, and civil society cannot adapt autonomously without reliable information and projections, especially since they should make some of their investment choices well before the effects of climate change are fully visible. A core responsibility of governments will be to see that it has access to, and disseminates domestically, good information on climate change. This will range from forecasts on the likely timing, extent and effects of climate change, to knowledge of drought and flood resistant crops and new crop planting techniques. Governments should provide these services given the public good nature of high quality climate information, noted in Chapters 18 and 19. The important role attributed to climate information is widely recognised. Tanzania's 1997 National

[24] In the jargon this is known as 'covariate risks' or 'correlated risks', as discussed in Chapter 5 and 19, Section 5.5 and 19.6 respectively.

Action Plan on Climate Change provides an example of a government focusing its efforts on raising awareness in the first two years of implementation.

Better information is a priority in many developing countries given the very low level of climate information currently available. The density of weather watch stations in Africa, for example, is eight times lower than the minimum level recommended by the World Meteorological Organisation, and reporting rates are the lowest in the world.[25] This low starting point indicates the size of the challenge compared to developed countries where government funded research programmes are already in place, such as UKCIP discussed in Chapter 19. Developing the necessary information is beyond the current capacity of many developing country governments. Many are not even able to monitor the climate, let alone forecast changes. Developing country governments will require international support in this area, as discussed in Chapter 26.

Effective communication of this information is also critical. Poor countries face barriers to free and easy communication such as:

- High illiteracy rates: in South Asia the female literacy rate is 46.3% and 53.2% in Sub-Saharan Africa, compared to 98.7% in developed countries;[26]
- Restricted access to electronic communication: while 70% of the population in North America are internet users, only 3.6% of Africans are, and only 10.8% of the population in Asia;[27]
- Inaccessibility of rural areas due to poor transport and road infrastructure.

Good development can go a long way in helping to overcome such barriers to effective information dissemination. However, it is important that governments take these issues into consideration in planning their communication strategies.

Government regulation can encourage private investment to take account of climate change. But its effectiveness will depend on the commitment and credibility of the government.

Land-use planning and performance standards can be important tools in encouraging private investment to consider the future risks and implications of climate change on their investment, as discussed in Chapters 18 and 19. However, the value of these interventions will be largely dictated by the commitment and credibility of the government. Poor governance is a problem in many developing countries, and indeed some developed countries, as demonstrated by the Corruption Perception Index.[28] This can lead to weak regulatory practices and poor enforcement of building standards. For example, while Iran adopted a seismic building code in 1989, legislation was not always enforced. As reported in the IFRC (2004), new buildings were sometimes certified as conforming to earthquake norms without thorough inspections being conducted, and laws were not in place to tackle municipalities that failed to retrofit infrastructure. Addressing problems with governance and weak enforcement will be crucial if regulation is to be fully effective.

---

[25] Washington et al (2004)

[26] UNESCO Institute for Statistics (2006) based on adult (15+) literacy rates on a regional basis, September 2006.

[27] Internet World Stats (2006)

[28] This is produced by Transparency International. This Index ranks more than 150 countries by their perceived levels of corruption, as determined by expert assessments and opinion surveys. http://www.transparency.org/policy_research/surveys_indices/cpi/2005

Unclear property rights and the illegality of much slum housing also pose major problems which need to be overcome by changes to ownership and property laws and stronger law enforcement. Without property rights and civil protection householders put off making necessary home improvements since authorities would then have an incentive to evict the occupants once the work was done and rent out the dwellings to others willing to pay for protection.[29] Development itself can help to overcome these constraints, by educating civil society, promoting transparency and institutional checks on power, and encouraging accountability. Governments can then help by providing better technical guidance on building standards and encouragement of monitoring and enforcement.

Governments should integrate adaptation into their development projects but may require support to overcome technical capacity constraints.

Developing countries should integrate adaptation into development policy, budgets, and planning.[30] This cannot be an add-on or an afterthought, since some degree of continuous adaptation will be required across many sectors and regions. Governments – working alongside donors, the private sector, and civil society – should ensure national policies, programmes and projects take account of climate change and the options for adaptation.

National planning and policies

The importance of integrating adaptation into development policy and process through national economic planning and budgetary processes is an important first step towards effective adaptation. The budget is an important process for identifying and funding development priorities. Adaptive activities should be integrated into the budget framework and relevant sectoral priorities to help ensure necessary actions receive adequate funding over the long term and are balanced against other competing priorities.[31] Yet there is little evidence of progress on this score so far. Recent (draft) analysis by the World Bank found that while most of the Poverty Reduction Strategy papers (PRSPs) reviewed established linkages to climate change, such as by highlighting vulnerabilities to climate risk factors and impacts on economic productivity, further in-depth discussion of the issue was rare. They found a similar story with the Country Assistance Strategies (CAS).[32] Developing countries face two key constraints in integrating climate change into broader national development planning:

- Institutional constraints: Governments face numerous constraints, including competing demands on scarce public resources. At present the adaptation process is generally channelled through the UNFCCC focal points, which are normally based in Ministries of the Environment. Such ministries usually have limited influence with other line ministries and with the Ministry of Finance. An integrated response requires activities led by a strong core ministry with overall responsibility such as Finance, Planning, Economic Affairs, and other line ministries. Climate change faces the same challenges that other crosscutting

[29] IFRC (2004)
[30] Burton (2006)
[31] Sperling (2003)
[32] Jimenez (2006)

issues, such as gender, HIV/AIDS, and rural livelihoods, have faced in the past. Given the importance of risk management in relation to climate change – and the potential impact on public sector investments – there are sound reasons for Finance and Economy Ministry engagement.

- Technical capacity: Many developing countries – particularly the poorest – have only recently begun preparing longer-term national development plans and budget frameworks. This planning capacity is essential for broader development, as well as for enabling the integration of climate issues, and is already being supported by many development programmes. This should also be supported by the process of preparing National Adaptation Plans of Action (NAPAs) in Least Developed Countries.[33] While NAPAs could help to fill this planning gap, it is essential that they be integrated within overall national planning. Otherwise they could become yet another issue without a strong championship ministry and therefore be ignored when budgets are prepared. So far, only five NAPAs have been completed, and there is no indication that any implementation has begun as a consequence of preparing a NAPA, so their effectiveness or otherwise is as yet untested.

Programmes and projects

At the programme and project level, climate change may reduce the efficiency with which development resources are invested and worsen outcomes. Hence the risks of climate change should be integrated into development programmes. This means, for example, using information related to climate-change risks in the design and construction of infrastructure and buildings. In addition to building the resilience of development programmes, integrating climate-change risks will also help to ensure action to achieve adaptation to climate change is consistent with action to reduce poverty. Several commentators have suggested how to incorporate climate change risks into their plans and programmes. Burton and van Aalst (2004), for example, have proposed a climate risk-screening tool for World Bank projects,[34] while the UNDP has compiled a series of technical papers to guide projects towards identification of appropriate adaptation strategies.[35] One crucial task will be for governments to manage public goods that may be sensitive to climate change through finance and investment decisions, for example by improving flood defences, public health and safety, and emergency planning and response. Some examples of adaptation in practice are given in Box 20.5 below.

In factoring climate-change risks into investment decisions, it will be important to make good use of information on the costs and benefits of various alternative

---

[33] The Least Developed Countries have received funding from the Least Developed Countries Fund (discussed in Chapter 26), implemented by the Global Environment Facility (GEF), to assist them in preparing these documents.

[34] This includes (i) a web-based knowledge tool that sets out the nature, magnitude and distribution of climate risks by country and region; and (ii) a routine project risk-screening tool modelled on the widespread practice of environmental impact assessment, where high-risk "hotspots" undergo a full risk assessment, while low and medium-risk projects undergo a vulnerability appraisal.

[35] The UNDP Adaptation policy framework is designed to be flexible so that those at an early stage of understanding can begin to assess vulnerability to climate variability and change, and those at a more advanced stage can begin to implement adaptation in practice. The overall approach embeds adaptation into key policies for development and places substantial emphasis on stakeholder engagement.

BOX 20.5 Adaptation in practice

(a) Climate resilience in the Pacific Islands

Several Pacific Islands are implementing climate risk management programmes:

- *Samoa*: community grants to strengthen coastal resilience and reconstruction of roads and bridges to cyclone-resilient standards. Such local initiatives may well be a fruitful approach since local people are usually able to identify more accurately points of vulnerability;

- *Tonga*: national programme to construct cyclone-resistant housing units and retrofit buildings to improved hazard standards;

- *Kiribati*: climate-proofing of major public infrastructure and promote effective water management;

- *Niue*: strengthening of early-warning system for cyclones, including satellite-phone back-up, solar-powered radios for isolated villages and email facilities. In addition, the government is promoting vanilla as a more resilient cash crop than taro that typically suffers heavy damage during cyclones.

Source: Bettencourt et al. (2006)

(b) Qinghai-Tibet Railway

The Qinghai-Tibet Railway crosses the Tibetan plateau with some 550 km of the railway resting on permafrost. About half of this permafrost is only 1°C to 2°C below freezing, and is therefore highly vulnerable to even moderate warming. Permafrost thawing could significantly affect the stability of the railway. To reduce these risks, design engineers have put in place a permafrost cooling system using crushed rocks. In the winter, the colder denser air above the rock layer will circulate downwards through the spaces between the rocks, forcing warmer air out and away from the ground. In the summer, the air will be warmer and lighter outside the rock layer, and the air within the rocks will cease to circulate, thus minimising the amount of heat absorbed by the permafrost. The technique could be applied to many types of infrastructure projects in permafrost zones around the world.

Source: Brown (2005)

(c) Adaptation of hydropower sector in Nepal

Glacier retreat and ice melt are adding to the size of Nepal's glacier lakes and increasing the risk of 'glacial lake outburst floods' (GLOFs), catastrophic discharges of large volumes of water following the breach of the natural dams that contain glacial lakes. The most significant flood occurred in 1985 when a surge of water and debris up to 15 metres high flooded down the Bhote Koshi and Dudh Koshi rivers for 90 km. The flood destroyed the almost-complete Namche Small Hydro Project, which had cost over $1 million. Much more attention is now being paid to the GLOF risks in Nepal and the likelihood that such risks will increase as a result of rising temperatures. Some adaptation options being considered include:

- Siting of hydropower facilities at low-risk locations (although this may only be feasible for new facilities);

- Early warning systems that can save lives far downstream (likely to cost around $1 million per basin);

- Design of hydropower infrastructure to limit vulnerability, such as powerhouse placed under ground;

- Direct reduction of risk through (i) siphoning or pumping water out of dangerous lakes, (ii) cutting a drainage channel, and (iii) taking flood control measures downstream.

Source: Agrawala (2005)

(d) Shanghai Heatwave Warning Systems

With a population of over 17 million, Shanghai is vulnerable to the effects of dangerous heat waves. The original heat wave warning system was triggered whenever temperatures in the city reached an arbitrary threshold of 35°C. The new system monitors a range of weather variables known to affect human susceptibility to the heat. For example, 'moist tropical' conditions are associated with the highest average temperatures and humidity and lead to the greatest increase in daily average mortality (35–63 excess deaths on top of a baseline of 222). The system can predict dangerous conditions up to two days in advance. Such a forecast triggers a series of activities by the Shanghai Municipal Health Bureau to reduce the population's vulnerability – media announcements on TV and radio, preparation of hospitals and public services, visits to the elderly in the city centre, and measures to ensure an adequate supply of water.

Source: Acclimatise (2006)

investments in terms of the damages avoided through adaptation and the benefits gained. An example is given in Box 20.6. These can mostly be at the project level, as in the case of retro-fitting buildings and flood defences. A key element determining the appropriate response will be the lifespan of the project and the options, for example, to retro-fit buildings and flood defences – noting that designing in adaptation at the beginning of the project can reduce the cost of retro-fitting.

BOX 20.6 Case-study of cost-benefit analysis for adaptation

Water supply in the Berg River Basin in South Africa

Runoff from the Berg River Basin provides a major source of water for Cape Town and the surrounding agricultural land. In the last 30 years, water consumption in Cape Town has increased three-fold, and is expected to continue to grow in the future, as a result of population growth (migration of households to the city from rural communities) and economic development. At the same time, climate models show that average annual run-off in the catchment could decrease by as much as 25% during the period 2010–2040 due to climate change. A dam for the basin was approved in 2004 to deal with these competing pressures, but the possible impacts of climate change were not taken into account. Similarly, arrangements for liberalising the market for water supply are also being discussed in order to provide an economically efficient and more resilient distribution of water.

A recent study has compared the net benefits of adjusting to development pressures and, additionally, adjusting to climate change under two strategies:

Strategy A: Constructing a storage reservoir to cope with development pressures and then adding capacity to cope with climate change.

Strategy B: Implementing water markets to cope with development pressures and then building a dam to cope with climate change.

Table showing present value estimates for costs and benefits of adjustments for increasing development pressures and climate change in the 2080s

Estimated benefit or cost measure	Strategy A	Strategy B
Development action (no climate change)	Construct dam, no water markets	Water markets, no dam
Net benefits of development action	15 billion	17 billion
Additional adaptation action (development + climate change)	Increase dam capacity, no water markets	Construct dam + water markets
Net benefits of adaptation (reduction in damages from adaptation minus costs of adapting)	0.2 billion	7 billion
Cost of not planning for climate change that does occur	−0.2 billion	−7 billion
Cost of planning for climate change that does not occur	−0.2 billion	−1 billion

Note: All monetary estimates are expressed in present values for constant Rand for the year 2000, discounting over 30 years at a real discount rate of 6%.

Both the dam and water market options individually have similarly large projected net present values, but adding the possibility of adaptation to climate change shows the benefits of adopting both simultaneously. Increasing the water storage capacity of the Berg Dam could have a significant benefit for welfare. The effect is particularly strong if efficient water markets are introduced (net benefit of 7 billion, discounted over 30 years). Under this flexible and economically efficient approach, the costs of not adapting to climate change that does occur are much greater than the costs of adapting to climate change that does not occur (−7 billion vs. −1 billion in the case of efficient markets).

Source: Callaway et al. (2006)

20.5 Adaptation costs in the developing world

Adaptation is projected to cost developing countries many billions of dollars a year, increasing pressure on development budgets.

Only a few credible estimates are now available of the costs of adaptation in developing countries, and these are highly speculative. In a world of rapid climate change, it is increasingly difficult to extrapolate future impacts from past patterns, so historical records are no longer reliable guides. Furthermore, the discussion above has shown that conceptually this is a difficult calculation to solve: adaptation is so broad and cross-cutting – affecting economic, social and environmental conditions, and vice versa – that it is difficult to attribute costs clearly and separately

Table 20.1 World Bank preliminary estimates of the added costs necessary to adapt investments in developing countries to climate-change risks

This table, based on World Bank analysis, examines the core flows of development finance and estimates the proportion of the investment that is sensitive to climate risk. An estimate of the additional cost to reduce that risk to account for climate change is given. The percentage figures relating to the estimated costs of adaptation require further research and revision.

Item	Amount per year	Estimated portion climate sensitive	Estimated costs of adaptation	Total per year (US$ 2000)
ODA and Concessional Finance	$100 bn	20%	5–20%	$1–4 bn
Foreign Direct Investment	$160 bn	10%	5–20%	$1–3 bn
Gross Domestic Investment	$1500 bn	2–10%	5–20%	$2–30 bn
Total International finance				**$2–7 bn**
Total Adaptation finance				**$4–37 bn**
Costs of additional impacts				**$40 bn (range $10–100 bn)**

Source: World Bank (2006a), updated through discussions with the World Bank

from those of general development finance. Adaptation should be undertaken at many levels at the same time, including at the household/community level, and many of these initiatives will be self-funded.

With these very important caveats, one can consider the range of estimates that is available. The most recent estimates come from the World Bank that show the additional costs of adaptation alone as $4–37 billion each year.[36] This includes only the cost of adapting investments to protect them from climate-change risks, and it is important to remember that there will be major impacts that are sure to occur even with adaptation.[37] The World Bank estimate is based on an examination of the current core flows of development finance, combined with very rough estimates of the proportion of those investments that is sensitive to climate risk and the additional cost to reduce that risk to account for climate change (5–20% as a very rough estimate).[38] See Table 20.1. While there is considerable debate as to the value of these figures, they provide a useful order-of-magnitude estimate and reinforce the importance of further research in this area.

Another source of information is the NAPAs, which five countries have completed so far. On the basis of these it is possible to get a preliminary indication of the funding required. The total estimated cost for these NAPAs is $133 million, averaging $25 million per country. Extrapolating up to the 50 Least Developed Countries suggests adaptation costs of $1.3billion for these (mostly small) countries alone, and for only the 'urgent and immediate' action that is required.

---

[36] World Bank (2006a) and subsequent revisions.
[37] As explained in chapter 18, adaptation will not fully insulate people or economies from climate change, rather it is a way of dampening the impacts. As such, there will still be residual costs.
[38] These estimates exclude flood risk and other categories for which no costs are available so can be considered an underestimate from this perspective

Information and knowledge about the additional costs of adaptation is very limited. This knowledge is essential in facilitating countries to integrate climate change risks and adaptation needs into their longer-term plans and budgets. More work is needed to arrive at more precise measures.

20.6   International assistance for adaptation

Just as individuals, communities, and firms in developing countries will require help from their governments to adapt efficiently to climate change, so these governments may need support from the international community. Chapter 26 will discuss how the international community can help to promote adaptation in developing countries. Nonetheless, it should be clear from the discussion above that adaptation will require coordinated efforts on many fronts. Donors and other international development partners should reorient their strategies to match national efforts and help to remove barriers to interventions that prove cost-effective once climate risk management is interpreted into development programmes. This would also help to mainstream adaptation into national development and planning processes and so promote sustainable development.

Equally, given that the most affected countries are often among the poorest, there is a real need for the international community to fully honour the commitments, made at Monterrey in 2002, the EU in June 2005, and at the G8 summit in Gleneagles in July 2005, to increase sharply the flows of aid to developing countries, with the EU confirming and setting a timetable to 0.7% of GDP for ODA. Chapter 26 explores in detail the question of international support for countries facing the challenge of adaptation.

20.7   Conclusion

The climate will continue to change over the next few decades, whatever the world manages to achieve on the mitigation side. But the costs of adaptation will rise exponentially if efforts to mitigate emissions are not successful. It is an unfortunate twist of fate that those affected most immediately and hardest are often the countries that contributed little to the problem and that are least able to afford the costs of adaptation. They can afford even less not to adapt, however. Adaptation efforts are already underway, but they must be accelerated. Much of the adaptation is and will have to be autonomous, driven by market forces and by the needs and devices of households and firms. Governments should assist this process.

This chapter argued that a first set of actions consists of policies that should already be high on each government's agenda, even in the absence of climate change. The first and best way for governments to accelerate adaptation is to promote development successfully. If individuals and communities are empowered by development and rendered less vulnerable overall, they will be better able to adapt to changes in their environment. Second, improving disaster preparedness and management saves lives, but it also promotes early and cost-effective adaptation to climate-change risks.

But, in addition, governments should adopt new policies targeted at the climate-change threat. One important new task for governments will be to provide firms and communities with high-quality information and tools for dealing with climate change. Governments will also have an important role in encouraging effective adaptation through the use of regulation. More generally, in light of the far-reaching implications of climate change, governments should integrate adaptation into their development projects and plans across the board. Investing in climate resilience has implications for each country's investments in natural, physical, human, technological, and social capital. There will be barriers and obstacles to these climate policies that will have to be taken into consideration – but development progress will help to address and overcome these constraints.

With all of these needs, the incremental investment costs of adaptation are projected to run into many billions of dollars a year for developing countries, including very poor countries that are already hard pressed to meet development goals. On top of those costs, these countries will have to bear the costs of the climate-change impacts that remain even after adaptation. Chapter 26 will return to the question of how the international community can best help developing countries adapt.

References

Acclimatise (2006): 'Review of adaptation to climate risks by cities', London: London Climate Chamge Partnership, available from http://www.london.gov.uk/climatechangepartnership/ adapting-jul06.jsp

Agrawala, S. (ed.) (2005): 'Bridge over troubled waters: Linking climate change and development', Paris: OECD.

Baer, P. (2006): 'Adaptation: who pays whom?', in pp. 131–153, W.N. Adger, J. Paavola, S. Huq, M.J. Mace (eds.), Fairness in Adaptation to Climate Change, Cambridge: MIT Press.

Benson, C. (1998): 'The cost of disasters. Development at risk?', in Natural disasters and the third world. J. Twigg (ed.) Oxford: Oxford Centre for Disaster Studies

Benson, C. and Clay, E. (2004): Understanding the economic and financial impacts of natural disasters, World Bank, Disaster Risk Management Series No. 4, 2004, Washington DC: World Bank.

Berkhout, F. (2005): 'Rationales for adaptation in EU climate change policies', Climate Policy 5: 377–391

Bettencourt, S., R. Croad, P. Freeman, et al. (2006): 'Not if but when: adapting to natural hazards in the Pacific Islands region', Washington DC: World Bank.

Brown, J.L. (2005): 'High-altitude railway designed to survive climate change', Civil Engineering 75: 28–38

Burton, I., (1996): 'The growth of adaptation capacity: practice and policy, in Adapting to Climate Change: An International Perspective J.B. Smith, N. Bhatti, G.V. Menzhulin, et al. (eds.)] New York: Springer-Verlag, pp. 55–67.

Burton, I., (1997): 'Vulnerability and adaptive response in the context of climate and climate change'. Climatic Change, 36 (1–2): 185–196.

Burton, I. (2006): 'Adapt and Thrive: Options for reducing the climate change adaptation deficit', Policy Options issue December 2005–January 2006 Global Warming – A Perfect Storm.

Burton, I. and van Aalst, M. (1999): 'Come hell or high water – integrating climate change vulnerability and adaptation into bank work', World Bank Environment Department Paper No 72, Washington DC: World Bank.

Callaway, J. M., D.B. Louw, J.C. Nkomo, J. C., et al. (2006): 'The Berg River dynamic spatial equilibrium model: A New Tool for Assessing the Benefits and Costs of Alternatives for Coping With Water Demand Growth, Climate Variability, and Climate Change in the Western Cape'. AIACC Working Paper No. 31. International START Secretariat, Washington, DC: AICC available from www.aiaccproject.org

Environmental Resources Management (2006): 'Natural disaster and disaster risk reduction measures- a desk review of costs and benefits', London: DIFID.

Hanemann, W. M. (2000): 'Adaptation and its measurement', Climatic Change 13: 571–581

Haughton, J. (2004): Global Warming – The Complete Briefing, third edition, available from http://www.london.gov.uk/climatechangepartnership/docs/adapting_to_climate_change.pdf

International Federation of the Red Cross and Red Crescent Societies (2002): 'World Disasters Report 2002'. Geneva: IFRC.

International Federation of the Red Cross and Red Crescent Societies (2003): 'World Disasters Report 2003'. Geneva: IFRC.

International Federation of the Red Cross and Red Crescent Societies (2004): 'World Disasters Report 2004'. Geneva: IFRC.

IMF Working Paper (2003a) Dealing with increased risk of natural disasters: Challenges and options, [Freeman, P.K., M. Keen and M. Mani], Washington: DC, IMF.

IMF (2003b): 'Fund Assistance for Countries Facing Exogenous Shocks'. Prepared by the Policy Development and Review Department (In consultation with the Area, Finance, and Fiscal Affairs Departments), Washington: DC, IMF.

Internet World Stats (2006): available from http://www.internetworldstats.com/stats.htm

Intergovernmental Panel on Climate Change (2001): 'Climate Change 2001: Working Group 1: Scientific Basis – Summary for Policymakers', Cambridge: Cambridge University Press.

Jimenez, R. (2006): How climate change is being addressed in strategic documents including CASs, PRSPs and Environmental Analytical Work. Consultant report prepared for the World Bank.

Jones, R. and R. Boer (2003): 'Assessing current climate risks', Adaptation Policy Framework: A Guide for Policies to Facilitate Adaptation to Climate Change, UNDP, in review, available from http://www.undp.org/cc/apf-outline.htm

Jones, R. and L. Mearns (2003): 'Assessing future climate risks', Adaptation Policy Framework: A Guide for Policies to Facilitate Adaptation to Climate Change, UNDP, in review, available from http://www.undp.org/cc/apf-outline.htm

Kleindorfer, P. R., and H. Kunreuther (2000): 'Managing catastrophe risk', Regulation, 23: 26–31, available from http://www.cato.org/pubs/regulation/regv23n4/kleindorfer.pdf

London Climate Change Partnership (2005): 'Adapting to climate change: a checklist for development'.

Mechler, R., J. Linnerooth-Bayer, and D. Peppiatt (2006): 'Micro insurance for natural disaster risks in developing countries: benefits, limitations and viability', ProVention/IIASA study, Geneva: Provention Consortium.

Mendelsohn, R. (2000) Efficient adaptation to climate change, Climatic Change 45: 583–600

Munich Re (2005): 'Topics Geo Annual review: Natural catastrophes 2005', Munich: Munich Re Group.

Nkomo, J.C., A. Nyong and K. Kulindwa (2006): 'The impacts of climate change in Africa', Report prepared for the Stern Review, available from http://www.sternreview.org.uk

Overseas Development Institute (2005) 'Aftershocks: natural disaster risk and economic development policy', ODI Briefing Paper November 2005

OECD (2004a): 'Statistical Annex of the 2003 Development Cooperation Report', Paris: OECD.

OECD (2004b): 'Natural Disasters and Adaptive Capacity' [J. Dayton-Johnson] Working Paper No 237, DEV/DOC(2004)06, Paris: OECD.

Pritchett, L., S. Sumarto, and A. Suryahai (2002): Targeted programmes in an economic crisis: Empirical findings from the experience of Indonesia. SMERU Working paper

ProVention Consortium (2005): 'Successful disaster prevention in LAC', available from http://www.proventionconsortium.org/goodpractices/

Roy, J. (2006): 'The Economics of Climate Change: A review of studies in the context of South Asia with a special focus on India'. Jadavpur University.

Schelling, T.C. (1992): 'Some Economics of Global Warming', the American Economic Review 82 (1): 1

Sen, A. (2004): 'A lecture on India: large and small', Public Lecture, December, 2004, India Habitat Centre, New Delhi

Sperling F. (ed). (2003): 'Poverty & climate change: reducing the vulnerability of the poor through adaptation', Washington, DC: AfDB, AsDB, DFID, Netherlands, EC, Germany, OECD, UNDP, UNEP and the World Bank (VARG).

Summers, L. H. (1992): 'Investing in all the people – educating women in developing countries', Vol. 45. Washington, DC: World Bank.

Tompkins, E. L. (2005): 'Planning for climate change in small islands: Insights from national hurricane preparedness in the Cayman Islands', Global Environmental Change, 15.

Tompkins, E. L., & Adger, N. W. (2005): 'Defining a response capacity to enhance climate change policy', Environmental Science & Policy, 8.

UNDP Reducing Disaster Risk (2004): A challenge for development

UNEP Global Environment Outlook 2003.

UNESCO Institute for Statistics (2006): 'Compendium on methods and tools to evaluate impacts of, and vulnerability and adaptation to, climate change', 2005, UNFCCC Secretariat, available from http://stats.uis.unesco.org/tableviewer/document.aspx?FileId=220

Venton, C and Venton, P. (2004). Disaster preparedness programmes in India. A cost benefit analysis. Humanitarian Practice Network, London: ODI.

Washington, R., M. Harrison, and D. Conway, (2004): 'African Climate Report: A report commissioned by the UK Government to review African climate science, policy and options for action', December 2004.

World Food Programme (2006): Progress Report on the Ethiopia Drought Insurance Pilot Project, WFP Executive Board Annual Session, :WFP

World Bank (2001): 'Engendering development through gender equality in rights, resources, and voice', volume 1. [King, E. M., and Mason, A. D], Washington, DC: World Bank.

World Bank (2004): 'Natural disasters: counting the cost'. Feature story 2nd March 2004, Washington, DC: World Bank.

World Bank (2005a): 'Scaling up micro insurance: the case of weather insurance for smallholders in India', Washington, DC: World Bank.

World Bank (2005b): 'Natural disaster hotspots: A Global Risk Analysis', IBRD, and Columbia University, Washington, DC: World Bank.

World Bank (2006a): 'Clean Energy & Development: Towards an Investment Framework' Annex K, Washington, DC: World Bank.

World Bank (2006b): 'Not if, but when: adapting to natural hazards in the Pacific Islands Region', a policy note, Washington, DC: World Bank.

VI | International Collective Action

Part VI of the Review considers the challenges of building and sustaining frameworks for international collective action on climate change.

It considers the various dimensions of action that will be required to reduce the risks of climate change: both for mitigation (including through carbon prices and markets, interventions to support low-carbon investment and technology diffusion, co-operation on technology development and deployment, and action to reverse deforestation), and for adaptation.

These dimensions of action are not independent. For example, a carbon price is essential to provide incentives for investment in low-carbon technology around the world, and can be strongly complemented by international co-operation to bring down the costs of new low-carbon technologies. The success of international co-operation on mitigation will determine the scale of action required for adaptation.

Part VI is structured as follows:

- Chapter 21 provides a framework for understanding international collective action, drawing on insights from game theory and international relations, and sets out an overview of existing international co-operation on climate change.
- Chapter 22 examines the challenge of creating a broadly comparable price for carbon around the world. It considers what can be learned from the implementation of the Kyoto Protocol, and looks at the scope for expanding and linking emissions trading schemes.
- Chapter 23 considers how the transition to a global low-carbon economy can be accelerated through action to promote the diffusion of technology and investment in low-carbon infrastructure in developing countries and economies in transition. It explores current arrangements including the Clean Development Mechanism and considers how flows of carbon finance can be transformed to respond to the scale of the challenge.
- Chapter 24 provides an analysis of how international co-operation can accelerate innovation in low-emission technologies and in technologies for adaptation.
- Chapter 25 considers the opportunities that exist to reverse the emissions from land use, and in particular the challenge of providing economic incentives to reduce deforestation.
- Chapter 26 examines how international arrangements for adaptation can support national efforts and contribute to an equitable international approach.
- Chapter 27 brings the Review to a conclusion, emphasising the importance of building and sustaining international collective action on climate change.

21 Framework for Understanding International Collective Action for Climate Change

KEY MESSAGES

Climate change mitigation raises the classic problem of the provision of a global public good. It shares some key characteristics with other environmental challenges that require the international management of common resources to avoid free riding.

International collective action is already taking place in a wide variety of forms, including multilateral, coordinated and parallel approaches.

- Multilateral frameworks such as the UNFCCC and Kyoto Protocol provide an essential foundation to build further co-operation.

- Partnerships, networks and organisations such as the International Energy Agency facilitate coordinated international action.

- Mutual understanding of domestic policy goals supports further action: the EU, China, and California are amongst those that have adopted strong mandatory initiatives that will reduce the growth of greenhouse gas emissions.

Stronger, more coordinated action is required to stabilise concentrations of greenhouse gases in the atmosphere. Successful efforts in many areas, including the protection of the ozone layer, have demonstrated that international co-operation can overcome issues of free riding. Insights from game theory help to inform the design of frameworks for international action.

Countries usually honour international commitments where they conform to shared notions of responsible behaviour, even through international law provides weak tools to enforce co-operation. **Existing multilateral frameworks can be enhanced by creating a shared understanding of long-term goals and responsible behaviour.**

The transparency and comparability of national action across a range of dimensions of effort are key to mutual understanding and recognition of what others are doing, as well as ensuring public accountability. Enhancing them will require a strong response from existing multilateral institutions, including those with expertise in monitoring economic policy.

Widespread public understanding of the climate change problem and support for action is growing rapidly. Public awareness and support is crucial for encouraging and sustaining co-operation.

21.1   Introduction

Climate change is one of the greatest challenges to international co-operation the world is currently facing. As we have described in the preceding Parts of this Review, the scale of the problem and consequences of failure to tackle it are immense. This Review has made a compelling case for action – on both mitigation and adaptation – demonstrating that the global economic costs of business as usual paths are likely to far outweigh the costs of taking action to reduce the risks. We have also explored some of the local and regional co-benefits that can act as incentives to take action. A wide range of policy tools for mitigation and adaptation are available to national governments. However, no two countries will face exactly the same situation in terms of impacts or the costs and benefits of action, and no country can take effective action to control the risks that they face alone. International collective action to tackle the problem is required because climate change is a global public good – countries can free-ride on each others' efforts – and because co-operative action will greatly reduce the costs of both mitigation and adaptation. The international collective response to the climate change problem required is therefore unique, both in terms of its complexity and depth.

This chapter sets out a framework for understanding the scale and type of international collective action required for climate change. The first section examines and applies theories and analyses of collective action that have been developed, pointing out both their insights and limitations. The next section reviews the current arrangements for action on climate change including multi-lateral, coordinated and parallel action, and initiatives by the private sector that go beyond international frameworks. The final section considers how to build on these initiatives to develop an international response at the much larger scale that is now required, and how to develop an effective and transparent approach to sustaining co-operation.

21.2   Understanding international collective action

Reducing the risks of climate change is the most important example of the provision of a global public good, as explained in Chapter 2. It is also in many ways the 'purest' example of a public good in that emissions of greenhouse gases (GHGs) from any one country have the same effect on the atmosphere as those from any other. Climate change also shares some key characteristics with other environmental challenges that require the international management of common resources, including the depletion of fisheries[1], the protection of the ozone layer, and with the provision of global public goods in other areas including health and development co-operation. While the impact of climate change is much larger in scale than any of these, there is much to be learnt from the experience of tackling these other problems.

Economists seek to understand the incentives relevant to situations that require collective action, and have studied the institutional arrangements that can facilitate co-operation. The study of collective action is concerned with

---

[1]   See, for example, Gissurarson (2000).

understanding how to overcome the market failures that lead to the under-provision of public goods where individuals or countries face an incentive to free-ride on the actions of others[2].

In *The Logic of Collective Action*, Olson (1965) argues that rational, self-interested individuals would not act to secure a common interest unless they were coerced, or induced to do so with incentives that were not available to those who did not participate. Collective action by independent sovereign nations is particularly challenging. In the area of climate change, there is no supranational authority to provide coercive sanctions[3], so co-operation requires that nations perceive sufficient benefits that they are willing to participate in international treaties or other arrangements, and share a common vision of responsible behaviour. They must also recognise that without their involvement, international collective action may fail.

Game theory is a tool that economists have used to study the challenges of collective action, especially the problems of provision of local and global public goods.

Game theory has been used to explore the underlying structure of some common problems. The Prisoner's Dilemma Game[4] has been used to explore a wide range of situations in which individuals act rationally in the light of their own situation and yet find themselves faced with an outcome that leaves them worse off than if they were able to co-operate.

The theory of collective action now recognises that many types of games are relevant, and in particular that strategic behaviour and repeated games provide a number of important insights for understanding how to promote international co-operation[5].

- Changing the structure of the incentives in the game can make co-operation more attractive. This can happen through increasing the shared understanding and awareness of the benefits of co-operation and making links to a wider range of benefits as well as through creating side payments (or, where costs of action are involved, sharing costs differently) to secure co-operation.
- Reciprocity plays a key role in situations where the players facing the prisoners' dilemma have the opportunity to play *repeated games* and remember the

---

[2] Wicksell K. (1896) identified the problem of free-riding. He showed that the voluntary provision of public goods would lead to undersupply, because all actors hope that others will bear the cost of provision, so do not contribute.

[3] In the area of international trade, for example, the rules-based World Trade Organisation exists and can exert coercive sanctions on countries. International trade – or rather, its liberalisation – has some public-good properties akin to action on climate change. The theory of comparative advantage suggests that the world as a whole can gain from the global reduction of trade barriers. However, countries may not wish to liberalise their markets fully and forswear tariffs, because of market power in international markets or distributional impacts. Impacts on the distribution of income can arise, for example, where the returns to capital and the returns to labour before liberalisation differ from the world average. There are also other potential barriers such as security - for example in food and energy production. Schelling (2002) suggests that countries are more willing to accept coercive sanctions in the area of international trade because it is a *detailed* system based on *reciprocity* - most sanctions tend to be bilateral and specific, so parties can retaliate and make penalties fit the crime. As we have noted in Chapter 2, the beneficiaries of action on climate change can't so easily organise themselves: today's poor as well as the generations as yet unborn.

[4] This is described in any standard microeconomic or game theory textbook, such as Gibbons (1992).

[5] See, for example, Sandler (2004).

BOX 21.1 Tragedy of the commons?

Hardin (1968) set out an example of how private incentives might be expected to operate in the absence of co-operation to manage a common environmental resource. In *The Tragedy of the Commons*, he showed that individual farmers had powerful short-term incentives to contribute to the overgrazing and destruction of common land.

The metaphor has been criticised as oversimplified. Ostrom (1990) demonstrated that many local communities can and do co-operate to manage common resources, from irrigation networks to forests. In an article reviewing the impact of Hardin's views, *The Struggle to Manage the Commons*, Dietz, Ostrom and Stern (2003) considered how global trends that drive environmental change limit the ability of local commitments to respond to those challenges.

Global environmental issues require choices to be made between clear and immediate local incentives and diffuse, long-term global benefits. These challenges cannot be resolved through local community action. They require co-operation between governments, as well as community involvement in local implementation.

previous choices of the other player. In particular, many players adopt a strategy of *conditional co-operation*, in which they contribute more to the provision of a public good the more others contribute[6].

- In repeated games, increasing the frequency of contact and transparency contributes to building co-operation, just as institutional structures and repeated negotiations do in international agreements[7].

- In repeated games, options for renegotiation of the rules at key stages play an important role[8]. Compliance mechanisms that rely on harsh punishments are hard to enforce, as they often have a detrimental effect on the punisher as well as the punished and create incentives for both the punisher and the defector to seek renegotiation in the event of a breach of co-operation[9].

- Reputation can play a significant role in influencing outcomes. A leader can create a positive dynamic by demonstrating a willingness to co-operate, and the actions of the leader have a strong influence on the beliefs that others in the game hold about the prospects for co-operation. It does not make a difference whether others in the game interpret these actions as 'rational' or 'irrational' – the point is they simply establish reputation[10].

Though extremely useful as a starting point for analysing international collective action, most of these theories tend to focus only on self-interest very narrowly defined, and so leave out perspectives on responsibility and ethical standards – for example, the views on what constitutes human decency that are expressed by the public. This does not mean the theories should be ignored – on the contrary, their conclusions are always imperative to implement correctly. However, a broader vision can acknowledge the important senses of community and shared endeavour that are evident in the history of many international frameworks for co-operation.

---

[6] See, for example, Sugden (1984); Joyce *et al* (1995); Fischbacher, Gachter and Fehr (2001).
[7] See, for example, Axelrod (1984).
[8] See, for example, Bernheim, and Ray (1989), Farrell and Maskin (1989).
[9] See, for example, Pecorino (1999).
[10] See, for example, Kreps *et al* (1982), Seabright (1993); Gaechter (2006).

Game theory has been used to try to identify key criteria for the design of frameworks for international collective action on climate change.

Arrangements for global collective action exist across a wide range of issues including international trade, health, development aid, terrorism and environmental protection. Sandler (2004) identified a number of conditions that would make it more or less likely that collective action would succeed in different circumstances. He found that international collective action was more likely to succeed where there was sufficient mutual self-interest (for example, international standards for telecommunications or aviation); in response to recognition of a shared threat (for example, increased co-operation on counter-terrorism in the immediate aftermath of 9/11), and where there was leadership by a dominant nation (for example, the role of the USA in securing agreement to protect the ozone layer). The barriers to action on climate change therefore included perceptions that country-specific costs of action dwarfed the benefits of action, and that was exacerbated by considerable uncertainty over the latter.

Barrett (2005) applied the lessons of collective action and game theory to an extensive review of over 190 arrangements for environmental co-operation – from the North Pacific Fur Seal Treaty to the Montreal Protocol on Ozone Depleting Substances. From this he concluded that the most successful treaties create a gain for all their parties, and sustain co-operation by changing the rules of the game – by restructuring the incentives for countries to participate and for parties to comply. Box 21.2 provides an example. Barrett suggested this requires a combination of carrots and sticks. Compensating payments may promote wide

---

BOX 21.2 Gaining cross-country participation to protect the ozone layer[11]

The Montreal Protocol on Substances that Deplete the Ozone Layer is often cited as an example of successful international co-operation. Just 24 countries signed the original Protocol in 1987, but as at October 2006, the Protocol has 74 ratifications, including the major developing countries. Emissions of most depleting substances have been brought under control. There are strong signs that the ozone layer will recover within the next 100 years.

Several factors contributed to the success of the Protocol. First, there was a high degree of scientific consensus and evidence that there was a problem that required urgent political action, and public opinion galvanised politicians. The Protocol thus established targets and timetables to phase out the use of ozone depleting chemicals, based on recommendations of expert panels including government and industry representatives. Second, although developing countries initial consumption of ozone depleting substances was low, it was growing fast. Developing countries participated because of the science, and because of the financial support provided for their transition to phase out of harmful substances – albeit at a slower pace than that for developed countries. However, the flows involved were not great, and were time-limited. Third, Montreal recognised the importance of stimulating and developing new technologies so that industry could use non-depleting alternatives, and providing access to technologies in developing countries. Finally, establishing groups of like-minded countries was useful in providing a forum to examine the complex issues involved in and consequences of taking action.

---

[11] Brenton (1994).

participation (for example because they distribute the gains from co-operation equally), while penalties, that are not too high to lack credibility, may deter non-participation and non-compliance.

21.3 Existing international arrangements for co-operation on climate change

International collective action to provide global public goods at the appropriate level can take place in a wide variety of ways, including specific binding treaties, arrangements embedded in other agreements, aspirational declarations, and participation in partnerships and regional coalitions. Formal multilateral agreements are at one end of a spectrum of co-operation, and can, if commitment is strong or enforcement mechanisms are credible, provide a high degree of assurance that countries will contribute to meeting shared goals. Other mechanisms allow for coordinated action even where there is no international legal instrument creating binding obligations. In some areas, where a number of actors perceive an advantage or a responsibility to adopting a leading position, parallel action is motivated by unilateral goals that may themselves be informed by an understanding of the magnitude of the climate change challenge.

The UN Framework Convention on Climate Change and the Kyoto Protocol embody the core principles of a multilateral response to climate change.

The international response to climate change dates back to 1979 when the first World Climate Conference highlighted concerns arising from the increased carbon dioxide in the atmosphere. In 1988 the UN General Assembly passed a resolution, proposed by Malta, in favour of the protection of the climate for present and future generations. In the same year, the World Meteorological Organisation and the United Nations Environment Programme jointly created the Intergovern-mental Panel on Climate Change (IPCC). The IPCC issued its First Assessment Report in 1990, confirming that climate change was a real concern and that human activities were likely to be contributing to it.

In recognition of the global nature of the problem, the United Nations Framework Convention on Climate Change (UNFCCC) was agreed at the Earth Summit in Rio de Janeiro in 1992. 189 countries, including all major developed and developing countries, have ratified the Convention[12]. The UNFCCC sets the overarching objective for multilateral action: to stabilise greenhouse gas (GHG) concentrations in the atmosphere at a level that avoids dangerous anthropogenic climate change. It also establishes key principles to guide the international response, in particular that countries should act consistently with their responsibility for climate change as well as their capacity to do so, and that developed countries should take the lead, given their historical contribution to greenhouse gas emissions. The Convention places a commitment to act on all countries. Whereas for developing countries this commitment is unquantified and linked to assistance from developed countries, the developed countries agreed to return greenhouse gas emissions to 1990 levels by 2000.

---

[12] As of October 2006.

The Kyoto Protocol, agreed in December 1997, set out an approach for binding international action and agreed specific commitments up to 2012. It entered into force in February 2005 and has been ratified by 162 countries[13]. However, the US and Australia have declined to join the Protocol, and the Canadian administration has signalled that it is likely to be unable to meet its commitments[14].

Climate change is becoming central to international economic relations, along with issues such as trade, development and energy security. A range of other institutions and arrangements support coordinated or parallel action on energy policy and land-use change.

Climate change is now a regular part of the agenda for G8 Summits, along with other aspects of international economic relations including trade and development. The Evian Summit in 2003 resulted in a statement on co-operation on various aspects of science and technology; at Gleneagles in 2005 leaders committed to an Action Plan for Climate Change, Clean Energy and Sustainable Development and launched a dialogue with other major economies; and at St Petersburg in 2006 the links between climate change and energy security were explored. Japan has asked for a report on progress from the Gleneagles Dialogue at its summit in 2008. G8 declarations are non-binding, but they have provided strong direction to a range of other international bodies (including the IFIs and the International Energy Agency (IEA)).

The IEA provides a forum for energy ministers from OECD member countries to debate energy policy and provides a wide range of technical information to support national policymaking. It now produces detailed analyses of the

---

BOX 21.3   Gleneagles dialogue on climate change, clean energy and sustainable development

The Gleneagles Dialogue is a process that brings together 20 countries with the greatest energy consumption, including the G8 and the major emerging economies of Brazil, China, India, Mexico and South Africa, and allows them to discuss informally innovative ideas and new measures to tackle climate change outside the formal negotiations under the UNFCCC. The Gleneagles Dialogue will also monitor the implementation of the Plan of Action, to ensure delivery of the commitments made by the G8 heads. To assist with the implementation of the Plan of Action, the G8 asked the IEA to develop and advise on alternative energy scenarios and strategies aimed at a 'clean, clever and competitive' energy future. In addition, the G8 have engaged with the World Bank and other international financial institutions to create a new investment framework for clean energy and development, including investment and financing.

The second Gleneagles Dialogue Ministerial meeting was held in Mexico in October 2006. The meeting saw progress on the Gleneagles Plan of Action (on which the Japanese Presidency of the G8 will receive a report in 2008); discussed the progression and debated the future direction of the work undertaken by the World Bank and other International Financial Institutions; considered how the IEA's programme of work can be utilised by governments; and debated the global economic implications of many of these policies.

---

[13] As of October 2006.
[14] Lessons from the experience gained from implementation of the Kyoto Protocol are considered in Chapter 22.

Table 21.1 Goals on climate change and clean energy adopted by 10 largest economies

Brazil	• National objective to increase the share of alternative renewable energy sources (biomass, wind and small hydro) to 10% by 2030 • Programmes to protect public forests from deforestation by designating some areas that must remain unaltered and others only for sustainable use

China	• The 11th Five Year Plan contains stringent national objectives including
	  • 20% reduction in energy intensity of GDP from 2005 to 2010
	  • 10% reduction in emission of air pollutants
	  • 15% of energy from renewables within the next ten years

France	• Kyoto Protocol commitment to cap GHG emissions at 1990 levels by the period 2008-2012 • National objective for 25% reduction from 1990 levels of GHGs by 2020 and fourfold reduction (75-80%) by 2050
Germany	• Kyoto Protocol commitment to reduce GHG emissions by 21% on 1990 levels by the period 2008-2012 • Offered to set a target of 40% reduction below 1990 levels by 2020 if EU accepts a 30% reduction target • National objective to supply 20% of electricity from renewable sources by 2020
India	• The 11th Five Year Plan contains mandatory and voluntary measures to increase efficiency in power generation and distribution, increase the use of nuclear power and renewable energy, and encourage mass transit programmes. • The Integrated Energy Policy estimates that these initiatives could reduce the GHG intensity of the economy by as much as one third.
Italy	• Kyoto Protocol commitment to reduce GHG emissions by 6.5% on 1990 levels by the period 2008-2012 • National objective to increase share of electricity from renewable resources to 20% by 2010
Japan	• Kyoto Protocol commitment to reduce GHG emissions by 6% on 1990 levels by the period 2008-2012 • National objective for 30% reduction in energy intensity of GDP from 2003 to 2030
Russian Federation	• Kyoto Protocol commitment to cap GHG emissions at 1990 levels by the period 2008-2012
United Kingdom	• Kyoto Protocol commitment to reduce GHG emissions by 12.5% on 1990 levels by the period 2008-2012 • National objectives to reduce CO_2 emissions by 20% on 1990 levels by 2010 and by 60% on 2000 levels by 2050
United States of America	• Voluntary federal objective to reduce GHG intensity level by 18% on 2002 levels by 2012 • California, the largest state, in the USA, has an objective to reduce CO_2 emissions by 80% on 1990 levels by 2050. • States in the North-East and mid-Atlantic have set up the Regional Greenhouse Gas Initiative to cut emissions to 2005 levels between 2009 and 2015, and by a further 10% between 2015 and 2018.

prospects for energy efficiency and technology to reduce greenhouse gas emissions from energy. Energy ministers at the IEA Ministerial in March 2005 considered the challenge of climate change and set out a vision of a "clean, clever and competitive" energy future. The International Energy Forum (IEF) also provides an opportunity to discuss energy policy responses to climate change, as it brings together oil producers including OPEC, and energy consumers including the IEA.

Climate change is also becoming increasingly important in the work of UN and other agencies (including the UN Environment Programme, and the UN Food and Agriculture Organisation) and partnerships (including. PROFOR, the collaborative programme on forests hosted by the World Bank) dealing with land use and agriculture.

In addition to formal multilateral arrangements, international partnerships launched in recent years allow interested governments, NGOs and private sector firms to co-operate in relevant areas. Some of these have been particularly successful at identifying opportunities for profitable action on climate change, including the Renewable Energy and Energy Efficiency Partnership and the Methane to Markets Partnership.

The Asia Pacific Partnership, launched in 2005, brings together energy, environment and foreign ministers and industry representatives from Australia, China, India, Japan, South Korea, and the USA – countries together responsible for around 50% of global GHG emissions, energy consumption, GDP and population. It has eight sectoral working groups, providing opportunities for networking and the development of joint public-private research and commercial projects for reducing greenhouse gas emissions. Other partnerships, such as the Carbon Sequestration Leadership Forum (CSLF) are focused on particular technologies, and will be discussed further in Chapter 24.

Many countries, regions, and cities have adopted approaches that complement and go beyond action under the multilateral framework.

National initiatives and policy measures designed to foster national and international co-operation in support of global environment issues are numerous, and rising in numbers. They can be found in countries at all stages of development. A comprehensive UNDP study (2005) found that more than half of these policy measures flow from national policy choices, while the others are undertaken in co-operation with multilateral organisations.

The majority of the world's largest economies now have goals in place to reduce carbon emissions, or to decrease energy intensity increase renewable energy and decrease deforestation. Countries have adopted a range of goals; if they can successfully deliver these, emissions will be reduced significantly below their 'business as usual' path. Table 21.1 summarises some of the relevant goals adopted by countries that account for around two thirds of the global economy and emissions.

Half the world's population lives in cities and many more travel into cities to work each day. By some estimates, urban areas account for 78% of carbon emissions from human activities[15]. Increasingly cities are taking initiatives aiming to reduce emissions. The Clinton Climate Initiative and the Large Cities Climate Leadership Group, a grouping of 22 of the largest cities in the world, have pledged

---

[15] http://www.epa.gov/oppeoee1/globalwarming/greenhouse/greenhouse16/vanguard.html

to reduce emissions and increase energy efficiency by creating a purchasing consortium to lower the prices of energy-saving products and accelerate their development. Cities in the developing world have also taken action, for example tackling local air pollution and congestion in ways that also have the effect of reducing greenhouse gas emissions.

International companies are taking a lead in demonstrating how profits can be increased while reducing emissions from industrial activities globally.

Multinational companies are accountable for their operations around the world, and a growing number of business leaders would now prefer to see a clear long-term international framework[16]. In many ways, large companies have longer time horizons than governments, and are making their own forecasts of where policy is likely to go, based in part on their views of current and future public opinion. For example, in an open letter to the British Prime Minister ahead of the G8 Summit, one group of business leaders said "We need to create a step-change in the development of low-carbon goods and services by rapidly scaling up our existing investments and starting to invest in new technologies. To achieve this, we need a strong policy framework that creates a long-term value for carbon emissions reductions and consistently supports and incentivises the development of new technologies."[17] The World Economic Forum has also convened a round table on climate change, which included businesses from around the world. A statement from the group urged G8 governments to "establish a long-term, market-based policy framework extending to 2030 that will give investors in climate change mitigation confidence in the long-term value of their investments"[18].

Businesses are motivated by opportunities to reduce costs from increased energy efficiency (as BP demonstrated through its introduction of an internal emissions trading scheme) and by intelligent forecasting of future markets – as for example with the development of hybrid cars by some auto manufacturers, the emphasis on low-carbon innovation in GE's Ecomagination campaign, and moves to explore non-fossil energy sources and carbon capture and storage by several major power and energy companies. We have discussed some of these incentives in Chapter 12. They are also motivated by opportunities to define and demonstrate responsible behaviour, including by protecting their staff and customers from the impacts of their emissions. Box 21.4 provides several examples.

Pressure from campaigners and stakeholders (including institutional investors and the general public) is also leading to increased board-level oversight of climate change risks. There have been several attempts to establish the legal liability of companies for their emissions, inspired by precedents including class action suits over tobacco and asbestos. Institutional investors are keen to see companies avoid being drawn into litigation. The US-based Ceres coalition of investors, environmental and public interest organisations regularly assesses the performance of companies in managing these and other direct and indirect risks from climate change[19]. In the UK, the Institutional Investors Group on Climate Change (representing investors with over $1 trillion in assets) has pledged to work with

[16] See, for example Browne (2004).
[17] http://www.cpi.cam.ac.uk/bep/clgcc/
[18] http://www.weforum.org/pdf/g8_climatechange.pdf
[19] http://www.ceres.org/pub/publication.php?pid=84

BOX 21.4 Visions for a zero carbon society - private sector leadership on climate change

A number of multinational companies in several sectors, including the automotive, power, energy intensive and financial industries, have begun to identify strategies for a zero-carbon society.

Toyota aim to build recyclable cars with zero emissions by minimising the environmental impact of vehicles over the lifecycle of a car. Energy use can be reduced through efficient manufacturing and production, engine types offer potential to reduce emissions from driving, and disposal at the end of life has been part of their vision of sustainable mobility.

In 2002, **Avis Europe** introduced a scheme to allow their car hire customers to offset carbon emissions, in partnership with the CarbonNeutral company (formerly Future Forests). They state that they have become 'carbon neutral' by 2005 by using their buildings more efficiently, recycling materials, and offsetting non-reducible emissions via tree planting and support of renewable energy and technology projects to reduce GHG emissions.

Vattenfall, an energy company that operates hydro, nuclear and coal generators has been developing and implementing three main CO_2-reducing measures: optimisation of existing technology to reduce emissions per unit of energy, increased use of non-CO_2 energy sources, and a long-term project to capture and permanently store CO_2 from fossil-fuel power plants.

Alcan has an ambition to become 'climate neutral' by no later than 2020 through the full life-cycle of its aluminium products. They have sought to increase energy efficiency through continued research and development in technology and process improvements, as well as reducing GHG emissions related to energy use, and pursuing the best energy mix from available energy resources and non carbon-based energy projects.

HSBC became the world's first major bank to become 'carbon neutral' in December 2005. To meet this goal, a Carbon Management Plan has been put in place which consists of three parts: reducing direct emissions, reducing the carbon intensity of the electricity used by buying from renewable sources where feasible, and offsetting the remaining CO_2 from the bank's own operations by buying emission reductions from 'green' projects.

governments and companies to promote a co-ordinated international response to climate change[20].

21.4 Building and sustaining coordinated global action on climate change

The scale of action required to reduce the risk of dangerous climate change requires both broad participation and high levels of ambition by all countries.

The existing international arrangements, national goals and business-led initiatives provide a strong foundation for action. Much has been learned in the last fifteen years, and there is growing international momentum to support moves to co-operation on a much greater scale. The UNFCCC Dialogue on Long-term

[20] http://www.iigcc.org/docs/PDF/Public/IIGCC_InvestorStatementonClimateChange.pdf

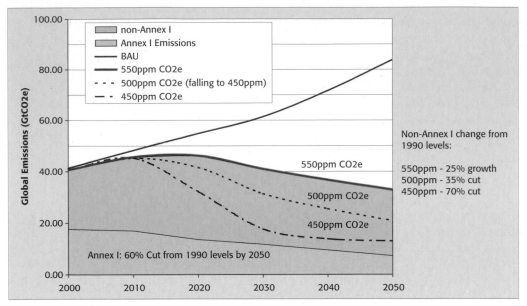

Figure 21.1 Emissions reductions in developed and developing countries, where developed countries take responsibility for cuts equal to 60% of their 1990 emissions by 2050

Action, the Kyoto Protocol discussions on the second commitment period, and a range of partnerships and initiatives provide room to explore a range of approaches.

We have argued in Chapter 13 of this Review that there is a strong case for stabilisation between 450-550ppm CO_2e. This would require very strong action to limit and reduce global emissions, starting now and continuing over the next 50-100 years. Robust, durable frameworks for international co-operation, based on a shared understanding of long-term goals, are required to meet this challenge.

It is essential that all major developed countries participate in this action. However, this will not be enough. Figures 21.1 and 21.2 demonstrate this by showing the extent of action that might be required globally for different possible stabilisation goals, given assumptions about emissions reductions by 2050 made by developed countries on their 1990 levels of emissions[21]. For example, even if developed countries reduce their emissions by 60% on their 1990 levels by 2050, depending on the overall stabilisation goal, the remaining emissions from developing countries could not exceed an increase of 25% on 1990 levels by 2050[22].

The distinction between developed countries taking responsibility for emissions reductions and making physical reductions within their borders is an important one. This is because the former can drive investment flows globally that can make it possible for developing countries to limit their emissions far below the levels they would otherwise be expected to reach.

---

[21] In Chapter 22, research is cited that, for developed countries, 60% to 90% cuts on 1990 GHG emissions are required to meet 450ppm and 550ppm CO_2e stabilisation goals respectively.

[22] This is in the context of the fact that developing countries' emissions as a whole have already increased substantially in recent years. GHG emissions in non-Annex I countries grew by 17% between 1990 and 2000, while they grew by 3% in Annex 1 countries over the same period.

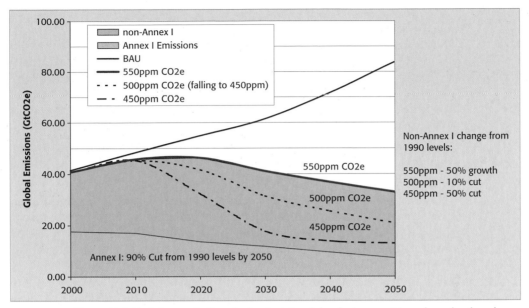

Figure 21.2 Emissions reductions in developed and developing countries, where developed countries take responsibility for cuts equal to 90% of their 1990 emissions by 2050

For example, were developed countries to take responsibility for reducing their emissions in 2050 by 90% on their 1990 levels, but put in place frameworks that allowed at least 50% of the investment in meeting these goals to take place outside their physical borders, they could meet the rest through investment in reducing carbon emissions in developing countries. This would mean, depending on the overall stabilisation goal, developing countries would still have to reduce the emissions within their physical borders in 2050 by around 50% on 1990 levels, but we calculate that they could also have flows of up to US$40 billion per year that could be directed towards helping achieve this[23]. Therefore, the more that developed countries commit to taking responsibility for, the more incentives could be provided for developing countries that take on commitments to limit or reduce emissions themselves.

It remains important that developing countries do take on commitments – in suitable forms and with the appropriate support. If the investment flows that are created by the rich countries take place only through the use of project mechanisms that allow them to offset their own commitments through action elsewhere, without any responsibility on the part of the recipient countries to take

---

[23] We calculate this with a very simple methodology that uses as a starting point the current value of CDM credits generated by an overall approximate 5% reduction in developing countries, and therefore assumes the difference between business as usual and emissions reduction paths remains stable up to 2050. We also assume that Annex I countries are currently meeting their reductions 50% domestically and 50% abroad, and a carbon price of $10/t CO_2. The UNFCCC Secretariat have used a different methodology to suggest that "100 billion dollars a year. . .would come about if half of the 60 to 80% reduction in emissions [by 2050] is met by industrialised countries through investment in developing countries".
http://unfccc.int/files/press/news_room/press_releases_and_advisories/application/pdf/20060919_riyadh_press_release_vs5.pdf

appropriate steps to constrain other sources of emissions themselves, there is a substantial risk of moral hazard[24].

Reductions on this scale are likely to be achieved only within frameworks that reduce the costs of action as far as possible, and that support an equitable distribution of effort. The following chapters will consider how global carbon markets can be mobilised to create the appropriate price signals and channel investment towards a low-carbon economy in both rich and poor countries, and how these frameworks apply to technology co-operation and reversing emissions from land use change.

The key challenge is to devise an agreement or a set of arrangements that attracts wide participation including all countries with significant sources of emissions, and achieves deep and lasting reductions in emissions from all sectors.

Countries are motivated to participate in international co-operation on climate change for a number of reasons, including the extent to which co-operation supports a range of short-term goals as well as the long-term goal of reducing the risks of climate change. For example, Chapter 12 discussed local co-benefits of mitigation.

Designing arrangements that are compatible with the underlying incentives of the participants is an effective way to ensure their continued adherence to the rules of the game and therefore a credible, lasting framework. Box 21.5 provides one illustration of the national short and medium term policy considerations that are relevant to international co-operation on climate change.

Box 21.5 Drivers for participating in international collective action on climate change

There are a number of drivers for participation in international collective action for both developed and developing countries. For example, an analysis of drivers for China's participation carried out by the Chinese Academy of Social Sciences (2006) shows a range of short and medium-term goals, including improving energy efficiency and financing the development and deployment of low carbon technology. Co-benefits include reducing air pollution and improvements to industrial structure, employment and regional development.

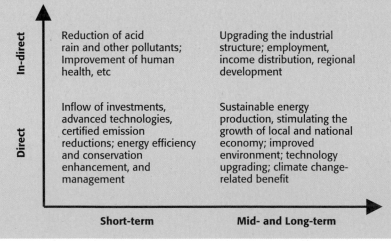

---

[24] "Offsetting" mechanisms include Kyoto's Clean Development Mechanism, which is introduced in Chapter 22 and discussed in more detail in Chapter 23. The offset credit is 'additional' if it represents a reduction that would not have otherwise happened under a business as usual path of emissions. Chapter 23 discusses how, in the absence of emissions reductions commitments, offsetting mechanisms can create moral hazard.

Shared notions of responsible and collaborative behaviour, within and outside governments, create the conditions in which countries honour international commitments.

The game theory that underpins analyses of international co-operation for global public goods tends to take as its starting point a narrow perspective of self-interest as the only motivation for action, distinguishing it from ethical approaches. In fact, these can be combined[25]. Although the key conclusions arising from these analyses are vital to examine, the creation of norms, and links to notions of responsible behaviour, are central to actions taken by governments[26]. Indeed, as we have noted, some game theory is moving beyond the traditional focus to examine the importance of reciprocity and reputation in solving collective action problems.

On many dimensions of international relations, governments make and respect international obligations because they are in line with perceptions of responsible and collaborative behaviour, and because domestic public opinion supports both the objectives and the mechanisms for achieving them.

Custom plays a very important role in international relations, and is often embodied in understandings and agreements that are not formally binding. These are often referred to as soft law. Environmental collective action provides numerous examples of the soft law approach and creation and recording of acceptable norms of behaviour between countries.

The principles set out in the non-binding 1972 Stockholm Declaration on the Human Environment were developed in numerous subsequent formal and informal agreements. They were picked up at the Earth Summit held in Rio de Janeiro in 1992. At Rio, world leaders signed conventions on climate change, biodiversity and desertification. They also adopted Agenda 21, a wide-ranging blueprint for action to achieve sustainable development worldwide. The Earth Summit concept of *think globally, act locally* inspired action from governments, community groups and individuals around the world. The Earth Summit was followed up at the World Summit on Sustainable Development in Johannesburg in 2002, where governments agreed a non-binding Plan of Implementation. This was supported by the launch of a large number of multi-stakeholder partnerships to take forward specific action. The UN Commission for Sustainable Development is currently reviewing the Johannesburg commitments on sustainable energy.

Soft law may allow countries to take on obligations that otherwise they would not. This is because non-binding instruments usually have an element of good faith that they will be adhered to by countries if possible, and may embody a desire to influence the development of state practices towards actual law making[27]. They can also be vehicles for focusing consensus on rules and principles and for mobilising a consistent, general response on the part of states. An example of this is 'tote-board diplomacy', whereby a collective standard for action is held up publicly, and countries that fail to agree are subject to collective pressure[28].

---

[25] For example, see Gauthier (1967).
[26] Some authors refer to this as the building of social capital, for example, Adger (2003); Dasgupta (2005).
[27] Birnie and Boyle (2002).
[28] Levy *et al*, 1992. The authors use the example of the 1979 Geneva Convention on Long-Range Transboundary Air Pollution, which created pressure on countries to tackle the problem of acid rain.

A collective sense of responsible behaviour and public acceptance of policy measures requires a shared understanding of action around the world. Governments also tend to look to the actions of neighbouring countries and key trade partners to benchmark the level of effort they are willing to make.

Co-operation across a broad range of issues including security and development can be sustained by norms of internationally responsible behaviour. Powerful statements stressing the importance of such behaviour in these contexts have been made by individual leaders, or expressed in a variety of non-binding international legal texts such as the declarations of the United Nations and communiqués from bodies such as the G8.

Collective action can be strengthened through actions taken at smaller, regional and national levels, for example, because "innovative rule evaders can learn how to get around a single type of rule more effectively than a multiplicity of rules-in-use."[29]. Therefore, codifying and passing commitments into domestic law can reinforce current and future commitments for action on a global public good. This sends a strong signal that a country is sincere in pledging action – and it means that reversing course becomes considerably more difficult and politically and legally challenging. Trust and credibility will be built especially when a country is seen to be taking real action to meet those commitments.

Formal compliance mechanisms have a role to play in managing specific and limited infractions of rules within international regimes. Agreed processes of adjustment may promote continued participation in a regime.

Where governments have set up a regime to take international action, compliance mechanisms can be used to maintain the credibility of that regime. The credibility of the regime will be damaged if rules of the regime are seen to be flouted, and this will quickly lead to a loss of support from other participants.

The existence of a compliance procedure may be sufficient to deter free-riding within the regime, provided that there is transparency, monitoring of actions, and, most importantly, there is pressure for the country concerned to remain part of the regime. However, participants can quit regimes. This means that for global public goods, formal compliance mechanisms are likely to only be effective for specific and limited infractions.

Chapter 14 discussed the issues for ensuring credibility of climate change mitigation policy on the national level[30]. National commitments, or sanctions applied in domestic law if those commitments are not met, may not be credible because governments can renege on their predecessors' commitments. This can also present a problem for international compliance[31]. We thus provided in Chapter 14 the rationale for short-term flexibility within an overall framework that has clear long-term goals in line with the scale of action required. The corresponding notion on the international level is that an international regime requires clear goals, and may require some form of adjustment of specific levels of effort to reach those goals over time to allow flexibility to respond to unforeseen circumstances. Adjustment could take account of economic growth, the underlying

---

[29] Dietz, Ostrom and Stern (2003); 1911.
[30] For example, see Helm *et al* (2004).
[31] See Aldy *et al* (2003). In particular, Schelling and Barrett propose regimes to take into account this issue.

carbon price in economies, the cost of low carbon technologies, or emissions reductions achieved. This, rather than automatic sanctions or punishment, may therefore create a way to respond to changing circumstances within one or a few countries without jeopardising the future of the entire framework.

It would be important that these rules were set, monitored and revised by a competent and credible international process, ideally a body independent of government ministries and influence in order to build credibility through reputation[1]. In the absence of such a body, representation of finance, external affairs and economic ministries in addition to environmental ministries would be important to obtain real buy-in to agreed rules.

Increasing the transparency and comparability of parallel national action is a significant challenge and will require a strong response from existing international institutions to enhance the coherence and cohesion of different policies.

Increasing understanding of action across different dimensions at different levels will build confidence amongst countries regarding the efforts of others and this could strengthen overall effort. Increasing information and monitoring may help to reduce free riding and improve accountability for the provision of public goods.

In the case of climate change, it is already clear that there are a number of dimensions of and a range of overlapping approaches to co-operation. Transparency and a shared understanding of action is required across all these dimensions, including on emissions reductions, the scope and level of carbon prices and policies, investment in innovation, parallel and coordinated approaches to standards and regulation, commitments to international co-operation on the deployment and diffusion of relevant technology, as well as international support for adaptation. The ways in which co-operation are assessed therefore have to be similarly broad, in the same way that the metrics used for organisational performance management have widened in recent years through use of approaches such as the balanced scorecard[32].

The task of benchmarking responsible action against other countries is made more complicated in the case of climate change by the competing priorities that can drive similar action. For example, the promotion of biofuels in Brazil, China and the US is often described as an energy security measure; in the EU, it is seen primarily as a response to climate change. Even more complex are the drivers for energy efficiency measures across countries. Therefore the definition of overall commitments for domestic climate change and energy policy also plays an important part in comparing efforts across countries.

The UNFCCC and Kyoto Protocol have already created a strong system for estimating and reviewing emissions according to standard guidelines[33]. Developed countries report emissions annually under this system. Formal national communications required from all countries also set out at a high level the policies and measures that are being implemented, but they are less frequent (every five years or so) and although there are agreed reporting guidelines, cross-country comparison is difficult.

---

[32] Kaplan and Norton (1996).
[33] The UNFCCC and Kyoto Protocol will be discussed in more length in Chapter 22.

Other initiatives can provide supplementary information. The G8 countries have agreed to provide annual updates in implementing the Gleneagles Plan of Action on Climate Change, Clean Energy and Sustainable Development, which covers areas including energy efficiency, cleaner power and the use of market-based instruments. The World Resources Institute has begun to develop an informal database of policy measures implemented in developing countries[34].

Transparency plays a key role in other areas of economic co-operation. The IMF, OECD, IEA, and many UN organisations systematically collect and compare data across countries on a wide range of economic policy issues[35]. It may be that a more systematic approach to monitoring economic policy relevant to climate change, including the explicit and implicit prices of carbon across the economy, would require the skills and expertise found in these institutions.

Global public concern and awareness about climate change are growing rapidly. They both influence and sustain international co-operation, national aspirations and private sector leadership on climate change.

As outlined in Chapter 17, individual preferences are subject to change, and public opinion across the world plays a very important role in sustaining co-operation on climate change. As on many other issues, public scrutiny of government policy matters. Public understanding of the challenge of climate change is essential to create the political space for governments to introduce and sustain the policies that are required to make the transition to a low carbon economy. International stakeholder pressure is also relevant, as a result of global investment flows and the responsibilities of multinational companies for their worldwide operations.

The public is influenced by the statements of, amongst others, politicians, scientists, Non-Governmental Organisations (NGOs), religious leaders and businesses, and by the presentation of the issues in the media. There has been a clear recent increase in public concern over climate change. Analysis of the incidence of references to climate change and global warming show that between 2003 and 2006, references in major newspapers doubled. International development NGOs and faith groups have increasingly become concerned about climate change. The UK's Stop Climate Chaos includes environmental and development NGOs as well as faith groups and trade unions. In the USA, a wide range of groups is campaigning on climate change issues. For example, the Evangelical Climate Initiative (ECI) released a statement signed by more than 85 evangelical leaders calling for action on climate change[36].

Pew Center polls on changing public attitudes around the world have sought to examine public attitudes to news stories. In a recent poll, awareness of climate change was high in the developed world, but in the developing countries sampled, awareness was generally lower than for a range of other issues. Clear majorities in most countries surveyed were concerned about the problem.

---

[34] This database is soon to be online at http://www.wri.org/climate/project_description2. cfm?pid=211.

[35] For example, the OECD regularly publishes Consumer and Producer Subsidy Equivalent statistics for the area of agriculture.

[36] http://www.christiansandclimate.org/statement.

BOX 21.6 Public attitudes to climate change around the world[37]

A poll by the Pew Center presented a snapshot of attitudes in 2006. Even in countries with limited formal participation in international action, at least half of the population now thinks that climate change matters a fair amount or a great deal.

Global Warming Concerns

	A great deal %	A fair amount %	Only a little/ Not at all %	DK %
United States	19	34	47	1
Great Britain	26	41	32	1
Spain	51	34	14	2
France	46	41	14	0
Germany	30	34	36	1
Russia	34	31	34	1
Indonesia	28	48	23	1
Egypt	24	51	23	1
Jordan	26	40	34	0
Turkey	41	29	23	8
Pakistan	31	25	39	5
Nigeria	45	33	20	2
Japan	66	27	7	0
India	65	20	13	2
China	20	41	37	2

Based on those who have heard about the "environmental problem of global warming

As the science of climate change is widely accepted, public attitudes will make it increasingly difficult for political leaders around the world to downplay the importance of serious action to respond to the challenge.

21.5 Conclusions

In this chapter we have examined the conditions for international collective action on climate change. We noted that extensive action has already begun on different levels – from the multilateral to the individual level, but that the scale of action now required demands a response on a much larger scale, involving all developed and developing countries in a collective endeavour to limit and reduce emissions.

[37] http://people-press.org/reports/display.php3?ReportID=280

Economic analysis can provide some guidance on the directions for effective, efficient and equitable frameworks for co-operation, and the following chapters will consider in more detail how to build key elements of international co-operation on climate change. These include carbon markets, support to developing countries in the transition to a low-carbon economy, international co-operation to accelerate innovation and to support the diffusion of energy efficient and low-carbon technologies, action to reverse emissions from land use change and forestry, and support for adaptation.

Each of these dimensions of action has its own specific challenges. An effective response to climate change requires co-operation in each area, supported by a shared understanding of long-term goals, and transparency about the contribution that each country is making towards them.

References

Adger, W. N., (2001): 'Social capital, collective action and adaptation to climate change'. Economic Geography, **79** (4): 387-404

Axelrod, R. (1985). 'The evolution of co-operation'. New York: Basic Books.

Aldy, J., S. Barrett and R.N. Stavins (2003): 'Thirteen plus one: a comparison of global climate policy architectures'. Discussion Paper. **26**, Washington, DC: Resources for the Future.

Barrett, S. (2005): 'Environment & statecraft: the strategy of environmental treaty making'. Oxford: Oxford University Press.

Brenton, T. (1994): 'The Greening of Machiavelli: the evolution of international environmental politics'. London: Earthscan Publications Ltd and Royal Institute of International Affairs.

Bernheim, D. B. and D. Ray (1989): 'Collective dynamic consistency in repeated games'. Games and Economic Behavior, **1**(4): 295-326

Birnie. P and A. Boyle (2002): 'International law and the environment', 2nd edn. Oxford: Oxford University Press.

Browne, J. (2004): 'Beyond Kyoto'. Foreign Affairs. **83**(4)

Chinese Academy of Social Sciences (2006): 'Understanding China's energy policy: economic growth and energy use, fuel diversity, energy/carbon intensity, and international co-operation'. Background Paper for the Stern Review on the Economics of Climate Change, available from
http://www.hm-treasury.gov.uk/media/5FB/FE/Climate_Change_CASS_final_report.pdf.

Dasgupta, P. (2005): 'Economics of social capital'. The Economic Record, **81**(1): 2-21

Dietz, T., E. Ostrom, and P.C. Stern (2003): 'The struggle to govern the commons'. Science, **302**: 1907-1912

Farrell, J. and E. Maskin (1989): 'Renegotiation in repeated games'. Games and Economic Behavior, **1**(4): 327-360

Fischbacher, U., S. Gaechter, and E. Fehr (2001): 'Are people conditionally co-operative? Evidence from a public goods experiment'. Economics Letters, **71**(3): 397-404

Gaechter, S. (2006): 'Conditional co-operation: behavioural regularities from the lab and the field and their policy implications', Discussion Paper, no. 3. Nottingham: The Centre for Decision Research and Experimental Economics, University of Nottingham.

Gauthier, D. (1967): 'Morality and advantage'. Philosophical Review, **76**: 460-475

Gibbons, R. (1992): 'Game theory for applied economists'. New Jersey: Princeton University Press.

Gissurarson, H. H. (2000): 'Property rights in marine resources: some new developments'. Hong Kong Center for Economic Research Letters, **60**

Hardin, G. (1968): 'The tragedy of the commons'. Science, **162**: 1243-1248

Helm D., C. Hepburn and R. Mash (2004): 'Time-inconsistent environmental policy and optimal delegation'. Conference Paper. London: Royal Economic Society.

Joyce, B., J. Dickhaut and K. McCabe K. (1995): 'Trust, reciprocity, and social history'. Games and Economic Behavior, Elsevier, **10** (1): 122-142

Kaplan, R.S. and D.P. Norton (1996): 'The balanced scorecard: translating strategy into action'. Boston, MA: Harvard Business School Press.

Kreps, D., P. Milgrom, J. Roberts, and R. Wilson (1982): 'Rational co-operation in the finitely repeated prisoners' dilemma'. Journal of Economic Theory, **27**: 245-52

Levy, M. A., R.O Keohane and P.M. Haas (1992): Institutions for the earth: promoting international environmental protection. Environment, 34(4): 12-17, 29-36

Olson, M (1965): 'The logic of collective action'. Cambridge, MA: Harvard University Press.

Ostrom, E. (1990): 'Governing the Commons: the evolution of institutions for collective action'. Cambridge: Cambridge University Press.

Pecorino, P. (1999): 'The effect of group size in public good provision in a repeated game setting.' Journal of Public Economics, 72: 121-134

Sandler, T. (2004): 'Global collective action'. Cambridge: Cambridge University Press.

Seabright, P. (1993): 'Managing local commons: Theoretical issues in incentive design'. Journal of Economic Perspectives, 7(4): 113-134

Schelling, T. C. (2002): 'What makes greenhouse sense?' Foreign Affairs. 81(3), pp. 2-9

Sugden, R. (1984): 'Reciprocity: the supply of public goods through voluntary contributions'. Economic Journal, 94(376): 772-787

Wicksell, K. (1896): 'Finanztheoretische Untersuchungen (Studies in the theory of public finance)'. Jena: Gustav Fisher.

22 Creating a Global Price for Carbon

KEY MESSAGES

A shared understanding of long-term goals must be at the centre of international frameworks to support large reductions in greenhouse gas emissions reductions around the world.

A broadly similar price of carbon is necessary to keep down the overall costs of making these reductions, and can be created through tax, trading or regulation. Creating a transparent and comparable carbon price signal around the world is an urgent challenge for international collective action.

Securing broad-based and sustained co-operation requires **an equitable distribution of effort across both developed and developing countries.** There is no single formula that captures all dimensions of equity, but calculations based on income, per capita emissions and historic responsibility all point to developed countries taking responsibility for emissions reductions of at least 60% from 1990 levels by 2050.

The Kyoto Protocol has established valuable institutions to underpin international emissions trading. There are strong reasons to build on and learn from this approach. There are also opportunities to use the UNFCCC dialogue and the review of the effectiveness of the Kyoto Protocol to explore ways to improve.

Private sector trading schemes are now at the heart of international flows of carbon finance. Linking and expanding regional and sectoral emissions trading schemes, including sub-national and voluntary schemes, requires greater international co-operation and the development of appropriate new institutional arrangements.

Common but differentiated responsibilities should be reflected in future international frameworks, including through a greater range of commitments and multi-stage approaches.

Carbon pricing and other measures should be extended to international aviation and shipping.

22.1 Introduction

At a national and regional level, as described in Chapter 14, approaches to mitigation include taxation, emissions trading and regulation. International collective action can build on these national approaches. As we have established in Chapter 23, such arrangements will be most successful if they take into account the underlying interests of the participants.

This chapter explains how international frameworks could be guided by long-term quantity goals and the corresponding global carbon price trajectory, and how they might also allow flexibility for national policy approaches.

The chapter considers how to build on and learn from the experience of the Kyoto Protocol so far. It also examines how the costs of mitigation can be minimised by international coordination and shared equitably, and the role of commitments and quota allocations. Finally we examine the challenges of expanding and linking regional and sectoral markets for carbon, and expanding carbon pricing to aviation and shipping.

22.2   Reducing the costs of mitigation through an efficient international framework

Very large reductions in greenhouse gas emissions are required around the world. A shared understanding of long-term goals, including for stabilisation of greenhouse gas concentrations in the atmosphere, is essential.

We set out in Chapter 14 the two key requirements for achieving efficiency for climate change mitigation. The first requirement is that greenhouse gas (GHG) emissions are reduced until the marginal cost of abatement[1] is equal to the marginal social cost of carbon (SCC) [2]. Defining the social cost of carbon requires a framework built around a shared understanding of long-term stabilisation goals.

A shared understanding of the scale of the challenge for both mitigation and adaptation can lead to a broad consensus on long-term goals for the stabilisation of GHGs in the atmosphere, as well as more medium-term considerations on appropriate pathways for global emissions, such as the depth of emissions reductions to be made by 2050. These goals can help to provide clarity and facilitate the development of national and international policies that minimise the costs and maximise the benefits of mitigation and adaptation. Policy-makers can then adjust national policy to operate in the context of a shared commitment to international collective action. Without this, there are risks that a series of fragmentary or short-term commitments would lead to inconsistent policies that raise the costs of action and fail to make a significant impact in reducing emissions.

It may not be essential to negotiate a single number for a long-term goal. As we have discussed in Chapter 21, declarations by political leaders and scientific and economic authorities can establish strong standards for responsible attitudes to

---

[1]   As we have emphasised throughout, risk and uncertainty are of the essence in climate change and we should really be speaking here in terms of mathematical expectations. But to avoid heavy language we keep it simple.

[2]   The social cost of carbon and carbon price discussed here are convenient short-hand for the social cost (and corresponding price) for each individual greenhouse gas. Their relative social costs, or 'exchange rate', depend on their relative global warming potential (GWP) over a given period and when that warming potential is effective, as the latter determines the economic valuation of the damage done. Suppose there were a gas with a life in the atmosphere one tenth that of CO_2 but with ten times the GWP while it is there. The social cost of that gas today would be less than the social cost of CO_2, because it would have its effect on the world while the total stock of greenhouse gases was lower on average, so that its marginal impact would be less in economic terms.

the climate. Recognition of the dangers associated with different stabilisation levels together with an understanding of what is feasible are likely to point to a fairly narrow range of goals for consideration. We argued in Chapter 13 that this range lies between 450ppm and 550ppm CO2e, given that the lower level could impose high adjustment costs in the near term for small gains given where we are now, and the upper level would substantially increase risks of very harmful impacts.

The scientific and economic evidence on climate change will continue to accumulate, including on the potential for dangerous climate change and future technologies. It is important that new information is reflected in international norms for climate protection, and that policy-makers are clear about how they will adjust their goals in the light of new evidence. The Intergovernmental Panel on Climate Change (IPCC) plays a vital part in assessing the scientific evidence and providing clear non-technical summaries that allow the issues to be widely debated. Long-term goals should be regularly revised in the light of the IPCC findings and other robust research.

A broadly similar global carbon price is an urgent challenge for international collective action. A global carbon price can, in theory, be created through internationally harmonised taxation or intergovernmental emissions trading, but neither is straightforward in practice.

The second requirement for efficiency discussed in Chapter 14 is that reductions in different countries are carried out as far as possible to the point where the marginal or incremental costs of further abatement across countries are just equal. Although the science tells us that the 'social cost' of emitting a tonne of GHGs is independent of where in the world it is emitted, there are currently significant differences in marginal abatement costs around the world, due to differences in rates of output and emissions growth, as well as differences in the structure of economies and energy sectors and levels of technical efficiency and differences in income. If the carbon price across countries is not broadly similar, there will be unexploited opportunities to abate an extra tonne of GHG more cheaply in one country compared with another, so the overall cost of abatement will be higher.

A similar carbon price around the world can be created in a number of ways, including through harmonised levels of net carbon taxes as part of national policy frameworks, intergovernmental emissions trading or expanding the use of private sector emissions trading; and/or using regulation to create an implicit price for carbon[3].

An internationally harmonized emissions tax – where all countries agree to set the same domestic carbon price across their economies – provides one model for an efficient approach to mitigation. Several analysts have argued that taxes have, on balance, advantages relative to quantitative limits at the international level[4].

A co-ordinated tax-based approach has the advantage that countries can take their tax decisions individually. It thus does not require elaborate structures and institutions, the construction of which can take time and effort. It allows compliance and monitoring to focus on the levels of net carbon tax in addition to monitoring of

---

[3] Therefore, when we refer to a 'carbon price' hereafter we mean an 'effective' carbon price that can be cumulatively generated by these sorts of instruments and schemes.

[4] These include Cooper (1998); Mckibben and Wilcoxen (2002); Pizer (2002); and Nordhaus (2005).

emissions. There are methodological challenges here, in untangling the multiple objectives of existing taxes, levels of direct and indirect subsidy applied and taking account of exchange rates. But they are not necessarily more complex than the existing monitoring of other policy areas carried out by institutions such as the International Monetary Fund (IMF), Organisation for Economic Co-operation and Development (OECD) or World Trade Organisation (WTO)[5].

Proponents of an internationally harmonised tax argue that it would also avoid difficulties associated with choosing baselines for trading. Efforts would be judged by the level of carbon tax rather than against an arbitrarily chosen historical base year of emissions. This would eliminate the asymmetry between early and late joiners, and remove the opportunity to create 'hot air'[6]. It would also avoid exceptionally large international transfers of wealth that could be generated by the initial allocation of emission rights under international trading regimes[7]. Under a tax-based approach, developing countries would retain all relevant tax revenue within their own borders. Crucially, any assistance from rich to poor countries would be made through direct public transfers tied to specific policy reform or programmes of action, and would be linked to the incremental cost of the action taken. This was the model for co-operation under the Montreal Protocol for Ozone Depleting Substances[8].

However, the international harmonisation of carbon taxes can be extremely difficult in practice. At a European level countries have previously failed to agree on a common carbon tax. Even the relatively homogenous group of four Scandinavian countries that sought to implement a uniform tax from the early 1990s ended up with a complex patchwork of partial application and exemptions between and within the countries[9]. Seeking an internationally uniform tax would preclude national discretion about ways of implementing environmental goals; and this may conflict with national sovereignty and the practical politics of domestic policy formation. There are also practical and political challenges in creating large-scale flows to poor countries, to support an equitable distribution of effort, through public budgets alone.

We argued in Chapter 14 that in the long-term, a global quantity constraint is the appropriate guide for policy-making. A global quantity constraint can be used to drive intergovernmental trading of emissions quotas, and this has already been adopted within the current multilateral framework, the Kyoto Protocol. Moreover, as we explained in Chapter 14, a key benefit of trading schemes for emissions quotas is that they allow the cost-effectiveness (via a common price) and distributional equity of action (via flows based on quota allocations) to be managed separately but simultaneously[10]. In a global and comprehensive system of quota trading, the initial allocation of national limits on emissions affects the

---

[5] Such as the OECD's Consumer and Producer Subsidy Equivalent statistics in the area of agriculture or the WTO's trade statistics.

[6] 'Hot air' can be described as quotas allocated to countries in excess of their requirements as a result of the negotiating process.

[7] Olmstead and Stavins (2006), p. 6 and Cooper (2001).

[8] We discussed the Montreal Protocol in Box 21.2.

[9] We illustrated the development of Norway's carbon tax in Chapter 15.

[10] This may not hold if there are high transactions costs, and/or participants (governments or firms) can exercise market power to influence the buying and selling of permits within a trading scheme (Olmstead and Stavins (2006), p. 5).

distributional equity of the scheme, but not the equilibrium distribution of emissions reductions, the market-determined carbon price or the costs of abatement[11]. Therefore these allocations represent the overall level of responsibility that each country undertakes, rather than the emissions reductions that are required to physically occur within its borders.

Nevertheless, some countries are currently unwilling to participate in intergovernmental emissions trading – including the USA and Australia, and there are real difficulties in enforcing quota allocations between governments under international law. The lessons of the Kyoto Protocol will be explored in more detail in Section 22.4 below.

In practice, a combination of approaches can achieve a similar price for carbon globally by building on existing national tax, trading and regulatory frameworks, but co-ordination is necessary.

Different sectors and countries have differing preferences, institutions and traditions. These affect the choices that governments make between policy instruments such as taxes, trading, regulation, and subsidies, and between mandatory and voluntary approaches. These issues were explored in Chapter 15. A key challenge for international frameworks is to allow for multilateral and parallel action in different countries, to manage and co-ordinate the interactions between different national approaches. This is because if policies adopted in different countries result in different effective carbon prices, the allocation of emission reductions will be inefficient.

The outcomes from using tax or trading schemes that create a price for carbon – such as their effectiveness in reducing domestic emissions – can also be influenced by their interaction with other instruments internationally, even if they are not explicitly linked. This is because, in theory, firms can relocate to different regions and market competition can eliminate high cost products[12]. For example, if one country chooses an emissions trading scheme and another a carbon tax, and if relocation is costless and there is perfect product market competition, arbitrage will occur so that the carbon price is capped by the tax rate[13]. However, the allocation of revenues will be determined by the quantity of allowances issued. This means that the country with the trading scheme has an incentive to increase the quantity of allowances to obtain more revenue – which can then be distributed to its firms or public. Overall, the environmental effectiveness of the instruments will be reduced.

Even if both countries choose to implement taxes, the tax base can make a difference. If taxes are levied on final goods on the basis of the emissions they produce (which is a relatively complex task), there is no incentive to relocate or benefits to competitors in other countries. However, if taxes are levied on domestic emissions, or on carbon content at the beginning of the supply chain, relocation and competition are more likely. In reality, as suggested in Chapter 11, these

---

[11] This statement abstracts from any 'income effects' that might shift demand patterns as a result of shifts in income or wealth associated with the allocation of limits. Olmstead and Stavins (2006).

[12] Tse (2006).

[13] It is possible for the carbon price to be below the tax rate if sufficiently many allowances are issued. This is unlikely in most cases.

kinds of impacts are likely to be substantially mitigated by costs of relocation and many other factors that influence the degree of competitiveness firms face – such as the degree of international exposure, price elasticity of demand for products, as well as market structure.

A uniform carbon price acts as a bedrock to efficient policy. But accommodating a range of dimensions of effort within international frameworks for mitigation is important.

We suggested some important caveats to the general conclusion on a single car-bon price in Part 4. For example, we acknowledged that a wide set of complemen-tary measures relating to the removal of subsidies, and removing behavioural barriers to energy efficiency can be useful. The process of managing the transition to a stable and predictable framework for carbon pricing may justify additional carefully targeted measures, for a specified duration, to overcome the numerous obstacles to the development and deployment of new low-carbon technologies. Moreover, given the contrast between short-term capital markets and the long-term nature of the climate problem, there may be a case for additional measures that could deter construction of long-lived carbon-intensive stock in favour of lower carbon options. We discuss these issues further in Chapters 23 and 24.

International frameworks designed to recognise and build on diverse national approaches require a shared understanding of long-term goals, and they must also allow countries to benchmark and compare action across a range of dimen-sions of effort. These include emissions reductions, the scope and level of carbon prices and policies, national investment in R&D and deployment sup-port, approaches to standards and regulation, commitments to international co-operation on the deployment and diffusion of relevant technology, as well as international support for adaptation.

22.3   Sharing the costs of mitigation

Securing broad-based and sustained participation in international co-operation to tackle climate change depends upon finding an approach widely understood as equitable.

As set out in Part III, any particular long-term quantity constraint can be met by different paths, and the costs involved will be kept down by increasing the flexi-bility about 'what, where and when' emissions are reduced. Scaling up action to reduce GHG emissions will require reductions to take place in both developed and developing countries. Given the ability to bear costs and historical responsi-bility for the stock of GHGs, equity requires that rich countries pay a greater share of the costs.

Frameworks for international collective action that recognise a global long-term quantity constraint on emissions must distribute responsibility for meeting the overall limit to nation states.

Both developed and developing countries can gain from mitigation policy, both because it will reduce the risks of dangerous climate change described in Part II and it because it can be designed to support the range of co-benefits described in

BOX 22.1 Empirical work shows that perceived fairness is important

It is important for any co-operation that those involved feel that the terms agreed are fair. An empirical demonstration of this idea is illustrated by the 'ultimatum game'. In the ultimatum game, 'a proposer' proposes to the other player, 'the receiver', how they should allocate $100. If the other player accepts, both parties divide the $100 as proposed by the proposer. If the receiver rejects the proposal, both parties receive nothing. Although it would be rational for the other player to accept low allocations rather than receive nothing, empirical experiments across different cultures have found that players consistently reject allocations below $30 because they believe they are unfair, while proposers tend to offer between $20 and $50[14].

Chapter 12. This does not mean that poor countries must bear the full costs of their participation. The incidence of imposing a global price of carbon is ultimately on the consumers of carbon-intensive goods and services, including consumers in rich countries who import those goods and services. Nevertheless, equity requires that poor countries should be compensated for some of the costs that they do bear. Emissions trading and similar mechanisms offer an effective route to achieving this.

In the case of climate change, a system of unco-ordinated national goals will not lead to an efficient or equitable distribution of effort. A major advantage of emissions trading schemes is that they enable efficiency and equity to be considered separately[15]. In the absence of trading, the allocation of responsibility for mitigation efforts requires considering efficiency and equity simultaneously.

The UNFCCC contains key principles for an equitable approach to sharing the costs of reducing global GHG emissions that remain relevant to further co-operation on climate change.

Concepts of equity suggest taking into account several aspects of a country's position or actions – which mostly complement each other[16]. The United Nations Framework Convention on Climate Change (UNFCCC) established that co-operation on climate change should recognise the 'common but differentiated responsibilities' of all countries, based upon their respective capabilities. This principle reflects several aspects of equity. First, it reflects the notion that, on the grounds of ability to pay, wealthier, more developed countries should support poorer countries in their efforts to adjust to climate change. Second, it acknowledges that the largest share of historic and current global emissions has originated in developed countries, and thereby applies historical responsibility or the 'polluter pays' principle[17]. Third, it accounts for the relative size of per capita emissions in developing countries and the requirement to allow their relative share of emissions to rise to accommodate their aspirations for growth and

---

[14] Güth et al. (1982).
[15] Rose and Stevens (1998) p. 336.
[16] Chapter 2 of this Review considers the issue of equity and climate change.
[17] See the Appendix to Chapter 2 for a discussion of the basis for this principle in terms of economic efficiency and jurisprudence.

poverty reduction (as recognised, for example, in the Millennium Development Goals (MDGs))[18]. Developed countries therefore took on a range of obligations under the Convention, including showing leadership in tackling their own emissions, transferring technology, supporting capacity building and financing the agreed incremental cost of emissions reductions in poorer nations, and supporting adaptation to the adverse impacts of climate change.

These three arguments all point to rich countries taking a greater share of the costs of mitigation, but they do not necessarily point to the same arrangements or rules for sharing those costs[19]. For example, the ability-to-pay approach suggests that the sharing of costs should be directly correlated to GDP or per capita GDP[20]. The 'growth-needs' approach applied simplistically suggests distribution on an equal per capita basis, whereas the historical approach might suggest that countries with similar economic circumstances have similar emissions rights and responsibilities.

There is no single formula that is likely to capture in a satisfactory way all relevant aspects of an equitable distribution of effort between countries across the various dimensions and criteria[21] – but the criteria tend to point in similar directions.

The correlation between income or wealth and current or past emissions is not exact, but it is strong. This means that equity criteria tend to lead to fairly similar policy approaches: as Ringius *et al* note, "we are in the fortunate situation that all the ...equity principles to a large extent point in the same direction"[22]. This can be demonstrated empirically.

Box 22.2 describes the work of Höhne (2006), who show that the impact of the methodology used to distribute initial mitigation obligations tends to be overridden by the powerful influence of the stabilisation goal on the level of effort required within an international framework for emissions reductions. The results indicate that emissions reductions of 60-90% on 1990 levels by developed countries would be required to meet a stabilisation range between 450 and 550ppm CO_2e.

In the end what matters is that total global effort matches the scale of the problem, that the parties perceive the distribution of effort to be fair, the accompanying goal of efficiency is not prejudiced, and public opinion across a wide range of countries is able to sustain co-operation on those terms over a long period.

---

[18] The Convention expressed this as "Recognizing the special difficulties of those countries, especially developing countries, whose economies are particularly dependent on fossil fuel production, use and exportation, as a consequence of action taken on limiting greenhouse gas emissions".
http://unfccc.int/essential_background/convention/background/items/2853.php.

[19] It is also possible to account for the distribution of the impacts of climate change under burden sharing. However, to avoid the implication that the victims of climate change should pay more because they will benefit most from mitigating climate change, we suggest it is probably the difference between those who bear the brunt of the impacts and their ability to pay to mitigate that should be taken account of. Hence, funding for adaptation to the impacts of climate change, is discussed separately in Chapter 26.

[20] Ringius *et al.* (2000) p. 10.

[21] Ashton and Wang (2003).

[22] Ringius *et al* (2000) p. 29.

BOX 22.2 The effect of stabilisation goals and allocation formulae

Höhne (2006) has compared the effect of the choice of stabilisation goal against different allocation methodologies on the distribution of quotas for emissions reductions between countries. They consider four allocation methodologies:

Convergence and contraction: Emissions in developed countries contract over time to allow emissions from developing countries to converge to a global equal per capita emissions level. This reflects the 'growth-needs' approach.

Common but differentiated convergence: Developed countries' per capita emissions converge to a low level. Developing countries' per capita emissions converge to the same level over the same time period – for example with no commitments or no-lose targets, but decrease after their per capita emissions are a certain percentage above or below the (time dependent) global average. This also reflects a combination of the 'growth-needs' and 'ability-to-pay' approaches.

Triptych: This takes into account differences in national circumstances relevant to emissions and emission reduction potentials. It was the model used for the EU's burden sharing agreement. It could be designed to reflect the 'growth-needs' approach, but it could equally compensate heavy emitters that might have difficulties in adjusting to mitigation policy.

Multi-stage approach: Countries would start at and move between different types and levels of commitment, depending on indices such as per capita emissions levels, income,

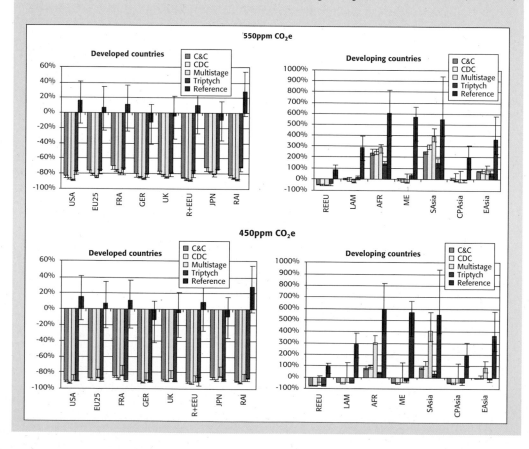

and so on. For example, here 4 stages are used: 1) no commitments; 2) incorporating climate change objectives within sustainable development policies, 3) commitments to moderate absolute limits on emissions – e.g. set above the starting year but below business as usual, and 4) absolute reduction limits.

The four graphs above show the results for both developed and developing countries or regions of 450ppm CO_2e and 550ppm CO_2e stabilisation goals combined with the four methods for sharing out the emissions reductions – here illustrated relative to 1990 levels alongside a reference scenario of business as usual emissions[23]. They do not incorporate international emissions trading. The results show that for developed countries, it is the overall stabilisation goal that is the main driver of the effort required – for all developed countries, action to meet a 450ppm CO_2e goal would require quotas to be set in line with a reduction in emissions of 70-90% on 1990 levels by 2050, and for a 550ppm CO_2e goal the reduction would be at least 60%. It is a similar story for the middle-income economies of Latin America, Central and East Asia and the Middle East, where all methodologies allow for a modest increase or very small decrease over current emissions by 2050. For Africa and South Asia, where both income and per capita emissions are currently very low, the allocation methodology makes a significant difference. Africa and South Asia have the greatest allocation under the methodologies that most closely relate to the 'ability-to-pay' equity criterion.

22.4   Putting efficiency and equity together: The experience of Kyoto

A global carbon price applied to emissions from all countries and sectors allows for efficient mitigation, and flows between countries allow for an equitable division of effort. Creating a framework that provides for both an efficient and equitable response is an urgent challenge for international collective action. This section explores how economic analysis might guide the development of such a framework for mitigation, starting with an evaluation of the current multilateral framework.

There is much to learn from the experience of implementing the Kyoto Protocol, and important opportunities to go beyond it in designing future international co-operation.

The Kyoto Protocol is an innovative attempt to apply emissions trading in the context of international collective action between sovereign states. Participating countries from Annex 1 (developed nations) have agreed to differentiated, legally binding commitments to reducing their overall emissions of a basket of six greenhouse gases by at least 5 per cent below 1990 levels over the first commitment period from 2008 to 2012. As such, an overall quota, or quantity ceiling, has emerged. Within their national limits, countries are free to choose how best to deliver emission reductions nationally.

The Protocol created flexible mechanisms to enable Annex 1 Parties to meet their commitments efficiently. International Emissions Trading (IET) allows trading of national quotas or allowances between countries. The Kyoto Protocol has provided the framework within which the EU has developed its cross-border

---

[23] Error bars show the spread using different reference scenarios.

private sector Emissions Trading Scheme (the EU ETS[24]), allowing over 11,000 energy-intensive installations in 25 countries to co-operate in reducing emissions.

Two further mechanisms, Joint Implementation (JI) and the Clean Development Mechanism (CDM), allow credits from emission reducing projects in one country to be used to meet another country's Kyoto commitment. Under JI, projects can be hosted in developed countries, and under CDM, in developing countries. Governments in Japan and Europe, for example, are expected to purchase CDM credits, and the EU ETS allows private sector participants to purchase credits generated from CDM and JI activities. In the period to 2012, projects generating credits for over 1 billion tons CO_2e are already in the pipeline, meaning the CDM is likely to provide between $5 and $15 billion in additional funding for mitigation in developing countries. CDM finance can also leverage new private and public investment, estimated at 6 to 8 times the amount of CDM finance[25].

The Protocol has also established the institutional basis for monitoring, reporting and verifying emissions, as detailed in Box 22.3. It also has a formal compliance

BOX 22.3 The institutions and processes set up under the Kyoto Protocol

- The Kyoto Protocol provides for detailed reporting and accounting for emissions and emissions allowance allocations within Annex I, and less onerous reporting and review obligations for non-Annex I parties.

- Prior to each 'commitment' period over which emissions reductions will be made, parties are required to submit initial reports establishing their 'Assigned Amount' – the emissions a country will be expected to emit over that period. If they exceed this they will have to purchase credits (allowances) from others that have emitted less than their assigned amount. Establishing an emissions inventory is crucial for this. International review teams review the reports and fix the amounts.

- Annex I parties must submit detailed annual emissions data on an annual basis in national inventory reports, with supplementary information on allowance holdings and transactions. Failure to submit annual reports and inaccuracy in reports can lead to suspension of eligibility to participate in the Kyoto mechanisms.

- Allowance holdings and transactions are monitored in real time by an electronic registry system comprising national registries, which are required to hold and record assigned amount information, as well as enforce detailed trading rules. Registries are linked to an international transaction log, which enforces transaction rules, and may suspend the operation of registries where consistent breaches of the rules have occurred. The CDM registry accounts for credits from projects in developing countries. Reports of the international transaction log are available to review teams in reviewing assigned amount information.

- At the end of the commitment period, following review of the inventory report for the final year, parties have a period of 100 days to ensure their assigned amount matches their emissions during the commitment period. Information on reconciliation, compilation of annual emissions and assigned amounts are forwarded to the compliance committee for final assessment.

---

[24] Discussed in detail in Chapter 15
[25] Ellis et al. (2004).

mechanism to discourage free-riding, containing three specific sanctions to be enforced by all Parties to the Protocol. First, there is a requirement to make up the amount required by the first commitment and incur a penalty of an additional 30% limit on top of their second commitment – this is essentially an interest rate on borrowing. Second, there is a requirement to develop a compliance plan of action – which provides an opportunity for international and national scrutiny of the adequacy of policy measures in place to identify ways of coming back into compliance in future periods. Third, there is suspension of eligibility for trading – which makes it harder for a country to meet its objectives in a cost-effective way, and may create difficulties for governments where businesses have invested in trading and parliamentary majorities are in favour of action to reduce emissions.

The Kyoto Protocol has been criticised on several grounds. However, Kyoto has, to its credit, established an aspiration to create a single global carbon price and implement equitable approaches to sharing the burden of action on climate change.

Criticisms of the multilateral approach adopted through Kyoto can be organised around three particular issues – incentive compatibility, the time horizons and ambition of commitments, and limited participation.

Analyses of international collective action, including those discussed in Chapter 21, point to the weakness of international law in enforcing obligations between sovereign states[26]. Governments can, if they choose, easily renege on their commitments, and they are more likely to do so if these commitments are not in line with widely adopted norms of international behaviour and with the commitments of key trading partners. International agreements that are not compatible with the underlying incentives of the participants are unlikely to succeed in creating significant changes in national action.

The Kyoto Protocol has a number of specific sanctions for non-compliance, but these are enforceable only where a government chooses to remain within the framework of the Protocol[27]. A country that exceeds its quota of emissions in the first commitment period can be suspended from eligibility for trading, and is required to make up its commitment and pay a penalty within the following commitment period. The suspension of eligibility to trade would be a significant concern for countries that wish to remain within the trading system and have a small variance from their limits to account for[28]. However, the second sanction creates an incentive for those countries that are not in compliance with their first phase limits to seek an alternative basis for any arrangements for future action[29]. Furthermore, the ratification threshold for the Kyoto Protocol is sufficiently high that a very small number of key countries can block the agreement of a second commitment period.

We discussed both the role of compliance mechanisms and how to build credibility in Chapter 21.

---

[26] For example, Victor (2001); Schelling (2002); and Barrett (in press).

[27] Alternative approaches to compliance were considered, such as the option of a compliance fund, but they also have drawbacks. See Wang and Wiser (2002) and Rolfe (2000).

[28] Going even further, Hovi & Kallbekken (2004) suggest that where a country may have a major role in supplying credits in the system, their suspension from trading would create perverse incentives, by raising the price of permits for the countries that must enforce the sanction. If the latter countries would suffer significant harm by doing so, suspension may not be credible.

[29] On the other hand others such as Rolfe (2000) have suggested the implied 30% interest rate on borrowing is low, so it is not a sufficient deterrent to non-compliance.

The second issue concerns the time horizons for action under the Kyoto Protocol. Stavins (2005) has recently repeated criticisms that the Protocol aims to do "too little, too fast"[30], aiming for excessively costly short-term reductions in emissions, without determining what should be done over longer timeframes – where there is more flexibility to make reductions in line with normal cycles of capital stock replacement. At the time the first commitment period for the Kyoto Protocol was set as 2008 to 2012, in 1997, it provided a 15-year window for action. However, the Protocol does not provide any guidance or formulae linking the action required in the first commitment period to an overall global quantity constraint or to long-term term timetable for emissions reductions. Coupled with the incentive compatibility problem described above, these issues mean that the Kyoto framework is not currently providing a sufficiently credible, long signal for countries or businesses to make long-term investments[31].

Finally, the Kyoto Protocol has been heavily criticised in some quarters for creating quantitative obligations only for the rich countries, without placing any constraints on emissions from the fast-growing emerging economies. The US and Australia have subsequently declined to ratify the Protocol, and a number of other countries are not taking strong steps to implement it. The developing countries did in fact take on obligations under the Kyoto Protocol, but these were unquantified and allowed climate change to be addressed as part of wider national policies on sustainable development. The CDM has been the mechanism by which non-Annex 1 countries have participated in formal action on climate change mitigation, but many non-Annex 1 countries already have policies in place – taxes, renewable energy and energy efficiency goals – that discourage carbon emissions that are not recognised as climate change commitments in the framework. Furthermore, the CDM has important limitations that are considered further in Chapter 23 – not least that credits are currently generated by offsetting against a business as usual baseline rather than by reductions below the baseline. Given the limited nature of participation in the first commitment period, the Kyoto Protocol has not in practice introduced a global price for carbon.

Nevertheless, the concepts underlying the Protocol – in particular, the aspiration to create a single, efficient carbon price across countries through the use of emissions trading and the recognition that mechanisms are required to make finance and technology available to poor countries on the basis of equity – are very valuable. These are elements to be strengthened within any future regime for action on climate change.

There are strong practical reasons to build on the achievements of Kyoto in the next round of negotiations, whilst exploring ways to learn from other approaches and to increase the breadth and depth of international co-operation for climate change.

The Kyoto Protocol can be seen as a first stepping-stone on the path to international co-operation on climate change, given political, economic and scientific realities[32]. The institutions, mechanisms and guidelines developed under Kyoto represent an enormous investment of negotiating capital. They reflect a fine

---

[30] Stavins (2005).
[31] Barrett (in press): p 6.
[32] Frankel (in press).

balance between the interests of over 130 countries. It is not obvious that starting from scratch with an entirely new approach would produce a more effective regime, and it could take many years for the shape of a new approach to emerge. Building on existing principles and established institutions, for example those described in Box 22.3, also helps to reduce uncertainty for investors about the intended direction of international climate policy, as well as to enhance trust between parties.

For countries that are willing to work within Kyoto, the institutions provide the framework within which to negotiate on future ambition that supports deep and liquid cross-border carbon markets. However, given the scale of action required to mitigate climate change, as we have emphasises throughout this Review and clearly demonstrated in Chapter 21, action taken by those countries that have signed up to Kyoto is necessary but is not sufficient. There are two aspects of the solution to this issue. First, as we have suggested in Chapter 21, transparent and comparable frameworks provide a way to benchmark a range of dimensions of effort between countries that prefer to work outside and within Kyoto. Second, it is important to build the kinds of institutions that enable Kyoto and non-Kyoto Parties as well as sub-sovereign bodies to engage in mitigation. We explore these types of institutions further below.

22.5   Building on national, regional and sectoral carbon markets

The scope for expanding private sector emissions trading markets is high, and can generate large flows globally.

Only a small portion of global emissions are currently covered by emissions trading schemes. The largest existing emissions trading scheme is the EU ETS. If trading expanded in future, for example, to cover the power and industrial sectors[33] in Australia, Canada, the EU, Japan and the USA, emissions trading would grow to 3 times the size of the current EU ETS. Expanding further to include all of the top 20 global emitters – a relatively small number of jurisdictions, which together account for almost 80% of global CO_2 emissions – would raise coverage by almost 5 times. This is shown in Figure 22.1[34].

An emissions trading market of the size of 5 times the current EU ETS would create allowances that could be worth between US$87 and US$350 billion[35]. These values are a function of the carbon price – which, as explained in Chapter 14, is determined by both marginal abatement costs in the covered sectors and the scarcity of allowances within schemes (i.e. the stringency of the overall cap on emissions within the scheme).

Expanding and linking regional emissions trading schemes globally will raise the scope for cost-effective emissions reductions.

As discussed in Chapter 15, an efficient and equitable framework for international collective action requires a broad, deep and liquid market for carbon, covering the

---

[33] These are the sectors currently covered by the EU ETS.
[34] This figure shows energy emissions only. We examine GHG emissions from land use change in Chapter 25.
[35] Assuming carbon prices of between $10 and $40. World Bank and IETA (2006).

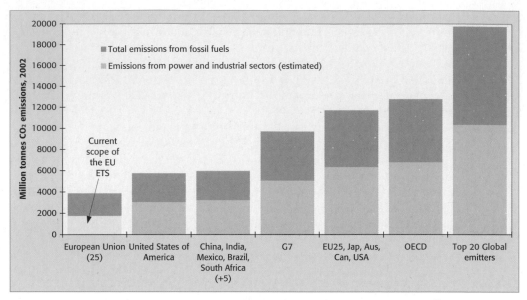

Figure 22.1 Scope of an international trading market in energy CO_2 emissions[36]

major emitters and operating with transparent rules. This emphasises the importance of an increase in the size and scope of emissions trading markets globally. This can occur when an existing scheme expands to incorporate new regions, through the merger of separate schemes, or through various approaches to linking, whereby several existing schemes may meet key criteria or develop harmonised rules for mutual compatibility.

Chapter 15 introduced several emissions trading schemes that have already been established or are planned in countries and regions across the globe. They vary in size, scope and characteristics. For example, the Chicago Climate Exchange (CCX) is a voluntary scheme. The proposed Regional Greenhouse Gas Initiative (RGGI) will only cover emissions from the power sector. The current UK Emissions Trading Scheme covers non-CO_2 and both direct and indirect CO_2 emissions. Some schemes may apply price caps, others may have differing penalties for compliance. The time periods for commitments also vary, often to reflect national circumstances. Creating a single scheme would entail considerable changes to harmonise these conditions.

Linking, although less efficient than a single global scheme, can nevertheless be very useful. For example, a small new scheme may see linking to an established scheme as a short-cut to establishing credibility and price stability. Links are already being made between existing schemes. For example, the EU ETS allows the use of project credits created by the Kyoto Protocol, and some non-Kyoto parties, including the CCX, also permit purchases of these credits. Box 22.4 describes another recent development.

The key issue for efficient markets when expanding and linking schemes is that caps are stringent and in line with shared international goals.

[36] Data taken from the World Resources Institute CAIT database.

BOX 22.4 UK-California announcement on climate change and clean energy collaboration

On 31 July 2006, the UK and California issued an announcement on climate change and clean energy. The mission statement includes a commitment to "evaluate and implement market-based mechanisms that spur innovation ... (and) evaluate the potential for linkages between our market-based mechanisms that will better enable the carbon markets to accelerate the transition to a low carbon economy".

California is currently developing specific proposals for a cap-and-trade scheme as part of its goal to reduce emissions 25% by 2020. The EU Linking Directive does not currently allow the EU ETS to be directly linked to schemes in countries that have not ratified the Kyoto Protocol or to sub-sovereign schemes. In the interim, one-way linking could occur through access to a common pool of offset credits from the Kyoto project mechanisms.

There are a number of policy issues that, although they may not have to be clarified in order to physically or feasibly link, tend to affect the desirability of linking, and therefore are important to overcome first[37]. The expansion or linking of trading schemes is particularly suited to situations when countries are willing to agree overall emissions limits as part of a negotiated international framework, since this encourages transparency and compatibility of emissions trading caps and provides the building blocks for key harmonisation criteria[38]. As Chapter 15 has suggested, the experience of implementing the EU ETS suggests that agreement on overall national emissions limits that are broader than the scope of the trading scheme allows governments considerable flexibility in determining the stringency of national allocations for sectors covered by emissions trading schemes. This can result in concerns about competitiveness and gaming that may undermine the effectiveness of the scheme. It could therefore be effective for international negotiations to focus directly on the stringency of emissions trading schemes.

In terms of harmonisation criteria, it is possible to link even if there are different *types* of emissions caps (such as absolute targets, or relative intensity targets[39]), safety valves, differing permitted use of offset credits, allocation methodologies, and differing financial penalties for non-compliance. However, such differences can make the environmental effectiveness of the schemes difficult to compare as well as lead to unintended transfers between countries. Significant shifts in exchange rates could also impact on the price of allowances, increasing volatility. There are solutions to these issues such as allocating *ex-post* rather than *ex-ante*, but these tend to increase the complexity and reduce the efficiency of schemes.

If expansion or linking is not well managed there may be negative impacts. For example, a scheme with an uncertain or unconstrained volume of allowances that can be purchased from outside the trading scheme's coverage over a relatively short time may cause price volatility. The process of linking schemes itself may cause price instability because of the introduction of uncertainty about the impacts of linking. Expansion and linking therefore require transparent negotiations and

---

[37] Ellis & Tirpak (in press).
[38] Blyth and Bosi (2004).
[39] These are discussed further below.

terms of agreement in advance of trading periods. This means new trading schemes should consider compatibility carefully, ideally mirroring, and influencing, as many of the features of existing schemes they wish to adjoin.

Sectoral approaches can introduce carbon pricing in sectors that are appropriate for early trading, to accelerate the movement towards global carbon markets, as well as overcome perceived competitiveness impacts.

Sectoral approaches can be used as a transition to introducing carbon markets throughout the global economy, and Chapter 15 has suggested some important reasons why certain sectors might be particularly suited to early trading. They can incorporate different levels of commitment and can be used at the multilateral or national level. Emissions intensities within sectors often vary greatly across the world, so a focus on transferring and deploying technology through sectoral approaches could reduce intensities relatively quickly, and could make it easier to fund the gap between technologies that developing countries can afford and existing cleaner technologies that the developed world is already adopting. Also, global coverage of particular sectors that are internationally exposed to competition and produce relatively homogenous products can reduce the impact of mitigation policy on competitiveness. Box 22.5 describes a global initiative already in place in the cement sector.

There are two important drawbacks to sectoral approaches. First, focusing on a few sectors may neglect emissions from other sectors that have lower abatement costs, thereby sacrificing 'where' flexibility. It may also lead to inefficiency by having different implicit carbon prices across sectors. This is more likely if just a few sectoral agreements are adopted. Second, there is potential for 'leakage' of emissions to sectors not included in such agreements if sectors are poorly defined, for example, if the agreements cover particular products but not their close substitutes. But even narrow coverage can make a large difference. For example, the Center for Clean Air Policy proposal for a sectoral scheme for power and industrial

---

BOX 22.5 Cement Sustainability Initiative[40]

Cement is one of the most energy-intensive industries. The World Business Council for Sustainable Development has developed the Cement Sustainability Initiative, with the participation of 17 companies with manufacturing facilities in Europe, the USA, India, SE Asia and Latin America. They are responsible for more than 50% of cement manufactured in the world outside China. Variations of energy use between countries shows clear scope for emissions reductions.

Through the CSI, the companies have developed common standards for monitoring and reporting CO_2 emissions, and pledged to set their own targets for reducing emissions per unit of output, and make progress reports available to the public. They have also developed guidelines to spread best practice throughout the industry. The CSI includes companies from countries not covered by targets under the Kyoto Protocol. Some have expressed strong support for a worldwide sectoral approach for their industry. Participation allows companies to explore how such a scheme would work.

---

[40] www.wbcsdcement.org.

emissions from the ten highest emitting developing countries would cover around 30% of developing countries emissions[41].

Several variants of sectoral approaches are possible, and include harmonised sectoral taxes and sectoral trading. The latter, as for other trading schemes, requires agreement of an initial goal or cap for the sector, with *ex-ante* provision of allowances at this cap, accompanied by a compliance mechanism to create a penalty for underachievement. The development of sectoral benchmarks – more generalised baselines or standards applicable to multiple projects in the same sector – can also be used to generate credits by sectors that beat performance against the agreed benchmarks. Sectoral approaches could also be designed around the phase-out of old technologies or phase-in of new, low-carbon or efficient technologies. Developing countries may be particularly interested in participating in such schemes where they offer an effective way to attract large-scale financing for sectoral reform, or incentives such as voluntary or no-lose targets.

A key issue is the degree of international negotiation that may be required to determine appropriate benchmarks, but sectoral agreements may offer the opportunity for firms in sectors to agree on emissions caps, taxes, benchmarks or standards amongst themselves. There are also methodological issues to consider, such as determining sector boundaries and baselines, but the approach itself can encourage development of relevant data and provide a step towards global sectoral trading. Some benchmarks for best available technologies in the electricity and industrial sectors have already been established by EU Member States for the purposes of the EU ETS, especially for new plant[42].

22.6   Building on common but differentiated responsibilities

Several types of commitment could be used to take into account equity concerns and widen participation in the international framework. Many are particularly applicable to developing countries.

In general, approaches to setting international emissions reductions obligations for trading schemes can be used to take account of countries' aspirations alongside key uncertainties. Emissions quotas can be set in relation to absolute emissions levels or per capita emissions levels, and these can be set in line with appropriately revised, credible long-term goals alongside rolling revision rules for flexibility. However, as explained in Section 22.4, and as the discussion in Box 22.2 illustrated, the methodology used to distribute emissions quotas has important implications for equity. Under a system based on trading of emissions permits, initial allocations reflect the level of responsibility that each country undertakes, rather than the actual emissions reductions required to be made by that country.

Pizer (2005) makes a case for emissions intensity targets indexed to economic growth. He suggests that relative or dynamic goals are more easily adjusted to levels that stop, slow or reverse emissions growth than absolute goals. As long as their limits are not revised, they can avoid penalising unexpectedly low economic

---

[41] Excluding emissions from land use, land use change and forestry. Schmidt et al. (2006).

[42] For example, UK benchmarks developed for over 20 categories of new entrants to Phase II of the EU ETS are available at: http://www.dti.gov.uk/energy/environment/euets/phase2/new-entrants/benchmarks-review/page29366.html.

growth and the decoupling of emissions from economic growth they aim at. Pizer also suggests that intensity targets are particularly suited to developing countries because they can alleviate concerns that economic growth will be stunted by taking on obligations to reduce emissions, and may reward middle income countries such as China that have high emissions intensity levels from which to descend.

There have been a number of proposals to build on equity considerations by taking into account developing countries' emissions reductions potentials, capacity to take action and development goals, and to provide positive incentives for their further participation in climate change mitigation.

As described in Box 22.2, a multi-stage or multi-track approach allows different types of participation depending on national circumstances[43]. Under these approaches, least developed countries would not be required to make reductions in their emissions in the near-term, but could be supported in making the transition to low carbon development paths either through direct financial flows, the use of flexible mechanisms, or allocations of quotas in excess of likely requirements. For middle-income and rich countries, a range of graduation criteria have been proposed that rely on indices including per capita income and emissions. Graduation criteria can allow countries to make the transition from, for example, project-based mechanisms to eligibility to participate in international emissions trading. This can also provide a useful compliance mechanism – for example, eligibility for project mechanisms could be withdrawn if a country does not introduce its own mandatory national policy frameworks for emissions trading once it has passed a graduation threshold[44].

Participation in emissions trading can also begin from 'no-lose' commitments. These are 'one way' commitments that provide a clear incentive for developing counties to make efforts to reduce their GHG emissions. They would allow developing countries to benefit from selling the emissions credits they generate for performance beyond an agreed limit (which could be either absolute or relative), but there would be no penalty for under-achievement. The concept could also be applied on a sectoral basis. However, it remains essential that some countries or sectors within the system have binding limits, in order to generate demand for surplus credits.

Positive recognition of developing country policies that generate emissions reductions alongside other goals may build trust.

The concept of giving formal recognition to sustainable development policy and measures (SD-PAMs) has attracted increasing attention from developing and developed countries alike. An SD-PAM would be a voluntary or mandatory commitment to implement a policy or measure that makes the development path of a country more sustainable, with the co-benefit of lowering GHG emissions, many of which were identified in Chapter 12. In this way it fits well with a development-centred approach to climate change mitigation[45].

SD-PAMs would increase the visibility of a wide variety of policies that are already being implemented in developing countries that tackle both sustainable development and climate change mitigation objectives, and this is something

---

[43] See Hohne (2006) and Den Elzen *et al.* (2006).
[44] Michelowa et al. (2005).
[45] See Winkler et al. (2002) and Bradley and Baumert (2005).

that has been missing from the international framework so far. The approach therefore provides a quantifiable alternative to emissions reductions obligations. Quantification of sustainable development and mitigation benefits of policies would help countries to identify future strategic opportunities for those PAMs that will reduce the growth of GHG emissions and meet their own national goals, as well as to compare effort across their peers. The World Resources Institute[46] has already begun to develop a database to record SD-PAMs. This might also facilitate international exchange of expertise and best practice, linking well to wider system of measures of effort suggested in Chapter 21.

Incentives to encourage the take up of SD-PAMs may be necessary, although that would intensify the importance of demonstrating that SD PAMs do provide emissions reductions over and above the emissions that would have occurred without the measure[47], as well as defining to whom they may apply, and making efficient links to existing carbon markets. SD-PAMs could also be a key method of combining and enhancing other funding sources that were previously devoted exclusively to climate or non-climate policies or measures, and attracting public as well as private investment.

There will be important issues to overcome before SD-PAMs are acceptable by developed and developing countries. Most importantly, numerous types of national policies could be covered by such an approach, and they could be complex. It would also be important to create a monitoring or review process to assess progress made against SD-PAM objectives. Pilot schemes would help clarify their applicability to key policy areas as well as the methodological issues.

22.7   Challenges of extending international co-operation to aviation and shipping

Extending the coverage of carbon pricing and other measures to international aviation will become increasingly important

Globally, international aviation emissions – defined as emissions from any aircraft leaving one country and landing in another – are about twice as great as domestic aviation emissions. As set out in Chapter 15, the impact of aviation on climate change is also higher than the impact of its CO_2 emissions alone. Aviation has negative local impacts on noise, local air quality, biodiversity, and local climate impacts, for which local policy interventions (such as regulation on noise levels) can be used.

However, there is currently no incentive to reduce international aviation emissions, as only emissions from domestic flights are currently allocated to any country within national emissions inventories. Furthermore, many large international markets are outside the current Kyoto obligations framework. However, the industry is growing fast, and people with lower incomes, especially in developed countries, are now able to travel globally due to low-cost flights. Many national

[46] The World Resource Institute has a work program to explore and define the SD-PAMs approach; look at specific SD-PAMs in detail; provide tools and analysis to assist those working on such policies and measures; and outreach activities to help policymakers incorporate SD-PAMs into international negotiations. A pilot database of SD-PAMS is available on-line at www.wri.org.
[47] I.e. some level of 'additionality'.

policy measures such as landing charges tend to be blunt instruments for cutting carbon emissions. However, differentiating them, for example, by length of flight or distance travelled, could improve their effects on reducing emissions.

International coordination on reducing emissions from aviation is important, for example, to avoid leakage of mitigation policies from travellers switching to different carriers, or air carriers changing their routes, or practices such as 'tankering' (i.e. carrying excess fuel on planes to avoid refuelling at airports where fuel taxes are levied). The UNFCCC has requested the International Civil Aviation Organisation (ICAO) to take action on aviation emissions, recognising that a global approach is essential. ICAO has established a Committee on Aviation Environmental Protection (CAEP), part of whose work plan relates to climate change emissions. Current tasks include developing guidance for states wishing to take forward emissions trading schemes, and developing a better understanding of the potential trade-offs between improvements in CO2 emissions and the effect on other environmental impacts. However, these measures do not, of themselves, regulate emissions.

The issue of aviation causing higher climate change impacts than simply that from its CO_2 emissions could be tackled by setting high carbon taxes on aviation. However, we noted the particular difficulty of co-ordinating international taxes in Chapter 15. The ICAO has recently endorsed the concept of an ETS for aviation, while the EU is currently developing a draft Directive to include aviation in the EU ETS. The EU Environment Council has suggested some preliminary guiding principles to be taken into account for its inclusion, so that it is a workable model that can be replicated worldwide. For example, coverage must be clear (options include domestic, intra-EU, all flights leaving or landing in the EU), trading entities should be air carriers and aircraft operators, and the allocation methodology should be harmonised at EU level. As suggested in Chapter 15, auctioning allowances would also raise revenue and increase the speed of adjustment to carbon markets. To account for the complete impacts of aviation within an ETS, some form of discounting could be used, analogous to the global warming potential factors that are used to convert GHG emissions to CO_2 equivalent emissions. Alternatively, combining emissions trading with a tax could provide extra revenue. This could provide strong incentives to innovate to reduce emissions within the sector, including in airframe efficiency, engine manufacture, airport operations, and air traffic management.

The international co-ordination of standards, including through voluntary approaches, is also an important measure. Existing international co-operation under the Advisory Council for Aeronautics Research in Europe (ACARE) requires new aircraft produced in 2020 to be 50% more fuel efficient per seat kilometre relative to their equivalents in 2000. As the target refers to new aircraft produced in 2020, it will take time for the fuel efficiency of the whole fleet to improve because of the long lifetime of aircraft. The ACARE target does provide some degree of challenge – in order to meet it, some technological breakthroughs will have to be achieved. The targets are broadly on track to being met. ACARE is an EU body, but the target is likely to have a significant impact on fuel efficiency internationally because aircraft manufacturers will want to keep up with fuel efficiency standards. In the US, the National Aeronautics and Space Administration (NASA) have set similar goals.

Complementary measures to trading and standard setting include co-operation on technology, sharing best practice in ground operations, and realising the potential to reduce emissions through enhanced air traffic management improvements.

Extending the coverage of carbon pricing to international shipping has been slow, but is likely to increase in momentum

Discussions on tacking the climate change impact of the international maritime industry are at a very early stage. The International Maritime Organisation (IMO) Assembly in December 2003 urged its Maritime Environmental Protection Committee (MEPC) to identify and develop the mechanism or mechanisms that can achieve the limitation or reduction of GHG emissions from international shipping, and asked for the evaluation of technical, operational and market-based solutions to limiting the GHG output of maritime transport.

The UK, under the lead of the domestic Maritime and Coastguard Agency (MCA), has been pushing the IMO to consider a full range of technical, methodological and market-based options for controlling maritime transport's emissions of GHGs, particularly CO_2. Discussions are continuing on the feasibility of the EU incorporating this sector into the EU ETS as a demonstration not only of the seriousness with which the EU views this issue but also of the effectiveness of emissions trading as a control measure.

22.8   Interactions with the international trade regime

The international trade regime offers one route to handle large disparities in levels of carbon pricing between major economies.

Some economists[48] have analysed the potential to use the international trade regime to respond to significant differences in the level of carbon prices applied in different economies. Countries could in theory impose a border tax on imports from countries with lower carbon prices – to correct for the under pricing of carbon in the country of origin. This could overcome carbon leakage or competitiveness concerns by reducing the incentive for domestic production to relocate abroad, and could increase the incentives for other countries to adopt similar measures to reduce GHG emissions. There is a clear logic here.

There has been a long-standing debate about whether border tax adjustments in response to carbon price differentials would be legal under World Trade Organisation (WTO) rules. Since the early 1980s, several cases have been brought to the General Agreement on Trade and Tariffs (GATT) and the WTO that have implications for environmental measures or human health-related measures[49]. In particular, the 1998 ruling on the 'shrimp-turtle' case[50] can be used to suggest that, as long as border adjustments or regulations on greenhouse gas intensity of the production process are carried out in a non-discriminatory way, they are likely to be permitted.

Adjustments to take account of carbon price differentials could also occur if exporter countries voluntarily impose export restraints within bilateral or multilateral agreements. For example, after the abolition of a global quota system,

---

[48] For example, Brack (1998), Frankel (2004) and Stiglitz (2006).
[49] They are listed and described at http://www.wto.org/english/tratop_e/envir_e/edis00_e.htm.
[50] United States—Import Prohibition of Certain Shrimp and Shrimp Products, WTO Doc. WT/DS58/R (panel report May 15, 1998), excerpted in 37 ILM 832 (1998); United States—Import Prohibition of Certain Shrimp and Shrimp Products, WTO Doc. WT/DS58/AB/R (Appellate Body Oct. 12, 1998), 38 ILM 118 (1999).

China had offered to raise its export tariffs and reduce export tax rebate rates to help manage the entry of their textiles into the EU and US markets. Under this arrangement, the revenues would have been paid to the Chinese government but EU and US producers would have been protected from high competition from abroad[51].

Notwithstanding the logic of trade measures, their potential misuse could have serious consequences for international relations and future co-operation.

As we have demonstrated in Chapter 12, the competitiveness impacts that underlie these arguments for adjustments should not be overplayed. Those findings also mean that, for many goods, given their cost structures, such border adjustments may not change patterns and trends of international trade significantly. However, border tariffs or similar measures to adjust for carbon price differentials could be undesirable for the following reasons:

- Barriers to trade are inefficient. The removal of trade barriers allows countries to develop comparative advantage in production. Therefore, even if effective, they are clearly second best to implementing a similar carbon price across the global economy.

- There would be technical challenges, whether border adjustments are set nationally or multilaterally, as the current structures of cross-border levies and subsidies are extremely complex.

- If the measures are effective, they could have detrimental effects on developing countries with high export dependency on carbon-intensive goods. In Chapter 23 we examine the transition to low-carbon economies in developing countries.

- The measures could become a pretext for other measures that are essentially protectionist and support inefficient industries. This has been the danger of imposing non-tariff barriers, such as phytosanitary standards, that can be used to deny entry of exports from developing countries into rich countries.

- Such measures could make it considerably more difficult to build the trust necessary for future international co-operation.

Nevertheless, there remains the risk that in the face of significant and long-running divergences in levels of carbon pricing across borders, industry will lobby for the implementation of these measures. Chapter 23 explores how the *removal* of trade barriers could be used to encourage mitigation, particularly in developing countries.

22.9   Conclusions

A broadly similar global carbon price is an essential element of international collective action to reduce greenhouse gas emissions. Creating this price signal,

---

[51] See Mueller and Sharma (2005), at http://www.scidev.net/content/opinions/eng/trade-tactic-could-unlock-climate-negotiations.cfm.

through international frameworks and through a range of regional and national policy instruments, is an urgent challenge.

The most important test for the international community will be to reflect the scale of action required sufficiently within their commitments. Approaches to equity can aid this process, but action from all countries is pressing.

Some elements of a potential future framework are becoming clear. The early formation and experience gained from the EU ETS, and the decisions by California and others to establish regional trading schemes strongly suggest that deep and liquid global carbon markets are likely to be at the core of future co-operation on climate change. Stronger international coordination as these schemes emerge, incorporating new sectors globally, will greatly increase their capacity to support an efficient and equitable response to climate change.

References

Ashton, J. and X. Wang (2003): 'Equity and climate: in principle and practice', in Beyond Kyoto: Advancing the international effort against climate change. Virginia: Pew Center on Global Climate Change, pp. 61–84.

Barrett, S. (in press): 'A multi-track climate treaty system', School of Advanced International Studies, Baltimore: Johns Hopkins University.

Blyth, W. and M. Bosi (2004): 'Linking non-EU domestic Emissions Trading Schemes with the EU Emissions Trading Scheme'. OECD/IEA Information Paper for the Annex I Expert Group on the UNFCCC, Paris: OECD.

Bradley, R. and K.A. Baumert (eds.) (2005): 'Growing in the greenhouse: protecting the climate by putting development first. Washington, DC: World Resources Institute.

Brack, D. (ed). (1998): 'Trade and environment: conflict or compatibility?', London: Earthscan Publications.

Cooper, R.N. (1998): 'Toward a real global warming treaty', Foreign Affairs, 77(2): 66–79

Cooper, R.N. (2001). 'The Kyoto Protocol: a flawed concept'. Fondazione Eni Enrico Mattei Noto di Lavoro 52

Den Elzen, M., M. Berk, P. Lucas, et al. (2006): 'Multi-stage: a rule-based evolution of future commitments under the Climate Change Convention'. International Environmental Agreements, 6(1): 1–28

Ellis, J. and D. Tirpak (in press): 'Linking GHG emissions trading systems and markets'. OECD/IEA Information Paper for the Annex I Expert Group on the UNFCCC, Paris: OECD.

Ellis, J., J. Corfee-Morlot and H. Winkler (2004): 'Taking stock of progress under the Clean Development Mechanism'. OECD/IEA Information Paper for the Annex I Expert Group on the UNFCCC, Paris: OECD.

Frankel, J. (2004): 'Kyoto and Geneva: Linkage of the climate change regime and the trade regime'. John F. Kennedy School of Government Faculty Research Working Papers Series, 42.

Frankel, J. (2005): 'Climate and trade: links between the Kyoto Protocol and WTO'. Environment, 47(7): 8–19

Frankel, J. (in press): 'Formulas for quantitative emissions targets', in Architectures for Agreement: Addressing Global Climate Change in the Post Kyoto World, (ed). J. Aldy and R. Stavins, Cambridge: Cambridge University Press.

Güth, W., R. Schmittberger, and B. Schwarze (1982): 'An experimental analysis of ultimatum bargaining', Journal of Economic Behaviour and Organization, 3(4): 367–388

Höhne, N., D. Phylipsen, S. Ullrich and K. Blok (2005): 'Options for the second commitment period of the Kyoto Protocol', Utrecht: ECOFYS.

Hovi, J. and S. Kallbekken (2004): 'The Price of Non-compliance with the Kyoto Protocol: The remarkable case of Norway', Working Paper 7, Oslo: Center for International Climate and Environmental Research.

Michelowa, A., S. Butzengeiger, and M. Jung (2005): 'Graduation and deepening: an ambitious post-2012 climate policy scenario'. International Environmental Agreements, 5(1): 25–46

McKibbin, W.J. and P.J. Wilcoxen (2002): 'The role of economics in climate change policy'. Journal of Economic Perspectives, 16(2): 107–129

Nordhaus, W.D. (2005): 'Life after Kyoto: alternative approaches to global warming policies'. NBER Working Paper, 11889.

Olmstead, S.M. and R.N. Stavins 2006: 'International policy architecture for the post-Kyoto era'. John F. Kennedy School of Government Faculty Research Working Papers Series, 9, Cambridge, MA: Harvard University.

Pizer, W.A. (2002): 'Combining price and quantity controls to mitigate global climate change'. Journal of Public Economics, 85 (3): 409–34

Ringius, L., A. Torvanger, A. Underdal (2000): 'Burden differentiation: fairness principles and proposals', Working Paper, 13, Oslo: Center for International Climate and Environmental Research.

Rolfe, C. (2000): 'The earth in balance: briefing notes for the november 2000 Climate Summit: Compliance', Note 2, Vancouver: West Coast Environmental Law Research Foundation.

Rose, A. and B. Stevens (1998): 'A dynamic analysis of fairness in global warming policy: Kyoto, Buenos Aires, and Beyond'. Journal of Applied Economics, 1(2): 329–362

Schelling, T.C. (2002):'What makes Greenhouse sense?', Foreign Affairs, 81(3): 2–9

Schmidt, J., N. Helme, J. Lee and M. Houdaschelt (2006): 'Sector-based approach to the post-2012 climate change policy architecture', Future Actions Dialogue Working paper, California: Center for Clean Air Policy.

Stavins, R. (2005): 'A better climate change agreement'. The Environmental Forum, January/February, p. 12, available from http://www.env-econ.net/stavins/Column_5.pdf

Stiglitz, J. (2006): 'Making Globalisation Work'. New York: W.W Norton.

Tse, M. (2006). 'A theoretical note on cross-border interactions between carbon abatement schemes'. Background Paper for the Stern Review on the Economics of Climate Change.

Victor, D.G. (1999): 'Enforcing international law: implications for an effective global warming regime',147, North Carolina: Duke University, Duke Environmental Law and Policy Forum.

Wang, X. and G. Wiser (2002): 'The Implementation and compliance regimes under the climate change convention and its Kyoto Protocol'. Review of European Community and International Environmental Law 11(2): 181–98

Winkler, H., R. Spalding-Fecher, S. Mwakasonda and O. Davidson (2002): 'Policies and measures for sustainable development', in Baumert et al. (eds.), Building on the Kyoto Protocol: Options for Protecting the Climate. Washington, DC: World Resources Institute.

World Bank and International Emissions Trading Association (2006): 'State and Trends of the Carbon Market 2006', Washington, DC : World Bank.

23 Supporting the Transition to a Low-Carbon Global Economy

KEY MESSAGES

Demand for energy and transportation is growing rapidly in many developing countries. The investment that takes place in the next 10–20 years could lock in very high emissions for the next half-century, or present an opportunity to move the world onto a more sustainable path. Investment in energy efficiency can reduce demand growth, and low-carbon technologies can further reduce the impact on climate change.

The transfer of technologies to developing countries by the private sector can be accelerated through national action and international co-operation.

Energy price and taxation reform will play an important role in improving the conditions for investment in more efficient and low-carbon technologies, as they can support other development priorities and encourage co-benefits from mitigation policies, including energy security and improved air quality.

Carbon pricing is essential to influence investment decisions in low-carbon technologies, including renewable energy and carbon capture and storage. The Clean Development Mechanism is currently the main formal channel for supporting low-carbon investment in developing countries, but in its existing form it has significant limitations.

The incremental costs of low-carbon investments in developing countries are likely to be at least $20-30 billion per year.

A transformation in the scale of and incentives for international carbon finance flows is required to support cost-effective reductions. This will require mechanisms that link carbon finance to policies and programmes rather than to individual projects, working within a context of national, regional or sectoral objectives for emissions reductions.

Long-term goals and early signals to provide continuity of carbon finance after 2012 are essential to deliver emissions reductions in developing countries.

There are opportunities now to build trust and to pilot new approaches to creating large-scale flows for investment in low-carbon development paths. The International Financial Institutions have an important role to play in accelerating this process, including through the creation of the Clean Energy Investment Framework.

The reduction of tariff and non-tariff barriers for low-carbon goods and services, including within the Doha Development Round of international trade negotiations, could provide further opportunities to accelerate the diffusion of key technologies.

23.1   Introduction

Shifting investment towards a low-carbon economy faces particular challenges in developing countries and economies in transition that will be explored in this chapter. Demand for energy is growing rapidly in many such countries. The choices made in the next 10–20 years on the levels of investment in end-use energy efficiency, the type of power generation systems, production processes and modes of transportation will affect greenhouse gas emissions for the next half-century. This chapter builds on the foundations of mitigation policy that are set out in Part IV to consider the key aspects of how best to assist developing countries to make the transition to a low-carbon economy.

This chapter first explores the context for investment decisions fast-growing emerging economies. There are significant requirements for investment in the energy sector, and finding resources to finance the incremental costs of investment in low-carbon technologies will be a challenge. There are also important financial, political and institutional barriers to clean energy investment in some developing countries and economies in transition.

Section 23.3 explores the role of national policy goals and reforms in making the transition to a low carbon economy. Energy price and taxation reform will play an important role in managing demand growth, as will improved end-use efficiency and facilitation of investment in more efficient and low-carbon technologies in several sectors. These reforms also support many other national objectives, because, as discussed in Chapter 12, mitigation policies have co-benefits such as improved air quality, increased access to modern energy services, and access to low carbon technologies. Many countries are already taking steps in these directions. It is in this context that assistance – financial, technical and so on, from both the public and private sector – can be enabled to facilitate a shift in the pattern of development.

As explained in Section 23.4, public policy also has a major influence in creating the conditions for the private sector to invest in and transfer low-carbon technologies (and the technologies relevant to adaptation) to developing countries. It is important to understand the various roles that the protection of intellectual property rights can play.

Section 23.5 discusses the essential role of lending and finance in supporting investment decisions in low-carbon technologies and energy efficiency, including through the Global Environment Facility. We consider what can be learned from the early experience of implementing the Clean Development Mechanism. Looking ahead, a transformation is required in these institutions to both generate and handle investment flows to enable developing countries to make the transition to a low-carbon economy. Section 23.6 examines the role of the World Bank and Regional Development Banks in creating frameworks to bring the issues discussed in this chapter together to ensure they complement each other. The chapter ends by examining the role that the international trade regime can play in supporting mitigation.

23.2   Understanding the context for energy sector investment

Demand for energy is growing rapidly in fast-growing emerging economies. The investment that takes place in the next 10–20 years could lock in very high greenhouse gas (GHG) emissions for the next half-century, or help move the world onto a more sustainable path.

Energy has a pivotal role in development – it helps promote access to better education, better health, increased productivity, enhanced competitiveness and improved economic growth. In many developing countries, under-investment in energy infrastructure is a brake on development[1]. The IEA (2006) has estimated that there are currently 1.6 billion people without access to energy (over a quarter of the world's population) and 2.5 billion using traditional biomass for cooking and heating[2]. Without new policies and financing, 1.4 billion people will remain without access to electricity by 2030.

In Chapter 12, we discussed the many co-benefits associated with reducing GHG emissions. Energy policy priorities in the developing world tend to be focused around facilitating economic growth and urbanisation; ensuring security of energy supply; providing access to energy; and reducing local and regional pollution from energy production and use[3]. These priorities can often lead to outcomes that reduce GHG emissions intensity – for example where there is a strong focus on energy efficiency, or when obsolete technologies are reduced or the use of carbon-intensive fuels is reduced. But there can also be conflicts, particularly where coal provides a cheap and readily available source of supply.

The IEA has identified a requirement for investment in the energy sector for developing countries of around $10 trillion to 2030[4]. This suggests that investment of around $165 billion per year is required from now to 2010 in the developing countries' electricity sectors alone, increasing at 3% per year through to 2030[5]. Out of this, $34 billion is required annually for energy access for poor people. This investment will come largely from national investment and from the private sector, and will depend to a large extent on the policy frameworks in place in the countries themselves.

There are financial, political and institutional barriers to encouraging clean energy in developing countries and economies in transition

Both the IEA and the World Bank note that the scale of actual current domestic and foreign investment is insufficient to meet these requirements. A large financing gap exists for investment in basic power sector infrastructure, in part because policy frameworks in the energy sector are not yet providing a sound environment for investment to take place. The World Bank estimates that there is a further significant gap, of around $20-30 billion per annum, to meet the incremental costs of low carbon investment in the power sector in developing countries.

There are strong pressures in fast growing economies to expand the supply of energy as quickly as possible. The implied returns to investment in the energy

---

[1] The World Bank estimates that in some countries under-investment in energy is reducing GDP growth by 1-4% per annum.

[2] See Chapter 12 and World Bank (2006b) for the effects of this on health.

[3] CCAP, 2006 and World Bank, 2006b.

[4] This figure is calculated as half of the IEA's (in press) total global capital investment estimate of US$20 trillion would be required to meet projected demand in the energy sector between 2005 and 2030 (of which around 57% would be required in the power sector). Proportion taken from IEA (2005).

[5] World Bank, 2006b.

sector mean it makes sense to expand generation capacity very quickly, often by using familiar capital stock and technology, and making use of domestic reserves of coal wherever possible, regardless of higher recurrent costs later through efficiency losses and local and regional environmental damage. These pressures have been particularly evident in China where power companies have been investing rapidly in new coal-fired power stations, but can also be seen in a range of other fast-growing economies. India and China's coal consumption is forecast to increase by 3% per year from 2004 to 2030, compared to an increase of 0.6% per annum for all OECD countries[6].

In addition, as Chapter 12 has noted, many developing countries subsidise their energy sectors – estimated at around $162 billion per year between 1995 and 1998. Many also have also built extended networks, and established fuel chains and users of dirty energy sources over time. For example, recent research from the Economic Commission for Latin America and the Caribbean shows that many of the countries in the region have had active fiscal policy to soften the impact on final consumer prices for scarce supplies of petrol and diesel, and do not differentiate between the polluting potential of these fuels[7]. Removing these distortions and pricing energy appropriately[8] could deliver long-term benefits for the climate and economy, but it requires careful management of any resulting redistribution of income between different parts of society.

These pressures are exacerbated by the difficulty faced by national governments and local authorities in enforcing environmental regulations or insisting on investments in untried technologies. These factors can slow down the introduction of more efficient technologies that are already cost-effective in developed countries, for example super-critical boilers for coal-fired power stations. In addition, low levels of capacity relative to demand means that it is difficult for operators to take plants off-line to make improvements to energy efficiency and delivery, given implications for local residents and industry. Hence, old and carbon-intensive infrastructure tends to be maintained in operation even where it would be cost-effective to upgrade it.

The following sections consider how international co-operation can support the achievement of ambitious national policy goals in the transition to a low-carbon economy, by creating an enabling environment for investment, accelerating transfer of relevant technologies to developing countries, and how carbon markets are beginning to create additional financial flows.

23.3   Improving the enabling environment for investment

There are a number of domestic barriers to investment and market development in clean energy technologies, many were identified in Section 23.2. The importance of these barriers will vary between countries, and according to the level of

---

[6] IEA (in press).
[7] Acquatella and Barcenas, 2005.
[8] In many cases the appropriate level is marginal cost of production, but the policy choice should depend on capacity and costs of outages, revenue constraints, and in some cases the incomes of the purchasers.

development of the country, the state of its financial sector, existing regulations and policies, as well as the availability of natural resources.

Many emerging economies are already engaged in a process of reforming the energy sector and introducing policies for sustainable transport, supporting national objectives for energy security, environmental quality, public finance and economic growth.

Taking action to reform the energy sector can be difficult, but as underlying distortions in energy prices and subsidies are removed, cost-effective efficient and low-carbon technologies will be taken up more widely, and there will be a stronger foundation for carbon markets to work more effectively. This can also increase the use of domestic capital as well as foreign domestic investment. An enhanced energy efficiency drive can also harness opportunities for significant gains by removing obsolete generation technologies, cutting losses in transmission, and enhancing positive impacts of removing carbon-intensive and locally polluting fuels. A case study commissioned by the World Bank (2006b) showed that an effective policy environment helped Vietnam to meet a sustained and rapid growth in demand for electricity.

Many developing countries are already advancing along these lines. In the 1990s, for example, China experienced rapid economic growth and a sustained fall in the energy intensity of its economy as it allowed prices to rise closer to market levels[9]. The 11th Five Year Plan seeks to continue this trend. The two key objectives are to double economic growth from 2000 to 2010 while reducing energy intensity 20% from 2006 to 2010. These objectives are supported by a wide range of policies, including the use of sales taxes to encourage the purchase of cars with smaller engines, and the use of regulation and other policies to encourage energy efficiency in the largest industrial enterprises (see Box 23.6). Chinese researchers have considered the extent to which reforms to energy taxation might contribute to this goal, as described in Box 23.1.

BOX 23.1 Modelling the potential impacts of energy taxation in China

China has now established a goal to reduce energy intensity by 20% between 2006 and 2010, reflecting concerns about energy security and air and water pollution. China has become increasingly reliant on oil imports (currently importing 43% of domestic oil consumption). Heavy reliance on the use of coal has caused high levels of air pollution[10]. Studies suggest that the economic costs of air pollution in China are between 2-7% of GDP, and that 16 of the 20 most polluted cities in the world are in China. China has also introduced legislation to promote energy conservation and the use of renewable energy, and is investing in a number of major national programmes to achieve the 20% energy intensity goal.

Research carried out for this Review[11] considered an illustrative example of how the introduction of energy taxation might support the delivery of China's energy, environmental

---

[9] CASS, 2006.
[10] Coal accounts for 70%, 90%, and 67% of total soot, sulphur dioxide, and nitrogen oxide emissions respectively (China Statistical Yearbook, 2005).
[11] CASS, 2006.

and social objectives, including lower air pollution and greater public resources for priorities such as education and health. The results indicated that:

- a flat tax of 50yuan/tonne coal equivalent (tce) on coal, oil, and natural gas would elicit a 6.3% reduction in energy demand (around 123 million tce) by 2010 compared with business as usual.

- variable tax rates of 120, 100 and 80 yuan/tce on coal, oil, and natural gas respectively to reflect the different carbon intensities of the fuels would result in an energy demand reduction of 16.2% (around 400 million tce) by 2030.

- the costs of introducing the tax was likely to be limited (0.4% of GDP in 2010 and 0.36% in 2030). This may be an overestimate because the calculations do not model the positive effects of reduced reliance on energy imports and the potential growth in environmentally friendly industries.

- the implementation of such tax rates might be expected to strengthen China's own public finances, raising approximately $11.6bn in 2010 and $31.5bn in 2030.

The Indian Planning Commission (2006) released a report on Integrated Energy Policy to contribute to its 11th Five Year Plan. This recommends a wide range of measures to increase competition in energy markets and allow energy prices to reflect market forces. It also recommends regulating prices to include environmental externalities, reduce losses in the power sector, and improve the transparency and targeting of subsidies. These reforms support the Indian government's goals of encouraging economic growth by reducing the cost of power and industrial energy intensity and extending access to electricity to all households by 2010. Such measures will also reduce ill health and mortality associated with indoor air pollution. As part of this strategy, the Indian Ministry of Power is working to remove market distortions caused by existing subsidies for kerosene in favour of less polluting, low-carbon home cooking systems based on solar and biomass technologies.

Specific local pollution control measures can also help control GHG emission growth. These policies are often designed and implemented by municipal rather than national authorities. For example, Mexico City has removed locally polluting carbon-intensive oil plants and replaced them with high-efficiency gas turbines. Likewise, Beijing has set up a plan to change industrial coal-fired boilers to natural gas and expand the use of natural gas in the grid in its effort to clean the city for the Olympics.

Long-term strategic planning is also essential to deliver the infrastructure for sustainable developments for the transport sector. The city of Curitiba in Brazil developed a plan to prevent urban sprawl and a high-capacity public bus system to keep total car use at 25% of that of comparable cities[12]. Similar proposals are advancing elsewhere. Bogotá, Colombia's capital city, has developed a methodology to account for the reduced emissions from implementing a Rapid Bus transit

---

[12] Michelowa and Michelowa, (2005): 22.

system to generate CDM credits from this project[13]. Cities in Mexico, Chile and Peru are planning to follow suit. Likewise, with World Bank support, Mexico has developed an umbrella program to expand new technology used for a Monterrey landfill-gas processing plant to other cities in the region.

Policies designed to support the deployment of new technologies such as feed-in tariffs and renewable portfolio standards, as described in Chapter 16, can also support investment, technology transfer and the formation of new national industries. Many developing countries have introduced such policies[14]. China and India have encouraged large-scale renewable deployment in recent years and now have respectively the largest and fifth largest renewable energy capacity worldwide[15].

The success of key developing countries in realising their current domestic energy and transport goals will play a part in limiting the growth of GHG emissions, and will facilitate further reductions over time. Notwithstanding the achievements so far, the goals that many of the large developing countries have set are ambitious, and there is much that international co-operation can do to support their implementation.

A number of international institutions and partnerships are focusing on increasing support for national policy reform to improve the environment for private sector investment and technology transfer.

There are a number of measures that governments can take to create a suitable investment climate for energy investment and the adoption of new technologies, such as[16]:

● Removal of broad-based energy subsidies and tariff barriers;
● Establishment of credible legal and regulatory frameworks;
● Creation of market-based approaches such as emissions trading, energy service companies, energy performance contracts, and credit guarantees;
● Information dissemination regarding energy savings and clean energy options;
● Including environmental costs in the price for energy services;
● Strengthening intellectual property rights;
● Developing product standards;
● Making markets more transparent.

It is important to involve the private sector in designing co-operation to enhance the climate for investment and technology transfer. The Renewable Energy and Energy Efficiency Partnership (REEEP), funded by a number of developed country governments, actively structures policy initiatives for clean energy markets and facilitates financing mechanisms for sustainable energy projects. REEEP provides opportunities for concerted collaboration among its partners, and has a bottom up approach to reflect local preferences, with the organisation playing a supportive role to the partners and members that run programmes rather than

---

[13] This has been with support from a Regional Development Bank – the Corporación Andina de Fomento. Also see Colombia's proposal at the Latin America Carbon Forum at http://www.latincarbon.com/docs/presentations/dia2/session2a/Presentaci%F3nMDLColombia-Ecuador.pdf
[14] REN21, 2005, p.20
[15] These are 2005 figures excluding large-scale hydropower. REN21, 2006, p. 6.
[16] World Bank, 2006a.

dictating approaches. This has proved popular and led to a diverse range of projects ranging from pure policy advice, such as compiling renewable energy legislation for Kazakhstan or devising clean energy policy and an action plan for Liberia, or more specific tasks such as promoting low energy buildings in China.[17]

The Asia Pacific Partnership, formed by Australia, China, India, Japan, South Korea and the US in 2005, takes a sectoral approach and, like the REEEP, focuses on the role of the private sector. The partnership includes a small amount of seed funding, but focuses on understanding the main drivers for investment in new technologies. Strong involvement of leading technology providers and investors provides a forum to explore practical steps to remove barriers to commercial cooperation on low carbon technologies. Over 90 private companies and industry groups and 150 senior representatives attended the inaugural ministerial meeting in January 2006. All eight sectoral task forces contain public and private sector members as equal participants rather than stakeholders.

The EU has its own partnerships on climate change and clean energy with China and India, as well as holding regular summits with the US, Canada, Russia and Latin America. Greater business involvement in these partnerships could provide an important channel for focusing on opportunities for profitable cooperation and priorities for policy intervention.

There are also opportunities to involve international lending institutions in identifying and advancing policy reform. This is discussed in Section 23.6.

23.4   Accelerating technology transfer to developing countries

Advances in technology play a key role in reducing the energy intensity of production in developed countries. The transfer of energy efficient and low-carbon technologies to developing countries allows developing countries to make similar progress.

The private sector drives significant transfers of relevant technology through markets, joint ventures, foreign direct investment and within policy frameworks such as the CDM. Governments have a role to play in creating the enabling environments for private sector transfers, and in setting the regulatory frameworks that govern international co-operation on intellectual property rights.

The creation of significant new national markets for a technology attracts foreign investors directly. For example, India's commitment to the expansion of wind power created the conditions for a successful joint venture between Vestas, the largest Danish wind turbine manufacturer, and India's RRB Consultants. This led to the creation of Vestas RRB, a wholly Indian owned company.

Joint ventures and licensing are a common entry vehicle for investment in emerging markets. There is some evidence that fear of competition and concerns relating to intellectual property rights may lead companies to offer older technologies[18] in such partnerships. However, the active role of the technology owner, particularly in the case of joint ventures, is likely to lead to effective technology

---

[17] http://www.reeep.org/index.cfm?articleid=33
[18] Saggi, 2000.

transfer since they have an incentive to ensure that the tacit knowledge[19] is also transferred to encourage effective use of the technology. Joint ventures are an effective long-term route to embed local firms into the learning network of transnational corporations[20].

Joint ventures played a particularly important role in China, where restrictions on Foreign Direct Investment (FDI) meant that between 1979 and 1997 the majority of FDI into China was in the form of joint ventures[21]. At the time there were conditions placed on the investment designed to spur technology transfer[22] that are no longer permissible following China's accession to the WTO. It is possible that these conditions reduced the overall supply of FDI, but they may have increased the quality of technology transfer in the FDI that did occur. The FDI to China had a significant impact on growth, especially through export growth[23].

The IPCC[24] conducted a study on the barriers that prevent the diffusion of key technologies relevant to climate change, and found that barriers arose at each stage of the process and varied by sectoral and regional context. The barriers included:

- Lack of information;
- Political and economic barriers such as lack of capital, high transaction costs, lack of full cost pricing, and trade and policy barriers;
- Lack of understanding of local needs;
- Business limitations, such as risk aversion in financial institutions; and
- Institutional limitations such as insufficient legal protection, and inadequate environmental codes and standards.

A recent report produced as part of a UK-India collaboration on the transfer of low-carbon energy technology[25] also explained that comprehensive technology transfer is much more than just hardware. It requires the transfer of skills and know-how for operation and maintenance and knowledge, expertise and experience for generating further innovation.

Barriers to technology transfer can be overcome through a combination of formal institutional mechanisms, measures to improve the enabling environment for private sector investment, and, where necessary, direct funding initiatives.

Formal co-operation on technology transfer can be built around any of the key stages in the technology transfer process. These stages were identified in the UK-India report as[26]:

- assessment of technology needs
- selection of technologies
- mechanism for technology import
- operating technology at design capacity

---

[19] Tacit knowledge is defined as knowledge that is not covered by the patent but embedded in skills and know-how.
[20] Buckley et al, 2006.
[21] OECD, 2000.
[22] Watson and Liu Xue, 2002.
[23] Graham and Wada, 2001.
[24] IPCC, 2000 and UNEP, 2001.
[25] SPRU, IDS and TERI (in press). Comprehensive literature review and five case studies.
[26] SPRU, IDS and TERI (in press) and Kathuria (2002).

- adapting technology to local conditions
- improving installed equipment
- development of technology

Different policy interventions maybe required at each stage depending on which functions private markets can successfully provide. Relevant policy interventions vary according to the nature of the technology, its stage of commercial development and the political and economic characteristics of both supplier and recipient countries.

In order to be sustainable, technology transfer must take place as part of a wider process of technological capacity building in developing countries. Building technological capacity relies on the transfer of skills, knowledge and expertise as well as hardware, especially if technologies are to be assimilated and developed further within recipient countries. Capacity building must be adapted to local circumstances, because there are many examples where a lack of technical, business or regulatory skills resulted in a failed attempt at technology transfer. A total package of human skills for technology transfer will also focus on creating improved and accessible competence in associated services, organisational know-how, and regulatory management, to strengthen and coordinate the networks through which stakeholders facilitate transfer.

The UNFCCC includes provisions on the transfer of technology to enable developing countries and economies in transition to mitigate greenhouse gas emissions and adapt to climate change. The UNFCCC Expert Group on Technology Transfer has recently completed a special report[27] that explored specific measures that can help develop technology flows across national borders, enhancing the technology framework under the UNFCCC. The key elements of the current approach to technology transfer include country-driven technology needs assessments; the provision of information through TT:Clear; a focus on understanding the aspects of national policy environments that facilitate private sector technology co-operation; and capacity building, for example, to help developing countries with project development process to meet lending criteria. The Special Climate Change Fund includes a provision for funding technology transfer. Intermediaries such as independent energy labs and foundations, such as the Energy Foundation[28], have played an important role identifying appropriate technologies.

A Technology Needs Assessment (TNA) is a country-driven activity that identifies the mitigation and adaptation technology priorities. It involves different stakeholders in a consultative process to identify the barriers to technology transfer and measures to address these barriers through sectoral analyses. It also examines regulatory options, fiscal and financial incentives and capacity building. More than 20 countries[29] have carried out assessments, including least developed countries, economies in transition and small island states (see Box 23.2 below for an example). For mitigation, key technologies identified included renewable energy for small-scale applications, such as biomass stoves; combined heat and power; and energy efficient appliances and building technologies such as compact fluorescent light

[27] http://unfccc.int/resource/docs/2006/sbsta/eng/inf04.pdf#search=%22FCCC%2FSBSTA%2F2006%2FINF.4%22
[28] http://www.efchina.org/home.cfm
[29] Synthesis Report on Technology Needs identified by Parties not included in Annex 1 to the Convention, SBSTA/2006/INF.1 available at http://ttclear.unfccc.int/ttclear/jsp/index.jsp.

BOX 23.2 Ghana's Technology needs assessment

Ghana submitted its TNA to the UNFCCC in 2003[30]. The assessment received major funding from the UNDP/GEF and technical support from the National Renewable Energy Laboratory in the US with funds from the Climate Technology Initiative and the US Department of Energy highlighting the role of international support and intermediaries.

The goal of the TNA is to communicate Ghana's climate change technology requirements by identifying a portfolio of technology development and transfer programmes that have the potential to reduce greenhouse gas emissions and contribute to Ghana's sustainable development. The assessment applied selection criteria to establish top priority technologies:

- Industrial energy improvements –demand side management including boiler efficiency enhancement

- Methane gas capture from landfill sites

- Use of bio-fuels (jatropha)

- Energy efficient lighting using Compact Fluorescent Lamps (CFLs)

Since the assessment, CFL promotion policies – including changes to Ghana's import tariffs, installation task forces and sales through employers and retail outlets – have led to a dramatic increase in adoption. This transformation in the lighting market has been sustainable and self-financing. An evaluation of the scheme shows it added US$10 million[31] to the Ghana Economy. Prior to the CFL support programme, lighting represented a third of energy consumption, and use of lighting also coincided with the peak consumption placing pressure on peak capacity. CFL promotion has reduced electricity consumption by around 6%, reducing the risk of a power crisis and demand for new generation capacity, and reducing the impact on consumers of a doubling of electricity price following reforms.

bulbs. For transport, traffic management and cleaner vehicles for public transport were most important. Institutional mechanisms and actions by intermediaries can help identify opportunities for private sector action.

The key barriers were identified as economic (including high upfront costs and incompatible prices, tariffs and subsidies), and lack of information about appropriate technology options. The Assessments have been followed up in various ways. Specific projects have been developed and presented to the GEF and to the UNFCCC workshop on innovative financing mechanisms. Some countries have used the results to make changes to their own development plans and enabling environments.

In many cases intellectual property rights are not the key barrier to transfer of technology.

Within international debates on climate change there has been a particular focus on the role of intellectual property rights (IPR) as a barrier to the international diffusion of technologies. In principle, patents that protect IPR and reward the

---

[30] For full report see http://ttclear.unfccc.int/ttclear/jsp/index.jsp.
[31] Benefits based on net present value calculated using a 25% discount rate (lower rates increase benefits). See http://www.oecd.org/dataoecd/37/53/34915266.ppt.

innovator are important as they provide an incentive to invest in developing new products. Weak IPR may deter domestic firms in developing countries from purchasing technologies as their competitors may be able to copy them without paying[32]. Companies with advanced technologies often cite insufficient IPR protection in developing countries as a barrier to technology transfer, and suggest stronger protection, for example by full implementation of the TRIPs[33] agreement, would help them deploy advanced technologies. Increasing the incentives for mitigation (for example by introducing a carbon price) increases the value of patents for low-carbon technologies and acts as a stimulus to investment in innovation in this area. The benefits of having an intellectual property (IP) regime do not imply that such rights should be increased without limit, especially if they reduce the beneficial effects of product market competition.

Patents can also be seen as creating a short-term monopoly and thus limiting efficient diffusion whilst the owner enjoys monopoly rents. From this point of view, patents on new products that could help developing countries to reduce their emissions or improve the resilience of their agriculture are inefficient – they make it more difficult to secure a global public good. IPR may have little impact on innovation and diffusion in countries without sufficient capacity to innovate, so could impose additional costs[34].

Company surveys indicate that patenting is the most important means of IP protection in only a few industries, such as pharmaceuticals and scientific equipment. A majority of companies in other industries make use of alternative protection methods. In an OECD report on innovation in the business sector[35], econometric estimates suggest that stronger IP protection has a substantial positive effect on patenting, but only a limited effect on R&D. Stronger patent regimes did help direct innovation towards patentable activities but such activities need not offer the greatest benefits for society as a whole. Other studies have found evidence that cross-country differences in patenting are positively related to cross-country differences in the strength of IP protection. However, others have suggested that the benefits of stronger IP protection are positive only when IP protection is initially weak. Most increases in patent claims in countries that have enhanced patent protection have been found to come from foreign residents, suggesting that strengthening patent protection, at least to some threshold level, can help to improve access to foreign ideas.[36] There is some evidence that a more robust IPR regime encourages transfer and that firms respond to changes in the stringency of IPR regimes. Different firms choose different modes of entry due to their relative sensitivity to protection. Firms with natural barriers to imitation tend to choose licensing, and vulnerable firms choose FDI, but stronger IPR may cause substitution between these modes. Not only is there an increase in FDI and licensing with stronger IPR, but also a change in the composition of technology transfer[37]. Another study[38] provides strong evidence that US multinationals respond to such

[32] Philibert and Podkanski, 2005.
[33] The agreement on Trade Related Intellectual Property Rights (TRIPs) is an international treaty administered by the World Trade Organisation which sets down minimum standards for most forms of intellectual property regulation within all WTO member countries.
[34] Falvey and Foster, 2006.
[35] OECD, 2005, pp. 39–42.
[36] Lerner, 2002.
[37] Nicholson, 2003.
[38] Branstetter et al, 2004.

changes in IPR regimes abroad by increasing technology transfers. The results of the study are however not sufficient to demonstrate that IPR reforms are welfare enhancing for the reforming countries.

In a series of case studies undertaken by the OECD, IPR did not appear to constitute an obstacle to technology transfer[39]. Some of the case studies found that there are many environmental technologies available that are not protected by patents, so IPR were not relevant to much of the volume of clean technology transfer. They also indicated that even when clean technologies were under patent, these patents were not a major concern either to importers or exporters. In general, exporters were willing to accept the risk of patent infringements, as by the time a process had been copied, it will have been overtaken. Importers of patented technologies did not generally find royalty fees to be a major obstacle, and were more concerned about other costs, such as that of capital investments in new plants and machinery[40].

IPR protection is just one issue in a complex process for technology transfer, and only a component of the cost of a technology and should not be overplayed. The level of tacit knowledge[41] not covered by the patent may prevent effective transfer rather than the IPR cost itself. Tacit knowledge ensures that transfer requires the co-operation of the IPR owner, and may mean that joint ventures and strategic programmes to enhance the capacity to manufacture and operate the equipment are the most effective means of accelerating the diffusion of key technologies.

There are also issues that arise in the case of advanced and dual use technologies such as nuclear power[42] and the advanced technology for gas turbines required in IGCC power stations. These are sensitive issues that require careful risk assessment, and can be resolved through proactive bilateral and multilateral diplomacy. Box 23.3 explores the case for public ownership of IPR.

BOX 23.3 Public ownership of IPR

In the pharmaceutical sector, production costs represent a small share of the price, so IPR provides an incentives during the costly research process. The demand for and impact of the drugs is predictable, so governments have a clearer understanding of the value of specific technologies, and have established channels to ensure that the drugs will reach those that need them. Public-private partnerships are useful in such settings, and may include:

- Purchasing commitments as an incentive for the development of new drugs[43]

- Voluntary buy-out of IPR for existing products, whereby governments agree a price with the IPR holder to buy all or limited rights to the IPR.

- Compulsory licensing approach whereby the government forces the holder of the IPR to grant use to the state or others. Usually, the holder does receive some royalties, either set by law or determined through some form of arbitration.

---

[39] OECD, 1992.
[40] Less and McMillan, 2005, p. 24.
[41] Tacit knowledge is defined as knowledge that is not covered by the patent but embedded in skills and know-how.
[42] See, for example, the recent US-India agreement on the use of civilian nuclear technology.
[43] Kremer and Glennerster, 2004.

For key mitigation technologies, such as electricity generation, IPR generally represents a much smaller component of cost due to the scale of the capital investments and running costs. A broad range of technological solutions is also available, so Governments will have difficulty in picking appropriate technologies and lack the information to negotiate a suitable price. Also, the tacit knowledge associated with using these technologies and challenge of re-engineering advanced energy technologies requires continued co-operation with the owners of the technology. This makes them less suitable for public funding of IPR or compulsory partnership. These factors all make public-private partnerships in this area, such as buying IPR rights for established technologies, problematic.

The development of new technologies, particularly those with significant public funding, will be more conducive to public IPR ownership. As these technologies would be collaboratively developed, the IPR could potentially enter into joint ownership by the partners involved with the aim of making the IPR available as a free or low cost public good. Some areas of adaptation, where there is a strong public good element, may also provide good reason to extend existing efforts to overcome IPR barriers, for example to deal with effects on health from climate change such as malaria.

23.5   International financial flows for energy efficient and low-carbon investment

Acting now, to ensure the current wave of investment in fast-growing economies incorporates energy efficient and low-carbon technology, will reduce the global cost of stabilising greenhouse gases in the atmosphere.

Private sector resources for energy sector investment far outweigh those available from governments and multilateral institutions, and public finance or loans can even be under-utilised in such countries. Middle-income countries, where the bulk of future GHG emissions growth is concentrated, have good access to capital from the private sector[44]. Public sector resources and flows of carbon finance provide an important lever to channel these larger flows of domestic and international private sector investment to energy efficient and low-carbon technologies.

The Global Environment Facility has a strong track record in financing programmes for energy efficiency and renewable energy, but is small relative to the scale of the challenge.

The main funding framework in the application of established low-carbon energy technologies is the Global Environment Facility (GEF)[45], working through its Implementing Agencies and with a range of multilateral and bilateral donors. Since its inception in 1991, the GEF has provided $6.2 billion in grants and generated over $20 billion in co-financing from other sources to support over 1,800

---

[44] Miller (2006) www.iddri.org/iddri/telecharge/climat/climat_dev_sept06/session_33/miller_finance.ppt.

[45] The Global Environment Facility (GEF) provides financial support through the World Bank, UNDP and UNEP to achieve the aims of the UN Framework Convention on Climate Change, Convention on Biological Diversity and the Stockholm Convention on Persistent Organic Pollutants.

projects that produce global environmental benefits[46] in 140 developing countries and economies in transition[47]. The GEF has financed the diffusion of energy efficient and renewable energy technologies, supported by wider investment in demonstration projects, local capacity building and institutional development. Projects to raise efficiency in a number of areas including boilers, lighting, and biomass stoves have delivered significant energy savings and related reductions in greenhouse gas emissions.

The World Bank has recently suggested that the GEF could play an enhanced role in encouraging technological learning and bringing down the cost of the low-carbon technologies that are most relevant to developing country priorities. Any increase should seek to overcome existing implementation challenges[48]. Current funds are small relative to the scale of the challenge. The GEF would require up to a two to three fold increase in current financing in order to ensure sustained market penetration of energy efficiency and renewable energy technologies over the next ten years. Financing a strategic, global programme to support the reduction in costs of pre-commercial, low GHG emitting technologies such as IGCC with CCS, solar thermal, or fuel cells would require more than a ten-fold increase[49]. This would in turn require significant changes in the GEF's institutional arrangements[50]. Whether it is through GEF or other institutional mechanisms, an expansion in the scale of funding is required if the deployment of low-carbon technologies is to be supported, and strong legal and regulatory environments and local partnerships are important in determining success. International efforts to develop low-carbon technologies are discussed in Chapter 24.

Lending can play an important role in supporting energy efficiency.

Financial institutions have a unique opportunity to encourage their clients to seek advice on the energy efficiency of proposed investments. By building this advice into the planning and financing stage of major investment in upgrades or new infrastructure, transaction costs can be greatly reduced. The European Bank for Reconstruction and Development has developed an effective business model for this, as described in Box 23.4.

The US Department of Energy is supporting the development of an International Energy Efficiency Project Financing Protocol as a method to accelerate the transformation of clean energy financial practices. This would provide standard methodologies and good practice guidelines for commercial lenders, especially to reduce the transaction costs associated with relatively small projects [51].

The Clean Development Mechanism provides an important channel for private sector participation in financing low-carbon investments in developing countries.

Under the UNFCCC and Kyoto Protocol, developing countries took on an unquantified responsibility to participate in action to limit the risks of climate change, in the context of their own priorities for economic and social development and

---

[46] Including benefits from reducing GHGs and other pollutants and increasing biodiversity.
[47] The GEF enjoys a 4:1 leverage ratio of total project funding to its initial contributions.
[48] Miller (2006) http://www.makemarketswork.com/client/makemarketswork/upload/Biography%20Miller.doc.
[49] World Bank, 2006b: 23.
[50] World Bank, 2006b.
[51] See http://www.evo-world.org/index.php?option=com_content&task=view&id=60&Itemid=148.

BOX 23.4 Lending for energy efficiency: the EBRD model

The European Bank for Reconstruction and Development has developed a successful business model to raise energy efficiency through financing industrial, SME, municipal infrastructure and power sector projects in transition economies. A dedicated energy efficiency team, operating at the core of the organisation, screens every new project proposal to identify potential energy efficiency financing opportunities. Comprehensive energy audits are provided to define the energy efficiency potential of a project and its financial return at the most relevant stage in the project lifecycle.

The EBRD is setting financial intermediation facilities across its regions of operations with local commercial banks to support energy efficiency investments in SMEs. Technical assistance is provided for market studies to assess the size, opportunities and constraints for the financing of SME energy efficiency projects and for project preparation and implementation support. The EBRD has signed energy efficiency credit lines with 11 banks in three countries targeting industrial SMEs, small renewable energy projects and the residential sector.

In addition, the EBRD has financed 35 industrial energy efficiency projects between 2002 and 2005 with €276 million of EBRD investment in energy efficiency components within a total project value of €1.45 billion. This has contributed to energy savings over 600,000 toe/year and to an estimated annual CO_2 reduction of 2.5 million tons. The Bank has financed 11 (largely municipally owned) district heating projects since 2001 with a total Bank investment of €265 million resulting in significant energy savings. It has also financed a portfolio of projects to improve the energy-efficiency of public transport vehicles and traffic management systems.

With the launch of its Sustainable Energy Initiative and the Multilateral Carbon Credit Fund in 2006, combined with the full integration of its energy efficiency activity across banking operations, the EBRD aims to step up its climate change mitigation investment to €1.5 billion during for the next three years[52].

poverty reduction. The Kyoto Protocol created a project-based mechanism – the Clean Development Mechanism (CDM) – to allow rich countries to use credits from investment in emissions reductions in poor countries to offset against their own emission reduction commitments[53].

The CDM has played an important role in building co-operation between the developed and developing parties to Kyoto, and it has helped to strengthen understanding of the main opportunities for abatement. It has also stimulated a strong private sector interest in climate change co-operation. Implementation has involved significant efforts at capacity building and project identification, both by bilateral government programmes[54] and the World Bank's Prototype Carbon Fund (PCF). A wide range of methods have been developed for crediting emissions reductions, ranging from industrial gases through energy efficiency to renewable energy projects.

The CDM in its current form is making only a small difference to investment in long-lived energy and transport infrastructure. Its role is limited by factors such as transaction costs, policy uncertainty, technology risk and other barriers.

[52] See http://www.ebrd.com/new/pressrel/2006/54may19.htm.
[53] See Grubb (1999) for a general introduction to the CDM.
[54] Such as Certified Emission Reduction Procurement Tender (CERUPT), a programme set up by the Netherlands to purchase greenhouse gas reductions through the CDM.

While a substantial international flow of funds is being generated through CDM[55], it falls significantly short of the scale and nature of incentives required to reduce future emissions in developing countries.

Around 35% of CDM credits in the current pipeline[56] come from 15 projects for industrial gases. Such projects are attractive because industrial gases have a very high global warming potential and thus generate a very large volume of emissions reductions compared to, for example, renewable energy projects[57]. There are still relatively few projects in many sectors that are important for the long-term reduction of GHG emissions. There has also been limited use of the CDM in the poorest countries, raising concerns about distributional equity of the CDM, and the appropriate mechanisms to tackle low-carbon infrastructure to support wider access to energy for poor people. There are a number of related reasons for these trends.

- The CDM provides funding on a project-by-project basis to offset against absolute reductions that would otherwise have been made by countries with commitments to reduce emissions under the Kyoto Protocol. For this reason, there are procedures involved in demonstrating additionality[58] on a case-by-case basis, which leads to high transaction costs.
- It has proved difficult, for example, to establish methodologies for energy efficiency in sectors dominated by small and medium-sized enterprises and for transport infrastructure and demand management[59], which may be more relevant to poorer countries.
- The CDM provides a funding stream on the basis of the carbon price, but does not necessarily cover the learning costs associated with the higher risks of using new technologies including advanced renewable energy technologies.
- Projects with longer payback periods may be affected by other capital market failures: where the benefits of long-term energy savings that occur beyond the standard pay-back period used in investment appraisal or are very heavily discounted both for time and uncertainty. This does not only happen with large projects – for example, this affects the uptake of small-scale solar technologies[60].

There are several proposals to streamline the CDM in its current form, including those described in Box 23.5.

The CDM plays a valuable role, but it has important limitations as a model for international co-operation in the longer term.

The CDM is explicitly designed to provide offsets to enable developed countries to meet their commitments more cheaply, while allowing developing countries to participate in carbon reduction and gain co-benefits from technology transfer. At the same time it allows the leveraging of investment in projects that meet local priorities for sustainable development. However, it does not represent additional

---

[55] Estimates as at October 2006 suggest that there are approximately 1.4billion CERs expected from projects up to 2012, valued at around $14billion (assuming a $10 price).
[56] As at October 2006.
[57] REIL, 2006: 9.
[58] Additionality is defined in the Marrakech accords: "A CDM project is additional if anthropogenic emissions of greenhouse gases by sources are reduced below those that would have occurred in the absence of the registered CDM project activity", This involves some difficulties in interpretation in practice.
[59] Browne *et al*, 2004.
[60] Philibert, 2006.

BOX 23.5 Proposals to streamline the CDM in its current form

Programmatic CDM was approved at the UNFCCC COP/MOP1 at Montreal in December 2005. It allows for specific programmes taking place in the context of national/regional policies to be credited. It can build upon national policies deployed by national or sub-national bodies to tackle both their own development objectives as well as reduce GHG emissions. Its main aim is to produce larger CDM projects with lower transaction costs. A programmatic approach to CDM can do so by aggregating smaller projects within a programme, for example incorporating reductions from households, small enterprises, rural electrification and transportation. These sectors cannot be tackled on an individual basis but can be tackled through an intentional government-led programme to facilitate reductions. Variants still being developed could boost incentives for developing countries to initiate such programmes.

Technology CDM would involve moving away from verification of project-specific information under the current CDM, towards a more principled or standardised approach to selection of eligible technologies and relevant baselines using technology standards. One variant of this approach is already possible under existing CDM rules, but is costly and complicated due to the need to determine appropriate technological benchmarks. A more streamlined approach, including prior crediting on the basis of an index of approved technologies, would enhance the attractiveness of the mechanism to investors, but at the cost of some environmental certainty – particularly if emissions reduction is about management performance as well as technology. Discounting or capping credits for these "wholesale" purchases might handle some of these concerns. This would require significant reform to the CDM modalities and procedures.

net emissions reductions over and above those required by developed country limits. Given the relative growth of emissions in both developed and developing countries, and the scale of the challenge represented by climate change, this approach can be seen as an important building block along the way to arrangements that support reductions on a much greater scale, rather than as the final shape of long-term structures for co-operation.

In particular, project-based carbon finance does not internalise the cost of the greenhouse gas externality for firms and consumers in the host country or for goods exported from the country. Project-based carbon finance acts as a form of subsidy; it reduces the emissions from a particular project, but it does not affect the demand for high carbon goods and services across the economy as a whole, so the overall level of emissions can remain high or increase. It also creates issues of moral hazard and gaming, where there are incentives to manipulate the system to increase the rewards received (or reduce the costs paid). For example, in the case of low-carbon investment, the implementation of second-best emissions reductions policies (such as increasing renewables within a subsidised power sector) may raise the costs of implementing first-best policies (such as removing subsidies). Both policies are important to implement in the long-term.

Improvements can also be made to carbon finance to raise the scope for emissions reductions programmes in the transport and buildings sectors. For example, complex decisions to channel resources to land-use planning, urban development, public transport and bicycle and pedestrian infrastructure are most important for sustainable transport use, as it is difficult to amend this infrastructure once in place. In many cases, this may suggest the use of non-uniform approaches in these sectors, including pilot approaches to carbon finance as well as direct funding – for example through bilateral assistance and GEF funds[61].

The transformation of carbon finance flows between developed and developing countries is required to support cost-effective reductions through policy and structural reform in developing countries. This, in turn, is likely to widen the scope of carbon finance to more regions and sectors and reduce global costs of mitigation.

Section 23.2 has demonstrated that large-scale flows are required to support the transition to a low carbon economy in developing countries. We provided illustrative calculations in Chapter 21 to demonstrate that large flows of carbon finance – up to around $40 billion a year – would be generated if developed countries were to *take responsibility for* significant emissions reductions to 2050 on 1990 levels, and if they were to meet a proportion of those through financing action in developing countries[62]. To reach long-term international goals, it would remain important for developing countries to take on their own commitments in suitable forms and with appropriate support. Investment flows could be directed to helping generate emissions reductions, for example, by financing the kinds of reforms suggested in Section 23.3. But this would also require a transformation of flows of carbon finance such as currently generated through the CDM.

The most cost-effective, large-scale emissions reductions are likely to be linked to strategic programmes, for example in supporting integrated programmes for urban transport and development, or in tackling a wholesale transition to lower carbon power generation including the retrofit of inefficient plants and the systematic use of carbon capture and storage. Programmes on this scale can take place only in the context of structural reforms and development policies implemented by national or regional governments. Investment in CDM projects tends to be directed towards countries where there is a strong enabling environment for private sector investment (for example, economic and political stability, liberalised markets, strong legal structures), and countries that have built up national capacity for using this source of funding[63]. This provides strong incentives for countries to develop such environments.

Useful lessons for broadening the scope of the CDM can be learnt from the proposal to use funds from intergovernmental emissions trading for programmes to reduce emissions in central Europe. Romania, Bulgaria and Hungary, for example, have all indicated a willingness to earmark funds from sales of their surplus allowances under the Kyoto first commitment period to emission reduction efforts, for example through programmes of building renovation. The countries would play the major role in identifying opportunities for these programmes and directing funds towards priority areas. The OECD/IEA and World Bank have examined these 'Green Investment Schemes'.[64]

Action at scale requires appropriate incentives across the economy. This implies moving carbon finance mechanisms closer to full emissions trading or to programmes that in other ways support the transition to carbon pricing in developing countries.

[61] Browne *et al*, 2004.
[62] Our methodology is described in Chapter 21.
[63] Fankhauser and Lavric, 2003.
[64] Blyth and Baron, 2003 and World Bank, 2004.

Carbon finance mechanisms could evolve to support the transition to full emissions trading in several ways or stages. One option is to design a policy-based CDM that would provide credits directly to developing country governments that introduce a policy relating to emissions reductions[65]. This approach could be used to provide incentives for emissions reductions in sectors that, for example, may not be immediately suited either to project-based CDM or to emissions trading, but where the early implementation of relevant policies could lead to long term emissions reductions. The policy reform could be credited using an estimate based on factors including volume of emissions sources affected, price elasticities and so on – for example to determine the impact of removing a subsidy. Where credits are granted without project-level monitoring and verification procedures, techniques including discounting, taxing or phasing of credits could be used to recognise uncertainty about final outcomes.

One challenge of policy-based approaches is that credits are likely to flow to the government while the costs of complying with the policy will fall on the private sector[66]. The design of policy-based schemes must therefore incorporate incentives for their implementation by the private sector. For example, revenues from credits could be used to compensate owners of inefficient facilities that would be closed down as part of an industrial restructuring policy, or could be used to encourage property developers and energy suppliers to introduce energy efficient lighting technologies or smart metering in new buildings.

Some sectoral crediting mechanisms and 'no-lose' commitments described in Chapter 22 would also move carbon finance in this direction. These approaches all require preparatory work, particularly regarding systems for data reporting and monitoring, and capacity building to enable firms to participate in the schemes. Some countries are already engaged in policies that would make it much easier to move in these directions; for example, China's programme to reduce energy use by its 1000 largest enterprises, described in Box 23.6. A number of international initiatives will also provide information to lay foundations for these approaches. For example, the IEA and World Bank have also announced co-operation to develop sector-specific benchmarks for energy efficiency for Brazil, China, India, Mexico and South Africa, as part of the Energy Investment Framework, to be discussed in Section 23.6[67].

Long-term goals and early signals to provide continuity of carbon finance after 2012 are essential to underpin emissions reduction policies in developing countries.

Debate on the future of the CDM is an important element of the international negotiations for co-operation on climate change beyond 2012. There is increasing interest, from governments and emissions trading schemes established inside and outside the Kyoto Protocol, in purchasing project-based credits from developing countries.

In the long-term, deep global reductions in GHG emissions will require that all countries with significant requirements for energy incorporate the externalities of using carbon into the structure of incentives in their own economies. This

---

[65] This proposal is in early stages. It was not approved following initial discussion at the UNFCCC COP/MOP in Montreal in December 2005.
[66] Michaelowa, 2005.
[67] World Bank, 2006a.

BOX 23.6 China's 1000 enterprises program

Industry accounts for approximately two thirds of total energy use in China. Improving industrial energy efficiency, in sectors such as iron and steel, is critical to delivering China's 11th five-Year Plan goal to reduce its energy use per unit of GDP by 20% between 2006 and 2010.

In March 2006, the Chinese government announced a program to manage and improve energy use among just over 1000 major energy consuming industrial firms and utilities that reportedly account for 47% of total industrial energy use. The program aims to save 70 Mtoe cumulatively over five years. This represents a major contribution towards the target of reducing overall energy intensity by 20% (which implies a reduction of approximately 170 Mtoe).

Under the scheme, each enterprise will have its energy use monitored and benchmarked against national and international market participants. Each will agree plans to deliver targets on the energy intensity of its outputs (such as average energy consumption per production unit). Those that meet or exceed their targets receive positive incentives, such as faster management promotion, while those that fail to deliver are publicly criticised as energy wasters.

China received assistance from the Energy Foundation to design the programme and seconded a member of staff from DEFRA for a year (partially sponsored by REEEP). Collaboration between the IEA and the Chinese administration may also assist delivery of the scheme, for example in developing indicators or statistics as part of the sector benchmarking process.

could take the form of full participation in international emissions trading, or could be achieved by a combination of domestic tax and regulation.

Long-term goals to underpin these developments are crucial. Ongoing research suggests that a lack of long-term goals and domestic policy frameworks could prevent carbon finance from facilitating the transition to a low-carbon future[68]. Therefore, early signals about the acceptability of particular types of credits from developing countries after 2012 in trading schemes worldwide could help to extend the role of carbon finance in advance of agreement on the final form of future mechanisms. This could include signals about the potential to reduce or remove the current restrictions in the EU ETS on the volume of project credits that can be used, and signals about the types of large-scale programmes that could become eligible for accelerated recognition. For example, the EU is examining changes to the ETS monitoring and verification methodology to incorporate carbon capture and storage, but a signal on whether and how CCS may be eligible for crediting under CDM could provide important incentives.

23.6 Developing an integrated approach to enhance investment in developing countries

The moves towards strong national goals, aspirations and policies described in Section 23.3 could provide a platform for enhanced co-operation based on

---

[68] Garibaldi, 2006.

international flows of carbon finance, public and private investment, risk guarantees and other instruments. And, as described in Sections 23.4 and 23.5, these flows can themselves be used to support the introduction of further domestic policies including energy market reforms and the use of new technologies. Therefore, channelling investment in developing countries towards energy efficient and low-carbon options requires an integrated approach.

The International Financial Institutions (IFIs) have an important role to play in accelerating this process. They work with national governments, providing technical assistance to set policy and institutional frameworks to create the right incentives in relevant sectors. They can help overcome capital market failures that lead to underinvestment in energy efficiency, and work with the private sector to increase the scale of low-carbon investment. Climate change is now a significant issue for economic growth and development and should be considered within country assistance strategies. The World Bank and Regional Development Banks (RDBs) are developing Clean Energy Investment Frameworks. The RDBs are working on specific initiatives or approaches to mitigation and adaptation that are likely to have resonance within their respective regions. These are described in Box 23.7.

These frameworks also provide the opportunity for IFIs to help facilitate the development of large-scale pilot programmes, for example to explore how the broadening of carbon finance or limited participation in emissions trading could be implemented in practice. This would require early agreement between developing countries willing to explore new approaches and developed countries with emissions trading schemes or other mechanisms to purchase credits that would be generated.

Combining carbon finance with public and private investment flows, risk guarantees and other financial instruments can support the deployment of emerging technologies.

Commercialising emerging technologies requires risk capital that is often unavailable in developing countries. Carbon finance alone may not be sufficient to fund incremental costs, and other types of support may be needed to make a project viable. Emerging technologies are perceived as higher risk and are thus less likely to attract domestic private investment or to receive export guarantees. There are significant opportunities for the IFIs to play a role in improving the pipeline of 'bankable' low-carbon projects that have risk profiles and business plans suitable for attracting private sector support, including through the use of public funding to improve project identification and the preparation of investment proposals. The use of financial and risk management instruments can reduce transaction costs, increase transparency and competitiveness of loan pricing, and share country and project risk.

Investment in the most advanced technologies may require a different approach. The IFIs are normally constrained by their procurement rules to purchase standard technologies rather than advanced technologies in their mainstream investment programmes. Initially, investors and managers in developing countries may require assistance including information and capacity building to use such technologies.

Public-private financing initiatives also have a role to play in reducing market place risks. The Johannesburg Renewable Energy Coalition (JREC) is made up of

BOX 23.7 The Clean Energy Investment Framework

At the G8 Summit in Gleneagles in 2005, the World Bank and the Regional Development Banks were asked to work with all their stakeholders to develop frameworks for investment in clean energy.

The approach presented by the World Bank at its Annual Meetings in September 2006 has three pillars: energy for development and access for the poor; transition to a low-carbon economy; and adaptation. The first two pillars of the framework focus on improving the coordination and coherence of existing sources of energy investment and risk management instruments from domestic and international capital markets as well as from the multilateral institutions. The framework will also combine financial and technical assistance to support developing countries on policy reform or sectoral initiatives, and help countries develop policies and enabling environments that are conducive to private sector investment.

Financing under the EIF is expected to include projects that accelerate the take up of technologies that enable more efficient and cleaner energy production and use, including the deployment of advanced super-critical coal-burning technologies in power stations and the introduction of more efficient operating practices and grid management and audits of energy-users to improve efficiency. The World Bank is examining vehicles for doubling concessional support to $4 billion per year in order to improve energy access for poor people. The Bank is also looking at how to increase the efficacy of its instruments and procedures (especially under its proposed Middle Income Strategy), as well as proposals to develop new instruments.

The EBRD has defined and is currently implementing the Sustainable Energy Initiative aimed at scaling up and accelerating the pace of investment in climate change mitigation projects in Central and Eastern Europe. Key target sectors include industry (both large corporates and SMEs), the power sector (including renewable energy) and the municipal infrastructure sector (including district heating, urban transport and solid waste).

The Asian Development Bank is focusing on both energy efficiency and transportation issues, and including additional carbon finance and adaptation components. Transportation is one of the largest causes of increased GHG emissions in Asia. The Inter American Development Bank is also developing a framework with four components: energy efficiency, renewable energy sources, biofuels, and adaptation. It also considers the development of carbon finance.

governments who have decided to co-operate actively on the promotion of renewable energy sources on the basis of concrete, ambitious and agreed objectives. The JREC Patient Capital Initiative[69] aims to develop an innovative public-private investment mechanism that creates and delivers risk capital to renewable energy project developers and entrepreneurs at affordable conditions. As part of this programme the European Commission is sponsoring the development of an innovative public-private financing mechanisms. The European Commission proposed Global Energy Efficiency and Renewable Energy Fund[70] in October 2006. It aims to contribute €80million over the next year, which, in addition to €20 million from other public and private sources, is expected to contribute to the financing of

[69] http://ec.europa.eu/environment/jrec/pdf/pci_summary_brochure_final.pdf
[70] http://europa.eu/rapid/pressReleasesAction.do?reference=IP/06/1329&format=HTML&aged=0&language=EN&guiLanguage=en

projects up to the value of €1 billion. It will lead to the creation of sub-finds that are tailored to developing countries and economies in transition in each region of the world, improving the access to clean, secure and affordable energy.

23.7   Enhancing trade in low-carbon goods and services

The incorporation of environmental benefits within the international trade regime could support some aspects of mitigation.

Co-operation within the international trade regime to account for the environmental benefits of traded goods can influence the extent to which mitigation is possible[71]. In a globalised, interdependent economy, the goods and services for effective mitigation and adaptation for climate change will often cross borders. Over and above the merits of wider liberalisation, there is a clear case for lowering tariffs on these goods. Increased trade allows effective and efficient mitigation or adaptation to climate change, and larger markets for these goods, allowing returns to scale and progression along learning curves and a contribution to global public goods. Reduced tariffs encouraged the adoption of energy efficient lighting in Ghana (see Box 23.2) and could help the development and dissemination of other technologies such as solar thermal technologies[72]. The reduction of subsidies for oil, coal and gas could also remove barriers to clean energy.

As part of the Doha Development round, which began in 2001, Ministers agreed to examine the reduction or, as appropriate, elimination of tariff, and non-tariff barriers to environmental goods and services. It would be important to establish broad principles over which goods should qualify taking into account climate change and other environmental effects. REIL (2006) suggest that in negotiations countries could identify a set of "positive green box" subsidies for clean energy that they would not challenge because of their positive environmental effects.

23.8   Conclusion

Many developing countries are already making efforts that will reduce their greenhouse-gas emissions in the long-term for many reasons, including local co-benefits. However, the challenge of building up and transforming institutions and mechanisms to handle large-scale low-carbon investment flows and to facilitate the diffusion of low-carbon technologies is now urgent. Long-term goals and supportive national policy environments will support the scaling up of these activities.

Actions outlined in each section of this chapter will complement actions taken elsewhere. Encouraging technology transfer and improving the enabling environment for investment will diminish the scale of the challenge for IFIs and carbon markets. Similarly increasing the scale of finance in low-carbon markets will encourage technology transfer and improve the environment for private sector investment. These will also build on the national actions outlined in Part IV of this Review.

Developing countries have a significant opportunity to work with the International Financial Institutions and with regions and countries that are willing to

---

[71] Border tax adjustments are discussed in Chapter 22.
[72] Philibert (2006b).

engage in emissions trading, to create large-scale programmes that will act as pilot schemes for new approaches and provide experience for negotiators to draw on for the future.

References

Acquatella, J. and A. Barcenas. (2005) (eds): Política fiscal y medio ambiente: Bases para una agenda común. Santiago: Comision Economica para America Latina.

Blyth, W. and R. Baron. (2003): 'Green investment schemes: options and issues. OECD/IEA Information Paper for the Annex I Expert Group on the UNFCCC, Paris: OECD.

Branstetter, L. G., R. Fisman, and C.F. Foley (2004): 'Do stronger intellectual property rights increase international technology transfer? Empirical evidence from US firm-level panel data', available from www.papers.ssrn/com

Browne, J., E. Sanhueza and S. Winkleman (2004). 'Getting on track: finding a path for transportation in the CDM', International Institute for Sustainable Development, Manitoba: Canada. Available at: http://www.iisd.org/climate/global/ctp.asp.

Buckley, P.J., J. Clegg, and C. Wang (2006): 'Inward FDI and host country productivity: evidence from China's electronics industry', Transnational Corporations/UNCTAD. vol. 15, available from http://www.unctad.org/en/docs/iteiit20061a2_en.pdf

Center for Clean Air Policy (2006): 'Barriers to increasing clean energy investment and consumption in Latin America and the Carribean'. Paper prepared for the Inter-American Development Bank.

China Statistical Yearbook (2005).

Chinese Academy of Social Sciences (2006): 'Understanding China's energy policy', Report commissioned by the Stern Review, available from
http://www.hm-treasury.gov.uk/media/5FB/FE/Climate_Change_CASS_final_report.pdf

Falvey, R. and N. Foster (2006): 'the role of intellectual property rights in technology transfer and economic growth: theory and evidence', Washington, DC: United Nations Industrial Development Organization available from http://exchange.unido.org/upload/3361_05-91453 e-book.pdf

Fankhauser, S. and L. Lavric (2003): 'The investment climate for climate investment: joint implementation in transition countries'. Climate Policy, 3: 417, 434

Garibaldi, J.A. (2006): 'A programmatic environment for resilient, low carbon economies', Presentation at Joint UK – Japan Research Meeting for Low Carbon Society. http://2050.nies.go.jp/200606workshop/presentations/4-2Garibaldi.pdf

Graham, E. M. and E. Wada (2001): 'Foreign direct investment in China: effects on growth and economic performance', IIE Working Paper Series, WP01-3, Washington, DC: Institute for International Economics, available from
http://www.iie.com/publications/wp/01-3.pdf

Grubb, M. (1999): 'The Kyoto Protocol: a guide and assessment'. London: Earthscan.

International Energy Agency (2005): 'World Energy Outlook 2005', Paris: OECD/IEA.

International Energy Agency (in press): 'World Energy Outlook 2006', Paris: OECD/IEA.

International Energy Agency (2006): 'Energy Technology Perspectives - Scenarios & Strategies to 2050', Paris: OECD/IEA.

Intergovernmental Panel on Climate Change (2000): Methodological and technological issues in technology transfer: a special report of the IPCC Working Group III. Cambridge: Cambridge University Press, available from
http://www.grida.no/climate/ipcc/tectran/

Kathuria (2002): 'Technology transfer for GHG reduction: A framework with application to India'. Technological Forecasting and Social Change, 69, pp. 405–430

Kremer, M. and R. Glennerster (2004): 'Strong Medicine: Creating Incentives for Pharmaceutical Research on Neglected Diseases', Princeton: Princeton University Press.

Lerner, J. (2002): '150 Years of Patent Protection', American Economic Review Papers and Proceedings, 92: 221–225

Less, T.C. and S. McMillan (2005): 'Achieving the successful transfer of environmentally sound technologies: trade-related aspects', OECD Trade and Environment Working Paper No. 2, Paris: OECD, available from https://www.oecd.org/dataoecd/44/20/35837552.pdf

Michaelowa, A. (2005): 'Climate Policy CDM: current status and possibilities for reform'. Paper No. 3, HWWI Research Programme on International Climate Policy.

Michelowa, A & K Michelowa (2005): 'Climate or development: Is ODA diverted from its original purpose?', Hamburg Institute of International Economics (HWWI) research paper no. 2, available from http://www.hm-treasury.gov.uk/media/071/4E/HWWIRP02.pdf

Nicholson, M.W. (2003): Intellectual property rights, internalization and technology transfer', FTC Bureau of Economics Working Paper No. 250, available from www.ftc.gov/be/workpapers/wp250.pdf

OECD (1992): Trade issues in the transfer of clean technologies, Paris: OECD.

OECD (2000): 'Main determinants and impacts of foreign direct investment in China's economy', Working papers on international investment 2000/4 , Paris: OECD, available from http://www.oecd.org/dataoecd/57/23/1922648.pdf

OECD (2005): Innovation in the business sector working paper 459 pages 39–42, available from http://www.olis.oecd.org/olis/2005doc.nsf/43bb6130e5e86e5fc12569fa005d004c/5d1216660b7d83dcc12570ce00322178/$FILE/JT00195405.PDF

Philibert, C. and Podkanski, J. (2005): 'International energy technology collaboration and climate change mitigation.' OECD/IEA Information Paper for the Annex I Expert Group on the UNFCCC, Paris: OECD.

Philibert, C. (in press): 'Barriers to the diffusion of solar thermal technologies'. OECD/IEA Information Paper for the Annex I Expert Group on the UNFCCC, Paris: OECD.

Renewable Energy Policy Network (REN21) (2005): 'Renewables 2005 Global Status Report'. Washington, DC: Worldwatch, available from http://www.ren21.net/globalstatusreport/RE2005_Global_Status_Report.pdf

REN21 Renewable Energy Policy Network (REN21) (2006): 'Renewables Global Status Report: 2006 update', Washington, DC: Worldwatch, available from http://www.ren21.net/globalstatusreport/download/RE_GSR_2006_Update.pdf

REN21 Renewable Energy and International Law Project, (2005): 'Post hearing submission to the International Trade Commission: World Trade Law and Renewable Energy: The case of non-tariff measures', available from http://www.reeep.org/media/downloadable_documents/c/0/REIL%20WTO%20paper%20-%20April%202006.pdf

Renewable Energy and International Law Project, (2006): 'The Clean Development Mechanism: Special Considerations for Renewable Energy Projects', available from http://www.yale.edu/envirocenter/renewableenergy/REIL_CDM_paper.pdf

Saggi, K. (2000): 'Trade, foreign direct investment, and international technology transfer: a survey', Policy Research Working Paper Series 2349, Washington, DC: The World Bank, available from http://www-wds.worldbank.org/servlet/WDSContentServer/WDSP/IB/2000/06/17/000094946_00061706080972/Rendered/PDF/multi_page.pdf

Science and Technology Policy Research (SPRU) and Institute of Development Studies (IDS) (University of Sussex) and The Energy and Resources Institute (TERI) (India), (in press): 'UK-India collaboration to identify the barriers to the transfer of low-carbon energy technology, UK and India Governments'.

United Nations Environment Programme (2001): 'Managing Technological Change: An explanatory summary of the IPCC Working Group III Special Report Methodological and Technological Issues in Technology Transfer, Division of Technology, Industry and Economics', Paris: UNEP, available from http://www.uneptie.org/energy/publications/pdfs/mantechchange_en.pdf

Watson, J. and Liu Xue (2002): 'Cleaner coal technology transfer: obstacles, opportunities and strategies for China' in D Runnalls et al (eds.) Trade and Sustainability: Challenges and Opportunities for China as a WTO Member Winnipeg: IISD, available from http://www.iisd.org/pdf/2002/cciced_trade_sus.pdf

World Bank. (2004): Options for designing a green investment scheme for Bulgaria. Report no. 29998.

World Bank. (2006a): 'An Investment Framework for clean energy and development: A progress report, Background Paper for the Development Committee Meeting, April 5th 2006, Washington, DC: World Bank.

World Bank. (2006b): An Investment Framework for clean energy and development: A progress report, Background Paper for the Development Committee Meeting, September 18th 2006, Washington, DC: World Bank.

24 Promoting Effective International Technology Co-operation

KEY MESSAGES

The private sector is the major driver of innovation and the diffusion of technologies around the world. But governments can help to promote international collaboration to overcome barriers to technology development. Technology co-operation enables the sharing of risks, rewards and progress of technology development and enables co-ordination of priorities.

Mutual recognition of the value contributed by country's investments in new technologies and innovation could usefully be built into international commitments.

International R&D co-operation can take many forms. Coherent, urgent and broadly based action requires international understanding and co-operation, embodied in a range of formal multilateral agreements and informal arrangements. Co-operation can focus on:

- Sharing knowledge and information, including between developed and developing countries

- Co-ordinating R&D priorities in different national programmes

- Pooling risk and reward for major investments in R&D, including demonstration projects

A global portfolio that emerges from individual national R&D priorities and deployment support may not be sufficiently diverse, and is likely to place too little weight on some technologies with global potential, such as biomass. International discussion and co-ordination of priorities for investment in R&D and early stage deployment could play an important role in developing a broadly-based portfolio of cost-effective abatement options.

A small number of technologies, including solar PV, CCS, bio-energy and hydrogen have been identified in international assessments as having significant global potential. **Dedicated international programmes could play a role in accelerating R&D in these areas.**

Both informal and formal co-ordination of deployment support can boost cost reductions by increasing the scale of new markets across borders. Transparency and information sharing have supported informal co-operation on renewable energy. Tradable deployment instruments could increase the effectiveness of support and allow greater co-ordination across borders. There is a strong case for greater international co-ordination of programmes to demonstrate carbon capture and storage technologies, and for international agreement on deployment.

International co-ordination of regulations and product standards can be a powerful way to encourage greater energy efficiency. It can raise their cost effectiveness, strengthen the incentives to innovate, improve transparency, and promote international trade.

24.1   Introduction

Co-operation to accelerate the development and diffusion of low-carbon technologies is likely to reduce the cost of achieving overall emission and stabilisation objectives. The benefits of developing cost-effective low-carbon technologies will be global but most costs will be incurred locally, including a significant proportion by the private sector.

This suggests that a combination of international and public-private co-operation may be required to increase the scale and effectiveness of investment in R&D[1] as outlined in Chapter 16.

An international approach to developing technologies can contribute to building trust and raising the overall ambition of action to tackle climate change. At the 2005 Gleneagles summit G8 leaders recognised the need for greater international co-operation and co-ordination of research and development of energy technologies[2]. At the same time, the Heads of Government of Brazil, Mexico, South Africa, China and India issued a joint statement looking to build a "new paradigm for international co-operation" in the future[3] including improved participation in R&D, international funding for technology transfer, and a concerted effort to address issues related to intellectual property rights (IPR).

Technology also has a vital role to play in adaptation. The development and diffusion of improved crop varieties, more efficient irrigation systems, and cultivation methods will reduce the costs of adapting to climate change in the agricultural sector. Improvements to design, materials and construction techniques can improve the resilience of infrastructure and urban development. Scientific and technological progress that improves the quality of climate predictions and weather forecasts will enable more effective adaptive responses to climate change. Some of these techniques are also relevant to mitigation – leading to lower emissions from rice cultivation[4], reduced energy use for space heating and cooling, for example.

This chapter explores the role of international co-operation on technology. The lessons apply for both adaptation and mitigation. Both formal multilateral action and a variety of arrangements to support co-ordinated or parallel action can play an important part in supporting co-operation. It looks at the role of international technology co-ordination (Section 24.2) and the models for R&D co-operation (24.3) and co-ordination of deployment programmes (24.4). In Section 24.5 it considers opportunities for greater international public-private co-operation at the commercialisation stage. Finally (24.6) it considers the role of global or regional co-ordination on regulation and standards.

---

[1]   Research and Development: In this chapter the term R&D will also cover the demonstration stage – Research, Development and Demonstration (R,D&D can be used for this but this can lead to confusion over the final D since some people use deployment or diffusion)
[2]   Gleneagles Plan of Action – Climate Change, Clean Energy and Sustainable Development http://www.fco.gov.uk/Files/kfile/PostG8_Gleneagles_CCChangePlanofAction.pdf
[3]   Joint Declaration of the Heads of State and/or Government of Brazil, China, India, Mexico and South Africa participating in the G8 Gleneagles Summit http://www.indianembassy.org/press_release/2005/July/5.htm
[4]   International Rice Research Institute (2006)

24.2   The role of international technology co-operation

The bulk of new technology development and commercialisation takes place within the private sector, which also spreads new technologies rapidly between countries.

In several cases developing countries have been able to "leapfrog" to advanced technologies – by installing mobile phone networks without ever developing systems of landlines, or in some cases by designing cities from the outset with mass rapid transit systems in mind. This may not be possible in some technologies where tacit knowledge[5] is important but, may occur in sectors where rigid infrastructure is yet to be built, such as building efficiency and combined heat and power.

Multinational companies use research bases around the world. Microsoft's research is strengthened by its operations in China[6] and India[7] to take advantage of local expertise. General Motors is collaborating with Shanghai Automotive to develop fuel cell cars on a commercial scale[8]. BP has begun a new programme of research on biofuels in India[9]. Co-operation between developing countries is also taking place through the private sector – including initiatives by Brazilian companies to market their biofuels technologies in Southern Africa. China has a number of highly competitive businesses exporting solar water heaters to other developing countries.

However, governments do have a role to play in sectors where the market under-provides new technologies. As outlined in Part IV, this requires governments to ensure that the private sector invests in developing and deploying low-emission technologies by creating a value for greenhouse gas emissions through pricing the externality. Additionally, in some climate sectors relevant to climate change, governments provide a significant proportion of R&D funds, and create markets through policy frameworks for deployment support. The central questions here are how to ensure that the combined international effort is sufficient relative to the scale and urgency required, and what types of co-operation and co-ordination are most useful.

Multilateral frameworks and joint funding arrangements have already supported technology development in other areas, and will be increasingly important for climate change technologies.

Formal co-operation on technology has supported advances ranging from basic science to space exploration and the launch of commercial satellite systems. There has been a growing debate over the importance of formal international agreements on technology co-operation as part of efforts to tackle climate change.

Carraro and Buchner[10] have suggested that technology could form an easier basis for international co-operation than carbon pricing, though ultimately a less effective one. Technology has some characteristics of a "club good" rather than a

---

[5] Much of the knowledge embodied in a technology is 'know-how' or 'gardeners craft' that is not codified

[6] http://research.microsoft.com/aboutmsr/labs/asia/default.aspx

[7] http://research.microsoft.com/aboutmsr/labs/india/default.aspx

[8] http://www.gm.com/company/gmability/adv_tech/400_fcv/fc_milestones.html

[9] http://www.bp.com/genericarticle.do?categoryId=2012968&contentId=7014607

[10] Carraro and Buchner (2004)

pure public good, in that despite the spillovers, some of the benefits of co-operation on innovation can be limited, for a time, to participants[11]. Benedick[12] has highlighted the importance of industry and government co-operation in identifying alternatives to the use of ozone depleting substances, and in developing appropriate timetables and safety valves for phasing out the polluting chemicals.

Barrett[13] examines the scope for international treaties focused on technology and R&D. In a recent paper he concluded that these are subject to the same underlying challenges for international collective action as those described in Chapter 21. But, he identified specific cases where formal international technology co-operation is important: where R&D can lead to breakthrough technologies that exhibit increasing returns to scale and where R&D co-operation might sustain a strong international response. Examples of technologies where formal co-operation may offer significant benefits include improved solar technologies, and the development of the infrastructure and networks required to support the use of hydrogen.

Informal arrangements can also play a valuable role in supporting co-ordinated or parallel action.

Co-operation on technology goes far beyond formal multilateral arrangements. Links between universities and research networks help to ensure that breakthroughs in basic research are widely available. Partnerships play a key role in bringing together smaller groups of public and private bodies to take a lead in developing particular technologies.

Recent IEA work[14] on the effectiveness of IEA and other technology partnerships highlights two key lessons. First, the involvement of a range of stakeholders, including the business community, is essential to the success of technology partnerships. Second, developing country participation is important, and not only from the point of view of building capacity and know-how. Increasingly, the wealth of scientific and technical expertise in developing countries means they have important contributions to make in their own right. A good example of this can be found in the case of biofuels (see Box 24.1), and in solar thermal technology.

International monitoring of R&D and deployment support should encourage greater recognition of national efforts to introduce relevant technologies as part of formal multilateral frameworks or informal arrangements for co-operation.

Data gathering and modelling by numerous institutions, particularly the IEA, enables policy makers to track technological progress. This can help to identify whether sufficient progress is being made and what further spending may achieve. It can also allow policy makers to check the balance of any support to ensure it is broadly proportionate to each technology's potential and stage of development.

National investment in technology is not currently recognised as a contribution to the objectives of the UNFCCC. Incorporating technology development

[11] Also in Neuhoff and Sellers (2006)
[12] Benedick (2001)
[13] Barrett (2006)
[14] IEA (2005a)

> **BOX 24.1 The Brazil-UK-Southern Africa biofuels taskforce**
>
> The aim of the project is to increase the production of biofuels[15] in Southern African countries using Brazilian technology. Brazil is the world's leading producer of biofuels (and the flexi-automobile engines which can use it) and a number of Southern African countries have the technical potential to produce sugar cane for local bioethanol production. There are considerable potential markets for bioethanol in Africa and globally.
>
> A technical feasibility study on the potential for bioethanol production in Southern Africa has now been completed and a taskforce of interested countries to undertake more detailed feasibility studies is being set up. The initial group of countries identified, in addition to Brazil, was South Africa, Mozambique, Zambia and Tanzania.
>
> This project has the potential to contribute to multiple aims in Southern Africa - rural development, added value to agricultural production, energy security, emissions reduction; and to enable South-South technology transfer from Brazil to Southern Africa. The objective is to more than double sugar cane production from around 0.7 to 1.5m hectares by 2020.

into the measurement of national commitments under the UNFCCC would have the advantage of recognising those countries that make a disproportionately large contribution towards developing new technologies. It is not possible to translate the impact of investing in innovation into resultant emission reductions so it is not possible to directly "trade-off" between the two. Thus international recognition of investment in innovation should be considered as part of a broader range of metrics over different dimensions of effort.

24.3   Models for R&D co-operation

International arrangements to support technologies for mitigation and adaptation could focus on the further development of a number of different types of co-operation, including

- Sharing knowledge and information
- Co-ordinating R&D priorities in different national programmes
- Pooling risk and reward for major investments in R&D, including demonstration projects

Sharing knowledge and information

Various arrangements can help to promote the positive spillovers of knowledge between technology programmes in order to speed the pace of innovation.

The IEA's Energy Technology Collaboration Programme includes more than 40 international collaborative energy research, development and demonstration projects known as Implementing Agreements. These enable experts from different countries to work together and share results, which are usually published for a wider audience.

[15] For more on biofuels see Boxes 9.5, 12.2 and 16.4 and Sections 12.6 and 16.3

Sharing information with developing countries who have not been strongly involved with these networks is important. It supports the development and transfer of technology as discussed in Section 23.4. The IEA has recently launched a further initiative on Networks of Expertise in Energy Technology (NEET) to encourage further co-operation with non-member countries. Conceived in response to a call from G8 leaders for more productive international partnerships for energy technology information exchange, IEA's NEET Initiative is set to play a catalytic role in promoting worldwide technology collaboration. It is linking existing energy R&D networks, bringing together policy-makers and stakeholders from the financial, business, research and other key sectors, in both IEA countries and major energy-consuming non-IEA emerging economies.

The challenge is not just creating new knowledge but ensuring that this knowledge is disseminated so it can be used effectively no matter where it originates from. This stimulates competition and reduces unnecessary duplication and ensures that other research efforts in both the public and private sector can benefit from the progress that is made.

Identification and co-ordination of research priorities

Competition plays an important role in driving innovation (see Section 16.1) but, international discussion, and to some extent co-ordination, can help to ensure R&D is directed towards the technologies that can make a significant global contribution to reducing greenhouse gas emissions. This is already happening to some extent, for example with hydrogen (see Box 24.2) and carbon capture and storage. However, as discussed below, there is scope to go further.

BOX 24.2 Partnerships can contribute to sharing knowledge and information

The International Partnership for the Hydrogen Economy[16], launched by the US in 2003 is an international institution dedicated to accelerating the transition to the hydrogen economy.

The IPHE provides a mechanism for partners to organize, co-ordinate and implement effective, efficient, and focused international research, development, demonstration and commercial utilization activities related to hydrogen and fuel cell technologies. The IPHE provides a forum for advancing policies, and common technical codes and standards that can accelerate the cost-effective transition to a hydrogen economy. It also educates and informs stakeholders and the general public on the benefits of, and challenges to, establishing the hydrogen economy.

It does not provide direct funding for research. However, it secures increased awareness and recognition of significant international collaborative research, development and demonstration projects. The strength of the IPHE is that it is a top-level political initiative – launched by Ministers – with high level official representation on its Steering Committee.

A global portfolio that emerges from individual national R&D priorities is likely to be unbalanced in respect to the global potential for different technologies.

As outlined in Chapter 16 the uncertainty and risks inherent in developing low-emission technologies are suited to a portfolio approach. National R&D policy

[16] http://www.iphe.net/

focuses on technologies where there is a compelling local need or a perceived first-mover advantage, in order to capture national benefits linked to lower cost energy, local health or agricultural priorities, and the development of new industries. The competitive and entrepreneurial energies motivated by seeking first-mover advantage have a powerful effect in spurring innovation. Nevertheless there are also disadvantages that policy can help overcome.

- Where a first-mover advantage exists, it is more likely to relate to products with significant economies of scale, production processes that are complicated and difficult to imitate, and strong export potential, including low transport costs.
- A policy focused only on first-mover advantage encourages countries to seek to reduce spillovers that would be beneficial to other countries, in the interest of their national industries.
- It can encourage a policy bias for local production rather than co-operation in developing manufacturing bases in other countries or using imported technology.
- It can bias the choice of technologies. Developed countries focusing on the technologies where they have comparative advantage, or where there are developed country applications, may fail to provide the technologies required for cost-effective reductions in the developing world, for example biomass and solar power.
- A fragmented approach is unlikely to create a sufficient market size to realise the learning potential of any technologies.

There is a wide range of models for international co-operation of research priorities in energy and transport technologies.

Extensive modelling work has provided an increasingly clear picture of the technologies that are likely to form part of the future energy portfolio[17]. This modelling often incorporates the cost uncertainty of future technologies and reflects this in the range of outcomes it delivers. There are further promising analytical tools being developed to aid understanding of a suitable global portfolio such as real options pricing[18]. Despite the inevitable uncertainty that surrounds such work, it provides a useful tool for policymakers to evaluate existing and planned policies and should be encouraged.

The G8 and OECD have both made efforts to identify international priorities for technology development. At the Evian Summit, G8 leaders issued an Action Plan on Science and Technology for Sustainable Development[19]. The Energy Research and Innovation Workshops hosted by the UK and Brazil fulfilled one of these commitments - delegates of energy policy and research experts from the G8 countries, Brazil, China, India, Mexico and South Africa have begun to meet annually to discuss how to facilitate co-operation in technology development amongst developed and developing countries[20]. It also led to the launch of international partnerships on specific technologies, including bio-energy (see Box 24.3), hydrogen and carbon

---

[17] For example IEA (2006)
[18] Pindyck and Dixit (1994)
[19] http://www.g8.fr/evian/english/navigation/2003_g8_summit/summit_documents/science_and_technology_for_sustainable_development_-_a_g8_action_plan.html
[20] The Energy Research and Innovation workshop held in Oxford 2005 and followed up in Brazil in September 2006 http://www.ukerc.ac.uk/content/view/75/67/

BOX 24.3 The launch of a Global Bio-energy Partnership responded to developing country priorities

The Global Bio-Energy Partnership[21], launched by Italy in May 2006 following the G8 meeting the previous year, focuses on the potential for the greater use of bio-energy. The involvement of developing countries is particularly important.

Biomass is widely used in developing countries as a source of domestic heat. Traditional biomass is a major source of indoor air pollution causing ill health and mortality (see Box 12.2). Biomass technologies could have a significant impact at the village and household level. Biomass also has the potential to form a significant part of mitigation in the power generation and transport sectors leading to export opportunities

The partnership will increase and facilitate an exchange of experiences and technologies not only North-South, but also South-South and South-North. The short and mid-term goals include the review of the current stakeholders network, knowledge and gaps in the understanding about bio-energy as well as the formulation of standard guidelines to measure the greenhouse gas emission reductions through the use of bio-fuels.

---

sequestration. This work could provide a platform for a more significant effort to accelerate these technologies.

The OECD Roundtable on Sustainable Development brought together scientists, heads of research councils and policymakers to undertake a full assessment of the current portfolio of research in energy technologies. The report, discussed by science and energy ministers from OECD and developing countries in June 2006, concluded[22] that the current portfolio is too small. It recommended that more attention should be given to funding research in:

- solar
- battery technologies
- carbon capture and storage

These technologies offer the prospect of substantial emissions savings because they have the potential to provide for a significant proportion of the market and all currently have limited public support.

These international assessments build on and complement existing national processes to allocate research funding and offer a model for further efforts at co-ordination of energy and transport priorities. A successful international model of R&D co-operation can be found in the case of the Consultative Group on International Agricultural Research (CGIAR) (see Box 24.4).

BOX 24.4 Lessons on R&D co-operation from CGIAR

A strong precedent exists for international collaboration on research and development in agriculture.

In the 1950s and 1960s a major concern was how to increase food supply given that the scope for increasing agricultural land area was becoming limited and the world's population was set to double by the end of the century. A major and successful effort was made to improve yields of agriculture research and extension, by bolstering both national research

---

[21] EC (2005a)
[22] OECD (2006) and Chairman's summary: http://www.oecd.org/dataoecd/38/59/37041713.pdf

stations facilitated by a network of international research centres, later brought together under the aegis of the CGIAR under the chairmanship of the World Bank.

The CGIAR was created in 1971; it now has more than 8,500 CGIAR scientists and staff working in over 100 countries. It draws together the work of national, international and regional organisations, the private sector and 15 international agricultural centres to mobilise agricultural science, promote agricultural growth, reduce poverty and protect the environment. It has an impressive record and can be expected to play a strong role in enabling the agricultural sector to adapt to the impacts of climate change through research on new crop varieties and farming methods. There is a good case for expanding this role to support mitigation and adaptation from the agricultural sector[23].

Several lessons from the experience of agriculture are relevant for an international programme in the development and use of low carbon technologies and practices. In the case of agriculture:

- There was a shared commitment among the sponsors;
- The programme evolved from an already extensive network of national research centres and supplemented and enhanced national efforts;
- It was based on real demonstration and R&D projects, and was not simply a 'talking shop';
- The efforts were not centred on one institution in one country, but divided across a set of institutions in several countries specializing on particular crops (rice, wheat, maize, agro-forestry and so forth) and livestock farming;
- There were good working links between the international and national centres of R&D;
- There were also good working links between the programme and the users (extension services and farmers), so that technology and knowledge could be rapidly diffused to those who would use it.

Pooling risk and reward for major investments in R&D

Co-operation can go beyond sharing information and co-ordinating of national priorities to include formal arrangements to spread the risk and cost of investing in new technologies.

The scale of some low-carbon technologies is too large for one country to take on alone. The classic example of this is nuclear fusion, where the benefits of a successful programme could be very large, but the technical challenges and scale of investment required are daunting.

The ITER[24] project to demonstrate the scientific and technical feasibility of nuclear fusion power is supported by European Union, Japan, China, India, the Republic of Korea, the Russian Federation and the USA each of which has committed to financing the projects $10 billion cost. The costs are shared amongst the participants: Europe will contribute 45.45%, and China, Japan, India, Korea, Russian Federation and the USA will contribute 9.09% each.

It can prove difficult to negotiate one-off projects where key questions of national interest arise. The start of the ITER project was delayed for several years as a result

---

[23] For more see Section 16.3 and Box 26.3
[24] http://www.iter.org/

of disagreements on its location. Where these problems can be overcome, however, the rewards can be appreciable. Discussions on a series of linked demonstration projects or for a number of different technologies could increase the opportunities to share the benefits of co-operation amongst the participants.

Traditionally OECD nations have been the primary focus of innovative investments. Arrangements that involve scientists and engineers from developing countries in the tasks of R&D in low carbon energy technologies and practices would have considerable economic merit. Already China and India are each graduating 250,000 engineers and scientists every year, as many as in the US and in the European Union combined. It is clear that a rich source of innovation is emerging and the traditional North to South view of technological progress is becoming outdated.

Dedicated international programmes could play a role in setting research priorities and sharing the costs of accelerating key technologies.

The number of technologies that have been proven viable and could potentially meet a large proportion of future energy needs, including those identified as part of the OECD assessment described above, is relatively small. An estimate of the learning cost of reducing the price of just one of these, solar PV, to the point of market competitiveness is €20 billion[25]. Costs of this scale provide a rationale for international co-operation (see Box 24.5 for an example of the costs and benefits of an ambitious international programme).

BOX 24.5 Illustrative estimate of the scale of costs and benefits of an international programme of R&D in clean energy[26]

The increases in R&D and deployment support outlined in Section 16.8[27] would probably be achieved mostly through national frameworks for supporting innovation. However, an international programme in fundamental R&D, support for demonstration projects and early stage deployment support could make a significant contribution to the global effort.

For example, a 20 year international programme to develop low carbon technology on a significant scale aggregating to perhaps 1-2 GW of electricity production per year, would require investment in the region of $6-10 billion per year. This would target technologies with significant potential for reducing greenhouse gas emissions where the nature of the costs and benefits of developing the technology benefit from action at an international scale. Around 50% of this cost could be leveraged through private investment, international offset programmes such as the CDM, and sales of the actual energy produced. Higher leverage rates would be achievable as the programme progressed and as conversion efficiencies and confidence in the industry improved. A key feature of such a programme could be involving scientists and engineers from developing regions which would deliver significant benefits. Such a programme could be built on existing international institutions or through collaboration between national programmes, and be perceived as part of international outreach and co-operation from developed countries.

---

[25] This is heavily dependent on the assumed learning rate. Source: Neuhoff (2005)

[26] Source: Dennis Anderson - Estimates from analysis undertaken as part of this review available at www.sternreview.org.uk

[27] Increase in public energy R&D of $10 billion and of deployment support of between $33 billion and $132 billion.

The positive externalities of such a programme would be substantial. The incremental costs of present programmes of investments in low carbon technologies (the cost beyond market dominant alternatives) in OECD countries amount to around $85 per tonne of CO_2 abated. But costs are declining and may become as low as $45 per tonne of CO_2 abated in 20 years time and $25 per tonne or less by 2050. Together the national and international programmes of R&D, plus the incentives provided by the more familiar instruments for encouraging innovation, are fundamental for such reductions to be achieved, and could reap worldwide benefits (as measured by consumers' plus producers' surpluses) of over $80 billion per year per gigatonne of carbon abatement.

There are other examples of countries pooling significant funds for R&D and investment in innovative new technologies, including the EU's R&D framework programme and the arrangements for public-private co-operation that have underpinned the Galileo satellite navigation system[28]. The European Commission is proposing that the model for European collaboration used in the Galileo project should now be rolled out as a new Community Instrument - the Joint Technology Initiative.

These initiatives, mainly resulting from the work of European Technology Platforms and covering one or a small number of selected aspects of research in their field, will combine private sector investment and national and European public funding, including grant funding from the Research Framework Programme and loan finance from the European Investment Bank. There is currently a proposal for a Joint Technology Initiative for hydrogen and fuel cells.

BOX 24.6 EU 7TH Framework Programme for R&D

Funding research and development at the EU level reduces the problem of spillovers and allows smaller countries to contribute to a large and diverse research portfolio. The EU has an R&D framework as part of the EU budget which will enter into its 7th programme, lasting 6 years and beginning in 2007, with a total fund of €48 billion (6% of the total EU budget). Of this, €5 billion will be spent on energy and environment issues.

EU research priorities are aligned using European Technology Platforms. These provide a framework for stakeholders, led by industry, to define research and development priorities, timeframes and action plans on a number of strategically important issues where achieving Europe's future growth, competitiveness and sustainability objectives are dependent upon major research and technological advances in the medium to long term.

Previous frameworks have invested in climate change research on:

- The science of climate change such as the impact on coastal zones[29] and adaptation[30];
- Technology development including wind turbines[31] and fuel cells[32].

[28] http://www.euractiv.com/en/science/galileo/article-117496
[29] http://ec.europa.eu/research/success/en/env/0069e.html
[30] http://ec.europa.eu/research/success/en/env/0336e.html
[31] http://ec.europa.eu/research/success/en/ene/0059e.html
[32] http://ec.europa.eu/research/success/en/ene/0265e.html

The 7th Programme's energy and environmental themes ensure that there is likely to be a greater emphasis on climate change research in the next programme and an intention to involve developing countries. The scale of investment required and the urgency suggests that this should be the case and the forthcoming fundamental review of the EU budget, which is to report in 2008/09, should consider the appropriate level of longer-term EU support in this area. Rebalancing within the EU budget, when combined with national and other international funding, could make a significant contribution to the increases set out in Section 16.8[33].

There is a strong case for greater international co-operation between national programmes to develop and demonstrate carbon capture and storage technologies.

Carbon Capture and Storage[34] (CCS) is a process that is yet to be deployed at full commercial scale in the power sector, so it remains at the demonstration stage of the innovation process. The IPCC special report on CCS suggested it would provide between 15% and 55% of the cumulative mitigation effort up to 2100. Failure to develop CCS would result in a narrower portfolio of low-carbon technologies and this would, on average, increase abatement costs. Recent IEA modelling shows that, without CCS, less abatement occurs at a higher cost as marginal abatement costs would increase by around 50%[35]. Modelling work undertaken for the Global Energy Technology Strategy programme showed that removing the option of CCS more than triples the cost of stabilisation for all concentration levels analysed.[36] This prominent role in future mitigation can be linked to the expected global growth of coal use.

The IPCC recently completed a special report[37] on the potential of CCS, providing an important assessment on key issues including the likely availability of geological storage sites. The Carbon Sequestration Leadership Forum[38] acts as a focal point for participating governments and industry to share updated information on national programmes and opportunities. A number of projects are under development, but so far, national governments have found it difficult to set up policy frameworks to cover the additional costs required for a full demonstration project.

A single CCS demonstration project costs several hundred million dollars over and above the cost of a standard power station. The IEA recommend that 10-15 such projects should be in place by 2015 at an estimated extra cost of $2.5 to $7.5 billion in order to demonstrate the commercial viability of the technology[39]. This is a dramatic increase on the $100 million that is currently spent on CCS R&D[40]. The 'lumpy' nature of CCS investments implies that it may be better for a limited number of countries to demonstrate CCS, but there are currently no arrangements for co-ordinating these efforts.

---

[33] Doubling of global public energy R&D from $10n billion to $20 billion.
[34] For more on CCS see Boxes 9.2 and 24.8.
[35] IEA (2006)
[36] GTSP(2005)
[37] IPCC(2005)
[38] www.cslf.org
[39] IEA (2006)
[40] Page 38, OECD (2006)

There have been several announcements from governments and the private sector on planned CCS projects. These include:

- the US Futuregen project[41] which is linked to the demonstration of IGCC coal generation technology
- BP's proposed project at Peterhead[42] which includes a 350MW hydrogen plant capturing 1.2 million tonnes of carbon each year; and RWE's feasibility study for a post-combustion techniques in a 1000MW coal plant in Tilbury; UK[43]
- A Japanese proposal to capture a sixth of all their emissions by 2020.
- Vattenfall's plan to build a 30 MW pilot coal plant in Germany. Construction has started and the plant will be in operation by mid 2008[44].
- A geological storage pilot project in the Otway Basin in Western Victoria[45] planned by a public-private research organisation in Australia. An LNG project[46]

BOX 24.7 Near-Zero Emissions Coal initiative in China

The EU agreement to develop a near-zero emissions coal plant in China is expected to lead to the construction of the first CCS project sited in a non-OECD country. This should create significant opportunities for learning. Undertaking this project as a joint venture encourages shared understanding of deploying CCS technology and reflects shared concerns over climate and energy security and the use of coal for power generation.

The Near-Zero Emissions Coal initiative was announced as part of the EU-China Partnership on Climate Change at the EU-China Summit in September 2005. It stated that the EU and China will aim "to develop and demonstrate in China and the EU advanced, near-zero emissions coal technology through carbon capture and storage" by 2020.

A Memorandum of Understanding (MoU) was signed between the UK and China on December 19th to detail specific UK funded action (Phase 1 Assessment). A complementary MoU was signed between China and the European Commission on February 20th 2006. This ambitious initiative will take place through a phased approach over several years allowing for the development, funding and implementation of the demonstration project:

Phase 1	Identifying early demonstration Opportunities	2006-2008
Phase 2	Define, Plan and Design a Demonstration Project	2009-2010
Phase 3	Construct and Operate a Demonstration Project	2011-2014+

The assessment of early opportunities for CCS demonstration under Phase 1 will begin in November 2006 with funding from the UK and the EU. The forecast investment of coal power stations in China provides a strong rationale for accelerating such a valuable project to create the option of more widespread deployment. Consideration should also be given to the case for demonstration projects in other developing countries with significant coal resources.

---

[41] http://www.fossil.energy.gov/programs/powersystems/futuregen/
[42] http://www.bpalternativenergy.com/liveassets/bp_internet/alternativenergy/next_generation_hydrogen_peterhead.html
[43] http://www.npowermediacentre.com/content/detail.asp?ReleaseID=676&NewsAreaID=2
[44] http://www.vattenfall.com/www/vf_com/vf_com/365787ourxc/366203opera/366779resea/366811co2-f/index.jsp
[45] http://www.co2crc.com.au/pilot/OBPP.html
[46] http://www.greenhouse.gov.au/challenge/members/chevron.html

Gorgon (North West Shelf), and the Stanwell ZeroGen IGCC-CCS project[47] are at the proposal stage.

● The EU has an initiative seeking to develop a CCS plant in China (see Box 24.7).

Building on these announcements, the enhanced co-ordination of national efforts could allow governments to allocate support to the demonstration of a range of different projects, and demonstration of different pre and post combustion carbon capture techniques from different generation plants[48], since the appropriate technology may vary according to local circumstances and fuel prices (see Box 24.8). One element that enhanced co-ordination could focus on is understanding the best way to make new plants "capture-ready", by building them in such a way that retrofitting CCS equipment is possible at a later date.

Governments should also develop legal, regulatory and policy frameworks to encourage deployment after demonstration. During the demonstration stage governments should simultaneously develop a regulation and policy framework, including the liability for any leaked CO_2 and reducing the probability of such an occurrence. Integrating this into policies such as emissions trading schemes and programmes to encourage renewables could have an important impact on deployment.

24.4   Co-ordinating deployment support

Chapter 16 estimated that the current level of deployment support should increase by 2 to 5 times to help deliver an appropriate portfolio of technologies. Understanding that others are taking significant measures to support technologies can encourage countries to increase their effort. Countries can also benefit from discussing effective policies and how to foster an appropriate portfolio of technologies, moving towards a common understanding of what this means. Most OECD and larger developing countries already have some sort of deployment support for low-carbon technologies, but they need to be increased to sufficient scale and ensure that potentially cost effective technologies are not ignored. International co-operation can complement national support strategies in enhancing investors' confidence for future markets, and thus encouraging innovative investments[49].

It is possible to conceive innovative policy structures to ensure that these goals are delivered. If the cost of developing technologies were not uncertain it would be possible to spread these globally in an equitable fashion. Given the inherent uncertainty, policymakers could agree a target level of deployment support and technology priorities and measure the contribution through the leaning cost incurred in each country (the cost beyond that of existing technologies within each country). However data on such costs may be hard to produce credibly and counterfactuals are unclear.

Informal sharing of experiences and, in some regions, co-ordination of deployment support appears to have provided an important boost to the use of renewable energy around the world.

Support for renewable energy sources is common throughout the OECD and in some non-OECD countries such as India and China. The structure and ambition

---

[47] http://www.zerogen.com.au/project/overview
[48] Integrated Gasification Combined Cycle and more traditional Pulverised Coal plants and dominant gas generators – Combined Cycle Gas Turbine generators.
[49] Neuhoff and Sellers (2006)

of this support varies greatly across countries and often within countries. There are now 41 states, provinces or countries with feed-in-policies (price support) and 38 with renewable portfolio standards (quantity targets) including many outside the OECD[50]. In addition, a number of countries use tax incentives to encourage the deployment of renewables. China applies a much lower rate of VAT to renewable energy technologies, and Mexico offers tax relief on clean energy R&D.

There is no formal co-ordination but the Bonn and Beijing Renewables Conferences and the REN21 network[51] have provided a powerful mechanism to gather and share information on different national approaches and to raise awareness of the scale of national efforts amongst policymakers and industry.

It is possible to make comparisons of the level of deployment support in different countries. This is easier for price support, mechanisms as the price is clearly evident. While recognising that other ancillary benefits may justify support it is possible to calculate the implicit carbon price for different policies. The price required to support a technology indicates the current cost of the technology and the degree to which it is a viable technology or a learning investment for the future. It is possible to calculate the cost of price support for new technologies in terms of carbon abatement (see Table 24.1). It is harder to estimate costs from quantity based targets, such as the renewable portfolio standards used in the US, as the price is bound up in the overall electricity price. However, it is possible to make an estimate using deployment figures and cost estimates. This allows comparison of the scale of effort (in terms of learning investments) in different countries.

Table 24.1 Implicit cost of carbon in existing deployment support[52]

Country	Application	Imputed carbon price, $ per tonne CO_2
Germany	Onshore wind	73
	Offshore wind	146
	Solar	1048
	Electricity from biomass	146
Austria	Wind	122
	Electricity from biomass	171
Spain	Wind	73
	Solar	804

More formal co-ordination of deployment support could include the use of internationally tradable policy instruments.

Currently, deployment policies such as renewables support mechanisms are implemented at the state or national level. However, learning depends on the overall global deployment, not where it takes place. The ability to trade obligations across borders would improve efficiency by ensuring that deployment takes place where it is cheapest to do so. The benefits from this may be significant where there are major

---

[50] Source: REN 21 (2006)
[51] REN 21 is the *Renewable Energy Policy Network for the 21st Century*. It is a global policy network that provides a forum for international leadership on renewable energy. More details of the conference are available at: http://www.renewables2004.de/en/2004/default.asp
[52] Source: Dennis Anderson paper available at www.sternreview.org

differences between countries in, for instance, the availability of a natural resource such as sunshine, or in lower labour or other costs. Such harmonisation has yet to be attempted. Even the 22 states in the US with renewable portfolio standards cannot co-operate across state boundaries to help reduce costs. An IEA study[53] identified that deployment of some technologies within non-OECD countries could prove much more cost-effective, particularly in the case of solar technologies. Where this is the case, countries could consider including financial support for deployment in developing countries towards national deployment targets.

Harmonising existing instruments may be very challenging in practice. Within the EU, for instance, countries use a mix of quantity instruments, similar to US state renewable portfolio standards, and price instruments, such as the German feed-in tariffs (see Box 16.7). However, the scope for cross-border links should certainly be considered when developing new policy. This could help improve the value-for-money of deployment support. The likely widespread introduction of deployment policies for CCS technology over the next 5-10 years offers an opportunity to look seriously at how these could be designed to take advantage of possible efficiency gains from international trading (see Box 24.8 below).

BOX 24.8 Options for supporting the deployment of carbon capture and storage

Carbon capture and storage technologies[54] have the significant advantage that their large-scale deployment could reconcile the continued use of fossil fuels over the medium to long term with the need for deep cuts in emissions. In the IEA's base-case, energy production doubles by 2050 with fossil fuels accounting for 85% of energy[55]. Coal use is forecast to grow in OECD countries, Russia, India, and China. The IEA forecast that without action a third of energy emissions will come from coal in 2030. Successfully stabilising emissions without CCS technology would require dramatic growth in other low-carbon technologies. The role CCS plays in avoiding these emissions will depend on the policy options that are chosen to support its deployment.

CCS is dependent on government intervention. Unlike other alternative generating technologies, CCS will always be more expensive than traditional fossil fuel[56] based alternatives, as it will always be cheaper to emit the CO_2 than to capture and store it. This is very similar to the problem of fitting flue gas desulphurisation equipment to tackle acid rain. This equipment is now widely used in OECD and developing countries, because it is recognised that the cost of using the technology is less than the cost of the externalities associated with sulphur dioxide emissions.

The economic viability of using CCS technology for power companies will reflect both the relative price of coal and natural gas and the level of the carbon price. Should the carbon price reach a sufficient level, with a credible expectation that it will remain there, widespread deployment of CCS can be expected. The choice of technology will also depend on the price of different fossil fuels, so if gas prices are high then coal will be chosen as shown in the figure below.

---

[53] IEA (2005b)

[54] For more see Box 9.2 and Section 24.3.

[55] IEA, 2006 – ACT MAP is a scenario in which includes CCS and where emissions are constrained to near current levels in 2050 following a technology 'push' for low-carbon technologies.

[56] Except perhaps under an extreme enhanced oil recovery scenario.

Impact of carbon and energy prices on CCS deployment[57]

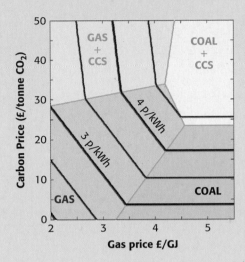

Alternatively, international agreement could focus on a regulatory approach to deployment. At the simplest level this would involve a commitment by participating countries to regulate that all new coal or fossil fuel electricity generation be fitted with CCS from a certain date. An example of this sort of regulation is the EU's Large Combustion Plant Directive, that places emission limit values on large plants with increasing stringency over time. It specifies different treatment depending on the age of the plant. It will ensure that by 2015 all European power stations conform to a common standard for air pollution emissions. For CCS, an agreement could set out a timetable for new plant to be capture-ready or to be fitted with CCS, and could establish differentiated responsibilities by giving more time or applying to a lower proportion of new plant in developing countries. The timing could be significant as mitigation costs will increase if significant investments are made in new capacity without, or precluding the addition of, carbon capture and storage technologies.

Renewable portfolio standards offer an alternative model for national or internationally co-ordinated policy instruments for the deployment of CCS. A CCS portfolio standard could require that a certain proportion of power supplied by generation companies is from plants fitted with CCS technologies[58]. This could begin with a very low proportion (e.g. 0.5%), consistent with the establishment by one or two operators in a market of demonstration plants. Other operators would share the risk of these projects through long-term contracts to purchase power from these plants to meet the CCS portfolio standard, and would pass the incremental cost through to all electricity consumers. Governments could set out a timetable for a strong increase in the level of the portfolio standard provided that the demonstration projects showed that key criteria could be met. This policy approach could include a tradable element to pool efforts across larger markets, minimise costs across regions or maintain differentiated responsibilities between countries at different stages of development.

---

[57] *Source*: Gibbins et al (2006) Coal price £1.4GJ 25 year plant life and a 10% Investment Rate of Return.
[58] As suggested Jaccard (2006)

24.5 The use of international public-private co-operation to support commercialisation

Finding niche markets where new technologies can benefit from market learning and building these into large-scale commercialisation opportunities is a key challenge for companies with promising low carbon technologies.

The private sector often succeeds in commercialising technologies, where the incentives are right, without intervention.

Partnerships between industry and academia can support the commercialisation of new research from universities, including across borders. The SETsquared Partnership[59] is a collaboration between four UK universities and two US universities to develop further their joint works, encouraging collaborative applied research and complimentary commercial ventures[60]. Together, the universities of the SETsquared Partnership represent the largest single source in the UK for academic knowledge transfer to the private sector as discussed in Section 16.5. This has led to the creation of many companies, for example, in marine energy. In the last 21/2 years, three SETsquared Partnership companies have achieved IPOs, with a total market capitalisation of £150 million.

Governments also play a role in supporting commercialisation, and could explore ways to extend this support across borders.

International co-operation between organisations such as the Carbon Trust could increase the access to international markets for technology developers. It is possible that a network of public-private investors could facilitate the creation of technology focused "commercialisation consortia", bringing together business participants and working to identify and overcome business, market and policy barriers to deployment.

Formal multilateral co-operation can also help in phasing out the use of emissions intensive products or processes for which a viable alternative exists, or in co-ordinating the introduction of infrastructure networks that are required to allow the adoption of a new low emissions technology.

There is a historical precedent for this approach with the Technology and Economic Assessment Panels (TEAPs) that were established to deliver reductions in CFC emissions following the Montreal Protocol. These played an important role in ensuring the roll-out of alternative technologies. This approach had the advantage of bringing government and business together to establish the technical feasibility of timetables for regulation. It built in some flexibility, with developing countries given more time to make the technological transition.

The scale and diverse range of sources of greenhouse gas emissions limits the applicability of the TEAP model in the case of the main greenhouse gases. It may be more relevant for setting limits on the creation of new sources of industrial gases with high global warming potential, such as Sulphur Hexafluoride (SF_6) and Perfluorocarbons (PFCs) (see Table 8.1).

It could also be relevant in the case of a major shift in transport fuels. Given the international market for vehicles, a global dialogue between vehicle manufacturers,

---

[59] http://www.setsquaredpartnership.co.uk/
[60] http://www.setsquaredpartnership.co.uk/news.cfm?item=59#viewing

fuel suppliers and infrastructure planners could help smooth a transition to a biofuel or hydrogen based system.

24.6 International co-ordination of performance standards, labels and endorsements

As outlined in Chapter 17, a range of failures and barriers in markets for energy efficiency determine that performance standards, labels and endorsements can complement or, occasionally, eliminate the need for, tax or trading instruments in order to elicit effective and efficient energy savings. In particular, such policies have the potential to drive demand for, and supply of, actions and investment to achieve energy savings. They can do this by: raising the visibility of energy costs; reducing uncertainty, complexity and transaction costs; inducing technological innovation; avoiding technology lock-in, for example where the credibility of carbon markets is still being established. They can also help in communicating policy intentions to global audiences.

International co-ordination of performance standards, labels and endorsements can reduce costs and increase their effectiveness, particularly in markets for highly traded goods.

As outlined in Chapter 17, careful appraisal, design, implementation and management of successful performance standards, labels and endorsements is important to their cost effectiveness. The locus of market intervention (for example national, regional or global) is one important factor affecting their cost effectiveness. There are many successful examples of these policies implemented by individual countries within a range of markets (see Boxes 17.2 and 17.5 for details). In addition, policy leadership by individual countries is generally welcomed. However, it is often desirable to co-ordinate the design and delivery of such policies across national boundaries, where they apply to markets for highly traded goods and services, in order to:

- *Influence conditions within larger markets:* create stronger incentives to innovate by influencing conditions within a larger market, and encouraging greater competition between manufacturers of efficient products;[61]
- *Increase transparency across markets:* improve the capacity of consumers, producers and vendors to compare the performance of products and components across different markets, and provide policy makers and utilities with better information about the capabilities and limits of particular technologies;
- *Reduce compliance costs:* decrease design and production costs for manufacturers arising from differences in national or regional compliance requirements. Co-ordinated standards, labels and endorsements can reduce policy design and management costs by employing economies of scale;
- *Removal of trade barriers:* international co-operation to harmonise or increase the compatibility of test protocols can discourage protectionism and enhance competition for international technology procurement contracts.

There are widespread opportunities to elicit greater energy savings in a more cost effective way through co-operation, for example on: the efficiency of electrical

---

[61] As markets and manufacturers move to comply with the new standards, the costs of product differentiation can create a tipping effect encouraging others to follow whether due to network effects, cost considerations (due to scale economies), or lock in. Barrett (2003) This occurred in the case of petrol where over 90% of the world petrol is now unleaded: http://www.unep.org/Documents.multilingual/Default.asp?DocumentID=277&ArticleID=3196

appliances, ICT[62] technologies (see Box 24.9 on stand-by power below) and power supplies, support for a more formal international Energy Star endorsement programme, as well as co-ordination of test and compliance protocols more generally.

There is considerable potential from energy efficiency policies implemented across the EU.

Policies implemented at the EU level to raise energy efficiency have the potential to be more efficient compared to subsidiary actions by individual states (although leadership from individual member states is welcomed). The EU Commission published a Green Paper[63] on Energy Efficiency which sets out proposals for delivering 20% energy savings by 2020. This builds on a suite of regulatory, information based and financing policies, as part of, for example, directives on: Eco-Design of Energy Using Products; Energy Performance of Buildings; Co-generation Energy End-Use Efficiency; and Energy Services.

The Energy Efficiency Action Plan adopted by the Commission in October 2006 represents an important opportunity to accelerate progress and set out ambitious action on energy efficiency. It has identified a number of priorities for action, in particular to: keep energy labelling up to date as well as set and progressively raise eco-design requirements for traded, energy using products and components (including on energy use). It also expands the scope of the Energy Performance of Buildings Directive to apply minimum performance standards for new and renovated buildings; and to build on existing on existing policies in relation to vehicle emissions.

BOX 24.9 Co-operation on Stand-by Power: The 1 Watt Initiative

Appliances and energy using consumer products are a major cause of growth in energy demand. They accounted for roughly two-thirds of the increase in electricity demand from buildings between 1973 and 1998 among IEA countries. Energy consumption used by appliances on stand-by mode is a major contributor.[64] In a typical Japanese or Danish household, for example, stand-by losses account for approximately 10% of total residential electricity consumption.[65]

International co-operation between policy-makers and stakeholders (including manufacturers and retailers) is necessary to reduce stand-by power related emissions (as well as those from the operating efficiencies of appliances). This is because the manufacturing, marketing and sales processes typically involve many countries. For example, a computer may be designed in the US, assembled in China using parts from Japan and Korea, and marketed and sold globally by a multinational company. As such, setting stand-by power use limits country by country would be unnecessarily difficult and costly.

The IEA launched the '1 Watt initiative' on the basis that more widespread use of existing power management technology could reduce total standby energy consumption by as much as 75% in some appliances and could form an important, cost-effective component of an overall global strategy to reduce greenhouse gas emissions. Countries including Australia have formally adopted the "1-watt plan" while others, including China, are seriously considering its adoption. In addition, the US now applies 1 Watt standards to federal procurement of energy using products (see Box 17.10). for further examples of driving efficiency through procurement.

---

[62] Information and Communications Technologies
[63] EC (2005b)
[64] One end-use metering campaign in 400 European households indicated that standby power now accounts for the largest potential energy saving among all non-thermal end-uses in the residential sector http://perso.club-internet.fr/sidler/index.html.
[65] IEA (2002)

The EU has a powerful role in shaping markets for automotive technologies, and its standards for vehicle exhaust emissions have been adopted in China and India. A voluntary agreement between manufacturers in the EU, Japan and Korea aims to reduce CO2 emissions to 140g per km across all passenger vehicles 1995 and 2008 (a cut of approximately 25% on 1995 levels). This agreement delivered reductions in CO2/km of approximately 12% between 1995 and 2004. Since then progress has slowed and the achievement of the 2008 target now appears unlikely, leading to the Commission to consider a stronger regulatory approach[66].

Harmonisation of test protocols could reduce costs and, where appropriate, provide a foundation for future consolidation of labels and standards.

Harmonisation of test protocols would bring reduced testing and compliance costs for manufacturers. It would also help consumers and manufacturers compare the performance of products and components across national boundaries; and, where necessary, provide a first step towards any future harmonisation of labels and standards. Successful harmonisation requires flexibility to account for regional and national differences in electricity, climate and local environments, product service features, and behavioural and product usage patterns.

Harmonisation of labels and standards can reduce costs but the cost effectiveness is likely to be greatest in markets where product characteristics and patterns of usage patterns vary least.

Harmonisation of labels and standards has the potential to deliver benefits in terms of increased transparency, reduced compliance and programme costs, and the promotion of innovation and growth. These opportunities are likely to be greatest for products in which characteristics and usage patterns vary least from country to country or region to region, for example, air conditioning units in South East Asia. However, significant barriers exist in certain product markets, for example in 'wet' goods (such as washing machines and dishwashers), in which regional and national differences in behavioural and product characteristics may mean the potential benefits for greater harmonisation are outweighed by the costs in terms of establishing tests, labels and standards at the lowest common national or regional denominator.

24.7   Conclusions

International technology co-operation can help speed the development and adoption of low-carbon technologies. It encourages the sharing of knowledge and information and the risks and rewards from major investments. It can also be used to monitor the pace of technological progress and the diversity of the portfolio of mitigation technologies being developed and ensure that investments are not disproportionately focused on particular technologies or regional interests.

This co-operation can take many forms with the complexities and uncertainties meaning that a range of approaches will be required in the future. Technology co-operation can build on existing experience and institutions though there may be some value in developing international programmes for research, demonstration and early stage deployment to complement national programmes.

[66] http://europa.eu/rapid/pressReleasesAction.do?reference=IP/06/1134&format=HTML&aged=0&language=EN&guiLanguage=en

References

For an exploration of the economic case for technology-based treaties see Barrett, 2006 "Climate treaties and 'breakthrough' technologies" and Carraro and Buchner, 2004 "Economic and Environmental Effectiveness of a Technology-based Climate Protocol" (full references below).

International co-operation by providing markets for low-carbon technologies and how this can be designed to encouraging market learning through is explored in Neuhoff and Sellers, 2006 "Mainstreaming new renewable energy technologies". For a broader exploration of technology co-operation and some case studies see IEA, 2005 "International Energy Technology Collaboration and climate change mitigation".

Barrett, S. (2003): 'Environmental Statecraft: The strategy of environmental treaty-making', Oxford: Oxford University Press.

Barrett, S. (2006): 'Climate treaties and 'breakthrough' technologies', American Economic Review, Papers and Proceedings **96**(2): 22-25

Benedick, R. (2001): 'Striking a New Deal on climate change', Issues in Science and Technology, available from http://www.issues.org/18.1/benedick.html

Carraro, C. and B. Buchner (2004): 'Economic and environmental effectiveness of a technology-based climate protocol', FEEM working paper, available from http://www.feem.it/NR/rdonlyres/05692881-5EB2-4EC4-9140-B05681635257/1135/6104.pdf

European Community (EC) (2005a): 'The support of electricity from renewable energy sources', Communication from the Commission, available from http://ec.europa.eu/energy/res/biomass_action_plan/doc/2005_12_07_comm_biomass_electricity_en.pdf

European Community (EC) (2005b): 'Doing more with less', Green paper, available from http://ec.europa.eu/energy/efficiency/doc/2005_06_green_paper_book_en.pdf

Gibbins, J. et al. (2006): 'Interim results from the UK Carbon Capture and Storage Consortium project', Paper presented at GHGT8, Trondheim, Norway, June 2006, available from www.ghgt8.no

Global Energy Technology Strategy Program (GTSP), (2005): 'Addressing climate change initial findings from an international public-private collaboration', available from http://www.pnl.gov/gtsp/docs/infind/cover.pdf

International Energy Agency (2000): 'Labels and Standards', Paris: OECD/IEA.

International Energy Agency (2002): 'Reducing standby power waste to less than 1 watt: A relevant global strategy that delivers', Paris: OECD/IEA.

International Energy Agency (2003): 'Cool appliances: Policy strategies for energy-efficient homes', Paris: OECD/IEA.

International Energy Agency (2005a): International Energy Technology Collaboration and climate change mitigation, Synthesis report, available from http://www.iea.org/Textbase/papers/2005/cp_synthesis.pdf

International Energy Agency (2005b): Deploying climate-friendly technologies through collaboration with developing countries, Paris: IEA/OECD, available from www.iea.org/Textbase/Papers/2005/Climate_Friendly_Tech.pdf

International Energy Agency (2006): 'Energy technology perspectives - scenarios & strategies to 2050', Paris: OECD/IEA.

International Rice Research Institute (2006): 'Climate change and rice cropping systems: Potential adaptation and mitigation strategies'. Philippines: IRRI.

Intergovernmental Panel on Climate Change (2005): IPCC Special Report on Carbon dioxide Capture and Storage, available from http://www.ipcc.ch/activity/ccsspm.pdf

Jaccard, M. (2006): 'Sustainable fossil fuels: the unusual suspect in the quest for clean and enduring energy', Cambridge: Cambridge University Press.

Neuhoff, K. (2005): 'Large-scale deployment of renewables for electricity generation', Oxford Review of Economic Policy, Oxford University Press, **21**(1): 88-110, Spring.

Neuhoff, K. and Sellers, R. (2006): 'Mainstreaming new renewable energy technologies', Electricity Policy Research Group Working paper, Cambridge http://www.electricitypolicy.org.uk/pubs/wp/eprg0606.pdf

OECD (2006): 'Do we have the right R&D priorities and programmes to support energy technologies of the future'. 18th Round Table on Sustainable Development background paper, available from http://www.oecd.org/dataoecd/47/9/37047380.pdf

Pindyck, A.K. and R.S. Dixit (1994): 'Investment under uncertainty', Princeton NJ: University Press.

REN 21 (2006): Renewables Global Status Report: 2006 update Washington, DC: Worldwatch, available from http://www.ren21.net/globalstatusreport/download/RE_GSR_2006_Update.pdf

25 Reversing Emissions from Land Use Change

KEY MESSAGES

Curbing deforestation is a highly cost-effective way of reducing greenhouse gas emissions and has the potential to offer significant reductions fairly quickly. It also helps preserve biodiversity and protect soil and water quality. Encouraging new forests, and enhancing the potential of soils to store carbon, offer further opportunities to reverse emissions from land use change.

Policies on deforestation should be shaped and led by the nation where the forests stand but there should be strong help from the international community, which benefits from their actions.

At a national level, establishing and enforcing clear property rights to forestland, and determining the rights and responsibilities of landowners, communities and loggers, is key to effective forest management. This should involve local communities, and take account of their interests and social structures, work with development goals and reinforce the process of protecting the forests.

Compensation from the international community should be provided and take account of the opportunity costs of alternative uses of the land, the costs of administering and enforcing protection, and managing the transition. Research carried out for this report indicates that the opportunity cost of forest protection in 8 countries responsible for 70 per cent of emissions from land use could be around $5 billion annually, initially, although over time marginal costs would rise.

Carbon markets could play an important role in providing such incentives in the longer term. But there are short-term risks of de-stabilising the crucial process of building strong carbon markets if deforestation is integrated without agreements that increase demand for emissions reductions, and an understanding of the scale of transfers likely to be involved.

Action to preserve the remaining areas of natural forest is urgent. Large-scale pilot schemes are required to explore effective approaches to combining national action and international support.

25.1   Introduction

The earth's vegetation and soils currently contain the equivalent of almost 7500 Gt CO_2[1], more carbon than that contained in all remaining oil stocks[2], and more than double the total amount of carbon currently accumulated in the atmosphere. The carbon presently locked up in forest ecosystems alone is greater than the amount of carbon in the atmosphere[3].

Plants and trees play a vital role in carbon sequestration. This is the natural process whereby living plants and trees remove carbon from the atmosphere through photosynthesis as they grow. Some of this is transferred to the soil through the roots and as leaves fall. But when soils are disturbed through ploughing or trees are cut down, the stored carbon oxidizes and escapes back into the atmosphere as CO_2.

Emissions from deforestation are very significant globally. Independent estimates of the annual emissions from deforestation more than 18% of global greenhouse gas emissions[4], greater than produced by the whole of the global transport sector[5]. These emissions could potentially be cut significantly fairly quickly – no new technology has to be developed – although considerable challenges have to be addressed, as discussed below.

While planting new trees is an excellent long-term policy, trees take decades to absorb the equivalent amount of carbon to that which is instantaneously released into the atmosphere when mature trees are cut down and burnt. Depending on the species, a tree may take 100 years to reach maturity, and much more land would have to be allocated for new forests to obtain the same amount of carbon absorption as would be released from burning an existing forest of mature trees. The biodiversity and other co-benefits of new forests are also likely to be much lower than those for natural forests. For these reasons, international support for action to protect existing forests should be kept distinct from the creation of new forest, through the latter is also important.

This chapter sets out the drivers of the release of emissions through deforestation, and how these can be reduced. It briefly addresses how atmospheric carbon can also be absorbed through changing agricultural methods, such as moving from deep ploughing to conservation tillage, and generally planting more trees and plants. It then discusses the international framework that can best support national programmes of action, the challenges that need to be overcome, and pilot schemes to start the process of taking action now and allow learning by doing.

---

[1] Prentice et al (2001)

[2] UNDP (2001) estimates this at 2400 Gt CO_2. Includes both conventional unconventional oil, known reserves and as yet undiscovered resources.

[3] Prentice et al (2001) gives ~4500 GtCO2 in forest ecosystems, compared with ~3000 GtCO2, the level with atmospheric concentration levels of 380ppm.

[4] Although all estimates suggest that land use emissions are significant, estimates of the scale of land use emissions vary. The WRI estimates used in this report estimate that emissions from deforestation are about 8 GtCO2 per year (see fig 25.1). This is within the range of the Third Assessment Report of IPCC which estimates emissions from land use change within the range equivalent to 2.2 to 9.9 GtCO2, with a central estimate of 6.2 $GtCO_2$. A fuller discussion setting out the range of estimates can be found in Baumert KA et al. (2005).

[5] CAIT, WRI. 2000 figures used.

25.2   Understanding deforestation

The drivers of deforestation are economic and challenging to reverse.

Action to prevent deforestation, as set out in Chapter 9, offers opportunities to reduce greenhouse gas emissions on a significant scale without much need for new technology except perhaps for monitoring. Action here can also bring significant national co-benefits in terms of local soil, water and climate protection, as well as opportunities for sustainable forest management and the protection of biodiversity and the livelihoods and rights of local communities.

As Figure 25.1 shows, deforestation is the main source of emissions from land use change. Harvesting leads to the release of CO_2 emissions, but growth absorbs CO_2. The difference between the two reflects the unsustainable exploitation of forest resources, such as timber from unsustainable logging[6]. Planting new trees[7] partially offsets emissions by absorbing CO_2.

The bulk of emissions from deforestation arise when the land is converted to agricultural production. Mature forests contain large stocks of carbon locked up within trees, vegetation and soils. Dense tropical forests have especially high carbon stocks per hectare. Conversion to agricultural land through slash and burn techniques releases most of this as CO_2. Burning is a cheaper way of clearing

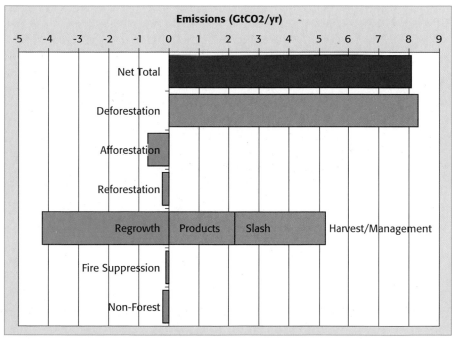

Figure 25.1 Sources of emissions from global land use change 2000

Source: Reproduced from Baumert et al (2005)

---

[6]   Although they are classified separately in this figure, unsustainable exploitation of a forest is similar to deforestation.

[7]   Reforestation (re-establishing former forests) and afforestation (establishing new forests).

land, releases CO_2 and leaves behind ash that gives a short-lived fertiliser effect to the newly cleared land.

As shown in Figure 25.2, the areas of globally significant forest most vulnerable to deforestation are mainly concentrated in tropical countries. The forces driving demand for additional agricultural land vary globally. In Africa, the main clearers are small-scale subsistence farmers. In South America, the drivers are large farming enterprises producing beef and soya for export. In South East Asia, the driver is a mixture of the two, with oil palm, coffee and construction timber the main products.

Logging, which is the process of harvesting large, valuable, mature trees mainly releases CO_2 from the cut trees and those damaged in gaining access to them. If logging is limited to valuable, single trees, forest recovery through re-growth can offset this over time. For these reasons, logging itself need not be a major driver of deforestation. Also if the timber is used for long-lived wooden products it actually conserves carbon during the product lifetime. Logging plays a greater role in specific cases such as Indonesia and elsewhere in South East Asia, where an unsustainable rate of logging is fuelled by the strong demand for timber from fast growing regional economies. The wider impact from logging is that building access roads, to bring in cutting equipment and take out the logs, makes forests more vulnerable to conversion to agricultural production. New logging access roads help to open up former closed regions and allow access to markets for agricultural products.

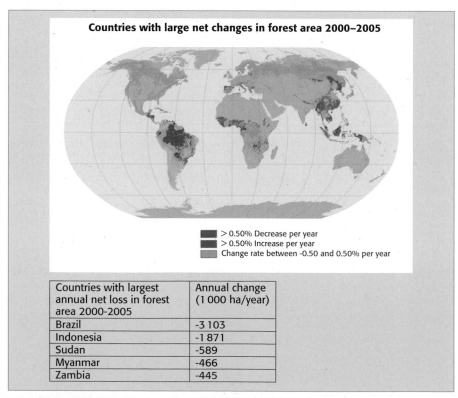

Figure 25.2 Deforestation is currently concentrated mainly in tropical areas

Source: FAO (2005a)

25.3   Changing economic incentives to reduce deforestation

Effective action to protect existing forests and encourage afforestation and reforestation requires changes to the structure of economic incentives that lead to
unsustainable logging and to the conversion of forestland to agriculture.

In Chapter 9 we summarised the findings of research into the direct costs of
reducing deforestation. These include net income from the sale of timber, the
opportunity costs of agricultural production, the costs of administering and enforcing forest protection, and some transitional costs.

Research commissioned by the Review, suggests that the direct yields from land
converted to farming, including proceeds from the sale of timber, are equivalent
to less than $1 per tonne of CO_2 in many areas currently losing forest, and usually
well below $5 per tonne[8]. The opportunity costs to national GDP would be some-
what higher, as these would include value added activities in country and export
tariffs. Other modelling studies, using alternative methodologies, have suggested
that, whilst there are significant opportunities to protect forests in some regions
at low costs, the marginal abatement cost curve could rise from low values up to
around $30 per tonne of CO_2[9] were deforestation to be eliminated completely.

Although the direct costs could be low at first, there are major institutional and
policy challenges that have to be overcome in achieving the transition away from
economic activities leading to deforestation towards those consistent with forest
conservation. This means that forest conservation and management projects, to be
successful, need to be part of a much wider, integrated resource management programme. Many countries have national forest programmes in place that increasingly take a broad inter-sectoral approach to the management and conservation
of forests. They espouse a participatory approach to policy formulation and planning, involving stakeholders at the local, sub-national and national levels. The
more developed of these programmes are closely linked to higher level policy and
planning frameworks, such as poverty reduction strategies, and provide a focus
for directing development assistance. Such programmes can be amended so that,
in a more targeted and effective way, they can tackle the main drivers to deforestation and unsustainable land use.

A recent World Bank study[10] of deforestation and related issues highlights two
key public policy challenges that forested countries face.

The first is to determine who has rights over the forest and what these rights
should be. The situation varies widely. In some countries, landowners clear forest
legally. Elsewhere, forests owned by the government are illegally encroached upon
by subsistence farmers, logging companies and agricultural businesses. Specific
circumstances require policies tailored to particular local and national conditions.
Over the last 20 years 26 tropical countries have experienced armed conflicts in
forested areas, and in some cases timber sales have financed the fighting[11].

---

[8] Grieg-Gran (2006), calculation assumes CO_2 levels per hectare of tropical forest preserved is
     500–750 t per hectare
[9] Sohngen (2006), Obersteiner (2006)
[10] At Loggerheads?: Agricultural Expansion, Poverty Reduction, and Environment in the Tropical
     Forests. Chapters 5 and 6 have comprehensive discussion of forest management policies. This
     section draws from the work of this report, and especially from the expertise of Ken Chomitz
     for which we are grateful.
[11] FAO 2005(b)

The second challenge concerns the social and economic decisions that national governments make about managing land use, including how to balance global and local environmental benefits with the opportunities for production of wood, food, fuel and fibres.

The World Bank study cites several examples of successful efforts to preserve forests and highlights some common themes. Reducing deforestation requires effective and capable institutions at the national, regional and local levels. Involvement of local communities is key to finding solutions that support local development goals.

Clarifying both property rights to forestland and the legal rights and responsibilities of landowners is a vital pre-requisite for effective policy and enforcement.

A lack of clear and enforceable property rights means that forests are often vulnerable to damage and destruction. Loggers can quickly exploit lack of clear ownership and their actions often open up the land for subsequent illegal conversion to farming. Historically there have been violent clashes between landless groups and large landowners, which stemmed from legal ambiguity, conflicting laws made both groups consider they had rightful claims to land and timber[12]. Clarity over boundaries and ownership, and the allocation of property rights regarded as just by local communities, will enhance the effectiveness of property rights in practice and strengthen the institutions required to support and enforce them.

BOX 25.1 Local and community ownership of forests

Latin America and South Asia have increasingly involved local communities in the ownership and stewardship of forests, and communities have often opted for more sustainable long-term programmes as a result. Another example is the Joint Forest Management Program in India. This has both improved forest regeneration and had a positive impact on livelihoods. Similarly in Guatemala 13 community concessions, almost all certified by the Forest Stewardship Council, have managed to combine highly profitable mahogany enterprises with deforestation rates lower than in protected or outside areas[13]. Other approaches have allowed local communities to benefit from timber revenues. This helps promote local support. In Cameroon, for example forest concessions were allocated through transparent auctions, with 50% of the royalties going to local communities.[14]

Land use planning has a key role to play in determining what kinds of activities are appropriate in forest areas: a complete ban on all activities may be justified in some areas, while in others, logging may be allowed subject to specific rights and duties. Logging concessions can be granted with conditions such as permissible extraction levels and sustainability requirements. Brazil has recently granted such contracts to private companies. The concessions run for 40 years, operations are required to meet key criteria for sustainability. The revenues have been used to set up and run the Brazilian Forest Service, which manages the concessions. In the first year of operation deforestation fell by an estimated 31%.

---

[12] Alston, Libecap and Mueller (2000)
[13] World Bank (2006) However deforestation is still present at a reduced rate.
[14] World Bank (2006)

There are many examples of perverse outcomes from poorly designed forestry policies, including policies that inadvertently create incentives for forests to be cleared illegally. For example, in one case, a tax on timber obtained from legally converting forestland, led to some farmers clearing the land by simply burning the forest[15]. More restrictive regimes for forest management have meant that in practice, it can be easier to get a permit for forest conversion than forest management.[16] This has led loggers to clear-cut and then abandon forest plots they would have been otherwise content to harvest selectively.

Rigorous enforcement of forest protection in one country without action to reduce demand for timber can displace logging to neighbouring countries. Following floods associated with deforestation in the upper reaches of the Yangtze River, China banned the logging of natural forest in 1998 and has greatly increased its own forest cover. However, timber imports from the Russian Far East, South East Asia and Africa have risen strongly since the ban has been enforced[17].

There are further challenges in institutional capacity, governance, and weak law enforcement. It is difficult to turn round entrenched systems of vested interests, although some countries are making significant efforts to do so. Indonesia is trying hard to improve governance, including tenure reform for judges and stricter law enforcement. Efforts to stem the trade in illegal merbau logs between Papau Province and China in 2005 resulted in an 83% drop in Chinese imports of this species[18].

Many frontier forests are remote and lack adequate communication facilities. This makes monitoring the forest difficult, and can cloak conflicts and resource grabs. However developments in remote sensing have started to improve real time monitoring for owners, the authorities and civil society.

Changing economic incentives and encouraging alternative economic activities are essential elements of sustainable forest management.

Competition for, and sometimes conflict over land use, reflects the many potential uses of the land, with changing values depending on the type of crop, world prices and other factors. Land-use planning forms part of the response but may have little impact in practice if land users face strong incentives for non-compliance. Planning that takes more account of the behaviour of those with claims on property, and which seeks popular support, may achieve greater success.

Poverty is often one of the key drivers for people who have little choice but to use forests unsustainably. It is important that the interests and livelihoods of those who would have gained income from converting forestland to agriculture are taken into account. Tackling the causes of poverty through an approach that offers local communities alternatives to deforestation is an important part of efforts to reinforce and sustain action. In the Philippines, conversion of lowland farms to labour intensive integrated rice production, tripled the employment of uplanders, and halved the rate of forest clearance by them[19]. Cameroon drew up a zoning plan on the basis of existing land use patterns, which is thought to have deterred conversion from forest to agriculture.

---

[15] Merry et al (2002)
[16] World Bank (2006)
[17] Chunquan et al (2004)
[18] Research in progress by CIFOR (Center for International Forestry Research)
[19] Shively and Pagiola (2004)

Many countries have set up protected areas, with the overall area increasing threefold over the past 30 years, while annual spending on protected areas in developing countries is estimated to have risen to $800m. The UN Global Environment Facility financed $3.6 billion of such projects during 1992–2002[20]. Potential areas are often chosen for biodiversity and national heritage value, and may not be at immediate risk of logging or conversion to agriculture. Experience has shown, that for Protected Areas to operate successfully, they need to be an integral part of a wider integrated natural resource management programmes, as otherwise the drivers that lead to deforestation cannot be addressed adequately.

However where people live in or close to forests, preserving the forest does not mean that it has to stay untouched. There are other ways of producing income from forests, and logging can also be carried out in a sustainable way. Estimates indicate that up to 5% of trees can be removed each year without risk to the forest[21]. Reduced impact logging, using known methods[22] can also reduce impacts to the soil from heavy logging machinery by 25% and preserve up to 50% of the carbon stored in the remaining vegetation.

Managing the tension between agricultural land use and forests.

Fluctuations in the rate of deforestation have occasionally been observed in response to global commodity prices. In Madagascar for example, deforestation increased sharply in response to higher maize prices[23], and in Brazil, increases in world prices for beef, soya beans and pig iron in 1999 greatly increased the incentive for deforestation. They contributed to a 33% rise in the rate of deforestation over the following five years.[24].

Opportunity costs of action essentially reflect the different returns on land depending on its use. The NPV of income[25] ranges from $2 per hectare for pastoral use to over $1000 for soya and oil palm, with one off returns of $236 to $1035 from selling timber. A study undertaken for the Stern report[26] estimates that these returns in 8 countries, responsible for 70% of emissions from land use, are $5 billion a year including one-off timber sales. This level of financial incentive would offset lost agricultural income to producers, although it would not reflect the full value chain within the country. Nor would it reflect the possible response of existing timber markets to reduced supply, given the current margin between producers and final market value, Nethertheless, the high carbon density of each hectare of forest that would be preserved (up to the equivalent of 1000t CO_2) suggests that reducing deforestation offers a major opportunity to reduce emissions at relatively low cost. Assuming a carbon price of $35–50, a hectare containing 500t CO_2, would be worth $17500–25000 in terms of the carbon contained if it were kept as forest, a large difference compared with the opportunity costs at the low end of the range.

Direct incentives can create a value for maintaining forest and form a key part of national programmes to reduce greenhouse gas emissions.

[20] World Bank (2006)
[21] C Kremen et al (2000)
[22] Priyadi, H et al (2006)
[23] Moser, Barrerr and Minton (2005), Minten and Meral (2005)
[24] Data from INPE (www.obt.inpe.br/prodes)
[25] These figures are calculated from income over 30 years, using a discount rate of 10%, except for Indonesia which uses 20%.
[26] Grieg-Gran, (2006)

BOX 25.2   Impact of avoided deforestation on availability of land for food production

The amount of potential agricultural production lost through better protecting forest, both within a country and globally, is likely in practice to be a small proportion of the existing farm output from converted former forest land. The level of output for any particular agricultural product is not fixed, and the additional output will in any case be small compared with total global agricultural output.

Completely eliminating deforestation in those countries covered in research carried out for the review would lead to an annual loss equal to 0.25% of land used globally for soybean production and 6% of land used for oil palm[27]. Depending on the elasticity of demand for products, this would be likely to have only a small impact upon commodity prices.

Much of the agricultural activity that currently takes place on converted forestland could be moved to other types of land, without a significant fall in productivity. For example, advancements in soil science have allowed farmers to grow soybeans and other crops in the infertile 'Cerrado' region of Brazil, a large area previously unusable by farmers. This has taken pressure off of the fertile Amazonian regions, whilst increasing overall agricultural production[28].

As set out in Chapter 2 of this Review, market failures can be corrected by adjusting prices to include the value of the externalities that are not fully captured by behaviour. Incentives that reflect the full benefits of forests to society could reduce the attractiveness of the potential income from agriculture on converted land. But transparent and legitimate ownership is vital for the success of any scheme that seeks to use incentives to protect forests by changing behaviour.

Several countries have successfully included incentive payments as part of their programmes to protect forests. In Costa Rica landowners can receive up to $45 a hectare per year if they if volunteer to maintain forests in the interests of carbon sequestration, biodiversity, hydrological protection and scenic beauty. Combined with other measures this has increased forest cover from 21% in 1977 to 51% in 2005, reducing rural poverty by benefiting 7000 families. Mexico has used similar payments involving payments of up to $28 a hectare a year to preserve forests, in a programme motivated by water scarcity and the need to raise water quality.

25.4   Project-based approaches to increasing carbon storage in land use

Protecting existing forest is the key to maintaining the large stocks of carbon contained in forests that are currently at risk. Action to protect these forests can be complemented by action to increase and store the uptake of atmospheric CO_2 in

---

[27] Calculations using Grieg-Gran (2006) and FAO Stat-available at http://faostat.fao.org
[28] The former Brazil Minister of Agriculture **H.E. Alysson Paolinelli** and former Technical Director of EMBRAPA Cerrado Research Center **Mr. Edson Lobato,** both of Brazil; and Washington Representative of the IRI Research Institute, **Dr. A. Colin McClung** of the United States were awarded the 2006 World Food Prize for their work in this area. http://www.worldfoodprize.org/press_room/2006/June/2006Laureates.html

Table 25.1 Countries with largest recent net gains in forest area

Countries with largest annual net gain in forest area 2000–2005	Annual change (1 000 ha/year)
China	4 058
Spain	296
Vietnam	241
United States	159
Italy	106

Source: FAO (2005a)

soils and trees. As with other types of mitigation, this can take place anywhere in the world, and produce the same benefits from reducing atmospheric carbon levels.

Planting new trees could be cost effective in many countries.

Forest cover can be increased in most areas of the world. Eight thousand years ago, 50% of the global land surface was covered by forest, compared with only 30% now. At modest carbon prices, there are potentially large areas of land in many countries where new forests could be planted, should the enabling environment be conducive. The costs of planting new forests depend on the value of an alternative land use and may be offset in the medium term by revenues from sustainable forest use. Reforestation (re-establishing former forests) and afforestation (establishing new forests) in marginal agricultural land and on abandoned land offer significant local benefits by reducing vulnerability to soil erosion and desertification.

Some countries already have programmes to encourage farmers to convert land and plant trees. For example China, as shown in Figure 25.2 and Table 25.1, in area terms has added forests at a rate equal to nearly half of global deforestation over the past 5 years. Measures include a programme that offers seedlings, cash and grain to farmers who retire marginal or steep, erosion-prone farmland and replant it with grass, fruit bearing trees or trees for timber. Under this plan 7m hectares of farmland was converted in the first 5 years. Vietnam is aiming to establish 3 million hectares of production forest, mainly via plantations, and 2 million hectares of protection forests by 2010. The programme has a strong focus on smallholder reforestation and allocation of forestland to private households, organizations and individuals. More than 1.4 million hectares have been allocated to 500000 families for periods up to 50 years.

An international framework for incentives for reforestation and afforestation is already in place for Parties to the Kyoto Protocol, see Box 25.3.

Changing agricultural practice can store carbon in soils and biomass.

As discussed in Chapter 9, cost effective carbon sequestration from agricultural land use change practices could amount to 1Gt of CO_2 in 2020. When soils are exposed to microbial activity, CO_2 emissions are released. These emissions can be reduced by disturbing the soil less, for example by using conservation tillage techniques and turning land into permanent set-aside.

BOX 25.3 Land use change in the Kyoto Protocol

Article 3, paragraph 3 of the Kyoto Protocol requires developed countries to account for afforestation, reforestation minus deforestation, since 1990 in meeting their commitments for the first commitment period. In other words they must take account of forestry activities that increase or decrease forest carbon stocks (or cause other greenhouse gas emissions) since the base year of the Protocol.

The Marrakesh Accords established that afforestation and reforestation would be eligible as project based activities for the CDM. By October 2006 no afforestation or reforestation projects had been registered by the CDM Executive Board, although one reforestation project was requesting registration and two reforestation projects were under consideration. Three afforestation and reforestation methodologies had been approved. Under Joint Implementation (JI), there was one afforestation project at the validation stage, to be hosted in Romania.

The agreement on forest activities has been criticised for its relative complexity, though this was regarded as necessary to reach agreement as the negotiations evolved over time. It is likely to be possible to simplify the inclusion of forestry in future.

Carbon emissions can also be reduced by improving the fertility of the soil because this increases the ability of the soil to sequester carbon, for example by using techniques known as conservation tillage, and by setting aside land to return to grassland. Techniques include planting particular crops and trees together to improve soil nutrient levels (agroforestry), erosion control, restoration, crop residue management and crop rotation.

Market based instruments can be used alongside agricultural extension activity to encourage biological carbon sequestration. The Chicago Climate Exchange[29] (CCX) allows participants (companies who have taken on voluntary commitments to reduce emissions) to purchase Carbon Financial Instruments from eligible projects. These eligible projects include reforestation, afforestation and soil carbon offsets. Soil carbon offsets are created through the use of conservation tillage and grass planting. There is a minimum four-year commitment to continuous no-till on enrolled areas. The projects must be enrolled through an intermediary registered with the CCX that serves an administrative and trading representative on behalf of multiple individual participants, known as an "Offset Aggregator". The first sale of an exchange of verified CO_2 offsets generated from agricultural soil sequestration took place in April 2005. By June 2006, approximately 350,000 acres of conservation tillage and grass plantings had been enrolled in Kansas, Nebraska, Iowa and Missouri.

Measures to enhance natural soil fertility and carbon sequestration potential can also have spin-off benefits in the form of reduced need for man-made fertilisers, reducing the need to deforest land, improved water quality and reduced power and fuel requirements to till land[30]. The Nhambita project in Mozambique, described in Box 25.4 provides an example of how these measures formed the basis of a carbon-offsetting project and also helped to reduce poverty.

[29] Source: www.chichagoclimateexchange.com
[30] International Soil Tillage Research Organization (ISTRO)

BOX 25.4   Sustainable agriculture and forestry project in Nhambita, Mozambique

The Nhambita Community project in Mozambique provides an example of the potential for a beneficial relationship between emissions reductions and poverty reduction. The natural habitat of the Gorongosa National Park was deforested and degraded during the country's 16 year civil war. The aim of the Nhambita project is to regenerate the environment, reduce CO_2 emissions and reduce poverty by incentivising local people to adopt sustainable agricultural and forestry practices. The following activities help to achieve these aims:-

- Agro-forestry is the practice of planting special types of trees and crops, such as the pigeon pea nitrogen fixing crop, to improve the fertility of the soil. This increases crop yields, reduces the need to use synthetic fertilisers that produce GHGs, and enhances the natural carbon absorption of the soil. It also saves emissions because by improving the soil fertility, the land can be farmed for longer and there will be no need to deforest other land to convert it to agriculture.

- Afforestation and planting other crops reduces GHG emissions as the biomass grows and sequesters carbon. Local people are paid to plant trees and crops appropriate to the local habitat and maintain the land. The Nhambita Community project has planted 150,000 trees over the last three years. The sustainable harvest of crops and trees provides a supply of fuel wood and other forest products.

- Forest fire fighting limits damage to crops and forest land. The Nhambita community has purchased mechanised fire fighting equipments and earns money for responding to forest fires.

To date there has been limited success in accrediting small-scale sustainable agriculture and forestry initiatives as CDM projects because the transaction costs are too great. The Nhambita community undertakes the sustainable practices described above under contract with Envirotrade, an organisation that brokers the carbon. The carbon credits from this project are independently verified, then purchased by organisations such as the Carbon Neutral Company on behalf of people who want to offset their emissions on a voluntary basis. The sustainable practices adopted by people in Nhambita are estimated to save 90 t CO_2 per hectare.

Source: Girling (2005) and Envirotrade[31].

25.5   International support for avoided deforestation

Existing international frameworks and processes relevant to deforestation include the United Nations Forum on Forests (UNFF), the International Timber Trade Organisation (ITTO) and initiatives on forest law enforcement, governance and trade (FLEG and FLEGT). There are also forest certification schemes that can be linked to procurement programmes and bilateral and multilateral initiatives.

However there are currently only limited international frameworks that focus upon reduced emissions from deforestation. Action to protect forest incurs costs, requires commitment of resources, and has to compete with other priorities. The

---

[31] www.envirotrade.co.uk

pressure for deforestation is greatest in a small number of developing countries, but all countries gain from preserving forests that provide global public goods.

Emissions from deforestation are within the Kyoto Protocol for Annex I countries, but non Annex I countries are where the vast majority of emissions take place. The Marrakesh accords rejected the inclusion of deforestation within CDM projects during the first commitment period, primarily because of concern about the risk that protecting forest in one project area would simply displace deforestation which would just take place elsewhere.

The scale of the problem is daunting. Without prompt action emissions from deforestation between 2008 and 2012 are expected to total 40Gt CO_2, which alone will raise atmospheric levels of CO_2 by ~2ppm, greater than the cumulative total of aviation emissions from the invention of the flying machine until at least 2025[32].

Taking action to protect forests is therefore too important to wait until the next commitment period. This means that pilot schemes outside the Kyoto Protocol are necessary. These need not be limited in scope - the more ambitious the reductions, the greater the benefit.

Currently, there are a number of schemes involving governments, companies, NGOs and individuals seeking to protect areas of rainforest. Examples include

Debt forgiveness in return for forest protection

Debt-for-nature swaps are designed to free up resources in debtor countries for conservation activities. The US Government has forgiven debt in exchange for forest protection in 10 countries under the 1998 Tropical Forest Conservation Act. A debt swap involves purchasing foreign debt at a discount and converting the debt into local currency to establish a Tropical Forest Fund, The fund then makes grants to local NGOs engaged in a variety of forest conservation activities. These include research on the protection and sustainable use of local plants and animals, development of sound forest management systems, training of local organizations in forest conservation management, and establishment and maintenance of protected areas. Signed agreements will generate over $100m over the next 10–25 years.

Using insurance markets to protect forest

Rather than increase premiums, insurance companies can reduce the cost of premiums payouts by improving forest management practice and selection of risk. This needs to be done in parallel with the realignment of forest insurers risk profile. For example the forestry insurance company, ForestRe proposes to use insurance premium criteria to reinforce the benefits from adopting a sustainable forest management system. As such, management is likely to reduce their risks of catastrophic loss, and their premiums will be reduced. It is also exploring linkages to ensure that sound environmental management (including reforestation and watershed management) is required to gain cover for large infrastructure projects, such as refurbishment of the Panama Canal.

---

[32] Calculation using IPPC data and IEA data and forecasts

International Finance to back national action

National action can be strengthened by the assistance of NGOs and International agencies. For example, the Amazon Regional Protected Area scheme, a collaboration between the Brazilian Government, the Global Environment Facility, the World Bank and the WWF has set up a project to create 18 million ha of Conservation Units. It includes areas where the forest is fully protected, and areas where sustainable exploitation is possible. Rights of indigenous people are respected and there is biodiversity monitoring and funding for protection of parks and reserves. Another example is the multi-stakeholder partnership proposed by the World Bank which is designed to bring together developing countries, industrialized countries, NGO's and the private sector. The partnership would implement and evaluate, on a prototype basis, incentive payments designed to reduce net deforestation rates in developing countries. The proposed partnership would integrate existing policies and programs for forest protection and management.

These initiatives offer the opportunity to learn what action is most effective, but they are not sufficient to ensure that forests are protected on a large scale.

Carbon markets could play an important part in providing incentives

Bringing deforestation into the broader multilateral mitigation framework would potentially allow trading of credits earned through preserving forests. The proposal by Papua New Guinea with other rainforest nations identifies a possible approach to integrating action to protect forests (see box 25.5)

In the long term, the main advantage of inclusion in a system of deep and liquid global markets for carbon is that this would support large-scale action. However any integration with the carbon market should be managed carefully since bringing

---

BOX 25.5 Compensated Reductions – Proposal by Papua New Guinea and Costa Rica

In the run up to the COP11 meeting in Montreal, Papua New Guinea (PNG) and Costa Rica, on behalf the Coalition of Rainforest Nations[33], led a move to reconsider approaches to "stimulate action to reduce emissions from deforestation". Their key proposal (commonly known as the PNG proposal) was to develop a mechanism to enable carbon saved through reduced deforestation in developing countries to be traded internationally.

Specifically, a country establishes a national baseline rate of deforestation (converted into carbon emissions) and negotiates a voluntary commitment (over a fixed commitment period) for reducing emissions below the baseline. Any reductions that are achieved below the baseline could then be sold under Kyoto or other carbon markets. No trading would be allowed if emissions were above the baseline in a commitment period.

The proposal has focused attention on how deforestation might be included, either as part of future commitments under the Protocol or under the Climate Change Convention itself. The proposal is now being reviewed by the UNFCCC's Subsidiary Body for Scientific and Technological Advice (SBSTA) to report back for COP13 in late 2007.

---

[33] Submission by the governments of Bolivia, Costa Rica, Nicaragua, and Papua New Guinea, supported by the Central African Republic, the Dominican Republic and the Solomon Islands. The Coalition currently consists of Bolivia, Central African Republic, Chile, DR Congo, Congo, Costa Rica, Fiji, Guatemala, Nicaragua, Panama, Papua New Guinea, Solomon Islands and Vanuatu

in a substantial tranche of new emission reductions, particularly if they are cheap to generate, could destabilise the carbon market. They could for example, represent a substantial disincentive on action to reduce emissions from long-lived energy and transport infrastructure unless national targets in participating countries were substantially increased.

Integration for the first commitment period in the Kyoto Protocol is in any case not possible under the existing agreement because the rules are already set. They do not include any provision in the CDM for reduced emissions from avoided deforestation. Beyond the first commitment period the level of commitments can be adjusted to accommodate the new reduction potential. In the longer term there are reasons to believe that the marginal costs of reducing deforestation will rise and that the technical challenges to include avoided deforestation in carbon markets can be overcome. Early crediting for the second commitment period could be a feature of pilot schemes discussed below.

Challenges to integrating deforestation into carbon markets

Looking beyond initiatives and project-based approaches in the longer term, there are good reasons to integrate action to reduce deforestation within carbon markets. This is challenging for a number of reasons.

Carbon measurement

Estimating carbon emissions to a uniform standard from forest preservation activities is more difficult than for energy-related projects. This is because the carbon content of forests varies significantly depending on the density, age and type of trees, and the soils. Detection of forest degradation, as opposed to actual deforestation, is particularly challenging. However, standard inventory methods have been developed by the IPCC and a combination of ground based and remote sensing methods is likely to be feasible. Brazil already uses advanced remote sensing methods, which are increasing in effectiveness while falling in cost. Such remote monitoring can also be used to monitor compliance.

Natural/accidental deforestation

Forests can be reduced through natural or accidental causes, such as fires or disease, causing unplanned fluctuations in emissions. Whilst inclusion with carbon markets would incentivise action to reduce the risks, the potential scale of events mean that that the markets would need to allow for this in some way. One approach would be to extend the period over which compliance was assessed, so as to average out fluctuations. The Chicago Climate Exchange[34] dealt with this for their Forestry Carbon Emissions Offsets by creating a carbon reserve pool of 20% of emissions to allow for catastrophic loss, released at the end of the programme. Losses could also be counted against future credits against the baseline or reference level. The way in which this issue is handled will affect credibility and could influence the price at which units are traded.

---

[34] See www.chicagoclimatex.com

Ensuring climate benefits

A key challenge is to ensure that emissions reductions are additional. The nature of the drivers of deforestation implies a substantial risk that, if small areas are protected, leakage to other areas could take place and overall emissions would not be reduced. The only way this can be overcome is to have projects over a large enough area to reduce this risk and induce a genuine change to behaviour of the people involved. This means a strategy for action will probably have to be adopted at a country level rather than relying only on local projects, and national baselines are a feature of the current proposed approaches from the Papua New Guinea and the Coalition of Rainforest Nations. The greater the international coverage, the lower the potential for leakage between countries.

Agreeing an equitable basis for participation and incentives

Setting baselines that are regarded as fair will be an important part of any future agreement to extend climate change agreements to include incentives to reduce deforestation, whether by emissions trading, a fund-based scheme or some other approach.

Determining the baseline of emissions from deforestation beyond which tradable credits would be earned will not be easy. Getting the level right may involve assessment of the historical trend and is a technical challenge given variability in deforestation rates year by year and lack of historical data in some countries. Setting a baseline incorrectly could lead to distortion in the level of effort.

As with the inclusion of any new sector, allocated limits would have to be re-examined to make sure they were appropriate, given the extended scope of the trading scheme and the limits and incentives adopted by new participants. Agreeing the terms under which countries can earn carbon credits will require consideration of the rate at which action can earn tradable credits. As discussed in Chapter 22, quota allocation must embody criteria of equity.

A particular challenge, when setting baselines, is how to treat countries that have already implemented policies to avoid deforestation such as China and Costa Rica. Focusing only on current deforestation would mean the countries currently removing forests most rapidly could benefit the most. Deforestation can occur at any time, and the potential returns from doing so, could rise if action elsewhere is successful. Potentially, as highlighted by Stiglitz[35], the combination of existing incentives in place to plant new forests, but no or insufficient incentives to preserve existing forests, could encourage perverse behaviour with forests being cut down, and then replanted. The result would be an increase in atmospheric carbon and a likely loss in biodiversity.

Under a global scheme, commitments by all countries to preserve natural forests and plant new forests could be rewarded appropriately. The design of a scheme should address the incentives so that the scheme is effective. Understanding and deciding upon the scale of transfers will be relevant to negotiations.

Finding agreement will need consideration by countries as to how to distribute available resources, and could prove challenging if a scheme were considered to channel excessive flows to a limited number of countries, or at the national level to

---

[35] Stiglitz (2006)

particular interest groups within countries. This might happen it a situation where the price of carbon was far higher than the cost of avoided deforestation. The difference might be considered rents or pure profits. Discounting and taxing credits offer options to handle the creation of excess rents.

Early action can reduce emissions significantly and allow learning to understand how to successfully address challenges arising from large-scale action.

International support for action by countries to prevent deforestation should start as soon as possible. Action starting with a few countries could start to turn the tide, and allow learning from the experience gained. In this way implementation can be used to refine and strengthen action as more countries choose to participate.

Since the rules for the first commitment period are already set, and do not include provision to credit reduced emission from deforestation, and there are difficulties with an immediate integration of deforestation into global markets, there is a need for pilot schemes. These pilot schemes will have to be separate from carbon markets in the first commitment period under the Kyoto Protocol, although the possibility for early crediting for the second commitment period exists.

The important step is to establish pilots to gain practical experience. Pilot schemes could be based on funds with voluntary contributions from developed countries, businesses and NGOs, This approach could also be an alternative to access to carbon markets for the longer term. Fund-based and market-based approaches largely share the preconditions just identified so it is not be necessary to make a final decision at the pilot stage. Practical experience will be needed for integration into global carbon markets or maintaining separate schemes.

Longer-term alternatives to inclusion in the carbon markets, by maintaining a separate but complimentary approach, offer the possibility of being more closely targeted on reducing deforestation and the issues associated with it. These alternatives might deliver savings more cheaply, depending on the long-term carbon price and the level of incentive required. These include:

Specialised funds

The advantage of specialised funds is that they can be targeted and directed to where they can provide most benefit. The stand-alone nature of protecting forests – there are few direct tradeoffs with other forms of mitigation -make it suitable for focused funds. A fund could work at country level, offering tailored support that provides resources at the outset of a programme and incentives to encourage success. It could also allow countries to generate resources for successfully tackling poverty and the other underlying drivers. The proposal by Brazil, see Box 25.6 could be developed into a specialised fund.

Targeting funding could allocate resources to individual country programmes depending on the opportunity costs faced, and could sharpen incentives. This could be better than a simple fixed global rate, which, depending on the level, could cost more overall or reduce the overall amount of action.

An example of a specialised fund for forests is the BioCarbon Fund, created in 2004 as a private sector trust managed by the World Bank. So far, the Fund is committed to a diversified portfolio of 23 projects worth $54m. Examples of the types of projects financed include, restoring forest ecosystems by connecting forest

BOX 25.6 Brazilian proposal of voluntary scheme[36]

At the UNFCCC Workshop in Rome in August 2006 Brazil proposed a scheme to offer positive incentives to developing countries that voluntarily reduce the greenhouse gas emissions from deforestation.

This would be a voluntary arrangement in the context of UNFCCC, that does not generate future obligations, and would not count towards emissions reductions commitments of Annex I countries. There would be a reference emission rate based upon previous deforestation rates, which would be periodically updated. This would allow annual or periodical emissions from deforestation to be compared to the reference level with standard values of carbon per hectare. Countries could earn credit, or debits (deducted from future incentives), with incentives distributed, according to the ratio of emissions reductions achieved.

This scheme has several elements in common with the Rainforest Coalition proposal - with the crucial difference that funding will be outside carbon markets. The proposal is that developed countries voluntarily share the cost of the scheme.

fragments with corridors, agroforestry projects, planting trees and improved forest management to enhance carbon storage.

Establishing separate markets for forest credits

A particular form of funding that could also be explored in the pilot phrase could be delivered through markets for biodiversity credits or deforestation credits. These credits would operate in a similar way to carbon credits, with demand coming in from those who wanted to invest in forestry projects linked to corporate social responsibility or other goals.

The credits could recognise a wider range of benefits than just avoided emissions. They could, for example, be based on the area of forest protected rather than complex measurement of carbon saved. If the credits were non-fungible with carbon finance, emissions reductions need not be the denomination, and it would not be necessary to look for parity with the global carbon price.

References

Alston, L.J., G.D. Lipecab and B. Mueller (2000): 'Land reform policies, the sources of violent conflict, and implication in deforestation in Brazilian Amazon', Journal of Environmental Economics and Management **39** (2): 162–188

Baumert, K.A., T. Herzog and J. Pershing (2005): Navigating the numbers: Greenhouse gas data and international climate policy'. Washington, DC: World Resources Institute.

Chunquan, Z., R. Taylor and F. Guoqiang (2004): 'China's wood market, trade and the environment', Washington, DC: WWF/Science Press USA Inc.

Food and Agriculture Organization of the United Nations (2005a): 'Global Forest Resources Assessment 2005: Progress towards sustainable forest management' Washington, DC: United Nations.

Food and Agriculture Organization of the United Nations (2005b): 'State of the world's forests', Washington, DC: United Nations.

[36] Presentation by Mr. Joao Paulo Ribeiro Capobianco to UNFCCC Workshop on Reducing Emissions from Deforestation in Developing Countries, Rome 30 Aug to 1 Sept 2006 "Positive incentives to reduce emissions from deforestation in developing countries: Views from Brazil".

Grieg-Gran, M. (2006): 'The cost of avoiding deforestation' – report prepared for Stern Review, International Institute for Environment and Development.

Girling, R. (2005): 'We're having a party', Sunday Times Magazine, July 3rd 2005.

Kremen, C., J. O. Niles, M. G. Dalton, et al. (2000): 'Economic incentives for rain forest conservation across scales', Science **288**(5472): 1828–1832, June.

F.D. Merry, P.E.Hildebrand, P. Pattie and D.R. Carter (2002): 'An analysis of land conversion from sustainable forestry to pasture: a case study in the Bolivian Lowlands'. Land Use Policy **19**: 207–215

Minten, B. and P. Méral (2005): 'International trade and environmental degradation: A case study on the loss of spiny forest cover in Madagascar', in Minten, Bart, eds.Trade, Liberalization, Rural Poverty, and the Environment: the Case of Madagascar, Washington, D.C: WWF.

Moser, C., C.B. Barrett, and B. Minten (2005): 'Missed opportunities and missing markets: Spatio-temporal arbitrage of rice in Madagascar'. SAGA Working Paper. 180. New York: Cornell and Clark Atlanta Universities.

Kindermann, G., M. Obersteiner and E. Rametsteiner (2006): 'Potentials and cost of avoided deforestation', presentation made at the Workshop on Reducing Emissions from Developing Countries, Bad Blumau, Austria, May.

Prentice et al. (2001): 'The carbon cycle and atmospheric CO_2', in Climate Change 2001: The Scientific Basis. Contribution of Working Group I to the Third Assessment Report of the Intergovernmental Panel on Climate Change [Houghton J. et al. (eds.)]. Cambridge: Cambridge University Press.

Priyadi, H., P. Gunarso, P. Sist and H. Dwiprabowo (2006): 'Reduced-impact logging (RIL) research and development in Malinau research forest, East Kalimantan: a challenge of RIL adoption', Proceedings ITTO-MoF Regional Workshop on RIL Implementation in Indonesia with Reference to Asia-Pacific Region: Review and Experiences, I. Gusti Made Tantra, E. Supriyanto (eds.). Bogor: ITTO and Center for Forestry Education and Training, Ministry of Forestry, pp 153–167.

Shively, G. and S. N. Pagiola (2004): 'Agricultural intensification, local labor markets, and deforestation in the Philippines', Environment and Development Economics **9**: 241–266

Sohngen, B. (2006): 'Cost and potential for generating carbon credits from reduced deforestation', presentation made at Workshop on Reducing Emissions from Developing Countries, Bad Blumau, Austria, May.

Stiglitz, J.E., (2006): Making globalization work', New York: WW Norton.

UNDP (2001): 'World Energy Assessment: Energy and the challenge of sustainability', New York, UNDP.

World Bank (2006): 'At Loggerheads? Agricultural expansion, poverty reduction, and environment in the tropical forests', Washington, DC: World Bank.

26 International Support for Adaptation

KEY MESSAGES

Adaptation efforts in developing countries must be accelerated. Adaptation is essential to manage the impacts of climate change that have already been locked into the climate system.

The poorest developing countries will be hit earliest and hardest by climate change, even though they have contributed little to causing the problem. The international community should support them in adapting to climate change. Without such support there are serious risks that development progress will be undermined.

Transfers to developing-country governments and civil society will be necessary to support adaptation. Additional costs to developing countries of adapting to climate change could run into tens of billions of dollars. **Donors and multilateral development institutions should mainstream and support adaptation across their assistance to developing countries.**

Public-private partnerships for climate-related insurance can help to support adaptation. At the household level, remittances are likely to have an important role in supporting autonomous adaptation.

The international community should also support adaptation through investment in global public goods, including:

- Improved monitoring and prediction of climate change;

- The development and deployment of drought- and flood-resistant crops;

- Methods to combat land degradation;

- Better modelling of impacts.

In addition, efforts should be increased to improve mechanisms for improving risk management and preparedness, disaster response and refugee resettlement.

The scale of the challenge makes it more urgent than ever for developed countries to honour their existing commitments – made in Monterrey 2002, and strengthened at the EU in June 2005 and at the G8 Gleneagles meeting in July 2005 – to double aid flows by 2010. Strong growth and development will enhance countries' ability to adapt.

Strong and early mitigation has a key role to play in limiting the long- run costs of adaptation. Without this, the costs of adaptation will rise dramatically.

26.1   Introduction

Adaptation is different from mitigation in two key respects: first, it will in most cases provide local benefits, and second, these benefits can be realized without long lead times (as discussed in Chapter 18). As a result, private actors – households, communities and firms – will carry out much adaptation on their own, without the active intervention of policy, in response to actual or expected climate change. People in even the smallest and poorest developing countries would benefit from any action they undertake to adapt their economies and societies in ways that make climate change less costly to them.

However, there are many barriers to effective adaptation ranging from a poverty-driven low adaptive capacity to market failures such as incomplete information. Government policy and support will therefore be critical in assisting and complementing individual responses, as set out in Part V. But governments in turn will require support from the international community. The poorest countries are the most vulnerable to the impacts of climate change *and* are particularly short of the resources required to manage a changing climate. The ethical foundations for this support were discussed in Chapter 2. Briefly they are (i) that common humanity points to support for the poorest members of the world community, and to efforts to build a more inclusive society, (ii) the historical responsibility of industrialised countries for the bulk of GHGs concentrations, and (iii) a common interest in avoiding the instabilities that could arise from the transfer of the dislocation of climate change.

The developed world should provide support for adaptation, including through existing aid delivery mechanisms for development and investment in global public goods. Under Article 4.8 and 4.9 of the UNFCCC, the Least Developed Countries are recognized as being among the most vulnerable to the adverse impacts of climate change, and all signatory countries are obligated to help developing countries adapt. Furthermore, many developed countries have acknowledged that there is a strong case for assistance. At the ninth Conference of the Parties (COP), Canada, the EU, Iceland, New Zealand, Norway and Switzerland reconfirmed an earlier pledge of $410 million by 2005 for the Special Climate Change Fund (SCCF) and the Least Developed Country Fund (LDCF).[1]

This chapter is divided into four broad issues that will require international collective action: honouring and improving current international commitments to assistance for development and, specifically, adaptation to climate change (Section 26.2); recognising and facilitating the role of international private financing for adaptation (Section 26.3); promoting and providing global public goods (Section 26.4); and improving international support for disaster risk reduction (Section 26.5).

---

[1] Nevertheless, many developing countries still believe too little is being done. For example, at Montreal in 2005 Bangladesh suggested a shift from the politics of aid to one of legal obligation where there could be 'compensation for damages due to unavoidable adverse impacts of climate change', and suggested that 'if voluntary obligations are not working then binding commitments might be necessary to secure adequate funds.'

26.2 International assistance for adaptation

The scale of the challenge posed by climate change and adaptation makes it more urgent than ever that donor countries honour their commitments – made in Monterrey 2002, and strengthened at the EU in June 2005 and at the G8 Gleneagles meeting in July 2005 – to double aid flows by 2010.

As Part V explained, autonomous adaptation may consist of a single farmer changing crop varieties or changing planting dates, to moving production or distribution facilities, or even leaving a country/region entirely. A major role of governments in tackling climate change will be to ensure that the private sector has the tools and incentives necessary to adapt autonomously. Helping people to build and develop their human capacity through investment in health and education, facilitating growth and diversification, and encouraging general development will be critical in supporting individual action to adapt. In addition, there will be an important role for Government in:

- Providing and disseminating information about climate change, and its likely impacts;
- Providing the additional services, and infrastructure investment that may be required to manage and prevent the impacts of climate change. For example, better water management, flood defences and agricultural extension services.

For developing countries, and especially the poorest developing countries, adaptation to climate change will substantially raise the costs of some investments, and may also require investments in new areas. These new demands will place pressure on already very scarce public resources. Meeting the Millennium Development Goals already requires international assistance to support action by developing countries. Climate change – and the need for adaptation – will pose an additional challenge for countries' growth and poverty reduction ambitions.

A major aspect of accelerating adaptation should be implementing good development practice. As Chapter 20 argued, many actions to promote growth and development should also help to reduce the vulnerability of developing countries to climate change and raise their ability and capacity to adapt. Scaling up development assistance will therefore be essential. And the developed country commitments to increase overall ODA – made at Monterrey in 2002, and reaffirmed at the G8 summit in Gleneagles in July 2005 – will therefore take on an even greater importance. The recent DFID White Paper on eliminating poverty summarises those historic commitments: donor countries pledged to "increase aid by $50 billion a year by 2010, with $25 billion of that to go to Africa; cancel debt worth another $50 billion; and provide AIDS treatment to all who need it by 2010".[2] (See Figure 26.1 below). ODA from DAC donors alone could double between 2004 and 2015 if the commitments and EU targets for 0.7% GDP in ODA by 2015 are met. So far, five DAC donors have met the 0.7 ODA/GNI ratio, and five others have announced timetables to meet this target.[3]

Recent increases in the efficiency of aid should make these flows more effective in helping recipient countries to tackle the additional challenge of adaptation. As

---

[2] UK Department for International Development (2006a)
[3] Additional ODA growth will come from non-DAC donors who are growing in importance.

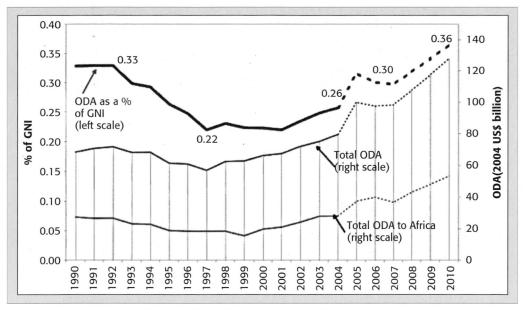

Figure 26.1 Scale of ODA if DAC donors honoured their commitments

Source: OECD (2005)

emphasized in the Commission for Africa report, three sets of factors have increased aid efficiency over the past decade or more: (i) improvements in policies, governance, and investment climate in recipient countries; (ii) aid allocations that have shifted more resources to countries that can use them well; and (iii) better quality of aid delivery.[4] In addition, the projected phase-in of aid increases over several years will also make it easier for recipients to use aid efficiently.

Looking to the future, and as set out in Part III, the international community should also recognise the crucial role of mitigation in limiting the potential damage from climate change. Without strong and early mitigation, the long-run costs of adaptation will rise sharply, and substantial additional resources will be necessary to finance this and to realise the internationally agreed poverty reduction goals.

To complement the broader increases in development budgets, a range of different funds have been developed under the UNFCCC to develop and integrate approaches to adaptation.

The main mechanisms for supporting adaptation are donor contributions to the Global Environment Facility (GEF) special funds for adaptation, the Adaptation Fund, and ODA and concessional lending of which a very small proportion (significantly less than 1%) is specifically focused on adaptation.[5] (See Box 26.1). World Bank estimates of the costs of adaptation in developing countries are in the tens of billions of dollars (discussed in Chapter 20). Contributions to dedicated adaptation funds are projected to amount to between $150–$300 million per year. In this context, the World Bank recently recognised the essential role of the

---

[4] Commission for Africa (2005). See Chapter 9 *Where will the money come from: Resources*
[5] World Bank (2006a)

International Financial Institutions in "ensuring that maximum impact is obtained from these funds by mainstreaming appropriate assessment and response to climate risk in the global development portfolio".[6]

International support to manage the effects of climate change will be significantly more effective if it fits with the rest of the international ODA architecture. This includes the Paris Declaration on Aid Effectiveness that focuses on the need to develop and reinforce national development plans, strategies and budget processes.[7]

BOX 26.1 Existing sources of dedicated funding for adaptation

A range of funding streams is available to support adaptation in developing countries:

GEF and associated funds

To help countries adapt to the adverse impacts of climate change, the Global Environment Facility (GEF) supports projects that reduce countries' vulnerability to climate change impacts and helps them build adaptive capacity. The GEF has adopted a three-stage approach to adaptation:

- Stage I: *planning* through studies to identify vulnerabilities, policy options, and capacity building.

- Stage II: *identifying measures to prepare* for adaptation and further capacity building.

- Stage III: *promoting measures to facilitate adaptation*, including insurance and other interventions.

GEF resources (established under the Climate Convention) include:

Least Developed Country Fund (LDCF): The GEF established the LDCF to address the extreme vulnerability and limited adaptive capacity of Least Developed Countries (LDCs). The LDCF initially supported preparation of National Adaptation Programmes of Action (NAPAs). To date, a majority of LDCs have received funds to prepare their NAPAs, many of which are now close to completion. The NAPAs conclude with a list of prioritized project profiles to be subsequently implemented with support from the LDCF. Pledges and contributions to the LDCF amount to $89 million as of April 2006.[8]

Special Climate Change Fund (SCCF): Adaptation activities to address the adverse impacts of climate change have top priority for funding under the SCCF, which is aimed at supporting activities in adaptation, technology transfer, economic diversification, and energy, transport, industry, agriculture, forestry, and waste management. The SCCF addresses the special needs of developing countries in long-term adaptation, with priorities given to health, agriculture, water and vulnerable ecosystems. To date, $45 million has been pledged in contributions to support adaptation and the transfer of technology.[9] There is currently a lack of agreement over the operational guidelines on economic diversification for this fund that has proved to be a constraint.[10] This issue relates to whether

---

[6] World Bank (2006a:46)

[7] Key principles include: ownership, alignment, harmonisation, managing for results, accountability and governance. www.oecd.org/dataoecd/11/41/34428351.pdf.

[8] World Bank data

[9] World Bank data, as of 25th September, 2006

[10] World Bank (2006a)

oil-producing countries should be compensated for lost revenues as a result of global agreement on reducing carbon emissions.

Neither fund is subject to the resource allocation framework of the main GEF Trust Fund and may receive between $100 million to $200 million per annum in donations.

Adaptation Fund

With the entry into force of the Kyoto principle, a 2% levy on most Clean Development Mechanism (CDM) transactions will be directed to an adaptation fund. The size of funding this will generate depends on both the extent to which the CDM is used and the carbon price (discussed in Chapter 23). The World Bank has estimated that the Adaptation Fund will generate funding in the range of $100–$500 million through to 2012. The priorities and management of the Adaptation Fund is still being negotiated.

Procedures for accessing international funding streams should be simple and transparent to ensure easy access by developing countries. Some commentators have suggested that the current adaptation funds should be unified and the process for access simplified to facilitate uptake by developing countries.[11] The role and demand for these funds should be kept under review to ensure that they are well placed to develop approaches to adaptation, are adequately resourced, and support the overall goal of ensuring that the pressures and risks posed by climate change are taken into account across all aspects of development.

New mechanisms to raise additional funding for development have also been proposed, with proposals for funding streams earmarked to particular activities, including adaptation.

A variety of additional mechanisms to scale up international funding for development have been proposed.[12] For example, the French government is introducing an air ticket tax linked to funding for HIV/AIDS. A number of specific suggestions have been made for mechanisms earmarked for adaptation. Box 26.2 summarises briefly some of those options.

While some of these options may have potential, they all suffer from the disadvantages common to all dedicated funds. Public finance principles would generally militate against the earmarking of revenues, on the grounds that it prevents efficient resource allocation across government. Dedicated funding sources could also make it harder to mainstream adaptation, if the funded activities are viewed as being outside the normal budgetary process. Given the far-reaching nature of the adaptation challenge, stand-alone funds and activities financed by supplementary levies and divorced from overall development budgets might make more difficult the task of integrating adaptation into the mainstream of development and its funding. Any additional funding for adaptation should therefore aim to feed into normal budgetary processes, and clearly within national development plans.

Donors should mainstream adaptation across their development programmes, to address the affects of climate change in all countries and sectors.

---

[11] For example Burton (2005), Huq (2006), Bouwer and Aerts (2006)
[12] Atkinson (2004)

BOX 26.2 Some alternatives for new dedicated funding streams for adaptation

A number of commentators have suggested possible dedicated financing mechanisms for adaptation in developing countries:

Levies on Joint Implementation Projects: As noted above, a 2% levy is applied on projects included within the CDM. This levy could apply also to Joint Implementation projects undertaken in transition countries. However, it should be noted that the existing levy has a perverse effect: while supplying funds for adaptation, the levy reduces the incentive for the private sector to invest in mitigation in developing countries and thus, ultimately, countries will have to adapt further.[13]

Adaptation levy: Some commentators have proposed the use of adaptation levies.[14] In particular, they suggest an air ticket levy may be particularly relevant given the low levels/exemptions from taxation from which it has benefited historically, and the projected growth in aviations emissions.[15] Such a levy could distinguish between short- and long-haul flights and classes of travel, and could be argued to have advantages on grounds of both equity (taxing "luxury" emissions rather than "survival" emissions) and efficiency (using a price instrument rather than quantity).[16] This type of levy would help to create disincentives to emit GHGs. The idea, which has been mooted by various commentators, has already been put into practice in the context of funding for health and education, among other sectors. The UK and France have recently made announcements in this area. France began collecting an air ticket levy in July 2006 and expects it to generate annual revenues of euros 200 million. They will hypothecate part of the duties raised to provide a long-term source of funding to an international drug purchase facility called UNITAID. The UK has an existing air ticket tax – the Air Passenger Duty – and some of the revenue from this will be allocated to the International Financing Facility for Immunisations (IFFIm).[17]

Auctioning of emissions permits: If auctioning were used to allocate some of the permits to emit GHGs, it would be theoretically possible to apportion a part of the auctioning revenue to help fund adaptation. There will, however, be many calls on the revenue that this generates. Finance Ministers will have to take decisions with regard to priorities, what will achieve the best value for money and the likely effects on the economy as a whole.

A new GDP-based levy on Annex 1 countries: Some commentators have suggested that a new levy on Annex 1 countries, set at a fixed percentage of GDP and allocated to adaptation, would be one way to give a clear funding commitment under the UNFCCC.[18] This option should be distinguished from using ODA increases, since this levy would provide a separate dedicated funding stream.

---

[13] This assumes that the CDM levy is kept – from an efficiency perspective it would be better to remove the levy from the CDM entirely.
[14] Mueller and Hepburn (2006)
[15] According to the IPCC (1999) this amounts to up to 15% of global emissions by 2050.
[16] Benito Mueller (2006)
[17] The IFFIm will use up-front long-term financial commitments from donors to provide additional resources more quickly and predictably.
[18] Bouwer & Aerts (2006)

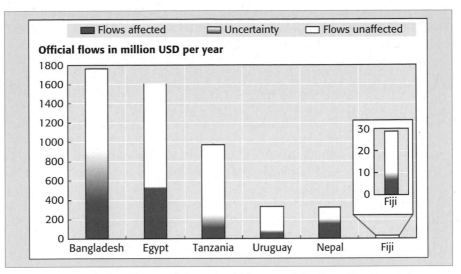

Figure 26.2 Annual official flows and share of activities potentially affected by climate change

Source: van Aalst and Agrawala (2005)

Chapter 20 discussed the importance of national governments integrating adaptation into their budgets and programmes. The same is true for donors – there is a role for the international community, including the development banks, in working with partner countries to promote a coherent response to climate change. A major aspect of accelerating adaptation should therefore be ensuring that development projects take account of climate change. An OECD analysis of ODA flows to six developing countries indicates that a significant portion of this aid is directed to activities potentially affected by climate risks, including climate change. Estimates range from as high as 50–65% of total national aid flows in Nepal, to 12–26% in Tanzania.[19] This is illustrated in Figure 26.2.

The international community has an important role in assisting countries as they develop their national development strategies (or poverty reduction strategies) to take account of adaptation across all government departments. Linked to this, the group of 50 LDCs have been asked to prepare National Adaptation Programmes of Action (NAPAs, discussed in Chapter 20). Effective NAPAs should help to ensure that national development strategies reflect adaptation priorities, and also help in the allocation of resources for adaptation. To date, five countries (Bangladesh, Bhutan, Malawi, Mauritania, and Samoa) have completed their NAPAs, and the costs of the priority projects they have identified total $133 million. Whilst NAPAs are useful in identifying funding priorities, it is important that the priorities they highlight are factored into broader national planning to ensure they are sustainable and effective – especially where they involve long-term investment decisions. For example, improving the resilience of drainage systems to the effects of climate change should be considered in the context of overall urban planning.

---

[19] van Aalst and Agrawala (2005)

26.3   The role of international private financing for adaptation

Private-sector financing for adaptation will come not only from domestic firms and households, but also potentially from international sources.

Remittances are the largest source of external financing in many developing countries. In 2005, remittance flows are estimated to have exceeded $233 billion globally, of which developing countries received $167 billion. Unrecorded flows amount to an additional 50% of the recorded flows.[20] In Ghana, for example, remittances account for 10–15% of national income compared with 3% from foreign investment, whilst in Bangladesh the wealth of the diaspora and the prevalence of migrant labour have led to remittances totalling $3.6billion in 2005, more than double ODA.[21] Remittances are especially important in times of crisis where they can provide very rapid and targeted financial assistance to those affected by climatic events and other crises. Banks and money transfer companies recorded sharp rises in remittances sent to the areas affected by the Pakistan earthquake and Asian tsunami immediately following those events, with increases of up to 400% in some cases. Because remittances usually accrue at the household level, they may be particularly important in funding autonomous adaptation of households.

Both private and public sector actions are needed to further unlock the potential of remittances to support adaptation. For example through making financial services, including remittance transfers, more accessible and better tailored for low-income senders and recipients. The public sector needs to ensure that favourable policies and legal environments are in place to encourage low value remittances to flow through licensed remittance providers (rather than informally), and that developing country payment systems are sufficiently well developed to distribute remittance flows efficiently and equitably to low income recipients too, who may not yet be banked with a country's largest banks.

Foreign direct investment (FDI) has also become important in many developing countries, particularly those in the upper middle-income category. While FDI flows will continue to be driven by the profit motive, they may – in some instances – also help to meet the incremental investment costs of adaptation. This may be the case, if, for example, the host country has regulatory requirements in place (such as building codes and standards for infrastructure). In such circumstances, foreign investors have the potential to demonstrate new ideas and technologies for dealing with and accelerating adaptation. The significance of FDI in facilitating and supporting adaptation will, however, vary between developing countries according to the scale of flows. Official flows, in the form of grants and loans, are much more significant for low-income countries, as demonstrated in Figure 26.3, and thus a higher priority area for integrating into development activities.[22]

Public-private partnerships, which harness the power of the market for public goals, are an attractive mechanism for supporting adaptation. Donors are beginning to use PPPs to promote the development and use of climate-related insurance markets in developing countries. There is great potential for expansion in this area.

---

[20] World Bank (2006b). Remittance flows are defined as the sum of workers' remittances, compensation of employees, and migrant transfers in the balance of payments statistics collected by the IMF.

[21] IMF (2005)

[22] van Aalst and Agrawala (2005)

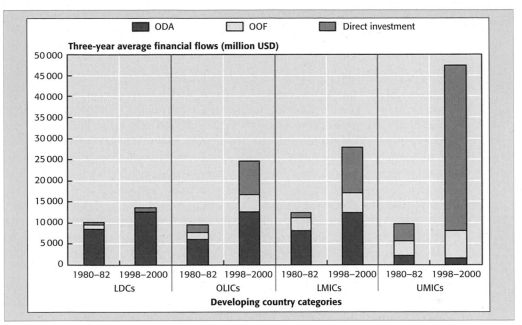

Figure 26.3 Official and private financial flows to developing countries (2000–04)

Source: van Aalst and Agrawal (2005)

It is crucial to develop insurance markets that can spread the growing climate-change risks, especially away from the most vulnerable households and countries. Part V discussed the importance of national-level action to develop such markets, but this action will require international support. Scale is crucial for insurance to be effective in reducing risk, because of the benefits of diversification across individuals and communities with uncorrelated risks (through re-insurance, for example). International risk-sharing mechanisms can also help in providing an element of subsidy for the poorest people and the poorest countries.

One approach to providing this international support is through public-private partnerships (PPP), which unite public institutions, private companies, and NGOs in an attempt to meet public goals by harnessing private efficiency and resources. A new example of such PPPs in the area of insurance is the Global Index Insurance Facility (GIIF), now being set up by the World Bank and the EU. This will help countries to access insurance markets for weather and natural disasters.

The GIIF will combine private and donor capital to support index-based insurance schemes (like weather derivatives) in developing countries. This risk-taking entity would originate, intermediate and underwrite indexable weather, disaster and commodity price risks in developing countries. The GIIF would lower the entry barrier to international insurance markets by pooling smaller transactions, thereby scaling up the transfer of risk from developing countries to those better able to carry these risks. At the local level the GIIF will promote capacity development of the financial sector. Current estimates are that annual risks totalling $0.2–$11.7 billion could be transferred to the market. A rough potential GIIF pipeline overview, based only on the projects led by the World Bank, suggests overall expected volumes of risk of $136 million in 2006, $214 million in 2007, and

$302 million in 2008.[23] Adoption of index-based insurance schemes will be more straightforward in those developing countries with relatively more sophisticated and deep financial systems (such as in South East Asia). The GIIF could help to stimulate adoption of insurance schemes in low-income countries, though may need to be supplemented with publicly-funded technical assistance.

One concern about using market-based insurance mechanisms to share risk is that the poorest households and countries will not be able to afford the premiums. Specific support to address weaknesses in developing countries' financial markets – for example, through technical assistance and capacity building – can help to tackle gaps in the domestic market. Precedents already exist for donor-supported insurance mechanisms; for example, the World Bank provides low-interest capital backup to the (public-private) Turkish Catastrophe Insurance Pool (TCIP) to make it affordable to property owners. Such initiatives can be on a local level (the Ethiopian weather derivatives, for example), a national level (as with the TCIP), or regional level (as has been proposed for the Caribbean states). Again, it is essential for any scheme to include incentives for participants to reduce their risks and, in the process, accelerate adaptation (as discussed in Chapters 19 and 20).

While this section has focused on PPPs supporting development of insurance markets, the PPP approach can be used elsewhere for adaptation as well. To date, most PPP efforts have been limited to mitigation activities to reduce GHGs. A key area in which to explore PPP would be the development of climate-resilient crops. Experience from previous publicly supported crop research demonstrates

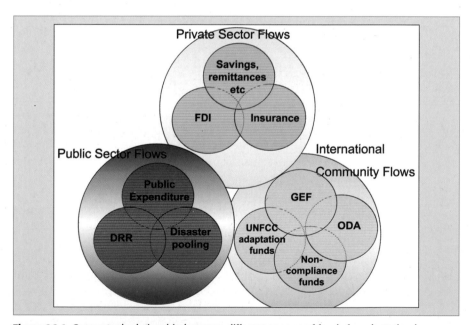

Figure 26.4 Conceptual relationship between different sources of funds for adaptation in developing countries at the national level

Source: Adapted from Bouwer & Aerts (2006)

[23] CRMG (2006)

the efficacy of this public-private approach. During the Green Revolution of the 1960s through 1980s, most crop research in wheat and rice particularly was financed by the public sector; now the majority is in the private sector. However, many advances are still prompted by publicly-funded research at universities and research institutions.

Figure 26.4 above summarizes current funding sources for adaptation from the public and private sectors and the international community.

26.4   Global public goods

In addition to providing financing directly to developing countries, the international community should invest in global public goods for adaptation.

Section 26.2 focused on mechanisms for direct international funding of the increased adaptation costs in developing countries. Given the arguments about mainstreaming, the key recommendation is for rich countries to deliver on their overall aid commitments. But there is much more that the international community can, indeed should, do to accelerate adaptation.

Ensuring global public goods (GPGs) are adequately financed will be especially important. While most adaptation measures will be at the individual, community, and country level, there are some global activities supporting adaptation where international co-ordination will be appropriate. These will tend to be characterised by benefits that can be shared widely at little cost, have economies of scale, and do not differ greatly across countries, so that the public good has international reach. Three important areas for global public good investment are discussed here:

- *Monitoring, forecasting, and researching climate change:* Adaptation will depend on comprehensive climate monitoring networks, and reliable scientific information and forecasts on climate change – a key global public good. Chapter 20 argued that developing-country governments should provide information to their own citizens but currently lack the capacity to do this, demonstrated by the shortage of weather watch stations. The international community should therefore support global, regional and national research and information systems on risk, including helping developing-country governments build adequate monitoring and dissemination programs at the national level. Priorities include measuring and forecasting climatic variability, regional and national floods, and geophysical hazards.[24] International networks of scientific organisations could enhance collaboration across national borders, such as the Global Climate Observation Systems. Following the Commission for Africa report, the G8 committed at Gleneagles in 2005 to help Africa obtain full benefit from the Global Climate Observing System with a view to developing fully operational regional climate centres in Africa. It is estimated that $200 million over 10 years is required for the Climate for Development in Africa programme; so far, very few pledges have been committed. As another example of possible GPG contributions in this area, the UK's

---

[24] Benson and Clay (2004)

Hadley Centre has developed a portable version of its Regional Climate Model, which is freely available for researchers in developing countries to run on standard computers.[25]

- *Research to improve crop resilience and reduce GHG emissions from agriculture:* The Consultative Group for International Agricultural Research (CGIAR) has proposed a new global challenge program that couples advances in agricultural science with research to mitigate climate change and adapt agriculture to its anticipated effects. That research could focus on development of rice varieties and water-management practices that reduce methane emissions; and crop varieties that resist higher temperatures, tolerate greater disease and insect pressures. They also need to withstand exposure to drought and excess water. Research is also needed into more efficient use of nitrogen fertilizers; simpler and more accurate ways to measure soil carbon; and farming systems that sequester carbon more effectively.[26] Such GPG investments have the potential for very high returns: evaluation research has estimated that the $7.1 billion (in 1990 US$) invested in CGIAR in the past has had a benefit-cost ratio of at least 9.0.[27] This type of research, particularly when coupled with the objective of strengthening national agricultural research systems, is highly valuable to developing countries. Box 26.3 describes the beneficial effects of research into improving rice plants and better use of fertiliser which enables positive adaptation by increasing rice yields in a changing climate. This is also an important example of an activity that combines both adaptation and mitigation benefits as the outcome contributes to a reduction in GHG emissions.

- *New methods to combat land degradation:* An important element of adaptation will be to prevent projected increases in the frequency of drought from leading to desertification. Approximately 2 billion people live in expanding drylands that currently cover 40% of the earth's surface. Protecting the biophysical foundations of agriculture – biodiversity, forests, livestock, soils, and water, are essential to combating the spread of desertification.[28] New techniques such as applying small amounts of fertiliser, or micro-dosing, increased grain yields by 30–50% in West Africa. Improved agro-forestry practices are helping regenerate nutrient-depleted soils in east Africa, while watershed programmes are reducing soil loss and increasing cropping intensity. Most adaptive practices will involve changes to farming or land management systems. Sometimes these systems can be transposed from elsewhere, others have to be developed and tested. This will require coherent programmes of information sharing, modelling of impacts, pilot programmes and extension services. Developing and testing such techniques is a global public good that would be a good focus for investments by the international community.

These global public goods are to some degree already funded internationally (for example, through the CGIAR or the World Bank), but they should be targeted

---

[25] http://precis.metoffice.com

[26] http://www.cgiar.org/impact/global/climate.html

[27] Under the plausible assumption the benefits will continue at present rates through 2011, the ratio rises to 17.3. Raitzer (2003)

[28] In recognition of the problem, the United Nations declared 2006 the International Year of Deserts and Desertification.

BOX 26.3 Adaptation and mitigation in rice production

Research into new rice plants could produce greater resistance to the changing climate and better grain quality. Wetland rice agriculture is also a major source of methane emissions due to anaerobic (without oxygen) decay of organic material caused by extended flooding periods. Higher yielding rice plants could utilise more carbon in its growth and hence reduce its emissions of methane. These higher yielding plants could also sequester more atmospheric CO_2 and utilize fossil fuel-based fertilisers more efficiently. New rice varieties could also yield higher revenues for rice farmers: for example, using one new rice variety, IR36, released in 1976 and planted on 11 million hectares in Asia in the 1980s, produced an additional 5 million tons of rice a year, boosting rice farmers' incomes by $1 billion.

Changes in fertiliser use can also have the dual benefit of reducing nitrogen oxide emissions from fertilisers and reducing indirect emissions from producing and transporting it. Rice plants can use the higher CO_2 concentrations in the atmosphere to their advantage by assimilating more carbon and using it to produce higher yields. However this CO_2 uptake effect can only be used when the plant has a sufficient nutrient supply. Site Specific Nutrient Management (SSNM) is an approach to application of fertilisers that uses the local characteristics of the land to determine how fertilisers should be applied. Balanced fertilisation, as developed under SSNM could improve nutrient supply using 30–40% less nitrogen fertiliser.

Initial evaluations of the use of SSNM in a large number of farmers fields in Asia finds significant environmental and financial benefits of SSNM over a range of fertiliser and rice prices. The costs associated with SSNM include additional time requirements for farmers' decision-making, but no significant up-front investment costs. In many rice growing countries fertilisers are subsidised, so lower use would also bring savings to the public finances: for example, in Indonesia the government spends $300 million on fertiliser subsidies and its minister of agriculture has requested a review of the subsidy level following roll-out of SSNM in the country.

Source: International Rice Research Institute (2006)

more directly at adapting to future climate-change challenges, in addition to responding to current problems. Given the extent of the inevitable climate change that is already on the way work on these GPGs should be intensified.

Investment in these global public goods should be scaled up; through existing mechanisms or through new instruments.

As already noted, for adaptation to climate change to be tackled effectively it should form an integral part of national development plans and budget planning. In addition, it is important to ensure the specific GPGs discussed above are funded fully. As such there may be a case for greater dedicated sources of funding to support these initiatives. This could be achieved either through existing mechanisms such as the GEF and the CGIAR, or through a new dedicated global fund and partnership.

Experience suggests that such dedicated funds can play a useful role where insufficient attention is being paid to an area, or where working across countries would add value.[29] These funds take advantage of returns to scale and collaboration in cases where action is urgently needed. Past efforts have had some success. A

[29] For a discussion of strengths and weaknesses of vertical funds, see UK DFID Practice Paper (2006b) How to work effectively with global funds and partnerships

recent review by the World Bank of 26 global funds (including the Prototype Carbon Fund and the Fund for the Implementation of the Montreal Protocol (MLF)) found that programmes delivering global public goods often add value, and rate well in their impacts on tackling the policy, institutional, infrastructural, and technological constraints that developing countries face.[30]

Effectiveness and efficiency suggests that the approach of choice should be built on existing mechanisms (such as the GEF). There are risks associated with a proliferation of vertical funds – in particular they can complicate efforts to co-ordinate aid and gain the full support of national governments.

26.5   Risk management and risk preparedness: responding to disasters and resettling refugees

More investment is required to manage and reduce the consequences of climate change.

Given the projected increase in frequency and intensity of climate-related disasters, the international community should support greater investment in managing and reducing the consequences of climate change through better risk management and preparedness, including improving mechanisms for refugee resettlement. This is especially important given that a recent World Bank report concludes: "[r]e-allocation is the primary fiscal response to natural disasters. Disasters have little impact on trends in total aid flows".[31]

Disaster risk reduction (DRR) includes the whole spectrum of prevention, preparedness, response and recovery. It focuses primarily on reducing the vulnerability of poor people by building capacity and livelihood resilience. DRR involves learning lessons from previous natural disasters, and working with governments at the local, national and regional levels to address the fundamental causes and consequences of the loss of lives and livelihoods. This includes:

- Reforming national disaster management agencies and establishing stronger co-ordination mechanisms between relevant line ministries;
- Linking community-level experience with national-level policy making;
- Strengthening building codes and land-use;
- Establishing well-resourced and prepared response systems with a focus on national and local capacity.

The key to successful DRR is ensuring it is integrated into development and humanitarian policy and planning. More effective financing for DRR should be based on country led approaches where national governments are accountable and committed to long-term investment.

While DRR will be essential in improving the resilience and capacity of poor people to manage a changing climate, it is impossible to avoid disasters altogether. Funding for humanitarian aid and improvement in the institutions and mechanisms for disaster recovery are critical. (See Parts II and V for a discussion of disaster recovery.) The international community has recognized the need for

---

[30] World Bank (2004)
[31] Benson and Clay (2004)

better, more integrated disaster-recovery systems that can react with greater agility, and has taken steps in that direction.

The disaster relief fund administered by the UN Office for the Co-ordination of Humanitarian Affairs has recently been renamed and re-launched as the Central Emergency Response Fund. The fund, launched in March 2006, has a target of $500 million (of which $222 million has been contributed so far).[32] UN agencies will be able to access these funds within 72 hours of a crisis. Individual agencies are also proposing to increase the sums that they can allocate to emergencies.[33] As discussed in Chapter 20, this is reactive adaptation funding; but climate change will bring more disasters to react to, even with investment in preventive measures. This funding will need to continue to rise significantly.

At the macroeconomic level, the IMF has recently introduced an exogenous shocks facility (ESF) that should help with recovery from natural disasters or commodity price shocks, or indeed any "event that has a significant negative impact on the economy and is beyond the control of the government". The ESF will become effective once the multilateral debt relief initiative is officially implemented. The IMF already has facilities to provide assistance to countries hit by certain types of shocks – those in post-conflict situations (Emergency Post-Conflict Assistance, or EPCA) and countries afflicted by natural disasters (Emergency Natural Disaster Assistance, or ENDA). Assistance is also provided under the Compensatory Financing Facility (CFF). These instruments have not been heavily used and the effectiveness of the ESF should therefore be monitored; but, in principle it is a sound idea, and the emphasis should be on ensuring it can work well and is co-ordinated with other facilities.

Even with strong and rapid action to manage the consequences of climate change through adaptation, in some cases the only effective adaptation response will be to migrate to higher land or safer areas with greater access to food and water. Adequate arrangements will be required in extreme cases where populations must be resettled, most notably in the case of the vulnerable small island states (see Part II for details). The United Nations Refugee Agency, United Nations Office for the Co-ordination of Humanitarian Affairs, and the International Organisation for Migration (UNHCR, OCHA, and IOM) should take on expanded roles for resettlement if others do not step forward to do so, given the permanent nature of such migration in response to climate change.

Recipient countries should develop reception and resettlement terms and strategies, with possible cost sharing across a broader range of countries on equity grounds. There are some very limited precedents from other organized resettlements of populations, often in forced circumstances. For example, when volcanic eruptions made much of Montserrat's housing uninhabitable in the 1990s, residents were given the option of moving to the UK or Antigua, and more than half of the population resettled. In that case, because Montserrat is a British overseas territory, responsibility for action was relatively clear. By contrast, in the future much of the resettlement may have to be across international borders, so arranging it and sharing costs will likely be much more complex.[34] Managing these resettlements will require not only funding, but also political will and co-operation.

---

[32] Note that this is not only for climate related disasters.
[33] For example, in 2006 UNICEF proposes to increase their Emergency Programme Fund ceiling from $25 million to $75 million per biennium.
[34] Commission for Africa (2005); UN Habitat

26.6 Conclusion

Reducing the vulnerability of poor people to climate variability and climate change should be the starting point for adaptation efforts in developing countries. Poverty limits the ability to cope with and recover from climate shocks — particularly when combined with other stresses, such as a high disease burden, land degradation, weak institutions, governance challenges and conflict. Poor people do adapt, but are constrained by limited additional resources.

If the international community is to continue its commitment to ambitious development aspirations, support to developing countries in adapting to climate change will be essential. The key mechanism for doing this will be following through delivery on commitments to scale up aid for development, since adaptation is a crosscutting challenge that will affect all aspects of development. Specifically, it is crucial that developed countries live up to the commitments they made at Monterrey in 2002, EU June in 2005 and the G8 Gleneagles meeting in 2005 and related recent international fora. And mainstreaming climate change into development priorities and measures will help ensure consistency between action to achieve adaptation to climate change and action for growth and poverty reduction, on all its dimensions.

The other major area for action is in providing global public goods (GPGs) for adaptation. This will require increased international co-operation and perhaps also dedicated funding sources for GPGs. Key GPGs include improved monitoring and prediction of climate change, better modelling of impacts, the provision of drought- and flood-resistant crops. It also requires planning approaches and infrastructure design better suited to a world of climate change. Further investment will also be required to improve mechanisms for improving risk management and preparedness, disaster response and refugee resettlement.

References

Adger W. N., (2005): 'Governing natural resources: institutional adaptation and resilience'. In F. Berkhout, M. Leach and I. Scoones (eds.) Negotiating Environmental Change: New Perspectives from Social Science. 193–208. Cheltenham: Edward Elgar.

Agrawala, S. ed. (2005): 'Bridge over troubled waters: Linking climate change and development', Paris: OECD.

Atkinson, A. B ed. (2004) 'New sources of development finance — UNU-WIDER Studies in Development Economics', Oxford: Oxford University Press.

Baer, P (2006:) 'Adaptation to Climate Change: Who Pays Whom?' in Fairness in Adaptation to Climate Change, eds. W.N. Adger et al. Cambridge, MA: MIT Press

Bals, C., I. Burton, S. Butzengeiger, Andrew Dlugolecki, Eugene Gurenko, Erik Hoekstra, Peter Höppe, Ritu Kumar, Joanne Linnerooth-Bayer, Reinhard Mechler, Koko Warner, MCII (2005): 'Insurance related options for adaptation to climate change, available from http://www.germanwatch.org/rio/c11insur.pdf

Benson, C. and E. Clay, (2004): 'Understanding the economic and financial impacts of natural disasters', World Bank, Disaster Risk Management Series No. 4.

Burton, I. (2006) `Adapt and Thrive: Options for Reducing the Climate Change Adaptation Deficit' in Policy Options issue December 2005-January 2006 Global Warming – A Perfect Storm available from http://www.irpp.org/po/archive/dec05/burton.pdf

Bouwer, L.M. and J.C. Aerts, March 2006H (2006): 'Financing climate change adaptation', Disasters 30: No 1.

Commission for Africa (2005): 'Our Common Interest – report of the Commission for Africa', London: Penguin.

CRMG/World Bank (2006): 'Global Index Insurance Facility (GIIF), Concept Note (Synopsis)', Washington, DC: Commodity Risk Management Group (CRMG), ARD, World Bank, available from http://www.proventionconsortium.org/themes/default/pdfs/GIIF_overview_Feb06.pdf

Environmental Resources Management (2006): 'Natural disaster and Disaster Risk Reduction Measures- a desk review of costs and benefits' London: DIFID.

Huq, S. (2006a): 'Adaptation funding after Montreal' a report for Tiempo Climate Newswatch, available at http://www.cru.uea.ac.uk/tiempo/newswatch/report060401.htm

IMF (2005): Finance & Development **42**(4), December.

IPCC (1999): 'Aviation and the Global Atmosphere. A Special Report of IPCC Working Groups I and III in collaboration with the Scientific Assessment Panel to the Montreal Protocol on Substances that Deplete the Ozone Layer', J.E. Penner et al. eds., Cambridge: Cambridge University Press.

International Rice Research Institute (2006): 'Climate change and rice cropping systems: Potential adaptation and mitigation strategies', Philippines: IRRI, available at www.sternreview.org.uk http://www.sternreview.org.uk

Mechler, R., J. Linnerooth-Bayer, D. Peppiatt (2006): ProVention/IIASA study, 'Micro insurance for natural disaster risks in developing countries: benefits, limitations and viability'.

Mueller, B. (2006): 'Adaptation Funding and the World Bank Investment Framework Initiative', Background Report prepared for the Gleneagles Dialogue Government Working Groups Mexico June 2006.

Mueller, B. (2006): 'Montreal 2005: What happened and what it means', Working Paper of the Oxford Institute of Energy Studies, EV35, Oxford: Oxford Institute for Energy Studies.

Mueller, B. and C. Hepburn (2006 forthcoming): 'An international air travel adaptation levy (IATAL): an outline proposal', need permission to quote

OECD (2005) http://www.oecd.org/dataoecd/57/30/35320618.pdf

OECD (2004) International Development Statistics Online Databases, www.oecd.org/dac/stats/idsonline.

Pew Centre On Global Climate Change (2005): 'International Climate Efforts Beyond 2012: Report Of The Climate Dialogue At Pocantico', Virginia: Pew Centre.

Raitzer, D. A (2003) *Benefit-Cost Meta-Analysis of Investment in the International Agricultural Research Centres of the CGIAR*. Report prepared on behalf of the CGIAR Standing Panel on Impact Assessment, Science Council Secretariat, Food and Agriculture Organization of the United Nations (FAO).

Sperling, F., and F. Szekely, (2005): 'Disaster risk management in a changing climate', Discussion Paper, prepared for the World Conference on Disaster Reduction on behalf of the Vulnerability and Adaptation Resource Group.

Sperling, F. (ed) (2003): 'Poverty & Climate Change: Reducing the Vulnerability of the Poor through Adaptation', by a consortium of 10 donors including the AfDB, AsDB, DFID, Netherlands, EC, Germany, OECD, UNDP, UNEP and the World Bank (VARG).

Tol, R.S.J, S. Fankhauser and J.B. Smith (1998): 'The scope for adaptation to climate change: what can we learn from the impact literature?', Global Environmental Change, 8, Number 2(15): 109–123.

UK Department for International Development (2006a): 'Eliminating world poverty: Making governance work for the poor', London: DIFID.

UK Department for International Development (2006b) 'Practice Paper: How to work effectively with global funds and partnership', London: DIFID.

UNCTAD (2005): 'World Investment Report – FDI from Developing and Transition Economies-Implications for Development', New York: United Nations.

Van Aalst, M. and Agrawala, S. (2005) "Analysis of Donor-supported Activities and national Plans" in Bridge Over Troubled Waters: Linking climate change and development, Agrawala, S. ed. Paris: OECD.

World Bank (2004): 'Addressing the challenges of globalisation: an independent evaluation of the World Banks approach to global programmes' Washington, DC: World Bank.

World Bank (2005): 'Scaling up micro insurance: the case of weather insurance for smallholders in India', Washington, DC: World Bank.

World Bank (2006a): 'An Investment Framework for Clean Energy & Development: A Progress Report', Washington, DC: World Bank.

World Bank (2006b): 'Global Development Finance: the Development potential of surging capital flows', Washington, DC: World Bank.

World Bank (2006c): 'World Development Indicators', Washington, DC: World Bank.

World Food Programme Executive Board Annual Session (2006): Progress Report on the Ethiopia Drought Insurance Pilot Project, Rome: WFP.

27 Conclusions: Building and Sustaining International Co-operation on Climate Change

KEY MESSAGES

- **Very strong reductions in carbon emissions are required** to reduce the risks of climate change. They are likely to provide benefits well in excess of the costs. Indeed the costs of not acting strongly are likely to be very high.

- **Action is urgent** since stocks of GHGs are rapidly approaching dangerous levels, there will be heavy investment in energy infrastructure that could lock in future emissions, and it will take time to develop technologies that deliver zero emissions at low cost.

- Without a clear perspective on **the long-term goals for stabilisation** of greenhouse gas concentrations in the atmosphere, it is unlikely that action will be sufficient to meet the objective.

- **Action must include mitigation, innovation and adaptation,** and there are many opportunities to start now, including where there are immediate benefits and where large-scale pilot programmes will generate valuable experience

- **Countries should agree a broad set of mutual responsibilities** to contribute to the overall goal of reducing the risks of climate change. These responsibilities should take account of costs and the ability to bear them, as well as starting points, prospects for growth and past histories.

- **The challenge now is to broaden and deepen participation across all the relevant dimensions of action** – including co-operation to create carbon prices and markets, to accelerate innovation and deployment of low-carbon technologies, to reverse emissions from land-use change and to help poor countries adapt to the worst impacts of climate change.

27.1 Introduction

This Review has considered the economics of climate change, and has come to some clear and strong conclusions.

That the science of climate change is robust, and that the risks of a "business as usual" path for climate change are very serious.

What happens in the next 10 or 20 years will have a profound effect on the climate in the second half of this century and in the next. Actions now and over the coming decades could create risks of major disruption to economic and social activity, on a scale similar to those associated with the great wars and the economic depression of the first half of the 20th century. And it will be difficult or impossible to reverse these changes.

Second, and in contrast, the costs of action – reducing greenhouse gas emissions to avoid the worst impacts of climate change – can be limited to around 1% of global GDP.

Third, prompt and strong action is, therefore, clearly warranted. Because climate change is a global problem, the response to it must be international. And it must be based on a shared vision of long-term goals and agreement on frameworks that will accelerate action over the next decade.

Fourth, the economics can provide a strong foundation for developing policy frameworks to guide action, reducing the costs by providing flexibility over how, when and where emissions are reduced. The costs of acting on climate change will be manageable if the right policy frameworks are in place. There are also benefits along the way, if policy is designed well, for energy security, environmental quality, health and access to energy for poor people. These policy frameworks must deliver on three fronts: creating a price for carbon, via, taxes, trading or regulation; promoting the development and deployment of new technologies; and deepening understanding of the problems, thus changing preferences and behaviour and overcoming market barriers that might inhibit action, notably on energy efficiency.

This final chapter considers the next steps that could be taken to bring about more effective and better co-ordinated international action on climate change.

The key building blocks for any collective action include

- Developing a shared understanding of the long-term goals for climate policy
- Building effective institutions for co-operation
- Creating the conditions for collective action

27.2 Developing a shared understanding of the long-term goals for climate policy

The voluntary nature of collective action means that each individual country has to be committed to playing their part in responding to the challenge. Commitment ultimately comes from the understanding that climate change is a serious and urgent issue, and that through co-operation the risks can be reduced to the benefit of all.

There is an urgent need for public and international debate on the appropriate range for stabilisation of greenhouse gases in the atmosphere. A broad consensus on the long-term goals for the stabilisation of greenhouse gases in the atmosphere, or for comparable measures including cumulative emissions over long time scales, would underpin a shared understanding of the scale of the challenge for both mitigation and adaptation. Without a long-term goal, there are grave risks that a series of fragmentary or short-term commitments would lead to inconsistent policies that would raise the costs of action and fail to make a significant impact in reducing emissions.

The IPCC plays a vital part in assessing the scientific evidence and providing clear non-technical summaries that allow the issues to be widely debated. Long-term goals should be regularly revised in the light of its findings, and other developments, particularly concerning the development of technologies.

An improved understanding of the likely impacts of climate change on each region and country, and the impacts on the most vulnerable, should inform the international response. More research is required on key regional weather systems including the impact on monsoon rains, and funding is essential to fill the gaps in the Global Climate Observation System including over Africa. It will also be very important to deepen understanding of the implications of sea level rise for vulnerable people in low-lying countries and small island states.

Shared assessments of the potential of technologies for mitigation and adaptation are also essential to guide policy-makers in developing effective approaches to co-ordinate increases in national and international support.

27.3   Building the institutions for effective co-operation

The current institutions for monitoring, reporting and verification of emissions, established under the UNFCCC and Kyoto Protocol are basically sound. They have laid important foundations and should form a key element of continuing co-operation. But they are just a beginning: the challenge now is to expand the scale of activities and put them on a secure footing for sustained and long-term action. In a number of dimensions this will require that the world advances strongly and develops and adapts to institutional structures and methods of collaboration.

The Kyoto Protocol has also established an effective basis for the registration of formal intergovernmental trading in emissions. The development of parallel regional emissions trading schemes, including some which are outside the Kyoto framework, presents a new set of challenges. Trading between these schemes requires further development of institutions and mechanisms.

A transformation of flows of carbon finance, linked to strong and effective national policy in developing countries, will be required to support the transition to a low-carbon global economy. Other sources of finance are also required to work alongside the carbon markets, including the Global Environment Facility and the range of instruments available to the IFIs. The IFIs can play a valuable role in accelerating the process: the establishment of a Clean Energy Investment Framework by the World Bank and the regional development banks offers significant potential to do this.

Both multilateral and co-ordinated action could be enhanced by building a stronger institutional base for monitoring and reporting policy action to reduce greenhouse gas emissions and support innovation. This could include developing an enhanced role for institutions such as the IMF, World Bank, OECD and IEA in monitoring and reporting on relevant policy implementation.

The challenges of mitigation and adaptation are becoming a core part of the management of the economy, and it is essential that economic and finance ministries develop their capacity to shape effective policy responses.

27.4   Creating the conditions for collective action

Effective action to reduce global emissions to a level consistent with the stabilisation of greenhouse gases in the atmosphere will require the broadest possible participation. Achieving effective and co-ordinated action on climate change will require international frameworks that allow countries to establish mutual responsibilities across the full range of dimensions of action.

But this does not mean that no action can begin in advance of agreement on the next phase of multilateral co-operation. Pilot programmes could and should begin early, building on the recent initiatives by the multilateral development banks to develop frameworks for investment in clean energy and energy efficiency. This process will depend on early signals from developed countries about the likely role of carbon finance mechanisms beyond 2012.

The negotiating process could be designed to support energetic and mutually reinforcing action, bringing forward increasingly ambitious responses as countries begin to make tentative offers. It may be helpful to begin a dialogue on the basis of pre-commitments: offers from countries which do not become binding unless reciprocal offers are made. The EU has already begun to do this: the European Council declared in March 2005 that it was ready to begin exploring with other developed countries the scope for targets in the range of 15-30% reduction of emissions by 2020.

Creating the conditions for collective action will require a step change in political leadership. The first commitment period of the Kyoto Protocol ends in 2012. This is already too short a time horizon for those who are making investment decisions in long-lived capital stock. Uncertainty on the international framework makes it more difficult for national policy-makers to give clear signals to investors. Agreement on the key elements of international frameworks for action should be an urgent priority for all areas of government policy – extending beyond the remit of environment ministries to include heads of state, foreign ministers and ministers of finance

Some of the elements of future international co-operation are becoming clear. At a minimum, they should include

- *Emissions trading:*   Expanding and linking the growing number of emissions trading schemes around the world are powerful ways to promote cost-effective reductions in emissions and to bring forward action in developing countries: strong targets in rich countries could drive flows amounting to tens of billions of dollars each year to support the transition to low-carbon development paths. And it is these decisions by private investors that will, over time, drive emissions down. Governments must create the frameworks but it will be largely the private sector that makes the investments. For them to act effectively the market signals must be credible.
- *Technology co-operation:*   Informal co-ordination as well as formal agreements can boost the effectiveness of investments in innovation around the world. Globally, support for energy R&D should at least double, and support for the deployment of new low-carbon technologies should increase up to five-fold. International co-operation on product standards is a powerful way to boost energy efficiency.
- *Action to reduce deforestation:*   The loss of natural forests around the world contributes more emissions each year than the transport sector. Curbing deforestation is a highly cost-effective way to reduce emissions; large-scale international pilot programmes to explore the best ways to do this could get underway very quickly.

- *Adaptation:*   The poorest countries are most vulnerable to climate change. It is essential that climate change be fully integrated into development policy, and that rich countries honour their pledges to increase support through overseas development assistance. International funding should also support improved regional information on climate change impacts and research into new crop varieties that will be more resilient to drought and flood.

27.5   Conclusions

This Review has focused on the economics of risk and uncertainty, using a wide range of economic tools to tackle the challenges of a global problem with profound long-term implications. Much more work is required, by scientists and economists, to tackle the analytical challenges and resolve some of the uncertainties across a broad front. But it is already very clear that the economic risks of inaction in the face of climate change are very severe.

There are ways to reduce the risks of climate change. With the right incentives, the private sector will respond and can deliver solutions. The stabilisation of greenhouse gas concentrations in the atmosphere is feasible, at significant but manageable costs. Delay would be costly and dangerous.

The policy tools exist to create the incentives required to change investment patterns and move the global economy onto a low-carbon path. This must go hand-in-hand with increased action to adapt to the impacts of the climate change that can no longer be avoided.

Above all, reducing the risks of climate change requires collective action. It requires co-operation between countries, through international frameworks that support the achievement of shared goals. It requires a partnership between the public and private sector, working with civil society and with individuals. It is still possible to avoid the worst impacts of climate change, through strong collective action starting from now.

Abbreviations and Acronyms

AAs	Assigned Amounts
ACT MAP	MAP Accelerated Technology Scenario (IEA – Energy Technology Perspectives publication 2006)
ADB	Asian Development Bank
AOSIS	Alliance Of Small Island States
AR (I-IV)	Assessment Report (first-fourth)
BAU	Business As Usual
bcm	Billion cubic meters (unit of volume, e.g. for gas)
BGE	Balanced Growth Equivalent
Bl	Barrel of oil
Boe	Barrels of oil equivalent
C	Carbon (to convert 1 tonne of C into CO_2, multiply by 44/12).
CH4	Methane (greenhouse gas)
CBA	Cost Benefit Analysis
CCGT	Combined Cycle Gas Turbine
CCS	Carbon Capture and Storage
CDM	Clean Development Mechanism
CER	Certified Emission Reduction
CFCs	Chloro Fluoro Carbons
CFL	Compact Fluorescent Lamp
CH_4	Methane
CGE	Computable General Equilibrium
CHP	Combined Heat and Power
CO_2	Carbon dioxide
CO_2e	CO_2 equivalent
CSD	Commission for Sustainable Development
dCHP	Decentralised Combined Heat and Power
DEFRA	UK Department for Environment, Food and Rural Affairs
DFID	UK Department for International Development
DFT	UK Department for Transport
DTI	UK Department for Trade and Industry
E	Exa: 10 to the power of 18
EBRD	European Bank for Reconstruction and Development
EC	European Commission
EIF	Energy Investment Framework
EIT	Economy In Transition
EMAS	Environmental Management Assistance System
EMF	Energy Modelling Forum (Stanford University)

EPA	Environmental Protection Agency
ESCO	Energy Service Company
ETS	Emission Trading System (EU)
EU	European Union
FCCC	Framework Convention on Climate Change (UN)
FDI	Foreign Direct Investment
FOB	Free on Board
G	Giga: 10 to the power of 9
GATT	General Agreement on Trade and Tariffs
GCM	Global Climate Model
GDP	Gross Domestic Product
GEF	Global Environment Facility
GHG	Greenhouse Gas
GMT	Global Mean Temperature
GNP	Gross National Product
GPG	Global Public Goods
GPP	Gross Primary Production
GtC	Gigatonne of Carbon
GWP	Global Warming Potential or Gross World Product
HDI	Human Development Index
HFC	Hydro Fluoro Carbon
HMT	UK Her Majesty's Treasury
HVAC	Heating, Ventilation and Air Conditioning
I-O	Input-Output
IA(M)	Integrated Assessment (Model)
IAEA	International Atomic Energy Agency
IEA	International Energy Agency
IET	International Emission Trading
IIASA	International Institute for Applied System Analysis
IMCP	Innovation Modelling Comparison Project
IMF	International Monetary Fund
IPCC	Intergovernmental Panel on Climate Change
IPR	Intellectual Property Rights
ISIC	International Standard Industrial Classification
ISO	International Standardisation Organization
IT	Information Technology
ITC	Induced Technical Change
J	Joule = Newton × meter (International Standard unit of energy)
JI	Joint Implementation
JRC	Joint Research Centre (EU)
K	Kilo: 10 to the power of 3
KWh	Kilowatt hour
Lbs	Pounds (unit of weight. 1 lbs = 0.454 kg)
LDC	Least Developed Country
LIC	Low Income Country
LNG	Liquid Natural Gas
LPG	Liquid Petroleum Gas

M	Mega: 10 to the power of 6
MAC	Marginal Abatement Cost
MERs	Market Exchange Rates
MDGs	Millennium Development Goals
MEA	Multilateral Environmental Agreements
MIC	Middle Income Country
MIT	Massachusetts Institute of Technology
Mtoe	Mega tonnes oil equivalent
N_2O	Nitrous oxide (greenhouse gas)
NACE	Nomenclature des Activités dans la Communauté Européenne (index of business activities in the EU)
NAFTA	North American Free Trade Agreement
NAP	National allocation plan
NCGGs	Non-Carbon Greenhouse Gases
NGO	Non-Governmental Organization
NIC	Newly Industrialized Country
NMHC	Non-Methane Hydro Carbon
NOx	Nitrogen oxides (local air pollutants)
ODA	Official Development Assistance
ODS	Ozone Depleting Substances
OECD	Organization for Economic Co-operation and Development
OPEC	Organisation of Petroleum Exporting Countries
P	Peta: 10 to the power of 15
PFC	Per Fluoro Carbon
ppm [v/w]	parts per million [by volume / weight]
PPP	Purchasing Power Parity
QELRCs	Quantified Emission Limitation or Reduction Commitments
R&D	Research and Development
RD&D	Research, Development and Demonstration
RFF	Resources for the Future
SAR	Second Assessment Report (by IPCC)
SBSTA	Subsidiary Body for Scientific and Technological Advice
SCC	Social Cost of Carbon
SD	Sustainable Development
SOx	Sulphur dioxide
SMEs	Small and Medium Enterprises
SRES	Special Report on Emissions Scenarios
SRLULUCF	Special Report on Land-Use, Land-Use Change and Forestry
T	Tera: 10 to the power of 12
TAR	Third Assessment Report (by IPCC)
tce	Tonnes of coal equivalent
THC	Thermohaline Circulation
Toe	Tonnes of oil equivalent. (Mtoe – mega tonnes of oil equivalent).
TPES	Total Primary Energy Supply
TWA	Tolerable Windows Approach

UK	United Kingdom
UN	United Nations
UNCED	UN Conference on Environment and Development
UNDP	UN Development Programme
UNEP	UN Environment Programme
UNFCCC	UN Framework Convention on Climate Change
US/USA	United States (of America)
US CCSP	United States Climate Change Science Programme
US EIA	United States Energy Information Administration
VAT	Value Added Tax
VOC	Volatile organic compound
W	Watt = Joule/second (International Standard unit of power)
WBCSD	World Business Council for Sustainable Development
WCED	World Commission on Environment and Development
WEC	World Energy Council
WEO	World Economic Outlook
WG (I–III)	Working Group (One to Three) of the IPCC
Wh	Watt hour
WHO	World Health Organization (UN)
WRI	World Resources Institute
WTA	Willingness To Accept compensation
WTO	World Trade Organization
WTP	Willingness To Pay
WWF	World Wildlife Fund

Postscript

The Review on the Economics of Climate Change, published on 30th October, has generated substantial interest and debate. We have now had the opportunity to present the Review to a wide range of audiences, including economists, scientists, business leaders and the international community, including the participants in the Nairobi Conference of the Parties to the UNFCCC, and to policy-makers at the European Commission and the African Union.

In this postscript, we offer some reflections in the light of the reactions and comments we have received in the first weeks since publication. In the main text, we have also taken the opportunity to correct any typographical errors found or which have been drawn to our attention. For example, revising the magnitudes of hurricane losses in Table 5.2. The discussion here follows the structure of the Review. The first issues concern the strength of the evidence base underpinning the recommendation of the Review that all countries should take urgent action to stabilise the concentration of greenhouse gases in the atmosphere at between 450–550 ppm CO_2e. The second set of issues concern the policy mechanisms that will support an effective, efficient and equitable approach to this action, and the importance of international co-operation to support adaptation to the adverse impacts of climate change.

The case for urgent action

Two key conclusions from our analyses of the science and economics of climate change provide important underpinning for the case for urgent action.

First, under a business-as-usual scenario, the stock of greenhouse gases could be more than treble pre-industrial (greater than 850 ppm CO_2e) by the end of the century. This may be a conservative estimate, for example, some IPCC scenarios suggest that the stock could be more than four times pre-industrial by 2100. Current scientific understanding suggests that a trebling of the stock would give at least a 50% risk of temperatures exceeding 5°C above pre-industrial levels during the following decades (Chapter 1).

In Part II of the report we brought together what can be said about impacts at high temperatures, based on the current state of the underlying science. This analysis has brought us to our second broad conclusion that the impacts of climate change across multiple dimensions are likely to be highly convex, with marginal damages that increase strongly as temperatures rise. Most impacts analyses focus on levels of warming of around 2–3°C above pre-industrial. Little is known about how the environment and human society will respond to larger increases in temperature. A warming of 5°C on a global scale would be well outside the experience of human civilisation, and would transform where we live and how we live our lives.

The analyses presented in Chapters 3–5 of the report demonstrate the great dangers of allowing temperatures to continue to rise, in terms of the environment, human health, and economic growth and development. Chapter 3 demonstrates that many of the impacts of climate change increase strongly in severity as temperatures rise. For example, the damage caused by hurricanes; the frequency of extreme events; and above a threshold, effects on agricultural production and heat-related mortality. Further, impacts can interact, bringing about rapid increases in damages at high temperatures: rising levels of pests in some areas may aggravate declines in agricultural production caused by heat or changes in water availability. In addition, current understanding suggests that at high levels of warming, the risks of major, irreversible changes to the climate, ecosystems and society are very real. These include physical changes, such as a collapse of ocean currents, and also the risk of major societal changes, such as mass migrations and political instability. Putting all these impacts together builds a strong picture of impacts rapidly rising with temperatures, with increasing damages for each marginal increase in temperature. High temperatures are likely to generate a hostile and extreme environment for human activity in many parts of the world.

It is the scale of these risks and an appreciation of the types and severity of damages involved that provide the *main case* for urgent and strong action to stabilize emissions below 550 ppm CO_2e, when one considers that the risks can be very substantially reduced by an expenditure of around 1% of GDP per year. Further, the costs of action to stabilise at any given level would rise rapidly if action were delayed.

In Chapter 6, as a complement to the disaggregated analysis, we investigated the role of formal economic modelling in providing an aggregate monetary estimate of the damages of climate change. We were explicit and clear about the severe limitations of such modelling, but we saw it as a perspective, which could provide some support, by adding structure, to an analysis of the case for action based on the disaggregated impacts.

As we made clear, the role of integrated assessment models is to give an illustration of the potential effects of climate change. Modelling of the economic impacts of climate change over very long time horizons cannot give precise results. The value of the approach is that it allows the investigation of the role of different specifications of model structure and ethical assumptions. The ethical judgements that have to be examined include those concerning how society should weight impacts on different generations. The impacts have been expressed in this Review using a technique that allows averaging over time, over risk and over country in a way that permits direct comparison with the costs of mitigation.

Two main modelling issues have been raised with us in discussions since the Review was published: first, concerns that the model we used may under-estimate the level of damages likely to be caused at different temperatures, particularly high temperatures, and second, concerns about the assumptions used in valuing or discounting the damages. The former is captured in the parameters of the function relating damages to temperature and the latter in the shape of the relationship of social utility to consumption and the pure time discount rate (see Chapter 2, its appendix and Chapter 6).

We have subsequently carried out sensitivity analysis on these issues, presented in a technical annex to this postscript. The sensitivity analysis allows us to explore the effect of different assumptions, but it does not change our overall conclusion, that climate change is likely to cause damages which are very severe and of much greater consequence than the costs of greatly reducing risks by strong reduction in emissions. In the report we calculated damages from business-as-usual which were equivalent to at least a 5% loss in consumption, based on a narrow definition of risks and impacts, and up to 20% if a broader range of risks and impacts are considered. The sensitivity analysis marginally reduces the lower end and increases the upper end. The only exception is where we use high pure time discounting rates, which are in our view implausible relative to most positions on ethical values, and take a very narrow view of impacts (i.e. excluding environment and health). Overall, unless the interests of future generations are heavily disregarded there is a very powerful case for strong mitigation.

Our estimates of damage from business-as-usual are higher than some previously published for the following sound reasons:

- We treat aversion to risk explicitly – this issue is all about risk and we invoke the economics of risk directly.
- We use the more recent literature, from the science, on the probabilities. This points to significant risks of temperature increases above 5°C under business-for-usual by the early part of the next century. Previous studies have focused on temperature increases of 2 or 3°C. The damages from 5°C would be very much higher – damages rise much faster than temperature.
- We adopt lower pure time discount rates than some earlier literature and thus, it was argued in Chapter 2 and its Appendix, the analysis gives future generations appropriate ethical weight. The effects of changing this assumption were set out clearly in Chapter 2 and its appendix, Chapter 6 and are explored in more detail in the Technical Annex to this postscript.
- We take account of the disproportionate impacts on poor regions, reflecting the fact that those in poverty will feel losses in consumption more keenly.

Few existing studies include all these factors, and as a result their estimates of the damages tend to be lower. One can compare these losses with the size of the losses from a recession, but climate impacts are actually more like an adverse supply-side shock than a large contraction in demand. And they are much more difficult to reverse. Our estimate in terms of per annum consumption losses (averaged over time, possible outcomes and across countries) of the costs of climate change can be interpreted as like a tax being levied each year, with the proceeds of the tax simply being poured down the drain. You could also think of it like an insurance premium – society would be willing to pay up to this amount to avoid the risks of climate change – in fact the actual cost of action to avoid climate change is much less, as Chapters 9 and 10 of the Review show, and as we will discuss again briefly below.

Our analysis leads us to the conclusion that the risks can be substantially reduced, but by no means eliminated, if concentrations of greenhouse gas emissions can be stabilised at 550 ppm CO_2e or below. The upper limit, 550 ppm CO_2e, is still a risky place to be. The analysis presented in the Review, based on an average of several models, suggests a 50:50 chance of a temperature increase above or below 3°C, and the Hadley Centre model predicts a 10% chance of exceeding 5°C even at this level (Chapter 8). Whilst the modelling of Chapter 6 and Part II of the Review, in general, brought together in Chapter 13, suggests that the damage from a 550 ppm CO_2 stabilisation level is far smaller than business as usual. Many people have suggested that this limit is too high. There is a strong case to examine whether it is possible to reduce these risks still further by reaching lower levels of stabilisation, and to keep this continually under review as policy-makers gain experience in managing the transition to a low-carbon economy.

The Review finds that the costs of bringing down the risks by stabilising at 500–550 ppm CO_2e are equivalent to around 1% of GDP, with a range of $+/-$ 3%. This range assumes that sensible policies are put in place and deliver the induced technological progress required. Some people have questioned whether the central estimate of 1% is too low, and others have suggested that while the overall level may be acceptable, the distribution of the costs may give rise to an unacceptable burden on some countries or sectors.

In response to the suggestion that the estimate of 1% is too low, it is worth noting a number of points. The figure of 1% is a central estimate within a range that is consistent with the literature, and that is therefore likely to be consistent with the review of the same literature currently being finalised by the Intergovernmental Panel on Climate Change for its Fourth Assessment Report. Achieving stabilisation at the lower end of the range of costs depends upon good policy frameworks, to bring forward appropriate low-carbon technologies and to provide flexibility in when, where and how emissions are reduced. The cost of 1% of GDP each year is certainly not trivial – expressed relative to current world GDP, it is equivalent to $350billion at market exchange rates or around $600 billion in terms of purchasing power parity. But this cost is manageable without slowing growth. An overall cost of around 1% of GDP to achieve stabilisation below 550 ppm CO_2e, as suggested here, would have an impact similar to a one-off 1% rise in price or cost indices. However, if investments in the next two or three decades were made in high-carbon infrastructure, it could cost far more than 1% subsequently to reduce the resulting emissions to levels consistent with stabilisation below 550 ppm CO_2.

As we made clear in Chapters 11 and 12, the costs of mitigation will not be evenly distributed across industry sectors. Carbon-intensive sectors will face higher costs, and it is right to consider the impacts of these costs on their competitiveness. Similarly, the costs of unabated climate change will fall more heavily on sectors that depend upon environmental resources, such as agriculture and tourism.

If all countries act in a broadly similar way, the impacts on competitiveness from action to mitigate climate change will be small for all of them. Where different policies are in place in different countries for mitigation, it is important to assess the increased carbon costs in the context of overall conditions for doing business in a particular country or region. For many industries, the impact of any higher energy costs associated with mitigation is very small in relation to the cost differentials of different wage rates between rich and poor countries or to transport costs over long distances. For a small number of internationally traded, carbon-intensive sectors, including aluminium and cement, it may make sense to develop specific sectoral arrangements that provide an international framework to support the efforts of those industries to upgrade their equipment and processes and reduce or offset their emissions. And it is important to recognise that the new technologies and investments will open up new economic opportunities.

While action is delayed, greenhouse gases in the atmosphere continue to accumulate, committing the world to greater impacts in the future or to the higher costs of bringing down flows of emissions more sharply to attain any stabilisation level. This cost of delay is a key element in the argument for urgent action.

Overall, we have heard three main arguments from those who do not support the conclusion that urgent action to reduce the risks of climate change, economically speaking, is a good deal. We suggest that all three are misplaced.

1   Some still deny the science of climate change.

There are legitimate debates over many particular details of the climate system, but it is no longer credible to doubt the underlying physical mechanisms associated with increases in greenhouse gases in the atmosphere, nor to doubt the importance of the natural carbon cycle and the potential for amplifying feedbacks that would be outside our control.

2   Some people accept the basic science, but still believe it is preferable to wait and see before taking significant action on mitigation. Some suggest a new technology will come along that will greatly reduce the costs of action, or that the changes will be such that future generations, with a higher capital stock available to them, will be able to adapt.

It is certainly true that for most countries, major transformational damages affecting the whole economy are not likely to be seen for several decades, or even a century or more – but if we wait until they appear and they are as difficult as we have reason to expect then we cannot go into reverse. Stocks of greenhouse gases are extremely difficult to reduce.

The range of human activity that gives rise to emissions is so broad, that there will be no single technology breakthrough that will bring about stabilisation. Further, technology development is not independent of the policy framework that is in place. The range of technologies required can only be brought forward by an appropriate policy framework.

Adaptation is necessary, but it is not the whole answer. The longer stocks of greenhouse gases are allowed to accumulate in the atmosphere, the greater the impacts to which we are committing the world. There are limits to adaptation at higher temperatures. Many of the effects could involve major dislocation, to whole nations and regions, with consequences that would be felt around the world. The only way to prevent very high future damages is to reduce greenhouse gas emissions today.

We should recognise the balance of risks. If the science is wrong and we invest 1% of GDP in reducing emissions for a few decades, then the main outcome is that we will have more technologies with real value for energy security, other types of risk and other types of pollution. However, if we do not invest the 1% and the science is right, then it is likely to be impossible to undo the severe damages that will follow. The argument that we should focus investment on other things, such as human capital, to increase growth and make the world more resilient to climate change is not convincing because of these irreversibilities and the scale and nature of the impacts. Similarly if we wait and see for 30 or 40 years then we are likely to go past the 550 ppm (CO_2e) that we argued would be a plausible upper limit. We might try to move rapidly from there but one cannot stop emissions in their tracks without great cost and disruption, if indeed it is feasible.

3   Some people prefer to place very low value on the future, or to put it another way, to place a very high value on near-term opportunities for consumption.

It is a key feature of the challenge of climate change that we must think long-term to understand the issues and to respond to them. It will always be possible to choose a pure time discount rate that makes the benefit of reducing future damages appear trivial.

In the Review, we do discount future damages for the likelihood that future generations will be richer than we are. But we apply only a low discounting to the future simply because it is the future (we account for the possibility of extinction). Choosing a high rate of pure time preference to analyse a long-term issue that affects the global environment is to make a profound ethical choice with, in this case, irreversible effects on future generations. It is as though a grandparent is saying to their grandchild, because you will live your life 50 years after mine, I place far less value on your well-being than I do on myself and my current neighbours, and therefore I am ready to take decisions with severe and irreversible implications for you. Nevertheless ethical choices appear different to different people and that is why in the technical appendix to this postscript we investigate different possible ethical positions concerning inequality and pure time discounting. The conclusion that strong mitigation is warranted is robust except where high pure time discounting is embraced.

An alternative view, associated with Bjorn Lamborg, that, it is agreed, places dealing with climate change low on the agenda, arises from comparing it with 'other ways' of spending public money and suggests that they have high social rates of return. There are important deficiencies in this approach. First, correcting an externality is a different policy question from spending public money.

Why would we refrain from a policy to correct an externality, a correction of a market failure because there are good ways to spend public money. Second, the argument as conveniently put takes little account of the severe risks of very high temperature increases from climate change, which we now know are possible, or indeed likely, under business-as-usual, and which cannot be reversed if they start to appear. Third, the costs of action for any given stabilisation level rise rapidly if action is delayed. Thus, this type of argument for low priority or for delay is completely unconvincing.

Responding to the challenge

We have also received comments and reactions to the policy issues discussed in the second half of the Review – the policy instruments to promote mitigation, approaches to adaptation, and the international framework.

Many people have welcomed the breadth of discussion on policy instruments, including the emphasis on the importance of all three strands of policy intervention – correcting the market failure on greenhouse gases, technology policy, and complementary measures to remove other barriers and to change perspectives on responsible behaviour. There has also been strong interest in the potential of each of tax, trading and regulation to play a role in the creation of a carbon price. We have been asked several times about the relative importance of each of these three approaches.

The answer to these questions must be guided by the principles of effectiveness (in terms of delivering greenhouse gas emission reductions), efficiency and equity. For different countries and different sectors, different approaches are likely to prove appropriate and effective. Many European countries have high fuel taxes, whereas in the USA regulation of vehicle efficiency standards has historically been more important. In the EU, emissions trading has from the outset taken the form of a mandatory cap and trade scheme, while in Japan and for some businesses in the USA, voluntary approaches are proving helpful in building up experience of using this instrument. For some areas, for example household appliances, labelling and standards are likely to bring about the fastest changes. Efficiency does not require that all these approaches be merged into one single scheme, but it does require that across countries and policies, a broadly similar price of carbon emerges. Otherwise, some sectors will be carrying a greater burden of emissions reductions when there are more cost-effective opportunities elsewhere. Equity does not mean that poorer countries should take no action to reflect the price of carbon in their own economies – otherwise producers and consumers in those countries will not see the signals that are required to support a transition to a low-carbon economy – but it does mean that these should be supported by rich countries in the process of managing the adjustments.

Emissions trading is particularly well suited to addressing both efficiency and equity across borders. If the rich countries set ambitious targets, consistent with the overall objective of achieving stabilisation between 450–550 ppm CO_2e, emissions trading will allow the private sector in those countries to seek out the most cost-effective opportunities to reduce emissions. Some of these opportunities will be at home – provided the signal is strong, credible and long-term, the carbon price will discourage further investment in high-carbon capital stock in rich countries. But many of the opportunities in the short term will be in developing countries, and trading can create substantial flows of carbon finance that will allow developing countries to avoid locking in new high-carbon infrastructure during the next few years, when substantial growth and investment is likely to take place. These flows must be supported by effective mechanisms, linked to national or sectoral policies and programmes to move away from carbon-intensive investment strategies. Such large-scale flows from the rich countries combined with strategic national or sectoral approaches in developing countries have the potential to transform the carbon intensity of the global economy, without capping national aspirations for growth and development in poor and rich countries. A project-by-project approach to such flows is very unlikely to be able to deliver the results required, either in terms of the effectiveness of emissions reductions or the potential scale of flows from rich to poor countries. Programme or policy-orientated schemes will be necessary to manage flows on a much larger scale.

Large-scale international flows of carbon finance will go a long way to addressing the issues of equity. However, the least developed countries have the fewest opportunities to benefit from private sector investment in emissions reductions. For some people, this suggests that a more equitable international framework would be based on equal per capita rights to emit. This view has some attractions but there are some practical and conceptual problems, which were discussed in Chapter 22. An alternative approach is to consider the challenges for the poorest countries directly. International co-operation can support access to clean, low-carbon energy for poor people, as demonstrated by the initiative of the World Bank and others in creating Clean Energy Investment Frameworks with a specific focus on energy access. The initiative on removing barriers to the use of the CDM in developing countries, launched at the Nairobi Conference of the Parties to the UNFCCC, also has the potential to increase the use of carbon finance in poorer countries. The underlying investment conditions for foreign and domestic private capital are fundamental to the success of such initiatives.

It is of great importance to move quickly on those actions and policies that can be rapidly agreed and implemented and to build the knowledge and trust that could arise from the experience. Fundamental to all of effectiveness, efficiency and equity and particularly to equity is strong ambition from the rich countries in terms of caps implemented and thus level of carbon price and potential financing flows to developing countries. From all three perspectives, implementing caps embodying ambitious reductions should be of high priority in rich countries. And it is crucial that trading schemes such as the EU ETS be long-term, to provide effective private sector signals, and open, so that as many countries as possible can be included, both from the perspective of efficiency and the building of international collaboration.

Equity also clearly points to support for adaptation to the adverse impacts of climate change. It remains very clear that adaptation to climate change is now both inevitable and very important. For developing countries, good adaptation and good development policy are very strongly intertwined, and it is right that climate change should now become central to national planning processes and to development assistance. International support for adaptation will come in large part through the delivery of the commitments made by rich countries to double aid by 2010 and the commitments made by many countries to meet the target of 0.7% of GNI by 2015. This will deliver an increase of hundreds of billions of dollars.

But there are limits to adaptation. Small island developing states threatened by sea level rise have fewer options to adapt. Sea defences are particularly costly for low-lying islands, and may do little to protect the tourism and fisheries that sustain the local economy. Development and diversification are still important strategies wherever possible, but ultimately the international community will have to find ways to support alternative responses, including the managed resettlement of some people for these states. This will bring many challenges, particularly for those people that must move. There will be much greater pressures if unabated climate change leads to sea level rise that threatens much larger populations in low-lying coastal areas.

Finally, some people have asked if it is really possible to create structures that will sustain co-operation and overcome the incentives for free-riding. Here, it is important to understand that public pressure for an effective response is growing in many countries, as people begin to realise the scale of the risks they and their successors face if no action is taken and as they see the wide range of initiatives by local governments, businesses and community groups that demonstrate that it is possible to do something about the problem. It is now more important than ever to build trust, through transparency and mutual understanding about the actions that different countries are taking, and to look for international mechanisms that build on and support national objectives, including by reducing costs and increasing the prospects for success.

Conclusions

Climate change presents a very serious challenge. The most severe damage will be felt in the future, often the far future, but decisions that we take now could lock in those damages.

The broad conclusion of our analysis is that urgent action should be taken to reduce the risk of committing the world to the real possibility of very high temperature increases. The next few years will be critical. Action is required now, if we are to stabilise somewhere in the range from 450–550 ppm CO_2e. Success will depend on continuity in the process of building carbon markets, and imagination and ambition in scaling up co-operation in areas such as technology and reducing deforestation.

The Review is intended as a contribution to the discussion. We welcome the debate that has been stimulated, and hope that further work will take place on all the issues raised by the Review, including those explored further in this postscript and its technical appendix on aggregate modelling.

Technical Annex to Postscript

Some commentators on the Review have focussed on particular technical issues associated with modelling the aggregated impacts of climate change.[1] Our estimates of damage from climate change derived from formal economic modelling are higher than many estimates in the literature, and there has rightly been strong interest in our underlying assumptions. This paper responds to some of the comments on the modelling we have received in the weeks since the publication of the report.

The questions concern both the model structure and the ethical judgements that are embodied in the evaluations. Investigating these questions allows us to use the models to clarify the roles of the different assumptions in a structured way. We did not present these results as part of Chapter 6, but we have subsequently carried out a sensitivity analysis in this area and the results are presented below. This Technical Annex can be seen, in part, as an annex to Chapter 6.

The Role of Integrated Assessment Models (IAMs)

Integrated assessment models attempt to summarise the impacts of climate change, usually in terms of aggregate gains or damages in terms of income. These models, on the basis of their assumptions, give an idea of the magnitude of risks, their evolution over time and sensitivity to emissions. As the Review makes clear, the role of IAMs is to give an illustration of the potential effects of climate change. Modelling of the economic impacts of climate change over long time-horizons cannot give precise results and is very sensitive to assumptions. Given the difficulty of modelling so far into the future, the models must be seen as highly speculative, but they do have the advantage of exploring the logic of assumptions.

Our results using IAMs complement our analyses of the overall risks and the disaggregated impacts of climate change. In the Review, we lay stronger emphasis on the disaggregated assessment of impacts, together with overall judgements on the riskiness of very high temperatures and of unknown territory in a context where greenhouse gas (GHG) concentrations and environmental damage are very difficult to reverse. The IAM analysis illustrates these risks but should not be seen as the first or most important argument in coming to an overall judgement concerning the importance of a strong reduction in GHG emissions.

---

[1] The comments have reached us in various ways – via remarks at seminars, e-mails and press comments. We focus here on the most commonly expressed concerns. We are particularly grateful to Partha Dasgupta and Bill Nordhaus for their comments.

It is important to recognise the limitations of IAMs. Expressing multi-dimensional impacts in terms of aggregated income losses masks the full environmental and human implications, which can be understood only through an analysis across several dimensions. In addition, in attempting to value these impacts in relation to a common income unit, IAMs add a degree of formality and precision, which can, from some perspectives, obscure rather than illuminate an overall assessment of the impacts. Existing IAMs rely heavily on literature that, in many cases, still excludes significant effects that have been explored only in the last few years, in particular the risks at high temperatures. The scientific literature has only recently been able to give probability distributions of temperatures associated with levels of greenhouse gas concentrations in the atmosphere. Crucially, this now allows more explicit analysis of the economics of risk and shows that the probability of temperature increases above 5°C under business-as-usual (BAU) may be high (above 50% in the most recent Hadley Centre estimates for some standard BAU emissions paths[2]).

The Review considers results from a range of IAMs and produces new results from one particular model: PAGE2002. The aim of this analysis was to provide an illustration of the scale of the potential impacts of climate change with an IAM that was updated to reflect recent probability estimates and incorporate the economics of risk (described below). These two features imply higher estimates than some previous literature but both are essential for a serious and up-to-date study of climate change. The economics is fundamentally about the economics of risk.

In addition we examined carefully the arguments for pure time discounting (see Chapter 2, its appendix, and below) and argued that whilst the growth arguments for discounting were sound (and included in the modelling) in this context, the ethical case for strong pure time discounting was weak. Lowering pure time discount rates raises estimates of losses.

Integrated Assessment Modelling in the Stern Review

In this section, we examine what shapes the outputs from the models, what innovations the Stern Review has made and what further innovations should be examined. There are four main elements: (i) the model structure; (ii) the underlying evidence; (iii) the issues being examined – here, particularly, the economics of risk; (iv) ethical judgements. We then provide a sensitivity analysis varying parameters relevant to the model structures and ethical judgements to cover issues raised with us by commentators. Finally we comment on directions for research in this area and the implications of the sensitivity analysis for the overall argument of the Review.

The model structure

The PAGE 2002 model was chosen for two reasons: (i) it is particularly convenient for examining risk; and (ii) it is designed to span the range of previous models. For example, the standard damage function of the model is designed to cover the range of estimates described in the IPCC Third Assessment Report (TAR, 2001)

---

[2] Discussed in Section 1.4 of the Review.

and the climate sensitivity range is consistent with the likely range given in that report. No changes were made to the core model structure for the Stern Review analyses in Chapter 6 of the report.

Scientific and other evidence

Through assessing the full range of possible outcomes based on current scientific evidence the results in the report go further than the majority of previous studies in attempting to quantify the impacts of climate change. This allows us to capture more fully the risks associated with higher temperatures. The 'baseline'-climate scenario of the model is designed to be consistent with probability distributions associated with the range of projections given in the IPCC TAR. The Stern Review builds on this by considering more recent scientific evidence pointing to greater risks of high temperatures due to additional feedbacks, such as weakening carbon sinks and increased natural methane releases. This is called the 'high'-climate scenario.

In addition to the more recent estimates of probabilities of different temperatures (used for both baseline and high climate), there is also an issue of how to evaluate consequences of different temperatures. As discussed in Part II, there are uncertainties here that can only be resolved once there are sufficiently good data. The G-ECON database is one project leading the way here (see Nordhaus, 2006a). The damage function of the PAGE2002 model is designed to capture stochastically the findings of other IAMs. For lower temperatures, a wide pool of published literature informs these damage estimates. However, as temperatures rise above around 3–4°C above pre-industrial, information becomes scarcer. Detailed empirical assessments of impacts at high temperatures are difficult to do because they take us far outside the range of human experience. Given that under a business-as-usual trajectory there is a significant risk of temperatures exceeding 5°C, more research is required to better understand the consequences of high temperatures.

In the PAGE2002 model, impacts are represented by a damage function that takes a simple form dependent on regional temperature[3] increases (T_R) and the damage exponent γ.

$$Damages \propto \left(\frac{T_R}{2.5}\right)^\gamma \tag{1}$$

The damage exponent is critical in determining the scale of the estimated impacts. In the standard model (as used in Chapter 6) this is defined by a triangular probability distribution, with minimum of 1, a mode of 1.3, and a maximum of 3. This range is based on results from several previous studies discussed in the IPCC Third Assessment Report. A value of 1.0 implies that damages are a linear function of global mean temperature. A value of 1.3 implies a weak convexity and 3 implies a stronger convexity. Figure PA.1 below demonstrates the dependence of damages

---

[3]   Note that T_R in the model is actually the 'vulnerable' temperature increase, as it is assumed that most regions can adapt to some degree of temperature rise. The regional temperature increase is dependent on the global temperature increase (a linear relationship) and the regional sulphate aerosol concentration.

at a given temperature on the damage exponent, relative to the damages at 2.5°C. For comparison, the global damage function from the DICE model is shown[4].

The disaggregated impacts analysis brought together in Part II suggests that the relationship between temperature and damages will be convex. Further, there are strong reasons to consider that the scale of impacts captured by the damage exponent of 1.3, the mode of the analysis in Chapter 6, does not adequately reflect the degree of convexity of likely damages.

Damages for many individual impacts rise steeply with temperature (see, for example, Table 3.1 and Box 3.1 in the Review). As well as the strong convexity that arises from individual effects there are also aggregate convexities that arise from their interaction. For example, most previous studies look only at the effects of average climate conditions. However, a 1°C increase in mean temperatures could lead to a ten-fold increase in the frequency of severe heat waves in some regions (Chapter 1). This will have knock-on effects, heightening damages (and strengthening convexities) in areas such as agriculture and health. The convexity of the aggregate damage function is supported by Nordhaus (2006a) using the new G-ECON database. This demonstrates a powerful cross-sectional relationship between temperature and output, as well as specific examples, such as the ninth power relationship between hurricane wind speed and damages (Nordhaus 2006b)[5]. The damages associated with such interactions between impacts have not been fully incorporated into previous aggregate analyses.

In addition, there is also the risk of major, irreversible changes in the climate, ecosystems and society (Chapter 3). As temperatures rise, these risks increase sharply. Some commentators have suggested to us forcefully that the types of risk associated with high temperatures as discussed in Part 2 of the Review are not well reflected in the formal modelling of Chapter 6. This is, in our view, a suggestion that is well founded.

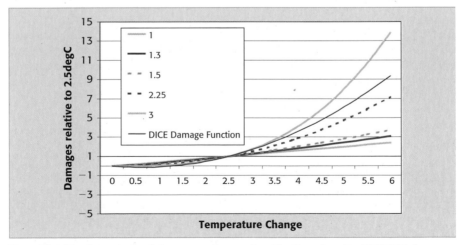

Figure PA.1 The dependence of damages on temperature. The lines show the PAGE2002 damages, as defined by the damage function in equation 1, for damage exponents (η) between 1 and 3

---

[4] See Warren et al. (2006)
[5] This is on top of the sharply increasing relationship between sea surface temperature and hurricane wind speed.

To test the sensitivity of the results to the damage exponent, the model is rerun with a new mode of 2.25. The lower bound of the range is increased to 1.5 and the upper bound is held constant. The range is chosen in the light of the proposed functional forms of the relationships illustrated in Box 3.1, analyses such as those just quoted, and the powerful reinforcing effect of combinations of these individual effects.

The economics of risk

Models and policy analyses are designed to investigate specific questions. In this case we have argued that the analysis of risks is crucial to the problem of climate change. Thus it is important that analyses are built around the economics of risk. For example, in the high-climate scenario with market impacts, risk of catastrophe and non-market impacts (Chapter 6), the 95th percentile estimate is a 35.2% loss in global per-capita GDP by 2200. This is not a statistical mean, but it is nevertheless a risk that few would want to ignore. Such risks can have a strong effect on welfare calculations, because they reduce consumption to levels where every marginal dollar or pound has a much greater value.

The Stern Review has adopted an expected-utility analysis, a standard tool in economics for working with risk. This is based on probability distributions of future outcomes that were not available in most previous analyses.

Ethical judgements and Discounting

In Chapter 2 and its appendix we examined a number of different ethical viewpoints. In the forward modelling of Chapter 6, with its very narrow view of outcomes in terms of monetary aggregates, we focused on a simple and standard framework in which discounted utility (as a function of consumption) of a generation is summed over time. We should also draw attention to a broader literature on sustainable development than referenced in Chapter 2 (a helpful analytical introduction and set of references is Dasgupta, 2001, and Arrow et al, 2003). We should also draw attention to an axiomatic approach to inter-temporal evaluations, which can lead to similar formulations, based on the work by Koopmans (1972). Simple aggregative modelling of the type used here usually precludes the relevant subtlety of evaluation.

Estimating the aggregate impacts of climate change requires us to consider the value of damages now compared with those in the future. For an evaluation of a marginal change of one unit at some time in the future, relative to a unit now, this is called the discount factor. Its rate of fall is the discount rate (see chapter 2 of the report and its appendix for a detailed discussion). Discount factors and rates depend on time and the path under examination. Discount factors and rates in the very aggregative models considered in the appendix to Chapter 2 and in Chapter 6 are shaped by two elements or questions:

1   How to take into account the fact that people are likely to be richer in the future.
2   Whether the future should be discounted simply because it is the future.

The first element appears in our modelling in a standard way. This is captured by the product of *elasticity of marginal utility of consumption*[6] η and the growth rate

---

[6]   This measures how fast the value of an increment in consumption falls as consumption rises, for example when it is equal to one, an extra unit to Person A, with three times the consumption of Person B, would have one third the value to that if the extra unit went to Person B. If the elasticity were equal to two, the extra unit would have one ninth of the value.

of consumption (See Chapter 2 and Appendix). In Chapter 6, we used an elasticity of marginal utility consumption of 1, in line with some empirical estimates.[7] For this case, the contribution to the discount rate at any time is equal to the rate of growth in consumption at that time on the path. Note that η has a dual role as both a parameter of inequality aversion and of relative risk aversion. Some previous studies have assumed that the discount rate at any point in time is independent of the scale of the impacts and of the path followed (the future growth trajectory). However, as climate change implies that strongly divergent paths for future growth are possible, the use of a single set of discount rates (over time) for all paths is inappropriate.

Such a value for the elasticity of marginal utility of consumption might be interpreted as implying a very high savings rate in some simple models. However, applying this type of framework to savings rates as a central object of analysis would require more focus on issues related to savings, for example, the lifetime of capital equipment, flexibility, uncertainty, relations and responsibilities within and across generations and so on. Similarly, arguments for high η would imply stronger preference for redistribution than is reflected in policy in many countries. That does not settle any argument about η but it does indicate that application of a simple theory and model structure focused on one issue applied directly to a second issue is likely to miss out much that is important for the second. Thus, arguments about implications for the second, while relevant, have to be handled with care. These ideas are discussed in the Appendix to Chapter 2.

The second component is captured by the *pure rate of time preference*. This requires a consideration of the ethical issues involved in comparing the incidence of costs and benefits between generations, some of which are very distant in time. We argued in the Review– in line with economists including Ramsey, Pigou, Solow and Sen – that the welfare of future generations should be treated on a par with our own. This means, for example, that we value impacts on our children and our grandchildren, which are a direct consequence of our own actions, as strongly as we value impacts on ourselves.

We argued that the primary justification for a positive rate of pure time preference in assessing the impacts of climate change is the possibility that the human race may be extinguished. As the possibility of this happening appears to be low, we assume a low rate of pure time preference of 0.1%, which corresponds with a 90% probability of humanity surviving a 100-year period, if the 'probability of existence' view of pure time discounting is invoked. Higher probabilities of survival would imply a still lower rate (see Table PA.1 below).

Table PA.1 Implication of pure time discount rate (δ) for probability of existence

	Probability of human race surviving 50 years	Probability of human race surviving 100 years	Probability of human race surviving 150 years
$\delta = 0.1$	0.95	0.91	0.86
0.5	0.78	0.61	0.47
1.0	0.61	0.37	0.22
1.5	0.47	0.23	0.11

[7] Pearce and Ulph (1999); Stern (1977).

Many previous studies have used higher pure rates of time preference. They have used rates similar to those often applied to the evaluation of project-based investments. However, in drawing such analogies much turns on the meaning of the uncertainty covered by the pure time discounting. In this respect, there are important differences between the kind of large-scale disinvestments in the environment involved in climate change and other types of long-term investment, e.g. a railway. In the railway example, we might think of pure time discounting as covering the possibility that the context would change in such a way that the investment would become irrelevant (e.g. the closure of the whole railway system). Or we might interpret pure time discounting as covering the possibility that the particular decision might be reversed in terms of non-renewal of the investment when it reaches the end of its life. These looser[8] but possible interpretations of pure time discounting in the project appraisal context apply to climate change only in a much weaker form. Climate change is long-term, severe and irreversible. Accumulated stocks of carbon cannot easily be reversed and we cannot opt for another planet. Thus, if these looser forms of interpretation of pure time discounting are introduced, they imply pure time discounting for other contexts than for climate change.

The analysis cannot avoid taking on directly the challenge of how to treat unrepresented generations. It is an ethical issue and cannot simply be derived from market behaviour. For example, Arrow (1995) and Samuelson and Nordhaus (2005) (and see references therein and in Dasgupta, 2005), rightly present the issue as 'prescriptive' rather than 'descriptive'. However, Arrow and Nordhaus come to different conclusions from those indicated here about the appropriate rate of pure time discounting. Some of those arguments were covered in the appendix to Chapter 2. See also the important discussion in Cline (1992).

The consequences of choosing a high pure time discount rate for evaluating the impacts of climate change should be very obvious and were emphasised in Chapter 2 of the Review and its appendix. They are clear from Table PA.1. For example, if the pure time discount rate is 1.5%, then benefits 50 years from now, for individuals who have exactly the same consumption, have a weight less than half that of now. In other words, a grandparent would tell a grandchild that simply because the latter's consumption flow came later (e.g. 50 years) in time than his or her own consumption flow it would be correct to assign a value of less than half to it in thinking about the consequences of actions today. In the case of climate change, this would mean that while we know the direct (stochastic) consequences of our actions today and whom they affect, we would nevertheless apply a very low weight to those consequences. Many people would find that ethical position very unattractive. It is hard to see why the logic should be any different from assessing externalities that affect members of the current generation. We must be transparent and clear. If you take little account of the interests of future generations you will care little about climate change. But ethical positions cannot be dictated by policy analysts, and sensitivity analysis of loss estimates to the rate of pure time preference is supplied below.

There are ways of thinking about the relationship between this and future generations in terms of implicit bargains rather than using an aggregate social

---

[8]   It is possible to argue that this type of risk should be embodied in the measurement of costs and benefits but it would play a similar computational role and this type of discounting and pure time preferences seems often to be combined.

welfare function as in Chapter 2, its appendix and Chapter 6. We might think that future generations would willingly accept a lower conventional capital stock (e.g. roads and railways) in exchange for a better climate. In that case the existing generation acting on their behalf would adjust its investment portfolio, without investing more, to invest in a better climate. These kinds of notions come in when we invoke the ideas of sustainable development (and see for example, Arrow, et al. 2003).

This formal modelling of Chapter 6 does not take into account the distribution of consumption across regions. In similar vein to a lower weighting for marginal increments to richer generations, increments in poorer regions should have a higher weighting than those in richer regions. Making such calculations was beyond the scope of this exercise, given the limited time available for analysis. Taking this regional approach would increase the climate change cost estimates, as illustrated in Section 6.2, so our decision to use a simpler global aggregation approach will bias our model toward lower cost estimates. How we might adjust for this was described in Chapter 6.

Other factors: growth rate and treatment of long time-scales

There are other aspects in the models used in the Stern Review that will affect the outcomes of the modelling exercise. We describe two briefly:

1   The baseline growth rate. A scenario with higher growth would be expected to generate greater emissions, but also have a reduced discount rate. The balance of these effects depends on the convexity of the damages function from emissions stocks and temperature change, and the elasticity of the marginal utility function.
2   The Stern Review, like other similar studies, is very conservative in its treatment of climate change after 2200. We assume that impacts post-2200 are equal to impacts in 2200. That is, we assume that the problem contains itself after this time. This assumption may lead us to underestimate the impacts of climate change.

Sensitivity Analysis

The above discussion and the comments we have received point to the importance of testing the sensitivity of the loss estimates to three key parameter choices in the model: the damage exponent γ, relevant to model structure, and the elasticity of the marginal utility of consumption, η, and the pure time discount rate, δ, relevant to ethical values.

We first consider changes in the damage exponent. Figure PA.2 shows the losses as a percentage of global GDP per capita for the scenarios above, with the standard range for the damage exponent [1, 1.3, 3] and the modified range [1.5, 2.25, 3] – see above. Note that this change to the model structure applies whichever ethical values are introduced.

Next we consider the effect of changes to the ethical values. The first table looks at the implications of changing the elasticity of the marginal utility of consumption, in combination with changes to the damage exponent.

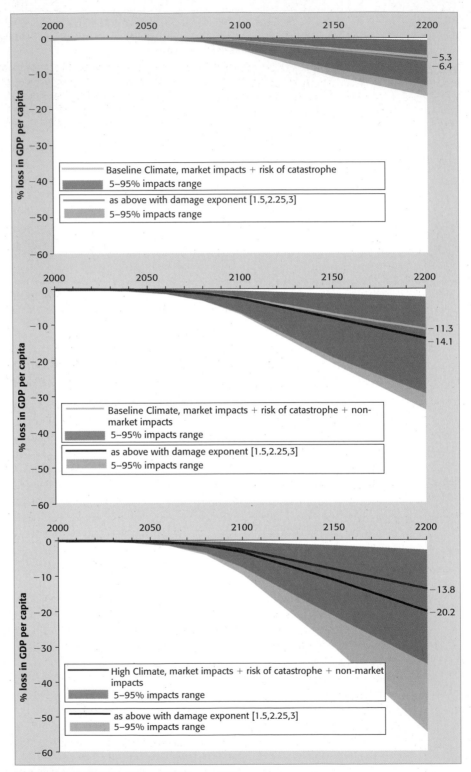

Figure PA.2 Percentage losses in GDP per capita

Table PA.2 Sensitivity analysis of estimates of the monetary cost of BAU climate change to the damage function exponent and the elasticity of the marginal utility of consumption, holding the pure time discount rate at 0.1% (*original Review estimate in italics*)

Damage function exponent	Elasticity of marginal utility of consumption	Baseline climate; market impacts + risk of catastrophe	Baseline climate; market impacts + risk of catastrophe + non-market impacts	High climate; market impacts + risk of catastrophe + non-market impacts
		Mean (5th percentile, 95th percentile)	Mean (5%, 95%)	Mean (5%, 95%)
Low range	*1.0*	*5.0 (0.6–12.4)*	*10.9 (2.2–27.4)*	*14.4 (2.7–32.6)*
	1.25	3.8 (0.6–9.6)	8.7 (2.2–21.7)	12.1 (2.7–26.0)
	1.5	2.9 (0.5–7.1)	6.5 (1.7–16.5)	10.2 (2.0–20.0)
High range	1.0	6.0 (0.8–15.5)	14.2 (2.8–32.2)	21.9 (3.7–51.6)
	1.25	4.6 (1.8–12.0)	11.3 (2.6–25.2)	18.2 (3.8–41.9)
	1.5	3.4 (0.3–9.0)	8.7 (1.8–19.2)	15.3 (2.8–33.1)

Table PA.2 presents results for three of the six scenarios originally reported in Chapter 6. They are:

1 Baseline climate; market impacts + risk of catastrophe;
2 Baseline climate; market impacts + risk of catastrophe + non-market impacts;
3 High climate; market impacts + risk of catastrophe + non-market impacts.

For a conservative scenario including baseline climate change and excluding non-market impacts on ecosystems and human health, increasing the elasticity of the marginal utility of consumption from 1.0 to 1.5 reduces the present value of the cost of BAU climate change from 5.0% to 2.9%. Using the same scenario, applying a higher probability distribution for the damage function exponent (with η constant at 1) increases the cost of BAU climate change from 5.0% to 6.0%.

We should note that higher values of η imply higher discount rates* via the growth effect. For example, a growth rate of 2% and an η of 1.5 would give a discount rate of 3%. And it should be noted in this modelling that we have not included declining discount rates in this modelling other than through the growth rate. There is a case for such a decline (see Appendix to Chapter 2 and references therein) and this would increase the loss estimates.

Although in the Review we have argued that it is preferable to value the impacts of climate change on health and the natural environment separately from its impacts on income, a comparison of the cost of mitigating climate change to the cost of BAU climate change, excluding non-market impacts gives a misleading signal. Interpreted literally, this would imply that these impacts have zero value. That would not be a tenable position. Zero is the most implausible of assumptions even though applying specific valuations raises difficult issues. The middle and third scenarios, which include non-market impacts, have a stronger claim on our attention. Increasing the elasticity of the marginal utility of consumption from 1.0 to 1.5 reduces the estimated cost of BAU climate change from 10.9% to

---

* Note that this is the discount rate to be applied to increments of consumption. This comes from the pure time discount rate, δ, the former is: $\eta(\dot{c}/c) + \delta$.

Table PA.3 Sensitivity analysis of estimates of the monetary cost of BAU climate change to the damage function and the pure time discount rate, holding the elasticity of the marginal utility of consumption at one (*original Review estimate in italics*)

Damage function exponent	Pure time discount rate (per cent)	Baseline climate; market impacts + risk of catastrophe	Baseline climate; market impacts + risk of catastrophe + non-market impacts	High climate; market impacts + risk of catastrophe + non-market impacts
		Mean (5th percentile, 95th percentile)	Mean (5%, 95%)	Mean (5%, 95%)
Low range	*0.1*	*5.0 (0.6–12.4)*	*10.9 (2.2–27.4)*	*14.4 (2.7–32.6)*
	0.5	3.6 (0.4–9.1)	8.1 (1.7–20.4)	10.6 (2.0–24.4)
	1.0	2.3 (0.4–5.8)	5.2 (1.2–13.2)	6.7 (1.3–16.0)
	1.5	1.4 (0.3–3.5)	3.3 (0.7–8.5)	4.2 (0.8–10.1)
High range	*0.1*	*6.0 (0.8–15.5)*	*14.2 (2.8–32.2)*	*21.9 (3.7–51.6)*
	0.5	4.3 (0.6–11.3)	10.2 (2.1–23.6)	15.8 (2.7–39.2)
	1.0	2.7 (0.4–7.2)	6.4 (1.4–15.5)	9.8 (1.7–25.6)
	1.5	1.7 (0.3–4.5)	4.0 (0.8–9.7)	5.9 (1.0–15.8)

6.5% (4.4 percentage points). On the other hand, applying a higher probability distribution for the damage function exponent increases the cost of BAU climate change from 10.9% to 14.2% (3.3 percentage points).

Substituting the high-climate scenario for the baseline-climate scenario, the probability distribution of the damage function exponent becomes the more important factor. Increasing this raises the cost of BAU climate change from 14.4% to 21.9% (7.5 percentage points). Increasing the elasticity of the marginal utility of consumption from 1.0 to 1.5 reduces the present value of the cost of BAU climate change from 14.4% to 10.2% (4.2 percentage points). We should note here that the elasticity of the marginal utility of consumption here plays a double role as an indication of (relative) risk aversion and of aversion to inequality. The former effect means that a high elasticity would increase damage estimates and the latter decrease them (via a stronger discounting). More sophisticated analysis could separate these effects.

Finally, we examine the sensitivity of loss estimates to the pure time discount rate, δ, presented in Table PA.3. The quantitative weighting following from different discount rates was presented in Table PA.1. The case where $\delta = 0.1\%$ was presented in Chapter 6 and is italicised in Table PA.3. As is intuitively clear, raising the pure time discount rate lowers loss estimates because the future is seen as less important. Nevertheless for all cases, even with the very high δ of 1.5% the loss estimates still exceed 1%, the estimated cost of strong mitigation. However, we would argue that even a pure time discount rate of 0.5% should be regarded as too high in this context, from an ethical or probability of extinction perspective (see Table PA.1 and related discussion).

Dr Chris Hope, the author of the PAGE2002 model, conducted a similar sensitivity analysis for the pure rate of time preference, which was published in the *Financial Times*[9] focusing on the social cost of carbon using the PAGE2002 model (baseline

[9] http://www.ft.com/cms/s/444ff4ae-783c-11db-be09-0000779e2340.html

climate scenario with non-market impacts and the standard damage function) and $\delta = 2$. This did not include the expected utility analysis used by the Review, but provides a useful comparison. Hope found that with the higher discount rate, the social cost of carbon is reduced by just over half to $40 (for the business-as-usual path). This is roughly consistent with the reductions outlined in Table PA.3.

Conclusions from Sensitivity Analysis

Where does this sensitivity analysis leave the overall case for strong mitigation as seen from the perspective of Chapter 6? First, let us re-emphasise that our first perspective on this argument was not Chapter 6, but the disaggregated analysis together with an overall assessment of risk. Formal modelling of the very simplistic kind carried out by IAMS, should not be the first claim on our attention in formulating policy. But pursuing the Chapter 6 approach, using the sensitivity analysis we can conclude that this perspective does provide a powerful argument for strong mitigation. For an analysis that takes account of non-market impacts all the calculations displayed give a loss estimate above 5% of consumption, except where the pure time discount rate is above 1%.

For the higher exponent on the damage function for temperature we find damages above the upper ranges provided in Chapter 6. Indeed even using the higher exponent on the marginal utility of consumption this statement remains true for the high climate case (for the pure time discount rate of 0.1% – see Table PA.2). We should recognise that the unitary value for the elasticity of the marginal utility of income together with $\delta = 0.1\%$ place stronger emphasis on later costs and benefits* than higher η or higher δ would imply. However, we have seen that provided δ is not extremely high (above 1%) the basic case from this approach for strong mitigation remains convincing, particularly when one takes account of higher damage exponents. And, in our view, the case for higher damage components in the context of the possibility of higher temperatures is convincing.

Many commentators have pointed to the importance of the pure discount rate. So did the Review, clearly and strongly, and it marshalled the arguments for the level chosen. On the other hand it is quite wrong as some have suggested, to argue that high losses from unabated climate change, relative to the costs of abatement, rest solely on this assumption. The sensitivity analysis demonstrates this clearly. Earlier authors who obtain lower damage costs do not take sufficient account of the most recent science linking probabilities of temperature increases to GHG concentration, and take insufficient account of the economics of risk.

The cost estimates presented here would increase still further if the model incorporated other important omitted effects. First, the welfare calculations fail to take into account distributional impacts, even though these impacts are potentially very important: poorer countries are likely to suffer the largest impacts. Second, the estimates here are conservative about damages post-2200. If they continued to rise after that then cost estimates would increase. Third, there may be greater risks to the climate from dynamic feedbacks and from heightened climate sensitivity beyond those included here. During the course of

---

* We note that, for these cases, increasing η increased the loss estimates, i.e. the risk aversion effect dominated the income distribution effect.

the Review, we examined the possibility that some of these factors could combine to produce significantly higher probabilities of large increases in temperature. The scientific evidence is not yet available to support any conclusions in this area, and we have not included the results of this work in the conclusions presented in the Review. This is an area where further scientific investigation would be very important as a basis for future economic analysis.**

We conclude with some brief remarks on possibilities of further research in this and related areas. We have already indicated our preference for a disaggregated approach to risk assessment in this area. Policy makers would (and probably should be) more convinced by a case which indicates the extent and seriousness of the risks involved in climate change, rather than aggregative results from speculative models that are highly sensitive to the assumptions built into them. Nevertheless these models do have a valuable supplementary role in the argument.

Thus, our first suggestion for further research is deeper investigation on the disaggregated effects of climate change. This should be oriented towards not only the 2–3°C range but also attempt to better understand the risks of 5°C and above, which we now know to be very serious possibilities under business-as-usual. This type of research would be important not only for understanding the case for strong mitigation but also be of great value in understanding what is necessary or advisable for adaptation.

This type of research would depend on high-resolution climate modeling which could provide much more detailed information on local impacts. This could and should be combined with detailed local studies, based on close local knowledge of possible implications of these climate changes.

At the same time, this type of high resolution *cum* local approach could be used, if sufficiently extensive, to inform global impact modeling. The work of Nordhaus (2006a) charts one very important line of investigation.

A second type of approach, building on the first, would be the development of the integrated assessment models to take more account of risk. Just one model was used here, chosen for convenience of use in stochastic analysis, and because it spanned a range of models. But other models should be used to develop different perspectives and in so doing test the robustness of our results.

In conclusion we should stress again that the analysis of the Review as a whole was always intended to be one contribution to a discussion. There have been, will be, and should be many more contributions.

References

Arrow, K, Dasgupta, P. and K-G Maler (2003): 'Evaluating projects and assessing sustainable development in imperfect economics', Environmental & Resource Economics, Vol 26 pp 647–685

Arrow, K.J. (1995): 'Inter-generational equity and the rate of discount in long-term social investment' paper at IEA World Congress (December), available at: www.econ.stanford.adu/faculty/workp/swp97005.htm

Cline, W.R. (1992): 'The Economics of Global Warming' Washington DC: Institute for International Economics.

** Technically, if consumption per head eventually grows at rate g and population is eventually constant, then convergence of the utility integral requires $(1 - \eta)g - \delta < 0$. Thus for $\eta = 1$ and $\delta > 0$, we have convergence but it is close to the borderline!

Dasgupta, P.S. (2001): 'Human Well-Being and the Natural Environment', Oxford University Press.

Dasgupta, P.S (2005): 'Three conceptions of intergenerational justice' in ed Mellor (2005).

Koopmans, T.C. (2972) 'Representation of reference orderings over time' in Maguire and Radner (1972)

McGuire, C.B. and R. Radner (eds) (1986) 'Decisions and Organization', North Holland, Amsterdam.

Mellow, H (ed) (2005): 'Ramsey's Legacy', Oxford University Press

Samuelson, P.A. and W.D. Nordhaus (2005): 'Economics', 8th Edition, New York: McGraw Hill.

Nordhaus, W. (2006a): 'Geography and macroeconomics: New data and new findings', PNAS, 103(10): 3510–3517

Nordhaus, W. (2006b): 'The economics of hurricanes in the United States, Prepared for Snowmass Workshop on Abrupt Climate Change', Snowmass: Annual Meetings of the American Economic Association, available from http://nordhaus.econ.yale.edu/hurricanes.pdf

Pearce, D.W. and A. Ulph (1999): 'A social discount rate for the United Kingdom' Economics and the Environment: Essays in Ecological Economics and Sustainable Development, D.W. Pearce, Cheltenham: Edward Elgar.

Stern, N. (1977): 'The marginal valuation of income', in Artis, M. and R. Nobay (eds.), Studies in Modern Economic Analysis, Oxford: Blackwell.

Warren, R. et al. (2006): 'Spotlighting Impacts Functions in Integrated Assessment Models', Norwich, Tyndall Centre for Climate Change Research Working Paper 91.

Index

Page numbers from figures, tables and boxes are in *italics*.
Page numbers from footnotes are accompanied with a suffix 'n'.